of

INTERPRETIVE LABORATORY DATA

THIRD EDITION

SECOND PRINTING

By

SEYMOUR BAKERMAN, MD., PH.D.

Revised by

PAUL BAKERMAN, M.D.

PAUL STRAUSBAUCH, M.D., PH.D.

Printed in the United States of America

Published by
Interpretive Laboratory Data, Inc.
Post Office Box 2250
Myrtle Beach, SC 29578
803-448-3055
0-700-ABC-BOOK

ISBN 0-945577-06-0

Paul Bakerman, M.D.
Associate Director
Pediatric Critical Care
Phoenix Children's Hospital
1111 E. McDowell Road
Phoenix, AZ 85006
(602) 239-3563

Paul Strausbauch, M.D., Ph.D.
Professor of Pathology
Department of Pathology and Laboratory Medicine
East Carolina University School of Medicine
Greenville, NC 27858
(919) 816-2809

The authors have extensively reviewed the information in this book to ensure that the recommendations are consistent with current practice. In an individual situation, these recommendations should not be considered absolute. Clinical practice is constantly evolving; current standards should be dictated by medical literature and specialists in various areas. This book is meant to provide information for teaching purposes. It is not meant to be a complete source, and other references should be consulted for more detailed information. It is the responsibility of the physician to assess a particular clinical situation and consult the appropriate experts in the field.

Reference range values are given for all tests. These values are usually those in general use or those employed in the authors' institution. In some instances, the source of the reference values is referenced and the reader may refer to the original publication. It must be remembered that reference values are dependant on many variables, including patient age, sex, specimen handling, analytical procedure, instrumentation, drug use, and patient population. Reference values for some substances have recently changed, especially those which employ monoclonal antibody technology. If there is any conflict between published procedures or reference ranges, refer to the instructions and values provided by your institution.

DEDICATION

This book is dedicated to the memory of Seymour Bakerman. Seymour Bakerman was an exceptional man; he was deeply loved by his family, students, staff and friends. As a person he approached life with overt enthusiasm that was infectious. He treated everyone equally, and with fairness. As a physician, Seymour Bakerman would want to be remembered as a teacher. When you walked into his classroom the first day, you knew immediately that something was different. He treated everyone with respect. He taught with organization, dedication, and enthusiasm, but at the same time always listened to the students to see if they were learning. The honor he cherished most was repeatedly winning the Golden Caduceus Award for outstanding teacher until it was renamed the Seymour Bakerman Award. As a family, we have continued the publication of this book as a tribute to Seymour Bakerman, teacher and physician; our father, husband and friend. He will always be remembered with love.

PREFACE

This book was started because there was a need for a clear, concise, pocket-sized paperback textbook on interpretation of laboratory data for the users of the laboratory and laboratory personnel.

The format in the text is uniform throughout: Name of test, specimen, reference range, method and interpretation. Usually, specimen volume is not given because of the wide variation in requirements from one method to another even within the same laboratory. The reference ranges are those that are generally acceptable but not necessarily applicable to all methods. The methods are selective and not all inclusive. The strength of this text is interpretation of the data.

The first and second editions of this book have been very popular with medical students and residents. Input from students and residents at East Carolina University School of Medicine and Phoenix Children's Hospital was helpful in the initial development of this book and in subsequent revisions.

PB: Dr. David Beyda made the initial suggestion that I revise this book, a project that looked insurmountable several years ago. I have received continued support from Dr. Beyda, as well as Dr. Paul Liu and Dr. David Tellez, an outstanding group of intensive care physicians with whom I work on a daily basis. Numerous attending physicians and pediatric residents at Phoenix Children's Hospital and Maricopa Medical Center have contributed in both general and specific ways to this text.

This work could not have been completed without tremendous support from my mother and the president of the publishing company, Winona Bakerman. Christina, Molly, Edward and Isaac provided psychological, physical, and intellectual support throughout this project. Beth and John Bakerman provided technical support.

PS: I first met Seymour Bakerman during my residency when he, among many others, taught me Clinical Pathology. Later as a faculty member in his department he continued to teach me and aid in my professional development. We taught together those years, both medical students and in postgraduate seminars. I now consider it a great privilege to aid in the revision of his book. Although there have been many changes, it still retains his character and the framework he built.

I must acknowledge the helpful questioning and inspiration of my fellow staff, the residents, and the medical students I have been associated with and taught over the years; the patience and support of my wife, Lynn, and family as I struggled over many hours of revisions; and the help and friendship of Winona Bakerman and her family.

Janine Jones Tripp spent countless hours above and beyond scheduled work time typing the text. Janine has ensured that each revision was as near perfection as possible, down to the smallest detail. Phyllis Broughton and Ruth Carson helped draw figures and provided technical support for this work.

Please contact us if you have any questions, comments or suggestions that would improve this text; our telephone numbers are (602) 239-3563. Paul Bakerman, M.D.; (919) 816-2809. Paul Strausbauch, M.D., Ph.D.

TABLE OF CONTENTS

Table of Contents (Cont)

Table of Contents (Cont)

Table of Contents (Cont)

Table of Contents (Cont)

Table of Contents (Cont)

PANELS

Table of Contents (Cont)

BLOOD COLLECTION TUBES

BLOOD COLLECTION TUBES

Stopper Color	Anticoagulant	Comment
Red	No Anticoagulant	Serum or clotted whole blood; serum must be separated from cells within 45 minutes of venipuncture. Most routine chemistries are done on serum. Most blood bank procedures: ABO; Rh; antibody screen and identification; direct and indirect antiglobulin tests
Red (Barrier Vacuum Tube)	None	Fill tube; invert once to accelerate clotting; allow to clot for 30 minutes; centrifuge. If sample is to be mailed, pour serum into separate serum vial. Barrier tubes are unacceptable for blood bank procedures.
Lavender (Purple)	Ethylenediamine Tetraacetate (EDTA)	Most hematologic procedures such as complete blood count; RBC, WBC, platelet counts and platelet function test, Hgb, Hct, red cell indices, differential; erythrocyte sedimentation rate(ESR); G-6-PD; Hgb electrophoresis; reticulocyte count; sickle cell preparation; CEA; renin (2 tubes on ice)
Blue	Citrate (must be full)	Prothrombin time(PT) and partial thromboplastin time(PTT); thrombin time(TT); factor assays (coagulation); fibrinogen level; G-6-PD assay (also 1 lavender top)
Green	Li Heparin	Blood gases (pH, PCO_2, HCO_3, base excess; PO_2, % Sat.) collected in a heparinized syringe, cap and place on ice; transport to laboratory immediately. Electrolytes; osmotic fragility; certain specific hematologic analyses, e.g., chromosomes; histocompatibility; ammonia (on ice); plasma hemoglobin
Grey	Potassium Oxalate, Sodium Fluoride	Blood glucose; fluoride exerts its action by inhibiting the enzyme system involved in glycolysis; lactate (on ice).
Yellow	ACD Solution	Blood group and type
Royal (Navy) Blue	None	Serum or clotted whole blood; special tube for trace metals.

The anticoagulants, EDTA, citrate and oxalate, act to prevent coagulation by removing calcium ions from the blood. Heparin acts by inactivating thrombin and thromboplastin. Do not use anticoagulant containing sodium and potassium for electrolyte determinations; use lithium or ammonium salt; do not use anticoagulant containing ammonium for ammonia determination.

ABO AND RH TYPE

SPECIMEN: Red top tube, separate cells from serum; do not use serum separation tubes. Include diagnosis, history of recent and past transfusions, pregnancy and drug therapy.
REFERENCE RANGE: See Interpretation
METHOD: Red Blood Cell Typing (Forward): Reaction of patient's red blood cells with known antiserum (anti-A, anti-B, Anti-A,B). Anti-A reacts strongly with A_1 but weakly or not at all with A_2, A_3, A_o or Am; anti-A,B reacts with Group A subgroups. The Rh test is performed with antiserum containing incomplete or blocking antibodies. (An incomplete blood group antibody does not react in saline). If the patient is Rh(D) positive, no further testing is done; if the patient is Rh negative, test for Rh variant and phenotyping. The Rh antigen is located on the red cell membrane only; the A and B antigens are located on the membrane of all cells.
Serum Typing (Reverse): ABO: Reaction of antibodies in patient's serum with red blood cells containing known surface antigens (known A_1, A_2, B and O cells).

The following conditions may interfere with typing: abnormal plasma proteins, cold autoagglutinins, positive direct Coombs test and some bacteria.
INTERPRETATION: ABO and Rh type is done on every ABO recipient before blood is issued for transfusion. The two antigens, A and B, on red blood cell membranes are responsible for the four blood groups. Cell type and serum antibody are:

Cell Antigen	Antibody in Serum
A	B
B	A
O	AB
AB	--

The majority of fatal transfusion reactions are due to ABO incompatibility.
Red Blood Cells: Reaction of patient's red blood cells with known antiserum (anti-A, anti-B, anti-A,B) is as follows:

Patient's Red Cells			ABO Blood				
Anti-A	plus Anti-B	Anti A,B	Blood Group	Whites	African American	Native American	Orientals
Negative	Negative	Negative	O	45	49	79	40
Positive	Negative	Positive	A	40	27	16	28
Negative	Positive	Positive	B	11	20	4	27
Positive	Positive	Positive	AB	4	4	<1	5

Some group A patients belong to subgroups of A; the approximate frequency of the major subgroups of A and B is given in the next Table (Sisson, J.A., Handbook of Clinical Pathology, J.B. Lippincott Company, Phila., 1976, pg. 345):

Major Subgroups of A and AB

Group A Subgroup	Approx. Frequency	Reactions with Anti-A_1 Serum	Reactions with Anti-AB	Approx. % with Anti-A_1 in Serum
A_1	78	Strong	Positive	None
A_2	22	Neg. or Weak	Positive	1-2%
A_3	Rare	Neg. or Weak	Positive	About 50%
A_o	Very Rare	Neg. or Weak	Positive	Over 50%
Am	Ultra Rare	Negative	Negative	Anti-H

Very young infants do not have their own alloagglutinins, that is, A cells, anti-B; B cells, anti-A. Those that are present are primarily passively transferred maternal antibodies. Elderly patients may not have alloagglutinins.
Rh: In contrast to the ABO blood group system, there are no isoagglutins (no anti-Rh) normally present in the serum of patients. 85% of the population is Rh(D) positive and 15 percent react negatively. Routine Rh typing for blood donors and recipients involves only the antigen $Rh_o(D)$.

ACETAMINOPHEN (TYLENOL, PARACETAMOL)

SPECIMEN: Red top tube, separate serum; or plasma (avoid heparin)
REFERENCE RANGE: Therapeutic range: 10-25mcg/mL (10-25mg/L); Time to Peak Plasma Level: 0.5-1.0 Hours; Half-Life: 2 to 4 hours; Time to Steady State: 10 to 20 hours; Toxic range: 100-250mcg/mL.
METHOD: Spectrophotometric technique (correct for salicylate). High pressure liquid chromatography. Ref: Sunshine, "Method Analytic Toxicology," C.R.C. Press, 1975, p. 14.
INTERPRETATION: Acetaminophen is the active ingredient of non-aspirin-containing analgesics. Acetaminophen is usually absorbed from the upper gastrointestinal tract; peak concentration occurs 1 hour after a therapeutic dose and 4 hours after an overdose. Unlike aspirin, acetaminophen has no a significant effect on platelet aggregation.

This drug may cause hepatic failure. Hepatic toxicity may appear 2-5 days after ingestion of a toxic dose. Most of the acetaminophen is normally metabolized in the liver to sulfate and glucuronide conjugates; a small amount is metabolized by the P-450 system to a potentially toxic intermediate, acetimidoquinone, which is metabolized by reaction with glutathione. Acetimidoquinone arylates vital nucleophilic macromolecules within hepatocytes, resulting in hepatic necrosis. Serious toxicity is likely to occur if the ingested dose is more than 140mg/kg. Tablets contain 80mg to 500 mg acetaminophen/tablet; fluid form may contain up to 2000mg acetaminophen/ fluid ounce. Blood specimens are collected about 4 hours and 8 hours after ingestion. A nomogram of plasma or serum acetaminophen concentration versus time since acetaminophen ingestion is shown in the next Figure; (Rumack BH, Peterson RG. Pediatr Supp. 1978; 62:898-903):

Nomogram of Plasma or Serum Acetaminophen Concentration versus Time Since Acetaminophen Ingestion

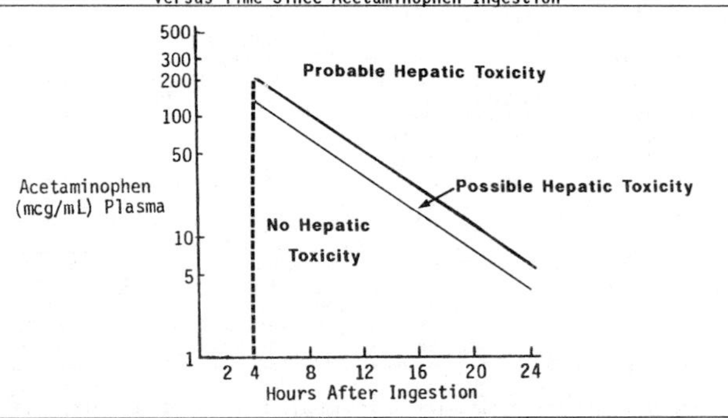

The half-life for acetaminophen elimination is 2 to 4 hours. The first sample is used to determine whether therapy with N-acetylcysteine (Mucomyst) should be started. Treatment should begin within 24 hours of ingestion. If the concentration of acetaminophen in the specimen collected 4 hours after ingestion is greater than 150mcg/mL, therapy should be begun using N-acetylcysteine in an effort to prevent the hepatic complications. N-acetylcysteine is believed to prevent hepatotoxicity because its sulfhydryl groups act as a glutathione substitute, binding to the metabolite. N-acetylcysteine (Mucomyst, Mead Johnson), available as a 20 percent solution, is administered orally, diluted 1:3 with a soft drink or grapefruit juice to mask its taste. N-acetylcysteine therapy should be initiated as soon as possible following acetaminophen ingestion and is still indicated at least 24 hours after ingestion (Smilkstein MJ. N Engl J Med. 1988; 319:1557-1562). The most common side effect of N-acetylcysteine therapy is vomiting; if vomiting occurs within one hour of administration of a dose, repeat dose. Liver function tests should be done to follow the patient. Contact the Rocky Mountain Poison Center, Denver, Colorado (1-800-525-6115) or other regional poison center for the most recent information on acetylcysteine therapy.

ACETONE, SERUM

SPECIMEN: Red top tube, separate serum; perform test immediately or refrigerate specimen; specimen should be free from visible hemolysis.
REFERENCE RANGE: Semi-Quantitative: The clinical laboratory usually expresses acetone content in terms of dilutions, e.g., 1/2, 1/4, 1/8, 1/16, etc., in order of increasing concentration. Quantitative: Negative (normal, 0.3 to 2.0mg/dL). To convert traditional units in mg/dL to international units in micromol/liter, multiply traditional units by 172.2.
METHOD: Semi-Quantitative: Nitroprusside (measures acetone and acetoacetate, not beta-hydroxybutyrate). Quantitative: Gas-liquid chromatography (specific for acetone).
INTERPRETATION: The assay of acetone is used as a measure of ketoacidosis. Acetone is formed from acetoacetate as shown in the next Figure:

Formation of Ketone Bodies

Acetoacetate (20 percent), $CH_3C(O)CH_2COOH$ —Spontaneously→ $CH_3C(O)CH_3$ Acetone (2 percent) + CO_2 Carbon Dioxide

Acetoacetate, $CH_3C(O)CH_2COOH$ —ß-hydroxybutyrate Dehydrogenase + NADH + H^+→ $CH_3CH(OH)CH_2COOH$ Beta-Hydroxybutyrate (78 percent) + NAD^+ Nicotinamide Adenine Dinucleotide

Acetone is formed in the conditions listed in the next Table:

Causes of Acetonemia

- Diabetes Mellitus, Uncontrolled
- Children:
  - Acute Febrile Illnesses
  - Toxic States with Vomiting or Diarrhea
- Alcoholism with Vomiting and Poor Food Intake
- Starvation, Prolonged and Some Weight Reducing Diets
- Isopropyl Alcohol (rubbing alcohol)
- Secondary to Acidosis in Von Gierke's Disease
- Stress

Ketosis occurs more frequently in the pregnant diabetic and develops at lower blood sugar levels (300-400 mg/dL range).

When ketone bodies accumulate, metabolic acidosis with anion gap develops. Beta-hydroxybutyrate is not assayed using nitroprusside. The extent of measured ketosis is dependent on the ratio of acetoacetate to beta-hydroxybutyrate. This ratio is low when a state of lactic acidosis coexists with ketoacidosis, because the reduced redox potential of lactic acidosis favors the production of beta-hyroxybutyrate. In this event, the level of ketosis as determined by the nitroprusside reaction appears to be inappropriately low for the degree of acidosis, and the additional diagnosis of lactic acidosis should be considered.

Furthermore, there is potential for confusion during the adequate therapy of diabetic ketoacidosis during which repeated measurements of serum ketones are performed. The failure of measured ketonemia to decline in the face of rising pH and falling blood sugar should not necessarily be cause for alarm. Beta-hyroxybutyrate levels fall rather rapidly, but this change is not detected by the nitroprusside reaction. In addition, beta-hydroxybutyrate can be metabolized into acetoacetate giving the false impression that the ketosis is worsening.

ACETONE, URINE

SPECIMEN: Random urine; specimen should be rejected if more than 3 hours old; false positives for semi-quantitative test are obtained with specimens containing bromosulphalein (BSP), L-dopa metabolites or phenylketones.
REFERENCE RANGE: Negative
METHOD: See ACETONE, SERUM
INTERPRETATION: Acetone is formed from acetoacetate as shown in the previous section, ACETONE, SERUM.

Acetone is formed in the conditions listed in the Table - See ACETONE, SERUM.

ACETYCHOLINE RECEPTOR ANTIBODY

SPECIMEN: Red top tube, separate serum
REFERENCE RANGE: ≤0.03 nanomoles/liter
METHOD: RIA
INTERPRETATION: Measurement of these antibodies are useful in the evaluation and follow up of patients with myasthenia gravis (MG). Myasthenia gravis is manifest, clinically, by weakness of skeletal muscle.

Muscular contraction is mediated by release of acetylcholine from motor nerve terminals and its binding to receptor proteins on muscle; contraction is terminated by the enzyme acetylcholinesterase which catalyzes the hydrolysis of acetylcholine to choline plus acetate. Myasthenia gravis is an autoimmune disease in which antibodies to acetylcholine receptor proteins reduce the acetylcholine sensitivity of the receptor.

The most common initial presentation of myasthenia gravis is ptosis and diplopia due to involvement of the extraocular muscles. In 20% of patients, symptoms are confined to the eye muscles (ocular MG). Patients with generalized MG may have abrupt or insidious onset of weakness of any muscle group, typically asymmetric distribution, and a fluctuating course (Seybold ME. JAMA. 1983; 250:2516-2521). Neonatal MG is caused by passively acquired maternal antibodies; congenital MG may also occur. Various subtypes of MG have been described, and MG has recently been reviewed (Linton DM, Philcox D. Disease-a-Month. 1990; 36:595-637).

Acetylcholine receptor antibody expression varies with clinical group and severity of MG: moderately severe or acutely severe 100%, chronic severe 89%, mild generalized 80%, ocular 50%, remission 24% (Tindall RSA. Ann Neurol. 1981; 10:437-447). Subtypes of acetylcholine receptor antibodies include binding, blocking, and modulating antibodies and these may be assayed individually (Specialty Laboratories, Inc., 800-421-7110). Patients with MG may have other autoimmune diseases including thyroiditis and systemic lupus erythematosus; other antibodies may be detected including antinuclear antibodies in 25-40% and rheumatoid factor, antithyroid, or other organ specific antibodies in 20% of patients (Zweiman B, Arnason BGW. JAMA. 1987; 258:2970-2973). MG and acetylcholine receptor antibodies have been associated with D-penicillamine therapy for rheumatoid arthritis, beta-blocking agents, and following bone marrow grafting; antibodies have also been reported in motor neuron disease (Mastalgia FL. Drugs. 1982; 24:304-321; Abbott RJ, et al. Lancet. 1986; i:906-907; Others).

Over 75% of myasthenia gravis patients have an abnormal thymus; 15% of these contain thymomas and the remainder are hyperplastic glands. C.T. scan and linear tomography may help to distinguish tumor from hyperplasia (Janssen RS, et al, Neurology. 1983; 33:534). Anti-striational antibodies may also be helpful in distinguishing MG with thymoma from those without. Anti-striational antibodies are found in >80% of MG patients with thymoma, and are absent in most without thymoma (Cikes N, et al. Mayo Clin Proc. 1988; 63:474-481).

ACETYLCHOLINESTERASE (see CHOLINESTERASE)

ACETYLSALICYLIC ACID (see SALICYLATE)

ACID MUCOPOLYSACCHARIDES (see MUCOPOLYSACCHARIDES)

ACID PHOSPHATASE, PROSTATIC

SPECIMEN: Red top tube, separate serum within 30 minutes; or lavender (EDTA) top tube; separate plasma within 30 minutes. Add 0.05mL, 20% acetic acid to 2.5mL serum or freeze within 45 minutes or add disodium citrate tablet.
REFERENCE RANGE: 0.1-0.8IU/L (variation with methodology); up to 10mg/mL by RIA.
METHOD: Over 20 isoenzymes with acid phosphatase (AcP) activities exist. Various substrates and assay conditions have been used to increase specificity. Thymophthalein monophosphatase is relatively specific for prostatic acid phosphatase; this substrate is not useful to assay for total acid phosphatase in Gaucher's disease. Another approach to increase sensitivity is to measure activity with and without tartrate. Prostatic AcP, band 2, is inhibited by tartrate. A third approach to increase sensitivity is to measure the prostatic AcP fraction by RIA. AcP activity is relatively unstable and requires careful preparation of the specimen.
INTERPRETATION: Serum AcP is used for the diagnosis of prostatic carcinoma and to monitor therapy in these patients. One might consider use of prostate specific antigen to substitute for this test in view of its greater sensitivity and specificity. The causes of elevation of prostatic serum or plasma AcP are given in the next Table:

Causes of Elevation of Prostatic Acid Phosphatase
Carcinoma of the Prostate
Prostatic Conditions Other than Carcinoma
Prostatic Palpation; Hyperplasia of the Prostate;
Prostatic infarction; Following Cystostomy;
Prostatic Surgery Including Biopsy
Non-prostatic Conditions

Note that the substrate, thymophthalein monophosphate is relatively insensitive to non-prostatic sources of AcP. However, elevated levels of AcP have been reported with this substrate in growing children, Gaucher's disease, leukemic malignancies, and with macro-acid phosphatase (AcP-immunoglobulin complexes). However, much of this lack of specificity may be avoided in the future by use of radioimmunological procedures.
Carcinoma of the Prostate: Serum AcP is elevated in 10%-20% of patients without metastases if carcinoma is within the gland; in 20%-40% of patients with metastases but without bone involvement and in 70%-90% of patients with bone involvement. A high level of AcP practically always means that the tumor is no longer confined to the prostate. The undifferentiated carcinomas may not produce AcP. AcP is not useful in the screening of healthy, asymptomatic men for occult prostatic cancer.

The metastatic lesions to bone tend to be osteoblastic; serum alkaline phosphatase is elevated in approximately 85% of patients with carcinoma of the prostate with skeletal metastases (Schwartz MF, Bodansky O. N.Y. Acad Sci. 1969; 166:775-793). Serum aspartate aminotransferase (AST) may be elevated in the serum of patients who have liver or other soft tissue metastases.

Serial determinations of AcP in patients with carcinoma of the prostate correlate with alterations of the clinical status. A decrease in the serum level of AcP indicates relative effectiveness of therapy; an increase indicates renewed activity of the metastases.

Measurement of prostatic AcP for the diagnosis of and management of prostatic carcinoma is rapidly being replaced by assay of prostate specific antigen (PSA). PSA measurements are much more sensitive than the older AcP assays. However, measurement of AcP may still be indicated to follow response to therapy, especially in those patients being treated with anti-androgen therapy (Rainwater LM, et al. Mayo Clin Proc. 1990; 65:1118-1126), since PSA requires androgen for expression, a requirement not needed for AcP expression. Some patients exhibit elevation of AcP with normal PSA levels.
Other Prostatic Conditions: Pearson et al obtained blood samples for enzymatic and RIA analysis for AcP before, five minutes, one hour and 24 hours after 30 seconds of prostatic massage in patients with prostatic carcinoma, prostatic hyperplasia and controls. AcP was elevated in 25 to 30 percent of subjects by both methods. It was concluded that specimens for AcP should be obtained before or at least 24 hours after prostatic examination (Pearson JC, et al. Urology. 1983; 21:37-41; Brawer MK. Urol Clin N Amer. 1990; 17:759-768). Serum prostatic AcP is slightly elevated in 10-30 percent of patients with hyperplasia of the prostate. Serum AcP usually increases markedly immediately following surgery; the enzyme returns to normal in 72 hours (Pearson JC, et al. already cited).

ACID PHOSPHATASE, TOTAL, PROSTATIC AND NON-PROSTATIC

SPECIMEN: Red top tube, separate serum; or lavender (EDTA) top tube, separate plasma. Separate from cells within 30 minutes. Add 0.05mL, 20% acetic acid to 2.5mL serum or freeze within 45 minutes or add disodium citrate tablet. Specimen rejected if hemolyzed, lipemic or icteric.

REFERENCE RANGE: Depends on method; in the newborn, serum acid phosphatase (AcP) is almost two times the upper limit of normal for an adult; during the growth period (up to age 13), serum AcP is approximately 1.5 times the upper limit for an adult; the upper limit for adult males is slightly greater than the adult level for females.

METHOD: P-nitrophenylphosphate; alpha-naphthyl phosphate or other substrate non-specific for AcP. These assays are often combined with tartrate inhibition. Prostatic AcP is inhibited by tartrate, while hairy cell, Gaucher cell, erythrocyte, and osteoclast AcP are resistant to tartrate inhibition.

INTERPRETATION: This assay is most appropriately used in patients with possible Gaucher's disease, Niemann-Pick disease and reticulo-endotheliosis (hairy-cell leukemia). Methods that are non-specific for AcP detect enzyme originating from prostate, osteoclasts of bone, red and white blood cells, platelets and liver.

The conditions associated with elevation of serum or plasma AcP are given in the next Table:

Conditions Associated with Prostatic and Non-Prostatic Acid Phosphatase
Carcinoma of the Prostate
Prostatic Conditions Other than Carcinoma
Elevation of Serum Bone Alkaline Phosphatase Isoenzyme Levels as Occur in Bone Disease
Liver Disease
Hematogenic Conditions:
Hemolysis, Thrombocytopenia, Myeloproliferative Diseases (Hairy Cell Leukemia)
Gaucher's Disease and Niemann-Pick Disease

Gaucher's Disease: Gaucher's disease is a rare lipid storage disease caused by a deficiency of the enzyme, glucocerebrosidase. In Gaucher's disease, serum AcP is elevated. However, the usual method of assay measures prostatic AcP and not the AcP found in patients with Gaucher's disease. It is therefore necessary to specifically indicate on the request form that assay of non-prostatic AcP is needed. This is also true for assays of AcP ordered to assess other non-prostatic conditions. See GLUCOCEREBROSIDASE.

ACID PHOSPHATASE, VAGINAL, EVIDENCE OF RECENT SEXUAL INTERCOURSE

SPECIMEN: An adequate amount of vaginal fluid is usually obtained by aspirating fluid from the vagina. A sample is collected by washing the vagina with a small amount, 2mL or less, of isotonic saline. Do not use cotton swabs because false positive results are obtained if the swab does not contain sufficient specimen. After the saline wash is obtained, centrifuge the specimen. Draw off the supernatant for acid phosphatase determination; high concentrations of acid phosphatase in vaginal samples may be accepted as proof that semen is present. Use the sediment for examination of spermatozoa; spermatozoa may be detected on swabs for as long as 48 hours. When spermatozoa cannot be identified, serological tests using antisemen sera may give a definite result. Seminal blood group antigens may be detectable for at least 24 hours. Reference: Editorial, Brit. Med. J. July 15, 1978; 154.

The specimens must be well identified and a clear chain of custody must be maintained. Specimens that are not assayed immediately for acid phosphatase may be stored at -20°C.

REFERENCE RANGE: 50U per sample or greater is considered "semen positive." Normal acid phosphatase activity in non-coital women is less than 10U/L (Lantz RK, Eisenberg RB. Clin. Chem. 1978; 24:486-488; Dahlke MB, et al. Am J Clin Pathol. 1977; 68:740-746). Acid phosphatase activity in the vagina remains relatively constant for about 14 hours; 40% of females are positive after 24 hours and 11% positive after 72 hours (Findley TP. Am J Clin Pathol. 1977; 68:238-242).

METHOD: Thymolphthalein monophosphate; this substrate is insensitive to non-prostatic sources of acid phosphatase.

INTERPRETATION: Assay of acid phosphatase is used to obtain evidence of recent sexual intercourse especially in those patients when it is no longer possible to detect spermatozoa. Detection of seminal fluid in the vagina means that recent sexual intercourse has taken place; detection of seminal fluid in other areas indicates sexual contact.

Acid Phosphatase may be detected when spermatozoa are absent as following vasectomy.

ACQUIRED IMMUNODEFICIENCY SYNDROME (AIDS) (See HIV TESTING)

ACTIVATED CLOTTING TIME (ACT)

Synonyms: Activated Coagulation Time; Ground Glass Clotting Time

SPECIMEN: Obtain 3 mL of blood in a plastic syringe. No anticoagulant. A gray stoppered vacutainer tube containing siliceous earth is also available. A clean venipuncture is required because contamination with tissue thromboplastin alters the clotting time.

REFERENCE RANGE: Approximately 150-200 seconds. The reference ranges vary with the laboratory. Commercial tubes will give shorter times.

METHOD: Timing of the test begins when one mL of the blood is placed in a glass tube which contains ground glass or celite which act as activator. It is preferable to perform the test at room temperature at the bedside. Tilt the tube at specific time intervals. The clotting time is the time interval from the start (when one mL of blood is placed in the glass tube) to the time that the blood has clotted.

INTERPRETATION: The activated clotting time is used to monitor heparin therapy. It is popular among cardiac surgeons and is useful in monitoring heparin therapy given during extracorporeal circulation. Although the partial thromboplastin time(PTT) is commonly used to assess heparin dosages, the PTT cannot be usually read at the high doses of heparin used during extracorporeal circulation. The ACT is used in this situation. Since the clotting time can be monitored at the bedside, heparin or its antagonists may be given to induce changes in the desired direction. High doses of heparin (cardiac surgery) are monitored with the ACT, usual dose of heparin (standard anticoagulation) are monitored with the PTT. Low doses (residual heparin effects) may be detected with the thrombin time test.

The ACT is not sensitive to deficiencies of blood coagulation factors except when these factors are very low (e.g., 1% of normal). Since it is also sensitive to effects other than heparin, it cannot be assumed that the prolonged ACT is due to heparin administration alone.

ACTIVATED PARTIAL THROMBOPLASTIN TIME
(see PARTIAL THROMBOPLASTIN TIME)

ACUTE ABDOMINAL PANEL

Laboratory tests for work-up of patients with an acute abdomen are listed in the next Table:

Acute Abdomen Panel
Complete Blood Count
Electrolytes, BUN, Creatinine
Urinalysis
Serum Amylase and Urine Amylase
Serum Bilirubin
Urine Porphobilinogen for Acute Intermittent Porphyria
Serum Glucose
Gram Stain of Vaginal Secretions
Hemoglobin Electrophoresis if Sickle Status in Black Unknown

Causes of acute abdominal pain by age are listed in the next Table (Jung PJ, Merrell RC. Gastroenterol Clin N Amer. 1988; 17:227-24; Hatch EI. Pediatr Clin N Amer. 1985; 32:1151-1164):

Causes of Acute Abdominal Pain by Age

Infants	3-11 Years
Intussusception	Appendicitis
Hirschsprung's Enterocolitis	Trauma
Strangulated Hernia	Meckel's Diverticulitis
Trauma (Abuse)	Pneumonia
Meckel's Diverticulitis	Bacterial Enterocolitis (Salmonella, Shigella, Campylobacter, Yersinia)
Bacterial Enterocolitis	Crohn's Disease
Pneumonitis	Pancreatitis
Pyelonephritis	Infected Mesenteric Cyst
Mesenteric Cysts	Ruptured Tumors
Testicular Torsion	Pyelonephritis
Pancreatitis	
Intestinal Obstruction-Volvulus	

Adolescent	Adult
Appendicitis	Appendicitis
Pelvic Inflammatory Disease	Peptic Ulcer Disease
Mittelschmerz	Cholecystitis
Crohn's Disease	Diverticulitis
Enterocolitis	Foreign Body Perforation
Peptic Ulcer Disease	Perforated GI Malignancy
Cholecystitis	Trauma
Pneumonia	Ectopic Pregnancy
Trauma	Bowel Obstruction or Perforation
Ectopic Pregnancy	Vascular Accident with Bowel Necrosis
Hematocolpos	Vascular Rupture (eg, Aneurysm)
Psychosomatic	

See also Bender JS. Med Clin N Amer. 1989; 73:1413-1422; Harberg FJ. Pediatr Annals. 18:169-178.

ACUTE GLOMERULONEPHRITIS PANEL

(Madaio MP, Harrington JT. N Engl J Med. 1983; 309:1299-1302).

Proteinuria, Quantitative (Usually 500mg to 3g/day)
Hematuria; Red-Cell Casts
↓
Acute Glomerulonephritis
C3 Complement and Hemolytic Complement

Low Serum Complement Level:

Systemic Diseases:
- Systemic Lupus Erythematosus (focal, approx.75%, diffuse, approx.90%)
- Subacute Bacterial Endocarditis (approx.90%)
- "Shunt" Nephritis (approx.90%)
- Cryoglobulinemia (approx.85%)

Renal Diseases:
- Acute Poststreptococcal Glomerulonephritis (approx.90%)
- Membranoproliferative Glomerulonephritis
  - Type I (approx.50-80%)
  - Type II (approx.80-90%)
- Renal Atheroembolization (Cosio FG, et al. Lancet. 1985; 2:118)

Normal Serum Complement Level:

Systemic Diseases:
- Polyarteritis Nodosa Group
- Hypersensitivity Vasculitis
- Wegener's Granulomatosis
- Henoch-Schonlein Purpura
- Goodpasture's Syndrome
- Visceral Abscess

Renal Diseases:
- IgG-IgA Nephropathy (Berger's)
- Idiopathic Rapidly Progressive Glomerulonephritis
- Anti-Glomerular Basement Membrane Disease (Goodpasture's)
- Immune-Complex Disease (Membranous Glomerulonephritis)
- Negative Immunofluorescence Findings (Minimal Change Disease)

Comment: Determine serum creatinine and creatinine clearance.

Nephrotic-range proteinuria plus low serum complement suggests lupus nephritis or membranoproliferative glomerulonephritis.

Obtain a renal biopsy if the information will substantially alter the specific treatment or overall management.

(1) Acute Glomerulonephritis with Low Serum Complement Level: The proposed mechanisms for low serum complement levels in acute glomerulonephritis are as follows: Immune complex formation: consumption of complement in the kidney and in other organs in some systemic diseases exceeds the production of complement components and results in a depression of the serum complement level. In other patients, antibodies form against complement components e.g. membranoproliferative glomerulonephritis. Individuals with genetic deficiencies have a higher incidence of glomerulonephritis than patients with normal complement levels. A diagnostic approach to the differential diagnosis of acute glomerulonephritis with low serum complement is given in the next Figure:

Differential Diagnosis of Acute Glomerulonephritis with Low Serum Complement

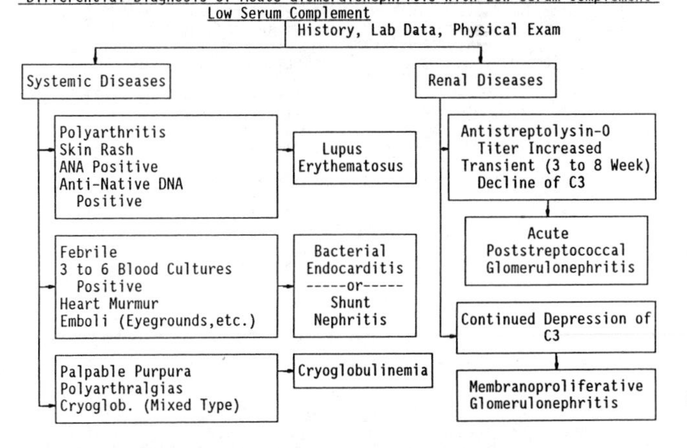

Acute Glomerulonephritis Panel (Cont)

Comments on Renal Diseases: Antistreptolysin-O-Titer: The antistreptolysin-O titer is elevated in up to 70 percent of patients with acute poststreptococcal glomerulonephritis and in approximately 20 percent of patients with membranoproliferative glomerulonephritis. Early use of penicillin prevents the antistreptolysin-O titer from rising. Many infectious agents, in addition to streptococcus, may cause postinfectious glomerulonephritis. The titer does not rise after cutaneous streptococcal infection.

Complement; Poststreptococcal Glomerulonephritis versus Membranoproliferative Glomerulonephritis:

	Complement			
Condition	C3	Alternate Pathway Activation	C3 Nephritic Factor	Classic Pathway Depressed C1,C4,C2
Poststreptococcal Glomerulonephritis	Transient Decline (3-6 wks.)	Yes	No	Normal or Slight
Type I Membranoproliferative Glomerulonephritis	Declines	No	Yes (Approx. 33 percent)	Yes
Type II Membranoproliferative Glomerulonephritis	Declines and Remains Depressed	Yes	Yes (Approx. 75 percent)	Normal or Slight

(2) Differential Diagnosis of Acute Glomerulonephritis with Normal Serum Complement Level: See discussion in article by Madaio MP, Harrington JT, already cited.

(3) Specific Laboratory Tests: Specific tests for glomerular disease are given in the next Table:

Tests for Glomerular Disease
Antistreptolysin-O Titer
Antinuclear Antibodies (ANA)
Anti-DNA Antibodies
Anti-Glomerular Basement Membrane (Anti-GBM)
Anti-Neutrophil Cytoplasmic Antibodies (ANCA)
Cryoglobulins
Bence-Jones Protein
Urine Protein Electrophoresis
Hepatitis-Associated Antigens
VDRL

Rapidly Progressive Glomerulonephritis (RPGN): RPGN refers to a rapid decline in renal function (weeks to months) associated with crescent formation. Specific serologic testing may be useful in RPGN as given in the following Table (Jennette JC, Falk RJ. Med Clin North Amer. 1990; 74(4):893-908):

Serologic Testing in Rapidly Progressive Glomerulonephritis
Anti-Neutrophil Cytoplasmic Antibodies (ANCA)
No Extra-renal Disease: Idiopathic Crescentic Glomerulonephritis
Pulmonary Necrotizing Granulomas: Wegener's Granulomatosis
Systemic Necrotizing Arteritis: Polyarteritis Nodosa
Anti-Glomerular Basement Membrane Antibodies
Lung Hemorrhage: Goodpasture's Syndrome
No Lung Hemorrhage: Anti-GBM Glomerulonephritis
Immune Complexes
Anti-DNA Antibodies: Lupus Glomerulonephritis
Anti-Streptococcal Antibodies: Post-Streptococcal Glomerulonephritis
Cryoglobulins: Cryoglobulinemic Glomerulonephritis

ACUTE MYOCARDIAL INFARCTION PANEL

BLOOD SPECIMENS: Red top tube, separate serum. Do not freeze samples for isoenzyme analysis. Obtain a minimum of three samples; at time of admission and at 12 and 24 hours after onset of symptoms of acute myocardial infarction. Some recommend samples at admission and at 8 and 16 hours if duration of symptoms is less than 24 hours.
INTERPRETATION: Diagnosis of acute myocardial infarct may be difficult. The WHO recommends that the diagnosis be based on two out of three of the criteria listed in the Table below:

Suggested Criteria for Diagnosis of Myocardial Infarct

Criteria	Sensitivity
Clinical Features	35%
EKG Changes	50-70%
Cardiac Enzymes	>95%

The changes in CPK, CPK isoenzymes, LDH and LDH isoenzymes are shown in the next Figure:

Serum Levels of Enzyme Following Uncomplicated Acute Myocardial Infarction

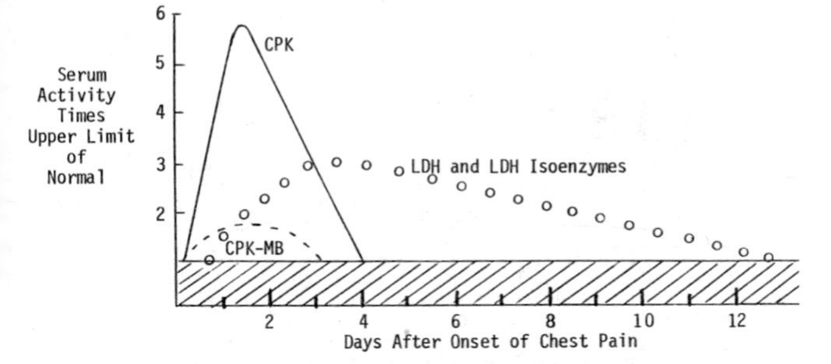

The changes in the serum enzymes, as shown in this Figure, are summarized in the next Table:

Changes in Serum Enzymes Following Acute Myocardial Infarction

Enzyme	Beginning Increase (Hours)	Maximum (Hours)	Return to Normal (Days)
Creatine Phosphokinase (CPK)	2-12	18-30	3-4
Creatine Phosphokinase Isoenzymes (CPK-MB)	2-12	12-36	2-3
Lactate Dehydrogenase (LDH)	8-12	36-72	8-12
Lactate Dehydrogenase Isoenzymes (LDH-1 and 2)	8-12	36-72	10-15

CPK begins to rise in 2 to 12 hours following onset of myocardial infarction; it reaches a maximum in 18 to 30 hours and it remains elevated for three to four days.
CPK-MB begins to rise in 2 to 12 hours, reaches its peak in 12 to 36 hours and returns to normal in 2-3 days. CPK-MB levels tend to peak at 12 hours with small or reperfused infarcts, and between 16 to 24 hours with large or non-reperfused infarcts.
LDH begins to rise in 8 to 12 hours, reaches a maximum in 2 to 3 days, and returns to normal in 8 to 12 days. In general, total LDH and LDH isoenzyme analysis are not as useful, sensitive or specific for diagnosis of myocardial pathology as the analysis of total CPK and CPK isoenzymes.
The LDH isoenzymes (LDH-1 and 2) begin to rise 8 to 12 hours following onset of myocardial infarction, reach a maximum in two to three days, and return to normal in ten to fifteen days. To diagnose acute myocardial infarct, one would like to observe a "flipped" LDH1/LDH2 ratio, i.e. LDH1 > LDH2. In general, total LDH and LDH isoenzyme analysis are not as useful, sensitive or specific for diagnosis of myocardial pathology as the analysis of total CPK and CPK isoenzymes.

Acute Myocardial Infarction Panel (Cont)

AST (SGOT): Serum Aspartate Aminotransferase (AST) begins to rise in 4 to 6 hours following onset of acute myocardial infarction, just slightly after CPK starts to rise; it reaches a maximum in one to two days, and it remains elevated for 4 to 6 days. AST has high sensitivity (about 95 percent) but low specificity for the diagnosis of myocardial infarction and is usually not ordered.

Serial CPKs are measured in the first 24 hours following symptoms. If the patient does not come to medical attention until much later, the CPK may have returned to the normal range. In such situations, measurement of LDH and LDH isoenzymes may be helpful.

CPK and LDH isoenzyme patterns 24 to 36 hours after an acute myocardial infarction are shown in the next Figure:

CPK and LDH Isoenzyme Patterns Following Acute Myocardial Infarction

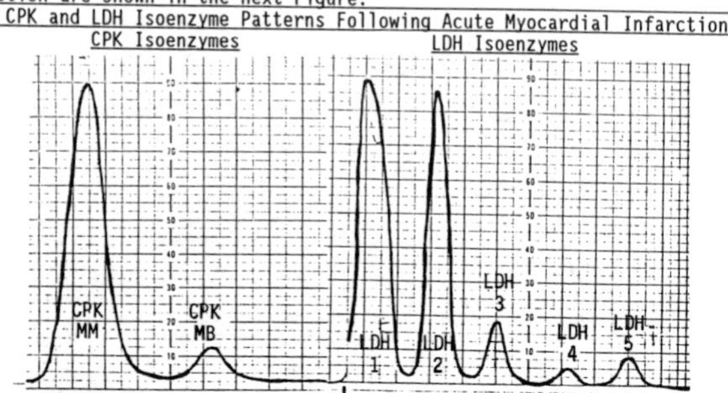

Following an acute MI, one would expect to see an elevation of the CPK-MB fraction in the first 6-48 hours after infarct. Changes in the LDH isoenzyme pattern should be evident after the first 24 hours. The diagnostic pattern is a LDH_1 greater than LDH_2, "flipped LDH."

Direct Current Countershock: Direct Current Countershock, which is used to convert cardiac arrhythmias to sinus rhythm, produces an increase in serum CPK in about 50 percent of patients; CPK-MM is elevated; occasionally, there is a mild elevation of CPK-MB. Tachycardia alone, or right ventricular temporary electrode pacing may raise the total CPK but not the CPK-MB (O'Neill PG, et al. Am Heart J. 1991; 122:709-714).

CPK-MB is highly sensitive and specific as compared to other serum enzyme determinations for the diagnosis of acute myocardial infarction as shown in the next Table (Wagner, et al. Circulation. 1973; 57:263-269):

Parameter Sensitivity and Specificity for Diagnosis of Acute Myocardial Infarction (Wagner et al, 1973)

Diagnostic Parameter	False-negative (%)	Sensitivity (%)	False-positive (%)	Specificity (%)
ECG	34	66	0	100
Total CPK	2	98	15	85
LDH1 > LDH2	10	90	5	95
CPK-MB	0	100	1	99

Mortality following myocardial infarction has been correlated with the level of serum enzymes as shown in the following Table:

Fifty Percent (50%) Mortality and Serum Enzyme Levels Following Acute Myocardial Infarction

Enzyme	Serum Enzyme Level Times Upper Limit of Normal	Serum Enzyme Levels in mIU
Aspartate Aminotransferase (SGOT)	7	(50)* 350
Lactate Dehydrogenase (LDH)	6	(200)* 1200
Creatine Phosphokinase (CPK)	15	(100)* 1500

*The numbers in parentheses are the upper limits of normal of the enzyme in mIU.

However, patients with a very small infarct may develop an arrhythmia and die and patients with large infarct may survive.

CPK in Thrombolytic Therapy: Measurement of serial CPK levels may be useful in the assessment of thrombolytic therapy (Hohnloser SH, et al. J Am Coll Cardiol. 1991; 18:44-49). Successful thrombolysis of a thrombosed vessel and reperfusion of an infarcted area results in a "washout" of released enzyme. In such a situation, elevated CPK levels appear earlier, reach higher levels, and peak earlier as compared to situations in which the vessel remains occluded and peak at about 10-12 hours after reperfusion.

In order to assess successful reperfusion measure CPK levels every 4 hours for 5 measurements. If information is required sooner in order to schedule surgical intervention, samples should be taken every hour for at least 3 hours. Such studies may be performed by measurement of total-CPK or CPK-MB.

Other Measurements: The development of other laboratory tests is in progress. Potentially attractive tests include measurement of myoglobin, troponin and myosin light chain. Myoglobin is released sooner from damaged myocardium than CPK, and hence its measurement would be an attractive indicator of myocardial damage. It is relatively non-specific due to its wide distribution. Troponin-T and troponin-I are cardiac specific. They begin to rise at the same time as CPK-MB but Troponin-T stays elevated for at least 10 days. Measurement of troponin could possibly replace isoenzyme studies in the diagnosis of AMI. Troponin levels have also been used to select a subgroup of patients with unstable angina at high risk for progression to myocardial infarct (Katus HA, et al. Circulation. 1991; 83:902-912; Hamm CW, et al. N Engl J Med. 1992; 327:146-150).

A number of other laboratory measurements have been employed in assessing risk of myocardial infarction. Elevated fibrinogen, >350mg/dL, in stroke patients may identify those at higher risk for subsequent cardiovascular events (Resck KL, et al. Ann Intern Med. 1992; 117:371-375). In one study elevated serum ferritin, >200 mcg/L, was associated with higher cardiac risk (Salonen JT, et al. Circulation. 1992; 86:803-811).

ACUTE PHASE REACTANTS

SPECIMEN: Red top tube, separate serum; except for fibrinogen-use blue (citrate) top tube, separate plasma.
REFERENCE RANGE: See below plus individual tests.
METHOD: See individual tests.
INTERPRETATION: Cellular injury and inflammation evoke synthesis of a heterogeneous group of proteins, the so-called acute phase reactants (Gewurz H. Hosp Pract. 1982; 67-81); some of these are listed in the next Table:

Acute Phase Reactants

Concentration may increase a hundredfold to a thousand fold:
- **C-Reactive Protein**: Normal plasma concentration: <0.5mg/dL; inflammation and host defense. Incremental changes are greater than for other acute phase reactants and hence the most sensitive. Synthesis induced by IL-1.

Concentration may increase twofold to threefold:
- **Alpha-1-Antitrypsin**: Normal plasma concentration: 200-400mg/dL; protease inhibitor especially neutrophil elastase.
- **Haptoglobin**: Normal plasma concentration: 40-180mg/dL; hemoglobin transport.
- **Fibrinogen**: Normal plasma concentration: 200-450mg/dL; coagulation.

Concentration may increase by about 50%:
- **Ceruloplasmin**: Normal plasma concentration: 15 to 60mg/dL; copper transport, free radical scavenger.
- **C-3 Complement**: Normal plasma concentration: 80-170mg/dL; modification of inflammation, host defense.

Similar indicators of inflammation include the ESR and plasma viscosity. Review: Brahn E, Scoville CD. "Biochemical Markers of Disease Activity." Bailliére's Clin Rheum. 1988; 2:153-183.

ADRENAL CORTEX PANEL

Tests of adrenal function and corticosteroid metabolism are usually performed to:

1) Evaluate adrenal insufficiency - primary (**Addison's**) or secondary to pituitary insufficiency
2) Evaluate hypercorticism - **Cushing's disease** (pituitary ACTH-producing adenoma), versus **Cushing's syndrome** (other etiologies)
3) Evaluate incidentally found adrenal adenomas
4) Evaluate congenital adrenal hyperplasia - inherited defects of steroid metabolism resulting in the **adrenogenital syndrome**
5) Evaluate hirsutism, menstrual irregularities, and infertility in women

Primary Adrenal Insufficiency: If you suspect that the patient has acute adrenal insufficiency, administer intravenous steroids immediately; then, initiate laboratory work-up. A basal plasma cortisol measurement is suggested as the initial test in evaluation of patients with suspected pituitary-adrenal insufficiency. A level above 300 nmol/L almost excludes ACTH-cortisol insufficiency while values below 100 nmol/L strongly suggest dysfunction. Use of a basal plasma cortisol screen is suggested to avoid more sophisticated and expensive testing (Hagg E, et al. Clin Endocrinol. 1987; 26:221-226).

Specific Tests for Primary Adrenal Insufficiency

Test	Primary Adrenal Insufficiency
Plasma Cortisol	Decreased
Urinary Free Cortisol	Decreased
Plasma ACTH	Increased
Urinary 17-OH CS, 17-KS, 17-KGS	Decreased
ACTH Infusion Test	Measure Blood Cortisol; No Change

17-OHCS = 17 Hydroxy Corticosteroids; 17-KS = 17 Ketosteroids; 17-KG = 17-Ketogenic Steroids.

An ACTH infusion test is used to distinguish primary adrenal insufficiency (Addison's) from secondary (pituitary). A Corticotrophin Releasing Hormone (CRH) stimulation test may be used to access pituitary reserve.

Other abnormal laboratory findings in primary adrenal insufficiency are shown in the next Table:

Other Abnormal Laboratory Findings in Adrenal Insufficiency

Test	Result
Serum Sodium (130mmol/liter or less)	Decreased
Potassium (6mmol/liter or more)	Increased
BUN (>25mg/dL)	Increased
Glucose (<70mg/dL)	Decreased
Hematocrit (>45%)	Increased
Eosinophils (normal 100 to 300/cu mm)	Increased

Cushing's Syndrome: The tests that are used to answer the question, "Does the patient have Cushing's syndrome?" are listed in the next Table:

Does the Patient Have Cushing's Syndrome? - Tests

Test	Result in Cushing's Syndrome
Plasma Cortisol	Increased
Diurnal Variation of Plasma Cortisol	Loss of Diurnal Variation
Urinary Free Cortisol (Best Test)	Increased
Urinary 17-OH CS, 17-KS, 17-KGS	Increased
Dexamethasone Screening Test (1mg at Midnight)	8 A.M. Cortisol greater than 5mcg/dL compatible with Cushing's
Dexamethasone Low Dose Test	Plasma Cortisol >5 mcg/dL at end of Second Day; 24 Hour Urinary 17-OH CS, 17-KGS, decreased

Measurement of urinary products of metabolism (17-OH CS, 17-KS, 17-KGS) have been largely replaced in many laboratories by measurement of plasma cortisol or urinary free cortisol. Reviews of diagnostic approaches in the patient with suspected Cushing's syndrome are given in: Carpenter PC. Endocrin Metab Clin N Amer. 1988; 17:445-472; Orth DN. N Engl J Med. 1991; 325:957-959. The diagnosis and treatment of Cushing's disease at Duke University for the years 1977-1982 have been reviewed (Burch WM. North Carolina Medical Journal. 1983; 44:293-296).

Other abnormal laboratory findings in Cushing's syndrome are shown in the next Table:

Abnormal Laboratory Findings in Cushing's Syndrome
Serum Sodium Increased and Serum Potassium Decreased
Hyperglycemia-Abnormal Glucose Tolerance Test in about 80% of Patients
Lymphocytes and Eosinophils Decreased
Tendency Toward Metabolic Alkalosis

Once a diagnosis of hypercorticism (Cushing's syndrome) has been made, the question is "What is the type of Cushing's syndrome?" There are a number of causes of Cushing's syndrome. These include: Cushing's disease (adrenal hyperplasia secondary to an ACTH producing pituitary adenoma), adrenal adenoma, adrenal carcinoma, ectopic ACTH producing non-pituitary neoplasm (often pulmonary small cell CA or carcinoid), iatrogenic or factitious administration of glucocorticoids, McCune-Albright Syndrome, micronodular adrenal disease, and gastric inhibitory polypeptide-dependant cortisol hypersecretion. This latter condition may be related to micronodular adrenal disease and was recently described as a form of hypercortisolemia that follows meals and that cannot be suppressed with dexamethasone (Lacroix A, et al. N Engl J Med. 1992; 327:974-980; Reznik Y, et al. ibid: 981-986). Laboratory findings similar to those found in mild hypercortisolism may be found in pseudo-Cushing's states. These include: severe obesity, alcoholism (chronic and withdrawal), depression, stress including surgery, renal failure, anorexia/bulimia, and glucocorticoid receptor resistance.

The usual problem is differentiation of bilateral adrenal hyperplasia secondary to a pituitary adenoma) from adrenal neoplasia, either adenoma or carcinoma. The tests that may be done are given in the next Table:

Differentiation of Adrenal Hyperplasia from Adenoma

Test	Hyperplasia	Adenoma
Plasma ACTH	Increased	Decreased
8mg/day of Dexamethasone	Suppression of Blood or Urinary Steroids	No Suppression of Blood or Urinary Steroids
Metyrapone Test	Increase in Urinary Steroids	No Change in Urinary Steroids
CRH Stimulation	Increase in Steroids	No Increase in Steroids

The results of these tests are not entirely specific. Over 50 percent of patients with Cushing's disease have ACTH levels within normal range (Besser GM, Edwards CRW. J Clin Endocrinol Metab. 1972; 1:451-490). ACTH-producing pituitary adenomas occasionally may not suppress with dexamethasone (Grossman AB, et al. Clin Endocrinol. 1988; 29:167-178). The metyrapone test may also not be conclusive. Measurement of petrosal sinus ACTH following CRH infusion is often useful in documenting and localizing pituitary adenomas (Perry RR, et al. Ann Surg. 1989; 210:59-68; Oldfield EC, et al. N Engl J Med. 1991; 325:897-905).

Ectopic ACTH-producing neoplasms such as pulmonary carcinoids may cause diagnostic problems. They sometimes mimic pituitary adenomas and suppress with dexamethasone. Likewise they are sometimes stimulated by CRH injection. CRH producing neoplasms are a rare cause of Cushing's syndrome. In such cases direct CRH measurements may be useful. Small adrenal cortical adenomas may be localized by imaging techniques employing CT or NP-59 scanning.

A low DHEA-S (<0.4mg/L or 1.1micromol/L) in a patient with Cushing's syndrome is said to be indicative of adrenal adenoma or pigmented nodular adrenal hyperplasia (Braithwaites SS, et al. Clin Chem. 1989; 35:2216-2219). Elevated DHEA-S concentrations are seen with adrenal carcinoma.

Incidentally Found Adrenal Masses: Use of contemporary imaging techniques has lead to the incidental discovery of numerous adrenal masses which typically have no clinical manifestations. Suggested protocols for the work-up of an incidentally discovered adrenal mass are given in: Ross NS, Aron DC. N Engl J Med. 1990; 323:1401-1405. Cautious removal of such masses is encouraged because these patients may have subclinical Cushing's syndrome and since the contralateral gland may be suppressed, removal may precipitate an Addisonian crisis (McLeod MK, et al. Ann Surg. 1990; 56:398-403; DeAtkine Jr D. NCMJ. 1992; 53:559-560; Huiras CM, et al. JAMA. 1989; 261:894-898). (cont)

Adrenal Cortex Panel (Cont)

Adrenal enlargement were found in approximately 150 patients who underwent abdominal CT (15,000 patients screened). The major causes of enlargement included metastases (20), non-functional adenomas (15), aldosterone secreting adenomas (4), pheochromocytomas (3), nodular hyperplasia due to MEN (3), primary carcinoma (4), corticosteroid producing adenoma (1) (Penn I, et al. Can J Surg. 1988; 31:105-109). In view of a fair number of pheochromocytomas and aldosterone secreting tumors that are incidentally discovered, it has been proposed that these patients undergo a screening test for pheochromocytoma (VMA, etc) and for primary hyperaldosteronism if hypertensive (check for low serum K^+)(Ross NS, Aron DC. N Engl J Med. 1990; 323:1401-1405). Screening for hypercortisolism is controversial (Ross and Aron, already cited) with some physicians finding a high rate of silent Cushing's syndrome and risk of post surgical Addisonian crisis following removal of the adenoma (Gross MD, et al. Ann Intern Med. 1988; 109:613-618; Caplan RH, Wickus GG. N Engl J Med. 1991; 324:1135).
Congenital Adrenal Hyperplasia results from the inheritance of enzyme defects which modify the metabolism of steroids, resulting in absence of cortisol production, release of feedback inhibition of ACTH production, elevation of ACTH, and resulting in adrenal cortical hyperplasia. In some instances, there is increase in sex steroids resulting in virilization of the female infant and precocious puberty in the male (adrenogenital syndrome). There are multiple enzymatic defects described but the principle ones are the 21-hydroxylase deficiency, 11-hydroxylase deficiency, and the 17-hydroxylase deficiency. Congenital adrenal hyperplasia has been reviewed in: New MI, Josso N. Endocrinol Metab Clin N Amer. 1988; 17:339-366; Cutler GB, Lane L. N Engl J Med. 1990; 323:1806-1813; and New MI. J Steroid Biochem. 1987; 27:1-7. The adrenal enzyme deficiencies are illustrated in the next Figure:

21-, 11- and 17-Hydroxylase Deficiencies in the Synthesis of Steroids

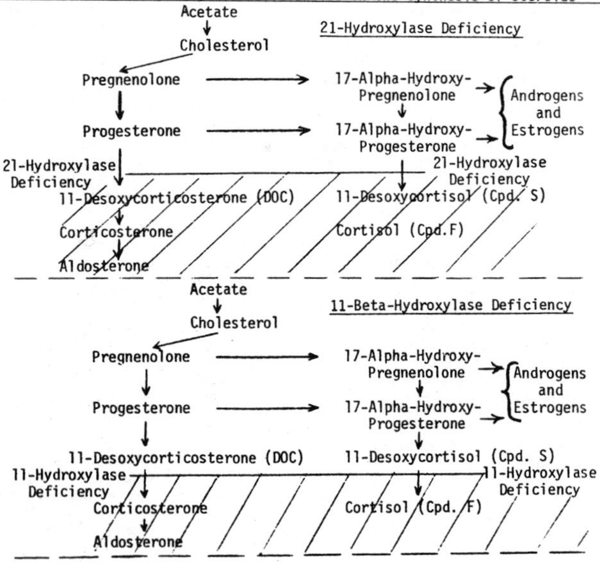

21-Hydroxylase Deficiency
Acetate
Cholesterol
Pregnenolone
17-Alpha-Hydroxy-Pregnenolone
Androgens and Estrogens
Progesterone
17-Alpha-Hydroxy-Progesterone
21-Hydroxylase Deficiency
21-Hydroxylase Deficiency
11-Desoxycorticosterone (DOC)
11-Desoxycortisol (Cpd. S)
Corticosterone
Cortisol (Cpd. F)
Aldosterone

11-Beta-Hydroxylase Deficiency
Acetate
Cholesterol
Pregnenolone
17-Alpha-Hydroxy-Pregnenolone
Androgens and Estrogens
Progesterone
17-Alpha-Hydroxy-Progesterone
11-Desoxycorticosterone (DOC)
11-Desoxycortisol (Cpd. S)
11-Hydroxylase Deficiency
11-Hydroxylase Deficiency
Corticosterone
Cortisol (Cpd. F)
Aldosterone

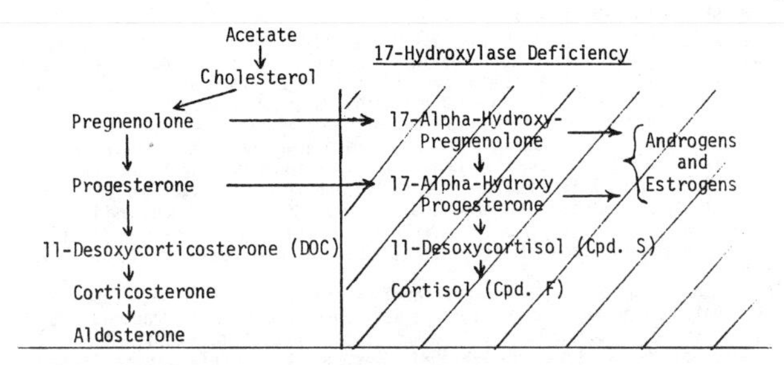

The classical findings associated with these syndromes are given below:

Sex Characteristics, Electrolytes and Blood Pressure in Adrenogenital Syndromes

Characteristic	Deficiency 21-Hydroxylase	11-Hydroxylase	17-Hydroxylase
Sex	Excess Sex Steroids	Excess Sex Steroids	Deficiency of Sex Steroids
Electrolytes:			
Sodium Retention	No	Yes	Yes
Sodium Loss	Yes	No	No
Potassium Loss	No	Yes	Yes
Hypertension	No	Yes	Yes

21-Hydroxylase Deficiency: The most common of the three conditions; the reported incidence varying from 1:5000 to 1:67000 births. Incidence is high in Native Americans in Alaska and now being screened for in newborns in 6 states. Since there is a lack of both aldosterone and of the mineralocorticoid 11-desoxycorticosterone (DOC), there is sodium loss, increased potassium, and hypotension. Blocking of the 21-hydroxylase pathway results in increased androgens. There are both mild and severe cases of the condition resulting in varying degrees of anomalous genitalia, hirsutism, lack of breast development, and failure to menstruate. A salt wasting form, a simple virilizing salt sparing form and a late onset non-classical form are recognized. First degree relatives of affected individuals often have polycystic ovaries (Hague WM, et al. Clin Endocrinol. 1990; 33:501-510). The phenotype may vary over time (Speiser PW, et al. N Engl J Med. 1991; 324:145-149). Laboratory findings include increased ACTH, urinary 17-ketosteroids, testosterone, androstenedione, progesterone, 17-OH progesterone, pregnanetriol, and serum potassium. Decreased levels of DOC, aldosterone, cortisol, urinary 17-OH corticosteroids, and serum sodium are found. Carriers are detected by giving ACTH and measuring increases in 17-OH progesterone or urinary pregnanetriol. Older testing relied on urinary pregnanetriol elevations as measured by 17-KGS. This is now being replaced by direct measurement of serum 17-OH progesterone.

11-Hydroxylase Deficiency: The metabolic block occurs at the hydroxylation of DOC step, resulting in an increase in this powerful mineralocorticoid. The patient has hypertension, sodium retention, and potassium depletion without increase in aldosterone. Sex steroids are increased. The female patient is masculinized. Prepubertal males may experience precocious development of the secondary sexual characteristic. 11-desoxycortisol is increased resulting in increased urinary 17-OH CS.

17-Hydroxylase Deficiency: The block in 17-hydroxylase activity results in a deficiency in sex steroids and all the 17-OH steroids. The picture is like that of primary hyperaldosteronism with increased sodium, decreased potassium, and metabolic alkalosis. Corticosterone is synthesized and this is able to partially feedback inhibit ACTH production like cortisol. (cont)

Adrenal Cortex Panel (Cont)

Evaluation of the Hirsute Woman: Sources of androgens include the adrenals, the gonads, or peripheral conversion of secreted precursors. DHEA-S is the major adrenal 17-KS while testosterone is the major circulating androgen. The differential diagnosis of hyperandrogenism includes polycystic ovary syndrome, congenital adrenal hyperplasia, "exaggerated adrenarche," prolactin, or GH-secreting adenomas, Cushing's syndrome, and virilizing tumors (Rosenfeld RL. Pediatr Clin N Amer. 1990; 37:1333-1358). The two basic tests in the workup of the patient with hirsutism are the DHEA-S and free testosterone, the former an indicator of an adrenal origin of the disorder (Bergfeld WF, Redmond GP. Dermatol Clin. 1987; 5:501-507). Occasionally a testosterone secreting tumor of the adrenal gland is noted (Matton JH, Phelan G. Surg Gynecol Obstet. 1987; 164:98-101).

In a survey of post-menarcheal women with hirsutism, late onset or attenuated 21-hydroxylase deficiency was found in only 1% of the patients (Motta P, et al. J Endocrinol Invest. 1988; 11:675-678; Chetkowski, et al. J Clin Endocrinol Metab. 1984; 58:595-598). However, others claim this disorder may account for 6-12% of hyperandrogenic hirsute women (Azziz R, et al. J Clin Endocrinol Metabol. 1989; 69:577-584; Chrousos GP, et al. Ann Intern Med. 1982; 96:143-148). The disorder is distinguished from polycystic ovary syndrome by the degree of elevation in 17OH-progesterone following ACTH stimulation.

ADRENOCORTICOTROPIC HORMONE (ACTH)

SPECIMEN: Fasting; collect blood in chilled plastic syringe containing heparin (1000 IU/mL). Transfer to ice-cooled green top tubes (heparin); place tubes in ice; centrifuge in a refrigerated centrifuge; freeze plasma.

REFERENCE RANGE: 0800 hours: 10-60pg/mL; 1600: 5-37pg/mL; 2400 hours: approximately 50% of A.M. value. Some advocate an afternoon ACTH measurement. In the patient with ACTH-dependent Cushing's syndrome: ACTH is above 4.5 pmol/L (>20 pg/mL) and cortisol >275 nmol/L (>10 g/dL). In the patient with ACTH-independent disease, cortisol values are similar but plasma ACTH is less than 1.1 pmol/L (<5 pg/mL) (Orth DN, N Engl J Med. 1991; 325:957-959). ACTH shows little variation with age or sex. To convert traditional units in pg/mL to international units in pmol/liter, multiply traditional units by 0.2202.

INTERPRETATION: ACTH blood levels are useful in differentiating the causes of Cushing's syndrome; in differentiating the causes of adrenal insufficiency; and in assisting in the diagnosis of the adrenogenital syndrome.

The secretion of ACTH in various conditions is illustrated in the next Figure (Catt KJ. The Lancet. 1970; 1275):

Secretion of ACTH in Adrenal Disorders

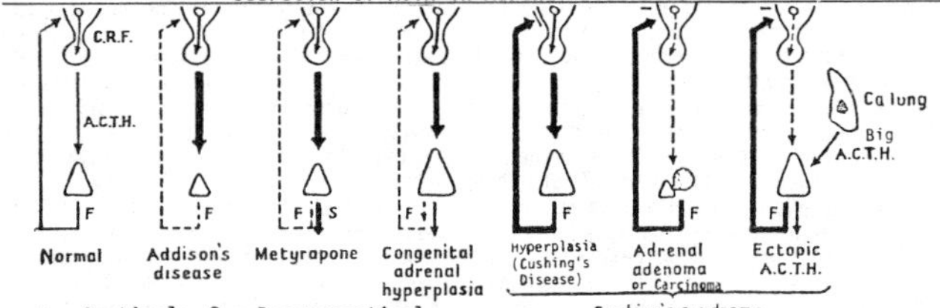

C.R.F.
A.C.T.H.
F
S
Ca lung
Big A.C.T.H.
Normal
Addison's disease
Metyrapone
Congenital adrenal hyperplasia
Hyperplasia (Cushing's Disease)
Adrenal adenoma or Carcinoma
Ectopic A.C.T.H.
Cushing's syndrome

F = Cortisol; S = Desoxycortisol

The change of ACTH and blood cortisol in adrenal disorders is summarized in the next Table:

Change in Plasma Cortisol and ACTH in Adrenal Disorders

Disorder	Plasma Cortisol	Plasma ACTH
Cushing's Syndrome:		
Adrenal Hyperplasia	Increased	Normal or Increased
Adrenal Adenoma or Carcinoma	Increased	Decreased
Nodular Hyperplasia	Increased	Decreased
Ectopic ACTH	Increased	Increased
Adrenal Insufficiency:		
Addison's Disease	Decreased	Increased
Pituitary Insufficiency	Decreased	Decreased
Other Conditions:		
Adrenogenital Syndromes	Decreased	Increased
Pseudo-Cushing's States	Increased	Increased

Normally, plasma ACTH undergoes a diurnal variation with levels in late P.M. being approximately 50% of the early A.M. levels. In Cushing's syndrome there is loss of diurnal variation of plasma ACTH.

In Cushing's syndrome due to adrenal hyperplasia (Cushing's disease/ pituitary-dependent adrenal hyperplasia) and ectopic ACTH, plasma cortisol is increased; however, over 50 percent of patients with Cushing's disease have ACTH levels within the normal range (Besser GM, et al. J Clin Endo Metab. 1972; 1:451-490). When Cushing's syndrome is due to adrenal adenoma or carcinoma or nodular hyperplasia, plasma cortisol is increased but ACTH is low. In multinodular hyperplasia of the adrenal, ACTH is usually low and cortisol production is often not suppressed by dexamethasone. These findings are now explained in part by the description of gastric inhibitory peptide stimulated cortisol production of multinodular hyperplastic adrenal lesions (Lacroix A, et al. N Engl J Med. 1992; 327:974-980; Reznik Y, et al. ibid: 981-986). In this condition, cortisol levels rise after food ingestion.

Pseudo-Cushing's states can be associated with some cases of severe obesity, alcoholism (chronic and in withdrawal), stress, renal failure, depression, primary glucocorticoid receptor resistance, and the McCune-Albright Syndrome.

There are three tests that modify the secretion of ACTH; these are the dexamethasone suppression test, the corticotrophin-releasing hormone (CRH) stimulation test, and the use of metyrapone. Dexamethasone is a fluorinated steroid which has about 30 times the potency of cortisol and suppresses ACTH in normal individuals; CRH stimulates release of ACTH from the pituitary; metyrapone inhibits the 11-hydroxylating enzyme responsible for the synthesis of cortisol from 11-deoxycortisol; these tests are discussed under the DEXAMETHASONE SUPPRESSION TEST, the ACTH: CORTICOTROPHIN RELEASING HORMONE (CRH) STIMULATION TEST, and METYRAPONE TEST.

In adrenal insufficiency due to Addison's disease, plasma ACTH is increased and plasma cortisol is decreased; ACTH infusion may be used to measure the functional reserve of the adrenal. In pituitary insufficiency, plasma ACTH may be decreased and plasma cortisol is decreased; metyrapone (metopirone) or CRH are used to measure pituitary reserve in patients with secondary adrenal insufficiency. Isolated ACTH deficiency is rare; there is only one report of isolated ACTH deficiency developing as a complication of post-partum hemorrhage (Stacpoole PW, et al. Am J Med. 1983; 74:905-908).

Pituitary ACTH reserve may be assessed by stimulation of ACTH release by stress testing with either glucagon or insulin administration (Little MD, et al. Clin Endocrinol. 1989; 31:527-533). Similar information may be obtained by a CRH stimulation test.

In the adrenogenital syndromes, the plasma cortisol is decreased and the plasma ACTH is increased. Various intermediates of steroid metabolism are increased.

ADRENOCORTICOTROPIC HORMONE (ACTH): CORTICOTROPIN-RELEASING HORMONE(CRH) STIMULATION TEST

SPECIMEN: Inject ovine corticotropin-releasing hormone at a dose of 100mcg or 1mcg per kg of body weight as an intravenous bolus injection. Collect blood at -15, 0, 5, 15, 30, 60, 90, 120, 150 and 180 minutes for measurement of ACTH and cortisol (see ACTH and CORTISOL assays for collection of blood specimens).
REFERENCE RANGE: See Interpretation
METHOD: RIA
INTERPRETATION: The corticotropin-releasing hormone (CRH) stimulation test may be useful in differentiating pituitary from ectopic causes of Cushing's syndrome. Following I.V. CRH, approximately 75% of patients with Cushing's disease (pituitary adenoma) develop a further increase in the already elevated levels of ACTH and cortisol (Grossman AB, et al. Clin Endocrinol. 1988; 29:167-178). Patients with ACTH-independent Cushing's syndrome or those with ectopic ACTH syndrome, who also have high basal plasma concentrations of ACTH and cortisol, usually have no ACTH or cortisol responses to CRH (Chrousos GP, et al. N Engl J Med. 1984: 310:622-626). However, a minority of approximately 10% of individuals with neoplasms producing ectopic ACTH will respond with a positive CRH stimulation test.

There are many potential uses of CRH: assessment of pituitary ACTH secretory capacity, differentiation between hypothalamic and pituitary causes of ACTH deficiency, evaluation of residual hypothalamic-pituitary functional abnormalities after various treatments for Cushing's syndrome, and possibly early detection of Cushing's disease or prediction of its recurrence (Orth DN. N Engl J Med. 1984; 310:649-651). The utility of this test has been increased by combining CRH administration with petrosal sinus catherization. ACTH samples drawn from both sinuses and a peripheral vein are compared. A central to peripheral ratio greater than 2:1 is indicative of Cushing's disease. Bilateral sampling may localize the abnormality to one side of the pituitary (Oldfield EH, et al. N Engl J Med. 1991; 325:897-905). The CRH test has also been used to evaluate pituitary suppression in patients on long term glucocorticoid therapy (Schlaghecke MD, et al. N Engl J Med. 1992; 326:226-230).

In a new approach, the CRH stimulation test has been combined with a course of dexamethasone suppression in order to distinguish between Cushing's syndrome and the mild hypercortisolism of pseudo-Cushing's sometimes seen in obesity, stress, alcoholism and anorexia (Yanovski JA, et al. JAMA. 1993; 269:2232-2238). The results of this initial study appear promising and the reader is referred to this article for details of the quite extensive protocol.

CRH-producing carcinoid tumors may occasionally result in Cushing's syndrome. In these instances, finding of an elevated blood CRH can lead to the correct diagnosis (Gerl H, et al. Dtsch Med Wochenschr. 1990; 115:332-336).

ADRENOCORTICOTROPIC HORMONE (ACTH) INFUSION TEST

Synonym: Cortrosyn Stimulation

PROCEDURE AND SPECIMENS (Screening Test): ACTH is given I.V. or I.M.; plasma cortisol and plasma aldosterone are measured. Obtain a blood specimen (green top vacutainer) for plasma cortisol and aldosterone assays (time 0); give 0.25mg of alpha 1-24 corticotropin (Cortrosyn) I.V. or I.M. (in patients less than 2 years old, give 0.125mg). An occasional allergic reaction has been reported to synthetic corticotropin. Collect blood at 30 and 60 minutes. The establishment of reference ranges for a number of steroid metabolites following ACTH administration has been reported (Grunwald K, et al. Gynecol-Endocrinol. 1990; 4:287-306). In many laboratories, a normal response at 60 minute poststimulation is two times the baseline value.

INTERPRETATION: This test is performed in order to establish the etiology of adrenal insufficiency: primary (Addison's) versus secondary (lack of pituitary ACTH). It may also be used in the evaluation of hirsute women with suspected defects in steroid metabolism (attenuated or late onset 21-hydrolase deficiency). The results for normal subjects and for patients with either primary Addison's disease or secondary adrenal insufficiency are given in the next Figures:

Effect of ACTH on Plasma Cortisol and Aldosterone in Normal Subjects and Patients with Primary (Addison's Disease) and Secondary Adrenal Insufficiency

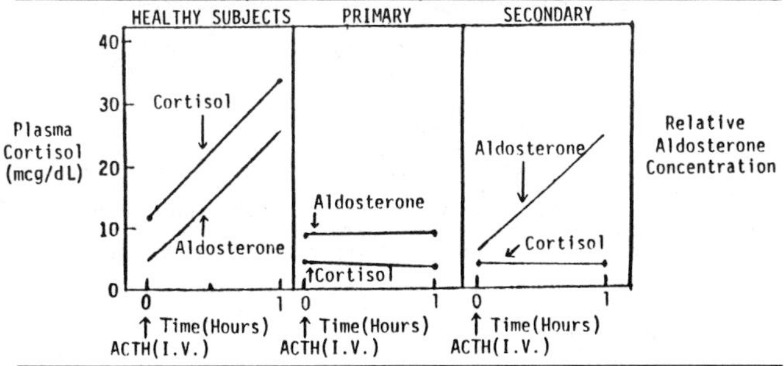

Normal: Following I.V. ACTH, a normal response is an increment of plasma cortisol greater than 10 mcg/dL and an increment of plasma aldosterone levels above control by at least 5 ng/dL.

Primary Adrenal Insufficiency: In primary adrenal insufficiency neither plasma cortisol nor plasma aldosterone increase following I.V, ACTH; the sensitivity is 100% and the specificity is 97% (Manu P, Howland T. Clin Chem. 1983; 29:1450-1451).

Secondary Adrenal Insufficiency: In secondary adrenal insufficiency, plasma cortisol shows no increase following I.V. ACTH; however, plasma aldosterone shows a normal increment.

21-Hydroxylase Deficiency: ACTH is given and serum 17OH-progesterone is measured. In severe 21-hydroxylase deficiency, 17OH-progesterone levels are >25ng/mL and an ACTH infusion test is probably not needed. The attenuated or late onset form is more problematic. 17OH-progesterone levels are typically 2-10ng/mL in both attenuated 21-hydroxylase deficiency, and polycystic ovary syndromes. Following ACTH stimulation, 17OH-progesterone levels are >10ng/mL while typically <3.5ng/mL in polycystic ovary syndromes (Burch Jr WM. NCMJ. 1992; 53:521). Testing should be done during the follicular phase (Gourmelen M, et al. Acta Endocrinol. 1979; 90:481-489; Azziz R, Zacur HA. J Clin Endocrinol Metabol. 1989; 69:557-584).

ALANINE AMINOTRANSFERASE (ALT)

Synonym: Serum Glutamic Pyruvate Transaminase (SGPT)
SPECIMEN: Red top tube, green (heparin) top tube may be used; serum stable for 3 days at room temp., 7 to 10 days at 4°C, and 6 months frozen.
Interference: excess hemolysis.
REFERENCE RANGE: Adult, 7-35U/liter; Newborn, 6 to 62U/liter decreasing to adult range in several months.
METHOD: Kinetic:

$$\text{Alpha-Ketoglutarate} + \text{Alanine} \xrightarrow{\text{ALT}} \text{Glutamate} + \text{Pyruvate}$$

$$\text{Pyruvate} + \text{NADH(340nm)} + H^{+} \xrightarrow{\text{LDH}} \text{Lactate} + NAD^{+}$$

INTERPRETATION: ALT is usually measured to evaluate liver disease. Distribution is limited to liver and kidney so that changes in level are a much more specific indicator of liver damage than the AST. Elevation of both AST and ALT is highly suggestive of liver disease. If the values are greater than 10x normal, severe hepatocellular damage is suspected. ALT is elevated in the diseases listed in the next Table:

Elevation of Alanine Aminotransferase (ALT)
Liver Disease
Congestive Heart Failure
Infectious Mononucleosis
Acute Myocardial Infarction
Acute Renal Infarction
Skeletal Muscle Disease
Acute Pancreatitis
Heparin Therapy

ALT is elevated in 60% of patients receiving heparin; the mean maximal increase is 3.6 times baseline value. The ALT returns to normal in 80% of patients after heparin therapy is discontinued and in 20% during therapy. In a retrospective study of samples of blood donors with an elevated ALT, 17% had detectable antibody to hepatitis C virus (Katlov WN, et al. Ann Intern Med. 1991; 115:882-884).

References: Rej R. "Aminotransferases in Disease," Clin Lab Med. 1989; 9:667-687; Landolphi DR, et al. "Interpretation of Liver Function Tests in Infectious Disease," Hosp Phys. Aug, 1990; 17-24.

ALBUMIN

SPECIMEN: Red top tube, separate serum.
REFERENCE RANGE: Range: Adult: 3.5-5.0g/dL; Newborn: 2.9-5.5g/dL; Child: 3.8-5.4g/dL. To convert traditional units in g/dL to international units in g/liter, multiply traditional units by 10.0.
METHOD: Bromcresol green(BCG) or bromcresol purple(BCP) dye binding method.
INTERPRETATION: Albumin is decreased in conditions listed in the next Table:

Decrease in Serum Albumin
Subacute and Chronic Debilitating Diseases
Liver Disease
Malabsorption
Malnutrition
Loss:
Renal Disease; Nephrotic Syndrome
Gastrointestinal Loss
Skin Loss
Third Degree Burns
Exfoliative Dermatitis
Dilution by I.V. Fluids
Genetic Variants such as Congenital Analbuminemia and Bis-Albuminemia

Albumin is synthesized in the liver; its half-life is 15 to 19 days. Serum albumin is dependent on albumin synthesis, degradation, loss, volume of distribution and exchange between intravascular and extravascular compartments.

The serum albumin level is considered a reliable index of severity and prognosis in patients with chronic hepatic disease; its main value lies in the follow-up therapy where improvement in the serum-albumin level is the best sign of successful medical treatment. Decreased serum albumin level within 48 hours of admission for an acute illness is a strong predictor of death, length of stay, and readmission in patients older than 40 years (Herrmann FR, et al. Arch Intern Med. 1992; 152:125-130).

Although serum albumin may identify a sicker population of patients, it is unclear whether albumin supplementation improves outcome. The indications for albumin supplementation in clinical practice are controversial (Kaminski MV, Haase TJ. Crit Care Clin. 1992; 8:311-321; Erstad BL, et al. Arch Intern Med. 1991; 151:901-911).

The only cause of increased albumin is dehydration; there is no naturally occurring hyperalbuminemia. Congenital variants such as bis-albuminemia may be diagnosed by high resolution serum protein electrophoresis.

ALBUMIN, URINE

Synonym: Microalbumin
SPECIMEN: First morning specimen preferred (Howey JE, et al. Clin Chem. 1987; 33:2034-2038). In some laboratories a timed specimen is used.
METHOD: RIA
REFERENCE RANGE: Normals: $\leq$ 11ng/min; At risk: above 30ng/min; or albumin: creatine ratio less than 2.6 (Cohen DL, et al. Diabetic Med. 1987; 4:437-440).
INTERPRETATION: Albuminuria is used as a means of accessing risk of nephropathy in patients with diabetes (Koven JC, et al. J Fam Pract. 1990; 31:505-510). In one study albuminuria was diagnosed in patients with an albumin of 15mg/L urine or an albumin: creatine ratio of 3.5mg/mmol (Microalbuminuria Collaborative Study Group. Diab Care. 1992; 15:495-501). Patients with Type I diabetes who excreted >0.2g urinary albumin/24 hours during the second decade of diabetes had a 100-fold greater risk of dying (macrovascular disease, end stage renal disease) as compared to those without proteinuria. This is also true of Type II diabetics (Andersen AR, et al. Diabetalogia. 1983; 25:496-501; Morgensen CE. N Engl J Med. 1984; 310:356-360). Cigarette smoking also increased the risk of albuminuria among subjects with type I diabetes (Chase HP, et al. JAMA. 1991; 265:614-617).

A variant of this test is to measure albumin excretion during exercise. Microalbuminuria during exercise is said to represent an early form of abnormal protein excretion (>30mcg/min) which can be reversed by improved glycemic control and may prevent progression of renal damage (Garg SK, et al. J Diabetic Complications. 1990; 4:154-158). Intensified treatment retards the microvascular complications of diabetes (Reichard P, et al. J Intern Med. 1991; 230:101-108).

ALCOHOL (see ETHANOL BLOOD)

ALCOHOLISM PANEL

The laboratory studies for patients with alcoholism are listed in the next Table (Wrenn KD, et al. Am J Med. 1991; 91:119-128):

Laboratory Studies - Alcoholism
Hematology
Complete Blood Count(CBC), Platelet Count
Blood Chemistry
Enzymes:
Aspartate Aminotransferase (AST or SGOT)
Alanine Aminotransferase (ALT or SGPT)
Bilirubin
Blood Urea Nitrogen(BUN)
Blood Glucose, Fasting
Hospital Setting:
Prothrombin Time
Amylase
Blood Alcohol Level
Lactate
Electrolytes, Na^+, K^+, Cl^-, CO_2 Content
Anion Gap
Arterial Blood Gases
Chemistry Panel to include:
Albumin
Total Protein
Inorg. Phosphorus
Magnesium
Creatine Phosphokinase(CPK)
Calcium
Uric Acid
Urinalysis
Microbiologic Studies:
Gonorrhea Culture:
Cervix, All Females
Urethral Discharge, Males
Pap Smear, Cervix, Females
Chest X-Ray

Complete Blood Count: Alcohol apparently has a direct toxic effect on bone marrow; low white cell counts and low platelets count may be observed in alcoholics. The changes in red blood cells that may be seen in alcoholics are given in the next Table:

Red Cell Changes in Alcoholics
Increased Mean Corpuscular Volume(MCV) without Folate Deficiency
Folate Deficiency
Sideroblastic Anemia
Iron Deficiency Anemia

Increased MCV without folate deficiency occurs in about 25 percent of alcoholics; these patients usually have a normoblastic bone marrow but occasionally have megaloblastic bone marrow. Alcohol has a direct toxic effect on the red cell.

The mechanism for low folate in chronic alcoholism is unknown but could be due to dietary deficiency, an effect of alcohol on folate absorption, a direct anti-folate effect of alcohol, enhanced utilization of folate as a co-factor in liver enzyme activity or a combination of these factors.

Sideroblastic anemia may be secondary to pyridoxine deficiency; pyridoxal phosphate is involved in the synthesis of delta-aminolevulinic acid, a precursor in the synthesis of heme.

Anemia in alcoholics may be due to iron deficiency; iron deficiency may occur from chronic blood loss following repeated hemorrhages from ruptured esophageal varices, gastritis or peptic ulcer.

Blood Chemistry: The sensitivity of laboratory tests in patients with alcoholism is given in the next Table (Morse RM, Hurt RD. JAMA. 1979; 242:2688-2690):

Sensitivity of Laboratory Tests in Patients with Alcoholism

Tests	Sensitivity(%)
Gamma Glutamyl Transpeptidase(GGTP)	63
Aspartate Aminotransferase (AST)	48
Triglycerides	22
Serum Alkaline Phosphatase (Alk. Phos.)	16
Bilirubin	13
Uric Acid	10

GGTP is the most sensitive enzyme used to detect liver damage from excessive alcohol intake. GGTP is situated on the smooth endoplasmic reticulum; alcohol causes microsomal proliferation (See GAMMA GLUTAMYL TRANSPEPTIDASE).

Increased serum triglycerides may represent endogenous release of triglycerides and increased hepatic synthesis. The lipoprotein electrophoretic pattern is that of secondary Type IV hyperlipidemia.

Hyperuricemia is secondary to increased lactate; lactate acid interferes with the renal excretion of uric acid. The hyperuricemia may precipitate attacks of gout.

Hypoglycemia may develop in those alcoholics who are drinking and not eating; liver glycogen stores may be depleted. Liver glycogen is depleted by a 72-hour fast.

Low BUN, (<5mg/dL) is found in advanced cirrhosis.

Arterial Blood Gases: Patients who have acute alcoholic intoxication, delirium tremens or chronic liver disease may develop disturbances as listed in the next Table:

Acid-Base Imbalance in Acute Alcoholic Intoxication

- Respiratory Alkalosis
- Metabolic Acidosis
  - Ketoacidosis
  - Lactic Acidosis
- Renal Tubular Acidosis

The most common disturbance is respiratory alkalosis.

Laboratory findings in alcoholic ketoacidosis are given in the next Table (Wrenn KD, et al. Am J Med. 1991; 91:119-128):

Laboratory Findings in Alcoholic Ketoacidosis

- Sodium Decreased (44%)
- Potassium Decreased (20%) or Increased (14%)
- Glucose Increased (34%) or Decreased (12%)
- Calcium Decreased (30%)
- Phosphate Increased (25%) or Decreased (25%)
- Magnesium Decreased (22%)
- Aspartate Aminotransferase (AST) Increased (89%)
- T. Bilirubin Increased (72%)
- Prothrombin Time Increased (11%)
- Albumin Decreased (4%)
- Total Protein Increased (22%)
- Amylase Increased (46%)
- Uric Acid Increased (63%)
- Creatine Phosphokinase Increased (60%)
- Hematology:
  - Hemoglobin Decreased (55%)
  - WBC Increased (30%) or Decreased (8%)
  - Platelets Decreased (17%)
- Acidosis:
  - Anion Gap Increased
  - Plasma Beta-Hydroxybutyrate Increased
  - Plasma Lactate Increased

(cont)

Alcoholism Panel (Cont)

Lactate: Lactic acidosis may occur in acute alcohol intoxication; lactic acidosis is usually not severe and is usually of short duration. The arterial pH is usually not less than 7.2 and the blood lactate level is usually not greater than five times the upper limit of normal. Spontaneous recovery usually occurs as alcohol is oxidized and the lactate ions are converted by the liver or kidney to glucose or are oxidized. The increased NADH/NAD ratio, secondary to the metabolism of alcohol, favors the conversion of pyruvate to lactate, as follows:

$$\text{Pyruvate} + \text{NADH} + H^{+} \xrightarrow[\text{(LDH)}]{\text{Lactate Dehydrogenase}} \text{Lactate} + \text{NAD}^{+}$$

Electrolytes: Serum potassium levels in patients during acute alcohol withdrawal are given in the next Table (Vetter WR, et al. Arch Intern Med. 1967; 120:536-541):

Serum Potassium Levels in Acute Alcohol Withdrawal

Range of Serum Potassium Concentrations	Percent of Patients Within Each Range
1.5 - 2.5	18
2.6 - 3.4	46
3.5 - 3.9	20
4.0 - 4.5	16

Eighteen percent of patients were severely depleted of potassium; 46 percent of patients were moderately depleted and 20 percent were low normal. There was a high incidence (45%) of cardiac arrhythmias in the moderately and severely potassium depleted groups and a low incidence (11%) in the normokalemic group.

Possible causes of potassium depletion include the following: low potassium content of inexpensive fortified wines and distilled liquors, poor dietary intake, and increased gastrointestinal loss of potassium through vomiting and diarrhea.

Other Chemistry Tests: The chemistry tests may include those tests listed in the next Table:

Chemistry Tests
Albumin
Total Protein
Inorg. Phosphorus
Magnesium
Creatine Phosphokinase(CPK)

Albumin: Albumin may be decreased in the alcoholic secondary to poor nutrition and/or liver disease. Albumin has a relatively slow turnover (half-life, 15 to 19 days) and does not respond rapidly to changes in diet or protein requirements.
Total Protein: Liver disease acts as a stimulus for the production of immunoglobulins; alcoholic liver disease tends to be associated with elevated IgA levels while other types of liver disease are not.
Inorganic Phosphorus: The hypophosphatemia in acute alcoholics is most likely caused by administration of glucose with a shift in phosphate from the extracellular space to the intracellular space.
Serum Magnesium: Hypomagnesemia is a common finding in chronic alcoholics, especially in delirium tremens. However, there is no clear clinical correlation between hypomagnesemia and the complications of alcoholism, e.g., delirium tremens or cirrhosis in the sense that alcoholics with normal serum magnesium levels may or may not develop these complications. The incidence of low serum magnesium in alcoholics is given in the next Table (Sullivan JF, et al. N.Y. Acad Sci. 1969; 162:947-962):

Incidence of Low Serum Magnesium

Diagnosis	Percentage with Low Magnesium
Alcoholism	30
Delirium Tremens	86
Post-Alcoholic Cirrhosis	37
Non-Alcoholic Patients	2

Creatine Phosphokinase(CPK): Alcohol has a direct toxic effect on skeletal muscle; increased serum CPK may be observed during hypophosphatemia during administration of glucose.

ALDOLASE

SPECIMEN: Red or green (heparin) top tube; separate serum or plasma immediately. Do not use hemolyzed specimen. Perform assay within 5 hours; otherwise, specimen can be stored for 5 days in the refrigerator or frozen for 2 weeks.

REFERENCE RANGE: 0-3 days: 4-24 IU/liter; 4 days-11 years: 2-12 IU/liter; 12 years-adult: 1-6 IU/liter.

METHOD: Aldolase is an enzyme occurring in the glycolytic pathway, glucose to pyruvate; it catalyses the conversion of fructose-1,6-diphosphate into dihydroxyacetone phosphate and glyceraldehyde-3-phosphate. The products are assayed using auxiliary and indicator reactions.

INTERPRETATION: Aldolase is increased in the serum in conditions listed in the next Table (Tormey WP. BJCP. 1990; 44(1):582-584):

Conditions Associated with Increased Serum Aldolase	
Skeletal Muscle Conditions:	
Duchenne's Muscular Dystrophy	Polymyalgia Rheumatica
Dermatomyositis	Fibromyalgia
Polymyositis	Viral Myositis
Limb-Girdle Dystrophy	Post-viral Myalgia
Myotonia Dystrophy	Non-specific Myalgia
Rhabdomyolysis	
Liver Disease:	
Hepatic Necrosis from any Cause, e.g., Viral Hepatitis	
Hepatotoxic Drugs	
Carcinoma Metastatic to the Liver	
Other Conditions:	
Acute Myocardial Infarction	Sensorimotor Neuropathy
Acute Pancreatitis	Anterior Horn Cell Disease
Prostatic Tumors	Hypothyroidism
Other Neoplasms (With Metastasis)	Hyperthyroidism (On Carbimazole)
Delirium Tremens	Acute Renal Failure
Injections of Drugs	Chronic Renal Failure
Rheumatoid Arthritis	Meningitis
Systemic Lupus Erythematosus	Peripheral Vascular Disease

The assay of creatine phosphokinase (CPK) has replaced that of aldolase for the evaluation of patients with skeletal muscle disorders; if CPK is within the normal range, aldolase should be measured. In acute viral hepatitis, the serum aldolase level tends to parallel the change in alanine aminotransferase (ALT).

ALDOSTERONE, SERUM

SPECIMEN: Fasting, green (heparin) top tube, separate plasma; or red top tube, separate serum; place specimen on ice and deliver to laboratory immediately; spin specimen and place 3mL of the plasma or serum in plastic vial and freeze (Renin often obtained at same time - see RENIN ACTIVITY, PLASMA).
REFERENCE RANGE: (on ad lib sodium intake) Supine: 3-10ng/dL; obtain specimen from recumbent patient in early A.M. prior to time that patient arises. Upright: (2-5 times supine) 5-30ng/dL; obtain specimen at 9:00 A.M. after patient is upright for two hours. On low sodium diet, serum aldosterone is 2-5 fold increase over ad lib sodium diet. Aldosterone is decreased in patients on heparin therapy. To convert traditional units in ng/dL to international units in pmol/liter, multiply traditional units by 27.74.
METHOD: RIA
INTERPRETATION: The usual reason for determining serum aldosterone is in the work-up of patients with hypertension for possible primary aldosteronism. The relative levels in serum and urine of aldosterone and renin in patients with primary and secondary hyperaldosteronism are shown in the next Table:

Aldosterone and Renin Levels in Primary and Secondary Aldosteronism

Aldosteronism	Aldosterone	Renin
Primary (60% Adenoma)	↑	↓
Secondary	↑	↑

The triad of polyuria, hypokalemia, hypertension tends to occur in primary aldosteronism; patients tend to have hypokalemic alkalosis. Clinical settings suggestive of primary hyperaldosteronism include spontaneous or diuretic-induced hypokalemia in a hypertensive patient, unexplained polyuria or nocturia in a hypertensive patient, and hypertension refractory to conventional therapy (Kannan CR. Disease-a-Month. 1988; 34(10):603-674).

The combination of elevated aldosterone secretory rate and low plasma renin activity is practically pathognomonic of primary aldosteronism.

Tests used to confirm the diagnosis of primary aldosteronism include the following (Melby JC. Endocrinol and Metab Clin N Amer. 1991; 20(2):247-255; Young WF, et al. Mayo Clin Proc. 1990; 65:96-110):

Tests Used to Detect Primary Aldosteronism

Test	Expected Value in Primary Aldosteronism	Remarks
Serum Sodium	Increased	Valueless
Serum Potassium	Decreased	False positives and false negatives
Hypokalemia with Inappropriate Kaliuresis	Urinary potassium >30mmol/24hr	Diuretic therapy is the most common cause of hypokalemia in patients with hypertension
Measurement of Plasma Renin Activity(PRA) after upright position. Diuretic (furosemide) added in some protocols	Low <1ng/mL/hr (Melby) <3ng/mL/hr (Young and Klee)	20% of patients with essential hypertensive have low PRA
Plasma Aldosterone (supine) at 8 A.M.	>15ng/dL	Diagnostic value debated
Urinary Aldosterone	>10-14mcg/24 hr	Must be adjusted for age
Urinary Aldosterone Metabolites:		
18-Monoglucuronide	>20mcg/24 hr	
Tetrahydroaldosterone	>65mcg/24 hr	
Measurement of Plasma Aldosterone concentration after I.V. of 2 L of normal saline over four hours	Increased >10ng/dL	False positive rate high in patients with bilateral idiopathic hyperaldosteronism

Tests Used to Detect Primary Aldosteronism (Cont.)		
Measurement of Plasma Aldosterone concentration after angiotensin-1-converting enzyme (captopril) inhibition	Increased >15ng/dL	
Measurement of Ratio of Plasma Aldosterone to Plasma Renin Activity (PAC/PRA)	Ratio above 20	Outpatient screening test; convenient, economical reliable outpatient screening test

Ratio of Plasma Aldosterone to Plasma Renin Activity (PAC/PRA Ratio): (Hiramatsu K, et al. Arch Intern Med. 1981; 141:1589-1593; Carey RM. Arch Intern Med. 1981; 141:1594; Muratani H, et al. Amer Heart J. 1986; 112(2):361-367): The PAC/PRA ratio is a convenient, economical and reliable outpatient screening test for primary aldosteronism. Unlike PRA and aldosterone concentration, the PAC/PRA ratio is not influenced by variations of sodium intake, diuretics, total body deficit of potassium or diurnal variation. The test is done as follows: Collect blood for the assay of plasma renin activity in EDTA; place immediately in plastic vial. Collect blood for assay of aldosterone in red stoppered vacutainer; separate serum and freeze serum. Aldosterone - (in ng/dL) PRA (in ng/mL/hr) ratio greater than 20 suggests primary aldosteronism.

The diagnosis of primary aldosteronism is based on the demonstration of hypokalemia with inappropriate kaliuresis, decreased plasma renin activity, nonsuppressible aldosterone level, and normal glucocorticoid levels (Young WF, et al. Mayo Clin Proc. 1990; 65:96-119). Different approaches to the diagnosis of primary aldosteronism have been described (Young WF, already cited. Melby JC, already cited).

In Bartter's syndrome, there is hyperplasia of the juxta-glomerular apparatus, increased renin, increased aldosterone, hypokalemia but without hypertension.

Four subgroups of patients with primary aldosteronism have been characterized; the causes and percent incidence are shown in the next Table (Carey RB. Arch Intern Med. 1981; 141:1594):

Subgroups of Patients with Primary Aldosteronism

Subgroup	Percent
Aldosterone Producing Adenoma (APA)	60
Bilateral Adrenal Hyperplasia (Idiopathic Hyperaldosteronism [IHA])	40
Glucocorticoid-Suppressible Hyperaldosteronism	Rare
Aldosterone-Producing Adrenocortical Carcinoma	Rare

Only patients with APA predictably respond to surgical therapy.

Hypoaldosteronism: Aldosteronism deficiency with intact glucocorticoid reserve is uncommon, causes of selective hypoaldosteronism are given in the following Table (Kannan CR. Disease-a-Month. 1988; 34(10):603-674):

Causes of Selective Aldosterone Deficiency

- Hyporeninemic Hypoaldosteronism
  - Interstitial Renal Disease
  - Decreased Prostacyclin I_2
  - Sympathetic Dysfunction
  - Impaired Prerenin to Renin Conversion
- Hyperreninemic Hypoaldosteronism
  - Addison's Disease
  - Enzyme Defects in Aldosterone Synthesis
  - Acquired Defects in Zona Glomerulosa
  - Critical Illness
- Pseudohypoaldosteronism
  - Tubular Resistance
- Miscellaneous
  - Heparin

The most common cause of selective hypoaldosteronism is hyporeninremic hypoaldosteronism (HHA). The components of this syndrome, include hyporeninemia, hypoaldosteronism, intact glucocorticoid reserve, hyperkalemia, metabolic acidosis with type IV RTA, and renal insufficiency. Typical patients have chronic renal failure, diabetes mellitus, and advanced age.

ALDOSTERONE, URINE

SPECIMEN: 24 hour urine; place 10 grams of boric acid as preservative into container prior to collection. Instruct the patient to void at 8:00 A.M. and discard the specimen. Then, collect all urine including the 8:00 A.M. specimen at the end of the 24 hour collection period; refrigerate urine during collection. Measure 24 hour volume and record on test request form. Mix urine and take the pH. Adjust the pH to 4.0 with glacial acetic acid. Obtain a 50mL aliquot and freeze.

REFERENCE RANGE: Concentration in urine is dependent on sodium intake; low sodium (10mmols): 20-80 mcg/24 hours; normal sodium diet (100-200mmols): 3-19 mcg/24 hours; high sodium diet (more than 200mmols): 2-12 mcg/24 hours; Prepubertal Children: normal sodium intake: 1-8 mcg/24 hours or 4-22mcg/g creatinine. To convert traditional units in mcg/24 hours to international units in nmol/day, multiply traditional units by 2.774.

METHOD: RIA

INTERPRETATION: The usual reason for determining urine aldosterone is in the work-up of patients with hypertension for possible primary aldosteronism. Elevations of the concentration of aldosterone in the urine are parallel to those of serum. For interpretation see ALDOSTERONE, SERUM.

ALKALINE PHOSPHATASE

SPECIMEN: Red top tube, separate serum; or green (heparin) top tube, separate plasma; do not freeze. Do not use samples anticoagulated with oxalate or citrate. Fasted samples preferred, meals promote increases in activity. Alkaline phosphatase activity increases on standing.

REFERENCE RANGE: Adult values: 39-117U/L.

The alkaline phosphatases (APs) vary with age and pregnancy. With age, the APs are elevated in growing children, decrease to an adult level, and then increase slightly in older people. The change of AP with age is shown in the next Figure:

Change of Alkaline Phosphatase with Age

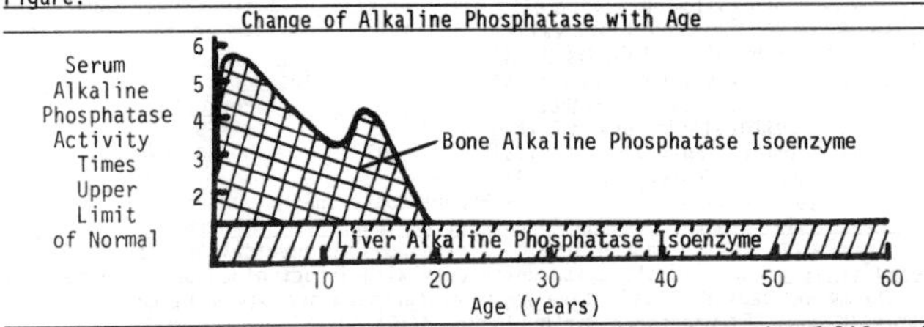

Alkaline phosphatase rises rapidly during the first few weeks of life and reaches values five to six times normal in about one month. Then, it decreases slowly until puberty and then there is another increase at puberty. This is followed by a decrease to adult level at 16-20 years of age. It remains constant until about age 40 to 50 and then it begins to rise to about ten percent above the adult level in older individuals.

The 95th percentile values for serum AP for males and females are given in the next Table (Meites S, Editor. Pediatric Clinical Chemistry. 1981; 72):

95th Percentile Limits for Alkaline Phosphatase

Age Group (Years)	Males	Females
Up to 1	477	442
1-2	333	415
2-5	291	341
5-7	291	341
7-10	316	354
10-13	362	393
13-15	-	322
15-17	365	-
15-18	-	122
17-19	176	-
18 and over	-	86
19 and over	126	-

During pregnancy AP values elevate to two times normal level, rise to three times normal level during labor and return to normal in three to four weeks. This increase in AP is principally derived from the placenta.

METHOD: Enzymatic/colorimetric.

INTERPRETATION: AP is principally measured to evaluate diseases of liver or bone. Conditions associated with elevated serum AP are listed in the next Table:

Conditions Associated with Elevated Serum Alkaline Phosphatase

Liver	(Osteoblastic) Bone Conditions
Cholestasis; Intrahepatic and Extrahepatic	Paget's Disease
Liver Disease	Metastatic Malignancy (Osteoblastic)
Congestive Heart Failure	Secondary Hyperparathyroidism
Infectious Mononucleosis	Renal Disease
Malignancy with Liver Metastasis	Osteomalacia
Acute Pancreatitis (occ.)	Malabsorption
Cytomegalovirus Infections	Rickets
	Primary Hyperparathyroidism (Late)
	Healing Fractures
	Hyperthyroidism

Non-Liver, Non-Bone:
- Pregnancy
- Sarcoidosis
- Gynecological Malignancies
- Amyloidosis
- Ulcerative Colitis
- Perforation of Bowel
- Pulmonary and Myocardial Infarction (Late)
- Sepsis

In a series of 110 patients with an isolated elevated serum alkaline phosphatase, the most common diagnoses were malignancy and acute or chronic congestive heart failure. Among the other patients with persistent elevations over a 12 month follow-up period diagnoses included phenytoin (dilantin) therapy (5 patients), congestive heart failure, hepatobiliary disease and metabolic bone disease (Lieberman D, Phillips D. J Clin Gastroenterol. 1990; 12:415-419).

Liver Disease: In liver disease, APs are elevated in both obstructive and hepatocellular processes. Highest levels of AP are found in primary biliary cirrhosis, cholestasis due to choledocholithiasis, cancer of the head of pancreas, drugs, and extensive metastatic or primary malignant disease of the liver. In partial obstructive disease, or in infiltrative disease the level of AP increases greater than the increase in bilirubin. In complete obstruction both increase. AP is slightly to moderately elevated in viral hepatitis, infectious mononucleosis and cirrhosis, both portal and post-necrotic. By contrast, in conditions with extensive hepatocellular damage, bilirubin is markedly elevated while AP is only slightly increased.

Bone Disease: The elevation of AP in bone conditions reflects increased osteoblastic activity of bone. In bone disease, AP is markedly elevated in osteitis deformans; moderately elevated in rickets, osteomalacia, hyperparathyroidism and metastatic bone disease; slightly elevated in healing fractures. Patients with chronic renal failure develop osteomalacia and elevated serum AP; during therapy with calcitriol, serum AP levels usually decrease, paralleling symptomatic response. Measurement of pretreatment serum alkaline phosphatase in patients with malignancies of bone may help in predicting treatment outcome with regard to relapse rate and mean survival (Bacci G, et al. Cancer. 1993; 71:1224-1230).

Other Conditions: Other conditions associated with an elevated AP include pregnancy (placental origin), various gynecological malignancies (Regan isoenzyme), GI conditions including inflammatory bowel disease, infarcts and ulcers (intestinal isoenzyme), organizing infarcts of heart and lung (AP from endothelium of granulation tissue), and infarcts of kidney, spleen and pancreas (AP isoenzymes from these respective tissues). Marked elevation of serum AP (4 to 10 times normal) and gamma glutamyl transpeptidase (GGTP) (5 to 13 times normal), little elevation of bilirubin (normal to 4 times normal) and minimal elevation of serum transaminases were found in patients with major systemic infections; the mechanism responsible for the marked elevation of AP and GGTP in infections is unknown (Fang MH, et al. Gastroenterology. 1980; 78:592-597).

(cont)

Alkaline Phosphatase (Cont)

Origin of Elevated AP: A common problem is to determine the source of an elevated AP- usually liver versus bone. The two approaches employed are to: 1) determine AP isoenzyme fractions, and 2) measure other enzymes that rise with liver disease but not bone disease. The latter indirect approach is the one most commonly followed and consists in measurement of gamma glutamyl transpeptidase, 5'-nucleotidase, or leucine amino-peptidase. Serum AP is elevated in 78 percent of patients with cancer and with metastases to the liver; gamma glutamyl transpeptidase (GGTP) is elevated in 97.3 percent of patients with metastases to the liver (Kim NK, et al. Clin Chem. 1977; 23:2034).

Decreases in AP: Conditions associated with decrease in serum alkaline phosphatase are malnutrition, hypophosphatasia, hypothyroidism, and pernicious anemia.

ALKALINE PHOSPHATASE ISOENZYMES

SPECIMEN: Red top tube, separate serum.

REFERENCE RANGE: There are at least five alkaline phosphatase (AP) isoenzymes; these are biliary, liver, bone, placental or Reagan isoenzyme and intestinal isoenzyme. Alkaline phosphatase isoenzymes are found in serum at different times in life as is shown in the next Table:

Predominant Alkaline Phosphatase Isoenzymes

Children	Adult	Pregnant Female
Bone	Liver Only	Liver
Liver		Placental

METHOD: Polyacrylamide gel electrophoresis. Other methods previously used to determine AP isoenzyme content include heat stability (bone AP relatively heat sensitive - "bone burns"), stability to urea, and differential inhibition of various isoenzymes by various amino acid derivatives or EDTA.

INTERPRETATION: AP isoenzymes are determined in order to differentiate the source of an elevated AP, usually due to bone or liver disease. AP isoenzyme fractionation has also been used to some extent in the evaluation of tumor markers (Reagan isoenzyme).

Liver Alkaline Phosphatase Isoenzyme: The activity of this isoenzyme remains relatively constant throughout our lifespan; it is the only isoenzyme that is present in the serum of all - young and old. One of the common clinical problems is to identify the source of an elevated AP. Elevation due to liver disease may be confirmed by documenting an elevation in the serum gamma glutamyl transpeptidase.

Bone Alkaline Phosphatase Isoenzyme: This isoenzyme is derived from osteoblasts of bone; it is found during the growth periods when bone osteoblastic activity is greatest. The increase of bone AP during the growth period is given in the previous section on ALKALINE PHOSPHATASE.

Placental Alkaline Phosphatase Isoenzyme: Alkaline phosphatase is produced by the placenta and appears in maternal blood. The change in AP during pregnancy and in the post-partum period is shown in the next Figure:

Change in Total Serum Alkaline Phosphatase in Pregnancy and in the Post-Partum Period

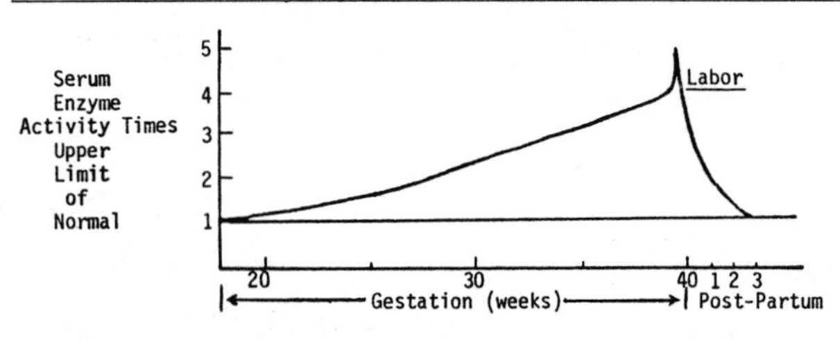

Measurements of maternal serum placental alkaline phosphatase are of no value in predicting fetal distress, the APGAR score at birth, or dysmaturity.

"Reagan" Alkaline Phosphatase Isoenzyme: The Reagan AP isoenzyme is an example of ectopic production of enzymes by cancers; it may be similar to ectopic production of hormones. The chemical, physical, and immunological properties of the Reagan AP isoenzyme are identical to that of the placental alkaline phosphatase isoenzyme.
Intestinal Alkaline Phosphatase Isoenzyme: The activity of this enzyme in the serum depends on three factors; these are blood group, secretory status and influence of a fatty meal. The relationship of serum intestinal alkaline phosphatase isoenzyme activity and blood group is as follows: O> B> AB> A. The serum activity of AP is higher in patients who are secretors. The activity is increased especially two to three hours following a fatty meal. The normal level of serum intestinal AP is up to 25 percent of the total AP.

ALPHA-1-ANTITRYPSIN

SPECIMEN: Red top tube; separate serum.
REFERENCE RANGE: 150-350mg/dL. To convert traditional units in mg/dL to international units in g/L, multiply traditional units by 0.01 (1.5-3.5g/L).
METHOD: Radial immunodiffusion.
INTERPRETATION: Alpha-1-antitrypsin deficiency is associated with chronic obstructive lung disease (emphysema) and less frequently with hepatic cirrhosis in infants and respiratory distress of the newborn. Alpha-1-antitrypsin is a protease inhibitor; it inhibits the action of the naturally occurring proteolytic enzymes, trypsin, elastase, chymotrypsin, collagenase, leukocytic proteases, plasmin and thrombin, which may be released during inflammatory reactions in the lung. In the absence of alpha-1-antitrypsin, inhibition of these enzymes does not take place, and these enzymes may digest pulmonary parenchyma.

A protocol for laboratory investigation of alpha-1-antitrypsin deficiency is given in the next Table:

Protocol for Laboratory Investigation of Alpha-1-Antitrypsin Deficiency
Serum Protein Electrophoresis; Carefully Examine Alpha-1-Band
Assay Alpha-1-Antitrypsin when Alpha-1-Band is Decreased or Absent
Perform Phenotyping if Indicated

Alpha-1-antitrypsin deficiency is inherited as an autosomal trait. The frequency of the heterozygote is about one in twenty and the frequency of the homozygote is about one in 1600.

Acquired decrease in alpha-1-antitrypsin occurs in the following conditions: nephrosis, malnutrition and cachexia.

Increase in alpha-1-antitrypsin occurs as an acute phase response to tissue necrosis and inflammation. (See ACUTE PHASE REACTANTS).

Alpha-1-antitrypsin may be spuriously elevated in heterozygous deficient patients during concurrent infection, pregnancy, estrogen therapy, steroid therapy and with exercise. Elevated levels may be seen in rheumatoid arthritis, bacterial infections, vasculitis and carcinomatosis.

ALPHA-1-ANTITRYPSIN PHENOTYPING

SPECIMEN: Red top tube, separate serum.
REFERENCE RANGE: Absence of homozygous ZZ; alpha-1-antitrypsin is usually associated with the Z state but other genotypes such as SS and SZ are also affected.
METHOD: Isoelectric focusing in polyacrylamide gels, Nephelometry.
INTERPRETATION: The Z allele is associated with cirrhosis or pulmonary emphysema. The frequency of protease inhibitor (Pc) (alpha-1-antitrypsin) types in the population of England and Wales is given in the next Table (Cook PJL. Postgrad Med J. 1974; 50:362-364):

Frequency of Protease Inhibitor (Pi) Types in England and Wales

Type	Frequency (%)
MM	86
MS	9
MZ	3 (Partial Enzyme Deficiency)
SS	0.25
SZ	0.2
ZZ	0.029 (2.9 out of 10,000)(Severe Enzyme Deficiency)

86% of the population is homozygote for MM and is normal. Genetic variants with at least one protease inhibitor Z (PiZ) allele have been associated with liver disease (Hodges JR, et al. N Engl J Med. 1981; 304:557-560). The ZZ allele is associated with severe alpha-1-antitrypsin deficiency; the MZ allele is associated with partial enzyme deficiency, the S allele may also be associated with deficiency.

The threshold protective level of alpha-1-antitrypsin is 11 micromol/L (approx. 80mg/dL). A highly purified alpha-1-antitrypsin standard has been developed to establish ranges for alpha-1-antitrypsin phenotypes as given in the following Table (Buist AS, et al. Am Rev Respir Dis. 1989; 140:1494-1497; Brantly ML, et al. Chest. 1991; 100:703-708):

Alpha-1-Antitrypsin Phenotype, Levels, and Risk of Emphysema

Phenotype	True Serum* Level(micromol/L)	Common Commercial Serum Levels(mg/dL)	Emphysema Risk
MM	20-53	150-350	Baseline
MS	18-52		Baseline
MZ	15-42	90-210	Baseline
SS	20-48	100-140	Baseline
SZ	10-23	75-120	Mild increase
ZZ	3.4-7.0	20-45	High risk
Null-Null	0	0	Very high risk

*5-95 percentile. The commercial standard overestimates the true level by 30-40%; the true serum level is more accurate because a pure alpha-1-antitrypsin standard is used (Ad Hoc Committee on Alpha-1-Antitrypsin Replacement Therapy of the Standards Committee, Canadian Thoracic Society. Can Med Assoc J. 1992; 146:841-844).

Replacement therapy with weekly infusions of human alpha-1-antitrypsin (Prolastin) is available and is generally reserved for patients 18 years of age or older with severe alpha-1-antitrypsin deficiency, serum levels less than 11 micromol/L and some abnormality of lung function (Buist AS, et al. Am Rev Respir Dis. 1989; 140:1494-1497).

ALPHA-FETOPROTEIN(AFP)

SPECIMEN: Red top tube, separate serum.
REFERENCE RANGE: <25ng/mL. To convert traditional units in ng/mL to international units in mcg/liter, multiply traditional units by 1.00.
METHOD: RIA, Enzyme Immunoassay.
INTERPRETATION: Serum alpha-fetoprotein level is useful in following the response to therapy of patients with hepatocellular carcinoma, non-seminiferous germ cell testicular tumors, certain ovarian tumors and in differential diagnosis of neonatal hepatitis versus biliary atresia in newborns and in maternal blood and amniotic fluid in certain fetal abnormalities (Adams MJ, et al. Am J Obstet and Gynecol. 1984; 148:241-254).
Hepatocellular Carcinoma: The blood level of AFP in 70% of patients with hepatocellular carcinoma is higher than 500ng/mL (upper limit <25ng/mL).
Testicular Tumors: Alpha-fetoprotein and human chorionic gonadotropin (HCG) in testicular tumors are given in the next Table:

Histopathologic Classification of Testicular Tumors, Alpha-Fetoprotein and Human Chorionic Gonadotropin (HCG)

Testicular Tumor	Alpha-FP	HCG	Frequency Of Markers	Other Markers
Germinal Cell Origin:				
Embryonal Carcinoma	+	-		
Embryonal Carcinoma with STGC	+	+		
Adult Type	+	+	88%	
Infantile Type (Yolk-Sac Tumor or Endodermal Sinus Tumor)	+	-		
Choriocarcinoma	-	+	100%	
Teratoma	-	-		
Compound Tumors				
Embryonal Carcinoma with Teratoma	+	-		
Any Combination of the Above	+	+	(if STGC present)	
Seminoma	-	-		NSE
Seminoma with STGC	-	+	7.7%	FSH
Nongerminal Cell Origin:				
Interstitial (Leydig) Cell Tumor; Sertoli Cell Tumor; Gonadal-Stroma Tumor; Compound Tumors				

STGC = Syncytiotrophoblastic Giant Cell; NSE = Neuron-Specific Enolase

The non-seminomatous germ cell tumors are associated with an elevated serum human chorionic gonadotropin (HCG) or alpha-fetoprotein (AFP) or both. Only 7.7% of the patients with seminomas showed elevated tumor marker in the serum and these were associated with STGC. Tumors of nongerminal cell origin were not associated with tumor markers. Anderson et al (Ann Intern Med. 1979; 90:373-385) reported that 85% to 90% of patients with nonseminomatous germ cell tumors had circulating tumor markers, e.g., AFP and/or HCG.
Ovarian Tumors: Alpha-fetoprotein is found in the serum of patients with germ cell tumors of the ovary; these are listed in the next Table:

Ovarian Tumors Secreting Alpha-Fetoprotein
Yolk Sac Tumor (Endodermal Sinus Tumor)
Embryonal Carcinoma
Mixed Germ Cell Tumor

Neonatal Hepatitis versus Biliary Atresia in Newborns: Serum alpha-fetoprotein is elevated in neonatal hepatitis but is usually normal in biliary atresia.
Alpha-Fetoprotein and Fetal Abnormalities: Alpha-fetoprotein increase or decrease is used to detect various fetal abnormalities. Alpha-Fetoprotein is the principal plasma protein of the fetus; this protein is detected in the amniotic fluid and maternal serum. Maternal serum alpha-fetoprotein is used as a screening test for congenital anomalies, chromosomal abnormalities, and adverse pregnancy outcomes.

Alpha Fetoprotein (Cont)

Conditions associated with an increase in alpha-fetoprotein (AFP) in maternal serum are listed in the following Table (Cunningham FG, Gilstrap LC. N Engl J Med. 1991; 325:55-57; Katz VL, et al. Obstet Gynecol Surv. 1990; 45:719-726; Burton BK. Obstet Gynecol. 1988; 72:709-713):

Conditions Associated with an Increase in Alpha-Fetoprotein in Maternal Serum
Myelomeningocele
Anencephaly
Omphalocele and Gastroschisis
Adverse Pregnancy Outcome: Low Birth Weight; Intrauterine Growth Retardation (IUGR); Premature Delivery; Placental Abruption; Intrauterine Fetal Death; Perinatal Death; Pre-eclampsia
Other Causes of Increased AFP: Renal Abnormalities (Including Congenital Nephrosis); Bowel Obstruction; Sacrococcygeal Teratoma; Hydrocephalus; Congenital Heart Disease; Esophageal Atresia; Tracheo-esophageal Fistula; Turner's Syndrome; Severe Rh Isoimmunization

The incidence of neural tube defects is about five per thousand births in England and six to eight per 10,000 births in the U.S.; about 3000 affected infants are born each year (Edmonds LD, James LM. MMWR. 1990; 39(SS-4):19-23). Approximately 15 percent of infants with myelomeningocele have other major anomalies, 10 percent will have chromosomal abnormalities, and 15 percent will have severe hydrocephalus (Luthy DA, et al. N Engl J Med. 1991; 324:662-666).

Conditions associated with a decrease in AFP in maternal serum are listed in the following Table (Burton BK. Obstet Gynecol. 1988; 72:709-713):

Conditions Associated with a Decrease in Alpha-Fetoprotein in Maternal Serum
Less Advanced Gestational Age
Nonpregnancy
Fetal Death
Hydatidiform Mole
Down's Syndrome
Maternal Factors: Increased Maternal Weight; Maternal Diabetes Mellitus

Maternal serum AFP is an imperfect screening test for Down's syndrome. The sensitivity of the test is low, with only one third of fetuses with Down's syndrome detected in women under the age of 35. The test is not specific, since all of the above listed conditions and other unknown factors may result in a positive test. If the risk is one in 270 or greater, counseling and amniocentesis is offered (DiMaio MSW, et al. N Engl J Med. 1987; 317:342-346). An investigational "triple test" utilizing unconjugated estriol, human chorionic gonadotropin as well as maternal serum AFP may increase the detection rate to 60% of affected pregnancies in women under the age of 35 (Cassidy SB (ed.), Mtn States Regional Genetics Services Network Newsletter. Spring 1990; 5).

Maternal Serum Alpha-Fetoprotein (MSAFP) Screening: In 1986, the American College of Obstetrics and Gynecologists recommended maternal serum alpha-fetoprotein (MSAFP) screening for all women. Measurements of MSAFP can identify 75 to 80 percent of pregnancies complicated by myelomeningocele. In practice, only 60 percent of pregnancies are screened and half of fetuses with neural tube defects identified (Hobbins JC. N Engl J Med. 1991; 324:690-691).

Screening for MSAFP is done as follows: Maternal serum is obtained between 16 and 18 weeks of gestation. AFP levels may be measured by various techniques and reported in various units, however, results should be reported as multiples of the median (MoM). Multiples of the median may be calculated as follows (Cassidy SB (ed.), Mtn States Regional Genetics Services Network Newsletter. Spring, 1990; 5):

$$\text{MoM} = \frac{\text{AFP raw value}}{\text{AFP median for gestational age}}$$

The distributions of maternal serum alpha-fetoprotein levels during the second trimester are given in the next Table (Tovey KC, Gerson M. British J Obstet Gynecol. 1979; 86:507-515):

Distributions of Maternal Serum Alpha-Fetoprotein Levels During the Second Trimester

Week of Gestation	Median(ng/mL)
15	20.4
16	30.9
17	30.2
18	33.9
19	35.5
20	45.7

About five percent of patients lie above 2.1 x median; about three percent of patients lie above 2.4 x median; about one percent of patients lie above 3.1 x median. The patient's MoM must be adjusted for maternal weight, insulin dependent diabetes and race. Weight adjustment is necessary because the amount of AFP produced and transferred into maternal serum is relatively fixed; AFP level will depend on maternal plasma volume which is determined by weight. Maternal weight adjustment factor is determined from the following Graph (Adams MJ, et al. Am J Obstet Gynecol. 1984; 148:241-254):

MoM Adjustment Factor for Maternal Weight

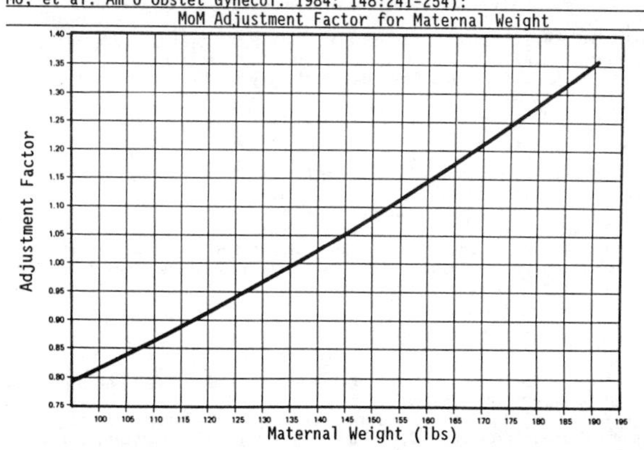

Diabetic women who are taking insulin prior to or during pregnancy have lower levels of alpha-fetoprotein. These women have a ten-fold higher incidence of neural tube defects. Although there is no consensus on the best correction factor to use, AFP levels are estimated to be 20 to 40 percent lower; correction factor should be 0.6 to 0.8. African Americans have median MSAFP levels that are 10 percent higher than Hispanics, Caucasians or Orientals. Correction can be made by multiplying the weight-adjusted MoM by 0.9 (Cassidy SB (ed.). Mtn States Regional Genetics Services Network Newsletter. Spring, 1990; 5).

Clinical interpretation and handling of abnormal MSAFP values varies considerably. General medical management principles are as follows: If the alpha-fetoprotein is elevated, a second serum specimen is obtained in 7 to 10 days following the first; if the result of the second test is elevated, ultrasound examination is done. Ultrasound is used to determine the gestation more exactly and potentially to identify abnormalities. If the serum AFP is elevated, amniocentesis is done. It has been suggested that amniocentesis may be unnecessary where high resolution ultrasonography and skilled operators are available (Nadel AS, et al. N Engl J Med. 1990; 323:557-561).

The concentration of AFP is amniotic fluid is 250 times greater than the maternal serum level; the concentration of AFP in fetal blood is 150 times that in amniotic fluid. When amniotic fluid alpha-fetoprotein is elevated, assay of acetylcholinesterase is helpful in confirming the diagnosis of an open neural tube defect; test for fetal blood is often performed as well since fetal blood contamination is a common cause of false elevations of amniotic fluid alpha-fetoprotein.

ALPHA-HYDROXYBUTYRATE DEHYDROGENASE

SPECIMEN: Red top tube, separate serum; reject if specimen hemolyzed.
REFERENCE RANGE: Up to 170 mU/mL; method dependent.
METHOD: Kinetic:

$$\text{Alpha-Oxobutyrate} + \text{NADH} + H^+ \xrightleftharpoons{HBD} \text{Alpha-Hydroxybutyrate} + \text{NAD}^+$$

INTERPRETATION: Alpha-hydroxybutyrate dehydrogenase (alpha-HBD) activity represents the activity of isoenzymes of LDH, primarily LDH-1. Conditions associated with an increase in alpha-HBD are given in the next Table:

Conditions Associated with Increase in Alpha-HBD
Acute Myocardial Infarction
Hemolytic Anemia
Megaloblastic Anemia
Progressive Muscular Dystrophy (Duchenne Type)
Cancer
Hemolyzed Samples

The elevation of alpha-HBD in disease reflects its tissue concentration, e.g., heart, erythrocytes and kidney; liver and skeletal muscle show little alpha-HBD.

Following acute myocardial infarction, alpha-HBD begins to increase in 10-12 hours, reaches a peak in 2-3 days and returns to normal in about two weeks. Currently in most laboratories the LDH-1 activity would be assessed by LDH isoenzyme fractionation.

AMIKACIN

SPECIMEN: Red top tube, separate serum and freeze. Do not use heparinized collection tubes. Obtain serum specimens as follows:
(1) 24-48 hours after starting therapy.
(2) 5 to 30 minutes before I.V. amikacin (trough).
(3) 30 minutes following completion of a 30 minute I.V. infusion (peak).

The following times should be recorded on the laboratory requisition form and on the patient's chart:

Trough Specimen Drawn _______ (Time)
Amikacin Started _______ (Time)
Amikacin Completed _______ (Time)
Peak Specimen Drawn _______ (Time)

Sampling is often done prior to attainment of steady state to help prevent toxicity and to ensure therapeutic efficacy.

High concentrations of beta-lactam antibiotics (penicillins, cephalosporins) inactivate aminoglycosides (gentamicin, streptomycin, amikacin, tobramycin and kanamycin); to reduce this interaction, specimens containing both classes of antibiotics should either be assayed immediately using a rapid method or stored frozen.

REFERENCE RANGE: Recommended Dose: 10-15mg/kg/day, I.M. or I.V.; Therapeutic Range: 15-25mcg/mL; Toxic Level, Peak: >30mcg/mL; Toxic Level, Trough: >5mcg/mL; Half-Life: 2-3 hours; Time to Steady State: Adults, 5-35 hours; Children, 3.5-15 hours; varies with dosage and age; may be significantly prolonged in patients with renal dysfunction.

METHOD: RIA; EMIT; SLFIA (Ames); Fluorescence Polarization (Abbott).

INTERPRETATION: Amikacin is an aminoglycoside antibiotic (gentamicin, streptomycin, amikacin, tobramycin and kanamycin-parenteral therapy) which is used frequently in hospitals to treat patients who have serious gram-negative bacterial infections, especially septicemia and some staphylococcal infections, untreatable with penicillins. Amikacin is the aminoglycoside of choice when gentamicin resistance is prevalent (Edson RS, Keys TF. Mayo Clin Proc. 1983; 58:99-102).

The three main toxic side effects of amikacin are ototoxicity, nephrotoxicity and neuromuscular blockage; it is important to control the dose given by monitoring peak and trough levels of the drug, particularly in patients with any degree of renal failure. As renal function declines, drug half-life increases.

To minimize risk of toxicity, it has been recommended that peak levels not exceed 30mcg/mL and that trough levels should fall below 5mcg/mL.

Amikacin is eliminated exclusively by renal excretion; excessive serum concentrations may occur and lead to further renal impairment. The renal damage is to the renal proximal tubules and is usually reversible if discovered early.

Ototoxicity is usually due to vestibular damage and is often not reversible.

AMINO ACID ANALYSIS

SPECIMEN: Urine: 24 hour collection preferred; the specimen should be kept refrigerated during collection; add 10mL of 6N HCl/liter to pH 2-3. A 50mL aliquot should be forwarded for analysis; note total 24 hour volume. Otherwise, random urine. Serum or Plasma: Red top tube, separate serum and freeze immediately; or green (heparin) top tube, separate plasma and freeze immediately. Specimens from amnionic fluid and CSF may be occasionally assayed for certain rare conditions. The qualitative thin layer technique is for screening for certain disorders. Not all amino acids will be separated by this technique.
REFERENCE RANGE: MetPath: Serum or Plasma: Ref. Levels(max.) in micromol/dL.

Amino Acid	Children	Adults	Amino Acid	Children	Adults
Taurine	11.5	14.0	Cystine	8.0	14.0
Aspartic Acid	2.5	5.0	Methionine	2.0	4.0
Hydroxyproline	--	--	Isoleucine	3.0	10.0
Threonine	9.5	25.0	Leucine	18.0	16.0
Serine	11.0	19.0	Tyrosine	7.0	9.0
Asparagine	2.0	5.0	Phenylalanine	6.0	12.0
Glutamic Acid	25.0	12.0	Beta-aminoisobutyric Acid	--	--
Glutamine	40.0	80.0	Tryptophan	--	--
Proline	45.0	44.0	Ornithine	9.0	13.0
Glycine	22.0	50.0	Lysine	15.0	26.0
Alanine	30.0	50.0	1-methyl-histidine	--	--
Citrulline	3.0	5.0	Histidine	9.0	12.0
Alpha-aminobutyric Acid	4.0	4.0	3-methyl-histidine	--	--
Valine	28.0	33.0	Arginine	9.0	15.0

METHOD: Quantitative: Gas-liquid chromatography; ion-exchange chromatography; HPLC chromatography; Qualitative: Thin-layer chromatography.
INTERPRETATION: This assay is used primarily in the evaluation of suspected inborn errors of amino acid metabolism. Conditions associated with increased amino acids in plasma and urine are given in the next Table (BioScience):

Amino Acids	PKU	Maple Syrup Urine Disease (MSUD)	Cystinuria	Homocystinuria	Hartnup	Arginino-Succinicaciduria	Histidinemia	Hyperprolinemia Type A	Citrullinuria
Leucine, Isoleucine		P;U			U				
Phenylalanine	P;U				U				
Valine, Methionine		P;U		P	U				P
Tryptophan, Beta-Amino Isobutyrate					U				
Tyrosine					U				
Proline		P						P;U	
Alanine, Ethanolamine					U		U		U
Threonine, Glutamate							U		U
Homocitrulline, Glycine, Serine, Hydroxyproline, Aspartic Acid, Glutamine, Citrulline		P			U	U		U	P;U
Homocystine, Asparagine				U					
Argininosuccinic Acid, Histidine, Arginine, Lysine Ornithine, Cystathionine, Cystine, Cysteine Hydroxylysine			U		U	U	P;U		U

P = Plasma
U = Urine

Generalized aminoaciduria is also found in Wilson's disease, Lowe's syndrome, galactosemia, cirrhosis of the liver and renal tubular abnormalities. Certain drugs: amphetamines, antihistamines, oral contraceptives and phenothiazines will interfere with assay.

AMINOPHYLLINE (see THEOPHYLLINE)

AMITRIPTYLINE (see TRICYCLIC ANTIDEPRESSANTS)

AMMONIA

SPECIMEN: Green (heparin) top tube or heparinized syringe or Lavender (EDTA) top tube; red top tube for serum may be used; the tube must be filled completely; kept tightly stoppered at all times. Specimen must be placed on ice immediately. Blood ammonia increases rapidly at room temperature. Test should be performed within 60 minutes of the venipuncture. Ammonia is stable in frozen plasma for several days at -20°C. Hemolyzed specimen should be rejected.
REFERENCE RANGE: Ammonia varies with age. (Units in micromol/liter):
Newborn, 65-105; 0-2 Weeks, 55-90; >1 Month, 20-50; Adult, 5-50. To convert international units in micromol/liter to traditional units in microgram/dL, multiply international units by 1.7.
METHOD:

$$NH_4^+ + \text{Alpha-ketoglutarate} + \text{NADPH (340nm)} \xrightarrow[\text{Dehydrogenase}]{\text{Glutamate}} \text{L-glutamate} + NADP^+ + H_2O$$

INTERPRETATION: The most common cause of elevated blood ammonia is severe liver disease; blood ammonia is also elevated in Reye's syndrome.

Normally, ammonia is produced in the intestine by bacterial action on protein. Ammonia is then transported through the portal venous blood to the liver. The hepatocytes metabolize ammonia to urea via the Krebs-Henseleit cycle:

$$\text{Ammonia} \xrightarrow{\text{Krebs-Henseleit cycle}} \text{Urea}$$

The formation of urea from protein is illustrated in the next Figure:

Formation of Urea from Protein

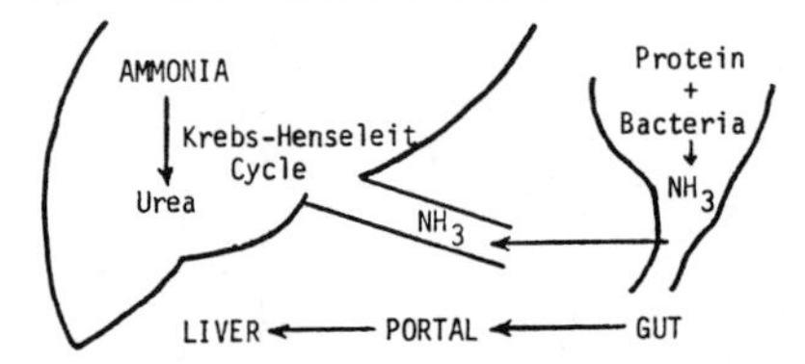

Other sources of ammonia are the kidney and muscle.

In liver disease, blood ammonia may be increased because of increase in collateral circulation which by-passes the liver and by failure of the diseased hepatocytes to metabolize ammonia to urea. The causes of elevated plasma ammonia are given in the next Table:

Causes of Elevated Plasma Ammonia

- Decreased Destruction of Plasma Ammonia
  - Impaired Capacity to Detoxify Normal Quantities of Ammonia
    - Inherited Metabolic Disorders
      - Urea Cycle
      - Organic Acidemias
      - Dibase Aminoacidopathies
    - Acquired Disorders
      - Liver Disease
      - Reye's Syndrome
  - Bypass of Liver (e.g. Cirrhosis)
- Increased Production of Plasma Ammonia
  - Urinary Tract Abnormalities
  - GI Bleed
- Other
  - Valproic Acid
  - Systemic Illness (Shock, CHF)
  - Perinatal Asphyxia

<u>Reye's Syndrome</u>: In Reye's syndrome, it has been suggested that peak plasma ammonia levels are predictive of disease severity and patient survival. When peak ammonia levels are less than 5 times normal, the survival rate was 100%, peak ammonia levels in excess of five times normal are associated with significant mortality. Peak ammonia levels occurred within four hours of admission in 88% of patients (Fitzgerald JF, et al. "The Prognostic Significance of Peak Ammonia Levels in Reye's Syndrome," Pediatrics. 1982; 70:997-999). The CDC case definition of Reye's Syndrome is given in the following Table:

CDC Definition of Reye's Syndrome
Noninflammatory Encephalopathy
Cerebrospinal Fluid ≤ 8 leukocytes per cubic mm (if available)
Cerebral edema without perivascular or meningeal inflammation
Hepatopathy by biopsy or a three-fold or greater rise of AST, ALT or ammonia
No other reasonable explanation for cerebral or hepatic abnormalities

The incidence of Reye's syndrome decreased dramatically between 1980 and 1985. This decrease coincided with the national decline in aspirin use as an antipyretic in children (Arrowsmith JB, et al. Pediatrics. 1987; 79:858-863). While some authors have continued to question the association between aspirin and Reye's syndrome (Orlowski JP, et al. Pediatrics. 1987; 80:638-642), other studies have continued to uphold this association (Forsyth BW, et al. JAMA. 1989; 261:2517-2524; Pinsky PF, et al. JAMA. 1988; 260:657-661; Hurwitz ES, et al. JAMA. 1987; 257:1905-1911).

AMNIOCENTESIS PANEL

Tests that are done on amniotic fluid are given in the next Table:

Tests on Amniotic Fluid
Bilirubin Amniotic Fluid: Evaluation of Hemolytic Disease of the Newborn
Lecithin/Sphingomyelin (L/S) Ratio and Phosphatidylglycerol (PG): Fetal Lung Maturity
Genetic Testing/Chromosomal Analysis

Indications for genetic amniocentesis are given in the following Table (Sullivan MM, Rawnsley BE, Hatch D. Check Sample, Cytopathology. 1982; 10:8):

Indications for Genetic Amniocentesis
Maternal Age > 35 Years
Previous Child with Chromosome Anomaly
Previous Child with Neural Tube Defect
Elevated Serum Alpha-Fetoprotein Screen
Previous Child or Known Carrier - Inherited Metabolic Disorder
Family History of Sex-Linked Recessive Disease

<u>Maternal Age >35 Years</u>: Children born to mothers over 35 years of age are at high risk for chromosomal abnormalities, particularly Down's syndrome. Down's syndrome occurs in 1 in 385 births to mothers age 35; the incidence of all chromosomal abnormalities is 1 in 202.

More than 90 percent of patients with Down's syndrome have trisomy 21; the appearance of chromosomes in trisomy 21 is shown in the next Figure:

Trisomy 21, Down's Syndrome

Down's Syndrome: Extra Chromosome, Group G, Trisomy 21

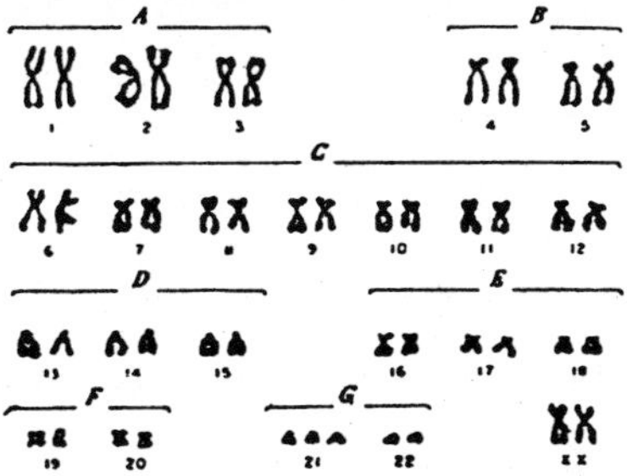

(cont)

Amniocentesis Panel (Cont)

The characteristics of Down's syndrome are as follows: General Features: Patients are mentally retarded and have short stature. They are hypotonic as newborns-later, they have hypermobility of joints. Patients have an increased incidence of leukemia. Cardiac: Increased incidence of congenital lesions of the heart, most commonly atrioventricular canal and ventricular septal defect. Head: flattening of occiput, fine sparse hair, epicanthic folds (a prolongation of a fold of skin of the upper eyelid over the angles of the eye), low nasal bridge, open mouth, large tongue, low-set ears, short neck. Hands and Feet: Hands have short metacarpals and phalanges; a typical pattern of palmer creases with one crease in fifth finger; feet have a wide gap between first and second toes.

Previous Child with Chromosome Anomaly: There is an increased incidence of chromosome anomalies in children whose mothers gave birth to a previous child with a chromosome abnormality.

Previous Child with Neural Tube Defect: If a mother has given birth to a child with a neural tube defect, the rates markedly increase in subsequent siblings.

Elevated Serum Alpha-Fetoprotein Screen: See ALPHA-FETOPROTEIN.

Previous Child or Known Carrier-Inherited Metabolic Disorder: Prenatal diagnosis is possible for numerous inherited metabolic disorders including disorders of lipid metabolism, disorders of carbohydrate or glycoprotein metabolism, disorders of mucopolysaccharide metabolism, mucolipidoses, disorders of amino acid metabolism and disorders of organic acid metabolism.

Family History of Sex-Linked Recessive Disease: Sex-linked conditions are listed in the next Table:

Sex-Linked Conditions
Color Blindness
Hemophilia A (Factor VIII Deficiency)
Hemophilia B (Factor IX Deficiency)
Muscular Dystrophy, Duchenne Type (1:36,000)
Glucose-6-Phosphate Dehydrogenase Deficiency
Rare Conditions:
Fabry's Disease
Lesch-Nyhan Syndrome (Deficiency of enzyme hypoxanthine-guanine phosphoribosyl transferase)
Diabetes Insipidus
Ectodermal Dysplasia, Anhidrotic Type
Ocular Albinism
X-Linked Ichthyosis
Agammaglobulinemia (Bruton Disease)
Wiskott-Aldrich Syndrome (an Immunodeficiency State)

References: D'Alton ME, DeCherney AH. "Prenatal Diagnosis." N Engl J Med. 1993; 328:114-120; Schwartz M, et al. "Amniotic Fluid and Advances in Prenatal Diagnosis." Clin Lab Med. 1985; 5:371-387.

AMNIOTIC FLUID ANALYSIS (see **BILIRUBIN, AMNIOTIC FLUID;** see **LECITHIN/SPHINGOMYELIN (L/S) RATIO)**

AMPHETAMINES, DEXTROAMPHETAMINE AND METHAMPHETAMINE ("SPEED", "ICE" OR "CRYSTAL")

SPECIMEN: Red top tube, separate serum; and/or 50mL random urine.
REFERENCE RANGE: Negative
METHOD: Thin layer chromatography (TLC) of urine; EMIT.
INTERPRETATION: Amphetamines are stimulants; stimulants increase alertness, reduce hunger and induce a feeling of well-being. They have been used to suppress appetite and to reduce fatigue.

The signs and symptoms of amphetamine toxicity primarily involve central nervous system, cardiovascular or neuromuscular effects as given in the following Table (Aaron CK. Emerg Med Clin N Amer. 1990; 8:513-526):

Signs and Symptoms of Amphetamine Toxicity

Organ System	Common	Uncommon
Central Nervous System	Agitation Anxiety Delusions Hallucinations Paranoia Seizures	Stroke-Ischemic or Hemorrhagic Hypertension Encephalopathy Due to Hypertension and Vasculitis Focal Neurologic Findings
Cardiovascular	Hypertension Sinus Tachycardia Bradycardia or AV Block Due to Alpha-Agonist	Tachyarrhythmias Cardiac Ischemia Myocardial Necrosis Systemic Vasculitis Mesenteric Ischemia
Neuromuscular	Hyperthermia Tremor Mydriasis	Rhabdomyolysis Muscular Rigidity

"Ice" is a smokable form of methamphetamine. "Designer" hallucinogenic amphetamines are chemically related to amphetamines and may not be detectable on routine screening. Nonprescription drugs may interact with screening tests such as TLC or immunoassay; medications which may produce false positive tests are given in the following Table (Schwartz RH. Arch Intern Med. 1988; 148:2407-2412):

False Positive Screening Tests for Amphetamines
Ephedrine Hydrochloride
Pseudoephedrine Hydrochloride (Sudafed)
Phenylpropanolamine Hydrochloride
Propylhexedrine
Phentermine Hydrochloride
Phenmetrazine Hydrochloride
Tenfluramine Hydrochloride
Some Theophylline-Containing Medications

AMYLASE, CLEARANCE

SPECIMEN: Red top tube, separate serum; and random urine specimen; timed urine specimen is not necessary.
REFERENCE RANGE: 5%; some laboratories have a value less than 5%.
METHOD: Serum and urine assays.
INTERPRETATION: Amylase clearance is increased with patients with acute pancreatitis. Amylase clearance has been useful in differentiating the causes of increased serum amylase, particularly, macroamylasemia. Use of amylase-creatinine clearance ratio to confirm or exclude the diagnosis of acute pancreatitis has limited value since any condition that increases renal tubular protein (e.g., amylase) excretion will increase this ratio (Eckfeldt JH, Levitt MD. Clinics Lab Med. 1989; 9:731-743).

Amylase clearance is the renal clearance of amylase, expressed as a percentage of creatinine clearance, and is given by the following equation:

$$\frac{\text{Amylase Clearance}}{\text{Creatinine Clearance}} = \frac{\frac{\text{Urine Amylase}}{\text{Serum Amylase}} \times \text{Urine Volume per Unit Time}}{\frac{\text{Urine Creatinine}}{\text{Serum Creatinine}} \times \text{Urine Volume per Unit Time}} \times 100$$

This equation simplifies to the following form:

$$\frac{\text{Amylase Clearance (\%)}}{\text{Creatinine Clearance}} = \frac{[\text{Urine Amylase}]}{[\text{Serum Amylase}]} \times \frac{[\text{Serum Creatinine}]}{[\text{Urine Creatinine}]} \times 100$$

As seen by inspection of the above equation, the clearance ratio, expressed as percentage, is calculated simply from the concentrations of amylase and creatinine in serum and urine samples obtained simultaneously. No timed collections are necessary.

In normal subjects, the value of the ratio amylase clearance/creatinine clearance is usually less than 4%. Increased amylase clearance occurs in patients with acute pancreatitis. The mean value in acute pancreatitis is about 3 times that of the normal value. An elevated ratio reflects defective proximal tubular reabsorption of amylase which occurs in virtually all patients with acute pancreatitis. Other conditions that are associated with acute defective tubular function include cardiopulmonary bypass, diabetic ketoacidosis, multiple myeloma, burns, aminoglycoside toxicity and renal failure; these conditions may also cause an elevation of the ratio.

The amylase clearance decreases over time during an acute episode of pancreatitis; the mean value for different time intervals post-onset of acute pancreatitis are as follows: Days 0-4, 6.8; days 5-8, 4.5; days 9-15, 4.
Macroamylasemia: In macroamylasemia, normal amylase is bound to an abnormal immunoglobulin or glycoprotein. Patients have an elevated serum amylase but amylase-creatinine clearance ratio is very low (<1%). Serum lipase is also normal in this condition. Macroamylasemia occurs in approximately 1% of the population and is generally considered to be a benign condition (Forsman RW. Clin Biochem. 1986; 19:250-253). Macroamylasemia is present in 25% of patients with systemic lupus erythematosus (Hasselbacher P, et al. British J Rheum. 1988; 27:198-201).

AMYLASE, SERUM

SPECIMEN: Red top tube, separate serum; or green top tube (heparin), separate plasma.
REFERENCE RANGE: Adult: 0-140U/dL. Practically no amylase activity is present in neonates; measurable enzyme activity is detected at approximately two months of age and increases slowly to adult values by the age of 1 year.
METHOD: Nephelometry; enzymatic
INTERPRETATION: Amylase is elevated in many conditions. Pancreatic isoamylase and lipase are more tissue specific for the pancreas and these tests are more sensitive and specific for the diagnosis of acute pancreatitis (Tietz NW. J Clin Chem Clin Biochem. 1988; 26:251-253). Causes of increased amylase are given in the next Table (Pieper-Bigelow C, et al. Gastroenterol Clin North Amer. 1990; 19:793-810):

Causes of Increased Serum Amylase
Pancreatic: Pancreatitis-Acute or Chronic; Pseudocyst; Pancreatic Ascites; Pancreatic Trauma; Endoscopic Retrograde Pancreatography Choledocholithiasis; Pancreatic Cancer; Early Cystic Fibrosis; Biliary Sludge
Salivary: Parotitis (e.g. mumps); Trauma; Surgery; Radiation; Calculi
Gut Diseases: Perforated Bowel; Mesenteric Infarct; Intestinal Obstruction; Appendicitis; Peritonitis
Liver Diseases: Hepatitis; Cirrhosis
Malignancies: Ectopic Production
Renal Failure
Female Genital Tract: Ruptured Ectopic Pregnancy; Fallopian or Ovarian Cysts; Salpingitis
Macroamylasemia
Miscellaneous: Alcoholism; Ketoacidosis; Nonketotic Acidosis; Extracorporeal Circulation; Pneumonia; Cerebral Trauma; Burns; Abdominal Aortic Aneurysm; Anorexia Nervosa; Bulimia; Drugs (Drug-Induced Pancreatitis - see Steinberg WM. Hosp Practice. 1985; 95-102).

Alcohol abuse and gallstone disease accounts for 70 percent of cases of acute pancreatitis in adults. Biliary sludge is a common cause of acute idiopathic pancreatitis (Lee SP, et al. N Engl J Med. 1992; 326:589-593). Evaluation of elevated serum amylase is given in the following Table (Panteghini M, Pagani F. Arch Pathol Lab Med. 1991; 115:355-358):

Evaluation of Elevated Serum Amylase

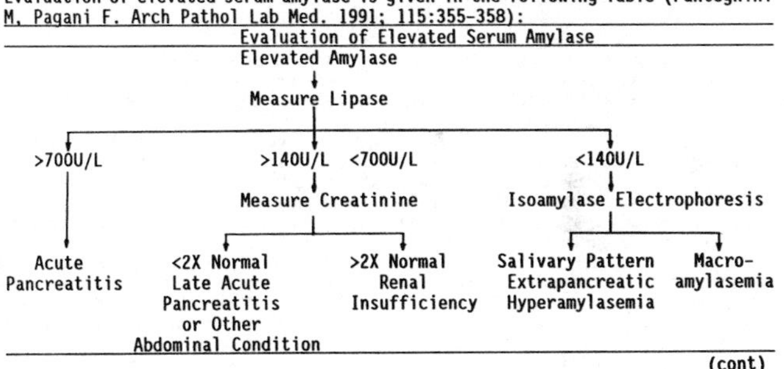

(cont)

Amylase, Serum (Cont)

The change in serum amylase, serum lipase and urine amylase in acute pancreatitis is illustrated in the next Figure:

Serum Amylase, Serum Lipase and Urine Amylase Following Acute Pancreatitis

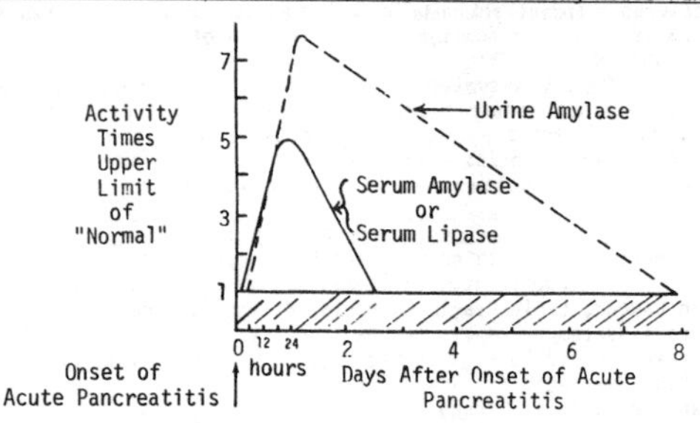

Following onset of acute pancreatitis, serum amylase begins to rise in 2 to 6 hours, reaches a maximum in 12 to 30 hours and remains elevated for 2 to 4 days. Urine amylase begins to rise 4 to 8 hours following onset of acute pancreatitis; this is several hours after the initial increase of serum amylase. Urine amylase remains elevated for 7 to 10 days, about five days after the serum amylase returns to normal. Serum lipase changes in a manner similar to that of serum amylase following onset of acute pancreatitis; the changes are summarized in the next Table:

Enzyme Changes in Acute Pancreatitis

Enzyme	Beginning of Increase (hrs.)	Maximum (hrs.)	Return to Normal (Days)
Serum Amylase	2-6	12-30	2-4
Urine Amylase	4-8	18-36	7-10
Serum Lipase	2-6	12-30	2-4

It is important to note that urine amylase is elevated when serum amylase is normal. This occurs because renal glomerular filtration for amylase is increased in acute pancreatitis; thus, amylase appears in the urine several days after the serum amylase returns to normal.

Pancreatitis in children is most commonly associated with systemic diseases (35%: Reye syndrome, sepsis/shock, hemolytic-uremic syndrome, viral infection, systemic lupus erythematosus). Other causes include idiopathic (25%), traumatic (15%), structural (10%: biliary stone, cyst, or stricture), metabolic (10%: hyperlipidemia, cystic fibrosis, hypercalcemia), drugs (3%), or familial (2%)(Weizman Z, Durie PR. J Pediatr. 1988; 113:24-29).

Pancreatic Pseudocyst: Pancreatic pseudocyst is the most common complication of pancreatitis, occurring in 2% to 10% of patients. Pseudocysts form as a result of alcoholic pancreatitis(60%), choledocholithiasis(20%), trauma(10%); blunt abdominal trauma is the most common cause of pseudocysts in children. Urine amylase is elevated in 60% of patients and serum amylase is elevated in 50% of patients. About one-third of patients have abnormal liver function tests, anemia, increased serum glucose, and elevated white blood cells.

Definitive diagnosis of pseudocyst is made by radiology: Plain film of the abdomen for calcification or mass lesion(40%); ultrasonography and contrast examination of the upper gastrointestinal tract(90%); CT Scan; IVP and contrast studies of the biliary tract (Van Landingham SB, Roberts JW, Hospital Medicine. 1984; 71-88).

Chronic Pancreatitis: A recent study demonstrated a significantly elevated risk of pancreatic cancer in patients with chronic pancreatitis (Lowenfels AB, et al. N Engl J Med. 1993; 328:1433-1437).

Reference: See excellent review of acute pancreatitis: Marshall JB. Arch Intern Med. 1993; 153:1185-1198.

ANAEROBIC CULTURE

SPECIMEN: The usual sources of anaerobic specimens are as follows: abscesses; body fluids: cerebrospinal fluid, pleural fluid, peritoneal fluid, pericardial fluid, synovial fluid, culdocentesis, amniotic fluid; tissues: surgical and autopsy specimens, placenta; wounds; sinus tracts; transtracheal aspirates; bone marrow; duodenal aspirates; middle ear aspirates; mastoid aspirates; eye; bronchial brushings.
INTERPRETATION: The relative incidence of anaerobic bacteria in various infections is given in the next Table (Feleke G, Forlenza S. Postgrad Med. 1991; 89:221-234):

Relative Incidence of Anaerobic Bacteria in Various Infections

Type of Infection	Incidence(%)	Predominant Isolates
Head and Neck		
Chronic Sinusitis	52	Oropharyngeal anaerobes
Chronic Suppurative Otitis Media	51	Oropharyngeal anaerobes
Neck Space Infection	100	Oropharyngeal anaerobes
Dental and Periodontal Abscess	92	Oropharyngeal anaerobes
Central Nervous System		
Brain Abscess	89	Oropharyngeal anaerobes
Pleura and Lung		
Aspiration Pneumonia	87	Oropharyngeal anaerobes
Lung Abscess	93	Oropharyngeal anaerobes
Empyema	76	Oropharyngeal anaerobes
Gastrointestinal Tract		
Intra-abdominal Infection	86	Bacteroides fragilis and
Liver Abscess	53	other colonic anaerobes
Female Genital Tract		
Salpingitis and Pelvic Peritonitis	56	Bacteroides fragilis and
Septic Abortion	63	other anaerobes in the
Pelvic Abscess	88	female genital tract
Skin and Soft Tissues		
Diabetic Foot Ulcers	81	B. fragilis, other Bacteroides
Decubitus Ulcers	63	species, Peptostreptococci
Bite-Wound Infection	51	Oropharyngeal anaerobes
Clostridial Myonecrosis	100	Clostridium perfringens
Synergistic Nonclostridial Myonecrosis	100	Bacteroides species and anaerobic Streptococci
Necrotizing Fasciitis	100	Clostridium and Bacteroides species

Materials which should not be routinely cultured for anaerobic bacteria (because anaerobes occur as normal flora) are as follows (Allen SD, Siders JA in Manual of Clinical Microbiology, 3rd ed. Am Soc Microbiology, Washington, DC. 1980): throat or nasopharyngeal swabs; gingival swabs; expectorated sputum; bronchoscopic specimens not collected by a protective double lumen catheter; gastric contents, small bowel contents, feces, rectal swabs, colocutaneous fistulae, and colostomy stomata; surface material from decubitus ulcers, swab samples of other surfaces, sinus tracts and eschars; material adjacent to skin or mucous membranes other than the above which have not been properly decontaminated; voided urine; vaginal or cervical swabs.

Some anaerobes are killed by contact with oxygen for only a few seconds. Great care should be taken to reduce contamination from adjacent surfaces.
Transport of Specimens: Porta-A-Cul or B-D anaerobic specimen collection tubes may be used for transport of specimens to the laboratory. A syringe may be used to collect and transport the specimens; expel all air, cap needle and transport to laboratory immediately.
Gram-Stain: Gram-stain is helpful for quality control in that preliminary information and may be obtained. Characteristics of some anaerobic bacteria with gram-stain are given in the next Table: (cont)

Anaerobic Culture (Cont)

Anaerobes and Gram-Stain

Characteristic	Anaerobe (Possible)
Gram Positive, Large, Broad Rods with Blunted Ends in Suspected Gas Gangrene	C. Perfringens
Gram Negative, Irregular Staining with Bipolar Staining; Specimen from Abscess	Bacteroides or Fusobacterium Species
Gram Negative, Filamentous Slim Rods with Tapered Ends	F. Nucleatum
Gram Positive, Cocci, Clusters and Chains, within Neutrophilic Exudate from Postoperative Intraabdominal Wound	Peptococcus Peptostreptococcus
Sulfur Granules from a Cervicofacial Lesion	Actinomycetes

The percentage distribution of anaerobic bacteria recovered from human clinical infections is given in the next Table (Allen SD, Siders JA in Manual of Clinical Microbiology. 34 ed. Am Soc Microbiology, Washington, DC. 1980; pg. 405):

Percentage Distribution of Anaerobic Bacteria

Group	Percent
Gram-Negative Nonsporeforming Bacilli	43
Bacteroides Fragilis Group	23
Gram-Positive Nonsporeforming Bacilli	23
Gram-Positive Cocci	21
Gram-Positive Sporeforming Bacilli	11
Clostridium Perfringens	5
C. Ramosum	2
C. Difficile	1
Gram-Negative Cocci	2

A survey of cases of bacteremia in England and Wales showed that 6.5 percent of all cases were due to anaerobic organisms, mainly Bacteroides (B. fragilis) species.

Culture Systems: Most low-volume laboratories rely on observed growth in the depths of thioglycollate broth tubes as the initial clue than an anaerobic bacterium may be present. There are different methods to generate an anaerobic atmosphere. A H_2-CO_2 generator (Gas Pak, BBL Microbiology Systems) may be used. Identification of anaerobes can be done using gas-liquid chromatography.

Treatment: Suggested empiric therapy for anaerobic and mixed aerobic/anaerobic infections is given in the following Table (Panichi G. Scand J Infect Dis. 1989; Suppl 62:47-51):

Empiric Antibiotic Therapy for Anaerobic Infections

Site	Empiric Antibiotic Therapy
CNS Infections	Benzylpenicillin or Second or Third Gen Cephalosporin plus Metronidazole or Chloramphenicol
Pleuropulmonary Infections	Benzylpenicillin or Second or Third Gen Cephalosporin plus Metronidazole or Clindamycin or Imipenem/Cilastatin or Aminopenicillin-Beta lactamase inhibitor combination
Intraabdominal Infections Female Genital Tract Infections	Imipenem/Cilastatin or Second or Third Gen Cephalosporin plus Metronidazole

Anaerobic infections have recently been reviewed (Finegold SM, et al. Infect Dis Clin N Amer. 1993; 7:257-275).

ANDROGEN PANEL [TESTOSTERONE, DEHYDROEPIANDOSTERONE-SULFATE (DHEA-S)]

SPECIMEN: Red top tube, separate and freeze serum.
REFERENCE RANGE: Testosterone: Normal reproductive females: 10-60ng/dL; Prepubertal children: <10ng/dL; Adult male: 350-800ng/dL.
Testosterone, Free: Normal reproductive females: 0.3-1.9ng/dL; Adult male: 9-30ng/dL.
Dehydroepiandosterone sulfate (DHEA-S): Normal reproductive females: 5-35ng/dL; Normal adult male: 15-55ng/dL.
METHOD: RIA
INTERPRETATION: Testosterone, free testosterone and DHEA-S aid in the evaluation of androgen deficiency or excess (hirsutism and/or virilism). The physiologically active testosterone is the free testosterone. The free testosterone gives a more accurate measure of the testosterone available to target cells. The total testosterone is influenced by the level of the testosterone binding protein; the testosterone binding protein is changed by the factors in a manner similar to those that change the concentration of thyroxine-binding proteins. Two conditions, hyperthyroidism and syndromes of androgen resistance are associated with increased total testosterone; however, free testosterone is normal. The clinical causes of hirsutism are given in the next Table:

Causes of Hirsutism

Common:
- Polycystic Ovary Syndrome

Rare:
- Adrenal Origin: Cushing's Syndrome; Late-Onset or Attenuated 21-Hydroxylase Deficiency (Congenital Adrenal Hyperplasia); Androgen-Secreting Tumors
- Ovarian Origin: Hilus Cell Tumor; Androblastoma; Teratoma
- Hypothyroidism
- Acromegaly
- Hyperprolatinemia

Drugs:
- Phenytoin; Diazoxide; Minoxidil; Menopausal Mixtures Containing Androgens, Corticosteroids (rarely)

Hirsutism in the female can be of adrenal or ovarian origin and may be caused by excess androgens. DHEA-S is a good indicator of adrenal function; testosterone is a good indicator of ovarian function. By measuring testosterone and DHEA-S, one can localize the lesion to the ovaries or the adrenal. Plasma testosterone over 200ng/dL is usually indicative of an ovarian abnormality.
Stein-Leventhal Syndrome (Polycystic Ovaries): The Stein-Leventhal Syndrome is the most common hormonal cause of hirsutism. Evidence of menstrual irregularity may point towards polycystic ovarian disease. Borderline elevations of the androgens, testosterone, and 17-ketosteroids are present; raised LH levels and increased LH/FSH ratios are consistent with this disease.

ANEMIA PANEL

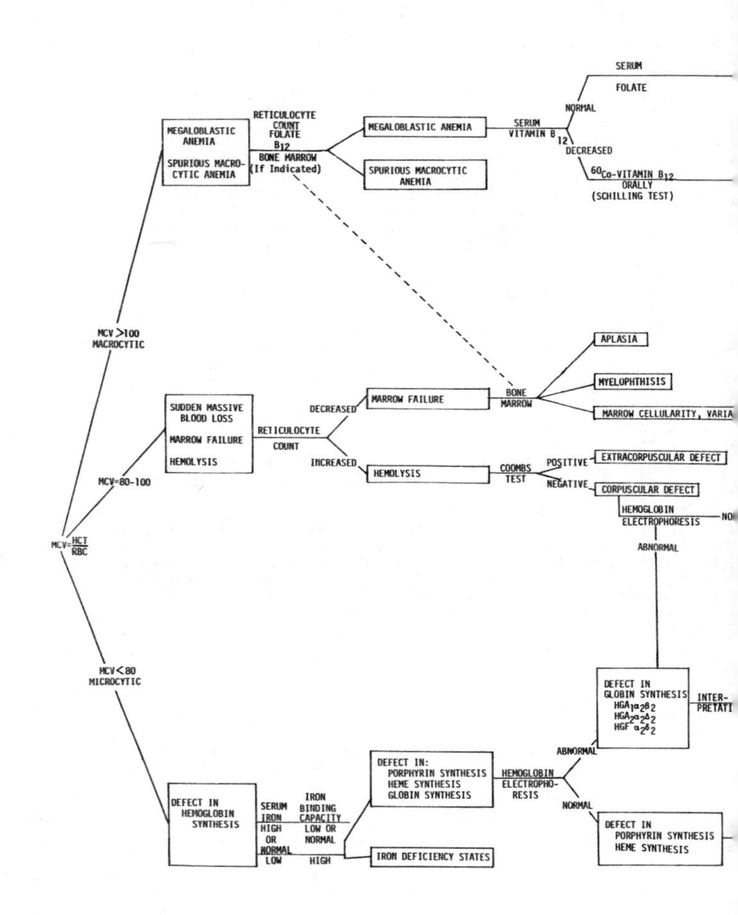

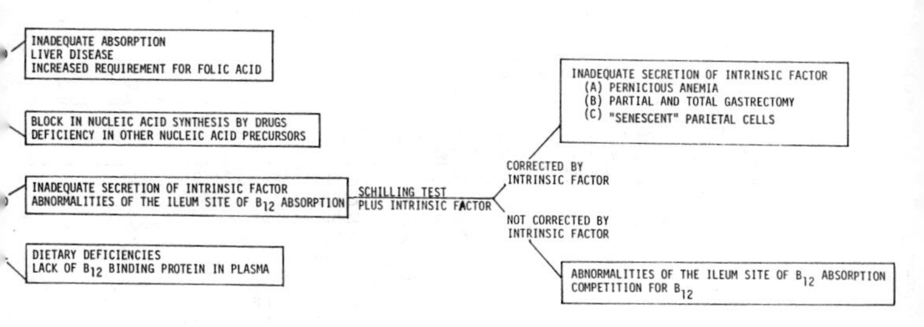
INADEQUATE ABSORPTION
LIVER DISEASE
INCREASED REQUIREMENT FOR FOLIC ACID
BLOCK IN NUCLEIC ACID SYNTHESIS BY DRUGS
DEFICIENCY IN OTHER NUCLEIC ACID PRECURSORS
INADEQUATE SECRETION OF INTRINSIC FACTOR
ABNORMALITIES OF THE ILEUM SITE OF B_{12} ABSORPTION
DIETARY DEFICIENCIES
LACK OF B_{12} BINDING PROTEIN IN PLASMA
SCHILLING TEST
PLUS INTRINSIC FACTOR
CORRECTED BY
INTRINSIC FACTOR
INADEQUATE SECRETION OF INTRINSIC FACTOR
(A) PERNICIOUS ANEMIA
(B) PARTIAL AND TOTAL GASTRECTOMY
(C) "SENESCENT" PARIETAL CELLS
NOT CORRECTED BY
INTRINSIC FACTOR
ABNORMALITIES OF THE ILEUM SITE OF B_{12} ABSORPTION
COMPETITION FOR B_{12}

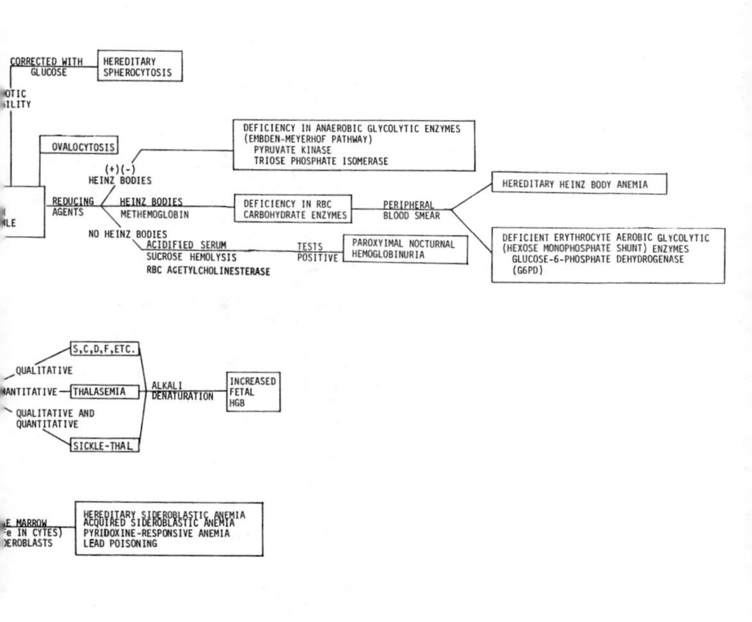
CORRECTED WITH
GLUCOSE
HEREDITARY
SPHEROCYTOSIS
OVALOCYTOSIS
DEFICIENCY IN ANAEROBIC GLYCOLYTIC ENZYMES
(EMBDEN-MEYERHOF PATHWAY)
PYRUVATE KINASE
TRIOSE PHOSPHATE ISOMERASE
(+)(-)
HEINZ BODIES
REDUCING
AGENTS
HEINZ BODIES
METHEMOGLOBIN
DEFICIENCY IN RBC
CARBOHYDRATE ENZYMES
PERIPHERAL
BLOOD SMEAR
HEREDITARY HEINZ BODY ANEMIA
DEFICIENT ERYTHROCYTE AEROBIC GLYCOLYTIC
(HEXOSE MONOPHOSPHATE SHUNT) ENZYMES
GLUCOSE-6-PHOSPHATE DEHYDROGENASE
(G6PD)
NO HEINZ BODIES
ACIDIFIED SERUM
SUCROSE HEMOLYSIS
RBC ACETYLCHOLINESTERASE
TESTS
POSITIVE
PAROXYIMAL NOCTURNAL
HEMOGLOBINURIA
S,C,D,F,ETC.
QUALITATIVE
THALASEMIA
QUALITATIVE AND
QUANTITATIVE
SICKLE-THAL
ALKALI
DENATURATION
INCREASED
FETAL
HGB
HEREDITARY SIDEROBLASTIC ANEMIA
ACQUIRED SIDEROBLASTIC ANEMIA
PYRIDOXINE-RESPONSIVE ANEMIA
LEAD POISONING

ANEMIA PANEL

The initial laboratory tests in differentiating causes of anemia are given in the next Table (Wallerstein RO. Consultant. August. 1980. 65-70):

Initial Laboratory Tests - Anemia
Hemoglobin, Hematocrit, Red Blood Count
Red Cell Indices (MCHC, MCV, MCH)
Peripheral Blood Film
Reticulocyte Count
Platelet Count
White Cell Count

BLOOD FILM: Morphologic alterations, characteristic of certain red blood cell conditions, are given in the next Table:

Morphologic Alterations of Red Blood Cells

Alteration	Differential Diagnosis
Spherocytes	Hereditary Spherocytosis Autoimmune Hemolytic Anemia Acute Alcoholism Hemoglobin C Disease Following Severe Burns Hemolytic Transfusion Reactions Severe Hypophosphatemia Acute Oxidant Injury in Hexose Monophosphate Shunt Defects Clostridium Welchii Septicemia
Oval Cells (Elliptocytes)	Megaloblastic Anemia Myelofibrosis Refractory Normoblasts Anemia Hereditary Elliptocytosis Thalassemia Syndromes
Fragmented Red Cells (Schistocytes)	Microangiopathic Hemolytic Anemias Thrombotic Thrombocytopenia Purpura(TTP) Hemolytic Uremic Syndrome(HUS) Disseminated Intravascular Coagulopathy(DIC) Secondary to Immune Mechanisms Giant Hemangioma Metastatic Carcinoma Malignant Hypertension Eclampsia Macroangiopathic Hemolytic Anemia Prosthetic Valve Replacement, etc.
Spur Cells (Acanthocytes)	Abetalipoproteinemia Cirrhosis/Hepatic Necrosis Pyruvate Kinase(PK) Deficiency Uremia Infantile Pyknocytosis
Stippling	Lead and Arsenic Poisoning Thalassemia Syndromes Sideroblastic Anemia Other Severe Anemias Unstable Hemoglobins Pyrimidine 5'-Nucleotidase Deficiency
Target Cells	Thalassemias Hemoglobin C, S Liver Disease

RED CELL INDICES: Diagnosis, derived from indexes, are shown in the next Table:

Indexes and Anemia

Index	Diagnosis
MCV>100	Megaloblastic Anemia: Folate or Vitamin B_{12} Deficiency Hemolytic Anemia (Elevated Reticulocyte Count) Liver Disease
MCV<80 and MCHC<30	Iron Deficiency Anemia or Thalassemia Minor
MCHC>36	Hereditary Spherocytosis

CLASSIFICATION OF ANEMIA: A classification of anemia is given in the next Table (Cohn LS. Diagnosis. Oct. 1983; 125-137):

Classification of Anemia

Classification	Index
Microcytic Hypochromic Anemia	MCV<80; MCHC<30
Normocytic Normochromic Anemia	MCV and MCHC, Normal
Macrocytic Anemia	MCV>100

Anemia is detected by decreased hemoglobin, hematocrit or erythrocyte count. Anemia is classified as follows: microcytic hypochromic anemia; normocytic normochromic anemia, e.g., hemolytic anemia; and macrocytic anemia.

MICROCYTIC HYPOCHROMIC ANEMIA: Microcytic hypochromic anemia is characterized by MCV<80 & MCHC<30. Causes of microcytic hypochromic anemia are given in the next Table:

Causes of Microcytic Hypochromic Anemia
Iron Deficiency Anemia
Thalassemia Syndromes
Defects in Porphyrin Synthesis
Hereditary Sideroblastic Anemia
Acquired Sideroblastic Anemia
Pyridoxine-Responsive Anemia
Lead Poisoning
Severe Protein Deficiency
Chronic Infection

Iron Deficiency Anemia: Causes of iron deficiency anemia are given in the next Table:

Causes of Iron Deficiency Anemia
Nutritional Iron Deficiency
Chronic Blood Loss, Usually G.I., Uterine, (e.g. Menstrual Blood Loss-Most Common Cause in Women), Hookworm
Achlorhydria and Gastrectomy
Defective Absorption e.g., Sprue or Steatorrhea; Billroth II Procedure for Peptic Ulcer Disease
Increased Demand, e.g., Pregnancy

Tests for iron deficiency anemia are given in the next Table:

Tests for Iron Deficiency Anemia

Test	Finding	Specimen
Peripheral Smear	Microcytic, Hypochromic	Fresh Capillary or Venous Blood; Fresh EDTA-Anticoagulated Blood may also be utilized
Red Cell Indices	Decreased	Lavender Top Tube
Serum Ferritin	Decreased	Red Top Tube
Zinc Protoporphyrin(ZPP)	Increased	Lavender Top Tube
Serum Iron	Decreased	Red Top Tube
Serum Iron Binding Capacity	Increased	Red Top Tube
Bone Marrow	Absence of Storage Iron	

(cont)

Anemia Panel (Cont)

Progressive Stages of Iron Deficiency: (1) Depletion of Iron Reserves: Bone marrow iron, dec.; serum ferritin, dec.; (2) Impaired Erythropoiesis: Serum iron, dec.; transferrin (total iron binding capacity) inc.; transferrin saturation, dec.; zinc protoporphyrin (ZPP), inc.; MCH, normal or dec.; MCV, normal or dec.; (3) Anemia: Hemoglobin, dec.; MCH, dec.; MCV, dec. (Finch CA. JAMA. 1984; 251:2004).

Iron Deficiency Anemia Versus Thalassemia Minor: Thalassemia minor and iron deficiency are associated with a marked decrease in the MCV. Patients with iron deficiency and beta-thalassemia trait may be differentiated using serum ferritin and MCV (Hershko C. Acta Haematol. 1979; 62:236-239) and zinc (erythrocyte) protoporphyrin(ZPP). The characteristics of iron deficiency anemia and thalassemia minor are given in the next Table:

Iron Deficiency Anemia and Thalassemia Minor

Parameter	Iron Deficiency Anemia	Thalassemia Minor
MCHC(gm/dL)	<32.5	Slightly Reduced
MCV(cubic microns)	<80	Usually significantly reduced below 80
MCV/RBC	>13	<13
Serum Ferritin	Decreased	Normal
Zinc Protoporphyrin(ZPP)	Increased	Normal

The anemia of thalassemia minor is always mild; hemoglobin is usually between 10 and 13 gm/dL. The red cell count may be higher than normal. The diagnosis is made by finding an elevated HgA_2 or HgF on hemoglobin electrophoresis. In iron deficiency anemia, the serum iron is low, serum iron binding capacity is elevated and the serum ferritin is low. There is reduced or no iron in the bone marrow.

The MCV/RBC is useful in differentiating iron deficiency anemia and thalassemia minor; MCV/RBC is >13 in iron deficiency anemia but <13 in thalassemia minor. In addition, the zinc protoporphyrin test(ZPP) is normal in thalassemia minor but elevated in iron deficiency anemia; serum ferritin is normal in thalassemia minor but decreased in iron deficiency anemia.

Anemia of Chronic Disease(ACD): See IRON, SERUM for differential diagnosis of iron deficiency anemia, ACD and chronic renal disease.

NORMOCYTIC NORMOCHROMIC ANEMIA: Normocytic, normochromic anemia is characterized by normal MCV (76-99) and normal MCHC (>32). The tests to be considered in normocytic, normochromic anemia are given in the next Table:

Tests in Normocytic, Normochromic (Hemolytic) Anemia
Complete Blood Count with Morphology
Reticulocyte Count
Direct Antiglobulin (Coombs) Test
Hemoglobin Electrophoresis
Bone Marrow - If Reticulocyte Count Low
Red Cell Survival
Osmotic Fragility
Serum Bilirubin, Total, Direct and Indirect
Serum LDH and Isoenzymes
Serum Haptoglobin
Urinalysis
Renal Function Testing
Blood Type and Cross-Match

The causes of normocytic, normochromic anemia are given in the next Table:

Causes of Normocytic, Normochromic Anemia
Sudden Massive Blood Loss
Hemolytic Anemia (Elevated Reticulocyte Count):
Acquired: (Morphologic Abnormalities)
Immune Mediated
Coombs Positive (Autoantibody)
Paroxysmal Nocturnal Hemoglobinuria(PNH)
Microangiopathic
Disseminated Intravascular Coagulopathy(DIC)
Hemolytic-Uremic Syndrome
Congenital:
Defect in Hemoglobin: (Morphologic Abnormalities)
Hemoglobinopathies
Unstable Hemoglobins
Defect of Red Blood Cell Membrane: (Morphologic Abnormalities)
Spherocytosis
Elliptocytosis
Stomatocytosis
Deficiencies of Red Blood Cell Enzymes: (Normal Morphology)
Embden-Myerhof Pathway, Most Common is Pyruvate Kinase(PK) Deficiency
Pentose Pathway, Most Common is Glucose-6-Phosphate Dehydrogenase Deficiency
Underproduction (Normal or Low Reticulocyte Count)
Pure Red Cell Aplasia
Drugs
Leukemia
Aplastic Anemia
Myelophthisis
Refractory Anemia
Chronic Disease, e.g. Infection
Renal Failure
Liver Disease (also associated with Macrocytosis)
Hypothyroidism (also associated with Macrocytosis)

Normocytic, normochromic red cells are seen with acute blood loss, hemolysis or bone marrow failure.

Reticulocyte Count: The reticulocyte count is useful in differentiating bone marrow failure from other causes of normocytic, normochromic anemia as illustrated in the next Figure:

Reticulocyte Count

Reticulocyte Count (Normal 25,000-75,000/ microliter; Adult, 0.5-2.0% Newborn, up to 5%)

- Low (0.5%) → Bone Marrow Failure
- Elevated (>2%) →
  - 200,000 to 500,000/microliter:
    - Hemolytic Anemia
    - Acute Blood Loss
    - Therapy for Iron and Vitamin B_{12} Deficiency
  - >1,000,000/microliter:
    - Autoimmune Hemolytic Anemia
    - Pyruvate Kinase Deficiency (rarely)

An absolute reticulocyte count is obtained by multiplying the percentage of reticulocytes by the red cell count (Normal reticulocyte count is 25,000 to 75,000 per microliter).(cont)

Anemia Panel (Cont)

Hemolytic Anemia: The usual findings in hemolytic anemia are: reticulocyte count, elevated; haptoglobin, decreased; serum indirect bilirubin, elevated; serum lactic dehydrogenase, elevated. Specific tests for hemolytic anemias are given in the next Table:

Specific Tests for Hemolytic Anemias

Condition	Test
Autoimmune Hemolytic Anemia	Direct Coombs Test
Hemoglobinopathy	Hemoglobin Electrophoresis(Hgb S, C, etc.)
Hereditary Spherocytosis	Osmotic Fragility, Autohemolysis
Hereditary Nonspherocytic Hemolytic Anemia	Glucose-6-Phosphate Dehydrogenase(G-6PD) Pyruvate Kinase(PK)
Paroxysmal Nocturnal Hemoglobinuria	(see PAROXYSMAL NOCTURNAL HEMOGLOBINURIA SCREEN)

Underproduction: Aplastic anemia, myelophthisis and malignancy are associated with bone marrow failure; bone marrow failure is secondary to replacement of normal myeloid tissue as listed in the next Table:

Bone Marrow Failure

Condition	Replacement
Aplastic Anemia	Fat
Myelofibrosis	Fibrotic Tissue
Malignancy	Multiple Myeloma, Leukemia or Lymphoma

The anemia of chronic disease, infection, inflammation or malignant disease is characterized by low serum iron, low TIBC, normal or increased serum ferritin and normal or increased bone marrow iron stores. Definitive diagnosis of bone marrow failure is made by bone marrow biopsy.

Coombs-Positive Hemolytic Anemia: Coombs-positive hemolytic anemia is detected by the direct Coombs test; this test detects antibody or complement on the surface of the red cells. Coombs-positive hemolytic anemias are divided into warm-antibody and cold-agglutinin disease.

Warm-antibody disease is generally due to IgG antibody, occasionally complement and sometimes both are present. Some of the causes of warm-antibody disease are given in the next Table:

Causes of Warm-Antibody Hemolytic Anemia
Drug-Induced, e.g., Alpha Methyl Dopa, Quinidine, Penicillin
Collagen Vascular Disease
Lymphoproliferative Disorders
Viral Infection
Autoimmune

Cold-agglutinin disease is generally due to IgM antibody, but IgM is generally not detected; however, a positive Coombs test is due to complement that remains on the cell surface after IgM has been eluted. This disease is identified by elevated cold agglutinin titer plus a positive Coombs test.

MACROCYTIC ANEMIA: Macrocytic anemia is characterized by MCV>100. The causes of macrocytosis are given in the next Table:

Causes of Macrocytosis

- Megaloblastic Anemia
  - Folate Deficiency
  - Vitamin B_{12} Deficiency
- Spurious Macrocytic Anemia: Aplastic; Myelofibrosis
- Macrocytosis
  - Reticulocytosis Associated with some Hemolytic Anemia and Blood Loss
  - Liver Disease
  - Hypothyroidism
  - Normal Newborn

Folate deficiency is much more common than vitamin B_{12} deficiency. The laboratory tests to be considered in macrocytic anemia are given in the next Table:

Laboratory Aids in Macrocytic Anemia

Test	Result
Screening Tests:	
Hematologic Tests:	
Red Cell Indices:	
MCV	Increased
MCHC	Normal
Blood Counts:	
Red Cell Count(RBC)	Decreased
White Cell Count(WBC)	Decreased
Platelets	Decreased
Hemoglobin and Hematocrit	Decreased
Peripheral Blood Smear	Hypersegmentation of the Nuclei of the Polymorphonuclear Leukocytes: Normal Segmentation is as Follows:
	Lobes: 2 — Percent of Polymorphs: 20 to 40
	Lobes: 3 — Percent of Polymorphs: 40 to 50
	Lobes: 4 — Percent of Polymorphs: 15 to 25
	There is an average of 3.42 segmentations per 100 cells; in megaloblastic anemia segmentation is increased. The differential diagnosis of hypersegmentation: Megaloblastic anemia; Congenital hypersegmentation; Patients with chronic renal disease.
Chemical Tests:	
Lactic Dehydrogenase(LDH)	Increased
LDH Isoenzymes	Increase in LDH-1 and LDH-2
Indirect Bilirubin	Increased (1.0 to 1.5mg/dL)
Serum Iron	Increased
Plasma Volume	Increased
Specific Tests:	
Folate, Serum	Decreased in Folic Acid Deficiency
B12	Decreased in Vitamin B12 Deficiency (Approximately 10% False +)
Other Tests in B_{12}Def.:	
Gastrin	Increased
Antibody to Intrinsic Factor	Present in about 2/3 of patients
Achlorhydria	Almost all patients with Pernicious Anemia

ANGIOTENSIN-1-CONVERTING ENZYME(ACE)

SPECIMEN: Red top tube, separate serum and freeze.
REFERENCE RANGE: Varies with methodology. This test is difficult to interpret in subjects less than age 20; these subjects have a markedly increased ACE.
METHOD: Radiochemical assay
INTERPRETATION: ACE is elevated in some patients with active sarcoidosis; it is usually not elevated in patients with other lung diseases such as fungal or tuberculous disease. The serum level of ACE in granulomatous diseases of unknown cause is given in the next Figure (Katz P, et al. Ann Intern Med. 1981; 94:359-360):

Serum ACE Activity in Granulomatous Disease of Unknown Cause

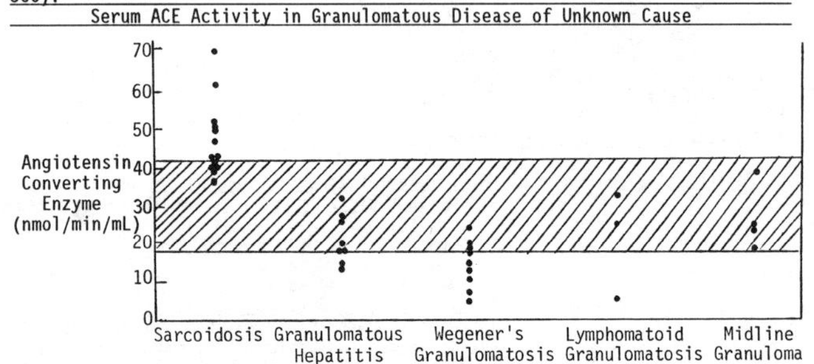

These data indicate that serum angiotensin-converting enzyme is useful in distinguishing sarcoidosis from these other granulomatous diseases.

The activity of serum ACE (SACE) reflects the stage of sarcoidosis. Sixty-seven percent of patients with Stage 1 disease (bilateral hilar adenopathy) had elevated SACE activity; 88 percent of patients with Stage 2 disease (bilateral hilar adenopathy plus parenchymal infiltrates) had elevated SACE activity; 95 percent of patients with Stage 3 disease (parenchymal infiltrates) had elevated SACE activity (Rohrbach MS, DeRemee RA. Clinical Laboratory Annual. 1981; 1). Serial SACE determinations are valuable for following the clinical course of sarcoidosis (Lawrence EC, et al. Am J Med. 1983; 74:747-756; Lufkin EG, et al. Mayo Clinic Proc. 1983; 58:447-451; Lieberman J, et al. Chest. 1983; 84:522-528).

SACE is produced by endothelial cells of blood vessels. In sarcoidosis, ACE values are increased in lymph nodes and it is suggested that the sarcoid granuloma particularly epithelioid cells and giant cells may be actively synthesizing ACE. The ACE level reflects the granuloma load in the body.

Sarcoidosis has an incidence of about 11 per 100,000 of population in the United States; in certain areas, an incidence of up to 70 per 100,000 has been reported. If affects about twice as many women as men and affects blacks about 10 to 17 times as commonly as whites. It has a mortality rate of 5 percent.

Other conditions are associated with elevated SACE as given in the following Table (Sharma OP. DM. 1990; 36:471-535; Lieberman J. Clinics in Lab Med. 1989; 9:745-755):

Conditions Associated with Elevated Serum ACE Level

Conditions Likely Confused with Sarcoidosis	Conditions Not Likely Confused with Sarcoidosis
Asbestosis	Coccidiomycosis
Berylliosis (75%)	Diabetes Mellitus (24-32%)
Granulomatous Hepatitis	Gaucher's Disease (100%)
Hypersensitivity Pneumonitis	Respiratory Distress Syndrome
Lennert's Lymphoma (75%)	Inflammatory Bowel Disease
Miliary Tuberculosis	Pulmonary Neoplasm
Primary Biliary Cirrhosis	Leprosy (53%)
Silicosis (21-42%)	Cirrhosis of the Liver (28.5%)
Histoplasmosis (16% of patients)	HIV Infection (Ouellette DR, et al. Arch Intern Med. 1992; 152:321-324)
	Hyperthyroidism (81%)
	ACE Inhibitors

Although SACE is not entirely specific for sarcoidosis, it is the most useful test available. Sensitivity for clinically active sarcoidosis is 90%, and specificity is 95% when measured on patients with various lung diseases (excluding patients with Gaucher's disease, hyperthyroidism, and leprosy) (Lieberman J. Clinics in Lab Med. 1989; 9:745-755).

ANION GAP, SERUM (Na^+, Cl^-, CO_2 CONTENT)

SPECIMEN: Red top tube, separate serum.
REFERENCE RANGE: Upper limit of normal: 15mmol/liter; borderline: 15-20mmol/liter; increased: >20mmol/liter; lower limit of normal: <5mmol/liter.
METHOD: Na^+ and K^+ by ion-specific electrodes of flame emission photometry; CO_2 content by electrode, colorimetrically or enzymatically.
INTERPRETATION: The causes of increased anion gap are given in the next Table:

Causes of Increased Anion Gap
Renal Failure (Phosphate and Sulfate Accumulation)
Ketoacidosis (Diabetic, Alcoholic, or Starvation)
Lactic Acidosis
Toxic Agents
Salicylates Poisoning (Salicylic Acid)
Ethylene Glycol (Oxalic Acid)
Methyl Alcohol (Formic Acid)
Paraldehyde (rarely)(Acetic Acid)
Propyl Alcohol

A mnemonic for anion gap acidosis is "A MUDPIE"; A=aspirin, M=methyl alcohol, U=Uremia, D=diabetic ketoacidosis, P=paraldehyde, I=idiopathic lactic acidosis, E=ethylene glycol.

Only four clinical conditions are associated with high anion gap metabolic acidosis; these are renal failure, ketoacidosis, lactic acidosis, and drugs or toxins. In the absence of renal failure or intoxication with drugs and toxins, an increase in anion gap is assumed to be due to ketoacids or lactate accumulation.

The anion gap is calculated by the equation:

$$\text{Anion Gap} = (Na^+) - [(Cl^-) + CO_2 \text{ Content}]$$

The normal anion gap is obtained by substituting the "normal" values for Na^+, Cl^- and CO_2 content into the above equation.

$$\text{Normal Anion Gap} = (140) - (103 + 25)$$
$$\text{Anion Gap} = 12\text{mmol/liter}$$

Anion gap has become a less sensitive test as a screen to detect hyperlactatemia. The most likely reason for this decrease in sensitivity is an upward shift in chloride normal values caused by wide application of chloride specific electrodes (e.g. ASTRA analyzers) which shifts the reference range for anion gap downward (to 3-11mmol/L)(Winter SD, et al. Arch Intern Med. 1990; 150:311-313). 60% of critically ill surgical patients with elevated lactic acid levels did not have an anion gap >16mmol/L (Iberti TJ, et al. Crit Care Med. 1990; 18:275-277).

The most common conditions associated with low anion gap are listed in the next Table:

Most Common Conditions Associated with Low Anion Gap
Multiple Myeloma
Hyponatremia
Hypoalbuminemia
Bromide Ingestion

Less common causes include other causes of increased, uncalculated cations such as hypercalcemia, hypermagnesemia, hyperkalemia, lithium intoxication, and polymyxin B administration.

ANION GAP, URINE

SPECIMEN: Urine
REFERENCE RANGE: Normal serum pH: value is positive (41 ± 9mmol/liter), Serum acidosis (pH <7.5): value is negative; a positive value indicates distal renal tubular acidosis.
METHOD: Na^+ and K^+ by ion-specific electrodes or flame emission photometry; CO_2 content by electrode, colorimetrically or enzymatically.
INTERPRETATION: Urinary anion gap may be useful in the evaluation of patients with hyperchloremic metabolic acidosis. A positive urine anion gap indicates distal renal tubular acidosis; measurement of plasma K^+ and urine pH may help in the diagnosis of hyperchloremic metabolic acidosis as given in the next Table (Battle DC, et al. N Engl J Med. 1988; 318:594-599):

Diagnosis of Hyperchloremic Metabolic Acidosis

Condition	Urine Anion Gap	Urine pH	Plasma K^+
Normal patient, exogenous ammonium chloride	Negative	<5.5	NL
GI Bicarbonate Loss	Negative	>5.5	NL-Low
Distal Renal Tubular Acidosis(RTA)			
Classic RTA	Positive	>5.5	NL-Low
Hyperkalemic Distal RTA	Positive	>5.5	Incr.
Primary Aldosterone Def.	Positive	<5.5	Incr.

In normal, non-acidotic patients, urine anion gap is positive; unmeasured anions (UA=bicarbonate, sulfate, phosphate, organic anions) exceed unmeasured cations (UC=ammonium, calcium, magnesium). Patients with hyperchloremic metabolic acidosis and normal renal compensatory mechanisms will increase ammonium excretion which decreases urine anion gap. Patients with distal renal tubular acidosis are unable to excrete ammonium.

The urinary anion gap is calculated by the following equations:

$$Cl^- + \text{Unmeasured Anions (UA)} = Na^+ + K^+ + \text{Unmeasured Cations (UC)}$$
$$UA - UC = Na^+ + K^+ - Cl^- = \text{urine anion gap}$$

ANTI-ACETYLCHOLINE RECEPTOR (see ACETYLCHOLINE RECEPTOR ANTIBODY

ANTIBODIES IN GRAVES' OPHTHALMOPATHY

SPECIMEN: Red top tube, separate serum.
REFERENCE RANGE: Negative
METHOD: ELISA (Enzyme-linked immunosorbent assay); Immunoblotting; Antibody-dependent cellular cytotoxicity (ADCC) assay. Commercial kits are not available; this test is not available in reference laboratories.
INTERPRETATION: Anatomically, patients with Graves' ophthalmopathy have abnormalities of extraocular muscles; muscle cells themselves are normal, but the muscles are enlarged by mucinous edema, proliferation of fibroblasts, and accumulation of lymphocytes. Clinical abnormalities may include proptosis due to enlarged muscles pushing the eye forward, periorbital and conjunctival edema due to compression of the orbital veins, diplopia due to restricted extraocular-muscle movement, and optic neuritis due to compression of the optic nerve (Utiger RD. N Engl J Med. 1992; 326:1772-1773).

Graves' ophthalmopathy occurs before, during or after thyroid dysfunction in 90 percent of cases; thus, the disorder occurs in the absence of any thyroid abnormality in 10 percent of cases. In patients with Graves' ophthalmopathy, 73 percent (40 to 55) had antibody against eye muscle antigen; tests were positive in 13 percent of patients with Hashimoto's thyroiditis, 2 of 12 patients with subacute thyroiditis, one of 20 patients with non-immunologic thyroid disease and none of 39 normal patients (Mengistu M, et al. Clin Exp Immunol. 1986; 65:19-27). In addition, patients with ophthalmopathy have been found to have antibodies that react with orbital fibroblasts (Bahn RS, et al. J Clin Endocrinol Metab. 1989; 69:622-628); these cells are stimulated to proliferate and produce collagen and glycosaminoglycans. Although the etiology of Graves' ophthalmopathy is unknown, antibodies to fibroblasts may be more important since eye muscle cells are normal.

Treatment of Graves' disease includes antithyroid drugs, thyroidectomy, or iodine-131. Compared to other therapy iodine-131 is more likely to be followed by exacerbation or development of Graves' ophthalmopathy (Tallstedt L, et al. N Engl J Med. 1992; 326:1733-1738).

ANTIBODY SCREEN (see COOMBS, INDIRECT)

ANTI-CARDIOLIPIN ANTIBODIES

Related Synonyms: Anti-phospholipid antibodies, lupus anticoagulant.
REFERENCE RANGE: Normals should be low or negative. Will vary between laboratories, but attempts have been made to standardize reference values in international workshops. Activity is expressed as GPL (IgG anti-phospholipid antibody) and MPL (IgM anti-phospholipid antibody). Values of 5-10 GPL and 3-10 MPL are 2SD above the mean in control blood bank sera. Results may be expressed in GPL/MPL units, or as low, medium and high. These correspond to: GPL-low <15, medium 15-80, high >80; and MPL-low <6, medium 6-50, high >50 (First and Second International Anti-Cardiolipin Standardization Workshop. Harris EN, et al. Clin Exp Immunol. 1987; 68:215-222; Harris EN. Am J Clin Pathol. 1990; 94:476-484).
METHOD: Methods and reagents vary, usually ELISA and Solid Phase RIA.
INTERPRETATION: Anti-cardiolipin antibodies are measured in the workup of patients with anti-phospholipid antibody syndrome. Anti-phospholipid antibodies are thought to be a subpopulation of anti-cardiolipin antibodies with lupus anticoagulant activity. These antibodies are found in some patients with SLE, thrombosis, recurrent late fetal loss, and thrombocytopenia. These antibodies may be detected in the ELISA/RIA type tests or they may be documented by clotting tests by the finding of a lupus anticoagulant (see LUPUS ANTICOAGULANT for further description). In general, the antibody tests are more sensitive and less specific. About 80% of patients with the lupus anticoagulant have anti-cardiolipin/anti-phospholipid antibody, while only 10-50% of patients with the anti-cardiolipin antibody have the lupus anticoagulant (reviewed by Lockskin MD. JAMA. 1992; 268:1451-1453). Patients may have serum containing the anti-cardiolipin/anti-phospholipid antibody without the associated clinical syndrome. Anti-cardiolipin antibodies and lupus anticoagulant activities may be physically separated (McNeil HP, et al. Br J Haematol. 1989; 73:506-513). Most work has concerned anti-cardiolipin antibodies of the IgG and IgM class. Recent studies suggests that anti-cardiolipin antibodies of the IgA class are important in patients with thrombocytopenia and vascular complications (Krilis SA. Clin Immunol. 1991; 11).

It appears that the antibody of interest binds to a complex of phospholipid and apolipoprotein H (beta 2 glycoprotein I). It is important to dilute the specimen while being tested with 10% serum (bovine) in order to preserve this cofactor.

Current feeling is that the lupus anticoagulant test is a better predictor of thrombotic complications, but that titers of anti-cardiolipin/anti-phospholipid antibody >40 GPL is more sensitive for identifying pregnancies at risk for fetal death (Derksen RH, et al. Ann Rheum Dis. 1988; 47:364-371; Lockskin MD. JAMA. 1992; 268:1451-1453).

Although anti-cardiolipin/anti-phospholipid antibody detection is less specific than the lupus anticoagulant detection it does have some advantages. These include: rapid screening of large numbers of specimens, ability to detect in serum or plasma, simpler transport and storage, quantifiable and readily standardized results, and ability to assay in the presence of concomitant anti-coagulation. The biological false-positive test for syphilis is not a good screening test for these antibodies. See LUPUS ANTICOAGULANT.

ANTI-CENTROMERE ANTIBODIES

SPECIMEN: Red top tube, separate serum.
REFERENCE RANGE: Negative
METHOD: Indirect immunofluorescent staining of metaphase cells (Fritzler MJ, Kinsella TD. Am J Med. 1980; 69:520-526); ELISA (Rothfield N, et al. Arthritis and Rheum. 1987; 30(12):1416-1419).
INTERPRETATION: Antibodies to centromeric chromatin are found in about 50 percent (range 19-96%: Soma Y, et al. Dermatologica. 1989; 178:16-19) of patients with the CREST syndrome: CREST is a mnemonic for the characteristics of this syndrome: Calcinosis cutis, Raynaud's phenomenon, Esophageal dysfunction, Sclerodactyly and Telangiectasia.

CREST is a variant of scleroderma or progressive systemic sclerosis (PSS). CREST patients differ from patients with PSS in that they do not have widespread involvement and the skin changes are confined to the hands and face.

Conditions associated with anti-centromere antibodies are listed in the next Table (Fritzler MJ, Kinsella TD. Am J Med. 1980; 69:520-526; Soma Y, et al. Dermatologica. 1989; 178:16-19; Powell FC, et al. Mayo Clin Proc. 1984; 59:700-706):

Anti-centromere Antibodies

Condition	Percent
CREST	50(19-96)
Raynaud's Disease	7-43
Systemic Sclerosis (Scleroderma)	9-22
Primary Biliary Cirrhosis	12
Sjögren's Syndrome	8
Systemic Lupus Erythematosus (SLE)	2-5
Mixed Connective Tissue Disease (MCTD)	0-7
Other: Myositis, Rheumatoid Arthritis, Chronic Active Hepatitis, Primary Pulmonary Hypertension	Rare

The titers of the anti-centromere antibodies are significantly less in patients with other diseases as compared to those with CREST (Tan EM. Hospital Practice. 1983; 79-84; Snaith ML. Brit Med J. 1983; 287:377-378).

Longitudinal follow-up of patients with anti-centromere antibodies (ACA) indicates that patients with Raynaud's Disease who test positive for ACA may later develop systemic sclerosis. Patients with other conditions who test positive for ACA are not at risk for developing systemic sclerosis. Patients with systemic sclerosis who are anti-centromere antibody positive generally have a more benign course than patients who are antibody negative (Takehara K, et al. Dermatologica. 1990; 181:202-206). Patients with CREST and positive anti-centromere antibodies have a better prognosis than CREST patients with anti-Scl-70 (anti-tropoisomerase I)(Steen VD, et al. Arthritis Rheum. 1988; 31:196-203). Anti-centromere antibodies are one of many antibodies that are associated with a speckled anti-nuclear antibody pattern. See also ANTI-NUCLEAR ANTIBODY (ANA).

ANTI-DNA

Synonym: Anti-double stranded DNA (anti-ds-DNA); Anti-native DNA (anti-n-DNA)
SPECIMEN: Red top tube, separate serum
REFERENCE RANGE: Depends on method used by reference laboratory; ≤ 1:10 in some laboratories.
METHOD: Crithidia immunofluorescent test, Farr Assay (ammonium sulfate precipitation), or membrane filter assay. The specificity and sensitivity of the immunofluorescent test was 100 percent and 67 percent respectively while the specificity and sensitivity of the Farr assay was 91 percent and 90 percent respectively (Burdash NM, et al. Ann Clin Lab Science. 1983; 13).
INTERPRETATION: This assay is useful in the evaluation of patients with SLE, predicting activity of disease, and monitoring therapy. High levels of antibody to native double-stranded deoxyribonucleic acid (anti-DNA) are found in patients with active systemic lupus erythematosus (SLE) but uncommonly in other diseases. Anti-DNA was elevated in about 70 percent of SLE patients when first seen and in 90 percent at some time during the clinical course (Weinstein A, et al. Am J Med. 1983; 74:206-216). There is good correlation between active SLE (particularly lupus nephritis) and elevated levels of anti-DNA antibodies; exacerbations of SLE are associated with a homogeneous or rim pattern on immunofluorescence. Inactive SLE usually is accompanied by low or absent serum levels of anti-DNA antibodies.

In individual patients, reappearance of anti-DNA antibodies in serum, or a rising titer of these antibodies, or persistence of elevated levels of anti-DNA antibodies, may correlate closely with subsequent exacerbation of disease; disappearance of the antibodies or a fall in titer may indicate remission.

Low levels of antibody to DNA are found in a number of other connective tissue diseases. In drug-induced lupus (hydralazine, procainamide), antibodies to DNA are absent.

The recommended approach to the diagnosis of a patient with suspected SLE is to first perform a test for anti-nuclear antibody. If positive, then perform additional tests such as the anti-DNA and anti-Smith (anti-Sm) tests. Both the anti-DNA and the older LE cell prep test probably give comparable information, however the anti-DNA is far superior and more specific (Clough JD, Calebrese LH. Cleve Clin Q. 1983; 50:59-68). The LE prep test is no longer offered by most hospitals.

ANTI-DEOXYRIBONUCLEASE-B (ANTI-DNase-B)

SPECIMEN: Red top tube, separate and refrigerate serum
REFERENCE RANGE: Preschool, <60 units; School age, <170 units; Adult, <85 units.
METHOD: The antigen, deoxyribonuclease, is incubated with patient serum dilutions, and a specific substrate (DNA methyl green) to the antigen is added to the mixture. The results are read as that dilution of patient's serum that inhibits the reaction of substrate and antigen. The end-point is a color change from green to decolorization.
INTERPRETATION: Elevation of DNase-B antibodies is especially evident in streptococcal pyodermal infections and acute glomerulonephritis, whereas the ASO response is weak. Detection of DNase-B antibodies is the most sensitive test for confirming postimpetigo nephritis. Streptococcal infections will result in a positive test for anti-DNase-B in 80 to 85 percent of patients. Anti-DNase-B titers rise slowly compared to ASO; peak levels occur at 4-8 weeks and persist for several months after streptococcal pharyngitis or impetigo (Ayoub EA. "Immune Response to Group A Streptococcal Infections." Pediatr Infect Dis J. 1991; 10:S15-S19).

ANTI-GLIADIN ANTIBODIES (see GLIADIN ANTIBODIES)

ANTI-GLOMERULAR BASEMENT MEMBRANE (GBM) ANTIBODY

SPECIMEN: Red top tube, separate serum and freeze.
REFERENCE RANGE: None detected
METHOD: RIA; EIA; IFA
INTERPRETATION: Anti-glomerular basement membrane antibody is present in the serum of patients with Goodpasture's syndrome (glomerulonephritis and pulmonary hemorrhage) or anti-basement membrane antibody-induced glomerulonephritis alone. These antibodies are directed towards a portion of the collagen type IV molecule. These antibodies tend to be present early in the course of the disease and last weeks to several years. Measurement of antibodies is useful in following the response to therapy. Some patients with lupus erythematosus have anti-glomerular basement membrane antibodies. Immunofluorescent studies on renal biopsies of patients with Goodpasture's syndrome demonstrate a linear fluorescent pattern.

Goodpasture's syndrome is a form of rapidly progressive glomerulonephritis. Full evaluation of this condition should also include assay for anti-neutrophil cytoplasmic antibodies (ANCA).

ANTI-HUMAN GLOBULIN TEST (see COOMBS, DIRECT)

ANTI-MICROBIAL SUSCEPTIBILITY (see SERUM BACTERICIDAL TITER)

ANTI-MITOCHONDRIAL ANTIBODIES (see MITOCHONDRIAL ANTIBODIES)

ANTI-NATIVE DNA (see ANTI-DNA)

ANTI-NEUTROPHIL CYTOPLASMIC ANTIGEN ANTIBODIES (ANCA)

SPECIMEN: Red top tube, serum.
METHOD: Immunofluorescence on tissue, EIA; ELISA; RIA.
REFERENCE RANGE: None detected; usually screen with 1:16 or 1:20 dilution of patient serum.
INTERPRETATION: Anti-neutrophil cytoplasmic antigen antibodies (ANCA) are measured in the diagnosis and management of Wegener's granulomatosis, periarteritis with glomerulonephritis (microscopic periarteritis), other vasculitis, idiopathic necrotizing and crescentic glomerulonephritis, and other renal diseases. The sensitivity and specificity of ANCA testing in multiple studies has been compiled (Goeken JA. J Clin Immunol. 1991; 11:161-174).

Staining of neutrophil preparations with positive ANCA patient serum will result in one of two patterns: c-ANCA or p-ANCA. The c-ANCA pattern exhibits a granular cytoplasmic pattern on ethanol fixed neutrophils and is probably due to antibodies against a 29 KD lysosomal proteinase termed serine proteinase 3 (Kallenberg CGM, et al. Immunol Today. 1991; 12:61-64). Antibodies against cationic protein 57 and cathepsin G may also be found (Goekin JA, already cited). The p-ANCA pattern exhibits a perinuclear distribution and is primarily due to antibodies directed against myeloperoxidase. Anti-elastase and anti-lactoferrin may also be detected in some of these patients. The p-ANCA pattern must be distinguished from anti-nuclear antibodies by the absence of antibodies to tissue sections of solid organs. The peri-nuclear distribution of the antigen appears to be due to a redistribution artifact during fixation. The immunofluorescent technique would appear to be the most reliable at present although there may be substantial day to day variation between results produced with different neutrophil preparations. Techniques such as ELISA appear to give a higher rate of false negatives.

Diseases associated with ANCA include: Wegener's granulomatosis (> 90% of patients)(mostly c-ANCA), idiopathic crescentic glomerulonephritis (80% of patients)(renal limited disease is mostly p-ANCA), microscopic periarteritis nodosa, Churg Strauss syndrome (50% of patients), and non-syndromic vasculitis (Jannette JC, Falk RJ. N Engl J Med. 1988; 319:1417). Functional asplenia and vasculitis has been reported in one patient with ANCA (Sunder-Plassmann G, et al. N Engl J Med. 1992; 327:437-438).

Measurement of ANCA levels is useful in following disease progression and response to therapy. Elevated levels of ANCA are associated with active disease and relapses, low or absent levels with inactive disease.

The combination of pulmonary hemorrhage and glomerulonephritis (Goodpasture's Syndrome) included atypical cases in which anti-glomerular basement membrane antibodies could not be detected. Glomerulonephritis with alveolar hemorrhage may now be divided serologically into three main groups partially based on ANCA: 1) ANCA-associated disease (e.g., Wegener's); 2) anti-glomerular basement membrane disease (e.g., Goodpasture's); 3) immune complex disease (e.g., SLE)(Jones DA, et al. NCMJ. 1990; 51:411-416).

The causes of diffuse pulmonary hemorrhage and glomerulonephritis are tabulated below:

Causes of Diffuse Pulmonary Hemorrhage and Glomerulonephritis
1. Anti-Glomerular Basement Membrane
Goodpasture's Syndrome
2. Immune Complex Disease
SLE
Henoch-Schönlein Purpura
MCTD
Cryoglobulinemia
3. Anti-Neutrophil Cytoplasmic Antibodies
Wegener's Granulomatosis
Microscopic Polyarteritis
Churg-Strauss Syndrome
Polyarteritis Nodosum

Adapted from: Discussion by Rosenblum ND. Case 16 - 1993. N Engl J Med. 1993; 328:1183-1190.

ANTI-NUCLEAR ANTIBODY (ANA)

Synonym: Fluorescent Anti-Nuclear Antibody (FANA)
SPECIMEN: Red top tube, separate serum. The serum should be stored at 2°C to 4°C if the specimen is to be analyzed within 24 hours; otherwise, the serum specimen should be frozen. Do not use grossly hemolyzed serum.
REFERENCE RANGE: Negative at 1:20 dilution; in some laboratories <1:40.
METHOD: In the fluorescent ANA test, the patient's serum is tested for the presence of antibody by detecting the binding of antibody to a cryostat-prepared nuclear-tissue preparation. In many laboratories a standardized preparation of cultured HEp-2 cells is used. The peroxidase method may be used to detect anti-nuclear antibodies, but patterns are not obtained. In an effort to simplify and standardize assays for ANA, procedures are being developed to perform these tests by ELISA techniques.
INTERPRETATION: This is a screening test for connective tissue diseases. A characteristic finding in the serum of patients with systemic connective tissue diseases is antibodies to nuclear antigens; these are referred to as anti-nuclear antibodies (ANA). The connective tissue diseases associated with the presence of anti-nuclear antibodies in the serum are given in the next Table:

Connective Tissue Diseases with Anti-Nuclear Antibodies
Systemic Lupus Erythematosus
Drug-Induced Lupus-Like
Procainamide
Hydralazine
Mixed Connective Tissue Disease (MCTD)
Sjögren Syndrome
Scleroderma
CREST Syndrome
Polymyositis-Dermatomyositis
Rheumatoid Arthritis

Lack of ANA essentially rules out SLE (over 96-99% positive). ANA's are found in approximately 3-4% of the non-diseases population. There are four fluorescent patterns by immunofluorescence; these are homogenous, peripheral, speckled and nucleolar patterns. These patterns reflect the presence of many types of anti-nuclear antibodies. The correlations of pattern, antibodies and connective tissue diseases are given in the next Table:

Correlations with Fluorescent Antinuclear Antibody Patterns

Pattern	Correlation
Peripheral	System Lupus Erythematosus
Homogenous	Systemic Lupus Erythematosus and Other C.T. Diseases
Speckled	Systemic Lupus Erythematosus
	Mixed C.T. Disease (MCTD)
	Sjögren Syndrome
	Polymyositis-Dermatomyositis
	Scleroderma
Nucleolar	Scleroderma
	Sjögren Syndrome

The peripheral pattern is relatively specific for lupus erythematosus; the nucleolar pattern is found in scleroderma and Sjögren syndrome. The speckled pattern is associated with a variety of nuclear antigens including antibodies to a specific nuclear antigen-the centromere of chromosome spreads (Burnham TK, Kleinsmith D'AM. Semin Arthr Rheum. 1983; 13:155-159) and RNP and non-histone proteins.

Some of the drugs that are associated with false positive ANA tests are as follows: p-aminosalicylic acid(PAS), carbamazepine(Tegretol), chlorpromazine (Thorazine), ethosuximide(Zarontin), griseofulvin, hydralazine(Apresoline), isoniazid(INH), mephenytoin(Mesantoin), methyldopa(Aldomet), penicillin, phenylbutazone(Butazolidin), phenytoin(Dilantin), hydantoin group, primidone (Mysoline), procainamide, propylthiouracil, trimethadione(Tridione). Also some drugs in the following categories: heavy metals, iodides, oral contraceptives, tetracyclines, thiazide diuretics and thiourea derivatives including sulfonamides.

ANTI-PHOSPHOLIPID ANTIBODIES (see ANTI-CARDIOLIPIN ANTIBODIES)

ANTI-RIBONUCLEAR PROTEIN (ANTI-RNP)

Synonyms: Anti-Nuclear RNP, Anti-Small Nuclear RNP, Anti-snRNP, Anti-UlsnRNP
SPECIMEN: Red top tube, separate serum as soon as possible and store at 2°C to 4°C if the specimen is to be analyzed within 24 hours; otherwise, the serum specimen should be frozen. Do not use grossly hemolyzed serum. If the specimen is sent by mail, the serum should be frozen.
REFERENCE RANGE: Negative
METHOD: Antibodies may be identified by EIA or immunodiffusion; titers are determined by hemagglutination. The Smith (Sm) antigen is not destroyed by ribonuclease while RNP is destroyed by ribonuclease; this is used to distinguish between antibodies to Sm and RNP in the hemagglutination test. Some of these tests are now employing genetically engineered antigens.
INTERPRETATION: Very high titers of antibody to nuclear RNP are found in mixed connective tissue disease(MCTD) occurring in 95% to 100% at titers ≥ 1:10,000 (Tan FM. Hospital Practice. 1983; 79-84).

Low titer antibody to nuclear RNP may also be present in systemic lupus erythematosus and progressive systemic sclerosis (scleroderma). The RNP antigens are one of a group of saline extractable antigens (Smith, SS-A/Ro, and SS-B/La are others). These antibodies are directed against small nuclear ribonuclear protein complexes consisting of small RNA's (labeled U_1, U_2, etc.) and associated polypeptides and small proteins. Both anti-RNP and anti-Smith antibodies are directed against these complexes but have different properties and clinical significance as tabulated below.

Comparison of RNP and Sm, Antigen, and Antibody

	Smith(Sm)	RNP
Reaction to RNAse	Stable Ag	Destroys Ag
Presence of Ab in SLE	20-30%	Low Titer
Presence of Ab in MCTD	No	High Titer

ANTI-SMITH(Sm) ANTIBODIES

SPECIMEN: Red top tube, separate serum.
REFERENCE RANGE: Negative
METHOD: Antibodies may be identified by EIA, immunodiffusion, or determined by hemagglutination of antigen coated, ribonuclease-treated cells. The Sm antigen is not destroyed by ribonuclease while RNP is destroyed by ribonuclease; this is used to distinguish between antibodies to Sm and RNP in the hemagglutination test.
INTERPRETATION: Antibodies to Sm antigen are useful in diagnosis of SLE. They are not useful in monitoring disease activity. Antibodies to Sm antigens are found almost exclusively in systemic lupus erythematosus (SLE), and are found in 20 to 30 percent of patients with SLE as determined by EIA. Clinical data indicate that there is an association between the presence of anti-Sm antibody with a higher incidence of vasculitis, resulting in peculiar visceral manifestations responding poorly to therapy (Beaufils M, et al. Am J Med. 1983; 74:201-205). Anti-Sm antibodies do not fluctuate much with disease activity.

ANTI-SMOOTH MUSCLE(ASM) ANTIBODY

SPECIMEN: Red top tube, separate serum and refrigerate
REFERENCE RANGE: Negative; positive at serum dilution of 1:10 or 1:20
METHOD: Indirect immunofluorescence technique. The patient's serum is diluted, usually 1:10, and added to fresh (frozen) tissue from mouse stomach or rat uterus or other appropriate tissue containing smooth muscle. The serum and tissue are incubated, and fluorescein-conjugated antiglobulin is added. The sections are examined by fluorescence microscopy.
INTERPRETATION: This test is useful in differentiating chronic active hepatitis(CAH) (lupoid hepatitis) from extrahepatic biliary obstruction, drug-induced liver disease, viral hepatitis and other conditions involving the liver. Conditions associated with increased anti-smooth muscle antibody are given in the next Table:

Increased Anti-Smooth Muscle Antibody

Condition	Percent Positive at 1:10 Dilution	Titer Suggestive of Diagnosis
Chronic Active Hepatitis(CAH) (Lupoid Hepatitis Type)	50-80	Generally, Titer Between 1:80 to 1:320 and Persists
Viral Hepatitis, (Acute) Viral, Infectious Mononucleosis	1-2	Titers Below 1:80 and Transient
Biliary Cirrhosis	0-50	Titers Between 1:10 to 1:40
Asthma	20	Low Titers
Normal Patients	3	Reference: Hawkins BR, et al. J Clin Immunol. 1979; 2:211-215.

Other conditions associated with anti-smooth muscle antibodies include uveitis, alopecia, primary pulmonary hypertension, CMV or Mycoplasma pneumonia infection, cancer, and alcoholic liver disease.

Anti-smooth muscle antibodies are a heterogeneous group of autoantibodies which react to at least two antigens - actin microfilaments and intermediate filaments. The lack of specificity of ASM antibodies limits the clinical usefulness of this test (Fusconi M, et al. J Immunol Methods. 1990; 130:1-8; Jorde R, et al. Acta Med Scand. 1987; 222:471-475).

Anti-Streptococcal Exozymes

ANTI-STREPTOCOCCAL EXOZYMES (see STREPTOZYME)

Anti-Streptococcal-O (ASO)

ANTI-STREPTOCOCCAL-O (ASO)

SPECIMEN: Red top tube, separate serum; a fasting specimen is preferred. Interferences: hemolysis, lipemia, contamination.

REFERENCE RANGE: Negative: <200IU/mL for school age children, <100IU/mL for preschool children and adults. A four-fold rise in ASO titer is consistent with an immunologic response to streptococcus.

METHOD: Latex Agglutination test

INTERPRETATION: A rise in titer of serum antibodies to streptolysin O (an oxygen labile hemolysin derived from group A streptococci) is a sensitive test and may be the single best test to document antecedent streptococcal infections. Increasing titers are seen in 80% of patients with untreated group A streptococcal pharyngitis; early use of penicillin prevents the ASO titer from rising (Madaio MP, Harrington, JT. N Engl J Med. 1983; 309:1299-1302). ASO titers are elevated in 85-90% of patients with acute rheumatic fever. ASO titer is elevated in 30-40% of patients with streptococcal pyoderma and 50% of patients with poststreptococcal glomerulonephritis. Anti-DNase B is a more sensitive indication of streptococcal impetigo or its renal sequelae (Kotylo PK, et al. Check Sample -Immunopathology. 1987; 11:1-5).

ASO titers are clinically useful only if serum is obtained at 2 to 3 week intervals. A marked rise in titer or a persistently elevated titer indicates that a focus of streptococcus infection or poststreptococcal sequelae is present. A rise in titer begins about one week after infection and peaks two to four weeks later. Evidence suggestive of a recent group A streptococcal infection is a four-fold or greater rise in titer between acute and convalescent phase sera. In the absence of complications or reinfection, the ASO titer will usually fall to preinfection levels within 6 to 12 months.

False positive ASO titers may occur if highly lipemic, old, or contaminated serum is used; false negative ASO titers may occur if the streptolysin O enzyme used in the assay is inactivated by exposure to oxygen (Ayoub EM. Pediatr Infect Dis J. 1991; 10:S15-S19).

ANTITHROMBIN III

Synonym: Heparin Co-Factor

SPECIMEN: Blue (citrate) top tube; separate plasma and freeze in a plastic vial immediately; serum activity unpredictably lowered and is approximately 1/3 less than that of plasma. Patient must be off heparin for at least 6 hours before drawing specimen.

REFERENCE RANGE: Immunologic: 17-30mg/dL; functional: approximately 70 to 145% compared to pooled control plasma. Newborn babies have about half the normal adult activity and the adult antithrombin III level is reached by six months of age (McDonald MM, et al. Thrombosis Haemostasis. 1982; 47:56-58).

METHOD: Immunologic: radial immunodiffusion (RID); Functional: different techniques; synthetic substrate using thiobenzl substrate or assay of inhibition of thrombin in presence of heparin.

INTERPRETATION: Antithrombin III is measured to assess thrombotic risk. Antithrombin III levels between 50 and 75% indicate a moderate risk for thrombosis and levels less than 50% indicate a significant risk of thrombosis.

Conditions in which there is a decrease in the level of antithrombin III are listed in the next Table:

Decrease in Antithrombin III
Familial Antithrombin III Deficiency (Autosomal Dominant)
Chronic Liver Disease (Severe)
Nephrotic Syndrome
Disseminated Intravascular Coagulation (DIC)
Fibrinolytic Disorders
I.V. Heparin for Greater Than 3 Days
Carcinoma
Acute Leukemia
Post-Surgical Trauma (Major)
Deep Venous Thrombosis
Thrombophlebitis
Asparaginase Therapy
Gram Negative Septicemia
Women on Contraceptive Pills
Pregnancy

Antithrombin III (AT-III) reacts with the negatively charged mucopolysaccharide anticoagulant, heparin. Heparin alone has minimal anticoagulant effects but when combined with antithrombin III, the inhibitory action of antithrombin III on coagulation enzymes results in the inhibition of thrombus propagation, e.g., venous thrombosis and prevention of pulmonary embolism. Factors Xa and thrombin are most sensitive to the inhibitory effects of the heparin-AT III complex.

Familial Antithrombin III Deficiency: Familial antithrombin III deficiency is rare (prevalence between 1 in 2000 and 1 in 5000). High-risk patients are as follows: Patients under the age of 35 who have had massive thrombosis after surgery or trauma; any thrombosis in the course of a minor illness; all patients with mesenteric thrombosis; individuals with a strong family history of thrombosis; women in whom thrombosis develops while they are taking oral contraceptives or during early pregnancy; infants born to AT-III deficient mothers; relatives of AT-III deficient patients; and patients difficult to heparinize (Editorial, The Lancet. 1983; 1021-1022). Patients with AT-III deficiency are resistant to anticoagulation with heparin, but may be anticoagulated with coumadin.

Immunological assays should be cautiously interpreted in screening for familial AT-III deficiency since there are individuals with normal levels of variant AT-III (by immunological tests), but which have reduced or absent activity.

Antithrombin III concentrations in plasma are elevated in normal survivors of myocardial infarction and in patients taking warfarin (Coumadin; Vitamin K antagonist) and depressed in those receiving heparin. It is significant if the antithrombin III level is low in a patient taking warfarin indicating that warfarin is not working effectively.

Ref.: Fareed J, et al. Seminars in Thrombosis and Hemostasis. 1982; 84:288.

ANTI-THYROGLOBULIN ANTIBODY (see THYROID ANTIBODIES)

ANTI-THYROID MICROSOMAL ANTIBODY (see THYROID ANTIBODIES)

APOLIPOPROTEIN A-I

SPECIMEN: Lavender(EDTA) top tube, separate plasma; red top tube, separate serum. Patient should be fasted (see below).
REFERENCE RANGE: Greater than 140 mg/dL.
METHOD: Immunological or turbidimetric technique.
INTERPRETATION: Apolipoprotein A-I is the major protein component of high density lipoprotein (HDL); the composition of HDL is given in the next Table:

Composition of High Density Lipoprotein (HDL)

Composition	Percent
Lipid	50
Phospholipids	40-50
Cholesterol	32
Triglycerides	10
Protein	50
Apolipoprotein A-I	65
Apolipoprotein A-II	30

HDL contains 50 percent protein and 50 percent lipid. Apolipoprotein A-I (65 percent) and apolipoprotein A-II (30 percent) are the major protein components of the high-density-lipoprotein (HDL). Apo A-I serves as an activator of lecithin cholesterol acyltransferase (LCAT) and thus serves an essential role in the mobilization of lipid from peripheral tissues.

Apolipoprotein A-I has been measured in both plasma and in serum (refer to your laboratory to determine which to use). HDL and its constituent proteins are usually measured on specimens from 12-14 hour fasted specimens. However, some have stated that serum is adequate for assessment of cardiac risk by Apo B/Apo A-I ratio. Apo B reflects the LDL level and Apo A-I the HDL level. The ratios and risk assessment table is reproduced below (Bates HM. Laboratory Management. 1989; 27:52).

Coronary Artery Disease Risk Assessment Based on Apo B/Apo A-1 Ratio

Risk	Men	Women
Lowest	<0.69	<0.47
Low	0.7-0.9	0.49-0.64
Moderate	0.95-1.22	0.66-0.9
High	1.25-1.48	0.92-1.15
Highest	>1.5	>1.17

Apolipoprotein A-I is said to be more useful than HDL-cholesterol for identifying patients with coronary-artery disease (Maciejko JJ, et al. N Engl J Med. 1983; 309:385-389). It is not known whether the apolipoprotein A-I level is useful as a screening test for predicting the future occurrence of coronary-artery disease in the general population (Blackburn H. N Engl J Med. 1983; 309:426-428).

In a recent report, although total HDL was important in predicting the risk of MI, detailed laboratory measurements of apoproteins A-I or A-II, or the subfractions HDL3 did not add to the diagnostic accuracy (Stamper MJ, et al. N Engl J Med. 1991; 325:373-381). Measurement of individual apolipoproteins has not been recommended by the National Cholesterol Education Program Expert Panel. JAMA. 1993; 269:3015-3023. Apo A-I is decreased in patients with Tangier disease (absent alpha lipoprotein).

See also APOLIPOPROTEINS; APOLIPOPROTEIN B; CARDIAC RISK ASSESSMENT.

APOLIPOPROTEIN B

SPECIMEN: Fasted (12-14 hours) serum, red top tube.
REFERENCE RANGE: 70-110mg/dL. Will vary by method, age, and sex.
METHOD: Immunological and turbidimetric techniques.
INTERPRETATION: Apolipoprotein B is the major apoprotein of LDL and VLDL. Following binding to the Apo B (LDL) receptor on the surface of hepatocytes, Apo B can regulate cholesterol synthesis by inhibition of HMG CoA reductase. Since fasting will affect lipoprotein levels, most laboratories request a fasted specimen for determinations. Results may also be reported as a ratio of apolipoprotein B/apolipoprotein A-I.

Elevated Apo B levels reflect increased LDL and associated increased myocardial risk. When measured in conjunction with Apo A-I levels, the Apo B/Apo AI ratio can be used to assess the level of risk (see APOLIPOPROTEINS and APOLIPOPROTEIN A-I). Low or absent levels of Apo B results in the condition abetalipoproteinemia (Rader DJ, Brewer HB. JAMA. 1993; 270:865-869. This is associated with low LDL levels (less than 50 mg/dL), and acanthosis on blood smears. These patients lack a microsomal triglyceride transfer protein (Wetteran JR, et al. Science. 1992; 258:999-1001).

Some workers claim that apoliprotein measurements are more accurate predicators of cardiac risk than conventional screening (Sniderman A, et al. Proc Nalt Acad Sci USA. 1980; 77:604-608). However others state that apolipoprotein A-1 and apolipoprotein B are surrogate measurements for HDL and LDL and do not present any advantages over total and HDL-cholesterol (Check WA. CAP Today. 1993; 717:1-32).

See also APOLIPOPROTEINS; APOLIPOPROTEIN A-I; CARDIAC RISK ASSESSMENT.

APOLIPOPROTEINS

SPECIMEN: Lavender(EDTA) top tube, separate plasma; or red top tube, separate serum. Patient should be fasted.
REFERENCE RANGE: Not clearly defined.
METHOD: Immunonephelometric Assay, Immunoturbidimetric Assay, RIA, Radial Immunodiffusion Assay. Standardization of apolipoprotein measurements has been undertaken by the Committee on Apoproteins of the International Federation of Clinical Chemistry. Test systems from 28 laboratories were evaluated. Calibration differences for apo A-I were reasonable (CV=7%) but for apoB, wide differences were observed (CV=19%). After uniform calibration, the CV improved from 19% to 6%. International reference materials are being developed (Marcovina SM, et al. Clin Chem. 1991; 37:1676-1682).
INTERPRETATION: Apolipoprotein A-I is the major protein component of high density lipoprotein (HDL); the composition of HDL is given in the next Table:

Composition of High Density Lipoprotein (HDL)

Composition	Percent
Lipid	50
Phospholipids	40-50
Cholesterol	32
Triglycerides	10
Protein	50
Apolipoprotein A-I	65
Apolipoprotein A-II	30

HDL contains 50 percent protein and 50 percent lipid. Apolipoprotein A-I (65 percent) and apolipoprotein A-II (30 percent) are the major protein components of the high-density-lipoprotein (HDL). HDL has two main subclasses - one is associated only with apolipoprotein A-I (LpA-I) and a second associated with both apolipoprotein A-I and apolipoprotein A-II (LpA-I:A-II). Antiatherogenic activity has been associated with LpA-I.

Apolipoprotein B-100 (ApoB) is the major protein component of both very-low-density lipoprotein (VLDL) and low-density lipoprotein (LDL). There is one molecule of ApoB per lipoprotein molecule; serum apoB levels reflect total VLDL and LDL concentration.

Apolipoprotein E (ApoE) is a constituent of VLDL and HDL.

Apolipoprotein levels may be more useful than HDL-cholesterol for identifying patients with coronary-artery disease. Results of selected recent studies are given in the following Table:

Apolipoprotein Levels in Coronary Artery Disease (CAD)

Study	Results
Kwiterovich PO, et al. Am J Cardiol. 1992; 69:1015-1021	203 patients, elective diagnostic coronary arteriography. Apo A-I and ApoB are better predictors of premature CAD than traditional risk factors.
Buring JE, et al. Circulation. 1992; 85:22-29	283 patients following first MI, 275 controls. Apo A-I and Apo A-II inversely related to MI and better predictors of risk than traditional risk factors.
Stampfer MJ, et al. N Engl J Med. 1991; 325:373-381	246 men with MI, 246 controls. Little or no predictive value for lipoprotein levels.
Reinhart RA, et al. Arch Intern Med. 1990; 150:1629-1633	502 patients elective diagnostic coronary arteriography. ApoB and ratio of Apo A-I/ApoB provide additional information above that given by traditional risk factors.
Kottke BA, et al. Mayo Clin Proc. 1986; 61:313-320	304 patients, 135 controls. Apo A-I and Apo A-II are better markers for CAD than traditional lipid levels.

The clinical utility of apoprotein levels is not known.

See also APOLIPOPROTEIN A-I; APOLIPOPROTEIN B; CARDIAC RISK ASSESSMENT.

APPENDICITIS, PANEL

The test panel for acute appendicitis is given in the next Table:

Test Panel for Acute Appendicitis
White Cell Count (WBC)
White Cell Differential for Percent and Band Neutrophils
C-Reactive Protein (CRP)
Tests to Rule Out Urinary Tract Infection and Ectopic Pregnancy

Automated methods for white cell differential, e.g., (cytochemical reaction, Technicon) may not separate band neutrophils from mature polymorphonuclear leukocytes.

The total white blood count (>10,500), and the percent neutrophils (>75%) have the highest sensitivities (81-84%) for the diagnosis of acute appendicitis. The sensitivity (positivity in disease) and specificity (negativity in subjects without the disease) of combinations of laboratory tests in the diagnosis of acute appendicitis are given in the next Table (Marchand A, et al. Am J Clin Pathol. 1983; 80:369-374):

Sensitivity and Specificity of Laboratory Tests in Diagnosis of Acute Appendicitis

Test Combinations					
WBC (Cells/cu mm)	Cytochem. Neut.	Manual bands	C-Reactive Protein (mg/dL)	Sensitivity Percent	Specificity Percent
>10,500	>75%	---	>1.2	97	42
>10,500	---	>11%	>1.2	100	47
>10,500	7,880/cu mm	---	>1.2	97	53
>10,500	---	1,150/cu mm	>1.2	100	47

These data indicate that when the results of three tests (WBC, neutrophils, CRP) are within reference range, the patient is unlikely to have acute appendicitis. However, increase in total white cell count is a late finding. The incidence of perforation in infants with acute appendicitis is very high, probably reflecting the difficulty in interpreting clinical signs and symptoms. Keep in mind that the morbidity and mortality of missing a case of acute appendicitis with subsequent rupture and abscess formation outweigh the morbidity and mortality associated with removal of a normal appendix. The value of laboratory testing may be to prompt observation in a patient with equivocal symptoms of appendicitis and normal laboratory studies (Hoffmann J, Rasmussen OO. Br J Surg. 1989; 76:774-779).

In adult men, a triad of nonoperative diseases (mesenteric adenitis, gastroenteritis and abdominal pain of unknown etiology) may be confused with acute appendicitis. Factors favoring appendicitis as the correct diagnosis are given in the following Table (Doherty GM, Lewis FR. Emerg Med Clin N Amer. 1989; 7:537-553):

Adult Men - Factors Favoring Appendicitis
Localizing Right Lower Quadrant Tenderness
Rebound Tenderness
WBC Elevated (>11,500)
Neutrophil Fraction Increased (>75%)

In ovulating women, pelvic inflammatory disease may be confused with acute appendicitis. Factors favoring pelvic inflammatory disease (PID) as the correct diagnosis are given in the following Table (Doherty GM, Lewis FR. Emerg Med Clin N Amer. 1989; 7:537-553):

Ovulating Women - Factors Favoring PID Versus Appendicitis
Absence of Anorexia, Nausea, and Vomiting
Duration of Pain > 2 Days
Onset Within 7 Days of Menstruation
History of Venereal Disease
Abdominal Tenderness Outside Right Lower Quadrant
Cervical Motion Tenderness
Bilateral Adnexal Tenderness

The diagnosis of appendicitis is more difficult in infants, young children and the elderly. Recognition of appendicitis is more difficult and patients may present later. The incidence of perforation is higher in patients under age 10 years or over age 60 (Scher KS, Coil JA. South Med J. 1980; 73:1561-1563).

APT TEST FOR FETAL HEMOGLOBIN

SPECIMEN: Bloody rectal discharge or vomitus from infant.
REFERENCE RANGE: Negative for fetal blood
METHOD: Mix specimen with an equal quantity of tap water; centrifuge or filter. Supernatant must have pink color to proceed. Add 1 part of 0.25 NaOH (1g NaOH, add water to 100mL) to 5 parts of supernatant.
Fetal Hemoglobin: Pink color persists over two minutes
Adult Hemoglobin: Pink color becomes yellow in two minutes or less.

A modification of the Apt test uses the same basic procedure described above. At exactly two minutes, the specimen is scanned in a spectrophotometer using a band width of 2nm and a scanning speed of 2000nm per minute between the wavelengths of 300 and 700nm. Adult hemoglobin loses peaks at 425, 541 and 575nm with alkaline denaturation (Crook M. Med Lab Sci. 1991; 48:346-347).
INTERPRETATION: This test is used to distinguish ingested maternal blood from gastrointestinal lesions of infants as a cause of bloody rectal discharge or bloody vomitus.

The theoretical basis for this test is the fact that fetal hemoglobin is relatively resistant to alkali denaturation as compared to adult hemoglobin (Apt L, Downey WS. J Pediatr. 1955; 47:6).

In rare cases, maternal blood may be relatively resistant to alkaline denaturation, such as hemoglobin variants (Ranier and Bethesda) and conditions associated with elevated fetal hemoglobin (Beta-thalassemia major and hereditary persistence of fetal hemoglobin)(Crook M, already cited).

ARSENIC, HAIR OR TOENAILS

SPECIMEN: Hair: Hair samples must have the root end and distal ends oriented and labeled for acute and subacute poisoning. Collect 0.5g of hair. Nails: Clippings at the end of toenails represent deposition of arsenic six months prior. Collect 0.5g of nail from all ten toenails.
REFERENCE RANGE: Hair: >1mcg/g(ppm); Nail: >2mcg/g(ppm). Reference range may vary somewhat depending on the laboratory. To convert traditional units in micrograms/g to international units in nmol/g, multiply traditional units by 13.35.
METHOD: Atomic absorption; anodic stripping voltammetry.
INTERPRETATION: Arsenic has a high affinity for the keratin in hair or nails. Hair: The earliest excess arsenic detectable in emerging hair (hair next to the root) appears two weeks after a dose of arsenic and may persist for months or years. Hair grows 1/2 inch per month (Weisaman W. "Laboratory Aids in Toxicological Problems," BioScience Laboratories).
Toenails: Toenails require from 6 to 9 months to grow so clippings obtained from the end represent deposition of arsenic six months prior.

Following exposure to arsenic, transverse white strips (Mees' lines) about 1mm wide and extending across the entire base of the nails appear in about two months; this band contains a very high concentration of arsenic.

ARSENIC, SERUM

SPECIMEN: Use Sarstedt syringes; draw blood into the syringe; cap the syringe; allow blood to clot and then centrifuge the specimen in the Sarstedt syringe; after centrifugation, pour the serum into a second Sarstedt syringe. Cap the syringe.
METHOD: Anodic stripping voltammetry; atomic absorption.
REFERENCE RANGE: <7.0mcg/dL. To convert traditional units in micrograms/dL to international units in micromoles/L, multiply traditional units by 0.1335.
INTERPRETATION: Arsenic combines with proteins, especially sulfhydryl groups; this affinity for intracellular proteins is responsible for the rapid removal of arsenic from the blood. Serum is useful only for diagnosis of acute arsenic poisoning. Arsenic clears the blood within four days after a significant dose. Blood is usually not the specimen of choice. See ARSENIC, URINE.

ARSENIC, URINE

SPECIMEN: Collect 24 hour urine in acid washed-containers. Use plastic containers (borosilicate, polyethylene or polypropylene); add 10% HCl solution to the container and allow to "soak" for 10 minutes; rinse with five volumes of tap water and then five volumes of deionized or distilled water. No preservative is needed. The patient should urinate at 8:00 A.M. and the urine is discarded. Then, urine is collected for 24 hours including the next day 8:00 A.M. specimen. Indicate 24 hour volume. A 50mL aliquot is used for analysis.
REFERENCE RANGE: Up to 20mcg/liter; high arsenic (seafood) diet: up to 100mcg/liter; industrial exposure: up to 200mcg/liter.
METHOD: Qualitative: Reinsch test; Quantitative: Atomic absorption; Anodic stripping voltammetry. Other methods are available.
INTERPRETATION: Arsenic is the most common acute heavy metal poisoning and is second only to lead as a chronic poison.

Urine is the preferred specimen from the living patient because high levels of arsenic persists in the urine for about a week after acute poisoning and up to about a month after chronic poisoning (McBay AJ. Clin Chem. 1973; 19:361-365). Urinary excretion of arsenic fluctuates and several negative urines are required to exclude arsenic poisoning. Arsenic combines with the proteins, especially sulfhydryl groups; this affinity for intracellular proteins is responsible for the rapid removal of arsenic from the blood. Blood is usually not the specimen of choice.
Acute: The acute symptoms relate to the central nervous, gastrointestinal and renal systems. There is headache, giddiness, dizziness, convulsions and coma; nausea, vomiting and profuse watery diarrhea ("rice-water" stools); shock and anuria.
Chronic: The chronic symptoms are multisystem: symmetrical hyperkeratosis of the hands and feet, symmetrical pigmentation, conjunctivitis, tracheitis, polyneuritis with sensory and motor involvement. Encephalopathy similar to Wernicke's syndrome and Korsakoff's psychosis may occur. Other organ involvement may include the GI tract (diarrhea), liver (hepatic damage), kidney (renal damage), blood (basophilic stippling, leukopenia with eosinophilia), and heart (prolonged QT interval).

Elemental arsenic is not toxic; inorganic trivalent arsenic trioxide is especially toxic; pentavalent arsenic is partially toxic. Inorganic arsenic is found in insecticides, herbicides, rodenticides, fruit sprays, moonshine, paint and smelters. Seafoods, particularly shellfish (oysters and mussels) are rich in arsenic and cause moderately elevated urinary arsenic levels; fortunately, the shellfish concentrate organic forms of arsenic which are apparently non-toxic.

2,3 Dimercaptopropanol (BAL, British anti-Lewisite) is used to treat arsenic poisoning; BAL is a reducing agent that converts arsenic to a less toxic form. The maximum excretion of arsenic in the urine occurs between three and four hours after intramuscular injection of BAL. Penicillamine, an oral drug, has been used successfully to treat acute and chronic arsenic poisoning. Hemodialysis removes arsenic from the circulation, and may be used in patients with concomitant renal failure. See review: Malachowski ME. "An Update on Arsenic." Clin Lab Med. 1990; 10:459-472.

ARTHRITIC (JOINT) PANEL

The diseases that involve joints are listed in the next Table:

Diseases Involving Joints
Osteoarthritis (Degenerative Joint Disease)
Traumatic Arthritis
Pseudogout
Gout
Rheumatic Fever
Acute Bacterial Arthritis
Tuberculous Arthritis
Rheumatoid Arthritis
Systemic Lupus Erythematosus

The laboratory tests for these conditions are as follows:

Laboratory Tests for Arthritis

Blood Tests:
- Complete Blood Count
- Erythrocyte Sedimentation Rate (ESR)
- Serum Uric Acid
- Immunologic Joint Disease:
  - Serum Protein Electrophoresis
  - Quantitation of Immunoglobulins
  - Serum Rheumatoid Factor (RF)
  - Serum Antinuclear Antibodies (ANA)
  - Serum Complement
  - Streptozyme and ASO Tests

Synovial Fluid Analysis (Hasselbacher P. "Arthrocentesis and Synovial Fluid Analysis." Primer on the Rheumatic Diseases, 9th ed., 1988; 55-60)
- Always:
  - Gross Appearance, Volume
  - Wet Preparation
  - Polarized Light Microscopy
  - WBC Concentration
- Usually:
  - WBC Differential Count
- When Indicated:
  - Gram Stain
  - Culture
- Tests of No, Limited, or Uncertain Value:
  - Protein
  - Glucose
  - Enzymes
  - Complement
  - Immune Complexes
  - Rheumatoid Factor or ANA
  - Viscosity
  - Mucin Clot
  - RBC Count

Mucin Clot (Ropes) Test: The mucin clot test reflects polymerization of hyaluronate. In the mucin clot test, a few drops of synovial fluid are added to 20mL of 5% acetic acid. Normally, a mucin clot forms within 1 minute; a poor clot indicates inflammation. This test was once one of the diagnostic criteria for rheumatoid arthritis. More accurate and objective tests are now available to test whether fluid is inflammatory, so this test is of limited value (Hasselbacher P. "Arthrocentesis and Synovial Fluid Analysis." Primer on the Rheumatic Diseases, 9th ed., 1988; 55-60). (cont)

Arthritic (Joint) Panel (Cont)

The diseases and useful serum tests (also skin test for T.B.) are given in the next Table:

Tests, Arthritis

Disease	Serum Tests
Osteoarthritis	None
Traumatic Arthritis	None
Pseudogout	None
Gout	Serum Uric Acid Usually Elevated
Rheumatic Fever	Immunological Response to Group A, Beta Hemolytic Streptococci; Tests: Streptozyme and Antistreptolysin O(ASO) Tests
Bacterial Arthritis	Blood Culture
Tuberculous Arthritis	T.B. Skin Test
Rheumatoid Arthritis	RF; ANA Positive, see Connective Disease Panel
Lupus Erythematosus	ANA; Anti-Double-Stranded DNA; Anti-Smith; Serum Complement

The most crucial decision in the investigation of joint fluid is to differentiate inflammatory from noninflammatory arthritis, and to identify cases of septic arthritis. Typical findings and diagnosis in noninflammatory, inflammatory, and septic arthritis are given in the following Table (Hoaglund FT, Maale G. Ortho Clin N Amer. 1979; 10:299-305; Hasselbacher P, already cited):

Diagnosis of Noninflammatory, Inflammatory, and Septic Arthritis

Test	Noninflammatory	Inflammatory	Septic
Clarity	Transparent	Translucent or Opaque	Opaque
Color	Straw, Yellow	Yellow or Yellow-Green	Variable
WBC/cumm	<2000	<100,000	>100,000
Poly's	<50%	50-90%	>95%
Diagnosis	Osteoarthritis	Rheumatoid Arthritis Reiter's Disease Partially Treated Septic Joint Fungal or Viral Infection Gout or Pseudogout Acute Rheumatic Fever	Bacterial Infection

There is considerable overlap of laboratory tests used to differentiate between noninflammatory, inflammatory and septic arthritis. Low WBC count (<2000/cubic mm) is seen in some patients with systemic inflammatory conditions such as systemic lupus erythematosus or scleroderma. Septic joints may have WBC counts less than 100,000 per cubic millimeter, particularly gonococcal or tuberculosis infection. Immunocompromised patients may not show a typical response to bacterial infection.

Acute Bacterial Arthritis: Only a few bacteria are responsible for most cases of bacterial arthritis, which is most often the result of bacteremia. Gonococcal infection is the most common cause of septic arthritis in young adults, but may occur at any age. The nongonococcal causes of septic arthritis with age of patient, are given in the next Table (Parker RH in Infectious Diseases, Hoeprich PD, Jordan MC, ed., JB Lippincott Company, Philadelphia, Fourth Ed., 1989; 1377):

Nongonococcal Causes of Bacterial Arthritis and Age

Organism	<2 yrs (%)	2-16 yrs (%)	>16 yrs (%)
Gram-Positive Cocci			
Staphylococcus Aureus	15	50	45
Streptococcus Pyogenes	2	2	4
Streptococcus Pneumoniae	10	5	5
Other Streptococcus Species	10	20	0
Gram-Negative Bacilli			
Enterobacteriaceae	10	10	20
Pseudomonas Species	5	3	5
Others			4
Fastidious Gram-Negatives			
Hemophilus Species*	45	5	5
Neisseria Meningitidis	3	2	
Anaerobes		3	2
Polymicrobial		1	10

*The incidence of Hemophilus influenza infection has decreased dramatically since initiation of routine vaccination against this organism.

The diagnosis of bacterial arthritis is particularly important. Tests used to diagnose bacterial arthritis are given in the following Table (Goldenberg DL, Reed JI. N Engl J Med. 1985; 312:764-771):

Diagnostic Tests for Bacterial Arthritis

Definitive or Highly Suggestive:
- Immediate Joint Fluid Aspiration Culture and Gram's Stain Positive
  - Leukocyte Count Markedly Elevated
- Blood Culture Positive
- Positive Cultures From Other Sites

Helpful But Not Definitive:
- Peripheral Blood Leukocytosis
- Elevated Erythrocyte Sedimentation Rate
- Antigen Determination
- Radionuclide Scan
- X-ray Films (e.g., Loosening of Prosthetic Joint)
- Response to Antibiotics (e.g., Disseminated Gonococcal Infection)

The presence of specific findings on examination of joint fluid may be highly suggestive of particular diseases, as listed in the following Table (Hasselbacher P, already cited):

Joint Findings Suggestive of Specific Diagnosis

Joint Findings	Diagnosis
Lupus Erythematosus(LE) Cells	SLE
Large Histiocytes with Ingested Neutrophils	Reiter's Disease
Rheumatoid Arthritis(RA) Cells (Neutrophils with Refractile Peripheral Inclusions)	Nonspecific
Sickled Red Cells	Sickle Hemoglobinopathy (Homo- or Heterozygous)
Crystals	Gout or Pseudogout

Crystals: Pseudogout: Using polarized light, the calcium pyrophosphate dihydrate crystals appear as rods, rectangles or rhomboids and weakly positive birefringent.

Gout: Crystals are seen in about 90% of patients during attacks of acute gouty arthritis and in about 75% between attacks. Using polarized light, monosodium urate crystals appear as birefringent rods or needles, strongly negative birefringent; the crystals appear yellow when parallel to the axis of the red compensator and blue when perpendicular to the axis. (Mnemonic: "U Pay Peb"; Monosodium urate crystals = Parallel, Yellow; Perpendicular, Blue. Turi M, personal communication). The pyrophosphate crystals of pseudogout have the opposite orientation. Note whether crystals are intracellular or extracellular. Intracellular crystals suggest urate as cause of acute arthritis. Incubate with uricase to confirm impression (McCarty DJ, Hollander JL. Ann Intern Med. 1961; 54:452).

ASCITIC FLUID ANALYSIS

SPECIMEN: Collect 3 tubes; purple (EDTA) for cell count; special vials for microbiological studies, aerobic and anaerobic; red or green (heparin) top tube for chemistries. Refrigerate; record volume and color. Culture, Gram stain and Ziehl-Neelsen staining should always be done on a centrifuged specimen.
If no fluid is aspirated then one liter of normal saline or Ringer's lactate (10 to 20mL/kg body weight) is infused over 15 to 20 minutes. After body movement to disperse fluid, the lavage fluid is siphoned back from the peritoneal cavity into the original container and examined.
REFERENCE RANGE: See below; normal volume, <50mL.
METHOD: The tests that should be done are given in the next Table:

Tests of Ascitic Fluid
Specific Gravity
Total Protein (Serum and Ascitic Fluid)
Lactate Dehydrogenase (LDH) of Serum and Ascitic Fluid
White Blood Cell Count and Differential (Vij D, et al. JAMA. 1983; 249:636-638)
Red Cell Count
Culture, Gram Stain, Ziehl-Neelsen Stain for Mycobacteria
Cytologic Examination
CEA for Tumor
Alpha-Fetoprotein (if Hepatoma or Endodermal Sinus Tumor of Ovary Suspected)
Amylase and Lipase (if Pancreatitis Suspected)
Glucose (Suspected Rheumatoid Disease or Infection)

If fluid is green, test for bilirubin; if positive, consider cholecystitis, pancreatitis, perforated intestine or gallbladder, or peptic ulcer.

If there is suspicion of bladder tap rather than ascitic fluid, measure blood urea nitrogen (BUN).

INTERPRETATION: The causes of ascites are given in the next Table (Krieg AF, Kjeldsberg CR, in Henry JB. Clinical Diagnosis and Management. W.B. Saunders, Phila., 1991; 18th edition, 467):

Causes of Ascites

Transudates	Exudates
Hepatic Cirrhosis	Malignancy: Metastatic Carcinoma; Lymphoma; Hepatoma; Mesothelioma
Congestive Heart Failure	Pancreatitis
Congestive Pericarditis	Infections
Hypoproteinemia (e.g., Nephrotic Syndrome)	Tuberculosis
	Primary Bacterial Peritonitis (May be Superimposed on Transudate)
Chylous Effusion	Secondary Bacterial Peritonitis (eg, Appendicitis, Intestinal Infarct)
Damage or Obstruction to Thoracic Duct, (e.g., Trauma, Lymphoma, Carcinoma, Tuberculosis, Parasitic Infestation)	Trauma
	Bile Peritonitis (e.g., Ruptured Gallbladder)

The most common cause of ascites is liver disease. Malignancy, congestive heart failure, tuberculosis and chronic pancreatitis are other important possibilities. Differentiation of transudate from exudate is given in the next Table (Boyer TD, et al. Arch Intern Med. 1978; 138:1103-1105; Editorial, Brit Med J. May 9, 1981; 282:1499):

Differentiation of Transudate from Exudate

Characteristic	Transudate	Exudate
Specific Gravity	<1.015	>1.018
Protein	<2 to 3gms/dL	>3gms/dL
Ascitic Fluid Protein = Serum Protein	<0.5	>0.5
Ascitic Fluid LDH = Serum LDH	<0.6	>0.6
WBC Count	<500/cu mm <25% Polymorphonuclear Nuclear Leukocytes	>500/cu mm (suggests Bacterial and Tuberculous Peritonitis)

ASCORBIC ACID (see VITAMIN C)

ASO (see ANTI-STREPTOCOCCAL-O)

ASPARTATE AMINOTRANSFERASE(AST)

Synonym: Serum Glutamic Oxalacetic Transaminase(SGOT)
SPECIMEN: Red top tube, separate serum; green (heparin) top tube may be used, separate plasma; stable for 3 to 4 days at room temp., 7 to 10 days at 4°C and 6 months frozen. Interference: hemolysis will give spuriously elevated results; turbidity and icterus may interfere.
REFERENCE RANGE: Adult, 0-35 U/liter; Newborn/Infant, 20-65 U/liter.
METHOD: Kinetic:

$$\text{Alpha-Ketoglutarate} + \text{Aspartate} \xrightarrow[\text{Pridoxal Phosphate}]{\text{AST}} \text{Glutamate} + \text{Oxalacetate}$$

$$\text{Oxalacetate} + \text{NADH} + \text{H}^+ \xrightarrow[\text{Dehydrogenase}]{\text{Malate}} \text{NAD}^+ + \text{Malate}$$

INTERPRETATION: AST is measured in the evaluation of conditions causing tissue damage. It is most commonly measured in acute myocardial infarct and liver disease. AST has a widespread tissue distribution. It is elevated in a wide variety of conditions as shown in the next Table:

Elevation of Aspartate Aminotransferase(AST)
Liver Diseases: Hepatitis; Cholestasis, Intrahepatic and Extrahepatic; Alcoholism; Drug Toxicity (eg, acetaminophen)
Heart: Acute Myocardial Infarction; Acute Myocarditis (any cause)
Skeletal Muscle: Skeletal Muscle Diseases; Trauma
Red Blood Cells: Hemolytic Anemia (Severe); Megaloblastic Anemia
Other: Malignancy; Infectious Mononucleosis; Congestive Heart Failure; Acute Renal Infarction; Acute Pulmonary Infarction; Acute Pancreatitis (Occ.); Tissue Necrosis; Third Degree Burns; Seizures; Eclampsia; Heparin Therapy

AST is primarily used to detect and monitor liver parenchymal disease. In general, the AST is more sensitive but less specific than the ALT in detecting liver disease. Very high values, over 500, may be found in liver disease, large necrotic tumors, congestive heart failure and shock. AST values are greatly increased in acute liver damage, e.g., viral hepatitis, and toxic damage. Moderate elevation occurs in cirrhosis, metastatic cancer, obstructive jaundice and infectious mononucleosis. There is slight to moderate elevation of serum levels in congestive hepatomegaly.

The AST level is elevated in almost 98 percent of patients with acute myocardial infarction. AST values rise and fall a few hours after the rise and fall of CPK.

The serum levels are elevated in muscle disease including certain forms of muscular dystrophy, crush injury and gangrene of muscle and in dermatomyositis.

Drugs which may cause serum elevations include oral contraceptives, acetaminophen, aspirin, isoniazid, codeine, cortisone and heparin. AST is elevated in 27% of patients receiving heparin; the mean maximal increase is 3.1 times baseline value. The AST returned to normal in 80% of patients after heparin therapy is discontinued and in 20% during therapy (Dukes GE, et al. Ann Intern Med. 1984; 100:646-650).

AST is decreased in patients with pyridoxine deficiency (B-6) and in patients with chronic dialysis.

References: Rej R. "Aminotransferases in Disease," Clin Lab Med. 1989; 9:667-687; Landolphi DR, et al. "Interpretation of Liver Function Tests in Infections Disease," Hosp Phys. Aug. 1990; 17-24.

ASPERGILLUS ANTIBODY SERUM

SPECIMEN: Red top tube, separate and then refrigerate serum. If titer is obtained, paired sera advisable to detect rising titers.
REFERENCE RANGE: Negative
METHOD: The immunodiffusion (ID; double diffusion) test using A. niger, A. fumigatus and A. flavus antigens is most often utilized. The formation of one to four lines of identify between patient's serum and reference serum versus A. antigen indicates a positive reaction. The I.D. test will detect precipitins in over 90 percent of patients with pulmonary aspergilloma and in about 70 percent of patients with allergic bronchopulmonary aspergillosis.

Serologic tests, other than immunodiffusion (ID) that are useful in confirming a diagnosis of noninvasive aspergillosis are as follows: complement fixation (CF), counterimmunodiffusion (CIE), enzyme-linked immunosorbent assay (ELISA) and passive hemagglutination assay (PHA).

IgE and IgG antibodies against A. fumigatus are measurable by radioimmunoassay (Wang JLF, et al. Am Rev Respir Dis. 1978; 117:917-927).
INTERPRETATION: Aspergilli are fungi, saprophytic (obtaining food by absorbing dissolved organic material) molds; the main species of clinical significance are A. fumigatus and A. flavus.

The predisposing condition to infection by aspergilli is the immunocompromised patient; this may occur in patients on immunosuppressive therapy and in patients with debilitating diseases. Up to 5 percent of patients with leukemia may become infected with aspergilli.
Allergic Bronchopulmonary Aspergillosis(ABPA): (Ricketti AJ, et al. Arch Intern Med. 1983; 143:1553-1557). The eight diagnostic criteria of ABPA are given in the next Table:

Diagnostic Criteria of Allergic Bronchopulmonary Aspergillosis (ABPA)
Laboratory Findings:
(1) Precipitating Antibodies Against A. Fumigatus Antigen
(2) Elevated Serum IgE and IgG Antibodies to A. Fumigatus
(3) Blood Eosinophilia (>1000/cu mm)
(4) Elevated Serum IgE Concentration
Clinical Findings:
(1) Episodic Bronchial Obstruction (Asthma)
(2) Immediate Skin Reaction to A. Fumigatus Antigen
(3) History of Pulmonary Infiltrates (Transient or Fixed)
(4) Central Bronchiectasis

Other findings in ABPA may include positive sputum culture of A. fumigatus, a history of expectoration of brown plugs and late (Arthus type) skin reactivity to intracutaneous testing with A. fumigatus antigen.

Patients with asthma and suspected ABPA should have sequential laboratory testing for ABPA as given in the following Table (Patterson R, et al. Arch Intern Med. 1986; 146:916-918):

Sequential Laboratory Testing for ABPA

Test	Result	Next Step
Skin Tests for AF:		
Cutaneous (Prick) Test	Positive	Serologic Studies
	Negative	Intradermal Test
Intradermal Test	Positive	Serologic Studies
	Negative	ABPA Excluded
Serologic Studies:		
Total Serum IgE (>2000ng/mL) Precipitins for AF (positive at 5x conc. serum)	Positive	Further Serologic Studies
Total Serum IgE (<1000ng/mL) Precipitins (negative)	Negative	ABPA Probably Excluded
Further Serologic Studies:		
IgE and IgG Indices	Positive (Both >2)	ABPA Diagnosed Perform Tomography
	Negative (Both <2)	Not Consistent with ABPA Consider Repeat
Tomography for Bronchiectasis:		
Location: Central		ABPA-Central Bronchiectasis
Normal		ABPA-Serologic

The IgE and IgG indices are calculated by dividing levels of specific IgE or IgG antibody in patient serum by the specific antibody levels found in a pool of sera from asthma patients.

The five stages in the natural history in patients with ABPA are as follows: acute, remission, recurrent exacerbation, corticosteroid-dependent asthma, and fibrosis (Patterson R, et al. Ann Intern Med. 1982; 96:286-291).

Conventional therapy for ABPA is high-dose systemic steroids; adjunctive therapy with itraconazole may be of benefit (Denning DW, et al. Chest. 1991; 100:813-819).

The serologic, radiologic, pathologic and clinical findings in ABPA has been reviewed (see Greenberger PA, Patterson R, Chest. 1987; 91:1655-1715).

Pulmonary Aspergilloma: The ID test will detect precipitins in over 90 percent of patients with pulmonary aspergilloma. This is a chronic infection generally arising in an old tubercular cavity or bronchial dilatation. The combination of caseated material and tightly branched radiating hyphae is called a "fungus" ball.

ASPIRIN (see SALICYLATE BLOOD)

ASTHMA PANEL

The diagnostic procedures in asthma are given in the next Table:

Diagnostic Procedures in Asthma

1. History
2. Physical Examination
3. Basic Laboratory Evaluation
   a. Chest X-Ray (Hyperinflation, Peribronchial Markings, Pneumonia, Atelectasis, Bronchiectasis)
   b. Complete Blood Count (Eosinophilia, Immunodeficiency, or Polycythemia)
   c. Spirometry
4. Optional as Indicated
   a. Allergen Skin Tests
   b. $Alpha_1$-Antitrypsin
   c. Blood Gases
   d. Complete Pulmonary Function Tests
   e. Electrocardiogram(ECG)
   f. Exercise Challenge
   g. Quantitative Immunoglobulins; IgG Subclasses
   h. Radioallergosorbent Test (RAST)
   i. Sinus Roentgenograms
   j. Nasal Smear for Eosinophils or Poly's
   k. Sputum Exam and Culture, AFB
   l. Sweat Test
   m. Trial of Allergen Avoidance
   n. Tuberculin Skin Test
   o. Stool for Ova and Parasites
5. Special Procedures
   a. Inhalation Challenge
   b. Occupational Challenge
   c. Oral Challenge
   d. Gastroesophageal Reflux Evaluation
6. Steroid Side Effects
   a. Growth Parameters (Height and Weight)
   b. Cushingoid Features, Striae, Acne
   c. Blood Pressure
   d. Electrolytes, Glucose
   e. A.M. Cortisol
   f. Serum Calcium, Phosphate, Alkaline Phosphatase, Vitamin D, Parathyroid Hormone
   g. 24-Hour Urine for Calcium and Phosphate
   h. Urinalysis
   i. Bone Age, Bone Density
   j. Ophthalmologic Exam for Posterior Subcapsular Cataracts or Glaucoma

(cont)

Asthma Panel (Cont)

The physiologic alterations in bronchial asthma are given in the next Table:

Physiologic Alterations in Bronchial Asthma

1. Increased airway resistance, decreased expiratory flow, (increased) bronchial smooth muscle tone, mucosal edema, mucus secretion).
2. Airway closure at higher than normal distending pressure (decreased vital capacity).
3. Increased lung volume (RV and FRC always)(TLC often).
4. Increased pulmonary artery pressure relative to pleural pressure (ECG evidence of right ventricular strain or cor pulmonale).
5. Increased negative pleural pressure resulting in marked respiratory variations in arterial pressure (pulsus paradoxus).
6. Increased ventilation-perfusion imbalance (decreased PO_2, increased deadspace, increased PCO_2 in severe cases).

The spirographic tracing of forced expiration is given in the next Figure:

Spirographic Tracing of Forced Expiration, Normal and Asthma

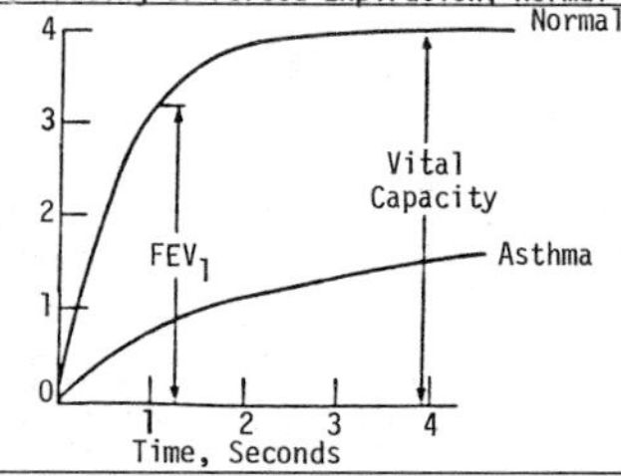

Volume in the spirometer is plotted against time. The vital capacity is represented by the total volume expired. One-second forced expiratory volume (FEV_1) is the volume expired during the first second.

Review: see Spector SL. "Asthma and Chronic Obstructive Lung Disease: A Pharmacologic Approach." Disease-a-Month. January 1991; 37.

AUTOHEMOLYSIS

SPECIMEN: Two green (heparin) top tubes

REFERENCE RANGE: Amount of hemolysis: No Sugar: 24 hours, 0.05-0.5%, 48 hours, 0.4-4.5%; With Sugar: 24 hours, 0-0.4%, 48 hours, 0.3-0.6%.

METHOD: When blood is incubated at 37°C, a small amount of lysis of the red blood cells occurs; glucose partially inhibits lysis.

INTERPRETATION: The autohemolysis test was classically used to aid in the diagnosis of hereditary spherocytosis, and to distinguish anemia due to a deficiency of glucose-6-phosphate dehydrogenase from that caused by a deficiency of enzymes of the glycolytic pathway, e.g. pyruvate kinase. Addition of glucose protects against hemolysis of the RBC of hereditary spherocytosis and G-6-PD deficient RBC, but has no protective effect on RBC with glycolytic enzyme defects. The results of the autohemolysis test are shown in the next Table:

Autohemolysis Test

Condition	No Glucose	Glucose
Hereditary Spherocytosis	Markedly Increased	Corrected
G-6-P-D Deficiency	Moderately Increased	Corrected
Pyruvate Kinase Deficiency	Markedly Increased	Not Corrected

Presently, enzyme defects are probably best diagnosed by measurement of individual enzyme activities.

BACTERIAL MENINGITIS ANTIGENS

SPECIMEN: One mL CSF (Other specimens are as follows: Red top tube, separate serum; 10mL urine; one mL pleural fluid; one mL of joint fluid); maintain sterility; store and forward to reference laboratory refrigerated or frozen.
REFERENCE RANGE: Negative
METHOD: Counterimmunoelectrophoresis(CIE), latex agglutination(LA) and ELISA. The sensitivity of ELISA methodology is greater than that of CIE.
CIE: The arrangement of the electrodes and the direction of flow of antibody and antigen are shown in the next Figure:

Counterimmunoelectrophoresis

(1) Place serum antibody into appropriate wells:

Well Well
Antibody Antigen

Antigens Assayed
Neisseria Meningitidis
Hemophilus Influenzae
Streptococcus Pneumoniae
Listeria Monocytogenes
Group B Streptococcus

(2) Electrophorese:

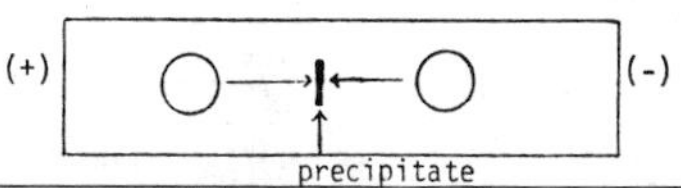

LA: Latex particles coated with appropriate antigens are available; the specimen (CSF, urine, etc.) is placed in contact with particles.
Rapid diagnostic tests are available for the following organisms (Feigin RD, et al. Pediatr Infect Dis J. 1992; 11:785-814):

Rapid Diagnostic Tests for Bacterial Pathogens
Group B Streptococci
N. Meningitidis Groups A,B,C,Y, and W135
H. Influenzae, Type b
S. Pneumoniae
L. Monocytogenes
E. coli, K1 strains
Klebsiella Pneumoniae
Pseudomonas Aeruginosa

INTERPRETATION: Rapid diagnostic tests may be useful to establish an etiology of bacterial meningitis, particularly in patients partially treated with antibiotics.
False positive reactions to group B streptococci may occur in some newborns with perineal or gastrointestinal tract colonization (Feigin RD, already cited).
E. coli K1 antigen is immunologically identical to the N. meningitidis group B antigen; these organisms cross react (McCracken GH, et al. Lancet. 1974; 2:246-250; Saez-Llorens X, McCracken GH. Infect Dis Clin N Amer. 1990; 4:623-644).
False positive H. influenza antigen detection may be seen with S. Aureus or E. coli K100 infection, asymptomatic carriers, and for up to 10 days after H. influenza vaccination (Salih MA, et al. Epidemiol Infect. 1989; 103:301-310; Cuevas LE. Ann Trop Med Parasitol. 1989; 83:375-379; Murphy TV, et al. Pediatr Res. 1989; 26:491-495).
Listeria monocytogenes antigen detection is relatively insensitive (McLauchlin J, Samuel D. Serodiagn Immunother Infect Dis. 1989; 3:17-26).(cont)

Bacterial Meningitis (Cont)

Bacteria Causing Meningitis: The organisms most commonly responsible for bacterial meningitis by age-group and suggested empiric therapy is suggested in the following Table (Tunkel AR, et al. Ann Intern Med. 1990; 112:610-623):

Usual Organisms and Empiric Therapy for Bacterial Meningitis

Age	Organisms	Therapy
0-4 weeks	E. coli Group B. Streptococci L. Monocytogenes	Ampicillin plus Third Generation Cephalosporin OR Ampicillin plus Aminoglycoside
4-12 weeks	E. coli Group B Streptococci L. Monocytogenes H. Influenzae S. Pneumoniae	Ampicillin plus Third Generation Cephalosporin
3 months to 18 years	H. Influenzae N. Meningitidis S. Pneumoniae	Third Generation Cephalosporin OR Ampicillin plus Chloramphenicol
18-50 years	S. Pneumoniae N. Meningitidis	Penicillin G OR Ampicillin
>50 years	S. Pneumoniae N. Meningitidis L. Monocytogenes Gram Negative Bacilli	Ampicillin plus Third Generation Cephalosporin

See also Greenlee JE. Infect Dis Clin N Amer. 1990; 4:583-598.

BARBITUATES

SPECIMEN: Red top tube, separate serum or 50mL random urine.
REFERENCE RANGE: The therapeutic ranges and toxic values for barbituates are given in the next Table:

Therapeutic Ranges and Toxic Values for Barbituates

Barbituates	Therapeutic Range(mcg/mL)	Toxic (mcg/mL)	Half-Life(hrs)
Short-Acting (3 to 6 hours)			
Pentobarbital (Nembutal)	1-5	coma >10	15-48
Secobarbital (Seconal)	1-5	coma >10	19-34
Intermediate-Acting (4 to 8 hours)			
Amobarbital (Amytal)	5-14	>30	8-42
Long-Acting (8 to 16 hours)			
Phenobarbital	15-40	>40	Adults: 50-120 Child: 40-70

METHOD: EMIT, gas-liquid chromatography, high-pressure liquid chromatography, RIA, thin-layer chromatography.
INTERPRETATION: Barbituates are used therapeutically to control seizures, manage elevated intracranial pressure, and to sedate patients. The symptoms of acute barbituate intoxication affect primarily the central nervous system and the cardiovascular system. The fatal serum concentration is greater in the long-acting barbituates than in the short-acting barbituates.

The depth and severity of coma in barbituate poisoning are graded; the grading system is given in the next Table:

Grading of Coma in Barbituate Poisoning

Grade	Characteristics		
	Deep Tendon Reflexes	Painful Stimuli	Vital Signs
I	Intact	Withdraws	
II	Intact	No Response	
III	Reduced or Absent	No Response	
IV	Absent	No Response	Respiratory Depression and/ or Circulatory Instability

A flat EEG and abnormal reflexes may occur during acute intoxication; these findings do not necessarily indicate structural damage to the CNS.

Barbituates are hypnotics and sedatives that have the following disadvantages: dependence, unnatural sleep, hangover and hepatic enzyme induction.

BENCE-JONES PROTEIN, URINE (LIGHT CHAIN EXCRETION)

SPECIMEN: 10mL of first morning specimen; refrigerate.

Determine concentration of protein in urine; if <100mg/24 hrs., then do not perform electrophoresis except in those patients with known monoclonal gammopathy. Dye binding determinations of protein such as urine dipsticks are more sensitive to albumin than to Bence-Jones protein. A urine protein electrophoresis is done prior to immunoelectrophoresis (IEP).

REFERENCE RANGE: None detected

METHOD: The urine is concentrated; then, electrophoresis is done followed by immunodiffusion.

INTERPRETATION: Urinary Bence-Jones proteins are determined in the work-up of patients with monoclonal gammopathies. The IEP is a test for qualitative identification of the light chains, kappa or lambda. Increased light chains (kappa or lambda chains) occur in malignant conditions such as myelomatosis, Waldenström's disease, lymphocytic leukemias and lymphomas and in non-malignant conditions, primary amyloidosis (Solomon A, N Engl J Med. 1976; 294:19) and very rarely in patients with benign monoclonal gammopathies (Dammacco F, Waldenström J. Acta Med Scand. 1968; 184:403). Light chains are excreted in the urine when cells acquire an imbalance in heavy and light chain synthesis and produce too many light chains. Bence-Jones proteins are found in 60 percent of patients with paraproteins. In about 15% of myelomatosis, only light chains are synthesized and released (see below). Light chains can pass into the glomerular filtrate because they have a relatively low molecular weight (25,000 daltons if monomers and 50,000 daltons if dimers); they do not usually accumulate in the serum except in renal failure.

Light chain disease is characterized as follows: (1) On serum protein electrophoresis, there is absence of M-spike but hypogammaglobulinemia is found as in multiple myeloma. (2) The peripheral smear fails to show rouleaux formation because of the absence of hypergammaglobulinemia. (3) Renal disease with elevated BUN and creatinine may be the initial laboratory findings. (4) 2.5% of cases are associated with amyloidosis. (5) Thrombocytopenia, anemia and leukopenia are absent. Only small amounts of Bence-Jones proteins are found in amyloidosis; about 60% are lambda light chains.

BENZODIAZEPINES (VALIUM, LIBRIUM AND SERAX)

SPECIMEN: Random urine

REFERENCE RANGE: Negative

METHOD: EMIT

INTERPRETATION: The benzodiazepines are used primarily as therapy for anxiety and insomnia and are the most commonly prescribed drugs for this purpose. Benzodiazepines are also used for sedation and as anticonvulsants. The drugs in this class are listed in the next Table (Pollack MH, Stern TA. J Intensive Care Med. 1993; 8:1-15):

Common Benzodiazepines

Drug	Half-Life (hours)	Usual Route
Diazepam (Valium)	20-100	PO, IV
Chlordiazepoxide (Librium)	5-30	PO, IV
Flurazepam (Dalmane)	40	PO
Clorazepate (Tranxane)	30-200	PO
Clonazepam (Klonopin)	15-50	PO
Lorazepam (Ativan)	10-20	PO, IV, IM
Oxazepam (Serax)	5-15	PO
Prazepam (Centrax)	50-80	PO
Temazepam (Restoril)	10-12	PO
Alprazolam (Xanax)	12-15	PO
Triazolam (Halcion)	1.5-3	PO
Midazolam (Versed)	1-12	IV, IM
Quazepam (Doral)	30-100	PO

Treatment of Anxiety: Anxiety disorders are the most common psychiatric disorders. Benzodiazepines are one class of drugs commonly used for anxiety disorders. The treatment of anxiety has been reviewed extensively (See Pollack MH, Stern TA. Already cited; Brown CS, et al. Arch Intern Med. 1991; 151:873-884; Ballenger JC. Arch Intern Med. 1991; 151:857-859; Cole JO. Med Clin N Amer. 1988; 72:815-830).(cont)

Benzodiazepines (Cont)

Long term therapy with benzodiazepines may lead to addiction. Patients may develop tolerance, and long term therapy leads to physical dependence; discontinuation of the drug leads to withdrawal. Long term benzodiazepine therapy must be tapered slowly over 6-8 weeks (Mackler SA, Schweizer E. Hosp Practice. September 30, 1992; 109-116).

Treatment of Insomnia: Triazolam (Halcion) is a short-acting benzodiazepine which has been used for treatment of insomnia. This drug is associated with significantly higher rates of confusion, amnesia, bizarre behavior, agitation, and hallucinations compared to temazepam (Restoril)(Wysowski DK, Barash D. Arch Intern Med. 1991; 151:2003-2008). Other authors have reported next-day memory impairment following triazolam administration (Bixler EO, et al. Lancet. 1991; 337:827-831). Elderly patients are particularly prone to sedation and other central nervous system effects and dosage in the elderly should be reduced by 50 percent if this drug is to be given (Greenblatt DJ, et al. N Engl J Med. 1991; 324:1691-1698).

BETA-GLUCOSIDASE (see GLUCOCEREBROSIDASE)

BETA-2-MICROGLOBULIN

SPECIMEN: Serum, CSF, and Urine; for serum specimen, use red top tube, separate serum and freeze; 1 mL CSF and freeze; 10mL of random urine, freeze.

REFERENCE RANGE: Serum: 10-30 yrs: 1.2-1.8mg/L; 31-50 yrs: 1.40-1.90mg/L; >50 yrs: 1.45-2.15mg/L; Cerebrospinal Fluid: 0.7-1.8mg/L; Urine: less than 0.3mg/L

METHOD: RIA, ELISA

INTERPRETATION: Beta-2-microglobulin has a molecular weight of 11,800; this protein is the light chain part of class-1 major histocompatibility (MHC) antigens. Due to its small size, it is 95% filtered at the glomerulus; after glomerular filtration, it is normally 99.9% reabsorbed. In tubular injury, urinary excretion of beta-2-microglobulin is increased and the urinary excretion of beta-2-microglobulin has been used as a marker of renal tubular cell injury. Beta-2-microglobulin is elevated in the urine in disorders affecting proximal tubule function such as acute tubular necrosis, interstitial nephritis, pyelonephritis, nephrotoxicity due to drugs (e.g. aminoglycosides or cisplatin) and transient increases are seen with trauma, sepsis, and pancreatitis. Any disease causing tubular dysfunction will cause elevated urinary beta-2-microglobulin including cadmium or mercury poisoning, Fanconi's syndrome, Wilson's disease, nephrocalcinosis, untreated congenital galactosemia, chronic potassium depletion, and cystinosis (Schardijn GHC, Statius Van Eps LW. Kidney International. 1987; 32:635-641; Hoekman K, et al. Neth J Med. 1985; 28:551-557).

Serum Beta-2-Microglobulin: Serum beta-2-microglobulin is elevated in any condition that decreases glomerular filtration rate (GFR). The gradual increase in serum levels with age is probably due to decreasing GFR. Patients with renal failure on dialysis have very high levels of beta-2-microglobulin. Serum levels may be used to detect diabetic nephropathy or renal transplant rejection.

Serum beta-2-microglobulin is elevated in malignancies such as Hodgkin's disease, non-Hodgkin's lymphoma, chronic lymphocytic leukemia, and multiple myeloma, and levels correlate with stage of disease or malignant cell mass in certain cases. Serum beta-2-microglobulin has prognostic significance in adult acute lymphocytic leukemia (Kantarjian HM, et al. Am J Med. 1992; 93:599-604).

Elevated beta-2 microglobulin levels have been detected in the serum of patients with HIV infection. Serum beta-2-microglobulin along with CD4 counts are useful markers for progression of HIV infected individuals to AIDS (Fahey JL, et al. N Engl J Med. 1990; 322:166-172; Lifson AR, et al. Lancet. 1992; 339:1436-1440; Hofmann B, et al. AIDS. 1990; 4:207-214).

CSF Beta-2-Microglobulin: The simultaneous measurement of CSF and serum level of beta-2-microglobulin is useful in detecting central nervous system involvement with leukemia or lymphoma. When the CSF level of beta-2-microglobulin was significantly higher than the serum level in patients with acute leukemia and lymphoma, the CNS was involved. This test is imperfect - there is considerable overlap with patients without CNS involvement (Ernerudh J, et al. Arch Neurol. 1987; 44:915-920; Oberg G, et al. Br J Haematol. 1987; 66:315-322).

Review: Bethea M, Forman DT. Clin Lab Sci. 1990; 20:163-168.

BETKE-KLEIHAUER (see FETAL HEMOGLOBIN STAIN)

BILIRUBIN, AMNIOTIC FLUID

SPECIMEN: 2mL amniotic fluid. The specimen is centrifuged for 10 minutes and the supernatant is removed; the specimen is kept in the dark prior to scan since bilirubin is unstable to light.

REFERENCE RANGE: See below

METHOD: Spectrophotometric scan of amniotic fluid between 300nm and 600nm. Bilirubin absorbs at 450nm. In order to obtain the absorbance at 450nm, draw a straight line between the absorbance at 550nm and 365nm. Draw a vertical line at 450nm from the top of the absorbance curve to intersect the straight line. The height from the curve to the straight line at 450nm is the absorbance and is called 450nm. A scan of amniotic fluid from 300nm to 600nm and clinical significance of absorbance is shown in the next Figure:

Spectrophotometric Scan of Amniotic Fluid and Clinical Significance of Absorbance Values

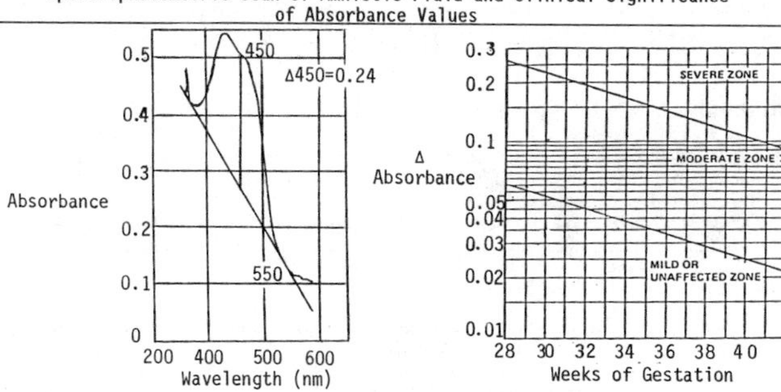

INTERPRETATION: The presence of fetal erythroblastosis fetalis in-utero is assessed by assay of amniotic fluid for bilirubin pigment; the amount of bilirubin pigment reflects the degree of hemolysis in the fetus. Erythroblastosis fetalis is due to Rh(-) mother and Rh(+) fetus, and occurs as follows: The mother, on exposure to Rh(+) fetal red blood cells, develops antibodies to the Rh(+) antigens; the antibodies pass from the maternal blood into the fetus. In the fetus, the antibodies combine with the Rh antigens of the fetal red blood cells; complement combines with antibody. Macrophages or lymphocytes attach to the Fc region of the immunoglobulin of the red blood cells. Sphering and then lysis of the red cells occurs; antibody-coated red cells may also be destroyed by polymorphonuclear leukocytes or by phagocytes. The pathogenesis of erythroblastosis fetalis is illustrated in the next Figure:

Lysis of Red Blood Cells in Erythroblastosis Fetalis

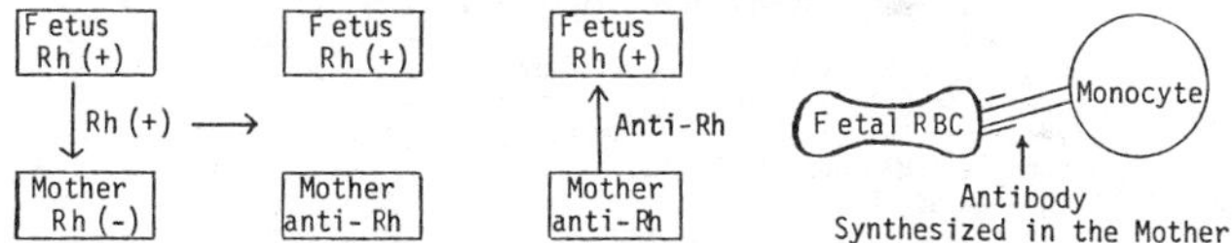

Diagnosis of erythroblastosis fetalis in the newborn is made as follows: mother, Rh(-); infant, Rh(+); Coombs test (+), then jaundice is due to Rh incompatibility. Anti-D is the most common antibody, but others producing disease include RhC, E, c and e, Kell, Kidd, and Duffy antibodies (Whittle MJ. Arch Dis Child. 1992; 67:65-68).

20-25 percent of patients with erythroblastosis fetalis develop hydrops fetalis. Severe hemolysis leads to hepatic erythropoiesis, hepatosplenomegaly, and congestive heart failure. Hepatic enlargement and dysfunction leads to portal hypertension and hypoalbuminemia. Patients develop ascites, pulmonary edema, pulmonary hypoplasia, and anasarca. Treatment in severe cases is intraperitoneal or intravascular fetal transfusion. (cont)

Bilirubin Amniotic Fluid (Cont)

Sources of error in amniotic fluid spectrophotometry includes contamination of the sample with blood, methemalbumin, or meconium, all of which absorb at or near 450nm leading to elevation of the peak. Other potential errors include exposure of the sample to light which destroys bilirubin, sampling of maternal urine which produces no 450nm peak and congenital anomalies such as anencephaly and obstructive upper gastrointestinal tract lesions which may cause a peak at 450nm. The spectrophotometric scan is not valid before 27 weeks gestation; severe hydrops fetalis often becomes manifest between 20 and 24 weeks. In such cases, ultrasound and fetal blood sampling may be helpful.

Rh(D) immune globulin (RhIg) became generally available in 1968. Since that time, the incidence of Rh hemolytic disease has decreased to 10.6 per 10,000 total births (Chavez GF, et al. JAMA. 1991; 265:3270-3274).

Review: Gibble JW, Ness PM. "Maternal Immunity to Red Cell Antigens and Fetal Transfusion." Clin Lab Med. 1993; 12:553-576.

BILIRUBIN, DIRECT (CONJUGATED, SOLUBLE)

SPECIMEN: Red top tube

REFERENCE RANGE: 0-0.2mg/dL. To convert from conventional units in mg/dL to international units in micromol/liter, multiply conventional units by 17.10.

METHOD: Reaction of bilirubin with diazonium salt

INTERPRETATION: Direct bilirubin is elevated in the conditions listed in the next Table:

Causes of Elevated Serum Direct (Conjugated) Bilirubin
Liver Disease:
Hepatocellular Disease
Biliary Tract Obstruction, Extrahepatic or Intrahepatic
Dubin-Johnson Syndrome
Rotor's Syndrome
Hepatic Storage Syndrome

Conjugated bilirubin is elevated in hepatocellular disease, biliary tract obstruction and some conditions involved in bilirubin metabolism.

Specific causes of direct (conjugated) hyperbilirubinemia in neonates in given in the following Table (Rosenthal P, Sinatra F. Pediatrics in Review. 1989; 11:79-86):

Causes of Neonatal Direct (Conjugated) Hyperbilirubinemia
Extrahepatic Obstruction
Biliary Atresia
Neonatal Hepatitis
Choledochal Cyst
Persistent Intrahepatic Cholestasis
Paucity of Intrahepatic Bile Ducts
Arteriohepatic Dysplasia
Benign Recurrent Intrahepatic Cholestasis
Byler Disease
Hereditary Cholestasis with Lymphedema
Acquired Intrahepatic Cholestasis
Infections: Bacterial Sepsis; Hepatitis B; Syphilis; Toxoplasmosis; Rubella; CMV; Herpes; Varicella; Echovirus; Coxsackie virus
Drug-Induced Cholestasis
Parenteral Nutrition Associated Cholestasis
Genetic and Metabolic Disorders
Carbohydrate Disorders; Galactosemia; Fructosemia; Glycogen Storage Disease Type IV
Amino Acid Disorders: Tyrosinemia
Lipid Disorders: Niemann-Pick; Gaucher; Wolman; Cholesterol Ester Storage Disease
Chromosomal Disorders: Trisomy 18; Down Syndrome
Miscellaneous: Alpha-1-Antitrypsin Deficiency; Neonatal Hypopituitarism; Cystic Fibrosis; Zellweger Cerebrohepatorenal Syndrome; Familial Hepatosteatosis

Physiologic jaundice does not produce conjugated (direct) bilirubin levels above 1.5-2.0mg/dL (Hammond KB, Wells R. Lab Med. 1983; 14:239-245).

Direct bilirubin is bilirubin conjugated with glucuronic acid at the smooth endoplasmic reticulum of the hepatic cell to form bilirubin-diglucuronide. This form of bilirubin is soluble in aqueous media. Healthy subjects have no detectable conjugated bilirubin; however, technical problems with available assays explain why laboratories detect some conjugated bilirubin. Direct conjugated bilirubin is the most sensitive test in the diagnosis of liver disease.

BILIRUBIN, TOTAL

SPECIMEN: Red top tube, separate serum

REFERENCE RANGE: 0.1 to 1.0mg/dL. To convert from conventional units in mg/dL to international units in micromol/liter, multiply conventional units by 17.10.

METHOD: Bilirubin + Diazotized Sulfanilic Acid ⟶ AzoDye (600nm)

INTERPRETATION: This test is useful in detecting and following liver disease and hemolytic disorders and is elevated in the conditions listed in the next Table:

Causes of Increased Serum Bilirubin
Cholestasis; Intrahepatic and Extrahepatic
Liver Disease
Congenital Hyperbilirubinemia:
Crigler-Najjar, Dubin-Johnson, Gilbert's Disease
Hemolytic Anemia
Reabsorbing Extravasated Blood
Malnutrition
Infection - Subacute and Chronic
Hyperthyroidism
NOTE: Increased with Exercise, Estrogens, Oral Contraceptives, Menstruation, Hemolysis, Fasting

The major source of bilirubin is hemoglobin catabolism resulting either from the destruction of adult circulating red blood cells or as a result of ineffective erythropoiesis within the bone marrow. In normal man, the sources of bilirubin are listed in the next Table (Berk PD, et al. J Lab Clin Med. 1976; 87:967):

Sources of Bilirubin

Source	Percent
Death of Senescent, Circulating Red Blood Cells	70
Ineffective Erythropoiesis	10
Other Principally Hepatic	20

BILIRUBIN, TOTAL (NEONATE)

SPECIMEN: 2 capillary tubes
REFERENCE RANGE: Newborn: Less than 12mg/dL decreasing to adult levels by end of first month. To convert from conventional units in mg/dL to international units in micromol/liter, multiply conventional units by 17.10.
METHOD: Neonatal Serum + PO_4^{-3} (buffer) ⟶ Absorption 452 nm.
Absorbance at 452nm is due to bilirubin and, if present, hemoglobin. At 540nm, bilirubin does not absorb while hemoglobin exhibits the same absorbance as it does at 452nm. The use of 540nm as the blanking wavelength, eliminates any hemoglobin contribution from the total absorbance at 452nm.
INTERPRETATION: Evaluation of hyperbilirubinemia in the neonate should begin with fractionation to rule out direct (conjugated) hyperbilirubinemia. See BILIRUBIN, DIRECT. The laboratory tests that should be considered in patients with significant jaundice are given in the following Table:

Laboratory Tests for Jaundiced Neonates
Serum Bilirubin, Direct and Total
Hemoglobin, Hematocrit
Reticulocyte Count
Peripheral Blood Smear
White Blood Cell Count, Platelet Count
Blood Type, Rh of Mother and Infant
Direct Coombs Test on Infant
Tests Useful in Specific Circumstances:
Sepsis Workup (Blood, Urine, CSF Cultures)
Prothrombin Time(PT)/Partial Thromboplastin Time(PTT)
Apt Test (Alkaline Denaturation of Hemoglobin in Emesis or Stool - Adult vs. Fetal Hgb)
Thyroid Screen

Common causes and characteristics of the conditions associated with indirect (unconjugated) hyperbilirubinemia in the newborn are given in the next Table:

Causes and Characteristics of the Conditions Associated with Indirect (Unconjugated) Hyperbilirubinemia in the Newborn

Causes	Characteristics
Physiological Jaundice	Onset 1-3 days; Peak 3-4 days; Peak level 6-12mg/dL; Return to normal 1.5-2 weeks; Incidence 50%
Red Cell Incompatibility	Rh Incompatibility: Rh(-) mother, Rh(+) fetus ABO Incompatibility: Mother's group O, Infant's group A or B Coombs Test Positive Onset within 24 hours
Breast-Feeding Jaundice	Early Onset: 3-4 days; Peak 4.5 days; Peak level 12-20mg/dL; Incidence 25% Late Onset: 4-5 days; Peak 10-15 days; Peak level 10-30mg/dL; Incidence 2-30% (Lascari AD. J Pediatr. 1986; 108:156-158)
Infection (sepsis)	Onset usually after 4th day
Other Causes: Hemorrhage, Red Blood Cell Defects, Swallowed Maternal Blood, Clotting Disorders, Placental Dysfunction, Infant of Diabetic Mother, Hypothyroidism, Intestinal Obstruction, Crigler-Najjar Syndrome, Lucey-Driscoll Syndrome	

Physiological Jaundice: Several factors contribute to the development of physiologic jaundice, as listed in the following Table (Rosenthal P, Sinatra F, already cited):

Factors Contributing to Physiologic Jaundice
Increased Bilirubin Load
Increased RBC Volume
Decreased RBC Survival
Increased Enterohepatic Circulation
Defective Hepatic Uptake
Defective Bilirubin Conjugation Due to Temporary Deficiency of UDP Glucuronyl Transferase
Defective Bilirubin Excretion

In full-term infants, the jaundice almost always appears after 24 hours of life and reaches a peak on the 3rd or 4th day; in preterm infants, it usually begins 48 hours after birth, peaks at day 5, and may last up to 2 weeks.
Red Cell Incompatibility: The main causes of red cell incompatibility jaundice are Rh incompatibility e.g., Rh(-) mother, Rh(+) fetus; and ABO incompatibility, the mother's blood is usually group O and the infant's group A or group B.
Breast-Feeding Jaundice: There is a strong association between breast-feeding and jaundice in healthy newborn infants. Caloric deprivation and increased enterohepatic circulation have been postulated as likely mechanisms for this phenomenon (Maisels MJ, Gifford K. Pediatr. 1986; 78:837-843).
Infection: The common causes of jaundice and infections in the newborn are septicemia and urinary tract infection; jaundice usually begins after the 4th day of life; in urinary tract infections, the jaundice is of hepatic origin. Infection should always be considered in infants with significant jaundice due to the morbidity and mortality associated with this diagnosis.
Other causes of hyperbilirubinemia occur less commonly, but may be clinically significant. Causes include hemorrhage (cephalhematoma), RBC defects (spherocytosis, elliptocytosis, G-6-PD deficiency, pyruvate kinase deficiency), swallowed maternal blood (see APT TEST), clotting disorders, placental dysfunction, infant of diabetic mother, hypothyroidism, intestinal obstruction, Crigler-Najjar syndrome, and Lucey-Driscoll syndrome.
Treatment: If the serum bilirubin concentration is greater than 15mg/dL in a healthy term infant, phototherapy may be needed. In ill or premature neonates, phototherapy may be initiated at a lower level. Many physicians begin phototherapy when bilirubin concentration is 5mg/dL less than the exchange transfusion level.

Phototherapy converts bilirubin into a colorless compound which does not have long-term effects on the infant. Exchange transfusion removes bilirubin by replacing the infant's blood with donor blood. Exchange transfusion is usually considered at a level of 20mg/dL in a term infant; patients with hemolysis and rapid rise in bilirubin (greater than 0.5mg/dL per hour, 10mg/dL at 24 hours, or 15mg/dL at 48 hours) will probably need exchange transfusion (Rosenthal P, Sinatra F, already cited). These are the "traditional" indications for exchange transfusion. Exchange transfusion is not necessarily a benign procedure and the risk of bilirubin toxicity at bilirubin levels above 20mg/dL is unknown. More recently, it has been suggested that bilirubin levels be kept below 23.4-29.2mg/dL (400-500micromol/L) in healthy term infants, and below 17.5-23.4mg/dL (300-400micromol/L) in neonates with hemolytic disease (Newman TB, Maisels MJ. Pediatr. 1992; 89:809-818; Series of editorials in Pediatrics. 1992; 89:819-833).
Bilirubin Encephalopathy: Theoretically, the level of free bilirubin in blood should be the most sensitive indicator of kernicterus, since only free bilirubin can directly affect the brain cells by diffusion across cell membranes; free bilirubin is soluble in the fat of cell membranes. It is thought that with increasing bilirubin, the bilirubin is increasingly bound to albumin until the three binding sites (one high affinity and two low affinity binding sites) are occupied; excess bilirubin appears free in the plasma and is then deposited in the brain. The relationship of free bilirubin is shown in the next Figure:

Free Bilirubin

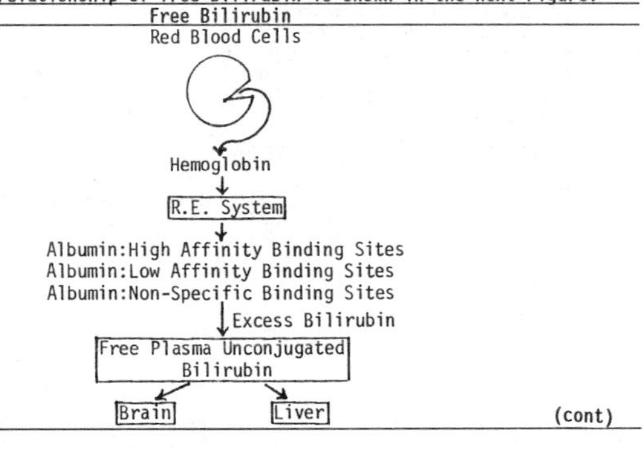

(cont)

Bilirubin Total (Neonate) (Cont)

Predictive Value for Neonatal Bilirubin Encephalopathy: The possible parameters that may yield information regarding predictive value for neonatal bilirubin encephalopathy are given in the next Table:

Possible Parameters for Neonatal Bilirubin Encephalopathy
Total Bilirubin
Bilirubin Loosely Bound to Albumin
Free Bilirubin
Binding Capacity of Albumin for Bilirubin
Bilirubin Bound to Albumin

Total serum bilirubin is the easiest parameter to measure. Kernicterus is frequent when total serum bilirubin exceeds 30mg/dL but rare when it is less than 20mg/dL. However, bilirubin encephalopathy may occur at values lower than 20mg/dL, especially in sick, low birth weight premature infants.

Free bilirubin, the value of which might be most useful to predict kernicterus, cannot be reliably estimated. A large number of tests have been investigated, and references to these may be found in the articles by Karp WB. Pediatrics. 1979; 64:361-368; Gitzelmann-Cumarasamy N, Kuenzle CC. Pediatrics. 1979; 64:3:375-377; Hammond KB, Wells R. Laboratory Medicine. 1983. 14:249-245.

BIOLOGICAL MARKERS FOR DEPRESSION

Biological markers for depression are listed in the next Table (Keffer JH. Am Soc Clin Pathologists, Check Sample, Clin Chem No. CC83-8. 1983; 23:8; Gold MS, et al. JAMA. 1981; 245:1562-1564):

Biological Markers for Depression
Dexamethasone Suppression Test
Thyrotropin-Releasing Hormone(TRH) Stimulation Test
Prolactin Response to Thyrotropin-Releasing Hormone(TRH)
3-Methoxy-4-Hydroxyphenylglycol(MHPG) Urine
Platelet Monamineoxidase(MAO) Activity
Phenylacetic Acid(PAA) Urine
Growth Hormone Release by Clonidine

Dexamethasone Suppression Test: The dexamethasone suppression test has been used to distinguish endogenous from nonendogenous depression (see DEXAMETHASONE SUPPRESSION TEST AS DIAGNOSTIC AID IN DEPRESSION)

Thyroid Releasing Hormone(TRH) Stimulation Test: The TRH stimulation test has been used to distinguish unipolar from bipolar depression. In this test, TRH is injected intravenously; serum specimens are obtained at 15 to 20 minute intervals for 1 hour (see THYROID STIMULATING HORMONE(TSH) FOLLOWING TRH). Agitated patients typically demonstrate a blunted TRH response, a reaction also characteristic for 77% of unipolar depressives; 17% of patients with bipolar depression demonstrate a blunted TRH response (Loosen PT, Prange AJ, Jr. Am J Psychiatry. 1982; 139:405-416; Kirstein L, et al. J Clin Psychiatry. 1982; 43:191-193; Sternbach HA, et al. JAMA. 1983; 249:1618-1620; Hein MD, Jackson IM. Gen Hosp Psychiatry. 1990; 12:232-244).

Prolactin Response to Thyrotropin-Releasing Hormone(TRH): see Judd LL, et al. Arch Gen Psychiatry. 1982; 39:1413-1416.

3-Methoxy-4 Hydroxyphenylglycol(MHPG) in Urine: The urinary level of MHPG reflects the central nervous system metabolism of norepinephrine and has been correlated with types of depression (Schildkraut JJ, et al. Psychopharma Bull. 1981; 17:90-91; Rosenbaum AH, et al. Am J Psych. 1983; 140:314-317).

Platelet Monamineoxidase(MAO) Activity: See Weinshilboum RM. Mayo Clin Proc. 1983; 58:319-330.

Clonidine Release of Growth Hormone: Depressed patients exhibit a blunted growth hormone response to clonidine due to α-2 adrenoceptor subsensitivity (Matussek N, et al. Psychiat Res. 1980; 2:25-36).

There are a number of medical conditions which are associated with psychiatric manifestations. These are listed in the next Table:

Conditions Associated with Psychiatric Manifestations
Hyperthyroidism and Hypothyroidism (Gold MS, et al. JAMA. 1981; 245:1919-1922)
Primary Hyperparathyroidism (Tibblin S, et al. Ann Surg. 1983; 197:135-138)
Cushing's Syndrome
Acute Intermittent Porphyria(AIP)
Vitamin B_{12} Deficiency (Spivak JL. Arch Intern Med. 1982; 142:2111-2114; Evans DL, et al. Am J Psychiatry. 1983; 140:218-221)
Resistance to Thyroid Hormone (Attention Deficit-Hyperactivity Disorder) (Hauser P, et al. N Engl J Med. 1993; 328:997-1001).

BLASTOMYCOSIS TESTING

SPECIMEN: Culture: Sputum, Urine, Pus; Antibody Tests: Serum, CSF.
REFERENCE RANGE: Negative
METHODS: Culture: Blastomyces dermatitidis may be detected by direct examination or by culture. This fungus is usually readily isolated and identified. Identification of blastomyces may be accomplished more rapidly by use of immunodiffusion testing for exoantigens A and K or by use of a DNA probe (Gen-Probe, San Diego). Antibody Tests: Various immunologic tests are available for detection of blastomyces antibodies. The sensitivity, specificity, positive predictive value and negative predictive value are given in the following Table (Kaufman L. CID. 1992; 14(Suppl 1):S23-S29):

Clinical Utility of Immunologic Tests for Blastomycosis

Test	Antigen	Sensitivity	Specificity	Positive Predictive Value	Negative Predictive Value
Complement Fixation	A	40%	100%	100%	81%
Immunodiffusion	A	65	100	100	88
Indirect Enzyme Immunoassay	A	80	98	94	93
Sandwich Enzyme Immunoassay	A	88	100	100	98
Western Blot	98-KD	88	92	82	95
Radioimmunoassay	120-KD	85	100	100	92

INTERPRETATION: Blastomyces dermatitidis is a fungus which causes pulmonary blastomycosis when inhaled. The fungus may form a solitary focus of consolidation and extend to regional lymph nodes; the fungi may then disseminate throughout the lung. Systemic dissemination may lead to infection in the skin, brain, bones and other tissues in the body; the skin is involved in 25 percent of cases. Direct examination and culture are the principal methods used to diagnose blastomycosis. In a recent review of 15 AIDS patients with blastomycosis, 6 patients had serologic testing by complement fixation or immunodiffusion - all tests were negative (Pappas PG, et al. Ann Intern Med. 1992; 116:847-853). Review of blastomycosis, see Bradsher RW. Infect Dis Clin N Amer. 1988; 2:877-898.

BLEEDING TIME

SPECIMEN: This test is done at the bedside. The test is not performed if the patient has taken aspirin or a nonsteroidal anti-inflammatory preparation within the 5 days, or if the platelet count is less than 50,000/microliter.
REFERENCE RANGE: 2-10 minutes; the reference interval for children is 2-8 minutes. A large number of drugs, diseases, physiologic factors, and test conditions may affect results of this test.
METHOD: (1) Place blood pressure cuff on the arm above the elbow and inflate to 40 mmHg. (2) Clean an area on the volar surface of the forearm which is free of visible veins. Use alcohol and allow to dry. (3) Incise skin with sterile blade. Use of a template device such as Simplate II permits the formation of a vertical incision of standard length and depth, leading to a more reproducible test. (4) With a round piece of filter paper, blot the blood every 30 seconds. Only the edge of the drop should touch the filter paper so as to not disturb the forming platelet aggregate. The blotted blood should turn from red, to straw-colored, to clear. The end point is taken at the straw to clear transition. Test results from a second incision should agree within 4 minutes. The test is continued for up to 20 minutes.
INTERPRETATION: The bleeding time is an in vivo test of hemostatic activity and is a test of the quantity and quality of platelet function, as well as by local tissue factors and other components of the coagulation mechanism.

The bleeding time is proportional to the platelet count below 100,000/ microliter. It is prolonged in the conditions listed in the next Table:

Prolongation of Bleeding Time
Low Platelet Count
Abnormal Platelet Function
Clotting Factor Deficiency: Von Willebrand's Disease or Syndrome; Hepatic Failure; DIC; Specific Factor Deficiency (Fibrinogen, V, VII, VIII, IX, X, XI, XII)
Systemic Illness: Amyloidosis; Anemia; Uremia
Drugs: Aspirin; Non-Steroidal Anti-Inflammatory Drugs; High Dose Penicillins and Cephalosporins; Heparin; Warfarin; Acute Alcohol Intoxication
Other: Diet (Fish Oils - Szechuan Purpura); Glanzmann's Thrombasthenia

When platelet counts are low, the expected bleeding time is calculated from the formula (Harker LA, Slichter SJ. N Engl J Med. 1972; 287:155).

$$\text{Bleeding Time} = 30.5 - \frac{\text{Platelet Count/Cubic mm}}{3850}$$

A bleeding time longer than that calculated from platelet numbers alone suggest defective function in addition to reduced number. A bleeding time shorter than that calculated suggests the presence of active young platelets, as in ITP. Many have experienced difficulty applying this equation and consider it unreliable.

Major indications for the performance of bleeding times are: (1) A diagnostic test for a bleeding disorder in a patient with a history of abnormal bleeding. (2) To monitor the progress of a patient being treated for abnormal bleeding function. (3) To predict those surgical patients who will most likely experience excessive bleeding. This last indication is most controversial, and recent studies show no predictive value for bleeding risk (Rodgers RPC, Levin J. Semin Thromb Hemost. 1990; 16:1; Lind SE. Blood. 1991; 77:2547-2552). The role of bleeding time determination in clinical practice is unclear (Review: Rodgers RPC, Levin J, already cited).

BLOOD COMPONENT THERAPY

The blood components available for transfusion and the indications for the use of these components are given in the next Table:

Blood Components and Indications

Blood Component	Indications
Red Cells	Chronic Anemia; Slow Blood Loss; Acute Blood Loss; Exchange Transfusions
Red Cells: Washed (Leukocyte and Plasma Poor) or Leukocyte Poor	Same as for Red Cells; Leukocyte Poor: Prevent Febrile Reaction Due to Leukocyte Antibodies; Plasma Poor: Prevent IgA Sensitization
Red Cells, Frozen-Thawed	Same as above plus Autotransfusions and Rare Blood Types.
Whole Blood	Massive Blood Loss
Plasma, Fresh Frozen	Treatment of Bleeding Patients with Multiple Factor Deficiencies
Albumin, 25% (25g/dL) Albumin, 5% (5g/dL)	Burns, Hypoalbuminemia Burns, Shock
Platelet Concentrates	Bleeding due to Thrombocytopenia or Platelet Function Abnormality
Granulocyte (Leukocyte) Concentrate	Granulocytopenia with Sepsis
Cryoprecipitate	Factor Deficiency: Hemophilia A, von Willebrand's Disease and Factor XIII Deficiency; Hypofibrinogenemic States
Factor VIII Concentrate	Hemophilia A

Red cell transfusions have largely superceded the use of whole blood. All blood products must be administered through a filter.

Preoperative Crossmatch Guidelines: For patients without preoperative complications undergoing elective surgery, the following guidelines are recommended. Blood is collected from the patient and testing is done for ABO, Rh(D) and antibody screen; preoperative crossmatch is not required if transfusion is not expected. If the patient should require emergency transfusion, blood of the same ABO group and Rh(D) type is selected; on initial spin, crossmatch is performed and the blood released in approximately 10-15 minutes. The standard crossmatch is completed while the patient is being transfused. This approach allows for a reduction in blood inventory and patient costs without sacrificing patient safety. If time allows, a standard crossmatch can be performed prior to transfusion.

RED CELLS: Red blood cells are prepared by centrifugation or sedimentation of a unit of whole blood from a hematocrit of 40% to a hematocrit of 60% to 80%. Red blood cells transfusions are indicated to increase oxygen-carrying capacity in anemia (e.g., chronic anemias, slow blood loss, acute blood loss) and for exchange transfusions.

Red blood cell transfusions are superior to whole blood transfusions for patients with cardiac disease, chronic anemia and for patients with liver or kidney disease who require restricted sodium or citrate intake.

A unit (hematocrit, 60 to 80%) which contains 200mL of red blood cells, should increase recipients hematocrit about 3% in a 70kg adult. Cells must be administered through a filter; a unit should not be transfused for longer than 4 hours. Fifty to 100mL of isotonic sodium chloride may be added to a unit of red blood cells to decrease viscosity and increase the transfusion rate. This step is unnecessary in ADSOL prepared units (see next section).

The side effects and hazards of red blood cell transfusion are like that of the transfusion of whole blood. **(cont)**

Blood Component Therapy (Cont)

ADSOL Red Blood Cells: ADSOL red blood cells is a preparation of red blood cells; ADSOL refers to ADditive SOLution system; ADSOL has the following advantages:

(1) This preparation has an improved red blood cell survival with extension of red blood cell shelf life to 42 days.

(2) The infusion rate is similar to whole blood; it is not necessary to add saline at the time of transfusion.

(3) When the product is prepared from the donor, the yield of platelets and plasma factors is increased when these components are separated.

The system is generically called Additive Solution System because it uses a second preservative solution in addition to the anticoagulant solution, citrate phosphate dextrose(CPD) needed for whole blood collection. The additive consists of adenine in 100mL of normal saline, with a small amount of citrate anticoagulant remaining. The plasma is separated from the red blood cells prior to the addition of the additive solution. Additive solutions differ; AS-1 and AS-5 contain mannitol as a red blood cell stabilizer, whereas AS-3 contains additional citrate and phosphate but does not contain mannitol. Patients with documented mannitol hypersensitivity should not receive AS-1 or AS-5 units.

Because of the small amount of mannitol (AS-1 unit contains 0.75g of mannitol) and the slightly increased volume per unit (60mL), rare patients with cardiac or renal failure may require volume reduction of the unit just prior to transfusion. In these patients, the order must be written as "volume reduced red blood cells." The ADSOL unit of red blood cells will be centrifuged in the Blood Bank just before transfusion and most of the liquid removed, resulting in a product with a hematocrit of about 80% which will not flow nearly as well as the ADSOL units. An order for packed cells without specifying "volume reduced" will be filled with the usual ADSOL RBC's.

A comparison of AS-1 RBC's with CPDA-1 RBC's and CPDA-1 whole blood is shown in the next Table:

Comparison of AS-1 RBC's with CPDA-1 RBC's and CPDA-1 Whole Blood

Parameters Measured After 35 Days Storage	AS-1 RBC's	CPDA-1 RBC's	CPDA-1 Whole Blood
Maximum Storage Period, Days (FDA)	42	35	35
Volume per Unit (mL)	310	250	510
Hematocrit, Percent	60	75	40
Viscosity, Relative to Whole Blood	1.2	2.6	1.0
Post-Transfusion Survival (Percent Recovery)	86	78	76
Supernatant Hgb (mg/dL)	112	900	43
Percent of RBC's Lysed During Storage	0.2	0.9	0.2
Supernatant Potassium (mmol/liter)	45	86	27
pH	6.70	6.58	6.72
ATP, Percent of Day Zero	74	52	62
Glucose, Percent of Day Zero	65	15	2

RED CELLS, LEUKOCYTE POOR:
Leukocyte Poor: Leukocyte removal helps to prevent febrile reactions. Recipients who have had multiple transfusions or pregnancies may develop antibodies to HLA or other antigens on leukocytes. Subsequent transfusions with cells carrying these antigens may cause a febrile (non-hemolytic) reaction. Leukocytes may be removed by centrifugation, filtration, or addition of sedimenting agents; the resulting unit contains fewer than 5 x 10^8 white cells (more than 95% removal). Leukocyte poor preparations should be given to patients who have repeated febrile reactions associated with transfusion and in patients that require frequent transfusions such as leukemia and aplastic anemia (Pisciotto PT (ed). Physicians Handbook, Amer Assoc of Blood Banks. 1989; 15-17). Leukocyte removal by filtration decreases the risk of transfusion-acquired CMV in newborn infants (Gilbert GL, et al. Lancet. 1989; 1228-1231).

RED CELLS, WASHED (LEUKOCYTE AND PLASMA-POOR):
Leukocyte Poor: After washing, less than 5 x 10^8 white cells remain and the unit contains more than 80% of the original red blood cells. Washing is less effective and more expensive than filtration or other means of leukocyte removal and should not generally be used for this purpose alone.
Plasma Poor: Three conditions in recipients are prevented by removing plasma; these are listed in the next Table:

Conditions Prevented by Using Plasma-Poor Red Cells

Component(s) Removed	Condition Prevented
IgA	Anaphylactic Reaction in IgA Deficient Patients: IgA deficient patients develop antibodies to IgA following exposure to IgA in a previous transfusion. Deglycerolized red cells is the component of choice for clinically significant antibodies to IgA.
Complement (C-3)	Hemolytic Episodes in PNH: Patients with paroxysmal nocturnal hemoglobinuria (PNH) have an acquired intrinsic defect in that their red blood cells bind more C-3 than do normal cells.
Electrolytes and Metabolic Products	Electrolyte imbalance

Each unit contains 185mL packed red blood cells in a 200-250mL volume. There are two disadvantages to the use of plasma poor red cells; one is the cost and the other is the shelf-life which is only 24 hours.

IRRADIATED RED CELLS: Certain patient populations are at risk for graft-versus-host disease (GVHD) if they receive unirradiated directed blood donations from HLA-homozygous or partially HLA-identical donors. Patients at risk include neonates and fetuses, patients with congenital immunodeficiency or immunosuppressive therapy (cancer chemotherapy or transplant immunosuppression), and in other situations. GVHD is caused by T-cells in transfused blood, proliferating and reacting against antigenic sites in the blood recipient. Irradiation destroys lymphocytes and prevents this reaction; granulocytes are not destroyed (Ferrara JLM, Deeg HL, N Engl J Med. 1991; 324:667-674).

DEGLYCEROLIZED RED CELLS (FROZEN-THAWED RED BLOOD CELLS): Red blood cells may be frozen for up to 3 years with glycerol. These preparations are particularly useful for autotransfusions, rare blood types, and clinically significant antibodies to IgA.

WHOLE BLOOD: (Volume 520 ± 45mL, Hct. 40%; smaller volumes for pediatric patients; administer through a filter)(Donor and recipient must be ABO identical crossmatch): Whole blood is given to patients who have had massive blood loss or who are bleeding profusely. Whole blood must be administered through a filter; transfusion should not exceed 4 hours/unit. In some centers, whole blood is no longer available; patient may receive a combination of packed red blood cells and fresh frozen plasma, as indicated.(cont)

Blood Component Therapy (Cont)

FRESH FROZEN PLASMA (SINGLE DONOR): Fresh frozen plasma, which may be stored up to one year, is used in the treatment of bleeding patients with multiple clotting factor deficiencies; fresh frozen plasma contains factors II, V, VII, VIII, IX, X, XI, and XIII. Stored whole blood also contains all of these factors but factors V and VIII are unstable in whole blood and are markedly decreased in activity by the second or third day of storage. Infusion of a unit of fresh frozen plasma causes all coagulation factors to rise about 8 percent; fibrinogen level rises about 13mg/dL. In children, transfusion of 10-15mL/kg raises levels about 20%. Fresh frozen plasma is stored at -20°C or below and requires 30 minutes to thaw in a 37°C water bath.

Fresh frozen plasma is used in the treatment of patients with conditions listed in the next Table:

Conditions Treated with Fresh Frozen Plasma
Disseminated Intravascular Coagulopathy (DIC)
Other Defibrination Syndromes
Liver Disease
Hemophilia and von Willebrand's When Specific Factors are Unavailable

The disadvantages of the use of fresh frozen plasma are as follows: excessive volume causing circulatory overload, possibility of hepatitis, allergic reactions with chills and fever and blood group antibodies; thus, fresh frozen plasma must be ABO-group specific with the recipient's red blood cells. Compatibility testing is not required.

Deficiency of factor VIII and fibrinogen (factor I) alone should be treated with cryoprecipite; factor VIII deficiency alone should be treated with purified factor VIII concentrates when therapy is clinically indicated.

Single donor plasma is used instead of pooled plasma because of the high incidence of hepatitis in pooled plasma. The plasma is administered through a filter.

ALBUMIN (25% and 5%): Albumin (25%, 25g/dL) is used in the treatment of burns and other causes of fluid loss to counteract fluid and sodium loss. The dosage is given so as to maintain the circulating plasma level of albumin at 2.5 ± 0.5g/dL (Tullis JL. JAMA. 1977; 237:460).

Albumin (5%, 5g/dL) is used in the treatment of burns and other causes of fluid loss. It may be used in the early treatment of shock due to hemorrhage until blood is available. The usual minimal effective adult dose is 250-500mL (Borucki DT. Blood Component Therapy, Am Ass Blood Banks, 3rd Ed., 1981).

PLATELET CONCENTRATES: Platelet concentrates are given to patients who are bleeding due to thrombocytopenia (usually <50,000) or due to abnormal platelet function, or prophylactically if platelet counts are below 10,000-20,000. One platelet concentrate increases the platelet count of a patient with a 5000mL blood volume by 5,000 per cu mm; the usual dose is six to eight units of platelet concentrates. In children, one pack per 6kg raises the platelet count by 50,000. Ideally, post-transfusion platelet counts should be obtained within one hour and after 18 hours to monitor response to transfusion and platelet survival.

ABO and Rh compatibility are desirable because of red blood cell contamination, but compatibility testing is not necessary.

Platelet concentrates are not given to patients with thrombocytopenia caused by accelerated platelet destruction, e.g., thrombocytopenic purpura (ITP), unless the patient has life-threatening hemorrhage.

Transfusion risks are hepatitis, immunization to HLA and red blood cell antigens, febrile reaction, allergic reactions with urticaria. Septic reactions occur in one per 4200 platelet transfusions; platelets are stored at room temperature and bacterial growth may occur; the incidence of sepsis is five times higher in platelets stored for five days compared to those stored for 4 days or less (Morrow JF, et al. JAMA. 1991; 266:555-558).

Advanced notification when possible is recommended due to short life span of platelets. Life span is 3-5 days depending on type of bag used for storage.

Platelet concentrates may be obtained by use of plateletpheresis; plateletpheresis is the centrifugal separation of platelets from whole blood either continuous or intermittent return of platelet-poor red blood cells and plasma to the donor. Plateletpheresis from one donor provides the equivalent of 6-8 units of platelet concentrate. Plateletpheresis involves some risk to the donor.

GRANULOCYTE (LEUKOCYTE) CONCENTRATE: Granulocyte transfusions have been given to patients who are granulocytopenic (<500 neutrophils/cu mm) and have infections, e.g., septicemia, that are not responsive to antibiotics or other forms of therapy. Bone marrow should show myeloid hypoplasia and the patient should have a reasonable chance for bone marrow recovery (Pisciotto PT (ed.). Physicians Handbook, Amer Assoc of Blood Banks. 1989; 19-20). Leukocytes are administered daily, usually for 5 days in succession, or until WBC rises above 500 granulocytes. Use of granulocyte transfusions is controversial.

Leukocyte concentrates must be ABO-group compatible and are administered through a blood filter. Transfusion risks include chills, fever, allergic reactions, viral hepatitis, CMV infection and in immunodeficient or immunosuppressed patients, graft versus host reaction. CMV seronegative granulocytes are recommended for immunosuppressed CMV-seronegative patients (e.g., bone marrow transplant recipients). The presence of red blood cells can result in hemolytic reaction.

Granulocyte concentrates may be obtained by use of leukapheresis; leukapheresis is the centrifugal separation of leukocytes from whole blood with either continuous or intermittent return of leukocyte-poor cells and plasma to the donor. Leukapheresis involves some risk to the donor.

CRYOPRECIPITATE: Cryoprecipitate is a source of Factors VIII:C (procoagulant activity), VIII:vWF (von Willebrand Factor), XIII and fibrinogen and is used to treat the conditions listed in the next Table:

Conditions Treated with Cryoprecipitate
Factor Deficiency
Hemophilia A (Factor VIII Deficiency)(Factor VIII Concentrate preferred)
von Willebrand's Disease (not responsive to DDAVP)
Factor XIII Deficiency
Hypofibrinogenemic States, e.g.,
Disseminated Intravascular Coagulopathy (obstetrical complications e.g., abruptio placenta and amniotic fluid embolism; endotoxin in bacterial sepsis; severe intravascular hemolysis; carcinomatosis)

Cryoprecipitate contains 150mg of fibrinogen, 80 to 100 units of Factor VIII, some Factor XIII and von Willebrand's factor; 1mL of normal plasma contains 1 unit of Factor VIII activity.

A patient with hemophilia A who is bleeding usually has a Factor VIII level of about 2 to 10 percent; the desired therapeutic level is 10 to 100 percent. Cryoprecipitate should be given every 12 hours to a severely bleeding hemophiliac or to a hemophiliac who is about to undergo surgery.

The dosage of cryoprecipitate required is calculated as follows:

$$\text{Number of bags of cryoprecipitate} = \frac{\text{desired factor VIII:C level(\%)} \times \text{Patients plasma volume(in mL)}}{\text{average units of Factor VIII:C per bag(min. 80)}}$$

In children, one bag per 6kg raises fibrinogen 75-100mg/dL and VIII:C by 30-40%. Side effects of cryoprecipitate include viral hepatitis, febrile and allergic reactions. Cryoprecipitate may be stored for 1 year at -20°C or below and requires 15 minutes to thaw in a 37°C water bath.

FACTOR VIII CONCENTRATE: Factor VIII is indicated for treatment of hemophilia A patients for the prevention and control of hemorrhagic episodes. In hemophilia, inherited as a sex-linked recessive, there is a deficiency of a part of the Factor VIII molecule, called Factor VIII:C. Factor VIII concentrate is obtained from a large number of donors, and recipients have developed acquired immunodeficiency syndrome (AIDS). There is also a high incidence of non-A, non-B hepatitis in patients receiving factor VIII concentrate (Andes WA. JAMA. 1983; 249:2331). These earlier clotting factor concentrates transmitted hepatitis viruses to nearly 100% and human immunodeficiency virus to 60-80% of patients with hemophilia. Newer preparations are much safer and involve heating the preparations for various time periods and under different conditions. Although more longitudinal data is needed, it appears that transmission of HIV has been virtually eliminated and that transmission of hepatitis has been greatly reduced. Hepatitis transmission has been curtailed by use of the hepatitis B vaccine and screening of blood for hepatitis C, as well as the availability of safer concentrate (Epstein JS, Fricke, WA. Arch Pathol Lab Med. 1990; 114:335-340).

(cont)

Blood Component Therapy (Cont)

TRANSFUSION REACTIONS: Transfusion reaction rates are given in the next Table (Goldfinger D, Lowe C. Transfusion. 1981; 21:277):

Transfusion Reaction Rates

Reaction	Rate (per thousand Units)
Febrile Nonhemolytic	3.0
Urticarial	1.6
Hemolytic, Delayed	0.7
Hemolytic, Immediate	0.06
	5.36

FEBRILE REACTIONS: Febrile reactions usually occur 1 to 2 hours after the transfusion has been started and lasts for several hours to 24 hours. A febrile reaction is defined as a 2 degree increase in body temperature or a one degree increase in temperature accompanied by shaking chills. The differential diagnosis of febrile reactions to blood transfusion products is given in the next Table:

Differential Diagnosis of Febrile Reactions to Blood Products

- Reactions to:
  - Leukocyte Antigens
  - Platelet Antigens
  - Plasma Proteins
- Hemolytic Reactions
- Bacterial Contamination
- Consider Causes Unrelated to Blood Products (eg, Drug Reaction)

In the most common cause of a febrile reaction, antibodies in the patient react with antigens present on donor leukocytes. Fever may result from complement activation or release of endogenous pyrogens such as interleukin-1.

Signs and symptoms of febrile reactions are as follows: febrile reactions, mild to severe; chills, nausea, vomiting, hypotension, cyanosis, tachycardia, transient leukopenia, chest pain and dyspnea.

Nurses should proceed as follows:

(1) Stop transfusion. Keep I.V. open with slow saline drip.
(2) Recheck patient identification on tie tag and patient arm band, and recheck blood unit numbers.
(3) Notify physician and blood bank
(4) Send clotted sample and blood container (without removing recipient set) to the blood bank and collect a post-transfusion urine specimen to send to the laboratory marked "Transfusion Reaction Specimen."

Physicians should proceed as follows:

(1) Evaluate patient for possible signs of hemolytic transfusion reaction.
(2) Ordinarily, order the transfusion discontinued.
(3) Administer antihistamine, antipyretics and possibly steroids as indicated.

Prevention: Consider washed (leukocyte poor) red cells for subsequent transfusion; use pretransfusion antihistamine in patients who have had previous febrile reactions.

ALLERGIC-ANAPHYLACTOID REACTIONS: Usually, these reactions are mild. Clinical signs and symptoms include pruritus, urticaria, occasional facial and periorbital edema; rarely bronchospasm and anaphylactoid shock. Urticaria is not seen with hemolytic reactions; therefore, a work-up for hemolysis is not necessary.

Usually, the transfusion is discontinued and an antihistamine is administered I.V. (e.g. diphenhydramine). If laryngospasm or bronchospasm occur, give epinephrine. If anaphylactoid shock occurs, establish an airway, give I.V. fluid, corticosteroids, and cardio-supportive drugs.

If the reaction is very mild, urticaria only, then slow transfusion rate and administer antihistamines I.V. If there is no clinical improvement, stop transfusion.

Prevention: Patients with a history of allergic transfusion reactions should receive oral diphenhydramine (Benadryl) 30 minutes prior to transfusion. Transfuse washed red blood cells instead of standard units. Plasma should be IgA deficient.

Causes: Most of the allergic reactions are probably caused by many different proteins.

Anaphylactoid transfusion reactions are rare; they most often occur in patients with IgA deficiency. About 1 in 850 individuals are deficient in IgA and about 25 percent have IgA antibodies. The diagnosis is made by finding anti-IgA in the recipient plasma.

HEMOLYTIC, DELAYED: These reactions usually occur 3-10 days after administration of red cells and are most commonly caused by antibody in the recipient reacting with donor red cell antigens. These reactions occur most often in patients who have had multiple transfusions or women previously sensitized to red cell antigens from pregnancy.

The signs and symptoms of delayed hemolytic reaction are: fever, jaundice and occasionally hemoglobinuria; lack of appropriate increase in hematocrit or hemoglobin. Obtain blood specimens as follows:

Blood Specimens for Delayed Hemolytic Reaction

Specimen	Assay
Purple (EDTA) or Blue (citrate) Top Tube	Direct Coombs Test
Red Top Tube	Antibody Screening Test
Red Top Tube	Antibody Titer
Purple (EDTA) Top Tube	Complete Blood Count (Hgb, Hct, RBC, Indices), Smear
Red Top Tube	Bilirubin, Haptoglobin

The goals of this work up are to: 1) note the drop in hemoglobin/hematocrit due to hemolysis, 2) note the increase in bilirubin, 3) screen pre- and post-transfusion serum for antibodies, and 4) document that at least one unit contained the antigen to which the patient developed antibodies. The blood bank must note the RBC antigens to which the patient is incompatible and keep this record on file.

The management of delayed hemolytic reaction is supportive; treatment of hypotension or renal failure is similar to acute hemolytic reaction. The usual responsibility of the blood bank is to search for antibody in the recipient.

HEMOLYTIC, IMMEDIATE: The most common cause of fatal hemolytic reactions is clerical errors. The most severe hemolytic reactions result from antibody in the recipient reacting with donor cells. Nonimmune hemolysis may occur if the blood is exposed to hypotonic solutions (e.g. 5% dextrose), hypertonic solutions (e.g. 50% dextrose), mechanical stress (e.g. cardiopulmonary bypass), freezing, overheating, or bacterial contamination (Piscioto PT (ed.). Physicians Handbook, Amer Assoc of Blood Banks. 1989; 79-80).

The features are as follows: fever, chills, hypotension, hemoglobinuria, oozing from incisional sites, nausea, anxiety, back pain, tachycardia, oliguria, shock, acute renal failure; death can occur. Guidelines for evaluation of a suspected immediate hemolytic reaction are as follows:

(1) Stop transfusion immediately; notify and send blood to blood bank; keep I.V. line open with saline.
(2) Recipient: Record and monitor (a) vital signs: temperature, pulse, blood pressure, and respirations; (b) urine output and fluid intake.
(3) Verify that recipient has received proper unit by checking donor numbers against recipient identification bracelet and transfusion request slip.
(4) Return complete I.V. set intact and unused portion of donor unit to blood bank:
    (a) Repeat ABO-Rh typing, antibody screen, compatibility testing
    (b) Bacterial culture
(5) Obtain blood specimens from recipient at a location away from the infusion site:
    (a) Blood Bank: (1 blue (citrate) and red top tube) Repeat ABO-Rh typing, antibody screen, direct Coombs test and donor compatibility testing.
    (b) Hematocrit: Lavender (EDTA) top tube
    (c) Blood Culture: Sterile yellow stoppered tube
(6) Urinalysis for blood assay

Blood Component Therapy (Cont)

Treatment: Treat shock with vasopressors and appropriate IV fluids. Monitor intake and output, maintain fluid balance. Maintain renal blood flow with IV furosemide (20-80mg) and/or mannitol. Test for disseminated intravascular coagulation (DIC). Administer appropriate compatible blood components as necessary, i.e., red blood cells, platelets, cryoprecipitate, fresh frozen plasma. Follow status of patient for possible renal failure requiring renal dialysis.

SIDE EFFECTS AND HAZARDS: Side effects associated with transfusion of blood are given in the next Table:

Side Effects Associated with Blood Transfusion

Side Effect	Comment
Hemolytic Transfusion Reactions	Incompatibility between donor red blood cells and recipient plasma or non-immune hemolysis.
Transmission of Infectious Disease	Viral hepatitis: Mostly non-A non-B non-C hepatitis. Other diseases that may be transmitted are malaria, brucellosis, babesiosis, trypanosomiasis, filariasis, Epstein-Barr virus, cytomegalovirus, toxoplasmosis, syphilis in the serological negative phase, Colorado tick fever, AIDS, other retrovirus (HIV-II, HTLV-1, HTLV-II). Bacterial contamination: Occurs rarely. Presents with fever (80%), chills (53%), hypotension (37%), and nausea or vomiting (26%), usually during transfusion (47%). Appearance of symptoms may be delayed (15 min-17 days). Mortality is 35% (Morduchowicz G, et al. Reviews of Infect Dis. 1991; 13:307-314).
Immunization of the Recipient	Recipients who have had transfusions or pregnancies may develop antibodies to HLA or other antigens on leukocytes, or platelets or red blood cells. Subsequent transfusions with cells carrying these antigens may cause a febrile reaction or an anamnestic reaction and a delayed hemolytic reaction (autoimmune hemolytic anemia-like) with a positive direct antiglobulin test.
Allergic Reactions	Urticaria occasionally accompanied by chills and fever; fever may be prevented in patients with known history by premedication with antihistamines. Anaphylactoid Reaction: About 1 in 500 individuals lack IgA and develop anti-IgA antibodies; on subsequent transfusion, an anaphylactoid reaction may occur; treat with epinephrine and corticosteroids.
Other Reactions:	Graft-vs-host disease; Circulatory Overload; Iron Overload; Metabolic Complications.

Trisodium citrate is an anticoagulant in blood; one unit of whole blood contains 17mEq of citrate; one unit of packed cells contain 5mEq of citrate. Following transfusions, citrate is metabolized to [HCO_3^-]; that is 17mEq of citrate is converted to 17mmol(mEq) of [HCO_3^-] and 5mEq of citrate is converted to 5mmol(mEq) of [HCO_3^-]. Normally, the [HCO_3^-] load is excreted by the kidney. However, there are certain conditions in which patients receive many units of blood and in which the excretion of [HCO_3^-] by the kidney is compromised; these are given in the next Table:

Blood Transfusions and Compromised Excretion of [HCO_3^-] by the Kidney

Conditions in Which Many Blood Transfusions May be Given	Compromised Excretion of [HCO3-] by the Kidney (Enhanced Renal HCO3- Reabsorption)
Shock	ECF Volume Contraction
Trauma	Potassium Depletion
Sepsis	Hypercapnia
Open Heart Surgery	Acute or Chronic Renal Disease
	Oliguria

An example of a consultation request for suspected transfusion reaction is given in the next Figure (McCord RG, Myhre BA. Laboratory Medicine. 1978; 9:39-46):

Consultation Request for Suspected Transfusion Reaction

__/__/__ ______ AM/PM ____________ ______________ mL
Date Time No. of Units Vol. Transfused
Clinical Diagnosis Prior to Transfusion ______________________

Donor No. of Last Unit Given ____________ ____________

PRE TRANSFUSION	POST TRANSFUSION
Temp.________	Temp.________
B/P ________	B/P ________
Pulse________	Pulse________

CHECK THOSE WHICH APPLY

______Pulse	______Chest Pain	______Back Pain	______Urine Output
______B/P	______Headache	______Jaundice	______Heat at I.V. Site
______Chills	______Dyspnea	______Rash	______Pain at I.V. Site
______Fever	______Hemoglobinuria	______Flushing	______Delirium
______Coma	______Nausea	______Pruritus	______Muscle Tenderness
______Syncope	______Vomiting	______Urticaria	______Petechiae

______Other (Specify)______________________

LIST DONOR NO. FROM ALL SUSPECTED UNITS IN THIS TRANSFUSION SERIES

Donor No.	Date	Time Started	Time Stopped	Amount Given	Whole Blood, Packed cells, Platelets, etc.	Reaction

Complete this portion of the form and return to the TRANSFUSION LABORATORY with post-reaction sample of the patient's urine and blood (clotted). Return the blood bags from all units in this transfusion series.

A standard report, used by residents in the Department of Pathology at UCLA School of Medicine, for febrile transfusion reactions and allergic transfusion reactions is as follows:

Standard Consultation Report for Febrile Transfusion Reactions

According to the above information and the chart entry of __/__/__, this age year old man-woman-child experienced chills and a fever (degrees F) during-shortly after the transfusion of a unit of packed cells-whole blood-platelets. There was no complaint of back pain or signs of dyspnea, shock or hemoglobinuria as is often the case in a hemolytic reaction. Furthermore, inspection of pre-transfusion and post-transfusion sera revealed that they were straw-colored, rather than pink or red as one sees with acute intravascular hemolysis. Repeat crossmatch and compatibility testing revealed that compatible blood was administered; vis., ABO, Rh type blood was given to a(n) ABO, Rh type recipient. In addition, antibody screening failed to demonstrate the presence of any unexpected antibodies. On the basis of these negative findings, we suspect that the patient experienced a FEBRILE TRANSFUSION REACTION due to pyrogenic material in the unit of blood he-she received. This material usually originates in the granulocytes contained in the blood. With regard to future transfusion of blood to this patient, please read the "Pathology Resident's Note on Febrile Transfusion Reactions" which has been taped into the progress notes.

The appropriate choice of underlined words is selected.

Blood Component Therapy (Cont)

Standard Consultation Report for Allergic Transfusion Reactions

According to the above information and the chart entry of _/_/_, this age year old man-woman-child experienced flushing and pruritus and urticaria during-shortly after the transfusion of a unit of packed cells-whole blood-plasma-cryoprecipitate. There was no complaint of back pain or signs of dyspnea, shock or hemoglobinuria as is often the case in hemolytic reaction. Furthermore, inspection of a pre-transfusion and post-transfusion sera revealed that they were straw-colored, rather than pink or red as one sees with acute intravascular hemolysis. Repeat crossmatch and compatibility testing revealed that compatible blood was administered; vis., ABO, Rh type blood was given to a(n) ABO, Rh type recipient. In addition, antibody screening failed to demonstrate the presence of any unexpected antibodies. On the basis of these negative findings, we suspect that the patient experienced an ALLERGIC TRANSFUSION REACTION due to allergenic material in the unit of blood he-she received. With regard to future transfusion of blood to this patient, please read the "Pathology Resident's Note on Allergic Transfusion Reactions" which has been taped into the progress notes.

Standard Note on the Causes and Prevention of Febrile Transfusion Reactions

Pathology Resident's Note on Febrile Transfusion Reactions

(If you have not already done so, please see the "Consultation Request for Suspected Transfusion Reaction", which has been inserted into the CONSULTATION section of this patient's chart).

Granulocyte pyrogenic material can be released in two ways. First, the recipient may have antigranulocyte antibodies (usually leukoagglutinins) in his blood. These antibodies can attach to receptor sites on the granulocyte membrane and, with the aid of the complement system, cause lysis of the granulocyte in vivo and concomitant release of pyrogenic substances. Secondly, stored refrigerated granulocytes normally undergo autolysis in vitro within 24 hours releasing substances which are pyrogenic. Hence, pyrogenic material can be present in stored blood in the absence of immune lysis.

The problem of febrile reactions due to granulocyte pyrogenic material* can be minimized by giving the patient washed packed erythrocytes. However, washing a fresh unit of packed red cells only removes about 80% of the granulocytes. The remaining cells (if the unit is used immediately) may be subject to immune lysis as described above. If the washed packed cells are stored, the remaining granulocytes will undergo autolysis and again release pyrogenic material.

Febrile reactions due to granulocytic pyrogenic material can be eliminated by ordering packed red cells which have been refrigerated or frozen. When ordering refrigerated packed red cells, be certain to request that the units be washed. The washing will remove the pyrogenic substances that has been released by granulocyte lysis. In addition, washing will remove any potassium or phosphate that has leaked out of the stored red cells, as well as any lactic acid generated by red cell metabolism. (Washing units of blood introduces the possibility of bacterial contamination, but we feel that the risk is minimal since Federal regulations require that blood be used within 24 hours after thawing and this should preclude significant bacterial growth). If frozen packed red cells are ordered, they will be routinely washed since washing is required to remove the cytoprotective agent, glycerol.

When the administration of whole blood is not vital, the BLOOD BANK recommends the use of frozen packed red cells for the following reasons:

1. These units are extensively washed to remove the glycerol. The pyrogenic material from the lysed granulocytes will be removed during the washing procedure.
2. The process of freezing and thawing destroys many granulocytes which would survive a normal washing procedure.
3. Glycerol is bacteriostatic, therefore greatly decreasing the chance of bacterial growth in the unit of blood.
4. The freezing process is believed to attenuate or even destroy hepatitis virus.

*Administration of platelets may also cause a febrile reaction, but this discussion is limited to correction of anemia only.

Standard Note of the Causes and Prevention of Allergic Transfusion Reactions

Pathology Resident's Note on Allergic Transfusion Reactions

(If you have not already done so, please see the "Consultation Request for Suspected Transfusion Reaction", which has been inserted into the CONSULTATION section of this patient's chart).

An allergic transfusion reaction is an example of an "immediate-type" hypersensitivity reaction. This kind of reaction is antibody-mediated and occurs within minutes of exposure to an antigen. (This is in contradistinction to the "delayed-type" hypersensitivity reaction which is cell-mediated and occurs within hours or days). Allergic transfusion reactions (flushing, pruritus, urticaria) are most commonly seen in people who are atopic or who have a congenital lack of IgA antibody; however, they may occur in anyone. Approximately 10% of the population are atopic and are sensitive to a variety of environmental allergens. They experience allergic reactions which are mediated by IgE antibodies. The other type of severe allergic transfusion reaction is the interaction between transfused IgA and anti-IgA antibodies in the recipient's plasma. In this case, the recipient has a congenital lack of IgA and has actually formed IgG antibody against the IgA in the transfused blood. About 1 in 500 recipients lack IgA and could theoretically be at risk.

Prevention of allergic transfusion reactions involves elimination of the offending allergen. When correction of anemia is the only hematologic problem facing the clinician, the BLOOD BANK recommends use of washed packed red cells or frozen reconstituted packed cells. The washing process should remove sufficient allergens (vis., environmental allergens or IgA antibody) to prevent an allergic reaction. In addition, washing removes any potassium or phosphate that has leaked out of the stored red cells, as well as any lactic acid generated by red cell metabolism. If frozen packed red cells are ordered, they will routinely be washed since washing is required to remove the cryoprotective agent, glycerol. (Washing units of blood introduces the possibility of bacterial contamination, but we feel that the risk is minimal since Federal regulations require that blood be used within 24 hours after thawing and this should preclude significant bacterial growth).

When it is absolutely necessary to administer whole blood, plasma or cryoprecipitate to a person who is known to have experienced mild allergic reactions, prophylactic administration of antihistamine should be considered. Furthermore, this patient has potential risk of development of laryngeal edema. Consequently, it is recommended that a physician be nearby for approximately one half hour post-transfusion; a nurse should continue surveillance for at least 12 hours post-transfusion.

When the administration of whole blood is not vital, the BLOOD BANK recommends the use of frozen packed red cells for the following reasons:

1. These units are extensively washed to remove the glycerol. The washing will remove any allergenic material in the unit. (Note: If a patient has ever experienced an anaphylactic reaction due to sensitivity to transfused IgA antibody, then transfusion of frozen red cells is absolutely essential, since routine washing does not remove enough plasma to eliminate the reaction).
2. Polymorphonuclear leukocyte breakdown products will also be washed out, thus removing the pyrogenic material responsible for febrile reactions.
3. Glycerol is bacteriostatic, therefore greatly decreasing the chance of bacterial growth in the unit of blood.
4. The freezing process is believed to attenuate or even destroy hepatitis virus.

Blood Component Therapy (Cont)

About two-thirds of all red blood cell transfusions are given in the perioperative period. In the past, it has been rigidly accepted that the criteria for perioperative red blood cell transfusion is a hemoglobin value of 100g/liter (10g/dL) or a hematocrit of less than 0.30 (30%). However, greater recognition of the risks of transfusion has alerted physicians to question and modify these criteria. The most important criteria for red blood cell transfusion is clinical judgement. Healthy patients with hemoglobin values of 100g/liter (10g/dL) or greater rarely require perioperative transfusion; those with hemoglobin values of less than 70g/liter (7g/dL) frequently require red blood cell transfusion. Some patients with chronic anemia, such as those with chronic renal failure, tolerate hemoglobin values of less than 70g/liter (7g/dL)(Consensus Conference, JAMA. 1988; 260:2700-2703).

ALTERNATIVES TO RED BLOOD CELL TRANSFUSION: Alternatives to red blood cell transfusion are given in the next Table (Consensus Conference. 1988; 260:2700-2703):

Alternatives to Red Blood Cell Transfusion
Autologous Red Blood Cell Transfusion
Intraoperative Blood Salvage
Modification of usual Criteria for Red Blood Cell Transfusion, e.g., Hemoglobin, 100g/liter (10g/dL); Hematocrit, 0.30 (30%)
Under Development: Modified Hemoglobin Solutions and improved Perfluorochemical Emulsions
Drugs: Desmopressin; deliberately induced Hypotension; Erythropoietin therapy

BLOOD GASES, ARTERIAL

Test Includes: pH, PCO_2, PO_2, Total CO_2, Bicarbonate, Base Excess, Oxygen Saturation

SPECIMEN FOR BLOOD GASES, ARTERIAL: Blood specimens are obtained from the radial, brachial, femoral or other artery. When blood is obtained from the radial artery, the patency of the ulnar artery should also be determined by the Allen test.

Use a 22 to 25 gauge, one to 1.5 inch disposable needle. Blood may be drawn into plastic disposable syringes which have better seals than glass syringes. A tourniquet is not used. Blood is collected using heparin as anticoagulant. Adequate volume for oxygen saturation, PO_2, PCO_2 and pH is obtained by using a 1 mL tuberculin syringe lubricated with heparin (100,000 units/mL). When the artery is penetrated, pulsations are noted when using a syringe. Multiple air bubbles and froth render the sample useless. After collection of the blood, the tip of the syringe should be sealed with an appropriate cap. After arterial puncture, pressure should be maintained on the arterial puncture site for at least five minutes.

Ice is used to preserve the samples if whole blood cannot be measured within 15 minutes; storage on ice will protect the specimen for 1 to 2 hours. If the specimen is not maintained on ice, glycolysis will occur and lactic acid will be produced; a fall in pH and a rise in PCO_2 will occur.

ARTERIAL pH:
REFERENCE RANGE: 7.36 to 7.44
METHOD: pH electrode
INTERPRETATION: The relationship of pH and $[H^+]$ is given in the next Table:

Relationship of pH and $[H^+]$	
pH	$[H^+]$ Concentration (nmol/liter)
6.8	160 (Four Times Normal)
7.1	80 (Twice Normal)
7.4	40 (Normal)
7.7	20 (Half Normal)

The pH is decreased in metabolic acidosis and respiratory acidosis; the pH is increased in metabolic alkalosis and respiratory alkalosis.

METABOLIC ACIDOSIS: In metabolic acidosis, HCO_3^- is decreased. Rule: A decrease in pH of 0.15 is the result of a decrease in bicarbonate of 10mmoles/liter. If compensation occurs, PCO_2 decreases. A diagnostic approach to the differential diagnosis of metabolic acidosis based on anion gap and serum potassium, is given in the next Figure (Narins RB. Diagnostic Dialog. 1981; 3:14-15):

Diagnostic Approach to Differential Diagnosis of Metabolic Acidosis

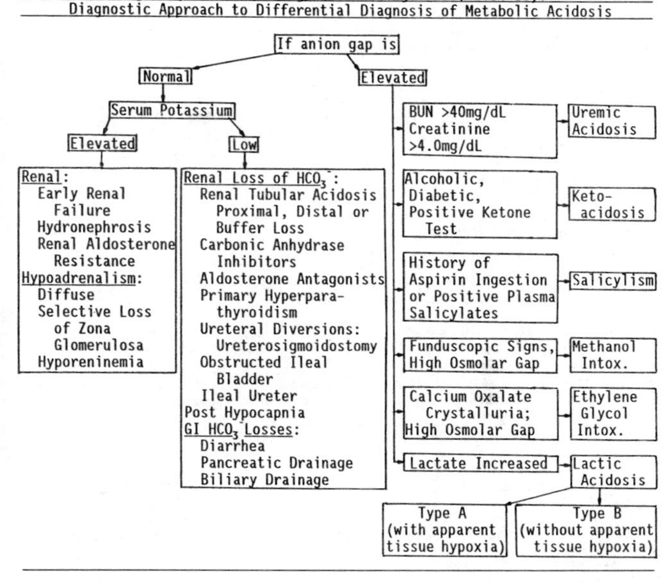

RESPIRATORY ACIDOSIS: In respiratory acidosis, PCO_2 is increased. Rule: An increase of PCO_2 of 10mmHg is associated with a decrease of pH of 0.08 units. If compensation occurs HCO_3^- is increased. Causes of acute respiratory acidosis are given in the next Table:

Causes of Acute Respiratory Acidosis

Condition Affecting Organ or System	Examples
Brain Depression	Sedative, Opiate, Other Drug and Anaesthetic Overdose Comatose States, e.g., Cerebrovascular Accidents Neurosurgery and Head Injury
Spinal Cord	Spinal Cord Trauma, Cervical Vertebral Fracture
Neuromuscular System Disorders	Respiratory Paralysis, Acute Neuromuscular Disease, Myasthenia Gravis, Poliomyelitis, Guillain-Barre Syndrome
Chest Wall Limitation	Rib Fracture with Flail Chest
Upper Airways	Airway Obstruction, e.g., Aspiration of Foreign Body, Croup, Epiglottitis, Laryngospasm
Lower Airways and Lungs	Acute Severe Pulmonary Edema Severe Pulmonary Infections, e.g., Severe Pneumonia Open Chest Wounds Prolonged Open Chest and Open Heart Operations Pneumothorax, Hemothorax, Atelectasis
Other	Mechanical Underventilation Abdominal Distention from Ascites and Peritonitis

Arterial Blood Gases (Cont)

Acute respiratory acidosis may occur when a patient with chronic pulmonary disease is given an excessive amount of oxygen. The respiratory center is normally very sensitive to changes in arterial PCO_2. When the PCO_2 rises above 65mmHg, the respiratory center becomes insensitive to PCO_2. Hypoxia then becomes the main stimulus to respiration (carotid and aortic bodies); these bodies respond to lowered arterial oxygen saturation.

Causes of chronic respiratory acidosis are given in the next Table:

Causes of Chronic Respiratory Acidosis

Severe Lung Disease
- Emphysema
- Chronic Obstructive Lung Disease, e.g., Chronic Bronchitis
- Chronic Asthma
- Bronchiectasis
- Pulmonary Fibrosis

Hypoventilation
- Extreme Obesity, i.e., Pickwickian Syndrome
- Chest Deformity: Kyphoscoliosis and Injury to Thoracic Cage

These patients also have low PO_2 and cyanosis may be present.

METABOLIC ALKALOSIS: In metabolic alkalosis, HCO_3^- is increased. Rule: An increase in pH of 0.15 is the result of an increase in bicarbonate of 10mmol/liter. If compensation occurs, PCO_2 is increased. Metabolic alkalosis is associated with conditions that cause a decrease in potassium and by conditions that cause a loss of hydrogen ions, by excess bicarbonate and by chloride losing diarrhea; these conditions are given in the next Table:

Causes of Metabolic Alkalosis

(1) Decrease of $[K^+]$

Urine Loss	G.I. Loss	Movement into the Intracellular Space
Diuretics: Thiazides; Loop Diuretics: Furosemide (Lasix); Ethacrynic Acid (Edecrine)	Vomiting	Diabetic Ketoacidosis (treated)
Antibiotics: Carbenicillin Amphotericin B	Nasogastric Suction	Familial Periodic Hypokalemic Paralysis
Magnesium Depletion	Pyloric Obstruction	**Other**
Increased Mineralocorticoid		Decreased K^+ Intake
Licorice Abuse		Acute Myeloid Leukemia
Bartter's Syndrome		

(2) Loss of $[H^+]$:
- (a) Gastrointestinal Tract: Vomiting or gastric aspiration
- (b) Kidney: Hypercalcemia: other than Primary Hyperparathyroidism

(3) Excess of $[HCO_3^-]$: Excessive $NaHCO_3$ or other alkaline salts

(4) Chloride Losing Diarrhea

Differential Diagnosis of Metabolic Alkalosis; (Harrington JT, Cohen JJ. N Engl J Med. 1975; 293:1241-1248). The most common causes of metabolic alkalosis are two-fold: (1) those associated with loss of hydrogen ion and extracellular fluid volume (chloride responsive type) and (2) those associated with pathologically excessive circulating levels of mineralocorticoid hormone (chloride resistant type). Measurement of urinary chloride is valuable in differentiation of causes of persistent metabolic acidosis, as given in the following Table:

Urinary Chloride in the Differential Diagnosis of Metabolic Alkalosis

Chloride Responsive	Chloride Resistant
(Urine Chloride $<$10mmol/liter)	(Urine Chloride $>$20mmol/liter)
Diuretic Therapy	Adrenal Disorders
Gastrointestinal Causes	Hyperaldosteronism
Vomiting	Cushing's Syndrome
Nasogastric Suction	Pituitary, Adrenal, Ectopic ACTH
Chloride-Wasting Diarrhea	Exogenous Steroid
Villous Adenoma-Colon	Gluco- or Mineralocorticoid
Rapid Correction of Chronic Hypercapnia	Licorice Ingestion (Glycyrrhizic acid has mineralocorticoid-like activity)
Carbenicillin	Carbenoxalone
Reduced Dietary Intake of Chloride	Bartter's Syndrome
	Alkali Ingestion

The most common causes of persistent metabolic alkalosis, e.g., loss of gastric juice and diuretic induced, are due to chloride depletion and are responsive to the administration of chloride adequate to replace body stores.

RESPIRATORY ALKALOSIS: In respiratory alkalosis, PCO_2 is decreased. Rule: A decrease of PCO_2 of 10mmHg is associated with an increase of pH of 0.08 units. If compensation occurs, HCO_3^- is decreased. Disorders which cause excessive pulmonary elimination of carbon dioxide result in respiratory alkalosis. Promptly, after the onset of hyperventilation, hydrogen ions from intracellular sources enter the extracellular fluid and bicarbonate moves into the red blood cells in exchange for chloride, thus minimizing extracellular alkalosis.

The causes of respiratory alkalosis are given under four headings:

Respiratory Center Stimulation
Hypermetabolic States
Artificial Ventilation
Mechanisms Unknown

In these patients, hyperventilation occurs with an increased tidal volume and increased alveolar ventilation.

The specific conditions causing respiratory alkalosis are given in the next Table:

Causes of Respiratory Alkalosis

(1) Respiratory Center Stimulation
- Central nervous system disease; Encephalitis; Brain stem lesions (cerebral-vascular accidents); Intracranial surgery
- Drugs: Salicylate intoxication (early stage); Infants of Heroin addicted mothers; Adults on withdrawal from Heroin; Paraldehyde intoxication; 2,4-Dinitrophenol intoxication
- Anxiety and hysteria (hyperventilation syn.)
- Hypoxia: Anemia from any causes; High altitude residents;
  - Congestive heart failure;
  - Pulmonary Disease (Pulmonary Fibrosis with Alveolar-capillary Block)
    - Pulmonary Emboli; Boeck's sarcoid; Beryllium granulomata; Asbestosis; Pulmonary scleroderma; Alveolar cell carcinoma; Diffuse metastatic carcinoma; Pulmonary fibrosis
  - Right to left shunt: Congenital heart disease; Pulmonary atelectasis
- Early septic shock
- Reflex hyperventilation: Pneumothorax; Pulmonary hypertension
- Pregnancy

(2) Hypermetabolic States: Thyrotoxicosis; Febrile States; Exercise
(3) Artificial Ventilation: Excessive ventilation; Mechanical overventilation using low PCO_2
(4) Others: Cirrhosis of the liver; Extracorpeal circulation; Beriberi; Transient phase following diabetic ketoacidosis; Alcohol intoxication--delirium tremens

ACID-BASE (MIXED STATES):
(Narins RG, Emmett M. Medicine. 1980; 59:161-187):
(a) Metabolic Alkalosis and Respiratory Alkalosis: Increased $[HCO_3^-]$, decreased PCO_2, Increased pH.
(1) Critically Ill Surgical Patients: The combination of metabolic alkalosis and respiratory alkalosis may occur in critically ill surgical patients. The causes of combined metabolic alkalosis and respiratory alkalosis are given in the next Table:

Causes of Combined Metabolic and Respiratory Alkalosis

Metabolic Alkalosis	Respiratory Alkalosis
Vomiting or Nasogastric Suction	Excessive Mechanical Ventilation
Massive Blood Transfusions	Hypoxemia
Lactated Ringer's Solution	Sepsis
High Dose Antacid	Hypotension
	Neurologic Damage
	Liver Disease
	Pain
	Drugs

Blood Gases, Arterial (Cont)

Blood Transfusions: Trisodium citrate is an anticoagulant in blood; one unit of whole blood contains 17mEq of citrate; one unit of packed cells contains 5mEq of citrate. Following transfusions, citrate is metabolized to [HCO_3^-]; 17mEq of citrate is converted to 17mmol(mEq) of [HCO_3^-] and 5mEq of citrate is converted to 5mmol(mEq) of [HCO_3^-]. Normally, the [HCO_3^-] load is excreted by the kidney. However, there are certain conditions in which patients receive many units of blood and in which the excretion of [HCO_3^-] by the kidney is compromised; these are given in the next Table:

Blood Transfusions and Compromised Excretion of [HCO_3^-] by the Kidney

Conditions in which many Blood Transfusions may be given	Compromised Excretion of [HCO_3^-] by the Kidney (Enhanced Renal HCO_3^- Reabsorption
Shock	ECF Volume Contraction
Trauma	Potassium Depletion
Sepsis	Hypercapnia
Open Heart Surgery	Acute of Chronic Renal Disease
	Oliguria

(b) Respiratory Acidosis and Metabolic Alkalosis: Increased [HCO_3^-], increased PCO_2, possibly normal pH. This combination is the most frequently encountered mixed acid-base disturbance. It occurs in patients with chronic hypercapnia (such as chronic obstructive pulmonary disease, COPD) who have received diuretics (such as the thiazides) or who had their PCO_2 abruptly lowered.

(c) Mixed Metabolic Acidosis and Respiratory Alkalosis: (See Narins RG, Emmett M. Medicine. 1980; 59:161-187: Gary NE, Resident and Staff Physician, 1981; 32-38s). This combination usually occurs in patients with uremia, lactic or ketoacidosis and liver or heart failure; bacteremic shock and salicylate intoxication may cause this combination.

(d) Mixed Metabolic Alkalosis and Acidosis: This combination may occur when there is a loss of bicarbonate due to diarrhea resulting in metabolic acidosis accompanied by vomiting resulting in metabolic alkalosis. Another cause of mixed metabolic alkalosis and acidosis is overzealous sodium bicarbonate treatment of acidosis. This pattern also occurs in infants with pyloric stenosis and hypovolemic shock due to vomiting.

PCO_2:

REFERENCE RANGE: Sea level, Men: 33 to 44mmHg (4.4kPa to 5.9kPa); Women: 32 to 42mmHg (4.12kPa to 5.59kPa); pregnancy: PCO_2 decreased; the normal adult respiratory volume of 5 liters per minute is increased in pregnancy to 10 liters per minute owing to the effects of progesterone which causes a chronic hyperventilation; during labor, the ventilatory rate may increase to 35 liters per minute with a further decrease of PCO_2 (Huch R, Huch A. Crit Care Med. 1981; 9:694-697).

One Mile Altitude: 34 to 38mmHg (4.52-5.05kPa)

Conversion of mmHg to kPa - multiply by 0.1333; kPa to mmHg - multiply by 7.52.

METHOD: The PCO_2 electrode consists of two basic components. There is a membrane permeable only to CO_2 gas between the solution to be measured and the pH electrode.

INTERPRETATION: PCO_2 is altered in the conditions listed in the next Table:

Alterations of PCO_2
Increased PCO_2:
Respiratory Acidosis
Compensation in Metabolic Alkalosis
Decreased PCO_2:
Respiratory Alkalosis
Compensation in Metabolic Acidosis

The causes of each of these conditions are given in the previous section.

Heparin may significantly lower PCO_2 measurements by dilution. This phenomenon occurs most frequently when small samples of blood are obtained (e.g., 0.2mL) in infants. This artifact is eliminated by using syringes containing lyophilized heparin (Zuerlein TJ. AJDC. 1990; 144:1287-1288).

BICARBONATE [HCO_3^-]:
REFERENCE RANGE: 20 to 26mmol/liter
METHOD: [HCO_3^-] is obtained by calculation by inserting the values of the pH and PCO_2 into the equation:

$$[H^+] = 24 \frac{PCO_2}{[HCO_3^-]}$$

INTERPRETATION: Bicarbonate [HCO_3^-] is altered in the conditions listed in the next Table:

Alteration of Bicarbonate (HCO_3^-)
Increased HCO_3^-:
Metabolic Alkalosis
Compensation in Respiratory Acidosis
Decreased HCO_3^-:
Metabolic Acidosis
Compensation in Respiratory Alkalosis

The causes of each of these conditions are given in the previous section.

Serum bicarbonate measurements are significantly decreased when vacutainer tubes are underfilled (Herr RD, Swanson T. Am J Clin Path. 1992; 97:213-216). Such an error could lead to misinterpretation of acid-base status and anion gap.

BASE EXCESS:
REFERENCE RANGE: -3 to +3mmol/liter
METHOD: Base excess is obtain directly from the blood gas analyzer from the PCO_2, pH and HCO_3^- values.
INTERPRETATION: Base excess suggests the presence of metabolic alkalosis; base deficit suggests metabolic acidosis. Base excess gives, in mmol/liter, the surplus of base or acid present in the blood or expressed in another way, base excess or deficit shows directly how many mmol/liter of acid or base excess must be added to "normalize" the pH of the blood.

P_aO_2, ARTERIAL:
SPECIMEN: 1mL arterial blood in heparinized syringe, capped and iced.
REFERENCE RANGE: Normal Adult: 75 to 105mmHg (10.0-14.0kPa) at sea level; 65 to 75mmHg (8.7-10.0kPa) at one mile altitude. The arterial blood gas values in mmHg in the fetus and first week of life are as follows; (Weisberg HF. Ann Clin Lab Sci. 1982; 12:245-253): Fetus: 16-24 (2.1-3.2); Birth, umbilical artery, 8-24 (1.1-3.2); 5-10 minutes, 33-75 (4.4-10.0); 20 minutes, 31-85 (4.1-11.3); 60 minutes, 38-83 (5.1-11.1); after 1 hour, 55-80 (7.3-10.7); one day, 54-95 (7.2-12.6); one week, 57-94 (7.6-12.5).
Pregnancy: (at term): 85-115 (11.3-15.3); during labor, the ventilatory rate can increase to as much as 35 liters per minutes, from the pre-pregnancy rate of 5 liters per minute, resulting in an increased PO_2. Conversion of mmHg to kPa - multiply by 0.1333; kPa to mmHg - multiply by 7.52.
METHOD: PO_2 electrode
INTERPRETATION: Conditions that are associated with decreased PO_2 and that require acute oxygen therapy are listed in the next Table:

Conditions Associated with Decreased PO_2 and That May Require Acute Oxygen Therapy
Carbon Monoxide Poisoning (give 100 percent Oxygen)
All Acute Pulmonary Disorders, e.g., Adult Respiratory Distress Syndrome; Pulmonary Edema; Severe Bronchial Asthma; Pulmonary Thromboembolism; Severe Pneumonia
Myocardial Infarction (Cardiogenic Shock)
Congestive Heart Failure
Drug Overdose, e.g., Barbiturate Poisoning
Head and Musculoskeletal Trauma, e.g., Crushed Chest Injury
Hepatic Failure; Acute Pancreatitis; Shock; Sepsis; Hemorrhage; Pneumonia

Blood Gases, Arterial (Cont)

Acute oxygen therapy, if prolonged, causes changes in alveolar structures (Davis WB, et al. N Engl J Med. 1983; 309:878-883).

In acute conditions, a low PO_2 is usually associated with a low or normal arterial carbon dioxide pressure.

Those conditions associated with hypoventilation and increased PCO_2 may be associated with hypoxia; these conditions were listed under the heading respiratory acidosis and are repeated here for convenience.

Causes of Acute Respiratory Acidosis

Condition Affecting Organ or System	Examples
Brain Depression	Sedative, Opiate, Other Drug and Anaesthetic Overdose Comatose States, e.g., Cerebrovascular Accidents Neurosurgery and Head Injury
Spinal Cord	Spinal Cord Trauma; Cervical Vertebral Fracture
Neuromuscular System Disorders	Respiratory Paralysis; Acute Neuromuscular Disease; Myasthenia Gravis; Poliomyelitis; Guillain-Barre Syndrome
Chest Wall Limitation	Rib Fracture with Flail Chest
Upper Airways	Airway Obstruction, e.g., Aspiration of Foreign Body, Croup, Epiglottitis, Laryngospasm
Lower Airways and Lungs	Acute Severe Pulmonary Edema Severe Pulmonary Infections, e.g., Severe Pneumonia Open Chest Wounds Prolonged Open Chest and Open Heart Operations Pneumothorax; Hemothorax; Atelectasis
Other	Mechanical Underventilation Abdominal Distention from Ascites and Peritonitis

The causes of chronic respiratory acidosis, which are associated with low PO_2 are given in the next Table:

Causes of Chronic Respiratory Acidosis Plus Low PO_2

- Severe Lung Disease
  - Emphysema
  - Chronic Obstructive Lung Disease, e.g., Chronic Bronchitis
  - Chronic Asthma
  - Bronchiectasis
  - Pulmonary Fibrosis
- Hypoventilation
  - Extreme Obesity, i.e., Pickwickian Syndrome
  - Chest Deformity: Kyphoscoliosis and Injury to Thoracic Cage

Patients who are mechanically ventilated may experience worsening oxygenation for a variety of reasons as given in the following Table (Glauser FL, et al. Am Rev Respir Dis. 1988; 138:458-465):

Worsening Oxygenation in Mechanically Ventilated Patients

- Disease Progression: Adult Respiratory Distress Syndrome; Pulmonary Edema; Pneumonia; Asthma; COPD
- Therapeutic Interventions or Procedures: Endotracheal Suctioning; Bronchoscopy; Chest Physiotherapy; Position Changes; Thoracentesis; Peritoneal Dialysis; Hemodialysis
- Medications: Vasodilators; Bronchodilators
- Ventilator-Related Problems: Endotracheal Tube Malfunction; Improper Ventilator Settings; Ventilator Circuit Malfunction
- New Problems: Pneumothorax; Atelectasis; Gastric Aspiration; Nosocomial Pneumonia; Pulmonary Emboli; Fluid Overload; Bronchospasm; Retained Secretions; Shock

Review of Acid Base: Preuss HG. "Fundamentals of Clinical Acid-Base Evaluation" Clin Lab Med. 1993; 13:103-116.

BLOOD GASES, CAPILLARY

Blood may be obtained from a vasodilated capillary bed (ear lobe, finger, or heel) for blood gas analysis. This technique is frequently used in pediatric patients due to the difficulty in obtaining arterial blood samples and the potential complications of arterial catheters.

Capillary PO_2 correlates poorly with arterial values. Arterial PO_2 values are consistently higher than capillary PO_2 values. Capillary pH and PCO_2 values may or may not correlate with arterial values (McLain et al. Arch Dis Child. 1988; 63:743-747; Courtney SE, et al. AJDC. 1990; 144:168-172). Capillary blood gases may be more prone to liquid heparin dilution with resultant lowering of PCO_2 due to the smaller sample volume utilized (Hutchinson AS, et al. BMJ. 1983; 287:1131-1132; Zuerlein TJ. AJDC. 1990; 144:1287-1288).

BLOOD GASES, VENOUS

SPECIMEN: Venous blood should be collected as follows: light tourniquet pressure for a short time; no movement of extremity; release tourniquet before sampling is complete.

REFERENCE RANGE: Normal values and ranges for venous and arterial blood gases and acid-base values are given in the next Table:

Gases and Acid-Base Values

Parameter	Venous	Arterial
pH	7.36(7.31-7.41)	7.40(7.36-7.44)
PCO_2(mmHg)	46(40-52)	40(34-45)
HCO_3(mEq/liter)	25(22-28)	24(22-26)
PO_2(mmHg)	40(30-50)	90(80-100)
O_2 Saturation(%)	75(60-85)	95 or Greater

The values for males are not significantly different than that of females.

INTERPRETATION: Venous pH is slightly less than that of arterial pH while that of venous PCO_2 is mildly greater than that of arterial PCO_2. There is no significant difference between venous and arterial HCO_3^-. Venous blood may be used in acid-base studies. However, in certain common clinical conditions, venous blood does not reflect whole body acid-base status; these conditions are congestive heart failure, shock, poor tissue perfusion, and in the newborn period. Venous blood taken from an extremity reflects the acid-base status of that extremity and not that of the whole body; this is especially true if there is poor circulation to that extremity. With stasis, PCO_2 increases 20 mm to 40 mm in one minute.

The oxygen content of venous blood is significantly less than that of arterial blood.

BONE PANEL

The tests done to reflect bone pathology are shown below:

Tests to Reflect Bone Pathology
Alkaline Phosphatase
Hydroxyproline
Serum Calcium
Serum Phosphorus
Vitamin D
Parathyroid Hormone

Bone alkaline phosphatase is derived from osteoblasts of bone and is elevated in osteoblastic lesions. Conditions associated with osteoblastic lesions of bone are given in the next Table:

Conditions Associated with Osteoblastic Lesions of Bone
Paget's Disease
Malignancy (Osteoblastic)
Secondary Hyperparathyroidism
Renal Disease
Osteomalacia
Malabsorption
Rickets
Primary Hyperparathyroidism
Healing Fractures
Hyperthyroidism

(cont)

Bone Panel (Cont)

Hydroxyproline is a measure of resorption of bone collagen; it is increased in the serum and urine in conditions listed below:

Causes of Increased Urinary Hydroxyproline
Growth Period
Bone Diseases:
Primary and Secondary Hyperparathyroidism
Paget's Disease
Myeloma
Osteoporosis
Osteomalacia/Rickets
Congenital Hypophosphatasia
Acromegaly
Rheumatoid Arthritis
Metastatic Neoplasms
Marfan's Syndrome
Scleroderma
Hyperthyroidism

Mnemonic for Tumors Metastatic to Bone: "P.T. Barnum Loves Kids"; P = Prostate; T = Thyroid; B = Breast; L = Lung; K = Kidney.

BOWEL INFARCTION PANEL

The diagnostic clues to bowel infarction are hemoconcentration, hyperamylasemia, hyperphosphatemia and metabolic acidosis (Jamieson WG, et al. Surg Gynecol Obstet. 1979; 148:334-338). However, laboratory tests are typically nonspecific and nondiagnostic. Laboratory findings in acute mesenteric ischemia are given in the following Table (Benjamin E, et al. Disease-a-Month. 1993; 39:129-212):

Laboratory Findings in Acute Mesenteric Ischemia
Leukocytosis
Hyperamylasemia
Hemoconcentration
Increased Hematocrit
Increased Blood Urea Nitrogen
Lactic Acidosis
Hyperkalemia
Hyperphosphatemia
Increased Alkaline Phosphatase
Increased Creatine Kinase
Increased Lactic Dehydrogenase
Increased Aspartate Aminotransferase (AST = SGOT)

Almost one-half of the patients have metabolic acidosis and almost one-half have respiratory alkalosis; a few patients have normal arterial pH values. The usual admitting medical diagnoses in bowel infarction are as follows: acute myocardial, upper gastrointestinal bleeding and sepsis.

The most common clinical symptoms are abdominal tenderness, pain, nausea and vomiting. Elderly patients may not present with typical symptoms (Finucane PM, et al. J Am Geriatr Soc. 1989; 37:355-358).

Review: See Benjamin E, et al. "Acute Mesenteric Ischemia: Pathophysiology, Diagnosis, and Treatment." Disease-a-Month. 1993; 39:129-212.

BREAST CANCER PROFILE

The tests in the following table are performed on tissue samples of breast tumors or their metastases and are considered useful in accessing prognosis and guiding therapy.

Tests Performed on Tissue
Estrogen/Progesterone Receptors
Cancer Antigen 15-3 (CA 15-3)
Cathepsin D
HER-2/neu Oncoprotein
Epidermal Growth Factor Receptor
DNA-Histogram (Ploidy/Cell Cycle Analysis)
Tumor Suppressor Gene P53
pS2 Protein
nm23

Those tests in bold may be found under their respective topic headings. Some of these tests have just been developed and their usefulness in patient care is still being evaluated. pS2 protein appears to be a marker of an intact estrogen system and may be useful in accessing which estrogen receptor positive patients will respond to hormone treatment (Schwartz LH, et al. Cancer Res. 1991; 51:624-628). Cathepsin D appears to be associated with invasiveness (Tandon A, et al. N Engl J Med. 1990; 322:297-302).

Other tests may be performed on the serum of patients with breast cancer to monitor their response to therapy. The principle serum tests performed on breast cancer patients is given in the next Table:

Serum Monitors in Breast Cancer Patients
Cancer Antigen 15-3 (CA 15-3)
Carcinoembryonic Antigen (CEA)
CA 549
Lipid-Associated Sialic Acid (LASA)
Gross Cystic Diseases Protein

Those tests in bold may be found under their respective topic headings.

BUN (see UREA NITROGEN)

CA-125; 15-3; 19-9 (see CANCER ANTIGENS)

CALCITONIN (THYROCALCITONIN)

SPECIMEN: Red top tube; separate serum, then freeze.
REFERENCE RANGE: Upper limit of normal varies with the sensitivity of the assay. After plasma extraction of monomeric calcitonin, normal calcitonin level is <10pg/mL when measured by sensitive RIA (Body JJ. Acta Clinica Belgica. 1992; 47:77-81). Normal calcitonin values at Massachusetts General Hospital are as follows: Male, 0-28pg/mL; Female, 0-20pg/mL (Jordan CD, et al. N Engl J Med. 1992; 327:718-724). To convert conventional units in pg/mL to international units in ng/liter, multiply conventional units by 1.0.
METHOD: RIA; Immunoradiometric assay.
INTERPRETATION: Serum calcitonin may be elevated in the conditions listed in the next Table:

Causes of Elevated Serum Calcitonin
Medullary Carcinoma of the Thyroid (10% of cases familial, usually associated with Multiple Endocrine Neoplasia, Type II Sipple Syndrome)
Zollinger-Ellison Syndrome
Pernicious Anemia
Pregnant Women at Term
Newborns
Associated with Some Cancers: Lung, Breast, Pancreas
Chronic Renal Failure

(cont)

Calcitonin (Cont)

Plasma calcitonin assays are useful to detect the presence of medullary carcinoma of the thyroid and in revealing the recurrence of treated medullary thyroid cancer. Very high serum calcitonin levels (>2000 pg/mL) are virtually diagnostic of medullary carcinoma of the thyroid. The familial form of medullary thyroid carcinoma is inherited as an autosomal dominant trait often associated with pheochromocytoma and parathyroid adenoma or hyperplasia (multiple endocrine adenomatosis, Type II). To detect familial medullary thyroid carcinoma, tests provocative of calcitonin secretion (infusion of calcium or pentagastrin or both) may be done. In these tests, calcium or pentagastrin are infused and blood samples are subsequently obtained for calcitonin assay. Elevated calcitonin levels are indicative of occult medullary thyroid carcinoma in individuals at risk (Review: Body JJ. Acta Clinica Belgica. 1992; 47:77-81).

Calcitonin has been used for treatment of hypercalcemia, Paget's disease, and osteoporosis. Calcitonin has been approved by the Food and Drug Administration for treatment of osteoporosis since 1984 but until recently was only available as an injectable medication. Calcitonin nasal spray is now available and more practical. Calcitonin is a potent inhibitor of bone resorption and is effective for the prevention of postmenopausal bone loss (See Review: Schneyer CR. MMJ. 1991; 40:469-473).

CALCIUM EXCRETION TEST

Synonyms: Calcium Load Profile; Calcium Tolerance Test

SPECIMEN: Specimens are taken at two times: fasting; and after a calcium test load. The test usually includes an initial serum sample (red top tube) for calcium, creatinine, uric acid and phosphorous; also two timed urines for volume, calcium, creatinine, uric acid, magnesium, and cyclic AMP. The protocol and substances measured may vary somewhat between laboratories.

Protocol:

1) The patient is fasted. High calcium meals should be avoided for one week (<400mg/day).
2) Upon arrival in early AM, blood collected for serum tests and patient voids. Discard urine.
3) Patient takes 600mL H_2O and urine over two hour period collected (fasting 2 hr urine).
4) Patient then ingests 1g calcium load with total of 600mL H_2O.
5) Collect urine excreted over next two hour period (2 hr urine, 1g calcium load).
6) From the urine volumes and substance concentrations, the total amount of substance excreted over a two hour period (fasted and 1g calcium load) is calculated in mg/2 hours.
7) The $\frac{\text{calcium}}{\text{creatinine}}$ ratio is then calculated.

References: Pak CYC, et al. N Engl J Med. 1975; 292:497-500; Broadus AE, et al. J Clin Endocrinol Metab. 1978; 47:751; Sakhall K, et al. Urology. 1979; 14:251-255.

REFERENCE RANGE: See Interpretation.

INTERPRETATION: The calcium excretion test is used as an aid in assessing the etiology of disorders of calcium metabolism and in the workup of renal stone disease. The primary use is in the diagnosis of hypercalciuria.

The causes of hypercalciuria are given in the next Table:

Causes of Hypercalciuria
Absorptive Hypercalciuria
Resorptive Hypercalciuria
Primary Hyperparathyroidism, Malignancy, Hyperthyroidism, Renal Tubular Acidosis
Renal Tubular Hypercalciuria

The expected serum calcium results and urine calcium/creatinine ratios in the fasted and calcium loaded state in various conditions is given in the next Table:

Expected Results of Calcium Excretion Test in Various Conditions

CONDITION	FASTING		POST 1 GRAM Ca^{++} Load	
	Serum Ca	Ca/Cr ratio	Serum Ca	Ca/Cr Ratio
Primary Hyperparathyroidism	↑	>0.11	↑	>0.20
Absorptive Hypercalciuria	N	<0.11	N	>0.20
Renal Tubular Hypercalciuria	N	>0.11	N	>0.20
Normocalciuric Nephrolithiasis	N	<0.11	N	>0.13
Normal	N	<0.11	N	<0.13

Absorptive Hypercalciuria: In absorptive hypercalciuria, there is normocalcemia, increased intestinal absorption of calcium and increased calcium in the urine. The primary abnormality is an increased intestinal absorption of calcium. Increased intestinal absorption of calcium increases the total body calcium and inhibits PTH leading to reduced reabsorption of filtered calcium by the renal tubule and increased urinary excretion. This pathogenic sequence is seen in an idiopathic form, vitamin D intoxication and in sarcoidosis. Absorptive hypercalciuria, type I, is associated with increased intestinal absorption of calcium and persists despite a low-calcium (200mg/day) diet.

Renal Tubular Hypercalciuria (Renal Leak Hypercalciuria): Calcium is filtered at the glomerulus; it is largely reabsorbed in the proximal tubule. In renal tubular hypercalciuria, a defect in calcium reabsorption of unknown cause occurs. Thus far, the pathogenesis of renal leak hypercalciuria (renal calcium leak) in all patients is idiopathic.

Pyrah has stated that hypercalcuria is either due to overabsorption from the GI tract or to renal leakage (Pyrah LN, in Renal Calculus, 1979, Chapter 8, Berlin, Springer-Verlog). However more recently, workers have questioned the utility of such a distinction and the ability of laboratory testing to discriminate between the two (Rose GA. Clin Chem Acta. 1986; 160:109-115; Thode J. Scand J Clin Lab Invest. 1990; Suppl. 197:1-45).

Primary Hyperparathyroidism: An oral calcium-load test has been said to be useful in patients with symptoms of hyperparathyroidism, but only minimal or intermittent elevations of PTH, serum calcium, or both. In this test intact PTH is measured at given intervals following ingestion of 1 gram of elemental calcium. Patients with primary hyperparathyroidism usually exhibit decreased PTH suppression compared to controls (Monchik JM, et al. Surg. 1992; 112:1103-1110).

Discussions of nephrolithiasis and hypercalcuria may be found in: Williams HE. N Engl J Med. 1974; 290:33-38; Coe FL, et al. N Engl J Med. 1992; 327:1141-1152.

CALCIUM, IONIZED (FREE)

SPECIMEN: Whole blood; Red top tube, separate serum; Green top tube, separate plasma.
REFERENCE RANGE: At pH 7.4, free calcium = 4.40 to 5.40mg/dL (1.10 to 1.35mmol/liter). To convert conventional units in mg/dL to international units in mmol/liter, multiply conventional units by 0.2495.
METHOD: Calcium specific electrode
INTERPRETATION: Ionized (free) calcium is abnormal in the same conditions that effect total calcium. Ionized calcium more accurately reflects biologically available calcium. Common causes of hypocalcemia and hypercalcemia are given in the following Table:

Common Causes of Elevated and Decreases Ionized Calcium

Elevated Ionized Calcium	Decreased Ionized Calcium
Hyperparathyroidism	Hypoparathyroidism
Ectopic PTH-Producing Tumors	Pseudohypoparathyroidism
Vitamin D Excess	Vitamin D Deficiency (Rickets)
Malignancies	Renal Failure
	Magnesium Deficiency
	Multiple ACD Blood Transfusions

Calcium circulates in the serum in three forms: free (47%), protein bound (43%), and diffusible calcium complexes (10%). About 80 percent of the calcium binding protein in serum is albumin; the remaining 20 percent is globulin. Calcium forms complexes with bicarbonate, citrate, phosphate, sulfate, and lactate. About 20 percent of the amino acids of albumin are glutamic and aspartic acids; binding of calcium to albumin probably occurs through these acidic amino acids. The free fraction is most important physiologically.

The serum forms of calcium are illustrated in the next Figure:

Serum Forms of Calcium

Albumin-Ca^{++} (43%)	Free-Ca^{++} (47%)	Complex-Ca^{++} (10%)
$-C(=O)-O^- \; Ca^{++}$	Ca^{++}	
$-C(=O)-O^- \; Ca^{++}$	Ca^{++}	HCO_3^-
$-C(=O)-O^- \; Ca^{++}$	Ca^{++}	Citrate, PO_3^{-3}, SO_4^{-2}
$-C(=O)-O^- \; Ca^{++}$	Ca^{++}	

Effects of pH on Free Calcium: Acidosis causes an increase in free calcium while alkalosis causes a decrease in free calcium; the reactions are given in the next chemical equations:

$$\text{Albumin} + Ca^{++} \underset{\text{Acidosis}}{\overset{\text{Alkalosis}}{\rightleftharpoons}} \text{Albumin-}Ca^{++} \text{ Complex}$$

The change in percent of free calcium with pH is shown in the next Figure:

Change in Percent of Free Calcium with pH

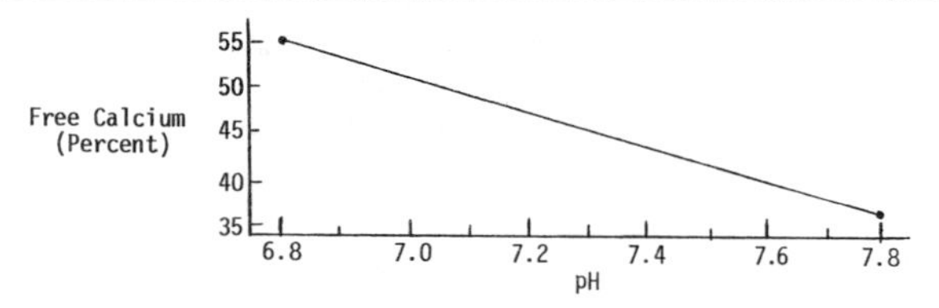

At a serum pH of 7.4, 47 percent of the total serum calcium is free calcium. As the pH of the serum increases, the percent of free calcium decreases. Acute respiratory alkalosis may precipitate tetany because of sudden decrease in free calcium.

Systemic acidosis has been associated with osteomalacia and rickets because low pH interferes with mineralization of newly formed bone and acidosis may interfere with the production of 1,25di(OH) cholecalciferol.

Inaccuracies Associated With Sampling: Several factors associated with sampling may influence ionized calcium level. The sample must be drawn anaerobically; the presence of air bubbles in the sample will allow loss of carbon dioxide, increase in pH, and decrease in measured ionized calcium. Forearm exercise with stasis during the blood drawing procedure may cause lactic acid accumulation which binds calcium leading to lowered ionized calcium. The use of liquid sodium heparin leads to decreased ionized calcium, partly due to dilution; dry heparin is preferred. Do not use anticoagulants such as citrate, EDTA or oxalate, since all will bind calcium leading to low ionized calcium concentrations. Sample pH should be determined along with ionized calcium (Forman DT, Lorenzo L. Ann Clin Laboratory Sci. 1991; 21:297-304).

Ionized Calcium Versus Total Calcium: Total calcium accurately reflects calcium hemostasis in most clinical situations, when acid/base status and protein concentrations are near normal. Ionized calcium may be preferable to total calcium in the following clinical situations (Forman DT, Lorenzo L, already cited; Bowers GN, et al. Clin Chem. 1986; 32:1437-1447):

Ionized Calcium Preferable to Total Calcium
Massive transfusion of blood or fresh frozen plasma containing citrate
Patients with hepatic or renal dysfunction
Neonatal hypocalcemia
Critically ill patients
Hyperparathyroidism when total calcium is normal

CALCIUM, TOTAL, SERUM

SPECIMEN: Red top tube

REFERENCE RANGE: 8.5 to 10.5mg/dL (2.1-2.6mmol/liter) in adults; premature: 7-10.0mg/dL, full term: 7.0-12.0mg/dL; children: 8.0-11.0mg/dL. Panic Values: less than 7mg/dL or greater than 12mg/dL. To convert conventional units in mg/dL to international units in mmol/liter, multiply conventional units by 0.2495.

METHOD: Reaction with cresolphthalein complexone

INTERPRETATION: HYPERCALCEMIA: The causes of hypercalcemia based on parathyroid status is given in the following Table (Bourke E, Delaney V. Clin Lab Med. 1993; 13:157-181):

Causes of Hypercalcemia Based on Parathyroid Status
Increased PTH:
Primary Hyperparathyroidism: Adenoma, Carcinoma, Hyperplasia (Including MEN Types 1, 2a)
Secondary Hyperparathyroidism Following Vitamin D Deficiency or Renal Osteodystrophy After Successful Renal Transplant
Tertiary Hyperparathyroidism: Chronic Renal Failure
Aluminum Bone Disease
Familial Hypocalciuric Hypercalcemia
Drugs: Lithium; Theophylline
Other: Neck Irradiation; Pheochromocytoma
Decreased or Normal PTH:
Malignancy: Myeloma; Breast Cancer; Lymphoma; Squamous Cell Cancers (Especially Lung)
Other Endocrine: Hyperthyroidism; Acromegaly; Addison's Disease
Hypervitaminosis D
Granulomatous Disorders
Iatrogenic: Hypervitaminosis A; Total Parenteral Nutrition; Milk Alkali Syndrome; Thiazides; Tamoxifen
Immobilization of Children and Adolescents
Hypercalcemia of Infancy

Mnemonic for Hypercalcemia: "VITAMINS TRAP"; V = Vitamins A and D; I = Immobilization; T = Thyrotoxicosis; A = Addison's Disease; M = Milk-Alkali Syndrome; I = Inflammatory Disorders; N = Neoplastic Disease; S = Sarcoidosis; T = Thiazides and Other Drugs; R = Rhabdomyolysis; A = AIDS; P = Paget's Disease, Parenteral Nutrition, Parathyroid Disease (Pont A. Endocrinol Metab Clin N Amer. 1989; 18:753-764).

Primary hyperparathyroidism, malignancy and drug-induced hypercalcemia are, by far, the most common causes of hypercalcemia. In the differential diagnosis of hypercalcemia, the rare causes, e.g., immobilization, Paget's disease and acute adrenal insufficiency, occur infrequently. Coma may be seen with levels above 13mg/dL; acidosis may intensify the effects of hypercalcemia. (cont)

Calcium, Total (Cont)

The most useful tests in the differential diagnosis of hypercalcemia are PTH, chloride, phosphorus, and alkaline phosphatase.

Primary Hyperparathyroidism: The diagnosis of primary hyperparathyroidism may be suggested by the following findings: increased serum calcium, decreased serum phosphorus, increased serum chloride, increased PTH, normal hematocrit; these findings will correctly classify primary hyperparathyroidism in 98% of the patients with hypercalcemias (Lafferty FW. Arch Intern Med. 1981; 141:1761-1766). Increased urinary calcium and cAMP are also seen.

Malignancy: The malignancies associated with hypercalcemia are shown in the next Table:

Tumors Associated with Hypercalcemia
Breast
Multiple Myeloma
Carcinoma of Squamous Lined Surfaces:
Lung (Squamous Cell)
Head and Neck
Esophagus
Cervix
Lung (Squamous Cell, Large Cell Anaplastic, Adenocarcinoma)
Prostate; Kidney; Lower Genitourinary Tract
Thyroid; Leukemia; Malignant Lymphoma; Melanoma; Chondrosarcoma

There are two predominant mechanisms by which tumors cause hypercalcemia - humoral hypercalcemia due to circulating hormones and local osteolytic hypercalcemia. Both mechanisms may be responsible for hypercalcemia in metastatic disease, and both mechanisms lead to hypercalcemia by increasing osteoclastic activity in bone. Not all metastatic bone tumors are associated with hypercalcemia.

At one time it was believed that some hypercalcemia of malignancy was due to ectopic PTH production. Now there is evidence that a PTH-related protein is produced by some malignant tissues. This PTH-related protein is produced by some malignancies, normal skin, placenta, pregnant uterus and lactating mammary gland. Elevations of PTH-related protein have been reported in cases of "humoral" hypercalcemia of malignancy, and during pregnancy and lactation (Lepe F, et al. N Engl J Med. 1993; 328:666-667). In such situations there is hypercalcemia, increased PTH-related protein, and decreased intact PTH.

Tests used to differentiate between primary hyperparathyroidism and malignancy-associated hypercalcemia are given in the following Table (Insogna KL. Endocrinol Metab Clin N Amer. 1989; 18:779-794):

Primary Hyperparathyroidism Versus Malignancy-Associated Hypercalcemia

	Primary Hyperparathyroidism	Humoral Hypercalcemia of Malignancy	Local Osteolytic Hypercalcemia
Serum Calcium	Incr	Incr	Incr
Serum Phosphorus	Decr	Decr	Variable
Urinary Calcium Excretion	Incr	Greatly Incr	Greatly Incr
Immunoreactive PTH	Greatly Incr	NL - Incr	NL - Decr
Nephrogenous cAMP	Incr	Incr	Decr
1,25-Dihydroxyvitamin D	Incr	Decr	Decr

Drug-Induced Hypercalcemia: The administration of thiazide diuretics to patients may result in increased serum calcium. The hypercalcemia is due to an increase in both total and ionized calcium. The serum calcium may begin to rise after 1 day of thiazides and increases gradually; the increase in serum calcium is usually moderate. Other drugs associated with hypercalcemia include lithium, theophylline, vitamin A or D, tamoxifen, or other anti-estrogen preparations.

Familial Benign Hypercalcemia (Familial Hypocalciuric Hypercalcemia): Familial benign hypercalcemia is frequently confused with primary hyperparathyroidism. Familial benign hypercalcemia is characterized by autosomal-dominant inheritance, asymptomatic life-long hypercalcemia, unexpected low urinary calcium excretion with a lower ratio of calcium clearance to creatinine clearance than in hyperparathyroidism and inappropriately "normal" levels of immunoreactive parathyroid hormone. All patients suspected of having primary hyperparathyroidism should routinely have 24-hour urinary calcium and renal calcium to creatinine clearance ratio determination to rule out familial benign hypercalcemia; a value below 0.01 for this ratio suggests familial benign hypercalcemia (Marx SJ, et al. Medicine. 1981; 60:397-412). High-resolution parathyroid ultrasonography frequently demonstrates enlarged parathyroid glands in patients with primary hyperparathyroidism; smaller glands are found in patients with familial benign hypercalcemia (Law WM, et al. Mayo Clin Proc. 1984; 59:153-155).

HYPOCALCEMIA: The causes of hypocalcemia based on parathyroid status is given in the following Table (Bourke E, Delaney V. Clin Lab Med. 1993; 13:157-181):

Causes of Hypocalcemia Based on Parathyroid Status
Decreased PTH:
Postsurgical: Accidental Parathyroidectomy During Thyroidectomy
DiGeorge Syndrome
Autoimmune
Infiltrative: Hemosiderosis; Wilson's Disease
Neoplastic: Metastatic
Iatrogenic: ^{131}I Treatment
Transient, Neonatal
Magnesium Deficiency (End-Organ Resistance)
Normal or Increased PTH:
Pseudohypoparathyroidism Types 1a, 1b, 1c, 2
Renal Failure
Vitamin D Disorders: Deficiency; Liver Disease; Anticonvulsants
"Hungry Bones": Post-Parathyroidectomy or Postrenal Transplant in Renal Osteodystrophy; Post-Thyroidectomy; Post-Vitamin D Replacement in Rickets; Increased Osteoblastic Activity (Metastases, Fluorosis); Decreased Osteoclastic Activity (Medullary Carcinoma of the Thyroid)
Drugs: Mithramycin; Biphosphonates; Calcitonin; Colchicine; Glutethimide; Estrogens; Ethylene Glycol Poisoning

Three of the conditions in the previous Table, renal failure, hypoparathyroidism, and pseudohypoparathyroidism, are associated with hyperphosphatemia as given in the next Table:

Conditions of Hyperphosphatemia and Hypocalcemia
Renal Failure
Hypoparathyroidism
Pseudohypoparathyroidism

Calcium in the "free" state is the important physiologically active form; calcium circulates in the serum in three forms: free (47%) protein-bound (43%), and diffusible calcium complexes (10%). Total serum calcium level cannot be properly interpreted without serum albumin level. Hypoalbuminemia is the most common cause of hypocalcemia. An equation for correction of serum calcium in patients with depressed albumin is:

Adjusted Calcium = Serum Calcium - Serum Albumin + 4.0

This equation corrects serum calcium so that serum calcium is increased by 1.0mg/dL for every 1.0g/dL that the albumin concentration is below 4.0g/dL.

Ionized calcium is normal in most cases of hypoalbuminemia; ionized calcium is a better test to assess calcium status in patients with hypoalbuminemia.

pH effects free calcium; acidosis (metabolic or respiratory) causes an increase in free calcium while alkalosis (metabolic or respiratory) causes a decrease in free calcium.

References: Bourke E, Delaney V. "Assessment of Hypocalcemia and Hypercalcemia." Clin Lab Med. 1993; 13:157-182; Zaloga GP. "Hypocalcemia in Critically Ill Patients." Crit Care Med. 1992; 20:251-262; Zaloga GP. "Hypocalcemic Crisis." Crit Care Clin. 1991; 7:191-200; Bilezikian JP. "Hypercalcemia." Disease-a-Month. 1988; 34:739-834; Marcus R (ed). "Hypercalcemia." Endocrinol and Metab Clin N Amer. 1989; 18:601-832.

CALCIUM, URINE

SPECIMEN: 24 hour urine collection. Instruct the patient on 24 hour urine collection as follows: Void at 8:00 A.M. and discard specimen. Collect all urine during the next 24 hours including the 8:00 A.M. specimen the next morning. Add 15mL of 6N HCl to pH 1 to 2. Refrigerate urine at 4°C.
REFERENCE RANGE: Varies with diet; low calcium, <50mg/24hr. Average diet, 50-150mg/24hr; high calcium, 100-300mg/24hrs. Pediatrics: Less than 4mg/kg/24hr. Urinary excretion of calcium varies greatly with its intake. To convert conventional units in mg/24hrs to international units in mmol/24hrs, multiply conventional units by 0.02495.
METHOD: Reaction with cresolphthalein complexone; atomic absorption
INTERPRETATION: Urine calcium is increased in the conditions listed in the next Table:

Increase in Urine Calcium
Primary Hyperparathyroidism (Increased in 2/3 of patients)
Vitamin D Toxicity
Hyperthyroidism
Renal Tubular Acidosis (Type I, Distal)
Osteolytic Conditions
Multiple Myeloma
Tumor Metastases
Paget's Disease
Sarcoidosis
Lymphoma
Drugs: Acetazolamide; Antacids; Protein Supplements; Furosemide; Ethacrynic Acid; Calcium Therapy; Corticosteroids
Osteogenesis Imperfecta (Chines A, et al. J Pediatr. 1991; 119:51-57)
Idiopathic

Many medical conditions associated with hypercalciuria are also associated with renal stone formation. Patients with hypercalciuria (urinary calcium greater than 200mg/24hrs) should have further evaluation including repeat urinary calcium following a calcium restricted diet and parathyroid hormone determination (DeVita MV, Zebetakis PM. Clin Lab Med. 1993; 13:225-234).
Pediatric Hypercalciuria: Causes of hypercalciuria are the same in children and adults. Neonates can develop renal calcifications due to hypercalciuria caused by chronic furosemide therapy. Patients may have residual glomerular and tubular dysfunction at one to two year follow-up. Urine calcium and calcium to creatinine ratio has been used to identify infants at risk for renal calcifications; abnormal urinary calcium to creatinine ratio is greater than 0.18-0.3 and abnormal urinary calcium greater than 4mg/kg/24hrs (Downing GJ, et al. J Pediatr. 1992; 120:599-604; Jacinto JS, et al. Pediatr. 1988; 81:31-35; Langman CB, Moore ES. Clin Pediatr. 1984; 23:135-137).
Familial Benign Hypercalcemia (Familial Hypocalciuric Hypercalcemia) vs. Primary Hyperparathyroidism: Familial benign hypercalcemia is frequently confused with primary hyperparathyroidism. Familial benign hypercalcemia is characterized by autosomal-dominant inheritance, asymptomatic life-long hypercalcemia, unexpected low urinary calcium excretion with a lower ratio of calcium clearance to creatinine clearance than in hyperparathyroidism and inappropriately "normal" levels of immunoreactive parathyroid hormone. All patients suspected of having primary hyperparathyroidism should routinely have 24-hour urinary calcium and renal calcium to creatinine clearance ratio determination to rule out familial benign hypercalcemia; a value below 0.01 for this ratio suggests familial benign hypercalcemia (Marx SJ, et al. Medicine. 1981; 60:397-412). High-resolution parathyroid ultrasonography frequently demonstrates enlarged parathyroid glands in patients with primary hyperparathyroidism; smaller glands are found in patients with familial benign hypercalcemia (Law WM, et al. Mayo Clin Proc. 1984; 59:153-155).

CALCULATIONS AND FORMULAS

SENSITIVITY: Sensitivity = Positivity in disease. Sensitivity indicates the probability of a positive test result when the disease is present and is given by the relationship:

$$\text{Sensitivity (\%)} = \frac{\text{True Positives (TP)}}{\text{Total Tested with Disease}} \times 100$$

$$= \frac{\text{True Positives (TP)}}{\text{True Positives(TP)} + \text{False Negatives(FN)}} \times 100$$

TP = True positives; the number of patients with the disease correctly classified by the test.

FN = False negatives; the number of patients with the disease misclassified by the test or abnormal test results in healthy patients.

The larger the number of false positives, the less the specificity.

SPECIFICITY: Specificity = Negativity in health. Specificity indicates the probability that the test result will be negative when the disease is not present and is given by the relationship:

$$\text{Specificity (\%)} = \frac{\text{True Negatives (TN)}}{\text{Total Tested Without Disease}}$$

$$= \frac{\text{True Negatives(TN)}}{\text{True Negatives(TN)} + \text{False Positives(FP)}} \times 100$$

TN = True negatives; the number of patients without the disease correctly classified by the test.

FP = False positives; the number of patients without the disease misclassified by the test or abnormal test results in healthy patients.

The larger the number of false positives, the less the specificity.

PREDICTIVE VALUE OF A POSITIVE TEST: Predictive value of a positive test indicates the probability that the disease is present when the test or procedure is positive. In other words, it reflects the probability that a positive test indicates disease. It is simply the number of true positive results expressed as a fraction of all positive test results, both true and false. It may be expressed as follows:

$$\text{Predictive Value(\%)} = \frac{\text{Subjects with a Positive Test Result}}{\text{Number of All Subjects with a Positive Test Result}} \times 100$$

$$= \frac{\text{True Positives(TP)}}{\text{True Positives(TP)} + \text{False Positives(FP)}} \times 100$$

The larger the number of false positives, the less the predictive value for a positive test.

PREVALENCE: $\text{Prevalence} = \frac{\text{Total with Disease}}{\text{Total Tests}}$

References: Griner PF, et al. Ann Intern Med. 1981; 94:553-600; Gottfried EL, Wager EA. Disease-A-Month, August, 1983.

PROBLEM: Calculate sensitivity, specificity, and predictive value of a positive test, and prevalence from the following information:

Enzyme Test Results in Coronary Intensive Care Unit

Enzyme Test	Patients with M.I.	Patients without M.I.	Total
Positive Test	450 (TP)	20 (FP)	470
Negative Test	50 (FN)	480 (TN)	530
	500	500	

(cont)

Calculations and Formulas (Cont)

Answer:

$$\text{Sensitivity} = \frac{TP}{TP + FN} \times 100 = \frac{450}{450 + 50} \times 100 = 90\%$$

$$\text{Specificity} = \frac{TN}{TN + FP} \times 100 = \frac{480}{480 + 20} \times 100 = 96\%$$

$$\text{Predictive Value of Positive Test} = \frac{TP}{TP + FP} \times 100 = \frac{450}{450 + 20} \times 100 = 96\%$$

$$\text{Prevalence} = \frac{TP + FN}{\text{Total Tests}} = \frac{500}{1000} \times 100 = 50\%$$

CORRELATION COEFFICIENT: The coefficient of correlation (r) ranges from +1 (perfect positive correlation) to 0 (no correlation) to -1 (perfect negative correlation).

Problem: Which one of the following correlation coefficients shows the strongest relationship between two variables? +0.85, +0.50, +1.25, -0.95, 0.00.

Answer: -0.95.

CREATININE CLEARANCE:

Creatinine clearance reflects glomerular filtration rate. The equation for the calculation of creatinine clearance is as follows:

$$\text{Clearance } \frac{mL}{min} = \frac{\text{Urine Volume (mL/min)} \times \text{Urine Creatinine}}{\text{Serum Creatinine}} \times \frac{1.73}{A}$$

where A=Body Surface Area

Creatinine clearances are usually performed on a 24 hour urine. There are (24 x 60) = 1440 minutes in each day. Hence the urine volume for 24 hours must be divided by 1440 to convert it to mL/min.

EXCRETION FRACTION OF FILTERED SODIUM:

This is a sensitive and specific test for acute tubular necrosis (Espinel CH, Gregory AW. Clin Nephrol. 1980; 13:73-77). This may be of use in differentiating prerenal azotemia from acute tubular necrosis. The measurements that are done are as follows: Serum and urine sodium, serum and urine creatinine; no timed specimen is necessary. The excretion fraction of filtered sodium is given by the equation:

$$\text{Excretion Fraction of Filtered Sodium} = \frac{\text{Urine Na}^+ \times \text{Plasma Creatinine}}{\text{Plasma Na}^+ \times \text{Urine Creatinine}} \times 100$$

AMYLASE CLEARANCE/CREATININE CLEARANCE:

Amylase clearance is increased in patients with acute pancreatitis but is normal in other causes of hyperamylasemia. Amylase clearance is useful in differentiating the causes of increased serum amylase; it is especially useful in confirming or excluding the diagnosis of acute pancreatitis. Amylase clearance is reduced in macroamylasemia. Amylase clearance is the renal clearance of amylase, expressed as a percentage of creatinine clearance, and is given by the equation:

$$\frac{\text{Amylase Clearance(\%)}}{\text{Creatinine Clearance}} = \frac{\frac{\text{Urine Amylase}}{\text{Serum Amylase}} \times \text{Urine Volume per Unit Time}}{\frac{\text{Urine Creatinine}}{\text{Serum Creatinine}} \times \text{Urine Volume per Unit Time}} \times 100$$

This equation simplifies to the form:

$$\frac{\text{Amylase Clearance (\%)}}{\text{Creatinine Clearance}} = \frac{\text{[Urine Amylase]}}{\text{[Serum Amylase]}} \times \frac{\text{[Serum Creatinine]}}{\text{[Urine Creatinine]}} \times 100$$

As seen by inspection of the above equation, the clearance ratio, expressed as percentage, is calculated simply from the concentrations of amylase and creatinine in serum and urine samples obtained simultaneously. No timed collections are necessary.

CORRECTION OF SERUM CALCIUM FOR DEPRESSED ALBUMIN:

Serum calcium level cannot be properly interpreted without serum albumin level; an equation for correction of serum calcium in patients with depressed albumin is:

Adjusted Calcium = Serum Calcium - Serum Albumin + 4.0

This equation corrects serum calcium so that serum calcium is increased by 1.0mg/dL for every 1.0g/dL that the albumin concentration is below 4.0g/dL.

CONVERSION OF BLOOD ETHANOL VALUES:

Blood ethanol values are reported in various units. Some state toxicology labs report the values in mg%. These units may be interconverted as follows:

0.10% Blood Alcohol = 0.10g/100g = 100mg%

THE ANION GAP:

The anion gap is a calculation of diagnostic convenience which represents the amount of unmeasured anions and cations in a clinical sample. It is based on the principle of electroneutrality - the total number of cations and anions in a solution must be equal. The commonly measured ions are Na^+, K^+, Cl^-, and HCO_3^- (or CO_2 content). From such measurements one calculates the anion gap:

Anion Gap = Measured Cations - Measured Anions

Anion Gap = (Na^+) - (Cl^- + CO_2 Content) K^+ usually not included.

Values: 5 - 15 mmol/liter, normal; 15-20 mmol/liter, borderline; over 20 mmol/liter, increased.

An increase in the anion gap is due to unmeasured anions such as phosphate, sulfate, lactate, acetoacetate, salicylate, formic acid, etc. An anion gap greater than 30 mmol/liter is usually due to an identifiable organic acidosis such as lactic acidosis or ketoacidosis.

In multiple myeloma, the strongly basic monoclonal immunoglobulin can sometimes serve as a source of positively charged cations and result in a decrease in the anion gap. The calculation of an anion gap can also be used as a quality control check to confirm the validity of a low serum sodium.

CONVERSION FORMULAS:

To convert drug concentrations to or from international units, determine the conversion factor(CF) using the following formula:

$$\text{Conversion Factor(CF)} = \frac{1000}{\text{Mol. Wt.}}$$

Traditional Units to International Units:

(microgram/mL) x C.F. = I.U.(micromol/liter)

International Units to Traditional Units:

I.U.(micromol/liter)/CF = (microgram/mL)

CANCER ANTIGENS (CA): 125; 15-3; 19-9

SPECIMEN: Red top tube, separate serum.
METHOD: Immunological; EIA
REFERENCE RANGE: CA 125: <35U/mL; CA 15-3: <25U/mL; CA 19-9 <70U/mL
INTERPRETATION: These cancer antigens are glycoproteins associated with malignant neoplasms. They are measured to aid in the management of patients with malignancies which produce these antigens in sufficient quantities that measurable levels of these materials can be found in the blood. None of the tests for these antigens are specific or sensitive enough for use as screening tests. The three tests described below have all been recently introduced and are finding use in selected situations.

CA 125

This antigen is most often followed in the patient with non-mucinous ovarian carcinoma. It may also be elevated in early pregnancy, endometriosis, and inflammatory disorders of peritoneum and abdominal organs. Levels are highest with serous, mixed and unclassified ovarian cancer, and lowest in mucinous tumors. Differences in survival correlate with preoperative and postoperative levels (Makar AP, et al. Obstet Gynecol. 1992; 79:1002-1010). It may also be elevated in persons with neoplasms of other gynecological sites (uterus, fallopian tube), pancreas, breast, lung and GI tract. CA 125 levels correlate with the course of disease in a high proportions of patients (Sevelda P, et al. Gynecol Oncol. 1991; 43:154-158). Virtually all patients with a CA 125 greater than 35U/mL prior to second-look operations have residual ovarian tumor (Makar A, et al. Gynecol Oncol. 1992; 45:323-328). Large scale screening for ovarian cancer with CA-125 alone is not recommended at this time (Helzlsoner KJ, et al. JAMA. 1993; 269:1123-1126; Einhorn N, et al. Obstet Gynecol. 1992; 80:14-18).

CA 15-3

This antigen is most often followed in the patient with breast cancer. Rising or falling levels of CA 15-3 correlate with progression or regression of the malignancy. CEA is another antigen often measured in the serum of breast cancer patients. The CA 15-3 may be elevated in some patients with lung, pancreatic and ovarian cancer, and also in patients with benign breast disease.

It has been observed that in approximately one-third of breast cancer patients, the CA 15-3 actually increases and "spikes" 14-60 days after starting antiestrogen therapy as a result of tumor cell lysis (Kreienberg R. Am J Clin Oncol. 1991; 14(suppl 2):556-561).

CA 19-9

This antigen is most often followed in the patient with pancreatic cancer and some patients with colon cancer (Steinberg W. Am J Gastroenterol. 1990; 85:350-355). Because CA 19-9 is related to the Lewis blood group, there is concern that this test may not be applicable in the Lewis A/Lewis B negative proportion of the United States population. The CA 19-9 may also be elevated in some colon, stomach, and hepatobiliary malignancies. CA 19-9 is also elevated in benign inflammatory disease of pancreas, gallbladder and liver. Elevated CEA may also be found in the patient with pancreatic carcinoma. The use of CA 19-9 versus CEA levels has been contrasted (Steinberg W, et al. Gastroenterology. 1986; 90:343-349). This is tabulated below:

Comparison of CA 19-9 and CEA in Disease

	CA 19-9	CEA
Pancreatic CA (Sensitivity)	81%	33-42%
Non-Pancreatic Tumors (% Positive)	22-67%	>50%
Benign GI Disease (% Positive)	14%	13%

Other

CA 195 might prove of use for patients with colon and other abdominal cancers. CA 549 may be of use in breast cancer patients. Other CA antigens are under investigation.

Review Article: Bates SE. Clinical Applications of Serum Tumor Markers. Ann Intern Med. 1991; 115:623-638.

CANCER MARKERS (see TUMOR MARKERS)

CANDIDA TESTING

SPECIMEN: Culture: Blood, Sputum, Urine, Body Fluids, Tissue, Intravascular Devices; Antibody Tests: Serum; Antigen Tests: Serum

REFERENCE RANGE: Negative

METHODS: Culture: Candida grows well on blood agar and frequently grows from cultures submitted for bacterial isolation. Lysis-centrifugation culture technique is more sensitive and fungal growth is detected sooner by this technique. Antibody Tests: Radioallergosorbent Test(RAST); Counter Immunoelectrophoresis(CIE); Double Diffusion(DD). Antigen Tests: Latex Particle Agglutination(LPA); Radioimmunoassay(RIA); Enzyme Immunoassay(EIA).

INTERPRETATION: Candidiasis is caused by several candida species; the common pathogen is Candida albicans.

Candida organisms normally occur in the mouth, gastrointestinal tract and the vagina of some normal individuals. Disease occurs in immunocompromised patients, especially those with depressed T-cell function (e.g. AIDS); such patients are prone to chronic mucocutaneous infection. Invasive candidiasis occurs when epithelial barriers are breached, particularly when patients are neutropenic or on broad spectrum antibiotics. Candidal endocarditis may occur following cardiac surgery, in drug addicts and in patients with long-term venous catheters. Young infants may develop oral candida (thrush) or candidal diaper rash. Vaginal candidiasis occurs in women taking oral contraceptives or during pregnancy.

Culture: Interpretation of a positive culture depends on the site being cultured. Candida colonization may be present on skin or mucous membranes. Positive cultures of sputum or urine must be interpreted with caution since samples are obtained through potentially colonized areas. Culture is the primary means by which candida infection is diagnosed.

Diagnosis of Candida infections of the skin and mucous membrane may be made by obtaining smears or scrapings on glass slides; add a drop of saline solution or 10 percent NaOH or KOH. Stains are usually not necessary; however, Gram, Wright, Ziehl-Neelson, Giemsa, Papanicolaou, periodic acid-Schiff(PAS) or methenamine silver method may be used.

Antibody Tests: The sensitivity of the tests for candida antibody is about 50 percent; antibody testing is not reliable for diagnosis of candidiasis.

Antigen Tests: Antigen tests detect candida cell wall antigens (e.g. mannans), cytoplasmic antigens (e.g. enolase), or metabolites (e.g. D-arabinitol). These tests are most useful if performed frequently (e.g. weekly) in high risk patients. As with antibody tests, antigen tests are not sufficiently sensitive to be clinically useful.

Review of candida testing, see Jones JM. "Laboratory Diagnosis of Invasive Candidiasis." Clin Microbiol Rev. 1990; 3:32-45.

CARBAMAZEPINE (TEGRETOL)

SPECIMEN: Red top tube, separate serum and refrigerate; or green (heparin) top tube, separate plasma and refrigerate. Reject if serum frozen on cells or left standing on cells for several days.
REFERENCE RANGE: Therapeutic: 2-10mcg/mL; Toxic: >12mcg/mL. Time to Obtain Serum Specimens (Steady State): 2-4 weeks (see dosage for initiation of therapy below); Half-Life: chronic dosing, adults and children, 5 to 27 hours.
METHOD: EMIT; HPLC; GLC; Fluorescence Polarization(Abbott); SLFIA(AMES).
INTERPRETATION: Carbamazepine is used as both an anticonvulsant and for the relief of pain associated with trigeminal neuralgia (tic douloureux); uses are given in the next Table:

Uses of Carbamazepine
Anticonvulsant
Complex Partial (Psychomotor or Temporal)
Tonic-Clonic (Grand Mal) Seizures
Mixed Seizure Patterns
Pain Associated with Trigeminal Neuralgia

Carbamazepine is used in the prophylactic management of partial seizures with complex symptomatology (psychomotor or temporal lobe seizures), generalized tonic-clonic (grand mal) seizures, and mixed seizure patterns. It has no effect on absence seizures (petit mal) or myoclonic seizures. Carbamazepine is used in the symptomatic treatment of pain associated with true trigeminal neuralgia (tic douloureux).

Carbamazepine is given by mouth and is slowly absorbed from the G.I. tract; it reaches a peak in the serum in 2 to 24 hours; about 3/4 of the drug is bound to albumin. Carbamazepine is metabolized (99%) by the liver. The clearance and half-life decrease with time so that it is advisable to initiate therapy gradually over two to four weeks; dose schedule and time to obtain blood specimens are as follows:

Dose Schedule and Blood Specimens

Age	Dose	Blood Specimens
Adults	Initially: 400mg/day in 2-4 divided doses Increase by 200mg/day every other day Maintenance: 600-1200 mg/day in 3 or 4 doses	Blood specimen obtained during week 3 or 4, several days after starting maintenance dose. Optimally, obtain blood specimen at start, middle and end (trough) of a dosing period; otherwise, obtain specimen at end of dosing period.
Children	Week 1: 5mg/kg/day (3 or 4 dose/day) Week 2: 10mg/kg/day (3 or 4 dose/day) Week 3: (Maint. Dose) 15-30mg/kg/day (3 or 4 dose/day)	Blood specimen obtained during weeks 3 or 4, several days after starting maintenance dose; optimally, obtain blood specimen at start, middle and end (trough) of a dosing period; otherwise, obtain specimen at end of dosing period.

Usual maximum adult dose is 1.2g daily, however, 1.6-2.4g daily is sometimes necessary.

There is variability in carbamazepine absorption and the trough concentration may not represent the lowest drug level during the dosing interval; therefore, multiple determinations may be necessary in some patients.

The time to peak concentration is about 3 hours after an oral dose when patients are on chronic therapy. The rate of clearance of carbamazepine in the presence of phenobarbital and phenytoin is increased. Clearance is decreased by erythromycin, calcium channel blockers, cimetidine, isoniazid and others. Carbamazepine accelerates the breakdown of estrogens necessitating higher estrogen doses in oral contraceptives. Metabolism of corticosteroids, theophylline, warfarin and haloperidol is increased (Review: "Carbamazepine Update". Lancet. 1989; 595-597).

Obtain pretreatment CBC. Minor hematologic changes (e.g., decreased leukocyte counts) are not uncommon and do not necessarily require discontinuation of therapy unless there is evidence of infection ("Carbamazepine Update". Lancet. 1989; 595-597). Aplastic anemia and agranulocytosis has been reported to occur in patients on carbamazepine; therefore, routine blood counts should be performed at 2-4 months intervals. Monitor hepatic toxicity (periodic liver function tests). Other side effects include rash, neuritis, drowsiness, dizziness, headache, tinnitus, diplopia, urinary retention and nausea. Mild asymptomatic hyponatremia is not uncommon; severe hyponatremia occurs rarely.

CARBON DIOXIDE CONTENT (CO_2 CONTENT)

SPECIMEN: Red top tube, separate serum
REFERENCE RANGE: 24-30mmol/liter (or mEq/L)
INTERPRETATION: The carbon dioxide content is the serum concentration of bicarbonate and carbonic acid as given by the following equation:

$$CO_2 \text{ Content} = [HCO_3^-] + [H_2CO_3]$$
$$[H_2CO_3] = PCO_2 \times 0.03$$

The $[HCO_3^-]$ is about 20 times the $[H_2CO_3]$; therefore the CO_2 content approximates the value of the $[HCO_3^-]$. CO_2 content is decreased in metabolic acidosis and increased in metabolic alkalosis. The CO_2 content is mildly increased in chronic respiratory acidosis with compensation. The CO_2 content is mildly decreased in chronic respiratory alkalosis with compensation.

Serum bicarbonate measurements are significantly decreased when vacutainer tubes are underfilled (Herr RD, Swanson T. Am J Clin Path. 1992; 97:213-216). This inaccuracy would also be reflected in the carbon dioxide content.

CARBON MONOXIDE (CARBOXYHEMOGLOBIN)

SPECIMEN: Grey (oxalate fluoride) top tube
REFERENCE RANGE: Rural Nonsmokers: 0.4 to 0.7%; Urban Nonsmokers: 1 to 2%; Smokers: 5 to 6%; Toxic: >15% Saturation. The half elimination time is 320 minutes breathing room air; it is 80 minutes when the patient is breathing 100 percent oxygen and 23 minutes with 100 percent oxygen at 3 atmospheres (Peterson JE, Stewart RD. Arch Environ Health. 1970; 21:165-175). To convert conventional units in % to international units, multiple conventional units by 0.01.
METHOD: Spectrophotometric (co-oxymeter)
INTERPRETATION: Carbon monoxide combines with hemoglobin as follows:

CO + Hemoglobin → CO Hemoglobin (Carboxy hemoglobin)

The correlations of CO in the atmosphere, saturation of CO in blood and symptoms, are given in the next Table:

Carbon Monoxide Levels and Symptoms

CO in Atmosphere (%)	% Saturation CO in Blood	Symptoms
0 to 0.01%	0 to 10%	None
0.01% to 0.1%	10% to 50%	Headache, Lethargy, Unconsciousness
0.1% to 1.00%	> 50%	Respiratory Failure and Death

Severe symptoms develop in almost all patients when CO saturation rises above 20% saturation. Levels above 40% saturation are fatal if not treated immediately. Metabolic acidosis is an ominous sign. If carbon monoxide poisoning is suspected, take a blood specimen and treat patients with 100% oxygen at a high flow rate.

Hyperbaric oxygen therapy may be indicated in certain cases: suggested guidelines for hyperbaric oxygen therapy in carbon monoxide poisoning are given in the following Table (Grim PS, et al. JAMA. 1990; 263(16):2216-2220):

Guidelines for Hyperbaric Oxygen Therapy in Carbon Monoxide Poisoning
Carboxyhemoglobin Level >25%; lower levels in children and pregnant women
History of Unconsciousness
Neuropsychiatric Abnormality
Cardiac Instability or Ischemia

Carbon monoxide is produced wherever organic matter is burned; cigarettes, automobile exhaust and fires are the most frequent sources of exposure to toxic gas. In a 10 year review of deaths in the United States (1979-1988), 56,133 death certificates indicated carbon monoxide as a contributing factor (suicides 25,889; severe burns of house fires 15,523; homicides 210; unintentional 11,547). Of the unintentional deaths, 57% were due to automobile exhaust, 83% of which were associated with stationary vehicles (Cobb N, Etzel RA. JAMA. 1991; 266(5):659-663). In children, riding in the back of the pickup trucks may result in carbon monoxide poisoning (Hampson NB and Norkool DM. JAMA. 1992; 267(4):538-540).

Carbon monoxide is tightly bound to hemoglobin; carbon monoxide has an avidity for hemoglobin more than 200 times that of oxygen. The major pathophysiologic disturbance in patients with carbon monoxide poisoning is the inhibition of oxygen kinetics by inhibiting oxygen transport via hemoglobin, oxygen delivery, and oxygen utilization of oxidative phosphorylation (Stewart P, et al. Crit Care Report. 1990; 1:359-364).

CARCINOEMBRYONIC ANTIGEN (CEA)

SPECIMEN: Lavender top tube (EDTA); separate plasma and place in plastic vial.
REFERENCE RANGE: Non-Smokers, <2.5ng/mL; Smokers, <4.0ng/mL.
METHOD: Enzyme-linked immunosorbent assay (ELISA), and RIA, Abbott; RIA, Roche. The antibodies in the Abbott and Roche kits measure different CEA antigen determinants and may give different values. Heparin may give spuriously elevated values with the Roche method.
INTERPRETATION: The CEA test is used to monitor response to therapy of patients following surgery and/or chemotherapy. It is of greatest use in patients with colon cancer, but may also be used in patients with breast, lung, or a variety of other epithelial tumors. It is not used as a screening test. Conditions that have been associated with an increase in serum concentration of carcinoembryonic antigen and the approximate sensitivities of CEA in these conditions are given in the next Table:

Conditions Associated with Increase in Serum Concentration of Carcinoembryonic Antigen

Carcinomas:	Sensitivities(%)	Non-Malignant Conditions:	Sensitivities(%)
Colon and Rectum	70-80	Pulmonary Emphysema	20-50
Pancreas	60-90	Active Ulcerative Colitis	10-25
Lung	65-75	Alcoholic Cirrhosis	25-70
Stomach	30-60	Cholecystitis	6-20
Breast	50-65	Rectal Polyps	4-20
Ovary	40	Benign Breast Disease	4-15
Other Carcinomas	20-50		

The CEA may also be elevated in a variety of GI disorders (hepatic, pancreatic, gastric, intestinal) and renal failure. Values are higher in smokers, men, and older individuals.

The CEA level is relatively high in patients with well differentiated and widely metastasized carcinoma of the colon and rectum as compared to poorly differentiated or localized disease (Goslin R, et al. Am J Med. 1981; 71:246-253).

CEA levels have been used as a prognostic factor in advanced colorectal carcinoma; patients with an initially normal level of CEA versus those with an abnormal level of CEA had median survivals of 23 and 9.2 months respectively (Kemeny N, Brawn DW. Am J Med. 1983; 74:786-794). Following successful surgery, CEA levels slowly drop, requiring up to 6 weeks to reach baseline values. The sensitivity of the CEA for recurrent disease is 56% (Meling GI, et al. Scand J Gastroenterol. 1992; 27:1061-1068). The effectiveness of the use of the CEA in monitoring recurrent disease following surgery has been recently questioned (Moertel CG, et al. JAMA. 1993; 270:943-947; Fletcher RH. ibid. 987-988). A contrasting view is given in: Chu DZ, et al. Arch Surg. 1991; 126:314-316.
CEA and Metastases: The CEA assay complements the use of liver scan for detecting liver metastases. In a study by McCarthy WH, Hoffer PB (JAMA. 1976; 236:1023-1027), 56 percent of patients who had liver metastases had an elevated CEA while 80 percent had a positive liver scan. If both CEA assays and liver scans gave positive results, the probability of metastases was virtually 100 percent and if both tests had negative results, the probability of liver involvement was approximately 1 percent.
CEA and Effusions: The CEA has been proposed as adjunct to cytology in the diagnosis of malignant pleural effusions (Pinto MM, et al. Arch Pathol Lab Med. 1992; 116:626-631).
CEA and Pneumocystis Pneumonia: Based on the fact the CEA is elevated in lung disease, CEA levels in HIV patients were studied. Patients with pneumocystis pneumonia had higher CEA levels than those who had other pulmonary disease and could be distinguished from those without lung disease. The level of CEA elevation also allowed the identification of patients with poor prognosis (Bedos JP, et al. Scand J Infect Dis. 1992; 24:309-315).

CARDIAC RISK ASSESSMENT

SPECIMEN: Fasting, 12-14 hours; red top tube, separate serum. Not all tests require a fasting specimen, but it is necessary for a complete assessment of risk as presently performed.

REFERENCE RANGE: See Interpretation. To convert conventional units in mg/dL to international units in mmol/L, multiply conventional units by 0.02586.

METHOD: See individual entries.

INTERPRETATION: The primary impetus to blood lipid measurement, lipoprotein analysis and other measurements is for the risk assessment of coronary artery disease and related events. There have been a number of ways in which this has been done.

Blood cholesterol screening: The Consensus Conference (JAMA. 1985; 253:2080-2086) established values for selecting adults by age at moderate and high risk or coronary heart disease:

Values for Selecting Adults at Moderate and High Risk Requiring Treatment

Age, Year	Moderate Risk mg/dL(mmol/L)	High Risk mg/dL(mmol/L)
20-29	> 200 (5.17)	>220 (5.69)
30-39	> 220 (5.69)	>240 (6.21)
≥ 40	≥ 240 (6.21)	≥260 (6.72)

This system takes into account the changing of cholesterol levels with age, but does not include the protective effect of HDL. A similar but easier to remember system is published by the National Cholesterol Education Program (Expert Panel. Arch Intern Med. 1988; 148:36-69):

Total Cholesterol Levels	
Desirable	<200 mg/dL (5.2mmol/L)
Borderline	200-239 mg/dL (5.2-6.2mmol/L)
High	>240 mg/dL (6.2mmol/L)

In this classification system, a low HDL (≤ 35 mg/dL) or elevated LDL (≥170 mg/dL), or additional risk factors such as smoking or hypertension should influence treatment of those patients with borderline levels. Screening may be done with unfasted specimens, but the subsequent HDL and LDL determinations must be done on fasted specimens. An updated report of cardiac risk assessment is given in: Expert Panel on Detection, Evaluation and Treatment of High Blood Cholesterol in Adults. JAMA. 1993; 269:3015-3023.

There is understandable concern that many individuals in the "desirable" or the "borderline" range may still be at high risk for coronary artery disease. In a study of individuals with total cholesterols less than 239mg/dL, 20% were considered at high risk because of elevated LDL, and 12% because of reduced HDL (Bush TL, Riedel D. Circulation. 1991; 83:1287-1293). Recently (June, 1993) the NCEP Treatment Panel added to the guidelines a recommendation that initial screening should include measurement of both total cholesterol and high density lipoprotein cholesterol.

Cardiac risk can also be ascertained by determination of the cholesterol/HDL ratio.

Cholesterol/HDL Ratio and Cardiac Risk

Risk	Percentile	Ratio-Men	Ratio-Women
Lowest	<25	<3.8	<2.9
Low	26-50	3.9-4.7	3.0-3.6
Moderate	51-75	4.8-5.9	3.7-4.6
High	76-89	6.0-6.9	4.7-5.6
Highest	>90	>7	>5.7

Recent studies employing the Cholesterol/HDL ratio can be found in: Luria MH, et al. Am J Cardiol. 1991; 67:31-36; Hong MK, et al. Am J Cardiol. 1991; 68:1646-1650. (cont)

Cardiac Risk Assessment (Cont)

With the development of immunochemical assays for apolipoproteins, it is possible to immunochemically measure a risk index. The apoprotein levels are less subject to recent diet. Dr. Harold Bates has recently published a table of ApoB/ApoAl based on over 500,000 subjects (Bates HM. Laboratory Management. 1989; 27:52). ApoB reflects the LDL level and ApoAl the HDL level. This ratios and risk assessment table is reproduced below.

Apo B/Apo A-I Ratio

Risk	Men	Women
Lowest	<0.69	<0.47
Low	0.7-0.9	0.49-0.64
Moderate	0.95-1.22	0.66-0.9
High	1.25-1.48	0.92-1.15
Highest	>1.5	>1.17

Similar risk assessment from apo A-I, apo B values has been published (Maciejko JJ, et al. Clin Chem. 1987; 33:2065-2069).

High Density Lipoprotein (HDL): It has been found that decreased levels of high density lipoprotein (HDL) in plasma lead to an increased risk in males of coronary heart disease. Among the various lipid risk factors, HDL cholesterol appears to have the strongest relationship to coronary heart disease. It may also be that raised levels exert a protective effect in that premenopausal women have HDL concentrations 30 to 60 percent higher than their male counterparts and subjects with familial hyper-alphalipoproteinemia have an above-average life expectancy.

The incidence of coronary heart disease by HDL cholesterol level is shown in the next Table (Gordon et al. Am J Med. 1977; 62:707-714):

Incidence of Coronary Heart Disease by HDL Cholesterol Level

HDL Cholesterol Level (mg/dL)	Incidence of Coronary Heart Disease	
	Men	Women
<25	18%	
25-34	10%	17%
35-44	10%	5%
45-54	5%	5%
55-64	6%	4%
65-74	2.5%	1.4%
75	0%	2%

Lower levels of HDL (<35mg/dL - 0.9 mmol/L) are associated with increased myocardial risk. The major causes of reduced serum HDL are listed below:

Major Causes of Reduced Serum HDL

- Cigarette Smoking
- Obesity
- Lack of Exercise
- Androgenic and Related Steroids
- B-adrenergic Blocking Agents
- Hypertriglyceridemia
- Genetic Factors
  - Primary Hypoalphalipoproteinemia

HDL can be fractionated into two measurable subfractions, HDL_2 and HDL_3. Alcohol intake appears to elevate HDL_3, apo A-I and apo A-II (Taskinen M-R, et al. Am Heart J. 1987; 113:458-463). Diet may elevate HDL by increasing the cholesterol content of the HDL while exercise may increase the number of HDL particles (Schwartz R. Metabolism. 1987; 36:165-171). Although HDL measurements are important in predicting myocardial risk, detailed laboratory measurement of HDL_2, HDL_3, or the associated apolipoproteins A-I or A-II do not add to the diagnostic accuracy (Stamper MJ, et al. N Engl J Med. 1991; 325:373-381).

Low Density Lipoprotein (LDL): A LDL greater than 170 mg/dL is associated with increased myocardial risk, even though the total cholesterol is in the desirable or borderline range. The LDL level is determined on a fasting sample and calculated from the values for total cholesterol, HDL and TG as follows:

$$LDL = \text{total cholesterol} - \left(\frac{TG}{5} + HDL\right)$$

Some prefer to use "6.25" in the denominator in place of "5".

Lipoprotein (a) Lp(a): Lipoprotein (a), [Lp(a)] is an independent risk factor for cardiac disease. Concentrations of Lp(a) greater than 30mg/dL are associated with a two-fold greater risk of developing coronary artery disease (Rader DJ, Brewer HB. JAMA. 1992; 267:1109-1112).

The distribution of plasma Lp(a) varies between various ethic groups and appears shifted into the upper ranges in the black population (Goldsmith MF. JAMA. 1992; 267:336-337; Uterman G. Science. 1989; 246:904-910). Dietary modification has little effect on Lp(a) concentration although vigorous exercise appears to be beneficial (Hellsten G, et al. Atherosclerosis. 1989; 75:93-94). The only drug reported to lower Lp(a) is nicotinic acid (Carlson, et al. J Intern Med. 1989; 226:271-276). See LIPOPROTEIN(a).

Triglycerides: At present the role of elevated TG as an independent risk factor for cardiovascular disease remains controversial. A NIH consensus conference concluded that current evidence did not allow one to conclude causality between high TG levels and CHD (NIH Consensus Conference. JAMA. 1993; 269:505-510). By contrast, publications suggesting such a link are: Bainton D, et al. Br Heart J. 1992; 68:60-66; Castelli WP. Am J Cardiol. 1992; 70:3H-9H; Manninen V, et al. Circulation. 1992; 85:37-45). In a few studies, TG appeared to contribute to risk in subgroups in association with low HDL and LDL (Criqui MH, et al. N Engl J Med. 1993; 328:1220-1225). See TRIGLYCERIDES.

Other Factors: Other non-lipoprotein conditions have been reported to be associated with increased risk for cardiac disease. The WBC count is positively associated with coronary heart disease (Kannel WB, et al. JAMA. 1992; 267:1253-1257). Hyperuricemia is significantly associated with elevated TG, body fat and myocardial infarct (Agamah ES, et al. J Lab Clin Med. 1991;118:241-249; Frohlich ED. JAMA. 1993; 270:378-379). Ubiquinone, a natural antioxidant found in LDL, is decreased in patients with ischemic heart disease (Hanoki Y, et al. N Engl J Med. 1991; 325:814-815). Hyperhomocysteinemia was found in a significant number of patients with cerebrovascular, peripheral vascular and cardiac disease (Clarke R, et al. N Engl J Med. 1991; 324:1149-1155; Stampfer MJ et al. JAMA. 1992; 268:877-881; Clarke R, et al. N Engl J Med. 1991; 324:1149-1155). Approximately 1-2% of the population is heterozygous for this condition. A high renin-sodium profile is possibly an independent predictor of MI in patients with hypertension (Alderman MH, et al. N Engl J Med. 1991; 324:1098-1104).

CAROTENE

SPECIMEN: Patient must be fasting and receive no vitamin supplement or foods containing vitamin A or carotene for 24 hours before testing for younger age group (0-6 mo.); 48 hours in the older group (older than 6 months). Red or green (heparin) top tube; protect from light by wrapping in foil; separate serum or plasma and continue to protect from light. The specimen may be stored in the freezer (-20°C) for four days.

REFERENCE RANGE:

Change of Serum Carotene with Age

Age	Range (mcg/dL)	Range (micromol/liter)
Infant	20-70	0.37-1.30
Child	40-130	0.74-2.42
Adult	60-200	1.12-3.72

To convert traditional units in mcg/dL to international units in micromol/liter, multiply traditional units by 0.0186 (Tietz NW. Textbook of Clinical Chemistry. WB Saunders Co., Phila. 1986).

METHOD: Spectrophotometric, HPLC (cont)

Carotene (Cont)

INTERPRETATION: Causes of increased and decreased serum carotene are given in the next Table:

Causes of Increased and Decreased Serum Carotene	
Increased	**Decreased**
Increased Carotenoids in the Diet (Carotenemia)	Liver Disease
Diabetes Mellitus	Fat Malabsorption Syndromes
Myxedema	Lack of Carotenoids in the Diet
Chronic Nephritis	High Fever
Nephrotic Syndrome	Protein-Energy Malnutrition
	Infections
	Stress
	Burns

The concentration of serum carotene is affected most by hepatic function.

Vitamin A is fat soluble and is absorbed with lipid; malabsorption of vitamin A may occur in any condition in which steatorrhea (pathological increase in stool fat) occurs such as pancreatic insufficiency, deficiency of bile and impairment of intestinal absorption.

Dietary beta carotene may exert a protective effect against cancers of epithelial origin. Protective effects have been most consistently reported for cancer of the lung, oral cavity, pharynx, larynx and cervix. Postulated mechanisms of this protective effect include enhancement of the immune response and anti-mutagenic actions of carotenoids (Bendich A. Proc Nutr Soc. 1991; 50:263-274; Mayne ST. Conn Med. 1990; 54:547-551; Basu TK, et al. Nutr Growth and Cancer. 1988; 217-228).

CATECHOLAMINES, FREE, URINE

SPECIMEN: 24 hour urine: There is a diurnal variation in catecholamines levels, hence the need for a 24 hour specimen. Instruct the patient to void at 8:00 A.M. and discard the specimen. Add 25mL of 6N HCl to container prior to collection. Then, collect all urine including the 8:00 A.M. specimen at the end of the 24 hour collection period. Caution patients against risk of skin burns. Refrigerate jug as each specimen is collected. Following collection, add 6N HCl to pH 1 to 2 (Do not use boric acid). Catecholamines are unstable in urine at alkaline pH and pH >3 is cause for rejection of specimen. Record 24 hour urine volume. In urine collected at a pH of less than 3 and stored at 4°C, the catecholamines are stable for at least one week. Forward 100mL aliquot of 24 hour urine for analysis.

Patient should be drug free if possible. Multiple drugs interfere and lists of these can be found in Stein PP, Black HR. Medicine. 1991; 70:46-66. Stress, exercise, caffeine and vitamin B-12 can also cause catecholamine elevation. Diet can interfere with some assays.

REFERENCE RANGE: Reference values vary with age and method. Children have lower values than adults. The more specific methods yield lower values.

Normal Values of Urinary Catecholamines	
Age	**Normal Values (micrograms/24 hours)**
Birth to 1 year	4-20
1 to 5 years	5-40
6 to 15 years	5-80
Over 15 years	20-100
Adults	Up to 100

Vorrhess ML. "Urinary Catecholamine Excretion by Health Children." Ped. 1967; 39:252-257. Some reference laboratories report these as the ratio: micrograms catecholamines/g creatinine.

METHOD: Measurement of urinary free catecholamines (UFC) by fluorometric methods was initially discouraged because of the poor specificity of these assays. Present techniques employ HPLC and are able to separately quantitate epinephrine (E), norepinephrine (NE), and dopamine (D). Some laboratories use chromatography followed by fluorescent assay. These measure only E and NE.

INTERPRETATION: Free catecholamines are elevated in patients with pheochromocytoma and neuroblastoma. Urinary catecholamines, E, NE and D, are excreted as the free amines or as the conjugated glucuronidase or sulfates. The measurement of only the free amines is the preferred test. Discriminate between this test and the presently disfavored total urine catecholamine measurement.

The ratio of E to NE is usually 0.2. A diagnosis of pheochromocytoma may be suspected if there is an elevated E to NE ratio even in the presence of a normal level of UFC (Brown J, et al. J Clin Pathol. 1966; 19:482-486). There was at one time a concern that UFC would not be diagnostic in patients with pheochromocytoma and intermittent hypertension. That no longer appears to be a concern. UFC are also elevated in patients with myasthenia gravis and progressive muscular dystrophy.

Fractioned 24 hour urine NE has a lower false positive rate than plasma NE, 2% vs. 5%, (Bravo E, et al. N Engl J Med. 1979; 301:682-686) and appears less subject to fluctuations (Duncan M, et al. N Engl J Med. 1988; 319:316-342).

For other tests for neoplasms of the adrenal medullary see: PHEOCHROMOCYTOMA SCREEN, METANEPHRINES, VMA AND CATECHOLAMINES, PLASMA.

CATECHOLAMINES, PLASMA, TOTAL OR FRACTIONATED

SPECIMEN: A successful measurement requires meticulous attention to detail in patient preparation and rapid processing of the specimen. The patient should be off drugs if possible and all acute stress should be avoided. Some laboratories request the patient be fasting (Bravo E, Gifford RJ. N Engl J Med. 1984; 311:1298-1303). Prepare the patient prior to obtaining the blood specimen as follows: Place a heparinized catheter into a vein and reassure the patient regarding the procedure. After the patient has remained in the supine position in nonstimulating surroundings for at least 30 minutes, draw blood through the indwelling catheter into a green top tube (heparin anticoagulant). Use minimum tourniquet time. Mix gently by inversion; do not shake. Immediately place tube in an ice bath. Centrifuge tube in the cold and separate the plasma; freeze plasma immediately. Some laboratories request an EDTA anticoagulant sample.
REFERENCE RANGE: For diagnosis of pheochromocytoma: ≤ 1000ng/L excludes the diagnosis; 1000-2000 ng/L is suspicious; >2000 ng/L is diagnostic (Bravo and Gifford, already cited).
METHOD: HPLC is able to give values for total plasma catecholamines or the individual amine: Epinephrine, Norepinephrine, and Dopamine.
INTERPRETATION: Because of the tedious patient preparation and specimen processing, this is usually not the initial assay chosen in the workup of suspected pheochromocytoma or neuroblastoma (Refer to PHEOCHROMOCYTOMA SCREEN). A urinary screening test (urine free catecholamines, Metanephrines, or VMA) is usually the first test ordered. There appears to be more false negatives with this test than the others (Stein PP, Black HR. Medicine. 1991; 70:46-66). Plasma catecholamines may be less sensitive in patients with paroxysms of episodic hypertension. Plasma catecholamines may also be measured in conjunction with a stimulating glucagon test, or suppressive clonidine test (see PHEOCHROMOCYTOMA SCREEN).

CATECHOLAMINES, TOTAL, URINE (INCLUDES UNCONJUGATED AND CONJUGATED)

Total urinary catecholamines are elevated in pheochromocytoma and neuroblastoma. However, fluorometric total catecholamine assays have a large reference range of normal which may obscure the diagnosis of a minimally secreting tumor and are strongly subject to interference by conjugated dietary catecholamines. Total unfractionated assay of urine catecholamine should be avoided because of these problems (Stein PP, Black HR. Medicine. 1991; 70:46-66). This assay should be replaced by the measurement of Urine Free Catecholamines. See CATECHOLAMINES, FREE, URINE.

CEREBROSPINAL FLUID (CSF)

SPECIMEN: 3 or more mL of CSF
REFERENCE RANGE: Cell count: lymphocytes and monocyte 0-10/cu mm; neutrophils and erythrocytes are absent; gram stain; cytology; CIE; latex agglutination; cultures; cryptococcus (India Ink Preparation); protein, 15-45 mg/dL; albumin/globulin ratio, 3:1; glucose, 50-85 mg/dL; chloride, 120-130mmol/liter; pressure (mm H_2O), 50-200; LDH <40; lactate, <20mg/dL.
PEDIATRIC REFERENCE RANGE: Pediatric reference ranges vary considerably with age. Diagnosis of meningitis should be based on clinical presentation and CSF composition including total WBC count, absolute neutrophil count (ANC ≥1/cu.mm.), glucose, protein, gram-stain and culture. Normal values for CSF cell counts for various ages has been compiled, as given in the following Table (Bonadio WA. Pediatr Infect Dis J. 1992; 11:423-432):

CSF Normal Cell Counts According to Age

Age	Total WBC(cu.mm.) Suggested ULN	Total WBC(cu.mm.) Mean(Range)	ANC Mean	% Neutrophils Mean(Range)
Premature Newborn	NR	9.0(0-29)	NR	7.0(0-66)
Term Newborn	22	8.2(0-22)	NR	61(NR)
0-4 Weeks	22*	11.0	0.40	2.2(0-15)
4-8 Weeks	15*	7.1	0.18	2.9(0-42)
>6 Weeks	5	2.3(NR)	0.68	NR(0-35)

*90th percentile (Bonandio WA, et al. Pediatr Infect Dis J. 1992; 11:589-591); NR = Not Reported; ULN = Upper Limit of Normal

Normal values for CSF glucose and protein for various ages has been compiled, as given in the following Table (Bonadio WA, already cited):

CSF Normal Chemistry Values According to Age

Age	Glucose(mg/dL)	CSF: Blood Glucose Ratio	Protein(mg/dL)
Premature Newborn	50(24-63)	0.74(0.55-1.05)	115(65-150)
Term Newborn	52(34-119)	0.81(0.44-2.48)	90(20-170)
0-4 Weeks	46(36-61)	NR (NR)	84(35-189)
4-8 Weeks	46(29-62)	NR (NR)	59(19-121)
>6 Weeks	61(45-65)	0.6 (NR)	28(20-45)

METHOD: Prior to lumbar puncture, examine eyegrounds for evidence of papilledema. Potential problems and complications of lumbar puncture are as follows: herniation of uncus or cerebellar tonsils; progression of paralysis with spinal tumor; hematoma in patients with clotting defects; meningitis in presence of sepsis; asphyxiation of infants due to excessive constraint of tracheal obstruction caused by pushing the head forward; and introduction of infection (Krieg AF. Clinical Diagnosis and Management. Henry JB, ed. Saunders, Philadelphia. 1979, 1:637).

In adults, lumbar puncture is performed between L3 and L4. The patient lies on his side with back flexed as much as possible to "open up" the space. mark the site for puncture, sterilize skin and anesthetize skin and underlying tissue. Place needle with stylet in place at midline, direct slightly cephalad and push slowly; at about 2-3 cm, remove stylet, check for flow of spinal fluid and before advancing further, replace stylet. If the needle strikes bone, withdraw and redirect. A "give" indicates that the needle is in the spinal canal.

In children, the technique is similar. The incidence of traumatic LP may be 20% in children and 36% in neonates (Schreiner RL, Kleiman MB. Develop Med Child Neurol. 1979; 21:483). The "Cincinnati method" has been suggested to minimize this risk. In this technique, the spinal needle and stylet are inserted until the epidermis and dermis are traversed. Subsequently, the stylet is withdrawn and the needle advanced until CSF is seen in the hub. After CSF is obtained, the stylet is reinserted prior to removal of the needle (Bonadio WA. Contemporary Pediatr. 1989; 109-116). Depth of LP is estimated to be 0.77cm + 2.56 x m^2 (Bonadio WA, et al. N Engl J Med. 1988; 319:952-953).

A manometric reading is obtained (opening pressure) when spinal fluid begins to flow. If the tap is bloody, drain 2 to 3 mL; if it clears, then a blood vessel was ruptured during puncture. Collect three tubes, each filled to 1 to 3 mL: one for microbiological studies (India Ink preparation for cryptococci, gram stain, latex agglutination tests, CIE, bacterial, fungal, and viral cultures); one for chemical determinations (protein, glucose, chloride, LDH, lactate); and one for cell count, differential and serology.

Differentiation of Traumatic Tap from Pathologic Bleeding: Traumatic tap is differentiated from pathologic bleeding by the following: The traumatic tap shows non-homogeneous mixing in the manometer; the traumatic tap shows visible clearing between the first and subsequent tubes and the erythrocyte count decreases in a similar manner; a very blood tap will clot on standing; xanthochromia (pink to orange to yellow color in the supernatant of centrifuged CSF) is found in pathologic bleeding. The specimen must be centrifuged within one hour after collection to avoid false positives.

INTERPRETATION: Typical CSF findings are shown in the next Table:

Typical CSF Findings in Some Conditions

Condition	Pressure mm CSF	Gross Appearance	Cells (cu. mm.)	Protein (mg/dL)	Glucose (mg/dL)
Normal	50-200	Clear, Colorless	0-10 Lymphs & Monocytes	Under 45	2/3 Blood glucose 50-80
Acute Bact. Meningitis	Increased 200-500	Turbid, May Clot	100-10,000; Polyps	50-500	Absent or very low
Tuberculous Meningitis	Increased 200-500	Turbid, Pellicle Common	10-500, chiefly Lymphs	10-300	Often under 40 (Average = 20)
Aseptic or Viral Meningitis	Increased 200-500	Clear or Slightly Turbid	10-500, chiefly Lymphs	45-200	Normal
Multiple Sclerosis	Normal	Clear, Colorless	Normal or 10-50 Lymphs	Normal or 45-100	Normal
Cerebral Thrombosis	Usually normal to slightly inc.	Clear	Usually Normal	Normal or 45-100	Normal
Cerebral Hemorrhage	Usually Normal	Bloody or Xanthochromic	Increased RBC	Increased 45-100	Normal
Subarachnoid Hemorrhage	Increased 200-500	Bloody or Xanthochromic	Increased RBC	Increased 50-1000	Normal
Brain Tumor	Usually Increased 200-500	Clear or Xanthochromic	Normal to 50 Lymphs	Normal or Slightly Increased	Normal or Slightly Decreased

Diagnosis of Infectious Diseases: There is no single test used to diagnose meningitis that is fully reliable. All tests must be viewed in clinical context. CSF testing has been extensively reviewed (see Lindquist L, et al. Eur J Clin Microbiol Infect Dis. 1988; 7:374-380).(cont)

Cerebrospinal Fluid (CSF) (Cont)

Tests of the cerebrospinal fluid (CSF) for the diagnosis of infectious diseases are given in the next Table:

Tests of the CSF for the Diagnosis of Infectious Diseases
Conventional Laboratory Methods:
Cellular, Protein and Glucose Content
Gram Stain
Culture with Isolation of Organisms
India Ink Preparation for Cryptococci
Detection of Specific Bacterial Antigens:
Latex Agglutination (LA)
Counterimmunoelectrophoresis (CIE)
Rapid and Non-specific Methods:
Lactic Acid
Lactate Dehydrogenase (LDH)
C-Reactive Protein (C-RP)
CSF Antibody Detection (Andiman WA. Pediatr Infect Dis J. 1991; 10:490-495)
Lyme Disease
Varicella-Zoster
Herpes Simplex
HIV-1

Bacterial and Viral Meningitis: The sensitivity and specificity of various CSF laboratory studies for bacterial meningitis versus viral meningitis are given in the next Table (Corrall CJ, et al. J Pediat. 1981; 99(3):365-369):

Sensitivity and Specificity (%)
Screening Tests for Bacterial Meningitis versus Viral Meningitis

Screening Test (CSF)	Sensitivity (%)	Specificity (%)
WBC >500 cells/cu mm	74	94
Polymorphonuclear Leukocytes >200 cells/cu mm	91	84
Protein >100 mg/dL	74	94
Glucose <40 mg/dL	78	100
Gram Stain	74	100
Lactic Acid >35 mg/dL(3.9 mmol/liter)	100	100
C-Reactive Protein	100	94
LDH >40	High	High

Early aseptic meningitis typically demonstrates a polymorphonuclear leukocyte predominance, which decreases over time: 0-12 hrs, 76% PMN's; 12-24 hrs, 57% PMN's; 24-36 hrs, 22% PMN's; more than 36 hrs, 12% PMN's (Amir J, et al. J Pediatr. 1991; 119:938-941).

Rapid and specific diagnosis of bacterial meningitis is done by latex agglutination (LA) tests and counterimmunoelectrophoresis (CIE) for assay of antigens of microorganisms; latex agglutination tests have largely replaced CIE since LA is more rapid, with better sensitivity and specificity. Latex agglutination may be positive in patients receiving HbOC (H. influenzae type b conjugate vaccine) for at least 7 days in concentrated urine, and 21 days in CSF (Darville T, et al. Pediatr Infect Dis J. 1992; 11:243-244). Latex agglutination tests are available for Neisseria meningitidis, Haemophilus influenza, Streptococcus pneumoniae and Group B streptococcus.

Bacteria Causing Meningitis: The organisms most commonly responsible for bacterial meningitis, by age-group, are given in the next Table:

Usual Organisms, by Age, in Bacterial Meningitis

Age	Organisms
Newborns	Gram Negative Bacilli, Group B Streptococci, Listeria Monocytogenes
Children, >2 months	Haemophilus Influenzae, Type b
Healthy Adults	Neisseria Meningitidis (Meningococci)
Elderly	Streptococcus Pneumoniae (Pneumococci)

Prior to 1987, the approximate incidence, in percent, of different microorganisms causing meningitis in the age range from two days to 10 years is as follows: H. influenzae, Type b (67%), N. meningitidis (16%), Str. pneumoniae (9%), Group B streptococci (5%), Listeria monocytogenes (4%). Introduction of H. influenza vaccine has produced a dramatic decrease in cases of H. Influenza Meningitis. Cases of bacterial meningitis at Children's Medical Center, Dallas, are given in the following Table (Nelson JD, McCracken GH. Bacterial Meningitis. Pediatr Infect Dis J. 1992; 11:661):

Bacterial Meningitis - Childrens Medical Center, Dallas

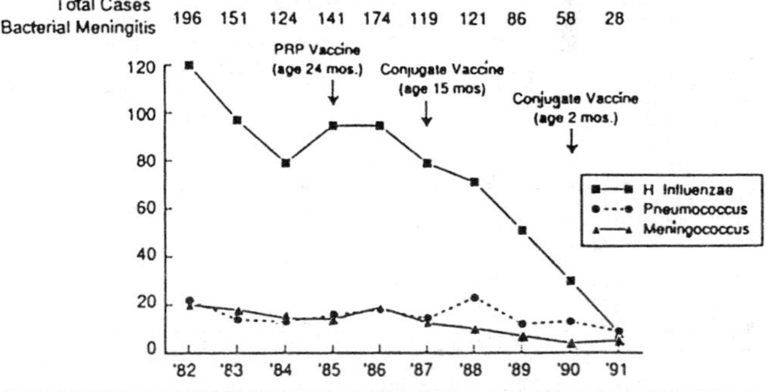

The approximate incidence in percent of different microorganisms associated with mortality with meningitis is as follows: Streptococcus pneumoniae (19%), H. influenza, type b (3%), Neisseria meningitidis (13%). Group B Streptococcus (12%), L. monocytogenes (22%)(Wenger JD, et al. J Infect Dis. 1990; 162:1316-1323). Bacteremia occurs in 40 to 42 percent of patients with meningitis.

The organisms that most commonly cause neonatal meningitis are Escherichia coli and Group B streptococci. In the first week of life, infections are likely to be caused by bacteria colonizing the maternal birth canal, and from organisms transmitted from people caring for the newborn infant (usually via the hands) and from organisms flourishing in the humidifying apparatus that is used in neonatal intensive care units.

The most common causes of meningitis in adults are meningococci and pneumococci. Bacterial meningitis may occur secondary to pneumonia, endocarditis, osteomyelitis or extension of infections of the sinuses and ears.

Opportunistic Infections of the CNS: There are specific conditions that have a striking association with infection of the CNS by particular organisms. Some of these conditions are associated with impaired immunity. This is largely the result of the widespread use of immunosuppressive drugs such as cytotoxic drugs and corticosteroids to treat malignant diseases and connective tissue disorders and after organ transplantation. These patients and those with congenital or acquired immune deficiency such as AIDS are particularly prone to serious infections. The microorganisms involved are many times not pathogenic to normal persons; these infections are termed opportunistic. Conditions that have a striking association with infection of the central nervous system by particular organisms are shown in the next Table:

Cerebrospinal Fluid (CSF) (Cont)

Conditions That Are Associated with Infection of the CNS by Particular Organisms

Condition	Organism
CNS Surgery	Gram-Negative Bacilli Staphylococcus Aureus or Epidermidis
Multiple Antibiotic Therapy in Seriously Ill Patients	Fungi
Leukopenia	Gram-Negative Bacilli Fungi
Leukemia, Lymphomas Cytotoxic Drug or Corticosteroids Administration	Viruses Fungi Listeria Nocardia
Treated Acute Lymphoblastic Leukemia in Childhood	Measles Virus
Splenectomy in Childhood and Adolescence	Streptococcus Species (Pneumococci) H. Influenzae
AIDS	Routine and Opportunistic Pathogens (Bacterial, Fungal, Viral, Tuberculous)

Culture: Spinal fluid should be cultured for bacteria, fungi, and tuberculosis. Neisseria meningitidis is cultured on chocolate and blood agar plates in a 3-5% CO_2 atmosphere. Incubate 2-3 mL to 37°C for 24 hours and incubate in a second set of cultures.

TUBERCULOUS MENINGITIS: The diagnosis of tuberculous meningitis requires demonstration of acid-fast organisms on smears of CSF or growth of mycobacteria in CSF cultures. The demonstration of T.B. on smears occurs in 25% of patients; growth on culture media may require 8 weeks.

CSF findings in tuberculous meningitis in pediatric patients, as well as adult patients with and without HIV infection are given in the following Table (Waecker NJ, Connor JD. Pediatr Infect Dis J. 1990; 9:539-543; Berenguer J, et al. N Engl J Med. 1992; 326:668-672):

CNS Findings in Tuberculous Meningitis

Age	WBC Count (cu mm)	Lymphocytes (%)	Protein (mg/dL)	Glucose (mg/dL)
Pediatrics	200(30-600)	75(20-100)	239(52-1324)	25(5-65)
Adult HIV-Infected	234(0-1200)	NR	52(10-300)	23(0-49)
Adult Non-HIV Infected	250(3-500)	NR	78(20-200)	20(7-52)

Mean (Range) are given. NR = not recorded.

ORGANISM-SPECIFIC ANTIBODY INDEX: This test measures immunoglobulins to specific organisms which may otherwise be difficult to document (e.g. Lyme Disease, Varicella-Zoster, Herpes Simplex and HIV-1). These tests are available in some hospital-based laboratories and by large commercial laboratories (Andiman WA. Pediatr Infect Dis J. 1991; 10:490-495).

MULTIPLE SCLEROSIS: See MYELIN BASIC PROTEIN and PROTEIN ELECTROPHORESIS.

PITUITARY TUMORS: See PITUITARY PANEL for hormones in the CSF in patients with pituitary tumor and suprasellar extension.

CEREBROSPINAL FLUID (CSF) PROTEIN

SPECIMEN: 1mL CSF
REFERENCE RANGE: Mean (range) are given. Premature Newborn: 115(65-150) mg/dL; Full-Term Newborn: 90(20-170) mg/dL; ≤1 month: 84(35-189) mg/dL; 4-8 weeks: 59(19-121) mg/dL; >6 weeks: 28(20-45) mg/dL (Bonadio WA. Pediatr Infect Dis J. 1992; 11:423-432) Adult Total: 15-45 mg/dL. To convert conventional units in mg/dL to international units in g/L, multiply conventional units by 0.01.

Percentage: Prealbumin, 7.0%; Albumin: 49%; Globulins: Alpha-1, 7.0%; Alpha-2, 8.6%; Beta: 19%; Gamma: 9.3%. The principal immunoglobulin in CSF is IgG. Normally, the concentration of CSF IgG is less than 8mg/dL while the concentration of IgA and IgM are less than 2mg/dL. Albumin is 26.0mg/dL or less.
METHOD: Trichloroacetic Acid precipitation of protein with light scattering at 340nm with DuPont ACA; nephelometric assay of albumin and IgG; electroimmunodiffusion; coomassie blue dye-binding method.
INTERPRETATION: An increase in CSF protein occurs in the conditions listed in the next Table:

Increase in CSF Protein
Multiple Sclerosis
Postinfective Polyneuropathy (e.g. Guillain-Barré Syndrome)
Spinal Block (e.g. Neoplasms in Spinal Canal)
Minor Degrees of Spinal Cord Compression e.g. Cervical Spondylosis; Recent Prolapse of an Intervertebral Disc
Meningitis (Bacterial, 100 to 500mg/dL; Viral, usually <100mg/dL)
Encephalitis
Cerebral Abscess
Cerebral Infarction
Neurosyphilis
Intracranial Venous Sinus Thrombosis

Multiple sclerosis is sometimes associated with CSF protein elevation, but this test lacks both sensitivity and specificity. Electrophoresis of concentrated CSF demonstrates oligoclonal banding (multiple monoclonal globulins, primarily IgG) in more than 90% of patients. Other causes of oligoclonal banding in CSF include encephalitis, meningitis, Guillian-Barré syndrome, polyneuritis, headache, stroke, tumor, etc (Hashimoto SA, Paty DW. Disease-a-Month. 1986; 32:518-589; Ebers GC, et al. Can J Neurol Sci. 1979; 7:275). Antibodies to cerebellar soluble lectin (a protein involved in central and peripheral nerve myelination) have been found in multiple sclerosis; specificity of the test is 85%, sensitivity is 93.5% (Zanetta J-P, et al. Lancet. 1990; 335:1482-1484).

CERULOPLASMIN

SPECIMEN: Red top tube; patient fasting; send frozen in plastic vial on dry ice.
REFERENCE RANGE:

Change of Serum Ceruloplasmin with Age	
Age	Serum Ceruloplasmin (mg/dL)
Newborn	2 - 15
Two	30 - 55
Ten	20 - 45
Young Adults	20 - 45
Adults	25 - 45

To convert conventional units in mg/dL to international units in mg/liter, multiply conventional units by 10.0.
METHOD: Ceruloplasmin catalyzes the oxidation of p-phenylenediamine by molecular oxygen with the formation of a blue-violet color.
INTERPRETATION: The causes of decreased serum ceruloplasmin are given in the next Table (Yarze et al. Am J Med. 1992; 92:643-654):

Causes of Decreased Serum Ceruloplasmin
Wilson's Disease
Homozygous (85-95%)
Heterozygous (10-20%)
Fulminant Hepatic Failure (Any Etiology)
Chronic Active Hepatitis in Children (25%)
Normal Neonates
Malnutrition
Intestinal Malabsorption
Renal Protein Loss
Menke's Kinky Hair Disease
Hereditary Hypoceruloplasminemia

Wilson's Disease (Hepatolenticular Degeneration): (Yarze JC, et al. "Wilson's Disease: Current Status," Am J Med. 1992; 92:643-654): Wilson's disease (hepatolenticular degeneration) is an uncommon (1:30,000) inborn error of copper metabolism inherited as an autosomal-recessive trait. In this disease, there is excessive deposition of copper in the liver, basal ganglia, cerebral cortex, kidney, and cornea. The clinical characteristics of 51 patients with Wilson's Disease at the time of diagnosis include the following (Stremmel W, et al. "Wilson's Disease: Clinical Presentation, Treatment, and Survival," Ann Intern Med. 1991; 115:720-726):

Clinical Presentation of Wilson's Disease (%)
Neurologic Manifestations (61%)
Dysarthria (51%); Tremor (47); Writing Difficulties (35); Ataxia (31); Asthenia (31); Hypersalivation (12); Psychiatric Symptoms (12); Nervousness (10); Headache (8); Hypomimia (8); Dizziness (4); Seizures (2)
Abdominal Manifestations (67%)
Hepatomegaly (49%); Splenomegaly (49); Abdominal Pain (41); Jaundice (14); Ascites (14); Gynecomastia (12); Esophageal Varices (10); Diarrhea (8); Vomiting (4)
Hematologic Manifestations (31%)
Thrombocytopenia (22%); Hemolysis (10); Anemia (8); Leukopenia (8)
Other Signs and Symptoms
Kayser-Fleischer Rings (67%); Arthralgia (10); Gigantism (10); Bone Deformities (4); Keratitis (4); Hypoparathyroidism (rare)(Carpenter TO, et al. N Engl J Med. 1983; 309:873-877).

Symptomatic patients usually show pigmented cornea rings and the presence of this ring is an important aid in diagnosis. The Kayser-Fleischer ring is a golden-brown, greenish, or brownish-yellow, or bronze discoloration in the zone of Descemet's membrane of the cornea.

The defect in Wilson's disease may be a lysosomal defect in hepatocytes leading to decreased biliary copper excretion. Almost all the circulating copper is normally bound to ceruloplasmin; the bound copper is known as the indirect-reacting copper. The remaining copper (2% to 5%) is loosely bound to serum albumin; this is known as the direct reacting fraction of serum copper. Patients with Wilson's disease have decreased total copper in their serum primarily because of the decrease in indirect reacting copper. This reflects a deficient level of ceruloplasmin. However, there is an increase in direct reacting copper which is loosely bound; this loosely bound copper is readily dissociated and deposited in tissue.

The laboratory diagnosis of Wilson's disease should be suspected in patients below the age of 30 who have chronic idiopathic hepatitis. The biochemical findings in Wilson's disease are listed in the next Table:

Biochemical Findings in Wilson's Disease
Decreased Ceruloplasmin in Serum
Decreased Total Copper in Serum
Decreased Indirect Reacting Copper
Increased Direct Reacting Copper (free serum copper)
Increased Copper in Urine
Increased Copper Deposited in Liver

The diagnosis of Wilson's disease is aided by the demonstration of decreased levels of the enzyme, ceruloplasmin, and copper in the serum and increased levels of copper in the urine.

A serum ceruloplasmin level of less than 20mg/dL, serum copper level less than 80mcg/dL, and a urinary copper level in excess of 100mcg/24 hour urine is compatible with the diagnosis of Wilson's disease (Stremmel W, et al, already cited).

The causes of elevated serum ceruloplasmin are given in the next Table:

Causes of Elevated Serum Ceruloplasmin
Oral Contraceptives
First Trimester of Pregnancy
Primary Biliary Cirrhosis
Infections (Acute Phase Reactant)

High serum levels of ceruloplasmin may cause a greenish cast to plasma.

CHLAMYDIA ANTIBODIES (PSITTACOSIS ANTIBODIES)

SPECIMEN: Red top tube, separate and refrigerate serum; obtain acute and convalescent serum samples; the convalescent serum specimen should be obtained 2-3 weeks after onset.

REFERENCE RANGE: Presence of antibody indicates chlamydial infection in the past. A fourfold or greater rise in antibody titer between acute and convalescent phase sera indicates recent infection.

METHOD: Complement fixation, immunofluorescence, or enzyme linked immunoassay (ELISA).

INTERPRETATION: (see CHLAMYDIA CULTURE) Serologic testing for chlamydial infections have significant limitations. Chlamydial sexually transmitted disease and respiratory infections are relatively common. The high prevalence of persistent chlamydial antibody limits the specificity of antibody testing for the diagnosis of acute chlamydia infection. Specificity is further limited since chlamydial antibodies may result from either respiratory or invasive genital infection (salpingitis, perihepatitis, proctitis, or epididymitis). Titers may not rise following superficial genital infection. Furthermore, many serologic tests do not differentiate between C. trachomatis, C. pneumonia (TWAR), and C. psittaci (Barnes RC. Clin Microbiol Rev. 1989; 2:119-136).

Complement Fixation: Complement fixation is generally useful in the diagnosis of lymphogranuloma venereum (LGV) and ornithosis (psittacosis) but is not sensitive enough to detect uncomplicated genital infections. C. pneumoniae infection may or may not result in a positive complement fixation (Ridgway GL, Taylor-Robinson D. J Clin Path. 1991; 44:1-5).

Immunofluorescence: Microimmunofluorescence (MIF) can identify IgG and IgM antibodies; antichlamydial IgM is very specific and sensitive for chlamydial pneumonitis and may be the diagnostic test of choice. This test is technically difficult to perform. Whole inclusion immunofluorescent (WIF) is sensitive only to chlamydiae in general and cannot differentiate between species (Barnes RC, already cited, 1989).

Enzyme-Linked Immunosorbent Assay (ELISA): ELISA tests may detect IgG or IgM antibodies. Like the WIF test, ELISA is sensitive to chlamydiae in general. Detection of chlamydial IgM by ELISA may be comparable to MIF for the diagnosis of chlamydia pneumonia in infants (Barnes RC, already cited, 1989).

CHLAMYDIA ANTIGEN, ENZYME IMMUNOASSAY

SPECIMEN: Place the conjunctival, nasopharyngeal, cervical, or urethral swab in a transport tube; the tube contains phosphate buffered saline to keep the organisms in solution.
REFERENCE RANGE: Negative
METHOD: Enzyme immunoassay (EIA-Abbott Lab). Since the C. trachomatis organism survives only inside living cells, a special solution is used to dissolve the cell membranes and liberate the organism; the assay is illustrated in the next figure:

Principle of the Abbott C. Trachomatis Enzyme Immunoassay

Release of C. Trachomatis from Cells

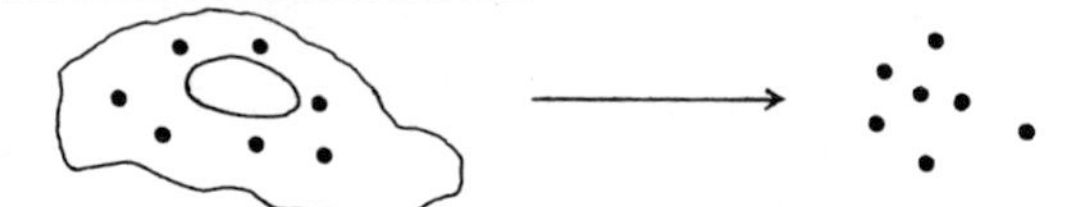

Reaction of C. Trachomatis with Beads

Reaction with Antibody

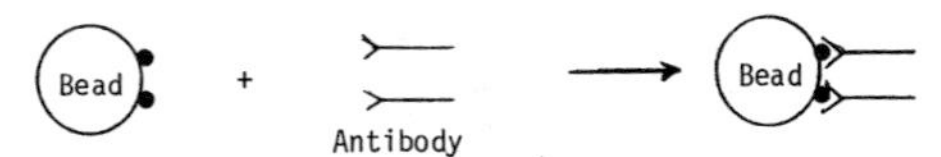

Reaction of Antibody with Enzyme-Labeled Antibody
E = Horseradish Peroxidase

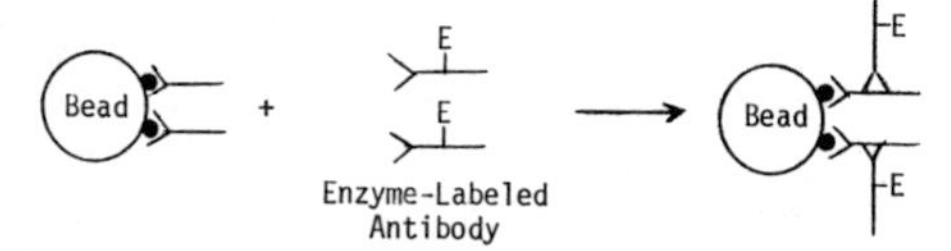

Detection
O-Phenylenediamine + H_2O_2 $\xrightarrow[\text{Peroxidase}]{\text{Horseradish}}$ Colored Product (492nm)

Variations on this technique are used in other assays.
INTERPRETATION: (See CHLAMYDIA CULTURE) Chlamydia trachomatis is the predominant venereal disease- causing organism in the United States.

The enzyme immunoassay test has a sensitivity range of 62 to 98 percent and a specificity range of 89 to 100 percent when used to detect chlamydia in cervical or urethral swabs in adult men and women (Review: Taylor-Robinson D, Thomas BJ. Genitourin Med. 1991; 67:256-266). This test is not specific enough for detection of chlamydial infection in prepubertal girls and is not recommended for evaluation of sexual abuse (Porder K, et al. Pediatr. Infect Dis J. 1989; 8:358-360). This test is acceptable for ocular or nasopharyngeal specimens, but not for rectal specimens (Hammerschlag MR, et al. J Clin Microbiol. 1987; 25:2306-2308; Rothburn MM, et al. Lancet. 1986; ii:982-983). This method will detect both live and dead organisms; therefore, dead organisms following successful therapy may yield a positive test result.

CHLAMYDIA ANTIGEN, FLUORESCENCE IMMUNOASSAY

SPECIMEN: A swab is taken of the endocervical, urethral, rectal, conjunctival, or nasopharyngeal area. The swab is rolled on a specially prepared slide; the slide is fixed in acetone or alcohol.
REFERENCE RANGE: Negative
METHOD: Fluorescence microscope is required. Organisms are detected by adding a fluorescein-labelled monoclonal antibody to the slide preparation. The slide preparation is incubated at room temperature or 37°C for 15 to 30 minutes and then rinsed, mounted and viewed with the fluorescence microscope. C. trachomatis organisms are stained green.
INTERPRETATION: (See CHLAMYDIA CULTURE) Chlamydia trachomatis is the predominant venereal disease - causing organism in the United States.

The sensitivity of this direct fluorescent antibody (DFA) test ranges from 68 to 100 percent; the specificity ranges from 82 to 100% (usually >95%). Interpretation of this test depends on the ability of the observer to detect small numbers of elementary bodies and differentiate between specific and non-specific staining (Taylor-Robinson D, Thomas BJ, Genitourin Med. 1991; 67:256-266). DFA testing has been used to detect C. trachomatis from nasopharyngeal swabs of infants with pneumonia and from conjunctival specimens of infants with conjunctivitis. DFA is not acceptable for evaluating rectal or vaginal specimens from prepubertal children with suspected sexual abuse (Review of pediatric implications: Hammerschlag MR. J Pediatr. 1989; 114:727-734).

Dead organisms may be detected by this method.

CHLAMYDIA CULTURE

SPECIMEN: Chlamydial organisms are labile in-vitro; use special transport media; the specimen must be kept cold; transport to laboratory as soon as possible. Use sterile swab to obtain specimen from suspected site. Plastic or metal-shafted swabs should be used, not wooden swabs. Adults: Urethra, cervix or eye. Infants: Nasopharynx, throat, sputum or eye. Extract swab into appropriate transport media, sucrose phosphate, without antibiotics. Discard swab and cap the vial containing transport media.
REFERENCE RANGE: Negative
METHOD: Because C. trachomatis is an obligate intracellular parasite, it cannot be grown on artificial media; tissue culture facilities are required. Cultivation of specimen in cycloheximide-treated McCoy cells (or HeLa229 cells) with subsequent staining with iodide at 48 to 72 hours postinoculation. Observe for inclusions present in the cytoplasm of the McCoy (or HeLa229) cells which are specific for C. trachomatis; C. psittaci inclusions do not stain with iodide; both C. trachomatis and C. psittaci inclusions stain with Giemsa stain.
INTERPRETATION: Chlamydiae are gram-negative intracellular bacteria; there are three species: C. trachomatis, C. pneumoniae, and C. psittaci. C. trachomatis is the causative organism in ocular and urogenital disease. C. pneumoniae (TWAR) is a newly recognized chlamydia species and is among the common causes of community-acquired pneumonia. C. psittaci causes respiratory, urogenital and systemic infections in a variety of animals; man is incidentally infected.
C. Trachomatis: C. trachomatis may be transmitted sexually; the type or stage is given in the next Table:

Chlamydia Trachomatis: Sexually Transmitted

Type or Stage	Comments
Urethritis or Cervicitis	"Urethral syndrome" with sterile pyuria in sexually active women
Pelvic Inflammatory Disease	Cervicitis may lead to invasion with endometritis and salpingitis and perihepatitis
Epididymitis	C. trachomatis is a major cause of epididymitis in young, sexually active men
Oculogenital Syndrome	Autoinoculation of the conjunctiva from a genital focus
Proctitis	Occurs in homosexual men practicing anal intercourse
Neonatal: Conjunctivitis Pneumonia	50-75% of infants born to infected mothers will be infected with C. trachomatis. These infants may have conjunctivitis, nasopharyngeal infection, pneumonia or asymptomatic infection of the vagina or rectum (Hammerschlag MR. J Pediatr. 1989; 114:727-734). Give erythromycin ophthalmic ointment to prevent conjunctivitis; neonatal pneumonia can be treated with systemic erythromycin for 14 days
Lymphogranuloma Venereum	Caused by certain types of C. trachomatis

The sexually transmitted diseases caused by C. trachomatis in men, women and infants are given in the next Table (Syva Monitor. 1984; 2:1-4):

Chlamydia Trachomatis in Men, Women and Infants

Men and Women	Women	Men	Infants
Lymphogranuloma Venereum	Cervicitis	Urethritis	Inclusion Conjunctivitis
Inclusion Conjunctivitis	Salpingitis	Epididymitis	Pneumonia
Otitis Media	Urethritis-Urethral Syndrome	Reiter's Syndrome (Suspected)	Rhinitis
Proctitis	Bartholinitis		Otitis Media
Pharyngitis (Suspected)	Perihepatitis		Vaginitis (Suspected)
	Endometritis		

C. trachomatis is treated with tetracycline and erythromycin. The treatment of C. trachomatis sexually transmitted infections is given in "1989 Sexually Transmitted Diseases Treatment Guidelines." MMWR. 1989; 38(S-8):1-40).

Venereal Disease: C. trachomatis is one of the most common genital infections in pregnant women. A retrospective analysis of a pregnant population suggests that eradication of C. trachomatis infection reduces the occurrence of premature rupture of membranes, premature contractions, premature delivery, small-for-gestational-age infants, and low birth weight (Cohen I, et al. JAMA. 1990; 263:3160-3163). It causes a spectrum of diseases like that of Neisseria gonorrhoeae with urethritis, cervicitis, epididymitis, proctitis, pelvic inflammatory disease, oculogenital syndrome and ophthalmia. C. trachomatis is the single most important cause of pelvic inflammatory disease and its resulting infertility. The estimated annual incidence of C. trachomatis venereal infections is three to four million; of the patients with Neisseria gonorrhoeae, about 20 percent of men and 30 percent of women are coinfected with C. trachomatis. The infection is asymptomatic in about 70 percent of women.

Diseases of Infancy: C. trachomatis is an important cause of neonatal infection; it can be transmitted to newborns by their passage through an infected birth canal. The rate of transmission from infected mother is 50 to 75 percent. The nasopharynx is the most frequent site of infection, and 30% of infants with nasopharyngeal infection develop pneumonia (Hammerschlag MR, et al. Pediatr Infect Dis. 1982; 1:395-401). Neonatal conjunctivitis develops in 20-50% of infants born to mothers with chlamydial infection. Erythromycin ophthalmic ointment prevents 80-93% of cases of conjunctivitis (Black-Payne C, et al. Pediatr Infect Dis. 1989; 8:491-498; Hammerschlag MR, et al. N Engl J Med. 1989; 320:769-772). Chlamydial conjunctivitis and pneumonia should be treated with systemic erythromycin. Chlamydial pneumonia usually presents at four to 11 weeks of age as an afebrile illness characterized by progressive tachypnea with rales, often with a staccato cough. Conjunctivitis occurs in half of these patients and is not helpful differentiating this type of neonatal pneumonia from others. Patients often have eosinophilia (>300/cubic mm); chest radiograph shows bilateral interstitial infiltrates with atelectasis and hyperinflation.

Diseases of Children: Perinatal maternal-infant transmission may result in chronic C. trachomatis infection. Infants may have positive conjunctival oropharyngeal or nasopharyngeal cultures for at least 28.5 months and positive rectal or vaginal cultures for at least a year (Bell TA, et al. JAMA. 1992; 267:400-402). Furthermore, C. trachomatis has been cultured from the vaginas of two sisters abused three years prior (Hammerschlag MR, et al. Pediatr Infect Dis. 1984; 3:100-104). Rectogenital chlamydia infection is associated with sexual abuse (Fuster CD, Neinstein LS. Pediatr. 1987; 79:235-238); Ingram DL, et al. Pediatr Infect Dis. 1984; 3:97-99); interpretation of positive cultures must be in clinical context.

C. trachomatis causes trachoma, the leading cause of preventable blindness in the world; almost all cases of trachoma are found in North America, the Mideast and Southeast Asia.

Chlamydia Pneumoniae: C. pneumoniae (TWAR) is a recently described chlamydia species that is a common cause of respiratory illness in children and adults. The spectrum of disease caused by C. pneumoniae includes asymptomatic infection, febrile "influenza-like" illness, bronchitis (often prolonged) and pneumonia which may be mild or severe. Pharyngitis and sinusitis often accompanies bronchitis and pneumonia. Sarcoidosis, myocarditis, and endocarditis have been serologically associated with TWAR infection (Grayston JT. Chest. 1989; 95:664-669). C. pneumoniae is responsible for causing 5-10% of cases of pneumonia in adults (Grayston JT, et al. J Infect Dis. 1990; 161:618-625).

Chronic C. pneumoniae infection may be a significant risk factor for the development of coronary heart disease (Saikku P, et al. Ann Intern Med. 1992; 116:273-278) and may cause wheezing, asthmatic bronchitis and asthma (Hahn DL, et al. JAMA. 1991; 266:225-230).

Chlamydia Culture (Cont)

Psittacosis: C. psittaci causes psittacosis, a pneumonia derived from infected birds.

Although culture is regarded as the "gold standard", it is estimated to be only 70-80% sensitive in detecting C. trachomatis from a single endocervical specimen (Lefebvre J, et al. J Clin Microbiol. 1988; 26:726-731). Other tests used to diagnose chlamydia infections are as follows (Barnes RC. Clin Microbiol Rev. 1989; 2:119-136):

Diagnostic Tests for Chlamydial Infections

Agent/Infection	Presumptive Test	Diagnostic Test
C. trachomatis		
Adult Oculogenital	Antigen Detection	Cell Culture
LGV	Complement Fixation Single Titer ≥ 1:32 with Clinical Findings	Cell Culture Isolate Typing
Neonatal Pneumonia	Pneumonia plus Detection of Agent by DFA or Culture	IgM Elevation Determined by Microimmuno-fluorescence or EIA
C. pneumonia		
TWAR (Respiratory)	No Accepted Standard; Micro-immunofluorescence Fourfold Rise to TWAR Antigen of High IgM Probably Diagnostic	Egg or Cell Culture
C. psittaci		
Ornithosis (Psittacosis)	Complement Fixation Single Titer ≥ 1:32	Complement Fixation Four-fold Rise Between Acute and Convalescent Titers

See also CHLAMYDIA ANTIBODIES and CHLAMYDIA ANTIGEN.

CHLORAMPHENICOL

SPECIMEN: Red top tube, separate serum

REFERENCE RANGE: Therapeutic: 10-20 mcg/mL; Toxic: >25 mcg/mL.

Half-Life: Adults and children: 1.5-5 hours;
Premature infants (one to two days old): 24-48 hours;
Premature infants (13 to 23 days old): 8-15 hours.

METHOD: HPLC; GLC

INTERPRETATION: Chloramphenicol is a broad-spectrum antibiotic which is used when less potentially dangerous drugs are ineffective or contraindicated.

The indications for the use of chloramphenicol are as follows: anaerobic infections, especially with Bacteroides fragilis; typhoid fever; Haemophilus influenzae meningitis; pneumococcal and meningococcal meningitis in patients allergic to penicillin; brain abscess; and rickettsial infections.

Chloramphenicol is the drug of choice for the treatment of severe cases of typhoid fever. Haemophilus Infections: Chloramphenicol may be used in conjunction with ampicillin for the initial treatment of meningitis, osteomyelitis, septic arthritis, cellulitis, epiglottitis, septicemia, or other serious infections caused by H. influenza although third generation cephalosporins are more commonly used. Rickettsial Infections: Tetracyclines are the drugs of choice for the treatment of Rocky Mountain spotted fever and other rickettsial infections; however, chloramphenicol is the drug of choice when tetracyclines cannot be used (Francke EL, Neu HC. Med Clin North Amer. 1987; 71:1155-1168).

Adverse Effects: Bone Marrow Depression: Bone-marrow depression is dose-related. Plasma levels >25mcg/mL are frequently associated with reversible bone marrow depression with decreased reticulocytes, decreased hemoglobin, decreased thrombocytes, decreased WBC's and increased serum iron. Anemia most often follows parenteral therapy, large doses, long duration of therapy or impaired drug elimination. Patients do not respond to iron or vitamin B_{12} therapy while receiving chloramphenicol. Complete recovery usually occurs within 1-2 weeks after drug discontinuation.

Gray Baby Syndrome: Fatal cardiovascular-respiratory collapse (gray baby syndrome) may develop in neonates given excessive doses and having plasma levels above 50-100mcg/mL. Large overdoses may induce a gray baby like syndrome in children and adults (Slaughter RL, et al. Clin Pharmacol Ther. 1980; 28:69-77).

Chloramphenicol is converted to chloramphenicol-glucuronide by glucuronyl transferase in the liver and then excreted rapidly (80 to 90 percent of the dose) in the urine by the kidney. In the neonate, conversion of chloramphenicol to the glucuronide is slow because the activity of the enzyme, glucuronyl transferase, is low; thus, chloramphenicol may accumulate. Chloramphenicol levels must be monitored, particularly in neonates and infants.

Aplastic Anemia: Aplastic anemia is rare, non-dose related, and fatal and can occur long after a short course of oral or parenteral therapy (Meissner HC, Smith AL. Pediatrics. 1979; 64:348-356).

Complete blood count should be obtained every two days to recognize reticulocytopenia, leukopenia, thrombocytopenia and anemia.

Chloramphenicol may increase the effects of many drugs that undergo hepatic metabolism (e.g., phenytoin and oral hypoglycemia agents). Effects of anticoagulants may be increased. Phenobarbital may increase chloramphenicol metabolism.

CHLORIDE, SERUM

SPECIMEN: Red top tube or green top tube (heparin)
REFERENCE RANGE: Premature: 95-100mmol/liter; full-term: 96-106mmol/liter; infant: 96-106mmol/liter; child and adult: 95-105mmol/liter. Conventional units (mEq/L) equal international units (mmol/L).
METHOD: Colorimetric measurement
INTERPRETATION: The causes of increased serum chloride are shown in the next Table:

Causes of Increased Serum Chloride (>106mmol/liter)
Hyperchloremic Metabolic Acidosis
Respiratory Alkalosis
Renal Disease:
Pyelonephritis
Polycystic Renal Disease
Obstructive Uropathy
Renal Tubular Acidosis
Severe Dehydration,
i.e., Diabetes Mellitus
Diabetes Insipidus
I.V. Saline

Increased Serum Chloride: Increased serum chloride occurs in hyperchloremic metabolic acidosis; anion gap is normal. In metabolic acidosis, $[HCO_3^-]$ is decreased; in hyperchloremic metabolic acidosis serum chloride increases as $[HCO_3^-]$ decreases in order to maintain electroneutrality. In respiratory alkalosis hydrogen ions from intracellular sources enter the extracellular fluid and bicarbonate moves into the red blood cells in exchange for chloride, thus minimizing extracellular alkalosis.

The causes of decreased serum chloride are shown in the next Table (Besunder JB, Smith PG. Crit Care Clin. 1991; 7:659-693):

Causes of Decreased Serum Chloride (<95mmol/liter)	
Renal Loss:	Other:
Loop Diuretics	Metabolic Alkalosis
Salt-losing Nephropathies	Respiratory Acidosis (chronic)
Bartter's Syndrome	Congestive Heart Failure
GI Loss:	Overhydration
Nasogastric Suction	Inappropriate ADH (SIADH)
Gastric Outlet Obstruction	Overtreatment with Hypotonic Solutions
(e.g. Pyloric Stenosis)	Dietary Deficiency
Zollinger-Ellison syndrome	Burns
Congenital Chloride-losing	Addison's Disease
Enteropathy	Perspiration (Cystic Fibrosis)
Secretory Diarrhea	

Decreased Serum Chloride: In metabolic alkalosis, $[HCO_3^-]$ is increased; in order to maintain electroneutrality, chloride decreases. In respiratory acidosis, chloride excretion is a necessary concomitant of renal compensation for respiratory acidosis. (cont)

Chloride, Serum (Cont)

References: Reviews of chloride homeostasis have recently been published: Koch SM, Taylor RW. "Chloride Ion in Intensive Care Medicine," Crit Care Med. 1992; 20:227-240; Besunder JB, Smith PG. "Toxic Effects of Electrolyte and Trace Mineral Administration in the Intensive Care Unit," Crit Care Clin. 1991; 7:659-693).

CHLORIDE, URINE

SPECIMEN: 10mL aliquot of 24 hour urine; preserved with 10mL of glacial acetic acid, indicate total volume.
REFERENCE RANGE: 110-250mmol/liter. Conventional units (mEq/L) equal international units (mmol/L).
METHOD: Colorimetric measurement
INTERPRETATION: The level of urinary chloride may be used in the differential diagnosis of metabolic alkalosis (Harrington JT, Cohen JJ. N Engl J Med. 1975; 293:1241-1248). The common causes of metabolic alkalosis are two-fold: (1) those associated with loss of hydrogen ion and extracellular fluid volume (chloride responsive type) and (2) those associated with pathologically excessive circulating levels of mineralocorticoid hormone (chloride resistant type). The measurement of urinary chloride is valuable in differentiation of causes of persistent metabolic alkalosis as given in the following Table:

Urinary Chloride in the Differential Diagnosis of Metabolic Alkalosis

Chloride Responsive (Urine Chloride <10mmol/liter)	Chloride Resistant (Urine Chloride >20mmol/liter)
Diuretic Therapy	Adrenal Disorders
Gastrointestinal Causes	Hyperaldosteronism
Vomiting	Cushing's Syndrome
Nasogastric Suction	Pituitary, Adrenal, Ectopic ACTH
Chloride-Wasting Diarrhea	Exogenous Steroid
Villous Adenoma-Colon	Gluco- or Mineralocorticoid
Rapid Correction of Chronic Hypercapnia	Licorice Ingestion (Glycyrrhizic acid has mineralocorticoid-like activity)
Carbenicillin	Carbenoxalone
Reduced Dietary Intake of Chloride	Barttler's Syndrome
	Alkali Ingestion

The most common causes of persistent metabolic alkalosis, e.g., loss of gastric juice and diuretic induced, are due to chloride depletion and are responsive to the administration of chloride adequate to replace body stores.

In hyperadrenocorticism, persistent alkalosis results not from the loss of chloride but from indirect stimulation of renal bicarbonate reabsorption.

CHOLECYSTITIS, ACUTE, PANEL

The test panel for acute cholecystitis is given in the next Table:

Test Panel for Acute Cholecystitis

Test	Test Result
Screening Tests:	
White Cell Count (WBC) and Differential	Leukocytosis with a Shift to the Left
Serum Bilirubin	Elevated (Mildly)
Serum Alkaline Phosphatase	Elevated (Mildly)
Aspartate Amino-Transferase (AST)	May be elevated
Serum Amylase	Elevated with Associated Pancreatitis
Confirmatory Tests:	
Flat Plate of Right Upper Quadrant of Abdomen	10 to 30 Percent of Stones are Radiopaque. Helpful to Rule Out Perforated Viscus (free air) or Right Lower Lobe Pneumonia.
Ultrasonography	Diagnostic Accuracy in Detection of Gallbladder Calculi Superior to Oral Cholangiography. Gives Information about Bile Duct Size, Hepatic Ducts and Pancreas. Does Not Determine Cystic Duct Obstruction.
Oral Cholangiography	Complements Ultrasonography
Cholescintigraphy	Preferred Method for Diagnosis of Acute Cholecystitis; I.V. Tc99m Diisopropyl Iminodiacetic Acid-Visualization of Common Bile Duct without Visualization of Gallbladder is characteristic of Acute Cholecystitis. Not Reliable with Prolonged Fasting, Liver Disease and Marked Hyperbilirubinemia
Computed Tomography	Selected Instances; Expensive, Insensitive

Refer to Radiological Examination of Gallbladder: Jacobson HG, Stern WZ. JAMA. 1983; 250:2977-2982; Dawes LG, Nahrwold DL. "Acute Cholecystitis: Update on Diagnosis and Treatment," J Crit Illness. 1992; 7:1409-1422.

Laboratory studies usually show an increase in white blood cells with a shift to the left. Serum bilirubin and alkaline phosphatase are mildly elevated. The presence of jaundice (bilirubin >3 mg/dL) suggests that stone or stones are present in the common bile duct. The Aspartate Aminotransferase (AST) may be elevated if there is ascending cholangitis; serum amylase and lipase should be checked for evidence of associated pancreatitis.

CHOLESTEROL, TOTAL

SPECIMEN: Red top tube, separate serum.

REFERENCE RANGE: Hypercholesterolemia values, for selecting men and women at moderate risk (75th to 90th percentiles) and high risk (>90th percentile) requiring treatment, are given in the next Table (Consensus Conference, JAMA. 1985; 253:2080-2086). The values given for the reference range are defendable, but understandably arbitrary. They were based on distribution of cholesterol values in the American population, which may or may not be either "normal" or ideal. To convert conventional units in mg/dL to international units in mmol/L, multiply conventional units by 0.02586.

Hypercholesterol Values for Selecting Men and Women at Risk for Treatment

Age (Years)	Moderate Risk mg/dL (mmol/L)	High Risk mg/dL (mmol/L)
20-29	>200 (5.17)	>220 (5.69)
30-39	>220 (5.69)	>240 (6.21)
40 and Over	>240 (6.21)	>260 (6.72)

Detailed tables containing cholesterol values are given in the appendix to: Expert Panel, National Cholesterol Education Program. Arch Intern Med. 1988; 148:36-69.

METHOD: Enzyme assay.

INTERPRETATION: Either a fasted or non-fasted sample is adequate for determination of total cholesterol. However, if a lipid or lipoprotein profile is also to be performed (determination of triglyceride, VLDL, LDL, HDL), a 12-14 hour fasted specimen is necessary (Cooper GR, et al. JAMA. 1992; 267:1652-1660).

Total cholesterol is an important risk factor for development of cardiovascular disease. The National Cholesterol Education Program has published cholesterol values to be used in selecting patients at risk and who require treatment (Expert Panel. Arch Intern Med. 1988; 148:36-69; Expert Panel. JAMA. 1993; 269:3015-3023). These values are tabulated below:

Desirable	<200 mg/dL (5.2mmol/L)
Borderline	200-239 mg/dL (5.2-6.2mmol/L)
High	>240 mg/dL (6.2mmol/L)

In this classification system, a low HDL (≤ 35 mg/dL) or elevated LDL (≥170 mg/dL), or additional risk factors such as smoking or hypertension should influence treatment of those patients with borderline levels. Screening may be done with unfasted specimens, but the subsequent HDL and LDL determinations must be done on fasted specimens. For each 1% reduction in cholesterol, it is estimated that there is a 2% reduction in risk of coronary artery disease (Lipid Research Clinics Program. JAMA. 1984; 251:365-374).

There is growing evidence that a low HDL-cholesterol imparts increased risk for CHD. Therefore a low HDL cholesterol (<35 mg/dL) is classified as a risk factor. This can be expressed as the total cholesterol/HDL ratio as seen below:

Cholesterol/HDL Ratio and Cardiac Risk

Risk	Percentile	Ratio-Men	Ratio-Women
Lowest	<25	<3.8	<2.9
Low	26-50	3.9-4.7	3.0-3.6
Moderate	51-75	4.8-5.9	3.7-4.6
High	76-89	6.0-6.9	4.7-5.6
Highest	>90	>7	>5.7

There is understandable concern that many individuals in the "desirable" or the "borderline" range may still be at high risk for coronary artery disease. In a study of individuals with total cholesterols less than 239mg/dL, 20% were considered at high risk because of elevated LDL, and 12% because of reduced HDL (Bush TL, Riedel D. Circulation. 1991; 83:1287-1293).

Total cholesterol levels fall after MI or cardiac surgery to reach a minimum in 4 to 6 days, and a return to baseline in 6 to 8 weeks. Some reports claim a serum cholesterol is accurate if performed in the first 24 hours following MI. There is approximately a 6% within-individual variation of total cholesterol measurements performed over a period of a year. Some of this variation is thought to be due to changes in posture, recent weight change, and alcohol consumption (Cooper, already cited). There is a gradual increase in cholesterol levels in males until age 50 and women until age 70 at which time the levels tend to plateau. Cholesterol screening in children is controversial (Newman TB, et al. JAMA. 1992; 267:100-101), and perhaps not accurate in accessing cardiac risk in geriatric populations (Garber AM, et al. Arch Intern Med. 1991; 151:1089-1095). The serum cholesterol is of value in cardiac risk factor assessment. This topic is discussed in detail under the heading: CARDIAC RISK ASSESSMENT.

The causes of elevated serum cholesterol are given in the next Table:

Causes of Elevated Serum Cholesterol
Lipoproteinemias (Types II, III, V)
Cholestasis, Intra- and Extrahepatic
Nephrotic Syndrome
Hypothyroidism
Oral Contraceptives
Normal Pregnancy
Acute Intermittent Porphyria (AIP)
Macroglobulinemia

A marked elevation of high density lipoprotein (HDL) cholesterol can cause the value for total cholesterol to exceed the upper limit of normal.

The causes of decreased serum cholesterol are given in the next Table:

Causes of Decreased Serum Cholesterol
Liver Disease
Malabsorption
Malnutrition
Hyperthyroidism
Anemia
Abetalipoproteinemia
Tangier's Disease

Abetalipoproteinemia: The demonstration of acanthocytes in fresh blood smears plus a serum cholesterol of less than 50mg/dL is virtually diagnostic of abetalipoproteinemia. The diagnosis can be confirmed by the absence (on electrophoresis) of all three lipoproteins that contain apolipoprotein B (LDL, VLDL and chylomicrons). Clinically, abetalipoproteinemia is manifested by steatorrhea, retinitis pigmentosa, and ataxic neuropathic disease. Much of this pathology is due to vitamin E deficiency. The mode of transmission is that of autosomal recessive with a gene frequency of 1 in 20,000. Abetalipoproteinemia is caused by a lack of a microsomal triglyceride transfer protein (Wetterau JR, et al. Science. 1992; 258:999-1001).

Tangier's Disease: This disease is characterized by reduced serum cholesterol and normal or raised triglycerides. There is a defect in the synthesis of HDL. In these patients there is marked accumulations of cholesterol in many tissues, e.g., spleen, lymph nodes, intestinal mucosa and blood vessels. The tonsils are orange or yellowish gray coloration. There is no HDL to mobilize the deposited lipid.

CHOLINESTERASE (ACETYL-)

SPECIMEN: Green (heparin) top tube or Lavender (EDTA) top tube; Amniotic Fluid
REFERENCE RANGE: Method dependent; units, U/liter
METHOD: Red blood cells are lysed and incubated with a buffered substrate containing acetylcholine:

$$\text{Acetylcholine} + H_2O \xrightarrow[\text{Cholinesterase}]{\text{Acetyl-}} \text{Acetate} + \text{Choline}$$

The acetic acid that is liberated decreases the pH of the solution; the rate of change of the pH is used as an index of acetylcholinesterase activity.
INTERPRETATION: Red cell cholinesterase is inhibited by insecticides; the toxic effect is detected longer in the red cells than it is in plasma or serum and may take 3 to 4 months to return to normal.

Acetylcholinesterase is found principally in the red blood cell, at motor end plates of skeletal muscle and in the central nervous system.

Decreased red blood cell cholinesterase is a useful adjunct in making the diagnosis of paroxysmal nocturnal hemoglobinuria(PNH).

Acetylcholinesterase is measured in amniotic fluid samples that have elevated alpha-fetoprotein levels. Acetylcholinesterase is not normally found in amniotic fluid; its presence is associated with neural tube defects. Minor blood contamination does not cause a false positive result.

CHOLINESTERASE (PSEUDO-)

SPECIMEN: Red or green (heparin) top tube; the specimens are stable at room temperature for 24 hours and for at least two weeks at 4°C.

REFERENCE RANGE: Cholinesterase activity is low at birth and for the first six months of life, increases to values 30% to 50% above adult values until about age five years, then gradually decreases until adult concentrations are reached at puberty (Hill JG in Pediatric Clinical Chemistry, Meites S, ed., 2nd edition, 1981; 153. Am Assoc Clin Chem. 1725 K. Street, N.W., Washington, DC 20006). Values depend on method. If possible, measure cholinesterase prior to anticipated exposure to pesticides.

METHOD: Hydrolysis of various choline esters including acetylcholine.

INTERPRETATION: There are two types of cholinesterase in the blood. There is true or acetylcholinesterase which is found principally in the red blood cell, at motor end plates of skeletal muscle and in the central nervous tissue; there is cholinesterase (pseudocholinesterase) which is found in plasma and in smooth muscle, liver and adipocytes.

Acetylcholinesterase has a well-defined physiological role at the motor end-plate; it acts on acetylcholine, which is an important mediator of nerve conduction.

The measurement of pseudocholinesterase is very useful in (a) monitoring exposure to organophosphate pesticides (b) identification of hypersensitivity to succinylcholine and (c) monitoring liver disease.

(a) Effect of Insecticides on Blood Cholinesterase Level: Cholinesterase activity levels can be used to detect the presence of organophosphate or carbamate insecticides; these insecticides inhibit the enzyme cholinesterase.

The reaction catalyzed by cholinesterase is as follows:

$$\text{Acetylcholine} + H_2O \xrightarrow{\text{Acetyl-Cholinesterase}} \text{Acetate} + \text{Choline}$$

The synthetic insecticides, e.g. organophosphates and carbamates, inhibit cholinesterase by combining with the enzyme; the organophosphates combine irreversibly with this enzyme. Both acetylcholinesterase and pseudocholinesterase are inhibited by organophosphate or carbamate insecticides. Pseudocholinesterase (serum) is depressed before acetylcholinesterase (red cells) but returns to normal before red cell cholinesterase. Serum pseudocholinesterase recovers in days to weeks whereas red cell cholinesterase levels may take 3 to 4 months to return to normal. Cholinesterase levels are useful in making a diagnosis of organophosphate or carbamate poisoning but have no relation to management.

When the action of cholinesterase in inhibited, acetylcholine accumulates. The signs and symptoms of poisoning by organophosphate or carbamate are similar to a continuous stimulation of the parasympathetic nervous system. Clinical features of organophosphate and carbamate poisoning are given in the following Table (Mack RB. Contemporary Pediatr. October 1985; 89-91):

Clinical Features of Organophosphate and Carbamate Poisoning

- Muscarinic:
  - SLUDGE (Salivation, Lacrimation, Urination, Defecation, Gastrointestinal Pain and Cramping, Emesis)
  - Miosis
  - Hyperactive Bowel Sounds
- Nicotinic:
  - Tachycardia
  - Hypertension
  - Muscle Symptoms (Fasciculations, Cramps, Fatigue, Loss of Deep Tendon Reflexes, Paralysis)
- Central Nervous System (Primarily Organophosphates, Not Carbamates):
  - Severe Headache
  - Tremor
  - Ataxia
  - Restlessness
  - Slurred Speech
  - General Weakness
  - Seizures
  - Coma
  - Cardiopulmonary Depression

(See Also: Sofer S, et al. Pediatr Emerg Care. 1989; 5:222-225; Zwiener RJ, Ginsburg CM. Pediatr. 1988; 81:121-126; Coye MJ, et al. Arch Intern Med. 1987; 147:438-442).

(b) Effect of the Muscle Relaxant Succinylcholine (Anectine) on Persons with Inherited Abnormality or Reduced Level of Cholinesterase Activity Secondary to Insecticides: Persons with an inherited abnormality in cholinesterase activity and those with reduced levels of cholinesterase activity secondary to synthetic insecticides may show marked sensitivity when exposed to the short-acting muscle relaxant, succinylcholine. Succinylcholine is used during tracheal intubation. Succinylcholine mediates its effects by its action on the myoneural junction; 90 to 95% of the drug is destroyed within a minute after injection. It is metabolized to the inactive form, succinylmonocholine, as shown in the next reaction (succinylcholine = succinyldicholine):

$$\underset{\text{(Active Form)}}{\text{Succinyldicholine}} + H_2O \xrightarrow[\text{(Myoneural Junction)}]{\text{Cholinesterase}} \underset{\text{(Inactive Form)}}{\text{Succinylmonocholine}} + \text{Choline}$$

When there is an abnormality of cholinesterase, the action of succinylcholine is prolonged; prolonged effects of succinylcholine may cause paralysis of respiratory muscles and apnea. However, patients with apnea lasting many hours are readily and safely maintained with mechanical ventilation.
(c) Liver Function: Pseudocholinesterase is decreased in patients with liver disease but it is not a reliable index of parenchymal liver cell damage.

Conditions that are associated with depressed pseudocholinesterase levels are listed in the next Table:

Depressed Pseudocholinesterase Levels
Insecticide Poisoning
Genetic Variants
Liver Disease
Metastatic Carcinoma
Malnutrition
Anemias
Acute Infections
Myocardial Infarction
Dermatomyositis
Pregnancy

Cholinesterase (Pseudo-)
Dibucaine/Fluoride Inhib.

CHOLINESTERASE (PSEUDO-), DIBUCAINE/FLUORIDE INHIBITION

SPECIMEN: Red top tube, separate and freeze serum; stable 6 months. The patient should not have taken muscle relaxant or anticholinergic drugs within 24 hours.
REFERENCE RANGE: Inhibition of pseudocholinesterase activity by dibucaine. Normal test value: 70-90% inhibition; Heterozygote: 30-60% inhibition; Homozygote: less than 30% inhibition.
METHOD: This test is done to detect homozygous or heterozygous "atypical" cholinesterase variant. The atypical variant is inhibited less strongly by positively charged inhibitors, e.g., dibucaine, than normal cholinesterase; dibucaine is a local anaesthetic.
INTERPRETATION: The atypical variant is autosomal recessively transmitted; the carrier rate is 1/20; the number of couples at risk is 1/400, and the incidence of the homozygote is 1/1600. If the level of inhibition is less than 70%, the patient should be considered to have an atypical pseudocholinesterase variant and the administration of succinylcholine or similar type drugs may pose a risk.

Other cholinesterase variants may be identified by fluoride inhibition. Still other variants may not be identified by either dibucaine or fluoride inhibition.

CHORIONIC VILLUS SAMPLING

SPECIMEN: Chorionic villus sampling (CVS) is usually performed between the 9th and 12th weeks of gestation. CVS may be performed through a transcervical catheter or transabdominal needle.

Transcervical CVS is performed with continuous ultrasonographic guidance. A catheter is inserted through the vagina and cervix, and into the placenta. Villi are aspirated. Transabdominal CVS is performed using a needle inserted through the abdominal wall and into the placenta under ultrasonographic guidance. Villi are aspirated. Villi are separated from maternal decidua using a dissecting microscope and transferred to culture medium (Hallack M, et al. Obstet Gynecol. 1992; 80:349-352).

INTERPRETATION: Chorionic villus sampling can provide a larger sample of fetal tissues than amniocentesis; results are obtained faster and the procedures can be done much earlier in a pregnancy.

Chorionic villi protrude from a membrane called the chorion which surrounds the developing fetus and becomes part of the placenta. The chorion is not an anatomical part of the fetus; however, it is fetal rather than maternal in origin.

The sample is used for chromosome, enzyme and other analyses. Amniotic fluid is not obtained; as such, amniotic fluid alpha-fetoprotein screening for neural tube defects is not possible.

The rate of spontaneous fetal loss through 28 weeks of pregnancy is 2.5 percent for transcervical CVS and 2.3 percent for transabdominal CVS. The chance of successful pregnancy with a surviving liveborn infant is 1.7 to 4.6 percent lower with CVS compared to second-trimester amniocentesis (MRC Working Party on the Evaluation of Chorionic Villus Sampling. Lancet. 1991; 337:1491-1499; Canadian Collaborative CVS-Amniocentesis Clinical Trial Group. Lancet. 1989; 1:1-6). Some studies have suggested an increased incidence of limb-reduction defects in patients undergoing CVS; other studies have failed to confirm this association (See review: D'Alton ME, DeCherney AH. "Prenatal Diagnosis," N Engl J Med. 1993; 328:114-120; see also editorials: Platt LD, Carlson DE. N Engl J Med. 1992; 327:636-638; Lancet. 1991; 337:1513-1515; Lilford RJ. BMJ. 1991; 303:936-937).

CLOSTRIDIUM DIFFICILE TESTING

SPECIMEN: C. difficile toxin testing, must send stool, not rectal swab; deliver immediately to laboratory in container with lid; refrigerate; if necessary, forward on dry ice to outside laboratory. C. difficile culture, swab is acceptable.

Clostridium Difficile Tests: Tests used to diagnose pseudomembranous colitis due to C. difficile are given in the following Table (Fekety R, Shah AB. JAMA. 1993; 269:71-75):

Clostridium Difficile Testing
Endoscopy and Radiology
Culture
Toxin Assays
Tissue Culture Cytotoxicity Assay
ELISA
Latex Agglutination

Endoscopy and Radiology: Flexible sigmoidoscopy and colonoscopy allows rapid and accurate diagnosis of pseudomembranous colitis (PMC); 95% of patients with PMC grow C. difficile (Gebhard RL, et al. Am J Med. 1985; 78:45-48). Computed tomography of the abdomen (thickened or edematous colonic wall with pericolonic inflammation) may be suggestive of PMC.

Culture: Clostridium difficile is readily cultured on selective media. Interpretation of a positive culture is problematic; one study demonstrated that 63% of patients with positive cultures were asymptomatic carriers. Clinical disease was associated with cytotoxin production (McFarland LV, et al. N Engl J Med. 1989; 320:204-210).

Toxin Assays: Tissue culture cytotoxicity assay is the diagnostic gold standard. Stool extract is inoculated into monolayer cell culture and observed for cytopathic effect. Toxin is identified by neutralization with C. sordelli antitoxin. C. difficile produces two toxins, cytotoxin and enterotoxin. Typically, cytotoxin is identified; cytotoxin production is usually symptomatic in adults but may be asymptomatic in infants (Fekety R, Shah AB, already cited).

ELISA and Latex agglutination tests are available to identify C. difficile. These tests are not as sensitive as culture or tissue culture cytotoxicity assay, but have good specificity (>90%). These tests may be useful as a rapid screening test, but must be interpreted in clinical context and confirmatory testing should also be performed.

INTERPRETATION: Pseudomembranous colitis (PMC), secondary to antibiotic use, has proved to be caused by an overgrowth of C. difficile, which produces an exotoxin that in turn leads to an enterocolitis with diarrhea.

Many antimicrobials have been shown to be capable of causing pseudomembranous colitis including ampicillin, cephalosporins, sulfamethoxazole and trimethoprim, tetracycline, and chloramphenicol. The clinical signs and symptoms usually develop within 4 weeks of antibiotic administration. Large-volume, watery diarrhea occurs, often with fever, leukocytosis, abdominal cramps and tenderness. Proctoscopic examination reveals a pseudomembrane (similar to diphtheritic membrane), composed of exudative, raised plaques with skip areas of edematous and hyperemic mucosa (Bartlett JG. Rev Infect Dis. 1979; 1:530-539).

Nosocomial transmission of C. difficile infection is very common; in one prospective study the transmission rate was 23%. Two-thirds of these patients were asymptomatic. Person-to-person spread occurs through direct contact between patients, through environmental or instrument contamination or on the hands of hospital personnel (McFarland LV, et al. "Nosocomial Acquisition of Clostridium Difficile Infection." N Engl J Med. 1989; 320:204-210). See also Fekety R, Shah AB. "Diagnosis and Treatment of Clostridium Difficile Colitis." JAMA. 1993; 269:71-75.

CLOT LYSIS TIME

SPECIMEN: Blue (citrate) top tube with liquid anticoagulant; place tube on ice and transport to coagulation laboratory immediately. Do not freeze.

REFERENCE RANGE: Intact clot for 48 hours

METHOD: One mL of whole blood is placed into each of two 10 x 75mm test tubes; #1 tube is allowed to clot in a 37°C water bath; #2 tube is allowed to clot in the refrigerator. A normal control is tested at the same time. Observe at intervals for degeneration or disappearance of the clot.

INTERPRETATION: With excessive fibrinolysis, fragmentation of clot occurs; this test is useful only in severe hyperfibrinolysis. Variants of this test have been used to monitor thrombolytic therapy. Related tests are the diluted whole blood clot lysis test and the euglobulin lysis test which are both more sensitive and require less time. see EUGLOBULIN LYSIS TIME.

CLOT RETRACTION TIME

SPECIMEN: Red top tube. Transport to lab within one hour.

REFERENCE RANGE: Clot begins to retract 30 to 60 minutes after collection.

METHOD: Clot retraction depends on intact platelets and the presence of divalent cations. The contractile protein of platelets, thrombasthenin, may be involved in this reaction. One mL of blood is placed into each of three 10 x 75mm test tubes and placed into a 37°C waterbath. The tubes are examined after two hours. If there is retraction in any of the tubes, the result is recorded as "positive." With normal blood, the clot begins to retract 30 to 60 minutes after collection.

INTERPRETATION: Clot retraction times are measured as an indicator of platelet dysfunction. The clot retraction time is a measure of platelet function; if there is poor platelet function, clot retraction will be poor or may fail to occur. Fibrinogen and hematocrit should be normal for valid test. Clot retraction is abnormal in the following conditions:

Abnormal Clot Retraction Time
Low Platelet Count
Aspirin Therapy
Thrombasthenia (Glanzmann's Disease)
Disseminated Intravascular Coagulation (DIC)
Polycythemia
Severe Hemophiliac States
Hypofibrinogenemia

CLOT UREA SOLUBILITY

Synonym: Factor XIII Screen
SPECIMEN: Blue (citrate) top tube, filled to capacity.
REFERENCE RANGE: Negative; in the absence of factor XIII, fibrin polymers will not become cross-linked and will be dissociated by urea into soluble fibrin monomers. Reported as positive (clot dissolves) or negative.
METHOD: Citrated plasma is recalcified and the fibrin clot is suspended in 5M urea solution at 37°C for 24 hours.
INTERPRETATION: This is a simple screening test for severe factor XIII (fibrin stabilizing factor) deficiency. Factor XIII deficiency is a rare inherited coagulation disorder manifested by umbilical cord bleeding at birth, delayed resorption of subcutaneous hematomas and poor wound healing.

A positive test result indicates severe homozygous factor XIII deficiency; a negative test does not exclude mild or moderately severe factor XIII deficiency since less than 5% of normal factor XIII levels are sufficient for hemostasis.

CLOTTING TIME

Synonym: Lee-White Clotting Time; Whole Blood Clotting Time
SPECIMEN: Obtain 3 mL of blood in a plastic syringe. No anticoagulant. A clean venipuncture is required because contamination with tissue thromboplastin can even cause a severe hemophiliac to have a normal clotting time.
REFERENCE RANGE: 5-10 minutes using 13x10 mm acid washed Pyrex glass tubes. The reference ranges vary with the laboratory.
METHOD: Timing of the test begins when one mL of the blood is placed in each of the three glass tubes (13x100mm acid washed Pyrex). If the tubes are kept at 37°C, clotting time is enhanced; this may obscure minimal abnormalities. It is preferable to perform the test at room temperature and in the room where the blood is drawn; transporting the tubes causes agitation of the specimen and temperature change. Tilt tube number 1 every 30 seconds until it clots. Then, tube number 2 is tilted until it clots. Then, tube 3 is tilted until it clots. The clotting time is the time interval from the start (when one mL of blood is placed in first glass tube) to the time that blood in the third tube has clotted. Technique may vary slightly between labs.
INTERPRETATION: This test has been used to monitor heparin therapy and the whole blood clotting system. A prolonged clotting time indicates that there is a clotting abnormality; a normal clotting time does not guarantee that clotting is normal. Patient should not receive heparin therapy within 3 hours of performing the test.

The Lee-White clotting time is not recommended for evaluating heparin activity because it lacks reproducibility and is no longer performed in many laboratories. The test that is recommended for monitoring heparin therapy is the partial thromboplastin time(PTT).

The sensitivity of the Lee-White clotting time is very low for detecting deficiencies of blood coagulation factors. The level of many factors must fall below 1% of normal before the clotting time is prolonged. The Lee-White clotting time is prolonged when fibrinogen falls under 50 mg/dL, prothrombin under 30% of normal and factors VIII and IX under 2% of normal. This test is completely insensitive to platelet factor 3 deficiency.

CLOTTING TIME, ACTIVATED (see ACTIVATED CLOTTING TIME)

COAGULATION FACTORS (see FACTORS, BLOOD CLOTTING)

COAGULATION, MIXING TEST

Prolongation of a coagulation test may be due to: 1) decreased activity (levels) of a coagulation factor, or 2) the presence of a circulating anticoagulant (heparin, anti-factor antibody, inhibitory peptides, inhibitory proteins). Mixing tests are employed to determine between the two possibilities.

In a mixing test, patient plasma and a normal plasma are mixed in defined ratios (1:1, 4:1, etc). If the prolongation of the coagulation test is corrected by the addition of normal plasma, it indicates a deficiency of one of the coagulation factors and specific factors tests can be run to confirm. By contrast, if addition of normal plasma fails to correct the prolongation of the coagulation test, it is evidence of the presence of a circulating anticoagulant. This is confirmed by specific testing for the suspected anticoagulant (heparin, lupus anticoagulant, specific factor antibody, fibrin split products, etc).

COAGULATION PANEL

SPECIMEN: Recommendations regarding specimen collection, transport, and storage for blood coagulation were revised by the National Committee for Clinical Laboratory Standards (document H21-A2. 1991; Villanova, Pa.). The optimal specimen consists of a ratio of 9 parts blood to 1 part sodium citrate anticoagulant (109 or 129mmol/L, trisodium citrate). A hematocrit above 55% (polycythemia) or below 20% (severe anemia) may lead to a spurious PTT or PT. In such cases the clotting volumes may be corrected by the formula Blood Volume to be added to 0.5mL anticoagulant equals 60/100-Hct x 4.5mL. If the specimen can be transported to the laboratory within one hour, room temperature transport is adequate. If more than one hour is necessary for arrival, transport on ice is recommended, however this is controversial because of: 1) impracticality; 2) "cold activation" in glass tubes resulting in a shortened PT; 3) neutralization of heparin by release of platelet factor 4. The PTT should be measured within 2 hours of arrival at the laboratory, the PT within 4 hours.

COAGULATION TESTS:

A list of some of the more common coagulation tests and the clinical indications for these tests are given in the next Table:

Coagulation Tests and Clinical Indications

Test	Principal Clinical Indications
Partial Thromboplastin Time(PTT) [Blue (Citrate) Top Tube]	Coagulation Screen Monitor Heparin Therapy Liver Disease DIC Factor Deficiency Lupus Anticoagulant Screen
Test is also called a Partial Thromboplastin Time, Activated (APTT)	
Prothrombin Time(PT) [Blue (Citrate) Top Tube]	Coagulation Screen Monitor Oral Anticoagulants Vitamin K Deficiency Factor Deficiency Liver Disease DIC
Activated Clotting Time (Plastic Syringe)	Monitor Heparin Therapy (Cardiac Surgery)
Antithrombin III [Blue (Citrate) Top Tube]	Thrombosis or Risk of Thrombosis
Bleeding Time (Bedside Test)	Platelet Function
Clot Lysis Time [Blue (citrate) top tube]	Severe Hyperfibrinolysis
Clot Retraction Time (Red Top)	Platelet Function

(cont)

Coagulation Panel (Cont)

Coagulation Tests and Clinical Indications (Cont.)	
Clot Urea Solubility [Blue (Citrate) Top Tube]	Severe Factor XIII Deficiency
Clotting Time, Lee-White (Plastic Syringe, Bedside)	Monitor Heparin Therapy (Obsolete) Severe Factor Deficiency (Obsolete)
Euglobulin Lysis Time [Blue (Citrate) Top Tube]	Monitor Thrombolysis Therapy Evaluation of Abnormal Fibrinolysis
Factor VIII [Blue (Citrate) Top Tube]	Classic Hemophilia von Willebrand's Disease DIC
Fibrin Split Products(FSP) (Red Top)	DIC
Fibrinogen [Blue (Citrate) Top Tube]	Bleeding Tendency DIC Monitor Thrombolysis Therapy Dysfibrogenemia
Lupus Anticoagulant [Blue (Citrate) Top Tube]	Thrombosis Recurrent Abortion
Plasminogen [Blue (Citrate) Top Tube]	DIC
Platelet Aggregation [Blue (Citrate) Top Tube]	Platelet Function von Willebrand's Disease
Platelet Count [Lavender (EDTA) Top Tube]	Bleeding Tendency DIC
Protein C [Blue (Citrate) Top Tube]	Thrombosis Risk
Protein S [Blue (Citrate) Top Tube]	Thrombosis Risk
Thrombin Time [Blue (Citrate) Top Tube]	Monitor Heparin Therapy DIC Fibrinogen Function
Thrombin Time, Protamine Sulfate Corrected [Blue (Citrate) Top Tube]	Fibrinolysis Monitor Heparin Therapy
Tourniquet Test (Bedside)	Vascular Abnormality Platelet Function

COAGULATION PROFILES:

COAGULATION PROFILES	
Pre-Operative Coagulation Panel	Prothrombin Time(PT) Partial Thromboplastin Time(PTT) Platelet Count Bleeding Time
Abnormal Bleeding Panel	Prothrombin Time(PT) Partial Thromboplastin Time(PTT) Platelet Count Bleeding Time Thrombin Time Fibrinogen Tourniquet Test Specific Factor Assay

COAGULATION PROFILES (Cont.)	
Easy Bruisability Panel	Prothrombin Time(PT) Partial Thromboplastin Time(PTT) Platelet Count Bleeding Time Tourniquet Test Platelet Function Tests Adhesiveness Aggregation Peripheral Smear for Megathrombocytes
Fibrinolysis Panel	Euglobulin Lysis Thrombin Time, Protamine Sulfate Corrected Fibrin Split Products D-dimer Test
Disseminated Intravascular Coagulation(DIC)	Prothrombin Time(PT) Partial Thromboplastin Time(PTT) Fibrinogen Platelet Count Fibrin Split Products(FSP) D-dimer Test Thrombin Time Euglobulin Lysis Time Peripheral Smear
Von Willebrand's Panel	Partial Thromboplastin Time(PTT) Factor VIII Panel Factor VIII Related Antigen Bleeding Time Platelet Count Platelet Aggregation, Ristocetin See VON WILLEBRAND'S DISEASE PANEL for discussion and further tests.
Thrombotic Tendency Panel	Partial Thromboplastin Time(PTT) Antithrombin III Lupus Anticoagulant Protein C Protein S
Qualitative Platelet Disorders	Platelet Count Platelet Factor 3 Bleeding Time Tourniquet Time Platelet Function Tests Adhesiveness Aggregation Clot Retraction Time

SCREENING TESTS IN CERTAIN COMMON DISORDERS OF COAGULATION (Cartwright GE. Diagnostic Laboratory Hematology, Grune and Stratton, New York, 1968, 4th Ed., 356; Aledort LM. Diagnosis. 1984; 87-90):

Screening Tests in Certain Common Disorders of Coagulation

Disorder	Bleeding Time	Prothrombin Time(PT)	Partial Thromboplastin Time(PTT)	Thrombin Time
Hemophilia	N	N	↑	N
von Willebrand's Disease	↑	N	↑	N
Deficiency of Vitamin K Dependent Factors	N	↑	Usually ↑	Usually N
Defibrination Syndrome	N or ↑	↑	N or ↑	↑

(cont)

Coagulation Panel (Cont)

Deficiency of Vitamin-K Dependent Factors: Deficiency of the vitamin-K dependent factors, prothrombin (II), factors VII, IX and X, occurs in coumadin toxicity, liver disease, malabsorption syndromes, hemorrhagic disease of newborns and vitamin K deficiency from any cause (e.g., antibiotic therapy).

Function and Hemostatic Tests

Function	Test
Vascular Integrity	Bleeding Time Tourniquet Test
Platelets	Platelet Count; Bleeding Time; Clot Retraction; Tourniquet Test; Platelet Aggregometry
Thrombosis	Activated Partial Thromboplastin Time(PTT); Prothrombin Time(PT); Fibrinogen; Whole Blood Clot Lysis; Antithrombin III; Protein S; Protein C; Lupus Anticoagulant
Fibrinolytic Studies	Euglobulin Lysis; Fibrin Split Products; Thrombin Time; D-dimer Test

Coagulation disorders are characterized by joint, soft tissue and organ bleeding. Defects in platelet function are characterized by purpura (petechiae and ecchymoses) of the skin and hemorrhage from oral mucous membranes and other tissues.

COCAINE ("CRACK", "ROCK", "SNOW")

SPECIMEN: Random urine specimen
REFERENCE RANGE: Negative
METHOD: Cocaine is rapidly metabolized and removed from the serum within hours. Most tests detect metabolites, benzoylecgonine and ecgonine methyl ester, which persist for up to 6 days. Screening tests and approximate levels of detection are as follows: thin-layer chromatography ($\geq$ 2000ng/mL); latex agglutination or enzyme immunoassay (300-1000ng/mL); radioimmunoassay (2-20ng/mL). Confirmatory tests: gas chromatography with or without mass spectrometry; high-performance liquid chromatography.
INTERPRETATION: Cocaine is a stimulant derived from the coca bush, a plant growing in Chile, Bolivia and Peru.

Freebased cocaine is a highly purified form known as "crack" or "rock" which has a lower melting point, making it amenable to smoking. Smoking produces a more intense, short-lived euphoria compared to the intranasal route.

Medical complications associated with cocaine abuse are listed in the following Table:

Medical Complications of Cocaine Abuse

Fetus (Volpe JJ. N Engl J Med. 1992; 327:399-407): Small For Gestational Age; Microcephaly; Prematurity; Abruptio Placentae; Perinatal Death; Dependence; Central Nervous System Abnormalities (Infarction, Hemorrhage, Teratogenic Effects)

Cardiovascular (Minor RL, et al. Ann Intern Med. 1991; 115:797-806): Myocardial Ischemia; Myocardial Infarction; Hypertensive Crisis; Cardiomyopathy; Myocarditis; Aortic Dissection and Rupture; Ventricular Arrhythmias; Atrial Arrhythmias; Asystole

Cerebrovascular (Levine SR, et al. N Engl J Med. 1990; 323:699-704): Cerebral Infarcts; Subarachnoid Hemorrhage; Intracerebral Hemorrhage; Transient Ischemic Attacks; Intraventricular Hemorrhage; Seizures

Psychiatric (Gawin FH, Ellinwood EH. N Engl J Med. 1988; 318:1173-1182): Stimulant Abstinence Syndrome (Crash, Withdrawal, Extinction); Psychiatric Comorbidity

Acute cocaine poisoning occurs in "body packers" or following massive intravenous infusion; "body packers" attempt to smuggle cocaine into the country in their gastrointestinal tracts wrapped in foil or rubber devices such as condoms. The drug packages may be successfully passed with purgation or surgery. The clinical findings in acute cocaine poisoning are as follows: acute agitation; diaphoresis; ventricular dysrhythmia with hypotension; grand mal seizures leading to lactic acidosis; severe respiratory acidosis due to hypoventilation and respiratory arrest and metabolic acidosis due to the seizures.

COCCIDIOIDES ANTIBODY, SERUM AND CSF

SPECIMEN: Red top tube, separate and then freeze serum. CSF specimens: 1.0mL of spinal fluid.
REFERENCE RANGE: Negative
METHOD: Complement fixation(CF); latex agglutination(LA); immunodiffusion(ID); tube precipitin(TP); counterimmunoelectrophoresis(CIE).

Antibody (IgG) is usually quantitated by CF. Titers of 1:2 and 1:4 are found early in the disease; titers of 1:16 to 1:32 are highly suggestive and over 1:32 are indicative of disease. Increasing titers are diagnostic.

INTERPRETATION: The change with time in skin test, complement fixation test(IgG) and tube precipitin(IgM) test following infection with Coccidioides immitis is shown in the next Figure (Penn RL, et al. Arch Intern Med. 1983; 143:1215-1220):

Skin Test, Complement Fixation Test (IgG) and Tube Precipitin (IgM) Test Following Infection with Coccidioides Immitis

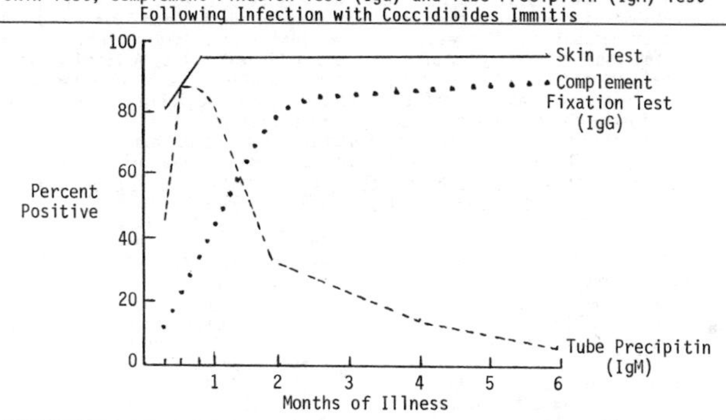

Skin Test: Positive coccioidin or spherulin skin test indicates current or prior infection.

Histopathology: Direct examination of samples demonstrating large mature spherules is diagnostic. In the absence of typical forms, immunofluorescence test for C. immitis may be useful.

IgM Tests: Immunodiffusion tube precipitin(IDTP) is sensitive and specific for acute infection. Latex agglutination(LA) is sensitive, but has a false-positive rate of 6%; positive tests must be confirmed by IDTP.

The TP(IgM) test is positive in about 90 percent of patients during the second and third weeks of illness.

IgG Tests: Immunodiffusion complement fixation(IDCF) or complement fixation(CF) may indicate recent or past infection. Titers of 1:8 are diagnostic. Titers of 1:2-1:8 may indicate early rise of IgG levels or cross reactivity; serial testing is necessary. Complement fixation titers are useful for monitoring disease progression or extension and response to therapy (Kaufman L. Clinical Infect Dis. 1992; 14(Suppl 1):S23-S29; Pappagianis D, Zimmer BL. Clin Microbiol Rev. 1990; 3:247-268).

Any titer of CF antibodies in CSF is indicative of meningeal infection.

Coccidioidomycosis is most prevalent in the Far West and Southwest of the United States and is particularly common in San Joaquin Valley ("valley fever" or "San Joaquin fever") of California.

Sixty percent of patients with coccidioidomycosis are asymptomatic. Two weeks following inhalation of the spores, the patient may develop symptoms; chest pain may be prominent. Within a day or two, a rash (usually erythema nodosum) may be seen. Infections usually involve the lungs but may progress to affect any or all organs of the body.

COLD AGGLUTININ TITER

SPECIMEN: Red top tube; keep specimen at room temperature and transport to the laboratory immediately. The blood should be incubated preferably at 37°C or alternately at room temperature and allowed to clot before the serum is separated. Do not refrigerate prior to separation of serum from red cells. Refrigerate serum.

REFERENCE RANGE: Titers less than 1:32 dilution.

METHOD: Two fold serial dilutions of serum are prepared. Add an aliquot of serum to human group O erythrocytes and incubate at 0 to 5°C for one hour. Positive test is agglutination of red blood cells. The agglutination is reversible when warmed to 37°C.

Patients with cold agglutinins develop anti-I or anti-i antibodies; these antibodies are usually of the IgM class, but may be IgG, IgA, or mixed IgM-IgG. These antibodies react with the I or i antigens on adult human Type O red cells at temperatures below 35°C, resulting in agglutination. Cold agglutinins react with Ii antigens on many types of normal and malignant cells. Antibodies may be monoclonal or polyclonal; monoclonal cold agglutinins are associated with lymphoreticular neoplasms whereas polyclonal cold agglutinins are usually associated with infection such as mycoplasma pneumonia.

Antibiotic therapy may interfere with antibody formation.

INTERPRETATION: Diseases associated with cold agglutinins are given in the following Table (Pruzanski W, Shumak KH. N Engl J Med. 1977; 297:583-589):

Diseases Associated with Cold Agglutinins

Polyclonal	Monoclonal
Mycoplasma Pneumonia	Cold Agglutinin Disease
Infectious Mononucleosis	Waldenstrom's Macroglobulinemia
Cytomegalovirus	Hodgkin Lymphoma
Listeriosis	Non-Hodgkin Lymphoma
Mumps Orchitis	Chronic Lymphocytic Leukemia
Subacute Bacterial Endocarditis	Chronic Myelogenous Leukemia
Syphilis	Kaposi's Sarcoma
Tropical Diseases (Malaria, Trypanosomiasis, Tropical Eosinophilia)	Multiple Myeloma
	Plasmacytoma
Collagen Vascular Diseases	Severe Combined Immunodeficiency (With B-lymphocyte Proliferation)
	Mycoplasma Pneumonia (Rare)

Mycoplasma Pneumoniae: A four-fold increase in antibody titer is found in 55 to 65 percent of patients with M. pneumoniae infections; 90 percent of patients with rising or falling titers are severely affected or have prolonged illness. In primary atypical pneumonia, cold agglutinins are detected one week after onset, peak at 12 to 25 days, and rapidly fall after day 30. The antibody is typically IgM directed against the I antigen. Hemolysis occurs rarely, 5-10 days following the acute infection. Culture and identification of this organism require three to four weeks. Culture is too slow and difficult to be clinically useful. Serologic testing is used to confirm the diagnosis of acute mycoplasma disease. Specific IgM and IgA levels are necessary for maximal detection of acute mycoplasma pneumonia infection - IgM alone does not detect all cases of acute infection (Sillis M. J Med Microbiol. 1990; 33:253-258). Specific IgG persists for months after mycoplasma infection.

Most cold agglutinins are autoagglutinins; these may cause problems in blood grouping or cross-matching.

Cold agglutinin disease(CAD): Cold agglutinin disease(CAD) is associated with a hemolytic process, characterized as follows: Red Cell Pathology: Clumping of red cells; apparent increase in MCV by some automated particle counters due to clumping; usually moderate normochromic normocytic anemia. Complement activation is the primary mechanism for intravascular hemolysis. Sedimentation Rate: Markedly increased at room temperature due to RBC agglutination. Other Findings in Acute Attacks of CAD: Hemoglobinuria; hemosiderinuria, haptoglobinemia; increased unconjugated bilirubin; increased lactic dehydrogenase; positive direct coombs.

Cold agglutination disease has recently been reviewed (Nydegger UE, et al. "Immunopathologic and Clinical Features of Hemolytic Anemia Due to Cold Agglutinins," Semin Hematol. 1991; 28:66-77).

COMA PANEL

The differential diagnosis of coma and stupor is given in the next Table (Nicholson DP. Med Clin of N Amer. 1983; 67:1279-1293):

Differential Diagnosis of Coma and Stupor

Central Nervous Disease:
- Intracranial Disease
- Injury
- Infection
- Subdural Hematoma

Metabolic:
- Diabetes: Hyperosmolar Coma; Diabetic Ketoacidosis
- Uremia
- Hepatic Encephalopathy
- Anoxic Encephalopathy (Include CO Poisoning and Inert Gas)
- Hypothermia
- Hypothyroidism (Myxedema Coma)

Electrolyte and Acid-Base Derangements:
- Hyponatremia
- Hypernatremia
- Hypercalcemia
- Hypo- or hyperkalemia
- Hypomagnesemia
- Respiratory Acidosis
- Respiratory Alkalosis

Drugs and Toxic Agents:
- Sedatives: Alcohol, Phencyclidine(PCP), Tricyclics and Opiates
- Metabolic Acidosis with Increased Anion Gap: Salicylates, Methyl Alcohol, Ethylene Glycol, Paraldehyde
- Enzyme Inhibitors: Heavy Metals, Organic Phosphates and Cyanide

Laboratory studies in coma of unknown cause are given in the next Table:

Laboratory Studies in Coma

Complete Blood Count(CBC)

Blood Chemistries
- Glucose
- Calcium
- Magnesium
- Liver Tests
  - AST
  - Bilirubin, Total and Direct
- Electrolytes: Sodium, Potassium, Chloride, CO_2 Content, Calculate Anion Gap
- Arterial Blood Gases: pH, PCO_2, PO_2
- Acetone
- Creatine Phosphokinase(CPK)
- Blood Urea Nitrogen(BUN)
- Creatinine
- Osmolality

Urinalysis, including Osmolality

Drug Screen (Toxicology)

Volatiles

- -

Chest X-Ray

Radiographic Examination of Any Injuries

Computed Tomography (CT) or Magnetic Resonance Imaging (MRI) for brain tumor, abscess, herniation, intracranial hemorrhage, or hydrocephalus

COMPLEMENT, C1Q BINDING ASSAY

SPECIMEN: Red top vacutainer, separate the serum and freeze immediately; ship frozen in dry ice.
REFERENCE RANGE: 0-13% Bound = Normal; 13-16% Bound = Borderline; >16% Bound = Abnormal (Scripps-BioScience Reference Laboratory).
METHOD: The Clq binding assay measures the binding sites on immune complexes available to radiolabeled Clq, a sub-unit of the first component of complement. If immune complexes are present, the complexes will bind the radioactive Clq; polyethylene glycol (PEG) is used to precipitate the complex. The percentage of total added Clq that is precipitated is calculated and expressed as "% Bound" (Scripps-BioScience Reference Laboratory)
INTERPRETATION: This is one of a number of methods used to measure circulating immune complexes. Some diseases reported to be associated with circulating immune complexes are given in the Table under IMMUNE COMPLEX PROFILE.

COMPLEMENT C3

SPECIMEN: Red top tube, separate serum; store frozen in plastic tubes. On standing at room temperature, C3 breaks down into split products which can give falsely high values.
REFERENCE RANGE: 83-177mg/dL. To convert conventional units in mg/dL to international units in g/liter, multiply conventional units by 0.01.
METHOD: Radial immunodiffusion; nephelometry.
INTERPRETATION: Measurement of C3 is useful in the diagnosis of certain disease states, especially renal, and in following the progress of a disease process. C3 is the component of complement found in highest concentration in plasma, and is the point at which the classical and the alternate pathways for complement activation converge. C3 is an acute phase reactant. C3 Complement is decreased in the serum in conditions listed in the next Table (Madaio MP, Harrington JT. N Engl J Med. 1983; 309:1299-1302):

Decreased C3 Complement in Serum
Systemic Diseases:
Systemic Lupus Erythematosus (Focal, approx. 75 percent; Diffuse, approx. 90 percent)
Subacute Bacterial Endocarditis (approx. 90 percent)
"Shunt" Nephritis (approx. 90 percent)
Cryoglobulinemia (approx. 85 percent)
Renal Diseases:
Acute Poststreptococcal Glomerulonephritis (approx. 90 percent)
Membranoproliferative Glomerulonephritis
Type I (approx. 50 to 80 percent)
Type II (approx. 80 to 90 percent)
Other:
Gram-Negative Septicemia
Fungal Diseases, such as Cryptococcal Septicemia

Conditions that cause acute nephritis, but associated with normal serum complement, are listed in the next Table (Madaio MP, Harrington JT. N Engl J Med. 1983; 309:1299-1302):

Conditions Causing Acute Nephritis but Associated with Normal Serum Complement
Systemic Diseases:
Polyarteritis Nodosa Group
Hypersensitivity Vasculitis
Wegener's Granulomatosis
Henoch-Schonlein Purpura
Goodpasture's Syndrome
Visceral Abscess
Renal Diseases:
IgG-IgA Nephropathy
Idiopathic Rapidly Progressive Glomerulonephritis
Anti-Glomerular Basement Membrane Disease
Immune-Complex Disease
Negative Immunofluorescence Findings

COMPLEMENT PROFILE, C3, C4, FACTOR B (C3PA)

SPECIMEN: Red top tube, separate serum and refrigerate serum; or lavender (purple) top tube, separate plasma and refrigerate. If the specimen is shipped, freeze and send frozen.
REFERENCE RANGE: C3, 83-177mg/dL; C4, 15-45mg/dL; factor B(C3PA), 17-40mg/dL.
METHOD: Immunodiffusion or immunoprecipitin or laser-nephelometric light scattering.
INTERPRETATION: Complement is the term for a system of sequentially reacting serum proteins which participates in pathogenic processes; the complement sequence leads to inflammatory injury. Complement pathways are illustrated in the next Figure:

Complement Pathways

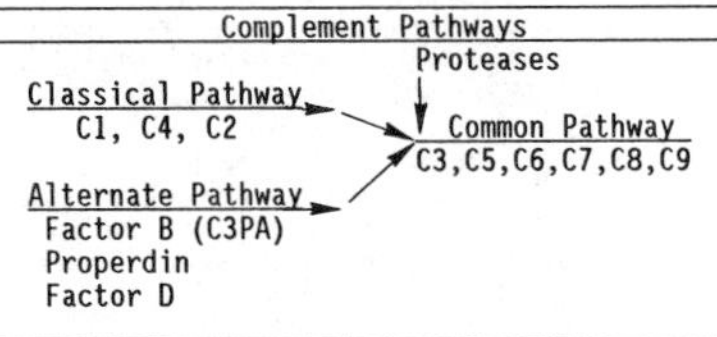

Complement may be activated through the classical pathway and/or the alternate pathway. Lowered C4 indicates activation by the classical pathway; lowered factor B (C3PA) indicates activation by the alternate pathway. The initial trigger for the classical pathway is interaction of immunoglobulin with C1q of the classical pathway. Causes of reduced serum complement, C3, C4, Factor B are given in the next Table:

Causes of Reduced Serum Complement, C3, C4, Factor B

C3	C4	C3PA	Mechanism	Diseases
↑	↑	↑	Acute phase reactants	Various infections and inflammation
↓	N	↓	Alternate pathway	Diffuse intravascular coagulation (DIC) Membranoproliferative glomerulonephritis (MPGN) (II), Gram-negative bacteremia, Snake venom, Cryptococcal sepsis, Post-Strep GN (late), Paroxysmal nocturnal hemoglobinemia (PNH)
↓	↓	N	Classical pathway	Bacterial endocarditis associated with GN, Acute GN (returns to normal), MPGN, "Immune complex disease", Falciparum malaria, Severe chronic liver disease
↓	↓	↓	Both pathways	Active SLE with renal involvement, Acute GN (returns to normal), MPGN, Post-Strep GN (later), "Immune Complex Disease"
↓	N	N	Proteolytic activation	Gram negative bacteremia with shock (prognostic value), Acute GN (returns to normal), Rheumatoid vasculitis, Temporal arteritis, Tissue damage, Hepatitis B, In vitro artifact
N	↓	N		Mixed cryoglobulinemia, Vivax malaria, Hereditary angioedema, Anaphylactoid purpura
N	N	↓		Nephrotic syndrome

Screening for complement component deficiencies and testing the integrity of the classical complement pathway may be assessed by measurement of the CH50 (See COMPLEMENT, TOTAL ACTIVITY).

COMPLEMENT, TOTAL ACTIVITY (CH50 OR CH100)

SPECIMEN: Red top tube; collect specimen on ice and separate serum from clot with minimum centrifugation time; refrigerate or freeze specimen. It is essential that serum be fresh or stored at -75°C for reliable results. Fluids other than blood, such as, pleural fluid and synovial fluid, should be treated in a similar manner, that is, maintain on ice and freeze.

REFERENCE RANGE: CH50 Test: Serum: 35-55 CH50 units/mL; synovial fluid: 20-40 CH50 units/mL; pleural fluid: 10-20 CH50 units/mL. CH100 Test: Serum: 58-114 units/mL.

METHOD: CH50 Test: This test uses sheep red blood cells coated with antibody. Addition of complement (serum or synovial fluid or pleural fluid) in the presence of Mg^{++} and Ca^{++} causes lysis of the coated cells. The amount of hemolysis is evaluated quantitatively and compared with known controls. CH100 Test: This is a radial diffusion test in agarose gel for quantitating hemolytic complement levels.

INTERPRETATION: These tests are done to detect complement deficiencies. The CH50 measures the integrity of the classical components C1-C9. It does not depend on alternate pathway components. It also does not measure the levels of the various complement components since many of these are present in excess amounts.

CONGENITAL INFECTIONS (see TORCH)

CONNECTIVE TISSUE DISEASE PANEL

Connective tissue diseases are listed in the next Table:

Connective Tissue Diseases
Rheumatoid Arthritis(RA)
Related and Associated Syndromes to R.A.
Felty's Syndrome
Sjogren's Syndrome
Juvenile Rheumatoid Arthritis (Still's Disease)
Reiter's Syndrome
Ankylosing Spondylitis (Marie-Strumpell)
Systemic Lupus Erythematosus(SLE)
Scleroderma and Progressive System Sclerosis(PSS)
CREST Syndrome
Polymyositis
Mixed Connective Tissue Disease(MCTD)

LABORATORY TESTS IN CONNECTIVE TISSUE DISEASES: The laboratory tests in connective tissue diseases may be classified as:

1) Non-immunological Laboratory Tests
2) Immunological Laboratory Tests

NON-IMMUNOLOGICAL LABORATORY TESTS: The non-immunological laboratory tests are those that are often done in screening; these tests are listed in the next Table:

Non-Immunological Laboratory Tests in Connective Tissue Diseases

Test	Remarks
Markers of Inflammation:	
Sedimentation Rate	Elevated
C-Reactive Protein	Elevated
White Blood Cell Count(WBC)	May be decreased May have lymphocytosis
Red Blood Cell Count(RBC)	Decreased
RBC Indices	Anemia, normocytic, normochromic
Platelets	Decreased
Coombs	Autoantibodies may be present
Renal Function Tests:	
BUN, Creatinine, Creatinine Clearance	May be abnormal with renal involvement

Markers of Inflammation: Fever, WBC, erythrocyte sedimentation rate(ESR), and C-reactive protein(CRP) increase with inflammation; fever, WBC, and ESR may be affected by conditions other than inflammation. However, CRP is said to be specific for inflammation and/or tissue necrosis. These patients often have a lymphocytosis. The white blood cell count may be decreased due to the presence of anti-leukocyte antibodies.

Red Blood Cell Count(RBC) and RBC Indices: The red blood cell count is often decreased possibly due to anti-erythrocyte antibodies or splenomegaly; normochromic, normocytic anemia is a common finding in connective tissue diseases.

Platelets: Platelets may be decreased due to anti-platelet antibodies or splenomegaly.

Renal Function Tests: The BUN and serum creatinine may be elevated in connective tissue diseases (except polymyositis) reflecting renal disease.

IMMUNOLOGICAL LABORATORY TESTS: The immunological laboratory tests in connective tissue diseases are listed in the next Table:

Immunological Tests in Connective Tissue Diseases
(1) Antinuclear Antibodies
(a) Fluorescence-Titer; Pattern
(b) Specific Antinuclear Antibodies
(2) Complement Profile
(3) Immune Complexes
(4) Rheumatoid Factor(RF)
(5) Cryoglobulins

(cont)

Connective Tissue Disease Panel (Cont)

RHEUMATOID ARTHRITIS: There are no laboratory tests that are specific for rheumatoid arthritis; the laboratory tests that should be considered in work-up of a patient for suspected rheumatoid arthritis are listed in the next Table:

Laboratory Profile in Rheumatoid Arthritis
Complete Blood Count(CBC): RBC; WBC; Indices; Platelets
Erythrocyte Sedimentation Rate(ESR); C-Reactive Protein
Rheumatoid Factor(RF)
Antinuclear Antibodies(ANA)
Examination of Synovial Fluid

Laboratory Findings in Rheumatoid Arthritis(RA): The laboratory findings in rheumatoid arthritis are listed in the next Table:

Laboratory Findings in Rheumatoid Arthritis(RA)

Finding	Remarks
Anemia	Normocytic, Normochromic
Lymphocytosis	25% of Patients
Erythrocyte Sedimentation Rate(ESR)	Elevated; Degree of Elevation Roughly Parallels Disease Activity
C-Reactive Protein	Elevated; Compliments the ESR
Rheumatoid Factor(RF)	Positive in 85% of Patients with Active Disease
Antinuclear Antibodies(ANA)	20% to 70% of Patients
Antibodies to Histones	20% of Patients
Synovial Fluid	Turbid, Poor Mucin Clot: Lysosomal Enzymes may Depolymerize Synovial Hyaluronate WBC's: 5000 to 20,000/cu mm, 2/3 pmn's Rheumatoid Factor(RF); Complement; Immune Complexes

Typically, RF is absent when the clinical signs and symptoms of rheumatoid arthritis first appear. Patients in which RF is demonstrable early in the course of rheumatoid arthritis have a greater risk of developing articular destruction and having sustained disabling disease. Juvenile rheumatoid arthritis is typically RF negative.

SYNDROMES RELATED TO OR ASSOCIATED WITH ARTHRITIS: Syndromes related to or associated with rheumatoid arthritis are listed in the next Table:

Syndromes and Arthritis

Condition	Remarks
(1) Felty's Syndrome	Rheumatoid Arthritis, splenomegaly and leukocytopenia; high RF titers
(2) Sjogren's Syndrome (Keratoconjunctivitis Sicca, Xerostomia)	May occur alone or associated with other connective tissue diseases. 90% occur in women, usually middle-aged. Keratoconjunctivitis sicca, pharyngitis sicca and parotid gland enlargement are the chief symptoms Heavy infiltration of lymphocytes in salivary, lacrimal and other secretory glands. Autoantibodies: RF, 90%; ANA, 70%; antisalivary, 60%-70% (Diagnostic); very high incidence of SS-A/Ro (occurs in 60-70% of patients), SS-B/La (occurs in 60% of patients).
(3) Juvenile Rheumatoid Arthritis (Still's Disease)	Pathology identical to Adult-Onset R.A. Low Frequency of RF (Seronegative RA) Three Clinical Forms: (a) Systemic - High fever, skin rash, lymphadenopathy, pleurisy, pericarditis. (b) Pauciarticular - Few large joints (c) Polyarticular - Multiple small joint involvement. Associated with HLA, B-27
(4) Reiter's Syndrome	Triad: Arthritis, Urethritis, Conjunctivitis Usually RF and ANA neg. Associated with HLA, B-27

Syndromes and Arthritis (Cont.)	
(5) Ankylosing Spondylitis (Marie-Strumpell)	Erosive sacroiliitis may progress to include entire vertebral column and may result in ankylosis and immobilization; shoulders, hips and small peripheral joints may be involved. Usually male disease. Associated with HLA, B-27

LUPUS ERYTHEMATOSUS: Lupus erythematosus is a disease of the connective tissue affecting about 500,000 people per year. Its cause is unknown. In this disease, the body forms auto-antibodies that attack healthy tissue. Criteria for diagnosis of systemic lupus erythematosus(SLE) are given in the next Table (JAMA, 1982; 48:622):

Criteria for Diagnosis of Systemic Lupus Erythematosus(SLE)

1. Malar Rash
2. Discoid Lupus
3. Photosensitivity
4. Oral Ulcers
5. Arthritis
6. Proteinuria Greater than 0.5 g/day, or Cellular Casts
7. Seizures or Psychosis
8. Pleuritis or Pericarditis
9. Hemolytic Anemia or Leukopenia or Lymphopenia or Thrombocytopenia
10. Antibody to DNA or Sm Antigen or the Presence of LE Cells or a Biologically False-Positive Serologic Test Result for Syphilis
11. Positive Fluorescence Antinuclear Antibody Test Results

A patient must have at least 4 of the 11 conditions to be classified as having SLE. The sensitivity is 96 percent and the specificity is 96 percent. Anti-nuclear antibodies are found in 99% of patients with SLE. Absence of ANA's essentially excludes a diagnosis of SLE. A peripheral (rim) pattern for antinuclear antibodies, an anti-DNA, or anti-Smith antibodies are relatively specific for SLE.

The laboratory tests results that should be considered in the work-up of patients with SLE are listed in the next Table:

Laboratory Profile in Systemic Lupus Erythematosus (SLE)

Non-Immunological Tests:
- Complete Blood Count(CBC): RBC; WBC; Indices
- Erythrocyte Sedimentation Rate(ESR)
- C-Reactive Protein
- Urinalysis
- Blood Chemistries (especially Total Protein, BUN and Creatinine)

Immunological Tests:
- Antinuclear Antibodies(ANA)
- Anti-Double-Stranded-DNA (anti-DNA)
- Anti-Smith Antibodies
- Complement Profile
- Immune Complexes

NON-IMMUNOLOGICAL TESTS: Hemolytic anemia (5%) or leucopenia (<4000/cu mm) (15%) or thrombocytopenia (<100,000/cu mm) (5%) or any combination of these may be found. Cell specific antibodies may be found in patients with lupus.

The incidence of renal disease in patients with lupus erythematosus is about 75%; the most common cause of death in these patients is renal failure. Renal lesions in lupus include focal glomerulonephritis, diffuse proliferative glomerulonephritis and membranous glomerulonephritis. Cellular casts develop in association with disease. Proteinuria occurs in about two-thirds of patients with systemic lupus erythematosus; proteinuria >3.5g/24 hr. occurs in about 20% of patients. (cont)

Connective Tissue Disease Panel (Cont)

IMMUNOLOGICAL TESTS: Antinuclear antibodies occur in lupus erythematosus. The rim or "ring" pattern is relatively specific for lupus. Antinuclear antibodies in SLE are shown in the next Table (Tan EM. Hospital Practice. 1983; 79-84):

Antinuclear Antibodies in Systemic Lupus Erythematosus(SLE)

Antibodies to:	Incidence
Native DNA	50-60% of Patients at Significant Titers
DNP	Up to 70% of Patients usually at liters of >1:10,000
Smith(Sm)	30% of Patients usually at liters of 1:40 to 1:640
Histones	60% of SLE and 95% of Drug-Induced Lupus Patients
SS-A/Ro	30-40%
SS-B/La	15%
RNP	30-40%
ANCA	<5%

Titers are by hemagglutination.

Anti-DNA: High levels of antibody to native double-stranded DNA are found in 40-60 percent of serum of patients with active system lupus erythematosus(SLE) but uncommonly in other diseases. There is good correlation between active SLE (particularly lupus nephritis) and elevated levels of anti-DNA antibodies; exacerbations of SLE are associated with a rim or a homogeneous pattern of immunofluorescence. Inactive SLE usually is accompanied by low or absent serum levels of anti-DNA antibodies. Low levels of antibody to double-stranded DNA are found in a number of other connective tissue diseases. See ANTI-DNA.

Anti-Sm Antibodies: Antibodies to Smith(Sm) antigens are considered almost specific of SLE, and they are found in 20 to 30 percent of patients with this disease. The presence of anti-Sm antibody is associated with a much higher incidence of vasculitis, resulting in peculiar visceral manifestations, which can be poorly responsive to therapy (Beufils M, et al. Am J Med. 1983; 74:201-216). There is one report of artifactual anti-HIV autoantibodies (ELISA and Western Blot, negative by PCR) induced in SLE (Jindal R, et al. N Engl J Med. 1993; 328:1281-1282). See ANTI-SMITH ANTIBODIES.

Serum Complement in Lupus Erythematosus: Serum complement levels are decreased in systemic lupus erythematosus; the components that are depressed involve both the Classical Pathway C-1, C-4, C-2, C-3 and the Common Pathway. The presence of high binding capacity of antibodies to native DNA and low C-3 was 100 percent correct in predicting the diagnosis of SLE (Weinstein A, et al. Am J Med. 1983; 74:206-216).

Other Immunological Tests in Systemic Lupus Erythematosus: Immunological tests other than those used for routine diagnostic purposes, are listed in the next Table:

Other Immunological Tests in Systemic Lupus Erythematosus
(1) L.E. Cell Phenomena (Obsolete)
(2) Rheumatoid Factor (30%)
(3) Biological False Positive - VDRL - (10-20%)
(4) Anti-Platelet Antibodies
(5) Anti-Thyroglobulin Antibodies
(6) Anti-Cytoplasmic Antibodies (Anti-Ribosomal)

Drug Induced Lupus Erythematosus: Drugs that can cause an illness with some of the features of SLE are shown in the next Table:

Drug Induced Lupus Erythematosus
Hydralazine (Apresoline) - Antihypertensive Agent
Procainamide - Used in Treatment of Cardiac Arrhythmias
Isoniazid; Practolol; Hydantoins; Chlorpromazine;
D-Penicillamine; Nitrofurantoin

There are differences between drug-induced disease and spontaneous SLE; some similarities and differences are listed in the next Table:

Similarities and Differences: Drug-Induced Disease Versus Spontaneous SLE

Systemic Lupus Erythematosus(SLE)	Drug-Induced Disease
Immune Complex Type Renal Disease is Common	Low Incidence of Immune Complex Type Renal Disease
Antibodies to DNA and Sm	Antibodies to DNA and Sm are Absent; ANA to Single-Stranded DNA and to RNP are present
Chronic - Primarily Females; Usually 20-40 Years of Age	Usually reversible - Male and Female; any Age

Procainamide: About 30% of patients who take therapeutic doses of procainamide for a prolonged time will develop a reversible lupus-like syndrome with arthralgias, myalgias, fever, pleuritis and pericarditis. Symptoms may develop as early as two weeks or as late as two years after starting treatment. The aromatic amino group of procainamide is important for the induction of procainamide lupus and acetylation of this amino group blocks this effect of procainamide (Kluger J, et al. Ann Intern Med. 1981; 95:18-23).

SCLERODERMA AND PROGRESSIVE SYSTEMIC SCLEROSIS (PSS):

The immunological findings in PSS are listed in the next Table:

Immunological Findings in PSS

Test	Remarks
Fluorescent Anti-Nuclear Antibody Patterns	Positive in almost 80% of patient but titers are usually low. Usually speckled pattern, but may be homogeneous or nucleolar; the nucleolar pattern is associated with PSS.
Specific Immunological Tests:	
Antibodies to Nucleolar RNA	Occurs in 40% to 50% of patients at titers varying from 1:100 to 1:1000 by immunofluorescence
Antibodies to Scl-70	Occurs in 20-40% of patients
Rheumatoid Factor	Present in about 25% of patients
Cryoglobulin	Immune Complex

Anti-Scl-70 antibodies (anti-topoisomerase I antibodies) are found in 40% or less of patients with systemic scleroderma. However, this antibody is relatively specific for this condition (Steen VD, et al. Arthritis Rheum. 1988; 31:196-203). Less than 20% of patients with limited scleroderma have this antibody. Approximately 3% of patients with primary biliary cirrhosis also have scleroderma.

CREST:

CREST is a mnemonic for the characteristics of this syndrome: Calcinosis cutis, Raynaud's phenomenon, Esophageal motility abnormalities, Sclerodactyly and Telangiectasia. CREST is a localized variant of scleroderma or progressive systemic sclerosis(PSS). CREST patients differ from patients with PSS in that they do not have widespread involvement and the skin changes are confined to the hands and face. Antibodies to centromeric chromatin (anti-centromere antibodies) are found in 80 to 90% of patients with the CREST syndrome. See ANTI-CENTROMERE ANTIBODIES.

POLYMYOSITIS:

The laboratory diagnosis is based on three tests; these are listed in the next Table:

Laboratory Diagnosis of Polymyositis

Test	Remarks
Muscle-Enzyme Studies	Serum CPK elevated
Electromyograph (EMG)	Involved muscles positive result
Muscle Biopsy	Pathognomonic. Initial finding is intense perivascular inflammation

Immunological Findings in Polymyositis: Some patients with polymyositis have findings similar to that observed in "connective tissue" diseases; these are given in the next Table:

Immunological Findings in Polymyositis

Finding	Remarks
Antinuclear Antibodies	Pattern variable; speckled positive in 30%
Anti-PM-1	Antibodies to PM-1 antigen occur in 50% of polymyositis and in 10% of dermatomyositis patients at low titers (Tan EM. Hospital Practice. 1983; 79-84).
Rheumatoid Factor	Positive in occasional patient

(cont)

Connective Tissue Disease Panel (Cont)

MIXED CONNECTIVE TISSUE DISEASE(MCTD):

Mixed connective tissue disease has a "mixed" clinical picture suggesting symptoms of systemic lupus erythematosus, scleroderma, and polymyositis. All of these patients have an unusually high titer of an antinuclear antibody with a specificity for a nuclear ribonucleoprotein(RNP) or anti-sn RNP antigen (anti-U1 RNP). Antibodies to RNP occur in 95%-100% of patients; there is absence of other antinuclear antibodies (Tan EM. Hospital Practice. 1983; 79-84). Patients are often responsive to corticosteroid therapy. Almost 3/4 of these patients have muscle pain, muscle tenderness and weakness with abnormal electromyograms consistent with inflammatory myositis.

The laboratory findings in MCTD are given in the next Table:

Laboratory Findings in MCTD

Finding	Remarks
Sedimentation rate	Elevated
Anemia	Moderate
Coombs Positive Hemolytic Anemia	Rare
Leukopenia	Moderate
Thrombocytopenia	Rare
CPK	Elevated when muscle disease present
Fluorescent Antinuclear Antibodies	Titers are High
Patterns	Low dilutions, mixed patterns; high dilutions, speckled pattern
RNP Antibody (anti-sn RNP)	High titers of anti-sn RNP antibody are found in MCTD (95%-100% of patients at titers >10,000) and SLE (low titer)
Rheumatoid Factor	Found in 50% of patients

COOMBS, DIRECT

Synonyms: Anti-Human Globulin Test, Direct Antiglobulin Test

SPECIMEN: Purple (EDTA) top tube, or 2 drops blood in saline micro-capillary tube; include history of recent and past pregnancy and drug therapy.

REFERENCE RANGE: Negative

METHOD: The anti-human globulin test is used to detect antibodies on red cells and is based on the principle that anti-human globulin antibodies induce agglutination of erythrocytes coated with globulins. The reaction is illustrated in the next Figure:

Coombs Test, Direct

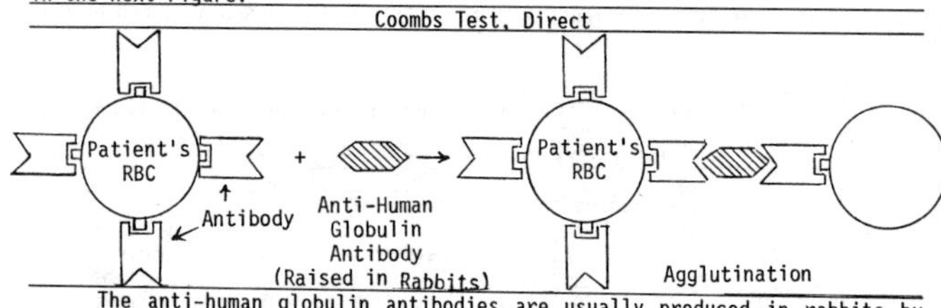

The anti-human globulin antibodies are usually produced in rabbits by immunization. These antibodies are directed against human immunoglobulins (mainly IgG) and complement (mainly C3).

INTERPRETATION: This test is used to answer the question: Are IgG and/or complement bound to red cell membrane? The direct Coombs test is positive in the following conditions:

Positive Direct Coombs Test
Hemolytic Disease of the Newborn (HDN)
Transfusion Reactions, Hemolytic
Autoimmune Hemolytic Anemias
Drug Induced: Methyl Dopa (Aldomet); Penicillin; Cephalosporins (Keflin); Quinidine, Insulin, Sulfonamides; Phenacetin
Warm Autoimmune Hemolytic Anemia
Cold Autoimmune Hemolytic Anemia
Paroxysmal Cold Hemoglobinuria

(cont)

Antibodies directed against human immunoglobulins (and occasionally complement) coating red blood cells are as follows: HDN, hemolytic transfusion reactions and antibodies induced by drugs, such as, methyl dopa (Aldomet), penicillin, cephalosporins (Keflin).

Antibodies directed against complement coating red blood cells are: Warm autoimmune hemolytic anemia; cold agglutinin disease; paroxysmal cold hemoglobinuria; antibodies induced by drugs, such as, quinidine, insulin, sulfonamides and phenacetin.

Hemolytic Disease of the Newborn (HDN): Classically, HDN is due to Rh(-) mother and Rh(+) fetus and occurs as follows: The mother, on exposure to Rh(+) fetal red blood cells, develops antibodies to the Rh(+) antigens; the antibodies pass from the maternal blood into the fetus. In the fetus, the antibodies combine with the Rh antigens of the fetal red blood cells; complement may combine with the antibody. Macrophages or lymphocytes attach to the Fc region of the immunoglobulin of the red blood cells. Sphering and then lysis of the red cell occurs; antibody-coated red cells may also be destroyed by polymorphonuclear leukocytes or by phagocytes. These effects are illustrated in the next Figure:

Lysis of Red Blood Cells in Erythroblastosis Fetalis

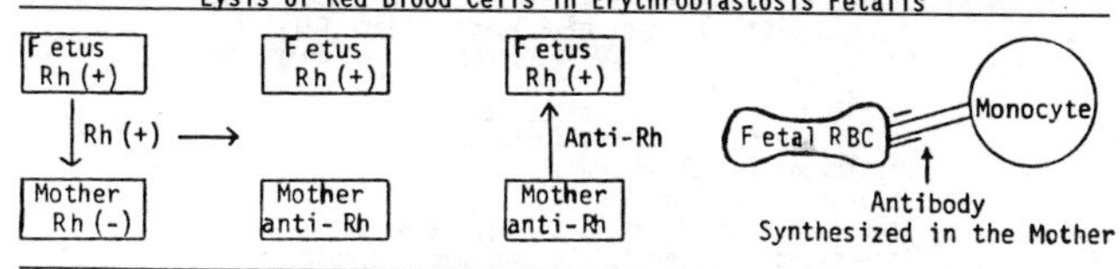

Diagnosis of erythroblastosis fetalis in the newborn is made as follows: mother, Rh(-); infant, Rh(+); Coombs test(+), the jaundice is due to Rh incompatibility.

Erythroblastosis fetalis can be prevented by passively administered antibody which can interfere with the induction of antibody formation. Other antigens can also be responsible for HDN. See HEMOLYTIC DISEASE OF THE NEWBORN (HDN) PANEL.

COOMBS, INDIRECT

Synonyms: Antibody Screen, Indirect Coombs

SPECIMEN: Red top tube, separate serum; do not use serum separator tube; include history of recent and past transfusions, pregnancy and drug therapy.

REFERENCE RANGE: Negative

METHOD: The indirect Coombs test is used to detect antibodies in sera; the sera, suspected of containing antibody, is reacted with red cells having known antigens on their surfaces. The reaction is illustrated in the next Figure:

Coombs Test, Indirect

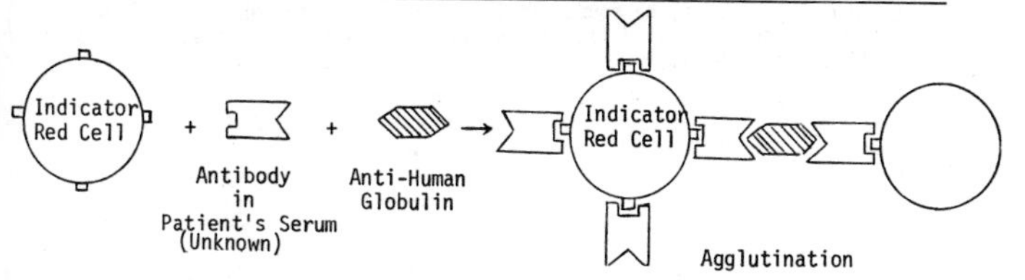

INTERPRETATION: This test is used to detect red cell antibodies in patient's serum in order to evaluate possible causes of hemolysis. This test is performed commonly when there are crossmatch/mismatch problems, transfusion reactions, and hemolytic diseases. This test is positive in the following conditions: antibody specific (previous transfusion) to red blood cell antigens, platelet antigens, leukocyte antigens, etc.; autoantibody, nonspecific in acquired hemolytic anemia.

COPPER, LIVER

SPECIMEN: Needle or wedge biopsy; refrigerate. Forward to reference laboratory in special metal-free plastic container; it is not necessary to refrigerate during shipment.
REFERENCE RANGE: 15-55mcg/gram (0.2-0.9micromol/gram) dry weight of liver. To convert conventional units in mcg/gram to international units in micromol/gram, multiply conventional units by 0.01574.
METHOD: Atomic absorption
INTERPRETATION: Copper, in liver, is increased in the conditions listed in the next Table:

Increase in Copper in the Liver
Hepatolenticular Degeneration (Wilson's Disease)
Primary Biliary Cirrhosis

In contrast to Wilson's disease, serum ceruloplasmin is elevated in primary biliary cirrhosis (Mistry P, Seymour CA, Q J Med. 1992; 82:185-196).

Characteristics of hepatolenticular degeneration (Wilson's Disease) are described (see COPPER, SERUM).

Primary Biliary Cirrhosis: The clinical features are listed in the next Table:

Clinical Features of Primary Biliary Cirrhosis
Middle-aged Females (20-80 years; peak 40-50 years)
Insidious Onset
Pruritis, Skin Pigmentation, Jaundice
Skin Xanthomata, Xanthelasmata
Hepatomegaly, Splenomegaly
Steatorrhea (Fat in Stool)
Signs and Symptoms of Portal Hypertension
Associated Diseases (Sicca Syndrome, Thyroiditis, etc.)
Death in Liver Failure

About 90 percent of the patients are women. The disease starts insidiously. Pruritis may be the presenting complaint, and usually precedes jaundice. Pruritis is usually attributed to raised bile acid concentrations. Hepatosplenomegaly is not usually a presenting finding. Steatorrhea is caused by defective bile acid excretion and thus poor fat and fat-soluble vitamin absorption leading to vitamin K deficiency and vitamin D deficiency. As the disease progresses, portal hypertension may develop.

Primary biliary cirrhosis occurs in association with other autoimmune conditions in more than 80 percent of cases, including rheumatoid arthritis, Sjögren's syndrome, scleroderma, renal tubular acidosis, autoimmune thyroiditis, and systemic lupus erythematosus.

Diagnosis of Primary Biliary Cirrhosis: Laboratory findings in primary biliary cirrhosis are those of biliary obstruction, e.g. serum alkaline phosphatase increased, serum bilirubin increased, and serum cholesterol increased. Serum IgM values are usually increased. This IgM is abnormally immunoreactive. Positive anti-mitochondrial antibodies occur in more than 90 percent of patients, but these antibodies do not appear to be the cause of primary biliary cirrhosis. Other autoantibodies may be found, including anti-nuclear, anti-smooth muscle, anti-thyroid, and rheumatoid factor (Review: Moreno-Otero R, et al. Med Clin N Amer. 1989; 73:911-929).

COPPER, SERUM

SPECIMEN: Red top tube, separate serum; refrigerate at 4°C.

REFERENCE RANGE:

AGE	Serum Copper(mcg/dL)
Newborn	10-27
Two	95-190
Ten	70-165
Young Adults	65-150
Adults	85-150

To convert conventional units in mcg/dL to international units in micromol/liter, multiply conventional units by 0.1574.

METHOD: Atomic absorption spectrophotometry

INTERPRETATION: Serum copper is decreased in the conditions listed in the next Table:

Causes of Decreased Serum Copper
Hepatolenticular Degeneration (Wilson's Disease)
Diarrhea
Malnutrition
Parenteral Nutrition with Cu Deficient Solutions
Malabsorption
Menke's Syndrome ("kinky" or "steely" hair syndrome)

Wilson's Disease (Hepatolenticular Degeneration): (Yarze JC, et al. "Wilson's Disease: Current Status," Am J Med. 1992; 92:643-654): Wilson's disease is an uncommon (1:30,000) inborn error of copper metabolism inherited as an autosomal-recessive trait. In this disease, there is excessive deposition of copper in the liver, basal ganglia, cerebral cortex, kidney, and cornea. The clinical characteristics of 51 patients with Wilson's Disease at the time of diagnosis include the following (Stremmel W, et al. "Wilson's Disease: Clinical Presentation, Treatment, and Survival," Ann Intern Med. 1991; 115:720-726):

Clinical Presentation of Wilson's Disease (%)
Neurologic Manifestations (61%)
Dysarthria (51%); Tremor (47); Writing Difficulties (35); Ataxia (31); Asthenia (31); Hypersalivation (12); Psychiatric Symptoms (12); Nervousness (10); Headache (8); Hypomimia (8); Dizziness (4); Seizures (2)
Abdominal Manifestations (67%)
Hepatomegaly (49%); Splenomegaly (49); Abdominal Pain (41); Jaundice (14); Ascites (14); Gynecomastia (12); Esophageal Varices (10); Diarrhea (8); Vomiting (4)
Hematologic Manifestations (31%)
Thrombocytopenia (22%); Hemolysis (10); Anemia (8); Leukopenia (8)
Other Signs and Symptoms
Kayser-Fleischer Rings (67%); Arthralgia (10); Gigantism (10); Bone Deformities (4); Keratitis (4); Hypoparathyroidism (rare)(Carpenter TO, et al. N Engl J Med. 1983; 309:873-877).

Symptomatic patients usually show pigmented cornea rings and the presence of this ring is an important aid in diagnosis. The Kayser-Fleischer ring is a golden-brown, greenish, or brownish-yellow, or bronze discoloration in the zone of Descemet's membrane of the cornea.

The defect in Wilson's disease may be a lysosomal defect in hepatocytes leading to decreased biliary copper excretion. Almost all the circulating copper is normally bound to ceruloplasmin; the bound copper is known as the indirect-reacting copper. The remaining copper (2% to 5%) is loosely bound to serum albumin; this is known as the direct reacting fraction of serum copper. Patients with Wilson's disease have decreased total copper in their serum primarily because of the decrease in indirect reacting copper. This reflects a deficient level of ceruloplasmin. However, there is an increase in direct reacting copper which is loosely bound; this loosely bound copper is readily dissociated and deposited in tissue. (cont)

Copper Serum (Cont)

The laboratory diagnosis of Wilson's disease should be suspected in patients below the age of 30 who have chronic idiopathic hepatitis. The biochemical findings in Wilson's disease are listed in the next Table:

Biochemical Findings in Wilson's Disease
Decreased Ceruloplasmin in Serum
Decreased Total Copper in Serum
Decreased Indirect Reacting Copper
Increased Direct Reacting Copper (free serum copper)
Increased Copper in Urine
Increased Copper Deposited in Liver

The diagnosis of Wilson's disease is aided by the demonstration of decreased levels of the enzyme, ceruloplasmin, and copper in the serum and increased levels of copper in the urine.

A serum ceruloplasmin level of less than 20mg/dL, serum copper level less than 80mcg/dL, and a urinary copper level in excess of 100mcg/24 hour urine is compatible with the diagnosis of Wilson's disease (Stremmel W, et al, already cited).

Copper is increased in the serum in the conditions listed in the next Table:

Causes of Increased Serum Copper
Spraying of Grapes with a Copper Sulfate Fungicide (Bordeau Mixture)
Dialysis Secondary to Use of Copper Tubing
Ingestion of Copper Sulfate
Copper in Drinking Water
Acute and Chronic Diseases
Malignant Diseases
Hemochromatosis
Primary Biliary Cirrhosis
Thyrotoxicosis
Various Infections
Patients on Oral Contraceptives
Patients Taking Estrogens
Pregnancy

COPPER, URINE

SPECIMEN: Collect 24 hour urine in acid washed-containers. Use plastic containers (borosilicate, polyethylene or polypropylene); add 10% HCl solution to the container and allow to "soak" for 10 minutes; rinse with five volumes of tap water and then five volumes of deionized or distilled water. The patient should urinate at 8:00 A.M. and the urine is discarded. Then, urine is collected for 24 hours including the next day 8:00 A.M. specimen. Indicate 24 hour volume. A 50mL aliquot is used for analysis.

REFERENCE RANGE: <40mcg/24 hour collection. To convert conventional units in mcg/24 hour to international units in micromol/day, multiply conventional units by 0.01574.

METHOD: Atomic absorption spectroscopy

INTERPRETATION: Copper is increased in the urine in the following conditions:

Causes of Increased Urine Copper
Wilson's Disease (Hepatolenticular Degeneration)
Primary Biliary Cirrhosis
Nephrotic Syndrome

A discussion of Wilson's disease is given under COPPER, SERUM.

A discussion of Primary Biliary Cirrhosis is given under COPPER, LIVER.

CORTICOTROPIN-RELEASING HORMONE (CRH) STIMULATION TEST **(see ADRENOCORTICOTROPIC HORMONE (ACTH): CORTICOTROPIN-RELEASING HORMONE (CRH) STIMULATION TEST**

CORTISOL, PLASMA

Synonyms: Compound F, Hydrocortisone
SPECIMEN: Green top tube (lithium heparin), separate plasma and freeze plasma, or red top tube, separate serum and freeze serum.
REFERENCE RANGE: A.M.: 7 to 25mcg/dL; P.M.: 2 to 9mcg/dL. To convert conventional units in mcg/dL to international units in nmol/liter, multiply conventional units by 27.59.
METHOD: RIA
INTERPRETATION: Serum or plasma cortisol levels are measured to diagnose disorders of the adrenal cortex, either states of adrenal hormone excess, adrenal hormone deficiency, increased or decreased pituitary function, or congenital abnormalities of steroid metabolism. The change in plasma cortisol in adrenal disorders is given in the next Table:

Change in Plasma Cortisol in Adrenal Disorders

Disorder	Plasma Cortisol
Cushing's Syndrome	Increased
Adrenal Insufficiency	Decreased
Secondary Adrenal Insufficiency	Decreased
Congenital Adrenal Hyperplasia	Decreased

Isolated measurement of a single cortisol value by itself is insufficient to diagnose Cushing's disease (hypercorticism due to excess ACTH production by, pituitary adenoma) or Cushing's syndrome (hypercorticism from a variety of causes). Mild hypercortisolism may be seen in pseudo-Cushing's states associated with a variety of disorders including severe obesity, alcoholism, depression, stress, renal failure, anorexia/bulimia, and glucocorticoid receptor resistance. The laboratory tests which may be used to distinguish between Cushing's syndrome and pseudo-Cushing's states are given in: Yanovski JA, et al. JAMA. 1993; 269:2232-2238. See also: ADRENAL CORTEX PANEL.
Diurnal Variation: In Cushing's syndrome there is loss of diurnal variation of plasma cortisol; elevated levels of plasma cortisol are maintained around the clock. Blood should be drawn at 8:00 A.M. and 4:00 P.M. to evaluate diurnal variation. The normal diurnal variation of plasma cortisol and the change in Cushing's syndrome are shown in the next Figure:

Normal Diurnal Variation of Plasma Cortisol and Change in Cushing's Syndrome

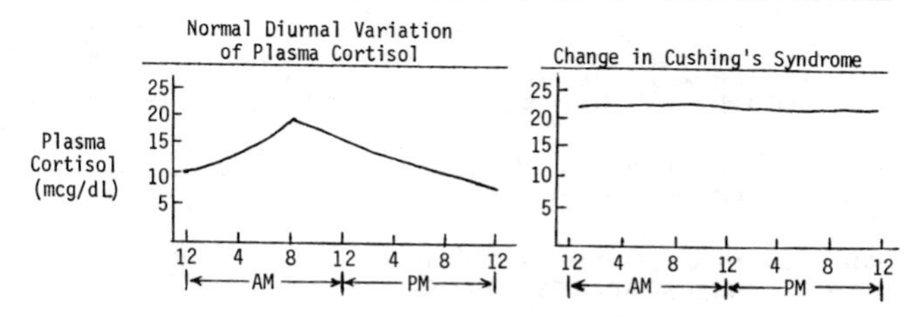

However, even a loss of diurnal variation in cortisol levels is not as sensitive in diagnosing Cushing's syndrome as other tests. Measurement of **urinary free cortisol** or an AM cortisol level following **Dexamethasone suppression** are more specific tests. Other tests performed to access adrenal function and to diagnose Cushing's syndrome such as 17-hydroxy corticosteroids, 17-ketosteroids and 17-ketogenic steroids are becoming replaced by more specific and less cumbersome assays.
Dexamethasone Suppression Test for Diagnosis of Cushing's Syndrome: Dexamethasone is a fluorinated steroid that has about 30 times the potency of cortisol. Dexamethasone, lmg, is taken orally at midnight and a fasting blood cortisol is obtained at 8:00 A.M.; a value of blood cortisol greater than 5mcg/dL is compatible with Cushing's syndrome. A negative test (<5mcg/dL) is found in less than 2 percent of the cases of Cushing's syndrome (Crapo L. Metabolism. 1979; 28:955-977). However, abnormal results (>5mcg/dL) are often found in patients under stress for any reason, including at least 30 percent of patients in a hospital who are acutely ill (Connolly CK, et al. Br Med J. 1968; 2:665-667).

(cont)

Cortisol Plasma (Cont)

ACTH Infusion Test: An ACTH infusion test is done to measure the functional reserve of the adrenal cortex. The manner in which the test may be done is as follows (Melby JC. N Engl J Med. 1971; 285:735-739): Obtain a blood specimen for cortisol assay; inject 0.25mg of alpha, 1-24, corticotropin (synthetic unit of ACTH), I.V.; obtain blood specimen at 0, 30, 60, 90 and 120 minutes for cortisol assay. The results are given in the next Figure (Melby, 1971):

Effect of ACTH Injected Intravenously on Plasma Cortisol Concentrations in Normal Subjects and Patients with Addison's Disease (Melby, 1971)

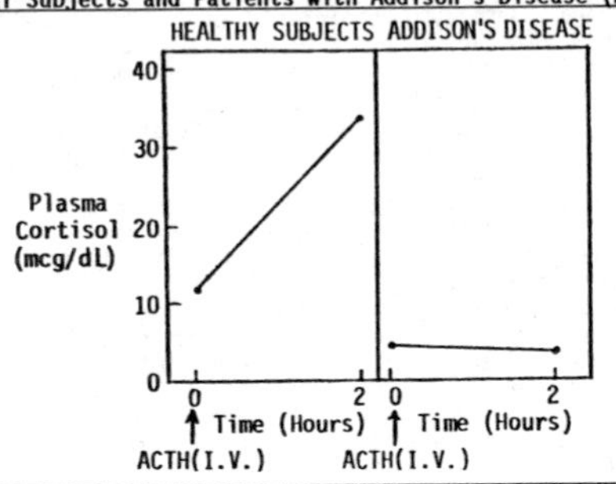

CORTISOL, URINE, FREE (UFC)

SPECIMEN: 20mL aliquot of 24 hour urine containing 10mL glacial acetic acid.
REFERENCE RANGE: 4 months-10 years: 35-176mcg/g creatinine; 11-20 years: 1-44mcg/g creatinine; adult: 13-60mcg/g creatinine.
METHOD: RIA
INTERPRETATION: Urinary free cortisol is the non-conjugated cortisol excreted in the urine; it reflects the concentration of the free cortisol in the serum, that is, cortisol not bound to transcortin. This test is used for initial screening in hospitalized patients suspected of having Cushing's syndrome. (See Hospital Practice, pgs. 37-43, June, 1983).

Measurement of UFC has largely replaced measurement of 17-hydroxy corticosteroids (17-OHCS) or 17-ketogenic steroids (17-KGS). Chromogenic materials do not interfere with it.

COUNTERIMMUNOELECTROPHORESIS(CIE)
(see BACTERIAL MENINGITIS ANTIGENS)

"C"-PEPTIDE, INSULIN (CONNECTING PEPTIDE, INSULIN)

SPECIMEN: Red top tube, separate serum; place serum in a separate plastic vial and freeze immediately.
REFERENCE RANGE: Children: 0.4-2.3ng/mL (fasting). Adult: 0.5-2.5ng/mL (fasting); 2.2-6.5ng/mL (2 hour postprandial). To convert from ng/mL to picomoles/mL, divide ng/mL by 3.56 (or multiply by 0.28). To convert from picomoles/mL to ng/mL, multiply picomoles/ml by 3.56.
METHOD: RIA
INTERPRETATION: The "C" peptide is the peptide that connects the A and B chains of insulin as illustrated in the next Figure:

Structure of Proinsulin

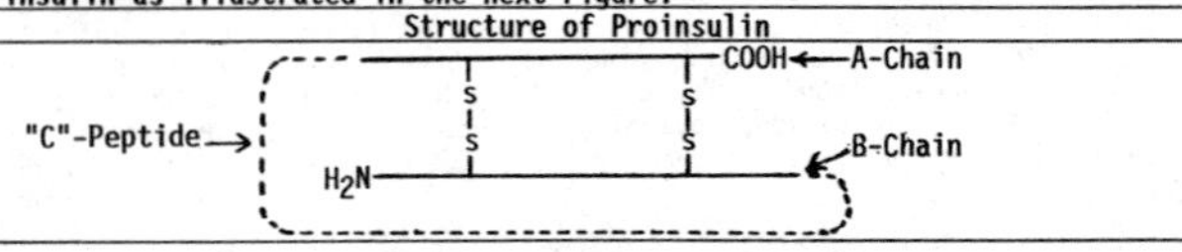

Assay of "C" peptide is done for the following reasons:
(1) Differential diagnosis of insulinoma versus factitious hypoglycemia: In insulinoma, "C" peptide and insulin are increased while in factitious hypoglycemia due to insulin injection, insulin levels are increased and "C" peptide levels, which reflect endogenous insulin secretion, are decreased. However, factitious hypoglycemia due to ingestion of drugs which promote insulin secretion will result in an elevated "C"-peptide. Urine tests for these drugs are necessary to establish the diagnosis in such a situation.
(2) Insulin Secretion in Diabetics: Almost all diabetics who receive non-human insulin develop serum antibodies after several months. Since commercial RIA insulin determinations employ heterologous antibodies, immunoassay of insulin in these individuals is virtually useless. Anti-insulin antibodies have also been reported in a large number of patients receiving preparations of the newer human recombinant insulin. The measurement of C-peptide is the best way to assess beta cell function. Since C-peptide and insulin are secreted in equimolar amounts, quantitation of C-peptide reflects insulin secretion; C-peptide does not contain antigenic determinants shared with the administered insulin; proinsulin does. Since the half-life of C-peptide is over three times longer than insulin, (14 minutes versus 4 minutes), normally there is more C-peptide than insulin in the blood.

Patients with insulin-dependent diabetes mellitus (type I) usually have no or very low concentrations of C-peptide; patients with non-insulin dependent diabetes mellitus (type II) tend to have normal or elevated C-peptide levels.

C-REACTIVE PROTEIN (CRP)

SPECIMEN: Red top tube, separate serum.
REFERENCE RANGE: None detected
METHOD: Latex agglutination, radial immunodiffusion (RID), electroimmunodiffusion (EID), nephelometry, enzyme-linked immunosorbent assays (ELISA).
INTERPRETATION: Measurement of C-reactive protein (CRP) is useful in non-specific screening for inflammatory and infectious disorders. CRP levels are sometime useful in discriminating between differential diagnoses and in monitoring the course of a disease process. CRP levels complement ESR and plasma viscosity measurements in response to inflammation. However, they may be particularly useful in the first 24 hours of the disease process, a time when the ESR and plasma viscosity are still normal. The "C" of CRP was derived from studies which demonstrated that serum from acutely ill individuals contained a substance that precipitated with the C-polysaccharide of the cell wall of pneumococci. CRP is synthesized in the liver in response to increased IL-1. Causes of an elevation in serum levels of "C"-reactive protein are listed in the next Table (Gewurz H. Hosp Pract. 1982; 67-81):

Uses and Causes of Elevation of Serum "C"-Reactive Protein
Screening for Organic Disease
General screening aid for inflammatory diseases, infections, and neoplastic diseases or tissue injury
Detection and Evaluation of Inflammatory Disorders
Rheumatoid arthritis
Seronegative arthritides (e.g., Reiter's syndrome)
Rheumatic fever
Vasculitic syndromes (e.g., hypersensitivity vasculitis)
Inflammatory bowel disease
Detection and Management of Infections
Neonatal infections
Postoperative infections
Intercurrent infections in leukemia
Bacterial infections in systemic lupus erythematosus
Pyelonephritis
Detection and Evaluation of Tissue Injury and Neoplasia
Myocardial infarction and embolism
Transplant rejection
Certain tumors (e.g., Burkitt's lymphoma)

(cont)

C-Reactive Protein (Cont)

Uses and Causes of Elevation of Serum "C"-Reactive Protein (Cont)

Aid in Differential Diagnosis
- Systemic lupus erythematosus versus rheumatoid arthritis and other arthritides (<SLE)
- Crohn's disease versus ulcerative colitis (< Crohn's)
- Pyelonephritis versus cystitis (> pyelonephritis)
- Bacterial versus viral infections (> bacterial)
- Acute bronchitis versus asthma (< asthma)

CRP is an acute reacting protein which may increase dramatically (up to a thousandfold) in inflammatory conditions; it is nonspecific. CRP levels follow the course of the acute phase. It can be used as a supplement to or complement to erythrocyte sedimentation rate (ESR) studies. The conditions that elevate ESR usually cause elevation of CRP but CRP levels more closely approximate the degree of ongoing tissue damage. CRP responds quickly to inflammation (6-10 hrs); it has a short half-life (5-7 hrs). CRP was found to be elevated in the serum of patients with bacterial meningitis but not in patients with viral meningitis or meningoencephalitis. CRP may be useful in monitoring the course of bacterial meningitis; levels characteristically return to normal within seven days in bacterial meningitis in the absence of complications (Peltola HO. Lancet. 1982; 1:980). The CRP has been used to distinguish between a variety of bacterial and viral disorders in children (Stanley TV, et al. NZ Med J. 1991; 104:138-139).

The rapid change in the serum level of CRP before and following cholecystectomy is shown in the next Figure (Gewurz H. Hosp Pract. 1982; 67-81):

Change in Serum Level of CRP Before and Following Cholecystectomy

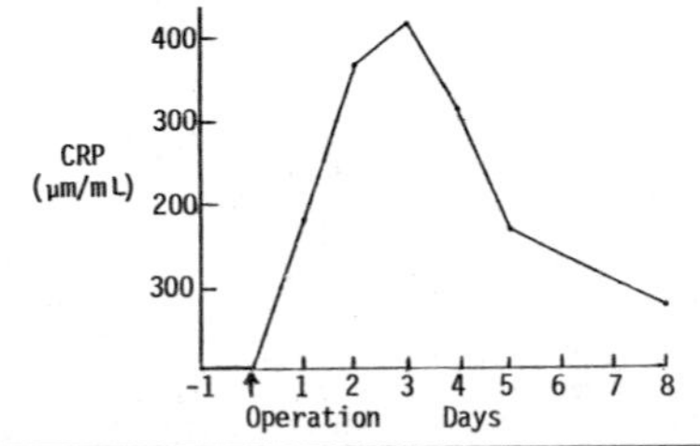

CRP in Acute Pyelonephritis and Acute Cystitis: CRP is increased in patients with pyelonephritis and normal in patients with cystitis as shown in the next Figure (Morley JJ, Kushner I. Ann N.Y. Acad Sci. 1982; 389:406-418):

Serum C-Reactive Protein (CRP) in Acute Pyelonephritis and Acute Cystitis

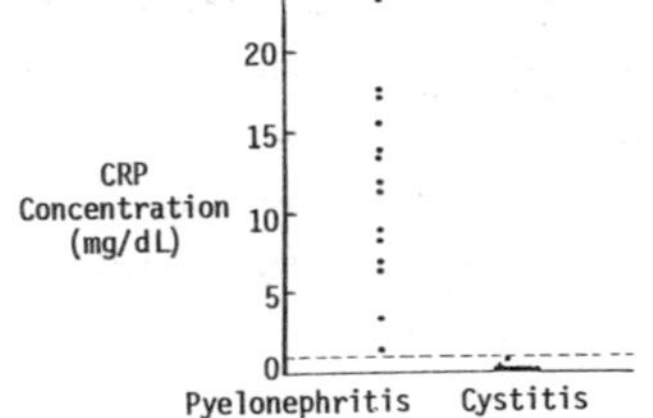

Serum CRP levels in patients with pyelonephritis are greater than 1mg/dL; serum CRP levels in patients with cystitis are less than 1mg/dL.

Children with pyelonephritis had serum levels of CRP greater than 2.5mg/dL; all children with cystitis had serum levels of CRP less than 1mg/dL (Jodal V, et al. Acta Pediatr Scand. 1975; 64:201-208).

Prediction of Chorioamnionitis in Premature Rupture of Membranes: The CRP has a sensitivity of 88 percent and a specificity of 96 percent for predicting chorioamnionitis (Hawrylyshyn P, et al. Am J Obstet Gynecol. 1983; 147:240-246).

CREATINE PHOSPHOKINASE (CPK)

SPECIMEN: Red top tube, separate serum; freeze if analysis delayed.
REFERENCE RANGE: Newborn: Up to 10x Adult Value, depending on trauma at birth; 4 Days: 3x Adult Value; >6 Weeks: Adult Value; Adult Male: 24-195 U/liter; Adult Female: 24-170 U/liter.
METHOD: CPK is assayed kinetically using two consecutive kinase reactions followed by a dehydrogenase reaction.
INTERPRETATION: CPK is usually measured because of suspected myocardial injury or infarction. It may also be assayed in the workup of suspected muscle disease. CPK is found in relatively high concentration in three tissues, the heart, skeletal muscle and the brain; conditions affecting these tissues that are associated with elevated CPK are listed in the next Table:

Conditions Affecting Heart and Skeletal Muscle Associated with Elevated Serum Creatine Phosphokinase

Heart
Acute Myocardial Infarction
Acute Myocarditis
Carbon Monoxide Poisoning

Skeletal Muscle	
Vigorous Exercise	Amyotrophic Lateral Sclerosis
External Cardioversion	Trauma
Tonic-Clonic Seizures	Surgery
Intramuscular Injections	Alcoholism
Skeletal Muscle Diseases; e.g., Polymyositis; Dermatomyositis; Rhabdomyolysis Muscular Dystrophies of all Types; Post-Polio Syndrome; Mycotoxic Drugs (Lovastatin)	Uremia
	Reye's Syndrome
	Rocky Mountain Spotted Fever
	Diabetic Ketoacidosis
	Head Injury
Malignant Hyperpyrexia (Hyperthermia)	Hypothyroidism
Severe Shock	Some Malignancies
Hypothermia	Labor and Delivery
Cerebrovascular Accidents	

CPK is elevated in the serum 3-4 hours following myocardial infarction, reaches its peak of activity between 18 and 30 hours and is followed by a sharp drop to normal levels by the third to fourth day. An intramuscular injection or injury prior to obtaining the serum may result in the elevation of the serum CPK; approximately one-third of patients develop elevation of serum CPK after intramuscular injection; elevation varies up to 10 times the upper limit of normal. Hypothyroidism is associated with a persistent CPK elevation without a CK-MB band. In some instance it may present as a cardiovascular event (LeMar HJ, et al. Am J Med. 1991; 91:549-552).

Serial CPK levels following thrombolytic therapy have been used to indicate reperfusion. Samples were drawn every 4 hours. If the CPK peaked "early", at or less than 12 hours post therapy, this was taken as indicating successful reperfusion (Hohnloser SH, et al. J Am Coll Cardiol. 1991; 18:44-49). However, other studies have claimed that these tests are not sufficiently accurate to reliably distinguish between reperfused and non-reperfused patients (Bosker HA, et al. Br Heart J. 1992; 67:150-154).

This enzyme is raised in progressive muscular dystrophy and other primary diseases (dystrophies, not atrophies) of skeletal muscle. It had been used to aid in screening of female carriers of muscular dystrophy, however, DNA genetic studies are more sensitive and accurate (Multicenter Study Group. JAMA. 1992; 267:2609-2615). Also, it may be elevated in cerebral infarcts. The enzyme is not elevated in liver disease or in pulmonary infarction. CPK of brain origin (CPK-BB) does not ordinarily appear in the serum even in patients with cerebrovascular accidents or head injury. It is CPK-MM of skeletal muscle origin that appears in the serum in these conditions. Occasionally, it may be necessary to distinguish the origin of elevated CPK, e.g., heart vs. skeletal muscle. This can be done by serum CPK isoenzyme studies. CPK tends to be decreased in hyperthyroidism. Measurement of CPK has been proposed to distinguish seizures from syncope in patients complaining of loss of consciousness (Libman MD, et al. J Gen Intern Med. 1991; 6:408-412).

CREATINE PHOSPHOKINASE ISOENZYMES (CPK ISOENZYMES)

SPECIMEN: Red top tube, separate serum; freeze if assay is not done within 48 hours. However, if LDH isoenzymes are desired on the same specimen, do not freeze, or save an unfrozen aliquot for LDH studies. The specimen should be rejected if there is gross hemolysis.

REFERENCE RANGE: There are more than 30 different commercial assays for CPK-MB determination. Each assay has it's own reference range. Values may be reported as a percentage of total or in absolute amounts. Representative values: CPK-MM: 94-100%; CPK-MB (Heart): 0-6%; CPK-BB (Brain): 0%; Normal: <10U/Liter; Borderline: 10-25U/Liter; Infarct: >25U/Liter; Normal <5.6ng/mL.

METHOD: Electrophoresis; immunochemical/kinetic. Historically, electrophoresis has been the usual method by which CPK-MB determinations have been made, and remains the "gold standard" to which other methods are compared. Immunochemical techniques are more useful for rapid testing and are more sensitive, but are less specific. It is estimated that approximately one-third of U.S. laboratories now perform CPK-MB by immunochemical assays. Much of the non-specificity of the immunochemical assay is related to its increased sensitivity.

INTERPRETATION: CPK isoenzymes are most often done to document an acute myocardial infarction. Optimally, a minimum of three serum specimens should be obtained for analysis for CPK-MB at admission and at 12 and 24 hours after onset of symptoms of acute myocardial infarction. Negative results of analysis for CPK-MB in samples obtained before 12 hours or after 24 hours should not be used to exclude the diagnosis of acute myocardial infarction. A total CPK value within the normal range is not a reliable screening test to exclude analysis of CPK-MB (Irvin RG, et al. Arch Intern Med. 1980; 140:329-334; Wagner GS. Arch Intern Med. 1980; 140:317-319; Fisher ML, et al. JAMA. 1983; 249:393-394).

Immunoinhibition kinetic assays may give elevated results and should be confirmed by electrophoresis or immunoassay.

Following acute myocardial infarction, CPK-MB begins to rise in 2 to 4 hours, reaches its peak in 12 to 24 hours and returns to normal in 24 to 36 hours. The change is CPK-MB is summarized in the next Table:

Change in CPK-MB Following Acute Myocardial Infarction

Beginning of Increase	Maximum	Return to Normal
2-4 hours	12 - 24 hours	24 - 36 hours

Irvin et al., already cited.

The causes of elevation of CPK-MB are given in the next Figure:

Causes of Elevation of CPK-MB

Pattern	Causes
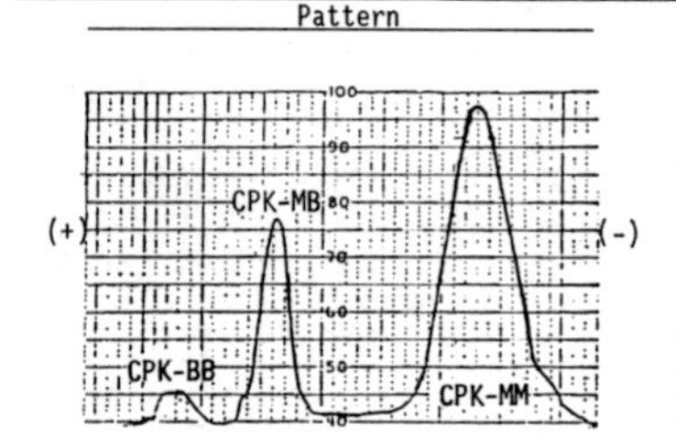	Acute Myocardial Infarction Myocarditis Acute Pericarditis with Myocardial Involvement Polymyositis Reye's Syndrome Malignant Hyperthermia Carbon Monoxide Poisoning Duchenne's Muscular Dystrophy Cardiomyopathy of Alcoholism Hypothyroidism Acute Myocarditis Cardiac Trauma Peripartum Period Cocaine Abuse Pulmonary Embolism

CPK-MB elevation is sometimes seen following pulmonary embolism. In one study, 8% of a series of patient with pulmonary embolism had an elevation of CPK-MB. The majority of these had echocardiograms demonstrating evidence of right ventricular infarction (Adams JE, et al. Chest. 1992; 101:1203-1206). Ventricular tachycardia alone should not raise the CPK or CPK-MB. External cardioversion will raise both CPK and CPK-MB. Right ventricular temporary electrode pacing will raise the CPK but not CPK-MB (O'Neill PG, et al. Am Heart J. 1991; 122:709-714).

There are efforts underway to increase the diagnostic sensitivity of CPK-MB isoenzyme and to shorten the time interval between onset of symptoms and time the CPK-MB test becomes diagnostically useful. In one approach, serial blood samples are taken at intervals (1-3 hours) and the rate of rise in CPK-MB determined.

Using such a protocol, studies indicate that patients with chest pain due to myocardial infarct can be differentiated from those with symptoms due to other causes (Marin MM, et al. Am Heart J. 1992; 123:354-361; Collinson PO, et al. J Clin Pathol. 1989; 42:1126-1131; Leung FY, et al. Ann Clin Biochem. 1991; 28:78-82).

Another approach to increase the sensitivity and early diagnostic capability of CPK isoenzyme studies is to assay for CPK isoforms. CPK-MB exists in tissues with a carboxy-terminal lysine. This tissue form is termed CPK-MB-2. Following release into the blood, blood proteases remove the carboxy terminal lysine to yield isoform CPK-MB-1. An increase in the proportion of isoform CPK-MB-2 to isoform CPK-MB-1 indicates recent myocardial injury. This may be reported in absolute quantities or as a ratio: CPK-MB-2/CPK-MB-1 or CPK-MB-2/total CPK. Changes in CPK-MB-2 can be detected earlier than changes in the usual CPK-MB (Prager NA, et al. J Am Coll Cardiol. 1992; 20:414-419; Puleo PR, et al. Circulation. 1990; 82:759-764).

CPK-MB is increased in conditions other than acute myocardial infarction. CPK-MB is elevated in the serum of patients with pericarditis when there is a concomitant underlying myocarditis (Marmor A, et al. Arch Intern Med. 1979; 139:819-820). About 15 to 20 percent of patients with polymyositis have involvement of cardiac muscle and increase in CPK-MB. Practically all patients with Duchenne's muscular dystrophy have elevation of CPK-MB and CPK-MM. Elevation of total CPK and CPK-MB may be seen in cocaine abuse (Schwartz JG, McAfee B. J Fam Pract. 1987; 6:27-39). An elevation of CPK and CPK-MB may be seen in marathon runners (Ohman EM, et al. Brit Med J. 1982; 285:1523-1526), however in one study the mean CPK-MB value was only 8.3% of the total CPK (Seigel A, et al. JAMA. 1981; 246:2049-2051). Ectopic production of CPK-MB was found in a patient with metastatic carcinoma of the colon (Annesley TM, et al. Am J Clin Pathol. 1983; 79:255-259). Elevation of CPK-MB is found in approximately 7-20% of patients on the first postpartum day following vaginal delivery (Leiserowitz GS, et al. J Reprod Med. 1992; 37:910-916; Satin AJ, et al. Am J Perinatal. 1992; 9:452-455).

Macro-CPK is formed from an immune complex consisting of autoantibody to CPK-MM and CPK-BB (macro-CK1) or to mitochondrial CPK (macro-CK2). On electrophoresis, macro-CPK usually migrates between CPK-MM and CPK-MB, and occasionally anodally to CPK-MM or with CPK-MM or CPK-BB. Macro-CPK leads to incorrect results when CPK-MB is separated by column, e.g., DuPont ACA (Pudek MR, et al. Clin Chem. 1982; 28:1400). Macro CPK can frequently interfere with immunoinhibition-type assays to CPK-MB (Wu AHB. Lab Man. 1985; 23(1):44-50; Wu AHB, Bowers GN. Clin Chem. 1982; 28:2017-2021).

Elevation of serum CPK-BB is uncommon; conditions associated with elevation of serum CPK-BB include infarction of the colon, Reye's syndrome, malignant hyperthermia, renal dialysis patients, certain malignancies (prostate, GI tract), biliary atresia, severe shock, hypothermia, cerebrovascular accidents and massive brain injury. It is reported to be raised in 90 percent of patients with metastatic carcinoma of the prostate (Silverman, et al. Clin Chem. 1979; 25:1432-1435).

CREATINE, SERUM

SPECIMEN: Red top tube, separate serum; freeze if analysis delayed
REFERENCE RANGE: Male: 0.17-0.5mg/dL; Female: 0.35-0.93mg/dL. Normal levels are higher in children and during pregnancy. To convert conventional units in mg/dL to international units in micromol/liter, multiply conventional units by 76.25.
METHOD: Creatine is converted to creatinine in acid solution; then creatinine is measured by the Jaffe reaction. The difference in the creatinine content before and after treatment with acid gives the creatine content.
INTERPRETATION: Creatine is formed primarily in the liver and is transported to the muscles where it is phosphorylated; creatine phosphate acts as a storage depot for muscle energy. Muscle contains 98 percent of the total body creatine pool, 60-70 percent is in the form of creatine phosphate. The muscle enzyme, creatine phosphokinase(CPK), catalyzes the reaction between creatine and creatine phosphate.

$$\text{Creatine} + \text{P} \underset{}{\overset{\text{CPK}}{\rightleftharpoons}} \text{Creatine Phosphate}$$

Creatine is filtered and actively reabsorbed by the kidney (Perrone RD, et al. Clin Chem. 1992; 38:1933-1953).

Creatine is increased in the conditions listed in the next Table:

Increased Serum Creatine
Skeletal Muscle Necrosis or Atrophy:
Trauma
Muscular Dystrophies
Poliomyelitis
Amyotrophic Lateral Sclerosis
Dermatomyositis
Myasthenia Gravis
Starvation
Endocrine Disorders:
Hyperthyroidism
Acromegaly
Other Conditions:
Acute Rheumatoid Arthritis
Systemic Lupus Erythematosus
Leukemia
Infections
Burns
Myocardial Infarction

This test has been largely replaced by CPK measurements to reflect skeletal muscle necrosis and dystrophies.

CREATINE, URINE

SPECIMEN: 24 hour urine, preserve with thymol or toluene; record 24 hour volume; obtain aliquot and freeze for prolonged storage.
REFERENCE RANGE: Male: 0-40mg/day; Female: 0-80mg/day; increased in pregnancy. To convert conventional units in mg/day to international units in micromol/day, multiply conventional units by 7.625.
METHOD: See SERUM CREATINE.
INTERPRETATION: Creatinuria is found in myopathies, myocardial infarction, infections, starvation, impaired carbohydrate metabolism and hyperthyroidism; See CREATINE, SERUM.

CREATININE CLEARANCE

SPECIMEN: Red top tube, separate serum and determine creatinine. Obtain 24 hour urine, record volume and measure urine creatinine. Instruct patient on 24 hour urine collection as follows: Void at 8:00 A.M. and discard specimen. Collect all urine during the next 24 hours including the 8:00 A.M. specimen the next morning. Keep specimen on ice during collection. Transport specimen to laboratory immediately. Measure 24 hour urine volume.

REFERENCE RANGE: Adult Male: 85-125mL/min; Adult Female: 75-115mL/min. Pediatric values for creatinine clearance are given in the following Table (Harriet Lane Handbook, 13th edition, 1993, 300):

Pediatric Values for Creatinine Clearance

Age		Creatinine Clearance
Newborn <24 hrs (27-43 weeks)		1.07 $\pm$ 0.12 mL/min/kg
Prematures >24 hrs		Unknown
5-7 days		50.6 $\pm$ 5.8mL/min/1.73M^2
1-2 months		64.6 $\pm$ 5.8
3-4 months		85.8 $\pm$ 4.8
5-8 months		87.7 $\pm$ 11.9
9-12 months		86.9 $\pm$ 8.5
12 months - adult	(Male)	124.0 $\pm$ 25.8
	(Female)	108.8 $\pm$ 13.5
Adults -	(Male)	105.0 $\pm$ 13.9
	(Female)	95.4 $\pm$ 8.0

METHOD: Alkaline picrate (Jaffé); Enzymatic

INTERPRETATION: Creatinine Clearance Calculated: Creatinine clearance reflects glomerular filtration rate. The equation for the calculation of creatinine clearance is as follows:

$$\text{Creatinine Clearance } \frac{mL}{min} = \frac{\text{Urine Volume (mL/min) x Urine Creatinine}}{\text{Serum Creatinine}} \times \frac{1.73}{A}$$

where A = body surface area.

The relationship between plasma creatinine and creatinine clearance is illustrated in the next Figure (Davidsohn I, Henry JB. Clinical Diagnosis by Laboratory Methods. Saunders. 1979, 140):

Relationship of Plasma Creatinine to Creatinine Clearance

Serum Creatinine (mg/dL)
8
4
2
1
15 30 60 120
Creatinine Clearance

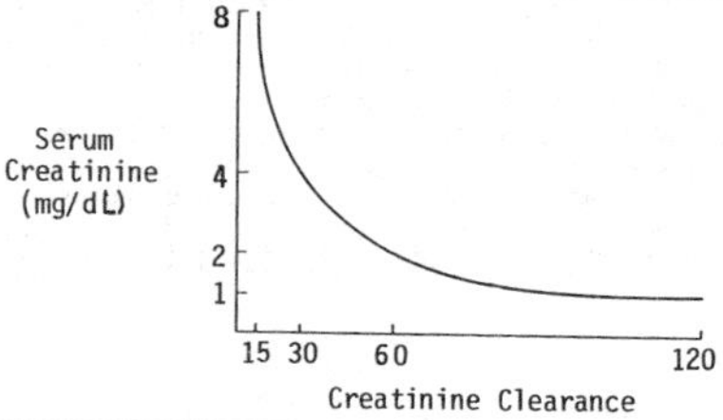

Serum creatinine increases as creatinine clearance decreases.

From this Figure, it can be seen that the serum creatinine level may be within normal range when the glomerular filtrate rate is significantly reduced. Normal creatinine clearance is 90mL/min to 130mL/min. Stages of renal failure may be arbitrarily related to creatinine clearance as follows: renal impairment, >50mL/min; renal insufficiency, 20mL/min to 50mL/min; renal failure, <20mL/min and as shown in the next Table:

Stages of Renal Failure and Creatinine Clearance

Stage	Creatinine Clearance
Renal Impairment	$\geq$50mL/min
Renal Insufficiency	20-50mL/min
Renal Failure, Uremia	5-20mL/min

The value of creatinine clearance is particularly useful when the dose of a drug depends on glomerular filtrate. (cont)

Creatinine Clearance (Cont)

Creatinine Clearance Estimates: Creatinine clearance may be estimated from the serum creatinine level alone by use of the following formula:

$$\text{Creatinine Clearance (Male)} = \frac{\text{wt } (140 - \text{Age})}{72 \text{ (Serum Creatinine)}}$$

$$\text{Creatinine Clearance (Female)} = \frac{\text{wt } (140 - \text{Age})}{72 \text{ (Serum Creatinine)}} \times 0.85$$

Serum creatinine may not reflect creatinine clearance or glomerular filtration rate. Use of equations is based on basic assumptions that creatinine values are at steady state (production equals excretion) and proportional to muscle mass, which is estimated from age, sex, and weight. Creatinine clearance is overestimated or underestimated by the conditions listed in the following Table (Perrone RD, et al. Clin Chem. 1992; 38:1933-1953):

Overestimation or Underestimation of CrCl by Serum Creatinine

CrCl Overestimated	CrCl Underestimated
Reduced Creatinine Production	Increased Creatinine Production
Cirrhosis	Weight-Lifting
Muscle Wasting	Anabolic Steroids
Malnutrition	Increased Creatinine Excretion
Reduced Creatinine Excretion	High Protein Diet
Vegetarian Diet	
Overestimation of Muscle Mass	
Obesity	
Edema	

In addition, drugs and illnesses that affect the creatinine assay itself may alter creatinine clearance estimates (see CREATININE, SERUM). When creatinine is collected for creatinine assay, errors in timing and incomplete specimen collection may lead to inaccurate estimates of creatinine clearance. Furthermore, drugs such as trimethoprim and cimetidine and disease states such as congestive heart failure, diabetes and dehydration may lead to decreased urinary creatinine and underestimation of glomerular filtration rate.

In pregnancy, there is an increase in renal blood flow and glomerular filtration rate (GFR). The normal values for BUN and serum creatinine are significantly lower; a serum creatinine concentration of 1.2mg/dL indicates an approximate 50 percent reduction in GFR in pregnant patient.

A two hour creatinine clearance test has been found to be useful in intensive care units when urine was collected using an indwelling Foley catheter. Accurate timing during the specimen collection is absolutely necessary (Wilson RF, Soullier G. Crit Care Med. 1980; 8:281-284).

Review: See Perrone RD, et al. Clin Chem. 1992; 38:1933-1953.

CREATININE, SERUM

SPECIMEN: Red top tube, separate serum.
REFERENCE RANGE: Varies with age and sex (Harriet Lane Handbook, 13th edition, 1993; 92):

Normal Values for Serum Creatinine

Age	Normal Range (mg/dL)
Newborn	0.3-1.0
Infant	0.2-0.4
Child	0.3-0.7
Adolescent	0.5-1.0
Adult	
Male	0.6-1.3
Female	0.5-1.2

To convert conventional units in mg/dL to international units in micromol/liter, multiply conventional units by 88.40.
METHOD: Alkaline picrate (Jaffé reaction) or Enzymatic Method
INTERPRETATION: Serum creatinine is increased in renal disease, muscle necrosis and hypovolemia.

Serum creatinine is useful in evaluation of glomerular function. However, the serum level is not sensitive to early renal damage and responds more slowly than the BUN to hemodialysis during treatment of renal failure. The serum creatinine together with serum BUN is used to differentiate pre-renal, renal and post-renal (obstructive) azotemia since an elevated urea with only slight to moderate elevation of creatinine suggests pre-renal or post-renal azotemia. The BUN/creatinine ratio in different conditions is shown in the next Table:

BUN/Creatinine Ratio

State	Ratio
Normal	10/1
Dehydration	15 to 20/1
Prerenal/Postrenal	> 10/1
Renal Disease	10/1 (BUN, Cr rise proportionately)

Excretion fraction of filtered sodium is the best test to differentiate prerenal from renal azotemia.

Creatinine is formed in muscles from creatine and creatine phosphate. Creatine is formed primarily in the liver and is transported to the muscles where it is phosphorylated; creatine phosphate acts as a storage depot for muscle energy. The muscle enzyme, creatine phosphokinase(CPK) catalyzes the reaction between creatine phosphate and creatine.

$$\text{Creatine phosphate} \xrightarrow{\text{CPK}} \text{Creatine} + \text{P}$$

The metabolic end product of creatine metabolism is creatinine, a cyclic anhydride of creatine. Determination of creatinine concentrations in serum is the most commonly used clinical method for measuring glomerular filtration rate. The plasma creatinine concentration is increased when the glomerular filtration is decreased. Serum creatinine varies with subjects age, body weight and sex.

Creatinine clearance may be estimated from the serum creatinine level alone by use of the following formula:

$$\text{Creatinine Clearance (Male)} = \frac{\text{Wt } (140 - \text{age})}{72 \text{ (serum creatinine)}}$$

$$\text{Creatinine Clearance (Female)} = \frac{\text{Wt } (140 - \text{age})}{72 \text{ (serum creatinine}} \times 0.85$$

This formula is based on calculations for normal males without renal or liver disease; wt is in kilograms, age in years, and serum creatinine in mg/dL. Creatinine clearance for females is based on the assumption that women have 15% less muscle mass then men.

Serum creatinine is low in subjects with relatively small muscle mass, cachectic patients, amputees, patients with muscle disease and some infants and children and older persons; older persons have a decreased muscle mass and a decreased rate of creatinine production. A serum creatinine level that would usually be considered normal does not rule out the presence of impaired renal function. (cont)

Creatinine, Serum (Cont)

Acute Renal Failure: The daily production rate of creatinine is approximately 15 to 30 mg per kg; with complete renal shut-down, the serum level will rise at a rate of 1 to 2 mg/dL per day. If the rate of rise is less, residual renal function is present. If the rate of rise is greater than 3 mg/dL per day, there is muscle disease, such as rhabdomyolysis or severe catabolism (Goldstein M. Med Clin N Amer. 1983; 67:1325-1341). The use of creatinine measurement in acute renal failure is subject to significant limitations. The rate of increase and final concentration of creatinine will depend on the severity of renal injury, rate of recovery of renal function, creatinine production, volume of distribution of creatinine, and extrarenal creatinine excretion.

Drug Interference in Creatinine Assays: Some cephalosporins, notably cefoxitin, cephalothin, cefazolin and cefamandole, falsely elevated creatinine values in the assay commonly used by hospital laboratories, that is, the alkaline picrate (Jaffe reaction); this interference is observed in all alkaline picrate methods including the widely used Beckman's kinetic Jaffe reaction (Steinback G, et al. Clin Chem. 1983; 29:1700-1701). At peak cefoxitin levels, this elevation may be 1.5 to 8.5 times the true serum creatinine. In patients with normal renal function, the serum creatinine assay is reliable two to four hours after the dose; in mild to moderate renal failure, six to eight hours. In severe renal failure, the serum creatinine determination is unreliable. To prevent interference during treatment with these cephalosporins, order serum creatinine assays to be drawn immediately prior to the dose. Creatinine clearance measurements are unreliable under all conditions while patients are receiving cefoxitin and the alkaline picrate method is used.

The enzymatic method for the determination of creatinine is not influenced by Cefoxitin (Steinbach G, et al. Clin Chem. 1983; 29:1700-1701). The antifungal agent 5-fluorocytosine (flucytosine) falsely elevates creatinine measured by this method by as much as 6mg/dL.

Diabetic Ketoacidosis and Creatinine: Falsely elevated serum creatinine values occur in diabetic ketoacidosis when the alkaline picrate reaction is used for assay of creatinine (Nanji AA, Campbell DJ. Clin Biochem. 1981; 14:91-93).

References:

Duarte CG, Preuss HG. Clin Lab Med. 1993; 13:33-52.
Perrone RD, et al. Clin Chem. 1992; 38:1933-1953.
Lyman JL. Emerg Med Clin N Amer. 1986; 4:223-233.

CRYOGLOBULINS, QUALITATIVE

SPECIMEN: Red top tube; ideally, at least 20mL of blood should be collected (to enhance detection of small amount of cryoprecipitate). Patients should be fasting since lipids interfere with the test by precipitating in the cold. Blood should be collected in tubes in warm water and transported to the laboratory immediately.

REFERENCE RANGE: Negative

METHOD: The specimen is allowed to clot at 37°C for one hour and separated in a warm centrifuge. Serum is removed and stored at 4°C for 3-7 days. The serum is examined daily for precipitate (Bloch KJ. N Engl J Med. 1992; 327:1521-1522). Positive cryglobulins are analyzed by immunoelectrophoresis.

INTERPRETATION: Cryoglobulins are serum globulins that precipitate at lower temperatures and redissolve upon warming to 37°C. The proteins in cryoglobulinemias are immunoglobulins and other proteins. Immunoproliferative, proliferative and autoimmune disorders are the most frequent diseases associated with cryoglobulinemia. Brouet JC, et al. (Am J Med. 1974; 57:775), classified patients with cryoglobulins into three types:

Type I - Isolated monoclonal immunoglobulins including IgG, IgM, IgA, and Bence-Jones.

Type II - Mixed cryoglobulins with a monoclonal component possessing antibody activity towards polyclonal IgG; combinations included IgM-IgG; IgG-IgG and IgA-IgG.

Type III - Mixed polyclonal cryoglobulins are composed of one or more classes of polyclonal immunoglobulins and sometimes non-immunoglobulin molecules such as complement or lipoprotein.

Laboratory Findings: Some laboratory findings in mixed cryoglobulinemia are given in the next Table (Gorevic PD. Am J Med. 1980; 69:287-308):

Laboratory Findings in Mixed Cryoglobulinemia
Increased Erythrocyte Sedimentation Rate(ESR)
Rheumatoid Factor
Serum Protein Electrophoresis: Increased Gamma Globulin
Increased Immunoglobulins(IgM or IgA or IgG)
Decreased Serum Complement(C4, C2, C3, CH50, CH100)
Anemia (Hematocrit <35%)
Liver Function Test Abnormalities: Elevated Serum Alkaline Phosphatase; Elevated Serum Transaminases (AST, ALT)
Proteinuria and Hematuria

Type I cryoglobulins are associated with multiple myeloma, Waldenström's macroglobulinemia and other lymphoproliferative disorders; rheumatoid factor activity is not associated.

Types II and III cryoglobulins are associated with viral, bacterial and parasitic infections, autoimmune disorders (systemic lupus erythematosus, Sjögren's syndrome, rheumatoid arthritis, and scleroderma) lymphoproliferative disorders, glomerulonephritis, and chronic liver disease. The term "essential mixed cryoglobulinemia" is used to refer to patients with primary Sjögren's syndrome and mixed cryoglobulinemia; other cases are classified as secondary mixed cryoglobulinemia (Bloch KJ, already cited). Hepatitis C virus RNA or antibodies has been found in a high percent of patients with Type II or Type III cryoglobulinemia (Angello V, et al. N Engl J Med. 1992; 327:1490-1495).

Clinical manifestations of Type I and some cases of Type II cryoglobulinemia include vascular obstruction causing distal ulceration and necrosis. Other cases of Type II cryoglobulinemia and most cases of Type III are manifest by acute recurrent purpura, Raynaud's phenomenon, arthralgias, glomerulonephritis, and neuropathy (Bloch KJ, already cited; Brouet J-C, et al. Am J Med. 1974; 57:775-788).

CRYPTOCOCCAL ANTIGEN TITER

SPECIMEN: Cerebrospinal fluid (CSF); or red top tube, separate serum; or urine.
REFERENCE RANGE: Negative
METHOD: Latex agglutination(LA)
INTERPRETATION: Latex agglutination testing is more than 90% sensitive for cryptococcal infection. Fluid samples should be boiled to eliminate rheumatoid factor which may cross react and controls to identify nonspecific agglutination reactions. False-positives occur with other infections (e.g. Trichosporon beigelii)(Perfect JR. Infect Dis Clin N Amer. 1989; 3:77-102).

A negative test does not exclude a diagnosis of cryptococcus; it may be necessary to examine more than one specimen. Occasionally patients who are serologically negative for cryptococcal antigen will have cutaneous lesions, sputums, or well-isolated pulmonary nodules that yield positive C. neoformans cultures. In cases of proven cryptococcal meningitis, LA has detected cryptococcal antigen in 91 percent of CSF and 99 percent of the serum samples (Chuck SL, Sande MA. N Engl J Med. 1989; 321:794-799). Prognostic significance of cryptococcal antigen and use of this test as an indication of response to therapy has yielded conflicting results.

Crytococcosis: Cryptococcus neoformans is by far the most common cryptococcal species responsible for human infection. The major environmental sources of cryptococcus are bird excreta and Eucalyptus camaldulensis (river red gum tree)(Ellis DH, Pfeiffer TJ. Lancet. 1990; 336:923-925).

Infection due to C. neoformans occurs in 7 percent of patients with AIDS.

Cryptococcus neoformans usually enters by the pulmonary route; it is the most common fungus to involve the CSF. Meningitis is the most common clinical manifestation of cryptococcal infection in patients with AIDS, occurring in 84 percent of patients. Meningeal cryptococcosis is a common initial manifestation of AIDS (Chuck SL, Sande MA, already cited).

Other Diagnostic Tests in Cryptococcosis: Cryptococci may be cultured from blood or other body fluid over 1 to 2 weeks. Histopathology may show encapsulated organisms. Cryptococcus neoformans antibody testing is not useful in the diagnosis or prognosis of cryptococcal infection (Penn RL, et al. Arch Intern Med. 1983; 143:1215-1220).

CULTURES

Recommended number of cultures, to detect or rule out the presence of a pathogen, is shown in the next Table (Fluornoy DJ. Clinical Microbiology Newsletter. 1982; No. 7, 4:50-51):

Culture Guidelines

Specimen	Organisms	Recommended Number of Cultures
Sputum (not saliva)	Acid Fast Bacilli	A series of 3-6 consecutive early a.m. specimens, one/day; duplicate specimens collected on same day are pooled.
	Bacteria	A series of 2 consecutive early a.m. specimens, one/day
	Fungi	Same as acid-fast bacilli
Blood	Bacteria	A total of 3-4 blood culture bottle sets (2 bottles/set, 5mL blood/bottle) collected in a 24 hr. period, 3 different venipunctures at different sites, preferably at 1 hr. intervals; accuracy is greater than 90%.
Urine	Bacteria	Women: 2 consecutive early a.m. clean-catch, midstream specimens, one/day; Men: 1 early a.m. clean-catch, midstream specimen; accuracy greater than 90%.
Stool	Bacteria	A series of 3 consecutive specimens, one/day
	Ova and Parasites	A series of 3 specimens collected 2-3 days apart, one/day; if these tests are negative, catharsis or sigmoidoscopy may be necessary.
Wound	Bacteria	1-2 specimens

If the pathogen is found in the first of a series of cultures, it may not be necessary to complete the series.

CYANIDE, BLOOD

SPECIMEN: Lavender(EDTA) top tube or grey (oxalate, fluoride) top tube. Use whole blood as specimen.
REFERENCE RANGE: ≤ 0.1-0.2 mcg/mL (mg/L); Potential toxic, >0.5 mcg/mL; Metabolic disturbances, >1.0 mcg/mL; Lethal, >3.0 mcg/mL. To convert conventional units in mcg/mL to international units in micromol/liter, multiply conventional units by 38.5.
METHOD: Colorimetric
INTERPRETATION: Cyanide acts by blocking cellular respiration by inactivating cytochrome oxidase. Oxygen cannot be utilized and venous blood may appear bright red. Cyanide causes a high anion gap metabolic acidosis due to the accumulation of lactate, and a narrow arteriovenous oxygen difference due to the blocked cellular oxygen utilization.

Well publicized events of cyanide poisoning have occurred. In 1978, mass poisoning occurred in Jonestown, Guyana, following ingestion of grape Kool-Aid mixed with cyanide. In 1982, cyanide was placed into the over-the-counter capsules of Tylenol and several people died following ingestion of these capsules. Fires involving synthetic polymers may result in toxic levels of cyanide in firemen and other people who may be exposed. Laetrile can yield hydrogen cyanide which may cause toxicity or death. Cyanide poisoning has been reported in children who ingest acetonitrile-containing cosmetics such as sculptured nail remover (Caravati EM, Litovitz TL. JAMA. 1988; 260:3470-3473).

Nitroprusside administration can cause cyanide poisoning and death. Infusion rates as low as 2 micrograms/kg/min can lead to potentially lethal cyanide levels; infusion rates of 10 micrograms/kg/min should not last more than 10 minutes. Cyanide level depends on rate of administration, total dose, hepatic function, renal function, and thiosulfate availability. Coadministration of nitroprusside with thiosulfate prevents significant rises in circulating cyanide concentrations. Each 100mg of sodium nitroprusside should be mixed with 1 gram of sodium thiosulfate.

References: Robin ED, McCauley R. Chest. 1992; 102:1842-1845; Curry SC, Arnold-Capell P. Crit Care Clin. 1991; 7:555-581; Holland DJ. Emerg Nursing. 1983; 9:138-140.

CYCLIC AMP, PLASMA, URINE, NEPHROGENOUS

SPECIMEN: Plasma and urinary cyclic AMP are measured; Nephrogenous cyclic AMP is calculated. Plasma Specimen: Lavender(EDTA) top tube; transport specimen to laboratory immediately; separate plasma and freeze immediately. Use this specimen for determination of plasma cAMP and creatinine. Do not perform determination of cyclic AMP, nephrogenous if the glomerular filtration rate is less than 25mL/min. Urinary Specimen: Timed two hour urine specimen; plastic container with no preservatives; measure volume and then freeze the urine specimen immediately. Use this specimen for determination of urinary cAMP and creatinine.

REFERENCE RANGE: Various references ranges have been reported (see Madvig P, et al. J Clin Endocrinol Metab. 1984; 58:480-487).

METHOD: RIA

INTERPRETATION: Nephrogenous cyclic AMP is calculated as follows:

Total Urinary cAMP (nmol/100mL GF) = Serum Creatinine (mg/dL) x Urinary cAMP (nmol/mg Creatinine)

Nephrogenous cAMP (nmol/100mL GF) = Total Urinary cAMP (nmol/100mL GF) - Plasma cAMP (nmol/dL); GF = Glomerular Filtrate

The cAMP found in the urine of normal persons is derived from two sources: 50% to 75% is derived by glomerular filtration of plasma and the remaining 25% to 50% is synthesized in the renal cortex of the kidney and excreted by renal tubular cells; this is illustrated in the next Figure (Broadus AE. Nephron. 1979; 23:136-141):

Renal Clearance and Sources of cAMP in Human Urine

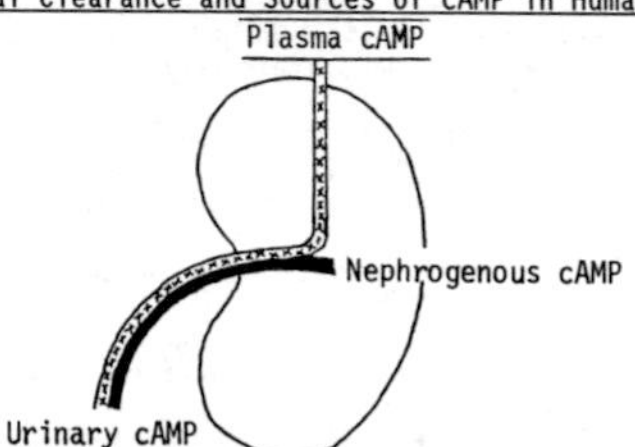

Parathyroid hormone stimulates adenyl cyclase in the renal cortex, thus converting ATP to cAMP. Measurement of urinary cyclic AMP is an index of parathyroid function.

Causes of increased and decreased renal cAMP are given in the next Table:

Causes of Increased and Decreased Renal cAMP

Increased Renal cAMP	Decreased Renal cAMP
Primary Hyperparathyroidism	Chronic Hypoparathyroidism
Some Patients with Vitamin D Deficiency and Osteomalacia	Pseudohypoparathyroidism
Patients with Calcium Urolithiasis and Hypercalciuria	Some Patients with Hypercalcemia other than Primary Hyperparathyroidism
Some Patients with Malignancy-Associated Hypercalcemia	Some Patients with Malignancy-Associated Hypercalcemia

(cont)

Cyclic AMP (Cont)

Vigorous exercise elevates plasma and urinary cAMP. Nephrogenous cAMP rises with age.

The assay of nephrogenous cAMP in response to intravenous or oral calcium administration has been used to diagnose patients with subtle primary hyperparathyroidism or intermittent hypercalcemia (Broadus AE. Nephron. 1979; 23:136-141). There is significant overlap between normal and hyperparathyroid patients which limits the clinical utility of this test. Hypercalcemia and elevated parathyroid hormone levels most accurately predicts primary hyperparathyroidism (Madvig P, et al., already cited).

Measurement of plasma and urinary cAMP after intravenous PTH is a reliable and sensitive test for the diagnosis of pseudohypoparathyroidism. Urinary cAMP is useful for the diagnosis of secondary hyperparathyroidism in patients with jejunoileal bypass. Nephrogenous cAMP may also be useful for predicting the risk of recurrent renal stone formation following a single renal stone.

Reference: Review of cAMP in clinical practice: Thode J. "Ionized Calcium and Cyclic AMP in Plasma and Urine. Biochemical Evaluation in Calcium Metabolic Disease," Scand J Clin Lab Invest Suppl. 1990; 197:1-45.

CYCLOSPORINE

SPECIMEN: Variable; purple top tube (whole blood) or red top tube (serum).

REFERENCE RANGE: Depends on the assay. Whole blood (HPLC) trough therapeutic: 100-300ng/mL (microgram/liter). Area under curve (AUC) monitoring involves 5 to 7 cyclosporine levels per dosing interval; this method may be superior to trough level testing, but is not in general use (Grevel J, et al. Ther Drug Monitor. 1989; 11:246-248).

METHODS: High Pressure Liquid Chromatography (HPLC); Radioimmunoassay (RIA); Fluorescence Polarization Immunoassay (FPIA); Affinity Column-Mediated Immunoassay (ACMIA); Enzyme-Multiplied Immunoassay Technique (EMIT); Chemiluminescence Immunoassay (CLI). (See reviews: Napoli KL, Kahan BD. Clin Lab Med. 1991; 11:671-691; Kivisto KT. Clin Pharmacokinet. 1992; 23:173-190).

INTERPRETATION: Cyclosporine is widely used as an immunosuppressant, with the ability to prolong graft survival by inhibiting delayed hypersensitivity without bone marrow suppression. It is used for the prevention of rejection in human kidney, liver, heart, heart-lung, and bone marrow transplantation.

Cyclosporine is a cyclic non-polar polypeptide consisting of 11 amino acids and is of fungal origin.

Cyclosporine selectively inhibits T-helper cell function and number. The drug depletes medullary thymocytes and splenic T-lymphocytes. There is a relative sparing of T-suppressor cells.

Cyclosporine reversibly inhibits the macrophage T-lymphocyte activation and impairs the production of interleukin-2, a growth factor essential for T-cell proliferation and differentiation into cytotoxic T-cells. The drug does not inhibit mature, proliferating T-cells.

Cyclosporine suppresses primary and, to a lesser degree, secondary antibody responses to T-cell-dependent antigens. The inhibition of antibody production to T-cell dependent antigens is a consequence of impaired T-helper cell function, which is required for B-cell growth and differentiation (Review: Kahan BD. N Engl J Med. 1989; 321:1725-1738).

Cyclosporine is administered orally or by IV infusion. Factors that affect the pharmacokinetics of cyclosporine are given in the following Table (Lindholm A. Ther Drug Monitor. 1991; 13:465-477; Rodighiero V. Clin Pharmacokinet. 1989; 16:27-37):

Factors Affecting the Pharmacokinetics of Cyclosporine
Absorption
Time After Transplant: Mechanism not completely understood.
Bile Flow and Bile Acids: Bile losses via T-tube dramatically decreases cyclosporine levels; after clamping the T-tube, levels increase twofold to sixfold.
Liver Function: Poor hepatic function causes decreased cyclosporine absorption.
Gastrointestinal Status: Impaired absorption is associated with diarrhea (pretransplant chemotherapy, acute graft-vs-host disease, infectious).

Factors Affecting the Pharmacokinetics of Cyclosporine (Cont)
Distribution
Hematocrit: Cyclosporine is distributed 50% in erythrocytes, 10% in leukocytes, and 40% in plasma. Cyclosporine concentration in mononuclear cells is 1000 times higher than in erythrocytes.
Lipoprotein Content: Cyclosporine binds to lipoproteins.
Elimination
Genetic Factors: Large interindividual variability in cyclosporine metabolism.
Age: Cyclosporine clearance in children is twice that of adults.
Drugs That Increase Cyclosporine Levels: Antibiotics (ketoconazole, erythromycin), Calcium Channel Blockers, Steroids.
Drugs That Decrease Cyclosporine Levels: Anticonvulsants (phenytoin, phenobarbital, carbamazepine), Antibiotics (rifampin, isoniazid, nafcillin).

Many problems exist with the cyclosporine testing procedure. Serum, plasma or whole blood may be tested. High performance liquid chromatography (HPLC) is considered the "gold standard" for cyclosporine testing, but this method is technically difficult and not without problems. Radioimmunoassay (RIA) is the most common technique for cyclosporine testing. Monoclonal antibodies used in this test recognize either just cyclosporine or cyclosporine and its metabolites. Therapeutic levels vary according to the specificity of the assay.

CYTOMEGALOVIRUS (CMV) TESTING

SPECIMEN: Direct Examination: Urine, Bronchoalveolar Lavage Fluid, Lung Biopsy Specimens; Culture: Urine, Blood, Bronchoalveolar Lavage Fluid, Saliva, Breast Milk, Semen, Cervical Secretions; Antibody Tests: Serum; Antigen Tests: Urine, Bronchoalveolar Lavage Fluid, Lung Biopsy Specimens.
REFERENCE RANGE: Negative; Positive culture, antibody or antigen tests must be interpreted in clinical context. No single positive test is diagnostic of acute CMV infection.
Antibody tests: The presence of IgM or a fourfold or greater rise in IgG titer suggests recent infection.
METHODS: Direct Examination; Microscopy; Culture: Direct inoculation into cell culture; Antibody Tests: Seroneutralization, Complement Fixation, Indirect Hemagglutination, Enzyme Immunoassay, Radioimmunoassay, Immunofluorescence, Others; Antigen Tests: ELISA, Immunofluorescence.
INTERPRETATION:
Direct Examination: A histologic study of stained urine sediments helps to make a presumptive diagnosis in 25 to 50 percent of symptomatic cases; the finding of single large red round intranuclear inclusions surrounded by a halo, giving the cell the appearance of an owl's eye, is distinctive for CMV infection. Electron microscopy has also been used.
Culture: Viral isolation is the "gold standard" for proof of CMV infection. The classic finding is cytopathic effect on tissue culture. CMV culture may become positive in two days but may take three weeks. Specimens should be refrigerated, not frozen. Sensitivity is increased by sending several specimens (typically, three separate urine samples are sent).

Positive CMV blood and urine cultures in HIV infected individuals correlate with degree of immunosuppression but have little diagnostic or prognostic utility; these cultures correlate poorly with clinical disease. Presence of viruria is highly nonspecific, but a negative urine makes the diagnosis of active CMV disease less likely (Zurlo JJ, et al. Ann Intern Med. 1993; 118:12-17).
Antigen Testing: CMV antigen detection in clinical specimens may be useful, particularly in conjunction with antibody testing and culture. The clinical usefulness of these tests has not been fully defined; one advantage of antigen tests is that results can be obtained quickly.
Antibody Testing: The presence of IgM may indicate current infection, but IgM antibodies may persist for years after initial exposure. IgG antibody titers fluctuate in normal individuals, but a four-fold or greater rise in titer between acute and convalescent sera is indicative of infection. (cont)

CMV Testing (Cont)

Congenital CMV: Congenital cytomegalovirus infection is probably the most common viral infection acquired in-utero by the fetus; in the United States, the incidence is one to two percent. CMV may also be transmitted to infants following blood transfusions; this transmission can be prevented by exclusive use of frozen deglycerolyzed or CMV-seronegative blood (Taylor B, et al. Pediatr Infect Dis. 1986; 5:188-191) or by blood filtration to remove leukocytes (Gilbert GL, et al. Lancet. June 3, 1989; 1228-1231).

If a woman who has had cytomegalovirus(CMV) infection has a recurrence of the infection during pregnancy, the chance that her baby will have a harmful congenital CMV infection is much less than it is if a woman has her first CMV infection during pregnancy. Preexisting maternal antibody to CMV before conception is predictive of much less damaging congenital CMV infection in the newborn (Fowler KB, et al. N Engl J Med. 1992; 326:663-667).

Unlike rubella, the presence of antibodies to CMV in the mother does not protect the fetus against infection. The absence of antibodies in the mother rules out congenital infection. IgG antibodies may be transferred from the mother to the infant by placental transfer; these antibodies will decay over 2 to 3 months. A rising (four-fold) or persistent titer of IgG antibodies in the infant's serum is suggestive of a congenital infection or a 1:64 titer or more is also suggestive of infection.

The presence of IgM antibodies in the infant indicates current infection since IgM antibodies, unlike IgG, do not cross the placental barrier. CMV IgM is neither sensitive nor specific for congenital infection. Many asymptomatic congenitally infected infants do not mount an IgM response.

Most infants with congenital CMV infections are asymptomatic or have mild hepatomegaly with moderately abnormal liver function tests and jaundice.

Postnatal and CMV Infections in Adults: The presence of IgM antibodies or a four-fold or greater rise in antibody titers between acute and convalescent sera is indicative of infection. About 80% of adults over the age of 35 have IgG antibodies.

Immunocompromised patients may have disseminated or fatal CMV and have no detectable antibody response.

False Positives: False positive IgM antibody may be present due to non-specific binding, anti-nuclear factor, and Epstein-Barr virus infections.

References: Galea G. Med Lab Sci. 1990; 47:297-303; Drew WL. Rev Infect Dis. 1988; 10(Suppl.3):S468-S476.

D-DIMER TEST

SPECIMEN: Blue top, green top or purple top tube: Citrated, heparinized or EDTA-treated plasma.

REFERENCE RANGE: Normal <0.5mcg/mL. Results are semi-quantitative and reported as <0.5mcg/mL; 0.5 to 1.0mcg/mL; or >1.0mcg/mL.

METHOD: Latex agglutination

INTERPRETATION: The D-dimer test may be useful in the diagnosis of thrombosis, DIC, hyperfibrinolytic coagulopathies, and monitoring fibrinolytic therapy. When used in conjunction with a test for fibrin split products (FSP), one can distinguish between degregation of fibrin versus fibrinogen. The FSP are composed of peptides A and B which are released from both fibrin and fibrinogen by thrombin action. By contrast, the D-dimer test measures crosslinked D-dimers which are released from factor XIIIa (plasmin) reacted fibrin alone (Hillyard, et al. Clin Chem. 1987; 33:1837-1840).

The D-dimer test is not subject to false positive results in the presence of heparin like the FSP test. In one study the D-dimer test was less sensitive (85 versus 100%) but more specific (97 versus 56%) than the FSP in the diagnosis of DIC (Carr, et al. Am J Clin Path. 1989; 91:280-287). The D-dimer test has also been used as an aid in the diagnosis of pulmonary embolism (Bounameaux H, et al. Lancet. 1991; 1:196-200).

DEHYDROEPIANDROSTERONE SULFATE(DHEA-S)

SPECIMEN: Red top tube, separate serum; freeze serum as soon as possible.
REFERENCE RANGE:

Change of DHEA-S with Age

Age	Traditional Units (ng/mL)	Conversion Factor	International Units (micromol/liter)
Newborn	1670-3640	0.002714	4.5-9.9
Children	100-600	"	0.3-1.6
Pre-Pubertal Male	2000-3350	"	5.4-9.1
Female(Premenopausal)	820-3380	"	2.2-9.2
Female(Postmenopausal)	100-610	"	0.3-1.7
Pregnancy(Term)	230-1170	"	0.6-3.2

METHOD: RIA
INTERPRETATION: The most common indication for measurement of DHEA-S is in the evaluation of patients with hirsutism. The workup of hirsutism in the female includes exclusion of an androgen secreting neoplasm and adrenal pathology. A minimal workup would include determination of serum testosterone, DHEA-S, and prolactin (Bailey-Pridham DD, Sanfillippo JS. Pediatr Clin North Am. 1989; 36:581-599). DHEA-S is a good indicator of adrenal function; testosterone is a good indicator of ovarian function. A marked elevation of DHEA-S (60-340mcg/dL) is usually an indication of an adrenal abnormality (tumor or enzymatic). DHEA-S measurements have largely replaced the more cumbersome and less accurate measurements of urinary 17-ketosteroids.

11-DEOXYCORTISOL (COMPOUND S)

SPECIMEN: Green(heparin) top tube; transport to laboratory immediately and separate plasma and freeze.
REFERENCE RANGE: 0-2mcg/dL; post-metyrapone: greater than 10mcg/dL. To convert traditional units in mcg/dL to international units in nmol/liter, multiply traditional units by 28.86.
METHOD: RIA
INTERPRETATION: Assay of 11-deoxycortisol (compound S) is used: 1) in the work-up of patients with possible 11-beta hydroxylase deficiency (congenital adrenal hyperplasia/adrenogenital syndrome); 2) in evaluating pituitary reserve following metyrapone (metopirone); and 3) in helping distinguish between hypercorticism due to a pituitary adenoma (Cushing's disease) from that due to an adrenal neoplasm by means of a metyrapone test. The causes of increased and decreased plasma 11-deoxycortisol (compound S) are listed in the next Table:

Causes of Changes in Plasma 11-Deoxycortisol

Increased:
- 11-Beta Hydroxylase Deficiency (Congenital Adrenal Hyperplasia)
- Post-Metyrapone
- Other: Eclampsia, Stress, Pancreatitis

Decreased:
- Hypofunction of Anterior Pituitary
- Addison's Disease
- 21-Hydroxylase Deficiency

The metabolic block in 11-beta-hydroxylase deficiency and in post-metyrapone are illustrated in the next Figure:

Metabolic Block: 11-Beta Hydroxylase Deficiency and Post-Metyrapone

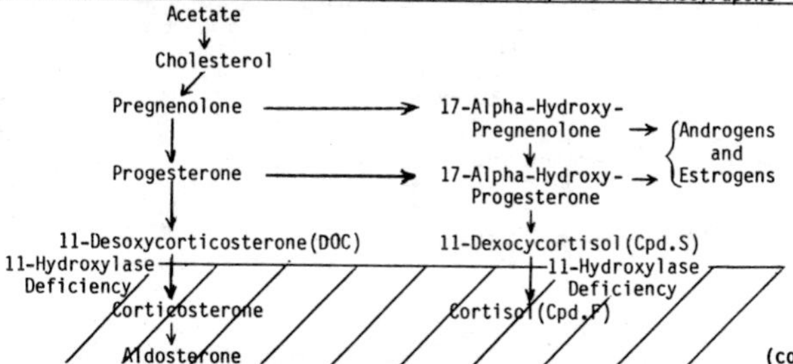

(cont)

11-Deoxycortisol (Compound S) (Cont)

The block in 11-beta-hydroxylase activity results in release of feedback inhibition by cortisol of ACTH production, and an increase in ACTH and 11-deoxycortisol production. There is no salt wasting in 11-hydroxylase deficiency as a result of the mineralocorticoid activity of 11-deoxycortisol.

The assay of plasma 11-deoxycortisol has largely replaced the measurement of urinary 17-hydroxy corticosteroids and 17-ketogenic steroids in the metyrapone test and workup of patients with suspected 11-hydroxylase deficiency (see METYRAPONE TEST).

DEPAKENE (see VALPROIC ACID)

DEXAMETHASONE SUPPRESSION TEST

SPECIMEN AND PROCEDURE: Dexamethasone is taken orally and the subsequent suppression of cortisol secretion, or lack of it is documented by measurement of blood cortisol or its urine metabolites. There are numerous variants of the dexamethasone suppression test with varying doses of dexamethasone administered, and differing sampling protocols, depending on the clinical question to be asked. The major variants of the Dexamethasone Suppression Test are:

1) Overnight screening test
2) Low dose suppression test
3) High dose suppression test
4) Modified overnight high dose suppression test
5) Overnight suppression test for depression
6) Corticotropin-releasing hormone stimulation following low-dose dexamethasone administration (combined dexamethasone - CRH test).

In all of these, a sample of blood, red top tube, or plasma, green (heparin) top tube is drawn. Sample is stored and shipped frozen. Alternatively, an aliquot of a 24 hour urine is assayed for steroid metabolites.

METHOD: RIA. Old test protocols relied on 24 hour urine measurements of steroid metabolites: 17-OHCS, 17-KGS and 17-KS. Newer assays such as plasma or serum cortisol, or urinary free cortisol are more convenient and accurate.

REFERENCE RANGE: Varies with type of test. See below.

INTERPRETATION:

Overnight Screening Test (Cushing's Syndrome): This test is performed to evaluate the possible presence of Cushing's syndrome (hypercorticism). The dexamethasone suppression test for the diagnosis of Cushing's syndrome is as follows: Dexamethasone, 1mg, is taken orally by the patient between 11:00 P.M. and midnight; a fasting blood cortisol is obtained the next morning at 8:00 A.M. In the normal individual, dexamethasone will suppress cortisol production to a blood value of less than 5mcg/dL.

A value of blood cortisol greater than 5mcg/dL is compatible with Cushing's syndrome. A negative test (<5mcg/dL) is found in less than 2 percent of the cases of Cushing's syndrome (Crapo L. Metabolism. 1979; 28:955-977). However, abnormal results (>5mcg/dL) are often found in patients under stress for any reason, including at least 30 percent of patients in a hospital who are acutely ill (Connolly CK, et al. Br Med J. 1968; 2:665-667). Conditions associated with mild hypercortisolism include stress, alcoholism (chronic or withdrawal), anorexia/bulimia, renal failure, depression and glucocorticoid receptor resistance. Urinary free cortisol is usually normal in the obese patient.

Alcoholism, especially associated with liver disease, may result in failure of suppression, and elevation of blood and urine cortisols. These values return to normal after a week of abstinence (Carpenter PC. Endocrin Metab Clin N Amer. 1988; 17:445-472). Increased rates of dexamethasone metabolism (seen with anticonvulsants) may produce results similar to that seen in Cushing's syndrome.

A small number of patients with Cushing's syndrome may exhibit a paradoxical dexamethasone suppression test. This may be due to a periodicity in cortisol production, or an inappropriately high dose of dexamethasone (adjust by basing dose on lean body mass).

Low Dose Suppression Test: Test is performed to establish a diagnosis of Cushing's syndrome. Give 0.5mg dexamethasone every 6 hours for two days. Measure plasma cortisol, urinary 17-hydroxy corticosteroids or urinary free cortisol on the second day of suppression. Blood cortisols should be suppressed below 5mcg/dL, urinary 17-hydroxycorticosteroids less than 4mg per day or 2mg/gm urinary creatinine, or a urinary free cortisol less than 20mcg per day. Alcoholism, stress, and obesity may confound this test.

High Dose Suppression Test: Used to establish the etiology of the hypercorticism once a diagnosis of Cushing's syndrome has been made. It is useful in distinguishing Cushing's Disease (adrenal hyperplasia secondary to an ACTH-producing pituitary adenoma) from other causes of Cushing's syndrome. A dose of 2mg dexamethasone is administered every 6 hours for 2 or 3 days. On the last day, urinary metabolites (17-hydroxy corticosteroids, 17-ketogenic steroids) or urinary free cortisol or blood cortisol are measured. Suppression of these derivatives by 50% is consistent with a diagnosis of Cushing's disease. A failure of suppression is consistent with adrenal neoplasia, either adenoma or carcinoma. Approximately 10% of pituitary adenomas fail to suppress with high dose dexamethasone. Occasionally these dexamethasone-resistant cases can be suppressed by cortisol (Carey RM. N Engl J Med. 1980; 302:275-279).

Interpretation of suppression tests is complicated by an occasional ectopic ACTH producing neoplasm which suppresses with high dose dexamethasone. The traditional high dose dexamethasone test will suppress pituitary function for a relatively long time. Hence, a metyrapone test cannot be performed until a few days following the high dose dexamethasone suppression test. It is best to schedule the metyrapone test before the dexamethasone test.

The traditional high dose dexamethasone suppression test has been modified to be performed on an overnight basis. Eight mg of dexamethasone are administered at 11:00 P.M. and blood cortisol is measured in the morning. The blood cortisol is suppressed by 50% in approximately 90% of patients with Cushing's disease (Tyrell JB, et al. Ann Intern Med. 1986; 140:180-186).

Dexamethasone Suppression Test in Depression: This is based on the fact that normals will suppress cortisol production for 24 hours following administration of 1mg dexamethasone at 11:00 P.M. Some patients with stress or depression will escape suppression by 4:00 P.M. or earlier. Further details are given in DEXAMETHASONE SUPPRESSION TEST AS DIAGNOSTIC AID IN DEPRESSION.

Combined Dexamethasone - CRH Test: This test has been proposed to accurately distinguish between Cushing's syndrome from pseudo-Cushing's states (conditions which may cause a mild hypercortisolism and may be confused with Cushing's syndrome). Pseudo-Cushing's states may include severe obesity, long-term alcoholism, alcohol withdrawal, stress, renal failure, anorexia/bulimia nervosa, depression, and primary glucocorticoid receptor resistance. In the combined dexamethasone suppression - CRH stimulation test, patients with pseudo-Cushing's syndrome exhibit a low cortisol as a result of combined suppression of ACTH levels and low CRH stimulation. By contrast, patients with true Cushing's would have higher cortisol and ACTH levels after dexamethasone and greater response to CRH. In the initial publication of this method, it appears that this technique is able to accurately distinguish Cushing's syndrome from pseudo-Cushing's states. the protocol is quite long and given in: Yanovski JA, et al. JAMA. 1993; 269:2232-2238.

DEXAMETHASONE SUPPRESSION TEST AS DIAGNOSTIC AID IN DEPRESSION

SPECIMEN: A protocol for the dexamethasone suppression test for endogenous depression is given in the next Table:

Protocol for the Dexamethasone Suppression Test on Inpatients

	Time	Specimen	Normal Value
Day 1	11:00 P.M.	Specimen for Blood Cortisol	<5mcg/dL
	Give one mg Dexamethasone (11:00 P.M. to 12:00 P.M.)		
Day 2	4:00 P.M.	Specimen for Blood Cortisol	<5mcg/dL
	11:00 P.M.	Specimen for Blood Cortisol	<5mcg/dL

On an outpatient basis, only a 4 P.M. post-dexamethasone serum cortisol is obtained. Collect blood in a green (heparin) top tube, separate plasma; or red top tube, separate serum; store serum or plasma frozen. Some workers suggest simultaneous measurement of dexamethasone to check patient compliance and aberrant clearance of the drug.

REFERENCE RANGE: Blood cortisol: <5mcg/dL

METHOD: RIA

INTERPRETATION: The dexamethasone suppression test is used to help predict which depressed patients will respond to medication or cognitive therapy. When 1mg of dexamethasone is given orally at 11:00 P.M. to 12:00 Midnight, normal subjects maintain suppressed plasma cortisol concentrations for at least 24 hours. Patients with endogenous depression may suppress their plasma cortisol concentrations temporarily but fail to maintain plasma cortisol suppression below 5mcg/dL for 24 hours. The abnormal escape of plasma cortisol concentrations usually occurs by 4 P.M. and may be observed as early as 8:00 A.M. Normalization of the results of the test occurs with clinical recovery. Studies have shown that the test will return to normal about 7 to 10 days after initiation of drug therapy. When the results of the test fail to normalize at discharge, patients are at high risk for early relapse.

This test should not be used as a screening test. It may be useful in questionable cases. Caution should be exercised with regard to the routine application of the dexamethasone suppression test for depression (Hirschfeld RMA, et al. JAMA. 1983; 250:2172-2175; Health and Public Policy Committee, American College of Physicians. Ann Intern Med. 1984; 100:307-308). There are contrary opinions about the clinical utility of the dexamethasone suppression test, and its use has been recently reviewed (Glassman AH, et al. Am J Psychiatry. 1987; 144:1253-1262; APA Task Force on Laboratory Tests in Psychiatry. Am J Psychiatry. 1987; 144:1253-1262). Some have suggested that the test not be used to determine a diagnosis or course of therapy, but that it can be used as a confirmatory test for depression (Arana GW. JAMA. 1991; 125:2253-2254).

Conditions and drugs that cause false positive and false negative results in the dexamethasone suppression test for endogenous depression are given in the next Table (Ritchie JC, Carroll BJ. Lab World. 1981; 24-29):

Conditions and Drugs Causing False Positive and False Negative Results in the Dexamethasone Suppression Test for Endogenous Depression

Conditions	Drugs
False Positive Tests	**False Positive Tests**
Major Physical Illness	Phenytoin
Trauma, Fever, Nausea, Dehydration	Barbiturates
Temporal Lobe Disease	Meprobamate
Pregnancy (or high dosage estrogens)	Methaqualone
Cushing's Disease	Glutethimide
Unstable Diabetes	Methyprylon
Malnutrition	Carbamazepine
Anorexia Nervosa (<80% ideal weight)	Reserpine (?)
Heavy Alcohol Use	
Alcohol Withdrawal (acute <10 days)	**False Negative Tests**
E.C.T. (on Post dex. day)	High-dose Benzodiazepines (>25 mg/day of diazepam)
Other Endocrine Disease?	Cyproheptadine
False Negative Tests	Synthetic Steroid Therapy
Addison's Disease	
Corticosteroid Therapy	
Hypopituitarism	

DIABETES INSIPIDUS (DI)

The clinical findings in diabetes insipidus are thirst and polyuria. The laboratory findings include hypernatremia, increased plasma osmolality, decreased urine sodium concentration, and increased urine volume. The diagnosis can be made by osmolality measurements over a 3 hour period during water restriction (Zerbe and Robertson. N Engl J Med. 1981; 305:1539). Clinical impressions can be confirmed by direct measurement of ADH or indirectly by the response to a trial of ADH (synthetic vasopressin).

This condition can result from hypothalamic/pituitary failure of ADH secretion (central or neurogenic DI) or lack of renal response to ADH (peripheral or nephrogenic DI). The partial or latent DI may occur as well. Nephrogenic cases of diabetes insipidus are much harder to threat. Causes of diabetes insipidus are listed in the next Table:

Causes of Diabetes Insipidus	
Neurogenic:	Congenital
	Tumors - Hypothalamic/Pituitary
	Trauma
	Meningitis
	Granulomas - TB, Sarcoid, Eosinophilic
	Lymphocytic Infundibuloneurohypophysitis
Nephrogenic:	Congenital
	Intrinsic Renal Disease
	Amyloidosis
	Multiple Myeloma
	Drugs
	Lithium
	Demeclocycline
	Electrolyte Imbalance
	Hypokalemia
	Hypocalcemia
Other:	Pregnancy (Increased Vasopressinase)

In general, patients with neurogenic DI respond to administration of synthetic ADH (1-deamino-8-D arginine vasopressin) while nephrogenic DI is associated with elevated levels of ADH and unresponsive to further administration of ADH. However, some cases of nephrogenic DI will respond to ADH if indomethacin is concurrently given (Stasior DS, et al. N Engl J Med. 1991; 324:850-851). Transient DI in pregnancy is rare, but has been ascribed to increased vasopressinase and lowered thirst threshold during pregnancy (Iwaski Y, et al. N Engl J Med. 1991; 324:522-526; Robinson AG, Amico JA. ibid. 556-558).

Approximately half the patients in one study of idiopathic neurogenic (central) DI had lymphocytic infundibuloneurohypophysitis as documented by MRI (Imura H, et al. N Engl J Med. 1993; 329:683-689). Some of the mechanisms of congenital DI are discussed by Lightman SL. N Engl J Med. 1993; 328:1562-1563.

DI is usually diagnosed by the failure of the patient to concentrate the urine on a water-deprivation test. The expected results of a 3 hour water deprivation test are given in the next Table:

Results of Water Deprivation

Condition	Plasma Osmolality		Urine Osmolality	
	Start	Finish	Start	Finish
Normal	Normal	Unchanged	≥ Plasma	Rises
DI	Increased	Rises (>300)	< Plasma	No Change
Compulsive H_2O Drinker	Low	Rises but (<300)	Low	Rises

Diagnostic impression and etiology may be confirmed with ADH measurements and clinical trials of ADH administration. In DI the hematocrit is usually above 50% and the sodium is usually greater than 150mmol/L. In hemoconcentration with sodium loss the sodium is usually less than normal.

DIABETIC KETOACIDOSIS (DKA) PANEL

Diabetic ketoacidosis (DKA) is a potentially lethal condition. DKA occurs at all ages, often in people previously unrecognized as diabetic and in insulin dependent diabetics. Disorders that accompany or precipitate DKA include bronchopneumonia, pyelonephritis, alcoholism, septicemia, gastroenteritis, myocardial infarction, pancreatitis and cerebrovascular disease.

Diagnosis of DKA: The diagnosis of DKA is made through determination of glucose, ketones, acid-base status, and electrolytes. The differential diagnosis of anion-gap metabolic acidosis includes: ketoacidosis: diabetes, alcohol, starvation, lactic acidosis; chronic renal failure; and drugs: salicylates, ethylene glycol, methanol and paraldehyde. The mean laboratory values in patients with DKA on admission to the hospital (Foster DW, McGarry JD. N Engl J Med. 1983; 309:159-169) and a schedule for monitoring laboratory tests for the first six hours are given in the next Table:

Mean Laboratory Values in Patients with Diabetic Ketoacidosis(DKA) on Hospital Admission and Schedule for Monitoring for First Six Hours

Laboratory Data	Average Concentration (Initial)	Schedule for Monitoring
Glucose (Lab)(80-120mg/dL)	500	Every Hour for 4-6 Hours
Glucose(Dextrostix;Chemstrip)	500	Every Hour for 4-6 Hours
Urine: Glucose	+3-4	Every 2 Hours
Ketones	+2-4	Every 4 Hours
Serum Osmolality(285-295mOsm/L)	313	Every Hour for 4-6 Hours
Arterial Blood Gases:	-	Every Hour for 4-6 Hours
pH (7.36-7.44)	7.1	
PCO_2 (32-45 mm Hg)	20-35	
HCO_3^- (24-29 mmol/L)	5-10	
Base Excess (-3 to +3)	-20	
Electrolytes:		Every Hour for 4-6 Hours
K^+ (3.5-5.0 mmol/L)	5.3	
Na^+ (135-145 mmol/L)	133	
Cl^- (95-110 mmol/L)	100	
CO_2 Content (24-30 mmol/L)	5-10	
Anion Gap: $[Na^+ - (Cl^- + HCO_3^-)]<15$mmol/L	25-35	Every Hour for 4-6 Hours
Ketones:		
Acetoacetate (<0.3mM)	4.0	Every Hour for 4-6 Hours
Acetone (<0.2mM)	4-5	
Beta-Hydroxybutyrate (<0.5mM)	12	
Lactic Acid (0.5-1.66mmol/L)	3.5	As Appropriate
BUN (5-20 mg/dL)	27	As Appropriate
Phosphorus: Adult: (2.3-4.3mg/dL)	4	Every 2 Hours
Urine Output		Continuously
Electrocardiogram(ECG)		Every 4 Hours
Pulmonary Capillary Wedge Pressures		As Appropriate

Other Tests: Vital signs (B.P., pulse and respirations) must be carefully monitored; the patient is often hypotensive and the pulse is increased. The respiratory rate is increased secondary to metabolic acidosis (Kussmaul respirations). Evaluation of weight loss may give some concept of dehydration. The state of mentation should be carefully monitored due to the risk of cerebral edema.

Other tests that should be considered initially include the following: blood and urine culture, urine analysis, serum creatinine, calcium and magnesium, chest x-ray, and complete blood count(CBC) with differential.

Specimens: Collect arterial blood for blood gases in a heparinized syringe (on ice). Use either arterial or venous blood for lactate; collect in a grey top (fluoride/oxalate) tube (on ice). Collect blood for culture in special tubes. Collect all other blood specimens in a red top tube. These specimens are collected every hour for four to six hours. Collect urine for culture and analysis.

Glucose: The average concentration of plasma glucose in patients with DKA on admission to the hospital is about 500 mg/dL; however, levels range from nearly normal to a thousand or more. Blood-glucose test strips (glucose oxidase) are used to determine blood glucose values in the physician's office, emergency room or at the bedside.

Urinalysis: Glucose and ketone bodies are measured in the urine by dip-stick methodology; this is done at the bedside.

Osmolality: The average serum osmolality is 310-315 mOsm/liter when the serum glucose is 500 mg/dL; the serum osmolality may be obtained by direct measurement or calculated from the formula:

$mOsm/kgH_2O = 2Na^+ + \frac{Glu}{18} + \frac{BUN}{2.8}$; Sodium (mmol/L), Glucose and BUN (mg/dL)

Each 100 mg/dL of glucose contributes 5.5 to the serum osmolality.

Arterial Blood Gases (pH, pCO_2, HCO_3): The average value for arterial blood pH is 7.1. At this pH, the HCO_3^- is 12 mmol/liter when the pCO_2 is 40 mmHg. The brainstem centers that regulate alveolar ventilation are stimulated by acidified extracellular fluid. Increased ventilation lowers pCO_2. Maximal respiratory compensation occurs within 24 to 36 hours. In metabolic acidosis with respiratory compensation (24 to 36 hours):

Expected pCO_2 (mmHg) = $(1.5\ HCO_3 + 8) \pm 2$

If HCO_3 = 10 mmol/liter, then pCO_2 = 23 (21 to 25) mmHg.

Base Excess: The base excess is about -17 when the blood pH is 7.1 and the pCO_2 is 40 mmHg; the base excess is -23 when the pH is 7.1 and the pCO_2 is 20 mmHg.

Electrolytes (K^+, Na^+, Cl^-, CO_2 Content):

Potassium: The serum potassium and electrocardiogram(ECG) are used to monitor serum potassium. The average value for serum potassium is 5.3 mmol/liter; serum potassium is normal or elevated in 80 to 90% of patients with DKA. The serum potassium does not reflect the total body potassium. In DKA, there is an osmotic diuresis with total body deficit of 3-5 mmol/kg and as much as 10 mmol/kg; however, there is a shift of K^+ from the cells as hydrogen ion enters the cell in exchange for intracellular K^+. A low-serum potassium indicates profound total body depletion and replacement (with K^+ as KCl or K phosphate) must be done as soon as possible.

Sodium: The average value for serum sodium is 133 mmol/liter; glucose draws water into the extracellular space and thus decreases the sodium concentration. Other causes of low sodium include vomiting, diarrhea, and dilution secondary to water intake. Serum sodium may be spuriously depressed due to severe hypertriglyceridemia; this occurs when the popular Beckman Astra or other electrolyte analyzers are used that dilute the serum sample prior to analysis.

Chloride: Serum chloride is usually within normal range; there is a normochloremic metabolic acidosis (anion gap).

Carbon Dioxide (CO_2) Content: The CO_2 content is the sum of the HCO^-_3 concentration plus the PCO_2; normally, the concentration of HCO^-_3 is twenty times that of the concentration of PCO_2. Thus, the CO_2 content reflects the concentration of the HCO^-_3.

Anion Gap: The anion gap is increased to 25 to 35 mmol/liter. The increase in the anion gap corresponds to the quantity of ketones present in plasma (usually 10 to 20 mmol/liter). The anion gap is calculated from the formula: $[Na^+ - (Cl^- + HCO^-_3)]$; when the average initial values for Na^+, Cl^- and HCO^-_3 found in DKA are substituted in this equation: [133 - (100 + 5)], a value of 28 mmol/liter is obtained.(cont)

Diabetic Ketoacidosis (DKA) Panel (Cont)

Ketone Bodies: Metabolic acidosis is caused primarily by the accumulation of ketone bodies, acetone and acetoacetate and beta-hydroxybutyrate. The nitroprusside test measures acetoacetate and acetone but not beta-hydroxybutyrate; nitroprusside tests are 15-20 times more sensitive to acetoacetate than to acetone.

Occasionally, the nitroprusside test is negative in the presence of severe ketoacidosis because ketone bodies are in the form of beta-hydroxybutyrate; this situation occurs in conditions associated with hypoxemia, i.e., hypovolemia; hypotension, low arterial pO_2 or association of DKA with alcoholism.

A screening test for ketone bodies in blood is done as follows: Dilute serum with an equal volume of water or saline consecutively, 1:1; 1:2; 1:4; etc. With the nitroprusside reaction, significant ketonemia is indicated if the serum ketones are positive as a dilution of 1:8 or greater.

Lactic Acid: About one-third of patients with DKA have lactic acidosis.

BUN and Creatinine: There is significant volume depletion in DKA with serum BUN in the range of 25 to 30 mg/dL; serum creatinine is usually normal; however, acetoacetic acid interferes with the assay of serum creatinine; at a serum acetoacetic acid concentration of 8 to 10 mmol/liter, the serum creatinine is falsely elevated by 3 to 4 mg/dL when measured by the alkaline picrate method.

Phosphorus: During DKA there is a shift of phosphorus from tissues to the extracellular compartment and excessive loss of phosphorus in the urine. The initial serum phosphorus level is usually normal but may be elevated or low. During therapy, the serum phosphorus falls progressively.

Prior to phosphorus replacement therapy, it is necessary to measure both the serum calcium level and BUN or creatinine to exclude the possibility of renal disease.

Electrocardiogram(ECG): The ECG is a helpful guide to potassium levels at the bedside, but is less reliable than direct measurements of potassium in the serum. In hypokalemia, there is a low T wave, the presence of a U wave and a depression of the ST segment. In hyperkalemia, there is a tall T wave, a wide QRS interval and decrease or absence of the P wave.

Pulmonary Capillary Wedge Pressure(PCWP): Pulmonary capillary wedge pressure is a useful and reliable indicator of left ventricular dynamics and pulmonary congestion (>21 mmHg). In severe DKA, PCWP may be increased.

Vital Signs (Blood Pressure, Pulse, Respirations): The degree of volume depletion may be gauged by evaluating vital signs; when volume depletion is severe, there is hypotension, marked tachycardia and altered mentation. Volume contraction can be life threatening, with development of myocardial infarction, stroke or irreversible shock from underperfusion.

Chest X-Ray: Chest X-ray is done to detect pulmonary infection and pulmonary congestion.

Other: The CBC demonstrates a leukocytosis which is associated with ketoacidosis.

Treatment of DKA Mellitus: Principles of treatment of DKA are listed in the next Table:

Principles of Treatment of Diabetic Ketoacidosis(DKA)
Correct Fluid and Electrolyte Disturbances
Give Sodium, Potassium and Glucose
Give Insulin to Restore and Maintain Intermediary Metabolism in a Normal State, e.g., Metabolize Ketone Bodies
Controversial: Sodium Bicarbonate and Phosphate Replacement

Fluid Therapy: It is estimated that a 70 kg man with DKA has a deficit of 5-7 liters of water, 300-450 mmol of sodium, and 200 to 400 mmol of potassium. In children, degree of dehydration should be estimated based on clinical signs and weight change.

Fluids are given to correct fluid deficits and lower the glucose level; the concentration of ketones and beta-hydroxybutyrate do not decrease when fluids are given without insulin. The physiologic goals of fluid therapy are normalization of tissue perfusion, improved renal perfusion, and lowering of hyperglycemia by hemodilution. Normal saline is recommended for acute volume resuscitation. Infusion rate should be adjusted according to clinical response. Urine output should be monitored closely; if output consistently exceeds intake, additional fluids may be necessary. Fluid therapy should replace the calculated fluid deficit over the course of therapy. Fluid deficit should be replaced over at least 24 hours, and preferably over 36-48 hours.

Insulin Therapy: Regular insulin is used. Insulin infusion rate is usually 0.1 units/kg/hr; under 5 years of age, insulin infusion is usually 0.06-0.08 units/kg/hr. Insulin bolus is recommended by some authors but not by others; usually recommended dose is 0.1-0.2 units/kg. Insulin is continued until ketosis is cleared.

The level of insulin in normal individuals is 5-40 mcU/mL in the fasting state. Infusion of regular insulin to raise the insulin level to 50 mcU/mL suppresses hepatic glucose output by 70 to 75 percent; at 100 mcU/mL, almost complete suppression of glucose output by the liver occurs and stimulation of peripheral glucose uptake is 70 to 75% of maximum. At plasma insulin levels of 50 to 100 mcU/mL, lipolysis is completely blocked. "Low dose" insulin, as described in this protocol, raises the plasma insulin level to 150 to 200 mcU/mL.

Glucose Levels During Therapy: As fluids and insulin are given, plasma glucose levels begin to fall. The expected rate of fall of plasma glucose is 75 to 100 mg/dL per hour. The plasma glucose level must not be allowed to fall below 250 mg/dL during the first four to six hours of insulin therapy. If this occurs, glucose should be infused to avoid hypoglycemia. It is necessary to continue insulin infusion even after glucose levels have fallen in order to eliminate ketone bodies; maintain glucose around 200 mg/dL. In general, hyperglycemia is corrected in 4 to 8 hours. Glucose levels must not fall too rapidly or cerebral edema and neurological complications will result.

Thirty to 45 minutes prior to the termination of the IV insulin infusion, subcutaneous insulin must be administered at an adequate dose to insure adequate coverage of glucose metabolism.

Potassium Therapy: As already mentioned, serum potassium is usually elevated initially despite total body potassium depletion. After therapy with fluids and insulin has started, potassium concentration in plasma decreases as potassium returns to the cell as acidosis is reversed. Administration of insulin in the presence of glucose will also facilitate potassium uptake. Potassium levels must be monitored frequently, since hypokalemia may precipitate cardiac arrhythmias.

Bicarbonate Therapy: Bicarbonate therapy is controversial. Authors that recommend bicarbonate generally reserve this therapy for patients with pH less than 6.9-7.1. Bicarbonate is administered slowly (usually mixed with IV fluids); the goal of therapy is to raise pH above 7.1.

Phosphate Therapy: Phosphate replacement is controversial. Total body phosphate may be markedly depleted in DKA, but there is no evidence that replacement improves outcome. Phosphate is necessary for formation of 2,3 diphosphoglycerate (2,3 DPG) in red blood cells, formation of adenosine triphosphate (ATP), and proper maintenance of neuromuscular tissue (including cardiac and respiratory muscles). Phosphate replacement may be in the form of potassium or sodium phosphate; serum calcium and BUN or creatinine should also be monitored.

Ketone Levels During Therapy: Plasma glucose levels fall before ketone bodies disappear; it is necessary to continue insulin infusion when acidosis or ketosis are still present even though glucose has fallen to 250 mg/dL. In general, ketosis is corrected in 10 to 20 hours. Ketonuria may persist for 24 to 48 hours after ketone bodies have cleared from the blood.

Chloride Levels During Therapy: Hyperchloremia normally develops during therapy and therapy using Ringer's lactate may be instituted.

Osmolality During Therapy: The fall in serum osmolality should be controlled at a rate not to exceed 10 mOsm/hour.

Monitoring Laboratory Data After First Four to Six Hours: Blood gases electrolytes, osmolality, ketones, and glucose should be assayed every 2-4 hours.

Complications: Complications of DKA are as follows: shock, ARDS, infection, arterial thrombosis and cerebral edema in children. Cerebral edema is usually fatal; if the patient survives, permanent brain damage may result.(cont)

Cerebral Edema: Cerebral edema is a devastating complication of therapy for DKA. Although the mechanisms are not completely understood, this condition may occur when plasma becomes hypotonic compared to the CNS resulting in fluid shifts into the brain and cerebral edema. Some degree of cerebral edema occurs as a consequence of DKA treatment in most if not all patients with severe DKA. Life-threatening cerebral edema may be associated with therapies that cause rapid decrease in serum osmolality such as excessive administration of free water leading to hyponatremia or rapid decline in serum glucose. Cautious fluid and electrolyte resuscitation is necessary to minimized the risk of this complication (Harris GD, et al. J Pediatr. 1990; 117:22-31; Rosenbloom AL. Diabetes Care. 1990; 13:22-33; Duck SC, Wyatt DT. J Pediatr. 1988; 113:10-14; Harris GD, et al. J Pediatr. 1988; 113:65-68).
General References: Kecskes SA. Pediatr Clin N Amer. 1993; 40:355-363; Bratton SL, Krane EJ. J Intens Care Med. 1992; 7:199-211; Cefalu WT. Crit Care Clin. 1991; 7:89-108; Sanson TH, Levine SN. Drugs. 1989; 38:289-300; Schatz DA, Rosenbloom AL. J Crit Illness. 1988; 3:31-45.

DIAZEPAM (VALIUM)

SPECIMEN: Red top tube, separate serum
REFERENCE RANGE: Therapeutic range: 100-1500ng/mL; Toxic Range: 3,000-14,000ng/mL
METHOD: Qualitative: Thin layer chromatography. Quantitative: Gas-liquid chromatography; Ultraviolet spectrometry.
INTERPRETATION: Diazepam (Valium) is a benzodiazepine and is used for the treatment of anxiety, insomnia, panic disorder and seizures. During the late 1970's and early 1980's, diazepam (Valium) was prescribed for oral administration more often than any other drug in the United States.

Physiologic dependence is likely to develop in patients taking benzodiazepines regularly for several months, and may be more likely in people who have been heavy users of alcohol or other sedatives (Woods JH, et al. JAMA. 1988; 260:3476-3480). Tight regulation of benzodiazepine prescription for anxiety or insomnia in New York state has reduced the number of prescriptions by half. The impact of the reduce prescribing rate on patient care has not been evaluated (Weintraub M, et al. JAMA. 1991; 266:2392-2397; Glass RM. JAMA. 1991; 266:2431-2433).

Diazepam acts rapidly when taken as a single dose; it is absorbed rapidly from the gastrointestinal tract, and since it is soluble in lipids, it quickly reaches therapeutic concentrations (serum therapeutic range, 100-1500ng/dL) in the central nervous system. Diazepam is metabolized in the liver by oxidation and has active metabolites; it is "long-acting" with a half-life greater than 24 hours. It may be more hazardous for the elderly and for patients with liver disease. In patients 65 years and older, risk of hip fracture was 70% higher in patients taking long half-life benzodiazepines compared to patients not taking psychotropic drugs, although this increased risk was not statistically significant (Ray WA, et al. JAMA. 1989; 262:3303-3307).

In the blood, diazepam is strongly bound to protein (85%-95%). Death is rare in overdose when diazepam is taken alone; when death occurs when diazepam is taken, the actual cause of death is more likely attributable to the other drugs ingested (Finkle BS, et al. JAMA. 1979; 242:429-434; Jatlow P, et al. Am J Clin Path. 1979; 72:571-577).

DIGOXIN

SPECIMEN: Red top tube; remove serum and freeze if greater than 24 hours; Reject specimen if not clotted. Blood specimen must be drawn 6-8 hours after administration of the last dose.

REFERENCE RANGE: Therapeutic range: 0.5-2.2ng/mL (0.6-2.8nmol/L); Toxic: >2.2ng/mL (>2.8nmol/L); Potentially toxic: >1.5ng/mL (>1.9nmol/L); in patients whose plasma potassium concentrations are below normal, plasma digoxin concentrations between 1.3 and 2.2ng/mL (1.7-2.8nmol/L) may be associated with toxicity. Half-Life: 36 hours (normal renal and hepatic function); 5.0 days (markedly reduced renal and hepatic function).

To convert conventional units in ng/mL or mcg/L to international units in nmol/L, multiply conventional units by 1.281.

The time to steady state is 1 to 2 weeks; the time to steady state may be prolonged with either renal or hepatic dysfunction. Therapeutic monitoring is done after steady state has been achieved.

METHOD: RIA; EMIT; Fluorescence Polarization (Abbott); HPLC. There is a natural substance in some patients that reacts like digoxin in several commercial kits used for assay of digoxin.

INTERPRETATION: Digoxin is given by mouth or intravenously. Digoxin is used in the treatment of congestive heart failure, atrial fibrillation and supraventricular tachyarrhythmias.

Congestive Heart Failure: When pathology of the left ventricular myocardium is involved, digitalis glycosides may be used. Digoxin is most useful in patients with dilated, failing hearts and impaired systolic function (S_3 gallop may be present). Digoxin is less useful in patients with elevated filling pressures due to decreased ventricular compliance but preserved systolic function (Review: Smith TW. N Engl J Med. 1988; 318:358-365). When the causes of congestive heart failure are pulmonary, endocrine, hypertension, mechanical or muscle replacement, digitalis will be slightly effective or ineffective.

The usual (oral) dose of digoxin is 0.25mg digoxin tablets; patients with reduced renal function take 0.125mg tablets or less. Firm recommendations about modification of dose of digoxin in patients with uremia have been devised (British National Formulary, London: British Medical Association and the Pharmaceutical Society of Great Britain 11, 1983).

Digoxin-Like Immunoreactive Substances (DLIS): Using radioimmunoassays, digoxin-like immunoreactivity has been found in the blood, urine, and amniotic fluid of patients not receiving cardiac glycosides. DLIS cross reacts with available radioimmunoassays. DLIS has been reported in neonates, older children, renal insufficiency, hepatic disease, hypertension, pregnancy, acromegaly, and agonal states. High levels of DLIS may be found in neonates, some children, pregnant women, and chronic renal failure (Friedman HS, et al. Chest. 1988; 94:1116-1117; Stone J, et al. J Pediatr. 1990; 117:321-325). In one study of 374 children less than 6 years of age not receiving digoxin, 27% had measurable DLIS levels and 27% of these patients had levels of 0.8-1.37ng/mL (1.0-1.8nmol/L)(Phelps SJ, et al. J Pediatr. 1987; 110:136-139).

Therapy is usually begun with or without a loading dose and therapeutic serum levels are maintained on a long-term basis after plasma-tissue equilibrium. The time course of distribution and elimination of a single oral dose of digoxin is shown in the next Figure (Fenster PE, Ewy GA. Resident and Staff Physician. 1981; 2S-15S):

Time Course of Distribution and Elimination of a Single Oral Dose of Digoxin

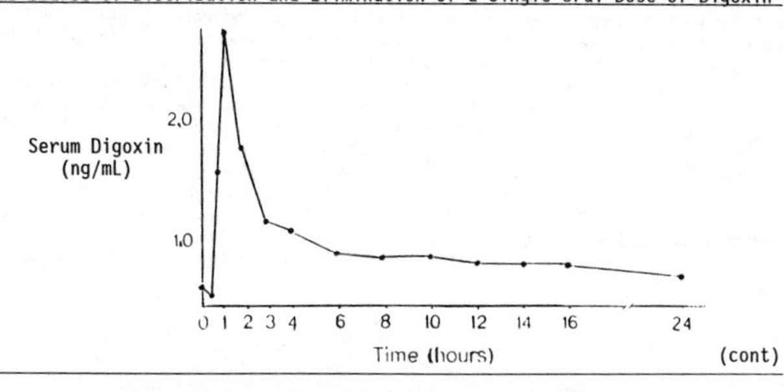

(cont)

Digoxin (Cont)

As with other drugs, after administration of a single dose, drug levels increase and then decline. The relatively high concentration of digoxin during the first six hours after oral dose reflects the relatively slow distribution of digoxin from the central compartment to the peripheral compartment. The serum level to determine peak activity should be obtained after the serum digoxin has had time to equilibrate with the tissues, that is, 8 to 24 hours after the oral dose. Digoxin is excreted mainly unchanged in the urine.

Toxicity: The most effective therapeutic dose of digoxin is a large fraction of the toxic dose, meaning that the therapeutic index is very low with only a very narrow difference between therapeutic and toxic dosages. Digoxin toxicity dropped from 14% to 6% after the assay of digoxin was introduced (Koch-Weser J. Therapeutic Drug Monitoring. 1981; 3:3-16).

The features of digitalis toxicity are given in the next Table (Anderson GJ. Geriatrics. 1980; 57-65):

Features of Digitalis Toxicity

Precipitating or Potentiating Factors:

Hypokalemia	Drug Interactions:
Hypercalcemia	Quinidine
Hypomagnesemia	Spironolactone
Hypoxia	Verapamil
Overdose	
Impaired excretion	
Severe heart disease	
Hypothyroidism	

Clinical Signs and Symptoms:

Anorexia	Abdominal pain
Nausea	Psychosis
Vomiting	Mesenteric venous occlusion
Diarrhea	Weight loss
Headache	Scotomas
Depression	Blue-yellow vision
Congestive heart failure	

Incidence of Cardiac Arrhythmias:

Ventricular arrhythmias	55-85%
Atrioventricular (AV) block	20-40%
Atrial arrhythmias	20-25%
Junctional arrhythmias	12-18%
Sinoatrial arrhythmias	5-15%
AV dissociation	10-20%

Clinical Features of Digoxin Toxicity: The systems that are most commonly involved are the gastrointestinal tract and the central nervous system.

ECG Manifestations of Digitalis Toxicity: Between 30% and 70% of patients with ECG manifestations of digitalis toxicity will not have any clinical signs or symptoms; ECG changes occur in 85% to 95% of patients with digitalis toxicity. The changes are listed in the previous table. Ventricular arrhythmias are the most prevalent ECG finding and are observed in about 60% of patients.

Precipitating or Potentiating Factors: Hypokalemia: Hypokalemia potentiates or precipitates digitalis toxicity because the potassium ion competitively inhibits the binding of cardiac glycosides to the cardiac membrane. Hypercalcemia: An increase in calcium concentration potentiates digitalis toxicity and induces abnormal automaticity through its effects on membranes.

Drug Interactions: Many drugs affect digoxin levels; interactions should be investigated prior to initiating any additional drug therapy in patients on digoxin. Quinidine increases digoxin level an average twofold. The effect begins in 24 hours and a new steady state is attained in 4 days. Quinidine increases digoxin levels by decreasing renal clearance and tissue binding of digoxin. Calcium channel blockers, especially verapamil, increase digoxin levels by up to 75%. Amiodarone increases digoxin levels by 100% or more. Erythromycin may also double digoxin levels. Spironolactone reduces elimination of digoxin.

Cholestyramine may bind digoxin in the gut. Kaolin and pectin reduces GI absorption.

Treatment: Restore body stores of potassium, digoxin toxicity is treated with lidocaine or phenytoin. More severe cases of digoxin toxicity can be treated with digoxin-specific Fab antibody fragments (Smith TW, et al. N Engl J Med. 1982; 307:1357-1362).
References: Smith TW. N Engl J Med. 1988; 318:358-365; Mooradian AD. Clin Pharmacokinetics 1988; 15:165-179; Michalko KJ, Bain L. Ther Drug Monitoring. 1987; 9:311-319.

DILANTIN (see PHENYTOIN)

DIPHTHERIA, TESTING

SPECIMEN: Culture: nasopharynx, throat. Obtain specimen before antimicrobial chemotherapy is started. Lesions should be cleaned to remove exudate before cultures are obtained. Depress the tongue to expose the pharynx. Use a sterile cotton or dacron swab. Commercially available sterile "Culturette" may be used. Rub vigorously over the posterior pharynx, tonsils, and tonsillar fossae; avoid the tongue, lips, and buccal mucosa. Serum: Antibody levels indicate degree of protection against Corynebacterium diphtheriae.
REFERENCE RANGE: Culture: negative; Antibody levels: protective, ≥ 0.1IU/mL; intermediate protection, <0.1 - ≥0.01IU/mL; no immunity, <0.01IU/mL.
METHOD: Culture on Loeffler or Pai medium. After 16-18 hours, methylene blue stain may be used for rapid presumptive diagnosis of diphtheria. Antibody testing: RIA; Hemagglutination.
INTERPRETATION: Infection by Corynebacterium diphtheriae is rare but should be suspected in patients who have pharyngitis but have not been immunized. Patients may not have a diphtheritic membrane. Even immunized populations may be at risk. Significant portions of the adult population lack immunity and in one study, 19% of children lacked immunity despite immunization (Christenson B, Bottiger M. Scand J Infect Dis. 1986; 18:227-233). Continued use of the tetanus and diphtheria toxoids(Td) booster in adults should help to prevent future outbreaks (Chen RT, et al. AJPH. 1985; 75:1393-1397).

DIRECT ANTIGLOBULIN TEST (see COOMBS, DIRECT)

DISOPYRAMIDE (NORPACE)

SPECIMEN: Red top tube, separate serum; or green (heparin) top tube, separate plasma. Spuriously low plasma concentrations have been reported when blood is collected in certain commercial rubber-stoppered collection tubes. Use all-glass tubes or test any commercial collecting tube thoroughly before use.
REFERENCE RANGE: Range: Therapeutic: 2.0-5.0mcg/mL. Time to Obtain First Serum Specimen (Steady State): 36-48 hours. Peak: 2-3 hours. Sampling Time: Just before next dose. Half-Life: 4 to 10 hours. Toxicity: >7mcg/mL.
METHOD: EMIT, GLC, HPLC
INTERPRETATION: Disopyramide is given by mouth. Disopyramide is used in the treatment of ventricular arrhythmias including ventricular premature beats and ventricular tachycardia. Disopyramide is a Type IA antiarrhythmic agent with properties similar to quinidine and procainamide. Disopyramide has a membrane depressant effect with decreased excitability, conduction, velocity, and contractility. The duration of the action potential is prolonged, as is the duration of the effect refractory period. Disopyramide has a negative inotropic effect.

Disopyramide level should be monitored 36-48 hours after initiating therapy; levels should be drawn as peak and trough.(cont)

Disopyrimide (Cont)

Dosage is modified in the conditions listed in the next Table:

Modification of Dosage of Disopyramide

Condition	Effect	Modification of Dosage
Renal Failure	Half-Life Longer Serum Conc. Higher	Decrease by increasing dosing interval
Acute Myocardial Infarction	Half-Life Longer; Volume of Distribution Higher	Decrease by decreasing maintenance dose. Not indicated for prophylaxis.
Congestive Heart Failure	Avoid disopyramide unless failure is caused by arrhythmias. Monitor closely. Avoid in moderate or severe failure.	
Hypokalemia		Higher Doses Are Necessary
Phenytoin	Induces metabolism	Higher Doses Are Necessary
Warfarin	Disopyramide potentiates	Warfarin dose modified

ADVERSE EFFECTS: Anticholinergic side effects are common (dry mouth, urinary hesitancy, etc). Negative inotropic effect. GI symptoms, syncope, Torsade des pointes may occur.

References: Willis PW. Angiology. 1987; 38:165-173; Siddoway LA, Woosley RL. Clin Pharmacokinetics. 1986; 11:214-222; Nestico PF, et al. Med Clin N Amer. 1984; 68:1295-1319).

DISSEMINATED INTRAVASCULAR COAGULOPATHY (DIC), TEST PROFILE

In DIC, the coagulation factors and platelets are consumed by intravascular coagulation. The laboratory tests and the test results in DIC are shown in the next Table modified from: (Corash L. Primary Care. 1980; 7:423-438):

Test Results in Disseminated Intravascular Coagulopathy(DIC)

Test	Test Result
Prothrombin Time(PT)	Prolonged
Partial Thromboplastin Time(PTT)	Prolonged
Thrombin Time	Prolonged
Plasma Fibrinogen	Decreased
Platelet Count	Decreased (<70,000/mm^3)
Fibrin Split Products(FSP)	Increased
D-Dimer Test	Increased
Euglobulin Lysis Time	Shortened
Peripheral Smear	Fragmented Red Cells

The PTT, PT, fibrinogen level and platelet count are screening tests; confirmatory tests are thrombin time, euglobulin lysis time and fibrin split product(FSP) titer. The combination of prolonged prothrombin time, hypofibrinogenemia, thrombocytopenia and abnormality of one of the confirmatory tests is strong evidence for the diagnosis.

Prothrombin Time(PT): The PT is prolonged due to decreases in the level of factors II, V, IX and fibrinogen.

Partial Thromboplastin Time: The PTT may be normal due to the presence of activated factors or prolonged due to depletion of factors.

Plasma Fibrinogen: Plasma fibrinogen is decreased because the catabolic rate exceeds the synthetic rate. However, fibrinogen is an acute reacting protein and may be elevated prior to the onset of DIC. Following onset of DIC, the level of fibrinogen may not fall below the normal range. This is also true in chronic DIC.

Platelet Count: Thrombocytopenia is found in DIC.

Fibrin Split Products(FSP): The excessive activation of thrombin leads to overactivation of the fibrinolytic system and to increased production of FSP. This is the test that tends to confirm the diagnosis of DIC although other conditions may give rise to FSP. Test for FSP cannot distinguish between degregation of fibrinogen versus fibrin. The D-dimer test can detect fibrinolysis alone.

Thrombin Time: The thrombin time is prolonged because of hypofibrinogenemia (usually at levels of fibrinogen less than 75 mg/dL) and because of circulating anticoagulants in the form of FSP.
Euglobulin Lysis Time: This is a test for increased fibrinolytic activity; the euglobulin lysis test is usually shortened to less than 120 minutes; however, this test is often not available in clinical laboratories.
Peripheral Smear: Beside thrombocytopenia, there is fragmentation of red blood cells which occurs in about 50 percent of patients. Other diseases may be associated with fragmented erythrocytes including thrombotic thrombocytopenia purpura, myelophthisic marrow disorder, acute leukemia and beta-thalassemia-hemoglobulin E disease (Visudhiphan S, et al. N Engl J Med. 1983; 309:113; Chaplinski TJ, personal communication).

Bacterial infection, particularly gram-negative septicemia, is the most common cause of DIC; other common causes are malignancy, trauma, shock, obstetrical problems, and surgery. Clinical conditions complicated by DIC are given in the next Table (Preston FE. Brit J Hosp Med. 1982; 129-137):

Clinical Conditions Associated with DIC

Mechanism	Clinical Condition
Infection	Bacterial especially Gram-Negative Sepsis and Meningococcemia, Viral, Protozoal, Rickettsial especially Rocky Mountain Spotted Fever
Neoplastic	Mucin-Secreting Adenocarcinoma, Acute Leukemia especially Promyelocytic Leukemia, Carcinoma of the Prostate, Lung and Other Organs
Tissue Damage	Trauma, Surgery especially Prostatic Surgery, Heat Stroke, Burns, Dissecting Aneurysms
Obstetric	Obstetrical Complications: Abruptio Placentae, Amniotic Fluid Embolism, Retained Fetal Products, Eclampsia
Immunological	Immune-Complex Disorders, Allograph Rejection, Incompatible Blood Transfusion, Anaphylaxis
Metabolic	Diabetic Ketoacidosis
Miscellaneous	Shock, Snake Bite, Cyanotic Congenital Heart Disease, Fat Embolism, Severe Liver Disease, Cavernous Hemangioma

Hemorrhage and thrombosis are the major criteria for beginning therapy. Thrombosis in arteries and veins of major vessels and small vessels of the digits, skin and organs occurs secondary to formation of fibrin clots. Hemorrhage occurs secondary to depletion of clotting factors and platelets. Therapy is first directed toward the underlying disease process. The fibrinogen concentration and platelet count are usually the first laboratory tests to show improvement. It is possible to determine if DIC is present in patients with liver disease by measuring factors V and VIII; decrease in both factors indicates DIC; decrease in factor V and normal factor VIII indicates liver disease only.

DNA HISTOGRAM (DNA PLOIDY/CELL CYCLE ANALYSIS)

Synonyms: DNA ploidy analysis, DNA Index, % S-phase, proliferation index, cell cycle analysis.
SPECIMEN: Variable depending on method. May be solid fresh or frozen tissue (0.5-1.0g), paraffin embedded formalin fixed tissue, touch preps hematological specimens (blood, bone marrow), lavage or brushing specimens from various sites. Contact laboratory for specific details and suitability of specimen.
REFERENCE RANGE: Normal diploid. Interpretation. Varies by type and site of specimen.
METHOD: Flow cytometry of tissue preparations or image analysis of slides and smears.
INTERPRETATION: This test is performed on neoplastic specimens in order to access prognosis and guide therapy. Analysis may be performed in different sections of the laboratory and may consist of multiple analysis depending on what is ordered.
DNA Ploidy Analysis (DNA Index): This measures the DNA content of the cell and determines the degree of aneuploidy. The DNA index (DI) is the ratio of the DNA content of the tumor cells to that of a normal diploid standard. The DNA aneuploid population is that population of tumor cells when the DI is neither 1.0 (diploid) or 2.0 (tetraploid).
% S-phase (proliferation index, cell cycle analysis): This measures the mitotic activity of the tumor cells. The % S-phase is the fraction of cells actively synthesizing DNA. In general, tumors with normal ploidy (diploid) have a better prognosis than the aneuploid tumors.
Breast Cancer: Prognosis of breast cancer is dependent on DNA content (Fallenins AG, et al. Cancer. 1988; 62:331-341). Women with node negative diploid breast cancer, but with a high S-phase fraction have a poorer prognosis.
Bladder Cancer: Employing both DNA ploidy analysis and cytology increases the detection rate of bladder cancer in the screening of symptomatic patients (Murphy WM, et al. J Urol. 1986; 136:815-819). DNA ploidy is useful in accessing the prognosis of bladder tumors (Blomjons ECM, et al. Am J Clin Path. 1989; 91:243-248).
Prostate Cancer: DNA ploidy analysis appears useful in accessing the malignant potential and prognosis of prostate lesions. In one study, no focal (stage Al) diploid tumors progressed (McIntire TL, et al. Am J Clin Path. 1988; 89:370-373). Patients with aneuploid tumors at all stages do poorly compared to diploid tumors.
Colon Cancer: A NIH Consensus Conference recommends stage II (Dukes B) colon cancer patients receive adjuvant therapy if their DNA content is aneuploid or S-phase fraction is high (NIH Consensus Conference. JAMA. 1990; 264:1444-1450). Adjuvant therapy is also recommended if the preoperative CEA is >5ng/mL or there is deletion of 17p or 18q.
Reference: See National Cancer Institute DNA Consensus Conference. Cytometry. 1993; 14:471-500 for a discussion of the application of DNA ploidy analysis on neoplasms of breast, bladder, prostate, colon and the hematopoietic system.

DRUG SCREEN (TOXICOLOGY)

SPECIMEN: 30mL of random urine; red top tube, separate serum; gastric aspirate.
REFERENCE RANGE: Negative
METHOD: Spot tests; Thin-Layer Chromatography (TLC); EMIT; Gas-Liquid Chromatography (GLC); High-Performance Liquid Chromatography (HPLC); Gas Chromatography-Mass Spectrometry (GC/MS).
INTERPRETATION: Drug screens are used for various purposes. Tests are useful in the emergency department for the evaluation of overdosed patients. Drug screens are also used to test for drugs of abuse in athletes and in various employment situations.

In the emergency department, toxicology screens have utility in various clinical situations, as given in the following Table (Osterloh JD. Emerg Med Clin N Amer. 1990; 8:693-723):

Clinical Situations in Which Toxicology Testing is Useful
"Rule-In" Value - Absence of a clear clinical picture
"Rule-Out" Value - Drugs are an unlikely cause and differential diagnosis is limited
Documentation that the working diagnosis was correct

The clinical presentation may suggest certain diagnoses; toxicologic causes associated with specific clinical signs and symptoms are given in the following Table (Osterloh HD, already cited):

Clinical Presentations and Toxicologic Causes

Presentation	Toxicologic Causes
Coma	Narcotics, Sedatives, Antipsychotics, Alcohol, Tricyclics, Benzodiazepines
Seizures	Theophylline, Tricyclics, Isoniazid, Stimulants, Camphor, Carbon Monoxide, Hypoglycemic Agents, Alcohol Withdrawal
Psychosis or Altered Mental Status	Anticholinergics, Stimulants, Withdrawal
Acidosis	Salicylates, Ethanol, Methanol, Ethylene Glycol, Cyanide, Drugs Causing Seizures
Respiratory Depression	Narcotics, Sedatives, Benzodiazepines
Pulmonary Edema	Salicylates, Narcotics, Iron, Paraquat
Arrhythmias	Tricyclics, Quinidine, Anticholinergics, Beta Blockers, Digoxin, Lithium, Antipsychotics, Organophosphates
Hypotension	Narcotics, Sedatives, Tricyclics, Antipsychotics, Beta Blockers, Theophylline, Iron
Hypertension	Cocaine, Amphetamines, Cyanide, Nicotine, Clonidine
Ataxia	Antiepileptics, Barbiturates, Alcohol, Lithium, Organomercury

Many drugs are detected by drug screens. Common drugs included on most toxicology screens are listed in the following Table (Osterloh JD, already cited):

Common Drugs Included on Most Toxicology Screens

Classification	Drug
Alcohols	Ethanol, methanol, isopropanol, acetone
Barbiturates or Sedatives	Phenobarbital, pentobarbital, secobarbital, amobarbital, butalbital, glutethimide, methaqualone
Anticonvulsants	Phenytoin, carbamazepine, primadone, phenobarbital
Benzodiazepines	Diazepam, alprazolam, temazepam, chlordiazepoxide
Antihistamines	Diphenhydramine, chlorpheniramine, brompheniramine, tripelennamine
Antidepressants	Amitriptyline, nortriptyline, doxepin, imipramine, desipramine, trazedone, amoxapine, maprotiline
Antipsychotics	Trifluoperazine (Stelazine), prochlorperazine (Compazine), chlorpromazine (Thorazine)
Stimulants	Amphetamine, methamphetamine, phenylpropanolamine, ephedrine, MDA, MDMA, cocaine, phencyclidine (PCP)
Narcotic Analgesics	Heroin, morphine, codeine, meperidine, pentazocine, propoxyphene, methadone
Other Analgesics	Salicylates, acetaminophen
Cardiovascular Drugs	Lidocaine, propranolol, metoprolol, quinidine, procainamide, verapamil
Others	Theophylline, caffeine, nicotine, oral hypoglycemics, strychnine

(cont)

Drug Screen (Cont)

Drugs that may not be detected by some emergency toxicology screens is given in the following Table (Osterloh JD, already cited):

Toxins Not Detected by Some Emergency Toxicology Screens

Problem	Drug
Too Polar	Antibiotics, diuretics, isoniazid, ethylene glycol, lithium
Too Non-Polar	Steroids, THC, digoxin
Too Non-Volatile	Plant and fungal alkaloids, some phenothiazines
Too Volatile	Aromatic and halogenated hydrocarbons, anesthetic gases, noxious gases (hydrogen sulfide, nitrogen dioxide, carbon monoxide)
Low Concentration	Clonidine, fentanyl, colchicine, ergot alkaloids, LSD, dioxin, digoxin, THC
Toxic Anions (Too Polar)	Thiocyanate, cyanide, fluoride, bromide, borate, nitrite
New Drugs	e.g., buspirone

A useful screening test for volatiles is serum osmolality. The contribution to serum osmolality of each 100 mg/dL of ethanol, ethylene glycol and methanol is given in the next Table:

Contribution to Serum Osmolality (Each 100mg/dL of Volatile Substance)

Substance	Contribution
Ethanol	21.7
Methanol	31.0
Ethylene Glycol	16.3

Thus, if serum osmolality is normal, the patient does not have ethanol, methanol or ethylene glycol toxicity.

Besides drug screen and serum osmolality, other useful laboratory tests are as follows: electrolytes, Na^+, K^+, Cl^-, CO_2 content; glucose, creatinine, urea, arterial blood gases, urinalysis, and liver transaminases.

Clinical symptoms by drug class are given in the following Table (Osterloh JD, already cited):

Symptoms of Overdosage by Drug Class

Drug Class	Clinical Symptoms
Narcotics	CNS depression; slowed respiratory rate; temperature normal or low; pinpoint pupils; deep tendon reflexes usually decreased
Alcohols or Barbiturates	CNS depression; ataxia; temperature usually decreased; deep tendon reflexes decreased; metabolic acidosis with alcohols and ethylene glycol except isopropanol
Anticholinergics*	Delirium; increased pulse; increased temperature; skin flushed, warm, pink; drug (no sweat); decreased bowel sounds; urinary retention; blurred vision; arrhythmias, prolonged QT
Stimulants	Acute psychosis; increased pulse, blood pressure, temperature, respiratory rate; agitation, dilated pupils; sweating; seizures
Antidepressants	Anticholinergic syndrome; hypotension; coma; seizures; widened QRS, QT; ventricular arrhythmias
Benzodiazepines	CNS depression; respiratory depression; deep tendon reflexes intact
Phenothiazines	Anticholinergic syndrome; decreased blood pressure and temperature; pinned pupils; rigidity, dystonia, torticollis; seizures
Salicylates	Abdominal pain; respiratory alkalosis (early); metabolic acidosis; shock; diaphoresis; hypoglycemia
Theophylline	Tachycardia; hypotension; hypokalemia; seizures
Iron	Abdominal pain; GI bleeding; hypotension; hypovolemia; acidosis; renal failure; cardiovascular collapse
Lithium	Tremor; chorea; abdominal pain; hyperreflexia; rigidity; seizures
Isoniazid	Metabolic acidosis; seizures; hepatitis
Oral Hypoglycemics	Hypoglycemia; coma; diaphoresis
Acetaminophen	Liver necrosis (late)
Beta Blockers	Bradycardia; hyperglycemia; hypotension with slowed conduction

*Anticholinergic symptoms are caused by various drugs including atropine, scopolamine, antihistamines, phenothiazines, tricyclics, quinidine, amantadine, and mushrooms.

Poison Control Centers: Information on treatment of poisons is available from poison control centers.

See review: Osterloh JD. "Utility and Reliability of Emergency Toxicologic Testing." Emerg Med Clin N Amer. 1990; 8:693-723.

Mandatory Drug Testing: Mandatory drug testing is based on the assumptions that there is a major drug abuse problem in the United States workforce, and that mandatory drug testing is a cost-effective method of addressing this problem. Guidelines were adopted by the National Institute on Drug Abuse (NIDA) which is a part of the Alcohol, Drug Abuse, and Mental Health Administration (ADAMHA), a division of the Department of Health and Human Services (DHHS).

Drug testing may be required in various situations, including pre-employment, prior to transfers or promotions, periodic, post-accident, reasonable cause, return-to-duty, follow-up, or random. Testing under the NIDA guidelines include the following components (Gerson B, Hahn R. Clin Lab Med. 1990; 10:517-529):

Components of NIDA Guidelines
Collect urine samples as described in NIDA guidelines
Initial screening by immunoassay in a NIDA certified laboratory
Confirm screen-positive results by gas chromatography/mass spectroscopy (GC/MS) in the same laboratory
Report all results to the medical review officer (MRO)

Concentration cutoffs for a positive report of substance abuse are given in the following Table (Gerson B, Hahn R, already cited):

Concentration Cutoffs for a Positive Drug Test

Drug or Class	Screen (ng/mL)	Confirm (ng/mL)
Marijuana Metabolites	100	15
Cocaine Metabolites	300	150
Opiate Metabolites	300	
Morphine		300
Codeine		300
6-Monoacetylmorphine		25
Phencyclidine (PCP)	25	25
Amphetamines	1000	
Amphetamine		500
Methamphetamine		500

Potential sources of error in testing urine samples for drugs of abuse are given in the following Table (Schwartz RH. Arch Intern Med. 1988; 148:2407-2412):

Sources of Error in Testing Urine Specimens for Drugs
Substitution of urine from a drug-free individual or of apple juice for the urine of the person being tested
Adulteration of the specimen (water, sodium chloride, vinegar, ammonia water, sodium hypochlorite, or soap)
Interference from nonprescription drugs or foods (especially a problem with amphetamine or opiates)
Collection or storage error
Technical error
Administrative error (improper labeling of specimens, inaccurate recording of results)

See also MacKenzie RG, et al. Pediatr Clin N Amer. 1987; 34:423-436.

D-XYLOSE (see XYLOSE)

ENDOCRINE EMERGENCIES

A list of conditions that may present as endocrine emergencies and the laboratory tests and findings are given in the next Table, as modified from Doggett P. Brit J Hosp Med. 1979; 38-43:

Conditions that May Present as Endocrine Emergencies and the Laboratory Tests and Findings

Condition	Laboratory Tests and Findings
Diabetes Mellitus	
Ketoacidosis	Serum Glucose; Usually in Range of 300 to 800 mg/dL Ketone Bodies; Elevated in Blood and Urine Blood Gases: pH: Usually Below 7.1 HCO_3: Usually Less Than 10 pCO_2: Usually Low Due to Compensation Other: Hemoglobin: Usually High due to Dehydration WBC's: May Be Raised Even in Absence of Infection
Hyperosmolar Coma	Plasma Glucose often 900 mg/dL or greater Plasma Osmolality Usually 350-400 mosmol/kg
Lactic Acidosis	Plasma Lactate is Very High Increased Anion Gap pH: Usually Below 7.0
Hypoglycemia	Blood Glucose is Low
Thyroid Gland:	
Myxedema Coma	Decreased T-4 and Free T-4; Elevated TSH Hyponatremia (Dilutional) Hypercholesterolemia Macrocytosis With or Without a Low Hemoglobin Acidosis: Increased pCO_2 and Hypoxia Secondary to Alveolar Hypoventilation
Thyroid Crisis	Elevated T-4 and Free T-4, decreased sTSH
Adrenal Gland:	
Addisonian Crisis	Low Plasma Cortisol and High Plasma ACTH Low Serum Sodium; High Serum Potassium
Iatrogenic Steroid Deficiency	
Acute Hypercortisolism	Hypokalemic Alkalosis Elevated Plasma Cortisol
Pheochromocytoma	Phentolamine Test (Use when Diagnosis has not previously been made); Increased Plasma and Urine Free Catecholamines; Metanephrines; VMA
Pituitary Gland and Hypothalamus:	
Hypopituitary Coma	Decreased Plasma T-4, Decreased Cortisol, Decreased TSH and Decreased ACTH Hypoglycemia Dilutional Hyponatremia
Diabetes Insipidus	Simultaneous Plasma and Urine Osmolality; Increased Plasma Osmolality (Usually Around 350 mosmol/kg) and Decreased Urine Osmolality (Usually Around 150 mosmol/kg) Large Urine Volume
Inappropriate ADH Activity	Simultaneous Plasma and Urine Osmolality Decreased Plasma Osmolality and Increased Urine Osmolality Dilutional Hyponatremia
Parathyroid Gland:	
Hypocalcemia	Serum Calcium <7 mg/dL (correct for albumin)
Hypercalcemia	Serum Calcium >13 mg/dL

EOSINOPHIL COUNT, TOTAL

SPECIMEN: Lavender(EDTA) top tube; deliver specimen to laboratory within one hour; stable up to 4 hours in the refrigerator.
REFERENCE RANGE: Newborn (<24 hours old): 20-850/cu mm; 1 day to 1 year old: 50-700/cu mm; 1 year to adult: 0-450/cu mm.
METHOD: Manual Method: Unopette (Becton, Dickinson and Co.) technique; the red cells are lysed and eosinophils only stain bright orange-red. Automatic Methods: (1) Selective cytochemical stains in a liquid milieu; sensing device measures light scatter and light absorption. (2) Stained blood smears; automated microscope with computerized morphologic and tinctorial criteria for cell identification. (3) Unstained cells are classified by phase microscopy on the basis of size and refractive index in a liquid milieu; classified on basis of size and density.
INTERPRETATION: The causes of increased blood eosinophils are listed in the next Table (Wolfe MS. Infect Dis Clin N Amer. 1992; 6:489-502; Nutman TB, et al. Allergy Proc. 1989; 10:33-46, 47-62):

Causes of Increased Blood Eosinophils
Parasitic Diseases: Trichinosis; Visceral Larva Migrans (Toxocara Canis or T. Cati); Ascaris Pneumonia; Strongyloides Filariasis; Schistosomiasis; Microfilariae (Causes Tropical Pulmonary Eosinophilia)
Allergic Diseases: Asthma; Seasonal Rhinitis (Hay Fever)
Skin Disorders: Atopic Dermatitis; Eczema; Acute Urticarial Reactions; Pemphigus; Dermatitis Herpetiformis
Pulmonary Eosinophilias: Loeffler's Syndrome; Pulmonary Infiltrate with Eosinophilia (PIE Syndrome); Hypersensitivity Pneumonitis [e.g. Allergic Bronchopulmonary Aspergillosis(ABPA)]; Tropical Pulmonary Eosinophilia (caused by Microfilariae)(Review: Umeki S, Arch Intern Med. 1992; 152: 1913-1919)
Collagen Vascular Diseases: Dermatomyositis; Progressive Systemic Sclerosis; Eosinophilic Fascitis; Hypersensitivity Vasculitis
Malignancies: Ovarian Carcinoma; Epidermoid Carcinoma (Cervix, Uterus, Penis, Lip, Tongue); Villous Carcinoma of the Bladder; Carcinoma of Lung; Metastasis from Vulvar or Penile Carcinoma; Adenocarcinoma of Colon
Immunodeficiency: Wiskott-Aldrich Syndrome; Hyperimmunoglobulinemia E; IgA Deficiency; Nezelof Syndrome
Hematologic Disorders: Polycythemia Vera; Pernicious Anemia; Myelofibrosis; Myeloid Metaplasia; Chronic Myelogenous Leukemia
Drugs: Arsenicals; Chlorpromazine (Phenothiazines); Gold; Iodides; Nitrofurantoin; Para-aminosalicylic Acid; Ampicillin; Phenytoin; Streptomycin; Sulfonamides
Hypereosinophilic Syndrome(HES) - Liesveld JL, Abboud CN. "State of the Art: The Hypereosinophilic Syndromes." Blood Reviews. 1991; 5:29-37).
Fungal Infections: Coccidioidomycosis; Histoplasmosis; Allergic Bronchopulmonary Aspergillosis (ABPA)
Other: Wegener's Granulomatosis; Eosinophilia-Myalgia Syndrome; Inflammatory Bowel Disease

EPSTEIN-BARR VIRUS(EBV) ANTIBODIES, SERUM

SPECIMEN: Red top tube, separate serum and refrigerate.
REFERENCE RANGE: Four antibodies to the Epstein-Barr virus are typically measured; these are antibodies to viral capsid antigen, (anti-VCA, IgM and IgG), antibody to early antigen, diffuse or restricted pattern [anti-EA(D) or anti-EA(R)] and antibody to Epstein-Barr nuclear antigen(EBNA). The reference ranges are as follows:

Antibodies to Epstein-Barr Virus

Antibody	Reference Range
Anti-Viral Capsid Antigen(VCA)IgM	< 1:10
Anti-Viral Capsid Antigen(VCA)IgG	< 1:10
Anti-Early Antigen (Anti-EA)	< 1:10
Anti-EB Virus Nuclear Antigen(EBNA)	< 1:5

METHOD: Indirect immunofluorescence for anti-VCA(IgG or IgM) and anti-EA; anti-EBNA by anti-complement immunofluorescence.
INTERPRETATION: Epstein-Barr virus(EBV) is the cause of infectious mononucleosis. Tests of antibodies to Epstein-Barr virus are recommended only when a screening procedure, such as Monospot Test or heterophil absorption are negative and the diagnosis of infectious mononucleosis is suspected. The Monospot test and heterophil tests are negative in about 10% of adult patients with infectious mononucleosis; the changes in the different antibodies in infectious mononucleosis are shown in the next Figure:

Changes in Heterophil Antibodies and Epstein-Barr Virus Antibodies During Course of Infectious Mononucleosis

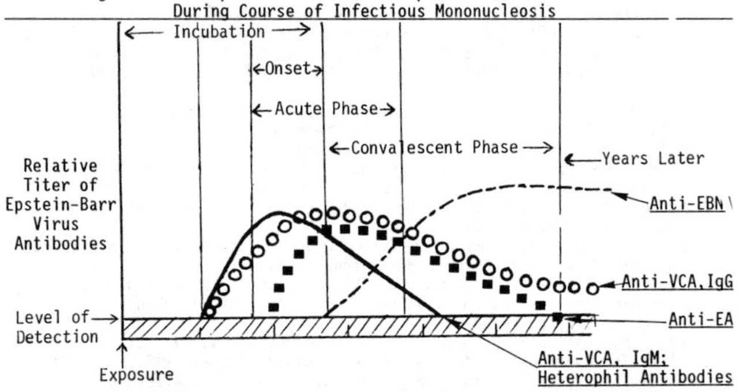

During the course of infectious mononucleosis, heterophil antibodies are detected in up to 90 percent of patients while antibodies to viral capsid antigen (anti-VCA, IgM and IgG), Anti-EA, and antibody to Epstein-Barr nuclear antigen (anti-EBNA) are detected in 100 percent of patients. The capsid is the protein coat that surrounds the nucleic acid, DNA or RNA, core of viruses.

Anti-VCA, IgM and anti-VCA, IgG and anti-EA reach high titers during the acute phase of infectious mononucleosis. Anti-VCA, IgM antibodies peak about the second week of illness and then, gradually disappear within one or two months, that is, soon after the disease subsides. Anti-VCA, IgG antibodies decrease to lower levels which persist for many years. Presence of anti-VCA, IgM, antibodies indicate recent, primary infection with EBV. A four-fold increase in anti-VCA, IgG antibodies is diagnostic of active disease; a single titer of 1:320 is strongly suggestive and a titer of 1:640 is definitive of active or recent EBV infection. Early antigens are proteins produced when latently infected cells are stimulated to produce virus. Immunofluorescent staining may be diffuse(D) or restricted(R). In most patients the response is diffuse, peaks at 3-4 weeks and persists for several months. IgG-EA is a marker for acute infection (Chetham MM, Roberts KB. Pediatr Ann. 1991; 20:206-213). Anti-EBNA antibodies may begin to appear in the fourth to eight week but are usually delayed for about six months after onset of illness. Anti-EB nuclear antibodies persist for life.

The specific tests for the diagnosis of infectious mononucleosis are given in the next Table (McCarthy LR. Clin Microbiol. Newsletter. 1984; 6:17-20):

Epstein-Barr Virus Specific Tests for the Diagnosis of Infectious Mononucleosis

EBV Test	Comment
IgM anti-VCA	Antibody formed early in course of disease, disappears in convalescent. Positive test indicates infection.
IgG anti-VCA	Antibody formed during acute illness; antibody remains elevated for prolonged duration. Fourfold increase in titer diagnostic. Formation of IgG may be delayed, and not present in early stage of acute infection.
IgG anti-EA	Antibody formed during acute illness; antibody peaks at 3-4 weeks, persists for months.
Anti-EBNA	Antibody formed during convalescence. Presence of antibody indicates past disease.
Virus Culture Throat	Positive during acute disease; intermittent/continuous shedding up to three months after disease.
Leukocyte Culture	Positive during acute disease. Cultures may be positive during convalescence.

The diagnosis of infectious mononucleosis is usually straight forward; the "classic" findings in infectious mononucleosis are as follows: malaise, fever, sore throat or pharyngitis, lymphadenopathy, mild to moderate hepatitis, splenomegaly, and positive Monospot or heterophil serology. Approximately 80% of patients present with pharyngitis and/or lymphadenopathy; the remaining 20% have fever without either pharyngitis or lymphadenopathy. About 50% of patients have splenomegaly.

The atypical lymphocytes are not diagnostic of infectious mononucleosis; they are also seen in other viral infections, such as hepatitis, mumps, rubella, rubeola, varicella, Herpes simplex, H. zoster, and cytomegalovirus infections, in drug-induced hepatitis and in toxoplasmosis. The atypical lymphocytes are T-cells.

The diagnosis of infectious mononucleosis may be difficult when the clinical features are atypical or absent and the Monospot and heterophil tests are negative. Serologic testing should be considered in patients with chronic or recurrent symptoms suggestive of EBV, patients with serious illness or acute neurologic disease of unknown etiology, evaluation of patients with suspected leukemia or other lymphoproliferative disorders, or suspected false positive Monospot or heterophil tests. Interpretation of EBV serology is given in the following Table (Sumaya CV. Pediatr Infect Dis. 1986; 5:337-342):

Interpretation of EBV Serology

Condition	IgM VCA	IgG VCA	IgG EA(D)	IgG EA(R)	Anti-EBNA
Susceptible	-	-	-	-	-
Acute Primary Infection(Adult)	+	+	+	-	-
Acute Primary Infection(Infant, Atypical, Asymptomatic)	+	+	-	+	-
Past Infection	-	+	-	-	+
Reactivation	+ or -	+	+ or	+	+

Evidence of Epstein-Barr virus(EBV) infection is present in almost every case of Burkitt's lymphoma in Africa; about 20% of sporadic or non-African Burkitt-like tumors are also associated with EBV. EBV is also associated with undifferentiated carcinoma of the nasopharynx, and a cytomegalovirus-like disease in renal transplant recipients and in post-transfusion syndrome. These conditions are associated with high levels of EBV antibodies.(cont)

Epstein-Barr Virus (Cont)

Complications of infectious mononucleosis are given in the following Table (Chetham MM, Roberts KB. Pediatr Ann. 1991; 20:206-213; Nelson JD. Pediatr Infect Dis. 1984; 3:S36-S37; Andiman WA. Pediatr Infect Dis. 1984; 3:198-203):

Complications of Infectious Mononucleosis

System	Complications
Neurologic	Aseptic Meningitis; Encephalitis; Transverse Myelitis; Guillain-Barré Syndrome; Optic Neuritis; Uveitis; Facial Nerve Palsy; Bell's Palsy; Seizures; Coma; Acute Cerebellar Ataxia
Cardiovascular	Myocarditis, Pericarditis
Respiratory	Upper Airway Obstruction Leading to Cor Pulmonale; Interstitial Pneumonia
Gastrointestinal	Hepatitis; Pancreatitis; Proctitis; Massive Hepatic Necrosis
Renal	Nephritis; Nephrosis
Hematologic	Hemolytic Anemia; Thrombocytopenia; Granulocytopenia; Pancytopenia; Disseminated Intravascular Coagulopathy
Immunologic	Fatal X-Linked Lymphoproliferative Disease

ERYTHROCYTE SEDIMENTATION RATE(ESR)

SPECIMEN: Lavender (EDTA) top tube

REFERENCE RANGE: Newborn: 0-2mm/hr; Neonates and children: 3-13mm/hr. Post-adolescent male (<40 years): 1-15mm/hr; Post-adolescent female (<40 years): 1-20mm/hr. The ESR (Westergren Method) varies with age; the maximum normal ESR at a given age is calculated using the formulas: Men: Age in years/2; Women: (Age in Years + 10)/2 (Miller A, et al. Br Med J. 1983; 286:266).

METHOD: The erythrocyte sedimentation rate (ESR) is the measurement, under standard conditions, of the rate of settling of erythrocytes in anticoagulated blood. Westergren sedimentation; Wintrobe; zeta method.

INTERPRETATION: The ESR is used as a marker of tissue inflammation, some causes of elevation of ESR are given in the next Table (Lascari AD. Pediatric Clinics of North America. 1972; 19:1113-1121):

Causes of Elevation of ESR

Infections	Collagen Diseases
Majority of bacterial infections	Rheumatic fever
Infectious hepatitis	Rheumatoid arthritis
Cat scratch disease	Lupus erythematosus
Post-perfusion syndrome	Dermatomyositis
Primary atypical pneumonia	Scleroderma
Tuberculosis	Systemic vasculitis
Secondary syphilis	Henoch-Schönlein purpura
Leptospirosis	Mediterranean fever
Systemic fungal infections	Renal
Hematologic and Neoplastic	Acute glomerulonephritis
Severe anemia	Chronic glomerulonephritis with renal failure
Leukemia	Nephrosis
Lymphoma	Pyelonephritis
Metastatic tumors	Hemolytic-uremic syndrome
Chronic granulomatous disease	Miscellaneous
Gastrointestinal	Hypothyroidism
Ulcerative colitis	Thyroiditis
Regional ileitis	Sarcoid
Acute pancreatitis	Infantile cortical hyperostosis
Lupoid hepatitis	Surgery, burns
Cholecystitis	Drug-hypersensitivity reactions
Peritonitis	

Many other conditions are associated with elevated sedimentation rate (see also Cunha BA. Hosp Phys. March 1989; 12-18).

The ESR has a relatively high sensitivity and low specificity. Changes in ESR may provide information concerning the activity of a disease process and serial readings may reflect therapeutic response. Normal results do <u>not</u> exclude serious illness.

An increased ESR is caused by enhanced erythrocyte aggregation; this is caused by increased levels of asymmetrical macro-molecules, principally fibrinogen and the globular proteins in the blood. Normally, red blood cells have a negative charge on their surface; the like charges on red cells cause these cells to repel each other. Plasma proteins, especially fibrinogen, tend to adhere to the red cell membrane and neutralize the surface charges and makes the cells more likely to aggregate forming stacks or rouleaux. The aggregated (stacked) cells have a higher ratio of mass to surface area than single cells and, therefore will fall out from the plasma more readily.

Conditions that alter ESR by interfering with the formation of rouleaux are as follows: acanthocytosis and poikilocytosis. Anemia may be associated with elevated ESR levels. In iron-deficiency anemia, there is reduction in the ability of smaller red cells to sediment; this may be compensated by the accelerating effect of an increased proportion of plasma.

The ESR is decreased in the conditions listed in the following Table (Cunha BA. Hosp Phys. March 1989; 12-18):

Causes of Low ESR (Approx. 0mm/hr)
High dose steroid or salicylate therapy
Severe anemia
Cachexia
Massive hepatic necrosis
Disseminated intravascular coagulation(DIC)
Polycythemia vera
Trichinosis
Chronic lymphocytic or myeloid leukemia
Hypofibrinogenemia
Macroglobulinemia (hyperviscosity syndrome)
"Chronic mononucleosis" syndrome

An ESR of 20 mm/hr in a patient with sickle cell anemia may equate to a value of 100 mm/hr in a normal individual and is a signal of possible acute infection (Thomas RE. Personal Communication).

In a study by Kirkeby AK and Leren P (Nordisk Medicin. 1952; 48:1193) of patients with ESRs of more than 100mm, 24% had pneumonia, 17% had chronic renal disease and 19% had malignancy (Zacharski LR, Kyle RA. JAMA. 1967; 202:264-266; Gibbs D. Brit J Hosp Med. May 1982; 493-496; Cunha BA. Diagnosis. Feb. 1983; 62-69).

ESTRADIOL

SPECIMEN: Red top tube, separate serum. The serum is stable at room temperature for 1 week, at refrigerator temperature for 1 month, in a frost-free freezer for 1 year and in a non-defrosting freezer for 3 years. Lavender(EDTA) top tube, separate plasma.

REFERENCE RANGE:

Reference Range for Estradiol

	pg/mL
Female	
Prepubertal, Normal	4.0-12.0
Ovulating, Normal	
Early Follicular	30-100
Late Follicular	100-400
Luteal Phase	50-150
Pregnant, Normal	Up to 35,000
Human Menopausal Gonadotropin Treatment: Therapeutic Range	350-750
Postmenopausal or Castrate	5.0-18.0
On Oral Contraceptives	Under 50
Male Normal	
Prepubertal	2.0-8.0
Adult	10-60

METHOD: RIA

INTERPRETATION: Estradiol is most often measured in the evaluation of amenorrhea; other clinical applications of serum estradiol measurements include precocious puberty in females; ovarian induction; and gynecomastia in males.

Amenorrhea: If estradiol is less than 30pg/mL, obtain serum FSH levels in order to differentiate primary from secondary ovarian failure.

Precocious Puberty in Females: Elevation of serum estradiol suggests estrogen secreting tumor or true precocious puberty associated with elevation of serum gonadotropins(FSH, LH) which are in the adult female range.

Ovulation Induction: Serum estradiol is measured during induction of ovulation (Blacker CM. Endocrinol Metab Clin N Amer. 1992; 21:57-84; Moghissi KS. Endocrinol Metab Clin N Amer. 1992; 21:39-55).

Gynecomastia in Males: Elevated estradiol levels have been observed in patients with cirrhosis and tumors of the liver and gonadal tumors.

ESTRIOL

SPECIMEN: Red top tube, separate serum and freeze; specimen must not remain at room temperature for more than one hour. Urine may also be tested.
REFERENCE RANGE: The change in estriol during pregnancy is shown in the next Figure (Specialty Laboratories, Inc. 800-421-4449):

Serum Estriol During Pregnancy (ng/mL)

Week	Unconjugated Estriol	Total Estriol	Week	Unconjugated Estriol	Total Estriol
18	2.4-7.2		31	5.8-17.2	24-125
19	2.7-8.0		32	6.3-19.3	26-140
20	3.0-9.0		33	6.8-20.6	29-155
21	3.3-10.1		34	7.4-22.3	33-170
22	3.8-11.2		35	8.0-24.2	39-200
23	4.2-12.5		36	8.9-26.3	42-220
24	4.4-13.2		37	9.8-30.1	48-250
25	4.4-13.3		38	11.5-35.0	55-295
26	4.4-13.4	18-85	39	13.3-39.6	65-320
27	4.6-13.4	19-90	40	14.2-43.0	75-340
28	4.6-13.5	19.5-97	41	15.0-44.0	85-340
29	5.0-14.2	21-105	42	15.0-44.0	100-325
30	5.5-15.8	22-115			

Urinary Estriol During Pregnancy

Week	Estriol (mg/24 hours)
32-34	6-27
35-37	8-39
38-40	11-44

Estriol increases until term, reaches a plateau at the 40th week of gestation.
METHOD: RIA
INTERPRETATION: Estriol is the main estrogen in the blood and urine of pregnant females. Estriol is synthesized in the placenta from fetal origin precursors. Estriol is conjugated in the mother's liver and excreted in the mother's urine.

Many factors influence estriol measurement. There are wide daily fluctuations and circadian variation in serum estriol. Conjugation and excretion of estriol are influenced by hepatic and renal functions; urinary estriol measurements depend on accurate collection of 24 hour urine specimens.

Estriol measurements (Total, Unconjugated, or Urinary) have been used to monitor fetal well being. Persistently low or rapidly falling levels suggest fetal distress. Generally, a baseline for each individual patient is defined; for example, the average or highest of three most recent estriol determinations may be used as baseline and a decline of this measurement by 40 percent or more may indicate fetal distress (Peters JB. Specialty Laboratories, Inc).

Estriol assay is not routinely used for predicting intrauterine growth retardation or for daily monitoring of insulin-dependent pregnant diabetic women because the number of false positive and false negative tests is unacceptably high (Miodovnik M, et al. Am J Perinatol. 1988; 5:327-333; Ray DA, et al. Am J Obstet Gynecol. 1986; 154:1257-1263; Evans JJ, et al. Clin Chem. 1984; 30:138-140). Ultrasonography, nonstress tests and contraction stress tests are typically used for monitoring these pregnancies.

Unconjugated estriol, in combination with maternal serum alpha-fetoprotein and human chorionic gonadotropin, may be a useful screening test for fetal Down's syndrome and other chromosomal anomalies (Haddow JE, et al. N Engl J Med. 1992; 327:588-593; Crandall BF, et al. Am J Obstet Gynecol. 1993; 168:1864-1869; Palomaki GE, et al. Prenatal Diagnosis. 1992; 12:925-930; Herrou M, et al. Prenatal Diagnosis. 1992; 12:887-892).

ESTROGEN/PROGESTERONE RECEPTORS

SPECIMEN: Depends on assay. The widely used dextran-coated charcoal technique (DCC) requires submission of 1g of trimmed, fat-free, fresh frozen tumor tissue. The more recently developed enzyme immunoassay (EIA) requires only 0.1-0.2g of similarly prepared tissue. The immunocytochemical assay (ICA) can be performed on frozen sections or on slides of paraffin embedded tissue.

The protocol for preparation of tissue for the DCC technique is as follows: Estrogen-receptors are very labile at room temperature; therefore, the tumor tissue should be packed in ice, without direct contact, as soon as it is removed. The tissue must not be placed in formalin. A biopsy specimen (at least one gram) should be placed in a container surrounded by ice in the operating room. "Normal" breast, to be used as control, should be obtained from an adjacent area. The specimens should be transported immediately to the laboratory where the pathologist should trim the tissue and obtain a small specimen for microscopic examination. The trimmed specimen should be weighed, and cut into small pieces. If the assay is not done on the same day the specimen should be immediately frozen in liquid nitrogen.

REFERENCE RANGE: Depends on laboratory and methodology. In some laboratories, values >3femtomol/mg cytosolic protein is evidence of estrogen receptor positivity. In other laboratories, 10-15femtomol/mg cytosolic protein is taken as the cutoff value.

ICA determinations are read by eye (qualitative) or image analysis (quantitative). These results are expressed as percent positive tumor cells.

METHODS: DCC is performed by incubating cellular extracts with radiolabelled estrogen and absorbing the estrogen-receptor complex on charcoal. EIA is performed by incubating cellular extracts with receptor specific antibody-enzyme complexes. ICA is performed by immunocytochemistry with specific anti-receptor antibodies on tissue sections.

The advantages of the various methods is as follows: The DCC method is the classical technique for measurement of receptor status and is the "gold standard" to which the other techniques are compared. EIA methods can be performed on less tissue and assay is not confounded by the presence of endogenous estrogen or anti-estrogen (Tamoxifen) therapy. ICA is performed on tissue sections so the actual tumor cells containing receptors are identified. It is amenable to very small pieces of tissues such as those obtained by fine needle aspiration biopsy and receptor status can be determined rectospectively on archived tissue. There is good to excellent correlation of receptor status results of tumors analyzed by these various techniques (Hanna W, Mober BG. Am J Clin Path. 1989; 91:182-186).

INTERPRETATION: Estrogen/progesterone receptor assays are performed on malignant breast tissue to assess prognosis and to determine which patients will most probably benefit from chemotherapy or hormonal manipulation. Estrogen/progesterone receptor assays have also been performed on tumors of the female genital tract.

The presence of estrogen or progesterone receptor is predictive of longer disease-free period and longer survival than patients who are receptor negative regardless of their treatment (Hubay CA, et al. Breast Ca Res Treat. 1981; 1:77-82). Determination of progesterone receptor is of equal or greater value for predicting disease free survival (Clark GM, et al. N Engl J Med. 1983; 309:1343-1347).

In general, receptor negative tumors do not respond to endocrine therapy, the majority of estrogen receptor positive tumors respond, and even greater response rates are achieved with tumors containing both estrogen and progesterone receptors. This is illustrated in the next Table:

Response to Endocrine Therapy	
Estrogen (-), Progesterone (-)	5-10%
Estrogen (+), Progesterone (-)	30%
Estrogen (+), Progesterone (+)	>75%

In another approach, the estrogen regulated proteins pS2 and cathepsin D can be detected immunocytochemically in histological sections. Expression of these products presumably indicate the presence of an intact estrogen receptor-response apparatus. Expression of pS2 is said to correlate with and predict the response to endocrine therapy. However, use of cathepsin D is controversial, those tumors with high expression of cathepsin D appear to have a worse prognosis, a behavior otherwise unexpected in estrogen receptor positive tumors.

ETHANOL (ALCOHOL) BLOOD

SPECIMEN: Do not use alcohol swab to clean venipuncture site; use iodine; visible hemolysis must be treated with 6% TCA. Red top tube, separate serum.
REFERENCE RANGE: Negative. To convert ethanol level in mg/dL to international units in mmol/L, multiply by 0.22.
METHOD: Enzymatic using alcohol dehydrogenase; gas-liquid chromatography.
INTERPRETATION: Blood alcohol level and probable toxic condition are given in the next Table (Smith MS. Ethanol. In Noji EK, Kelen GD, Goessel TK (eds): Manual of Toxicologic Emergencies. 1989; 250):

Toxic Effects of Ethyl Alcohol in Nontolerant Individuals

Blood Alcohol Level (mg/dL)	Clinical Condition
<50	Limited muscular incoordination; Driving not seriously impaired.
50-100	Incoordination; Driving increasingly dangerous.
100-150	Mood, personality, and behavior changes; Driving is dangerous (legally drunk in most states).
150-200	Prolonged reaction time; Driving is very dangerous (legally drunk in all states).
200-300	Nausea, Vomiting, Diplopia, Marked Ataxia
300-400	Hypothermia, Dysarthria, Amnesia
400-700	Coma, Respiratory Failure, Death

Ethanol is absorbed through the stomach wall (20-25%) and the remaining is absorbed in the small intestines. The peak level in the blood usually occurs about 1 hour after ingestion; food in the stomach decreases the rate of absorption. Once the blood alcohol has peaked, the hourly elimination from blood varies from 4 to 40 mg/dL/hour; the female elimination rate is about 1.25 times that for the males.

Ethanol distributes itself between the blood and gas phase of the lungs (according to Henry's Law: the amount of gas dissolved is proportional to the pressure of the gas) and the ratio of the concentrations of alcohol in the blood and the alveolar air is a constant (k = 2100). Thus, the alcohol concentration of the alveolar air is an accurate reflection of the concentration of alcohol in the blood. Actually, breath alcohol determinations tend to underestimate the blood concentration of ethanol (Lovell WS. Science. 1972; 178:264-272).

Measurement of serum osmolality may be used as a screening test for alcohol in that each 100mg/dL contributes 22mosm/liter to the serum osmolality. If the normal serum osmolality is 290mosm/liter, 100mg/dL of ethanol in the serum will yield a serum osmolality of 312mosm/liter. Other alcohols contribute to serum osmolality as follows: methanol 34mosm/liter per 100mg/dL; ethylene glycol 20mosm/liter per 100mg/dL; acetone 18mosm/liter per 100mg/dL and isopropanol 17mosm/liter per 100mg/dL (Burkhart KK, Kulig KW. Emerg Med Clin N Amer. 1990; 8:913-928). A laboratory decision tree to differentiate an unknown alcohol, based on serum osmolality is given in the next Figure (Glasser L, et al. Am J Clin Pathol. 1973; 60:695-699):

Laboratory Decision Tree for Differentiation of an Unknown Alcohol

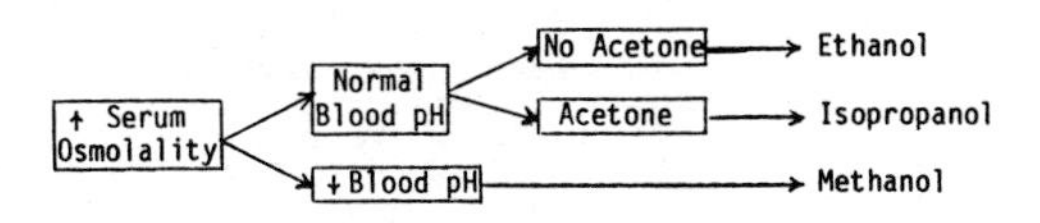

About 95 percent of the alcohol absorbed into the blood is oxidized in the liver to acetaldehyde and then to acetic acid.(cont)

Ethanol (Cont)

There are two laboratory tests that serve as "markers" for alcoholism; these are the enzyme gamma-glutamyl-transpeptidase (GGTP) which tends to be elevated in the alcoholic and mean corpuscular volume (MCV) of red blood cells which also tends to be elevated even without folate deficiency. Hyperbilirubinemia and hypoprothrombinemia are associated with high morbidity and mortality from alcoholic liver disease (See Review: Groover JR. Emerg Med Clin N Amer. 1990; 8:887-902).

The laboratory tests that tend to be abnormal in acute alcoholism are as follows: increased triglycerides; acid-base changes: respiratory alkalosis, metabolic acidosis; hypokalemia; hypomagnesemia; hypophosphatemia; hyperuricemia; and hypoglycemia.

Patients presenting to the Emergency Department with acute intoxication often provide limited history. Suggested evaluation for the most common clinical presentations related to acute alcohol intoxication are given in the following Table (Marco CA, Kelen GD. Emerg Med Clin N Amer. 1990; 8:731-748):

Evaluation of Common Clinical Presentations of Acute Intoxication

Presentation	Evaluation	Diagnostic Studies
Altered Mental Status	History Vital Signs Cervical Spine Precautions Papillary Response, Nystagmus Neck Examination Trauma Evaluation Neurologic Evaluation	Electrolytes, Glucose Calcium, Magnesium, Phosphate Blood Alcohol Level Ammonia Arterial Blood Gas Consider CT Scan Consider Lumbar Puncture
Chest Pain	History Vital Signs Cardiopulmonary Examination	Chest Radiograph ECG Consider Arterial Blood Gas Consider Ventilation Perfusion Scan
Abdominal Pain	History Vital Signs Cardiopulmonary Examination Abdominal Examination Rectal Examination	Electrolytes Amylase Bilirubin Liver Enzymes Prothrombin Time Consider Nasogastric Tube Consider Ultrasound or CT Scan

Urine samples have an alcohol content 1.3 times that of blood. Saliva testing is also available (Schwartz RH, et al. Ann Emerg Med. 1989; 18:1001-1003).

ETHOSUXIMIDE (ZARONTIN)

SPECIMEN: Red top tube, separate serum; or green (heparin) top tube, separate plasma. Reject: hemolyzed specimen or serum frozen on cells.
REFERENCE RANGE: Therapeutic: 40-100mcg/mL. Time to Obtain First Serum Specimen (Steady State): Children: 6 days, Adults: 12 days. Trough: Immediately prior to the next oral dose. Time to Peak Concentration: 2-4 hours after an oral capsule; less for syrup. Half-Life: Children: 30 hours, Adults: 60 hours. Toxic: >100mcg/mL.
METHOD: EMIT, GLC, HPLC
INTERPRETATION: Ethosuximide is given by mouth; it is the drug of choice in the treatment of patients with absence (petit mal) seizures; absence seizures are characterized by transient episodes of loss of awareness without convulsive movements. Ethosuximide is ineffective against generalized tonic-clonic seizures. Guidelines for therapy and serum specimen monitoring are given in the next Table:

Guidelines for Therapy and Monitoring

Therapy	Monitoring
Children (<11 years): 15-40mg/kg/day Adults: 15-30mg/kg/day	Children: 6 days Adults: 12 days Maintenance Monitoring: 4 to 6 month intervals provided therapeutic response is good

Around puberty, a lower dose is usually required to achieve the same serum level.

Ethosuximide is oxidized to inactive form in the liver; 10-20% is excreted unchanged by the kidneys.

Adverse Effects: Frequent dose-related side effects are nausea, vomiting, lethargy, hiccups, drowsiness and dizziness; these may diminish with time. Occasionally, Parkinson-like symptoms, photophobia and agitation are observed. Rarely, a lupus-like syndrome, Stevens-Johnson syndrome, leukopenia, thrombocytopenia, aplastic anemia and ataxia are seen.

Ethosuximide may increase the frequency of grand mal seizures in patients with mixed type seizures (Pellock JM, Pediatr Clin N Amer. 1989; 36:435-448).

ETHYLENE GLYCOL, SERUM OR URINE

SPECIMEN: Red top tube or random urine; store serum or urine at 4°C.

REFERENCE RANGE: Normally not present in serum or urine.

Toxicity: If sampled shortly after ingestion, ethylene glycol >3mmol/L (20mg/dL) in serum. To convert conventional units in mg/dL to international units in mmol/L, multiply by 6.2.

METHOD: Gas-liquid chromatography; Spectrophotometry assay.

INTERPRETATION: Ethylene glycol is the major ingredient in antifreeze. Small amounts of ethylene glycol can be ingested without ill effects; the minimum lethal dose of ethylene glycol is approximately 100mL in adults. Survival has been reported with massive ethylene glycol ingestion with initial level as high as 146.1 mmol/L (Curtin L, et al. Arch Intern Med. 1992; 152:1311-1313).

Blood and urine levels of ethylene glycol during hemodialysis and therapy with oral ethanol are given in the next Figure (Peterson CD, et al. N Engl J Med. 1981; 304:21-23):

Blood and Urine Levels of Ethylene Glycol

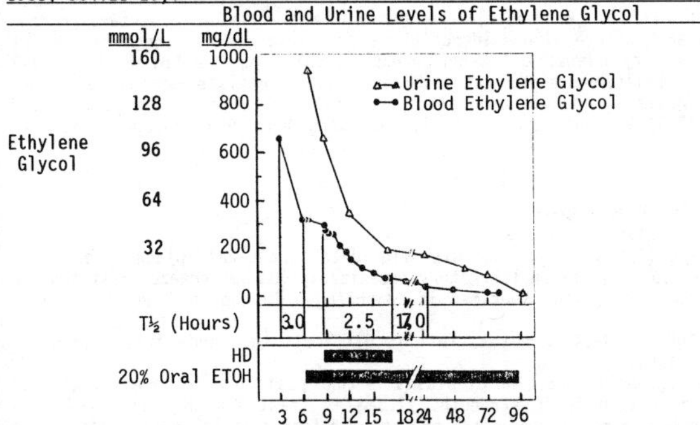

The half-lives of ethylene glycol are as follows: No therapy, 3.0 hours; hemodialysis, 2.5 hours; oral ethanol, 17.0 hours. Ethanol is used to treat ethylene glycol toxicity; ethanol competes with ethylene glycol for metabolic sites in the liver.

Most laboratories do not offer gas-liquid chromatography on a "stat" basis. However, a metabolic acidosis with an increased anion gap and a substantial discrepancy between the measured and calculated osmolality exists for ethylene glycol ingestion.

The increase in serum osmolality for each 16mmol/L (100mg/dL) of ethylene glycol is 16.3mosm/liter; the normal range for serum osmolality is 285-295mosm/liter. Therefore, 16mmol/L (100mg/dL) of ethylene glycol will yield a serum osmolality of about 306mosm/liter. (cont)

Ethylene Glycol (Cont)

The metabolism of ethylene glycol may help to explain acidosis, and acute renal failure. The pathway for metabolism of ethylene glycol is shown in the next Figure (Levinsky NG. Discussant in Case Records of the Mass. Gen. Hosp. N Engl J Med. 1979; 301:650-657):

Metabolism of Ethylene Glycol

```
                              ethylene glycol
                                  ↓ alcohol dehydrogenase
                              glycolaldehyde
                                  ↓
                            ↙ glycolic acid
              glycine ←           ↓
             ↙              ↖ glyoxylic acid
      hippurate                   ↓
                              oxalic acid
```

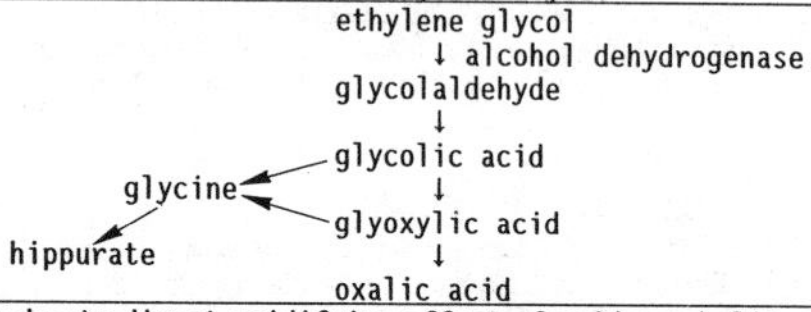

Acidosis is due to direct acidifying effect of acid metabolites of ethylene glycol and due to the toxic effects of these metabolites resulting in lactic acid production. Oxalate and hippurate crystals may be seen in the renal collecting system. Appearance of oxalate crystals in the urine strongly suggests the diagnosis. Antifreeze usually contains sodium fluorescein to detect radiator leaks. This compound will fluoresce using a Wood's Lamp - urine, gastric aspirate or skin can be screened (Winter ML, et al. Ann Emerg Med. 1990; 19:663-667).

The clinical symptoms of ethylene glycol toxicity are as follows: During the first stage (30 minutes to 12 hours), CNS symptoms predominate; in the first hour or so, the patients may appear drunk; coma and convulsions, hypertension and leukocytosis may follow. Other findings include anion gap acidosis, increased osmolal gap, and oxalate may bind calcium to produce hypocalcemia and tetany. The second stage (12 to 24 hours) the patient may develop hypertension and cardiopulmonary failure with evidence of cardiac enlargement, congestive heart failure, pulmonary edema, and bronchopneumonia. During the third stage (2-3 days), renal failure due to intratubular calcium oxalate deposition. Late findings (1-2 weeks) include cranial nerve deficits (Palmer BF, et al. Amer J Med. 1989; 87:91-92; Spillane L, et al. Ann Emerg Med. 1991; 20:208-210).

EUGLOBULIN LYSIS TIME

Synonym: Euglobulin Clot Lysis

SPECIMEN: Blue (citrate) top tube with liquid anticoagulant; place tube on ice and transport to coagulation laboratory immediately. Do not freeze. Test must be performed immediately. The specimen is centrifuged in the cold and the plasma removed.

REFERENCE RANGE: 120-240 minutes. Test is difficult to standardize. Times may vary between labs.

METHOD: The euglobulin fraction of plasma is that portion which precipitates at low pH and decreased ionic strength. Plasma is treated with acetic acid and refrigerated for 15 minutes or until precipitate forms; it is centrifuged in the cold. The precipitate (euglobulins) contain fibrinogen, plasminogen, plasminogen activator and plasmin. The supernatant is decanted and saline-borate buffer is added to dissolve the precipitate. Thrombin is added to the solution and the resultant clot is incubated at 37°C and checked after 10, 20 and 30 minutes and every subsequent half-hour period. Time of lysis is recorded.

INTERPRETATION: This test is used to screen for increased fibrinolytic activity. A shortened lysis time indicates activation of the fibrinolytic system. Shortened lysis time may also result from dysfibrinogenemia and fibrin split products. The euglobulin lysis time has been used during cardio-bypass surgery since the acid precipitation step separates inhibitors including heparin. This test has also been used in monitoring urokinase and streptokinase therapy (Schafer AI. Patient Care. 1982; 87-115). This test is mainly a measure of plasminogen and plasminogen activator. Plasminogen is proteolytically activated to plasmin; plasmin attacks fibrinogen and fibrin to produce fibrin split products(FSP).

EXCRETION FRACTION OF FILTERED SODIUM

SPECIMEN: Urine: 10mL aliquot of random urine specimen; no preservative is necessary; this urine specimen will be used for assay of sodium and creatinine. Blood: Red top tube, separate serum; this specimen will be used for the assay of sodium and creatinine. Green top tube (plasma) may be used.
REFERENCE RANGE: See interpretation below.
METHOD: Flame photometry or ion-specific electrodes for sodium; alkaline picrate is the usual method for creatinine.
INTERPRETATION: This is a sensitive and specific test for acute tubular necrosis (Espinel CH, Gregory AW. Clin Nephrol. 1980; 13:73-77). The measurements that are done are as follows: Serum and urine sodium, serum and urine creatinine; no timed specimen is necessary. The excretion fraction of filtered sodium is given by the next equation:

$$\text{Excretion Fraction of Filtered Sodium } (FE_{Na}) = \frac{\text{Urine } Na^+ \times \text{Plasma Creatinine}}{\text{Plasma } Na^+ \times \text{Urine Creatinine}} \times 100$$

The results of the test for excretion fraction of the filtered sodium (FE_{Na} Test) for patients with acute tubular necrosis and those with prerenal azotemia are illustrated in the next Figure:

Excretion Fraction of Filtered Sodium (FE_{Na} Test) for Patients with Acute Tubular Necrosis Compared to Patients with Prerenal Azotemia

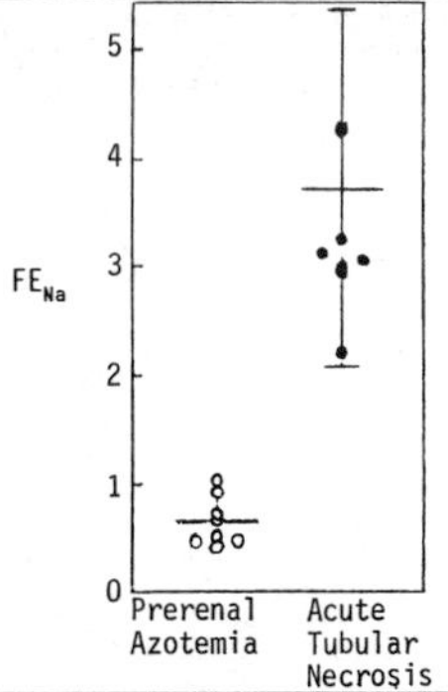

The two conditions are clearly distinguished by the FE_{Na} values. Patients with acute tubular necrosis present with an FE_{Na} of more than 2. Patients with prerenal azotemia present with an FE_{Na} of less than 1. Levels of FE_{Na} in different conditions are shown in the next Table (Espinel CH. Clin Lab Med. 1993; 13:89-102):

Levels of FE_{Na} in Different Conditions

Low FE_{Na} (<1)	High FE_{Na} (>1)
Prerenal Azotemia	Acute Tubular Necrosis (Oliguric, Nonoliguric, Post Renal Transplant)
Acute Glomerulonephritis and Vasculitis	Urinary Obstruction
Hepatorenal Syndrome	Chronic Uremia
Renal Transplant Rejection	Diuretics

In acute glomerulonephritis and hepatorenal syndrome, the renal tubule reabsorbs sodium avidly.

In acute tubular necrosis, functioning nephron units excrete a large fraction of the filtered sodium. In chronic uremia, sodium hemostasis is maintained by a reduction in tubular sodium reabsorption. Inhibition of sodium reabsorption by diuretics results in a high FE_{Na}.

EYE CULTURE AND SENSITIVITY

SPECIMEN: Lid and Conjunctiva: sterile cotton-tipped applicator moistened in thioglycollate media; Corneal cultures: Kimura platinum spatula; Anterior Chamber Fluid: Needle aspiration; Vitreous Fluid: Needle aspiration or vitrectomy.
REFERENCE RANGE: Normal flora depends on the site. The most common organisms are Staphylococcus epidermidis or S. aureus.
METHOD: Usual culture techniques.
INTERPRETATION: The indications for cultures or scrapings are given in the following Table (Brinser JH. Infect Dis Clin N Amer. 1992; 6:769-775):

Indications for Obtaining Cultures or Scraping	
Required:	Conjunctivitis (Neonatal; Hyperacute; Membranous; Severe)
	Corneal Ulcers - not obviously herpetic
	Endophthalmitis (Postoperative; Traumatic; Endogenous)
Helpful:	Conjunctivitis (Chronic; Unilateral; Infectious; Eczematous; Vernal; Atopic)
	Keratoconjunctivitis Sicca
	Blepharitis
	Lacrimal Infections
	Preoperatively in mononuclear patients with an ocular prosthesis

The most common ocular infection is conjunctivitis. When infection spreads to the lid or cornea, the infection is called blepharoconjunctivitis or keratoconjunctivitis. Conjunctivitis may be bacterial, viral or allergic; differentiation of the types of conjunctivitis is given in the following Table (Syed NA, Hyndiuk RA. Infect Dis Clin N Amer. 1992; 6:789-805):

Differentiating Features of Bacterial, Viral, and Allergic Conjunctivitis

Signs and Symptoms	Bacterial	Viral	Allergic
Discharge	Purulent Mild-Severe	Watery Mild-Mod	White, Ropy Mild
Injection	Severe	Mod	Mild
Preauricular Nodes	Rare	Usual	None
Smear	PMN's Bacteria	Lymphs	Eosinophils
Pathogens	H. Influenzae S. Pneumoniae Staph Aureus Staph Epidermidis N. Gonorrhoeae Chlamydia	Adenovirus Herpes Simplex Varicella-Zoster Coxsackievirus Epstein-Barr Virus Influenza	

Blurred vision, pain, photophobia, corneal haze, pupillary changes, and high or low intraocular pressure are signs and symptoms of keratitis (bacterial or viral), iritis or acute glaucoma. See Barza M, Baum J [ed.] "Ocular Infections." Infect Dis Clin N Amer. December 1992; 6(4).

FACTOR VIII

Synonyms: VIII:C, antihemophilic factor, AHF
SPECIMEN: Blue (citrate) top tube; tube must be filled to capacity. If multiple studies are being done, draw coagulation studies last; otherwise, draw 1-2mL into another vacutainer, discard and then collect factor assay. This procedure avoids contamination of specimen with tissue thromboplastins. Separate plasma and refrigerate. Sample must be frozen if determination is not done immediately.
REFERENCE RANGE: 50-150 percent of normal activity.
METHOD: Patient sample is mixed with a plasma deficient in factor VIII. PTT reagents are added and a conventional PTT is run. Assay results are compared with graphed results of a assay performed with a control of limiting dilutions of normal plasma (100% activity) mixed with the plasma deficient in factor VIII.
INTERPRETATION: This assay measures the clotting ability (function) of factor VIII. This assay is used to aid in the diagnosis of hemophilia A, von Willebrand disease, acquired deficiencies of factor VIII, the response to factor VIII preparations, and the quality control of factor VIII preparations. Factor VIII is decreased in the following conditions:

Causes of Decreased Factor VIII
Inherited
Hemophilia A
von Willebrand's Disease
Acquired
DIC
Autoantibodies to Factor VII
(Autoimmune States, Drugs, Late Pregnancy)

Hemophilia A is a sex linked disorder. Deficiency of factor VIII can be diagnosed in the newborn, since adult levels of this factor are reached at the time of birth. It is important to determine factor VIII levels in hemophilia A since there are both mild (30%) and severe (<1%) forms of the disease. Levels of factor VIII will determine the amount of blood component to be administered. Infusion of factor VIII to a level of 10-30% will usually stop bleeding (Hilgartner MW: in Hilgartner MW, Pochedly C (eds). Hemophilia in the Child and Adult. New York, NY: Raven Press, 1989; 1-26). During orthopedic procedures, 100% activity should be maintained. Approximately 10-15% of severe hemophiliacs develop inhibitors to factor VIII. see FACTOR VIII INHIBITOR.

Increased levels of factor VIII are seen in inflammatory states (acute phase reactant). Patients with von Willebrand's disease show an "overresponse" of factor VIII levels following transfusion (levels in excess of the amount of factor VIII administered).

FACTOR VIII INHIBITOR

Synonym: Bethesda inhibitor
SPECIMEN: Blue (citrate) top tube.
METHOD: Factor VIII in normal plasma is inhibited by dilutions of patient plasma containing the inhibitor. A 2 hour preincubation period is followed by addition of the reagents and running of a PTT assay. Inhibitory activity of the patient sample is compared to that of normal plasma and the amount of inhibitor is calculated from a standard curve. The amount of inhibitor is reported in Bethesda Units. One Bethesda Unit will inhibit 50% of the factor VIII activity in normal plasma.
REFERENCE RANGE: None present
INTERPRETATION: Factor VIII inhibitors are most commonly found in patients with severe hemophilia A. This assay is usually used to document the presence of these inhibitors and to titer their levels prior to surgery or to follow the response to plasma exchange.

FACTORS, BLOOD CLOTTING

Assay of the following blood clotting factors are commercially available:

Clotting Factor Assays
Factor I, Fibrinogen
Factor V, Labile Factor/Proaccelerin
Factor VII, Stable Factor/Proconvertin
Factor VIII, Antihemophilia Factor
Factor VIII Antigen, Von Willebrand Antigen
Factor IX, Antihemophilia Factor B/Christmas Factor
Factor X, Stuart Factor
Factor XI, Antihemophilia Factor C/Plasma Thromboplastin Antecedent
Factor XII, Hageman Factor
Factor XIII, Fibrin Stabilizing Factor
Factor XIV, Protein C Antigen
Protein S

The majority of these assays measure functional activity in various clotting tests employing factor deficient plasma. Factor VIII and Factor VIII Antigen assays are employed in the diagnosis and management of hemophilia A and von Willebrand patients and are described under their respective headings. Factor IX assays are used in the diagnosis and management of hemophilia B patients. Fibrinogen levels and factor XIV (protein C) and protein S assays are occasionally useful in patients with coagulation problems. The other factor assays are of limited use in general clinical practice and performance is limited to the more specialized coagulation laboratories.

FAT, FECES (see FECAL FAT)

FAT MALABSORPTION PANEL

The causes of fat malabsorption are given in the next Table:

Causes of Fat Malabsorption

(1) Deficiency of Pancreatic Digestive Enzymes:
- Chronic Pancreatitis
- Cystic Fibrosis
- Pancreatic Carcinoma
- Pancreatic Resection

(2) Impairment of Intestinal Absorption:
- Celiac Disease (Coeliac Disease, Non-Tropical Sprue, Gluten-Sensitive Enteropathy)
- Rare Causes:
  - Tropical Sprue
  - Abetalipoproteinemia
  - Lymphangiectasis
  - Intestinal Lipodystrophy
  - Amyloidosis
  - Lymphoma
  - Surgical Loss of Functional Bowel

(3) Deficiency of Bile:
- Extrahepatic Bile Duct Obstruction
- Intrahepatic Disease
- Cholecystocolonic Fistula

The causes of steatorrhea in 47 elderly patients are given in the next Table (Price HL, et al. Brit Med J. 1977; 1:1582-1584):

Causes of Steatorrhea in 47 Elderly Patients

Cause	Age <65	Age >65	Total
Celiac Disease	12	4	16
Pancreatic Insufficiency			
Carcinoma	4	0	4
Other	3	7	10
Postgastrectomy	6	2	8
Jejunal Diverticula	1	1	2
Tropical Sprue	0	2	2

The common causes of steatorrhea are celiac disease and pancreatic insufficiency; other causes of steatorrhea are "collagen" disease, diabetes mellitus, scleroderma and Whipple's disease; all these occurred in patients less than age 65. Patients with conditions associated with deficiency of bile were not included in the population examined by the investigators.

Tests for Malabsorption and Steatorrhea: Tests for malabsorption are as follows:

(a) Fecal Fat (72 Hour Collection): More than 7g/24 hours represents steatorrhea. (See FECAL FAT)

(b) Microscopic Examination of the Stool Fat: Sudan III or Oil Red O are used to stain fat globules; the sensitivity is 72% and the specificity is 95% (Bin TL, et al. J Clin Pathol. 1983; 36:1362-1366).

(c) Trypsin, Immunoreactive, Serum: Serum immunoreactive trypsin is decreased in pancreatic insufficiency with associated steatorrhea; serum immunoreactive trypsin is normal in pancreatic insufficiency without steatorrhea and in steatorrhea with normal pancreatic function (Jacobson DG. N Engl J Med. May 17, 1984; 310:1307-1309). (See TRYPSIN)

(d) Serum Carotene: (See CAROTENE)

(e) Schilling Test: (See SCHILLING TEST)

(f) D-Xylose Absorption Test: (See XYLOSE)

(g) Sweat Electrolytes for Cystic Fibrosis: (see SWEAT TEST)

(h) Intestinal Biopsy for Intestinal Disorders

(i) Trypsin in Stool

(j) Glucose Tolerance Test for Pancreatic Deficiency

(k) Breath Test for Lactose Intolerance: (See LACTASE)

Laboratory Findings in Malabsorption: Laboratory findings reflecting malabsorption are given in the next Table:

Laboratory Findings Reflecting Malabsorption

Fat Malabsorption
- Vitamin D Deficiency
  - Serum Calcium Decreased
  - Secondary Hyperparathyroidism, Alkaline Phosphatase Increased, (Osteomalacia)
- Serum Cholesterol Decreased

Folic Acid Deficiency
- Megaloblastic Anemia
- Lactate Dehydrogenase (LDH) Increased

Protein Malabsorption
- Total Protein Decreased
- Albumin Decreased

Vitamin K Deficiency
- Prolonged Prothrombin Time (PT)

Vitamin A Deficiency
- Serum Carotene Decreased
- Retinol Binding Protein Decreased

Deficiency of fat soluble vitamins (A, D, E and K) may accompany malabsorption.

FEBRILE AGGLUTININS (TYPHOID, PARATYPHOID, BRUCELLA)

SPECIMEN: Collect acute phase specimen. Use a red top tube, separate serum and freeze serum. Then, collect convalescent-phase sera, ten to fourteen days later. Use red top tube, separate serum and freeze.
REFERENCE RANGE: Less than fourfold increase in titer between the acute and convalescent serum titers.
METHOD: Slide and tube agglutination tests
INTERPRETATION: Tests for febrile agglutinins have been widely used in the past in screening patients with fever of unknown origin (FUO). However, the need for these screening tests has diminished significantly in recent years for the following reasons: sensitivity and specificity often low; improved understanding of these diseases; decreased incidence of these diseases; improved ability to culture the causative organisms; and greater sensitivity and specificity of selected antibody tests. The antigens commonly used in this test to detect antibodies are listed in the next Table:

Antigens Commonly Used in Tests for Febrile Agglutinins

Infections	Antigens
Salmonella	H Antigens: S. typhi d; S. paratyphi a,b,c O Antigens: Salmonella A,B,C,D and E
Rickettsial Infection	Proteus OXK, OX2, OX19
Brucellosis	Brucella

Salmonella Infections: This test is useful when organisms in parenteral tissue invade the blood stream and stimulate the production of antibody; as a corollary, if blood invasion does not occur, there will be no antibody production.

The antibody response that characteristically occurs in salmonella infection is as follows: Antibodies to O Antigen: Titers of O agglutinins in patients with Salmonella disease are elevated in 50% of patients by the end of the first week and in 90 to 95% of patients by the fourth week peaking in the sixth week and falling or disappearing within six to 12 months. Antibodies to H Antigen: Titers of H agglutinins in patients with Salmonella disease are detected later, peak later, and remain elevated for several years. Antibody to H antigen may be detected for years following immunization.

Specificity is poor in that antigens of Salmonellae are shared by Enterobacteriaceae, and there may be nonspecific stimulation of Salmonella agglutinins by febrile disease of other cause.

The most appropriate test for Salmonellae infection is culture.(cont)

Febrile Agglutinins (Cont)

Rickettsial Infection: The basis of this test is that Proteus antigens are shared by R. rickettsii, R. prowazekii, R. mooseri and R. tsutsugamushi and infections by these Rickettsial agents often elicit antibodies that react with Proteus antigens. This is a very insensitive test in that infection by Rickettsial agents may not elicit antibodies that react with Proteus antigens and antibodies may be produced by other infections, e.g. Proteus, spirochetal, brucellosis, tularemia and others.

Brucellosis: This test is more sensitive than the two previous tests but is less specific in that other conditions produce antibodies that react with Brucella antigen.

Reference: Fuchs PC. Medical Laboratory Observer. 1983; 15-17.

Febrile agglutinin panels may include tests for Francisella Tularensis (Tularemia), Q-fever, Leptospirosis, Scrub Typhus and Murine Typhus.

Specific tests should be ordered when a particular disease is suspected rather than as a panel.

FECAL EXAMINATION FOR OCCULT BLOOD

SPECIMEN: Hemoccult slide test for occult blood (Smith Kline Corp.):

(1) Collect a very small stool specimen on tip of wooden applicator.
(2) Apply thin smear of specimen inside the circle.
(3) Close cover; dispose of applicator.
(4) Open perforated window in back of slide.
(5) Apply two or three drops of developing solution to slide opposite specimen.
(6) Read results after 30 seconds.
(7) Dried samples may be rehydrated with a drop of deionized water.

REFERENCE RANGE: Positive: Presence of blue color. Negative: No detectable blue color. Hemoccult begins to turn positive in the presence of about 5.0mg of hemoglobin per gram of feces. Some specimens may turn from positive to negative after storage for two days.

METHOD: A stool specimen is applied to guaiac-impregnated filter paper. In the presence of hemoglobin and hydrogen peroxide the reaction occurs as follows:

$$\text{Guaiac} + H_2O_2 \xrightarrow{\text{Heme}} \text{Blue Color} + H_2O$$

Guaiac is the chromogen, orthotolidine, and is converted to blue oxidized orthotolidine by peroxidation. The heme moiety of hemoglobin has pseudoperoxidase activity and catalyzes the oxidation of guaiac by hydrogen peroxide to a blue color.

Methods, other than the guaiac procedure, have been developed (Welch CL, Young DS. Clin Chem. 1983; 29:2022-2025; Schwartz S, et al. Clin Chem. 1983; 12:2061-2067).

INTERPRETATION: Colorectal cancer is one of the most common tumors; it is the second leading cause of cancer death in females and the third leading cause of cancer death in males in the United States. One hundred and fifty thousand new cases are diagnosed yearly. The test for occult blood in the feces is done to help to detect colorectal cancer.

Trials of fecal occult blood screening for the detection of colorectal cancer have been variable; two recent studies have suggested that screening reduces mortality from colorectal cancer (Mandel JS, et al. N Engl J Med. 1993; 328:1365-1371; Selby JV, et al. Ann Intern Med. 1993; 118:1-6).

Various foods and drugs can yield inaccurate results for fecal occult blood testing. False-negative tests occur with vitamin C. False-positive tests occur with excess dietary meat or plant peroxidases (horseradish, yellow turnips, apples, oranges, and bananas). GI bleeding may be induced by alcohol, aspirin, steroids, non-steroidal anti-inflammatory drugs, iron therapy and others.

FECAL FAT

SPECIMEN: Fresh random stool (semi-quantitative) or 72 hour stool (quantitative)
REFERENCE RANGE: 1-7 grams/24 hours (3.5-25mmol/day). To convert conventional units in g/day to international units in mmol/day, multiply conventional units by 3.515.
METHOD: Semi-Quantitative: Sudan III or Oil Red O. This test has a sensitivity of 72% and specificity of 95% (Bin TL, et al. J Clin Pathol. 1983; 36:1362-1366). Sudan stain is more specific for the detection of triglyceride and fatty acid in a matrix; phospholipid and cholesteryl ester are not stained with sudan dye, but are detected as part of a 72 hour stool determination (Khouri MR, et al. Gastroenterology. 1989; 96:421-427). Methods for stool examination for fat have been described in detail (Sondheimer JM. Contemp Pediatr. Feb. 1990; 63-82). Quantitative: Saponification, acidification, extraction of lipids with petroleum ether, vaporization of ether and weigh residue.
INTERPRETATION: This test is done in the work-up of patients for possible steatorrhea; steatorrhea refers to a pathological increase in stool fat. The major causes of steatorrhea are listed in the next Table:

Causes of Steatorrhea
(1) Deficiency of Pancreatic Digestive Enzymes:
Chronic Pancreatitis
Cystic Fibrosis
Pancreatic Carcinoma
Pancreatic Resection
(2) Deficiency of Bile in Intestinal Lumen:
Extrahepatic Bile Duct Obstruction
Intrahepatic Disease, e.g., Primary Biliary Cirrhosis
Cholecystocolonic Fistula
Intestinal Stasis Syndrome
(3) Impairment of Intestinal Absorption:
Celiac Disease (Nontropical Sprue, Gluten-Sensitive Enteropathy)
Postgastrectomy
Tropical Sprue
Lymphangiectasis
Intestinal Lipodystrophy
Amyloidosis
Lymphoma
Surgical Loss of Functional Bowel

The major causes of steatorrhea are: (1) deficiency of pancreatic digestive enzymes, (2) deficiency of bile and (3) impairment of intestinal absorption.

FECAL LEUKOCYTES (see STOOL LEUKOCYTES)

FERRIC CHLORIDE TEST FOR AMINOACIDURIA

SPECIMEN: Random urine; an early morning specimen is preferred to reduce variations due to diet. If not analyzed immediately, the urine should be acidified to a pH less than 4; if necessary, freeze at -20°C for no longer than one week.
REFERENCE RANGE: Negative, that is, no color change.
METHOD: Add 10% $FeCl_3$ by drops to 1-2mL of urine.
INTERPRETATION: The ferric chloride test is non-specific and gives color reactions with metabolites from amino acid disorders, other metabolites and drugs as shown in the next Table:

Ferric Chloride Test in Urine

Disorder and Urinary Product	Color Change
Phenylketonuria (PKU)	
Phenylpyruvic Acid	Blue or Blue-Green, Fades to Yellow
Tyrosinemia	
p-Hydroxyphenylpyruvic Acid	Green, Fades in Seconds
Alkaptonuria	
Homogentisic Acid	Blue or Green, Fades Quickly
Maple Syrup Urine Disease	
Alpha-Ketoisovaleric Acid	Blue
Alpha-Ketoisocaproic Acid	Blue
Alpha-Keto-Beta-Methyl Valeric Acid	Blue
Histidinemia	
Imidazole Pyruvic Acid	Green or Blue-Green
Diabetics, Alcoholics, Starvation	
Acetoacetic Acid	Red or Red Brown
Other Products:	
Bilirubin	Blue-Green
Ortho-Hydroxyphenyl Acetic Acid	Mauve
Ortho-Hydroxyphenyl Pyruvic Acid	Red
Alpha-Ketobutyric Acid	Purple; Fades to Red Brown
Pyruvic Acid	Deep Gold-Yellow or Green
Xanthurenic Acid	Deep Green, Later Brown
Drugs:	
Salicylates	Stable Purple
Aminosalicylic Acid	Red-Brown
Phenothiazine Derivatives	Purple Pink
Antipyrines and Acetophenetidines	Red
Phenol Derivatives	Violet
Cyanates	Red

The ferric chloride test has been essentially replaced by more specific tests for inborn errors of metabolism, drugs, etc.

FERRITIN

SPECIMEN: Red top tube, separate serum.
REFERENCE RANGE: Females, 20-120ng/mL; Males, 20-300ng/mL; Newborn, 25-200ng/mL; 1 month, 200-600ng/mL; 2-5 months, 50-200ng/mL; 6 months-15 years, 7-142ng/mL. Borderline low, 10-20ng/mL; Iron deficiency, less than 10-12ng/mL; Iron overload, more than 400ng/mL. To convert conventional units in ng/mL to international units in mcg/liter, multiply conventional units by 1.00.
METHOD: RIA or enzyme immunoassay
INTERPRETATION: The conditions in which serum ferritin measurements are of value are given in the next Table (Skikne BS, Cook JD. Laboratory Management. May 1981; 31-35):

Conditions in Which Serum Ferritin Measurements Are of Value
Detection of Iron Deficiency
Detection of Response and/or Endpoint to Oral Iron Therapy
Differentiation of Anemia of Chronic Disease from Iron Deficiency Anemia
Monitoring Iron Status of Patients with Chronic Renal Disease
Detection of Iron Overload
Monitoring the Rate of Iron Accumulation in Iron Overload
Monitoring the Response to Iron Chelation Therapy in Iron Overload
Determination of Iron Status of a Population and Response to Iron Fortification

The causes of altered serum ferritin are given in the next Table:

Causes of Altered Serum Ferritin	
Decreased	Increased
Iron Deficiency	Iron Overload:
	Hemochromatosis
	Transfusion Hemosiderosis
	Gaucher Disease
	Acute Hepatitis
	Inflammation
	Chronic Liver Disease

Iron Deficiency Anemia: The first biochemical change in iron deficiency is a low serum ferritin level; this occurs before iron is decreased and before morphologic abnormalities appear in red blood cells. Iron deficiency and no other disease is associated with a serum ferritin level less than 10ng/mL.

Serum ferritin level is helpful in differentiating between iron deficiency anemia and anemia of chronic inflammation, infection or chronic disease. In iron deficiency anemia and in anemia associated with chronic conditions, serum ferritin is low (below 10ng/mL). However, in anemia associated with infection and malignancy, serum ferritin is above 10ng/mL; a serum ferritin level lower than 50ng/mL in a patient with obvious inflammatory disease is a strong indicator of iron deficiency.

In the presence of liver disease, serum ferritin is increased; in iron deficiency anemia plus liver disease, serum ferritin may be normal.

In patients with iron deficiency anemia who are being treated with iron orally, serum ferritin measurements are useful in monitoring the response to therapy and in determining the time when iron should be discontinued; a normal hemoglobin does not necessarily indicate that the body iron stores have been replenished. Serum ferritin assays should be done at 3 to 4 week intervals until serum ferritin rises above 50ng/mL (body iron stores of about 400mg) (Skikne BS, Cook JD. Laboratory Management. 1981; 31-35).

Serum ferritin reflects iron stores if the patient does not have liver disease or an acute inflammatory reaction that may produce spurious elevation of serum ferritin concentration; 1.0ng/mL of serum ferritin is equivalent to 8mg of storage iron. Total iron stores are 0.7 to 1.5g. About 15 to 20 percent of iron is stored within ferritin. Serum ferritin does not reflect bone marrow stores, examples: patients treated with iron; dialysis patients; uremia with depression of erythropoiesis.
Hemochromatosis: Serum ferritin is usually elevated in patients with hemochromatosis. During the development of hemochromatosis, iron is confined mainly to liver parenchymal cells and in the absence of cellular damage, relatively little ferritin is released.

Not all patients with hemochromatosis have elevated serum ferritin; in a study of homozygous patients with hemochromatosis 23 of 32 patients had elevated serum ferritin (Edwards CQ, et al. Ann Intern Med. 1980; 93:519-525). (cont)

Transferrin saturation is a more sensitive test for hemochromatosis than is serum ferritin (Edwards CQ, et al, already cited). The three tests that are used in the diagnosis of hemochromatosis are given in the next Table (Gollan JL. Patient Care. Sept. 30, 1982; 102-119):

Tests in Diagnosis of Hemochromatosis

Test	Abnormal Value	False Positive Frequency	False Negative Frequency
Serum Iron	>175mcg/dL	10%	24%
Transferrin Saturation	>50%	33%	0%
Serum Ferritin	>250ng/mL	2%	3%

A suggested protocol for hemochromatosis screening and treatment is given in the following Table (Edwards CQ, Kushner JP. N Engl J Med. 1993; 328:1616-1620):

Protocol for Hemochromatosis Screening and Treatment

Measure Transferrin Saturation
If >60% in men or >50% in women

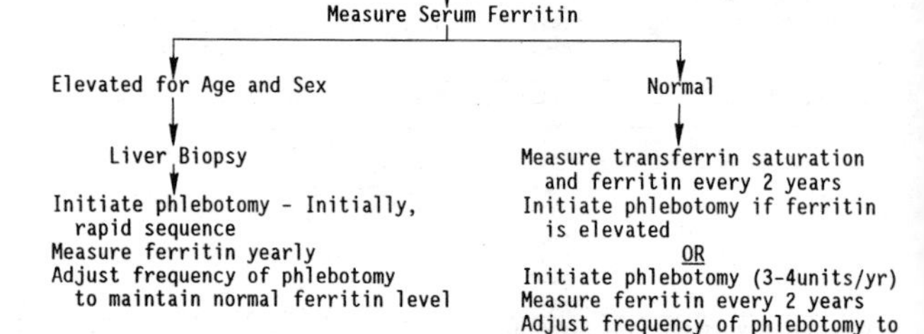

Characteristics of Ferritin: Ferritin is the main storage molecule for iron. It is present in highest concentration in the reticuloendothelial cells of the liver and spleen and in the erythroblasts of bone marrow. It consists of two basic constituents - a protein shell (apoferritin) and a central core of iron in the ferric form existing as heterogenous ferric oxyhydroxide micelles with some phosphate.

FETAL HEMOGLOBIN (see HEMOGLOBIN F)

FETAL HEMOGLOBIN IN MATERNAL BLOOD; SCREEN: THE ROSETTE METHOD

SPECIMEN: Purple (EDTA) top tube; for best results, examine fresh specimen immediately.

REFERENCE RANGE: Negative

METHOD: Ortho Diagnostics: Fetal Screen: A suspension of maternal red blood cells is incubated with anti-Rho(D) typing serum and then washed and incubated with an Rho(D)-positive indictor red blood cell in combination with a potentiating medium. If any Rho(D)-positive (fetal) red blood cells are present in the maternal sample, rosettes will be formed. Rosetted red blood cells are quickly and easily detected microscopically.

INTERPRETATION: This is a screening test for fetal-maternal hemorrhage. All Rho(D)-negative women who give birth to Rho(D)-positive babies should be tested for fetal-maternal hemorrhage. This method will detect the presence of 0.3% Rho(D)-positive cells (fetal-maternal hemorrhage of approximately 7.5mL of fetal red cells). The test is negative in individuals who have less than 0.1% Rho(D)-positive cells (2.5mL of fetal-maternal hemorrhage) and may be positive or negative between 2.5 and 7.5mL of fetal-maternal hemorrhage.

False negative results may occur in the presence of ABO incompatibility between mother and fetus. False positive results may occur if maternal red cells have a positive direct antiglobulin (Coombs) test.

Quantitation of a positive screen should be done by the Betke-Kleihauer stain (see next section).

FETAL HEMOGLOBIN STAIN (BETKE-KLEIHAUER STAIN)

SPECIMEN: Blood: Purple(EDTA) top tube; for best results, examine fresh specimen immediately; otherwise refrigerate. Amniotic Fluid: No preservative; send to laboratory for analysis immediately; do not refrigerate.
REFERENCE RANGE: Negative. No fetal cells in maternal blood.
METHOD: Dilute blood sample with saline, 1:1; if hematocrit is low, saline dilution is not necessary. Prepare thin film of the diluted capillary or fresh EDTA blood. Air dry slides and fix with 80% alcohol in order to precipitate hemoglobin within red cells. Treat slides with citric acid-phosphate buffer, pH 3.2; at this pH adult hemoglobins are soluble and are dissolved out of the red cells while HgbF remains precipitated and is next stained with eosin. The cells that contained HgbA appear as "ghosts" while the HgbF cells stain red. Five hundred cells are counted under high dry magnification.
INTERPRETATION: This test is most often used as a quantitative measure of Rho(D) cells in the maternal circulation; use the Fetal Hemoglobin Test as a screen. There are three different patterns; these are illustrated in the next Figure:

Patterns of HgbF and HgbA in Red Cells and Examples

HgF HgA 500 Cells

Mother's Blood: Fetal-maternal hemorrhage; 50-90% of fetal RBC's contain HgbF. Collect maternal blood immediately following delivery if newborn has anemia of newborn or when a mother is Rh negative or Du-negative. In the case of a full-term delivery, the red blood cells of the infant must be Rho(D) positive and the direct antiglobulin test (for anti-Rho(D)) negative for the mother to be a candidate for RhIG. The amount of fetal blood that has escaped into the maternal circulation is roughly calculated using the formula:

$$\text{Fetal Blood (mL)} = \%\ \text{HgbF cells} \times 50;\ \%\ \text{HgbF} = \frac{\text{No. of Fetal Cells}}{\text{No. of Maternal Cells}} \times 100$$

$$\text{Vials of Rho(D) Immune Globulin Needed (RhIG)} = \frac{\text{mL of Fetal Blood}}{30}$$

One vial (300 microgram) of RhIG will suppress the immunization of 15mL of fetal red blood cells or 30mL of whole fetal blood hemorrhage; two vials will suppress the immunization of 2 x 15mL or 30mL, etc. Because quantitation by this method is imprecise and the consequences of undertreatment are serious, RhIG dosage should be overestimated to provide a margin of safety. One suggested method is to calculate the number of RhIG vials round to the nearest whole number and give an additional vial (Technical Manual, Tenth Edition, 1990, Am Assoc Blood Banks, 1117 North 19th Street, Suite 600, Arlington, VA 22209).

- -

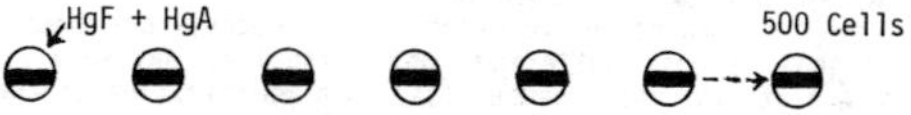

HgF + HgA 500 Cells

Even distribution of HgbA + HgbF (Heterozygous Form)
Hereditary Persistence of HgbF (2 Alpha Chains, 2 Gamma Chains):
Homozygous Form: 100 percent HgbF
Heterozygous Form: Hemoglobin F (15 to 35%) and is evenly distributed throughout the red cells.

- -

Uneven Distribution of HgbA + HgbF: Thalassemia minor and major, HgbS-S disease, Fanconi Anemia and hereditary spherocytosis.

FIBRIN SPLIT PRODUCTS (FSP)

Synonyms: Fibrin degradation products (FDP); Fibrin Breakdown Products (FBP)
SPECIMEN: Obtain specimen before starting heparin therapy; collect blood in dry syringe; inject blood into FSP sample tube. Immediately mix by inverting the tube gently several times; the blood will clot firmly within a few seconds. Do not shake the tube, transport to laboratory as soon as possible. Do not put specimen on ice. The FSP sample tube contains thrombin and proteolytic inhibitor to inhibit plasmin activity.
REFERENCE RANGE: <10mcg/mL. No agglutination at 1:5 dilution.
METHOD: Anti-fibrinogen fragments D and E antibody bound to latex particles (Burroughs-Wellcome) are aggregated in the presence of peptides A and B.
INTERPRETATION: Fibrin split products are measured to detect in vivo fibrinolytic activity. However this assay will not distinguish between fibrinogenolysis and fibrinolysis. Likewise, it will not distinguish between lysis of pathological and physiological thrombi. Elevated FSP are found in a number of conditions including DIC, therapeutic thrombolysis (streptokinase, urokinase), thrombosis, pulmonary embolus, Thrombotic thrombocytopenic purpura (TTP), hemolytic uremic syndrome, liver disease, and recent surgery. Falsely elevated levels may be encountered if the sample does not clot fully (e.g., heparin therapy), or if rheumatoid factor is present.

Other useful tests may include RBC morphology, PTT, PT, thrombin time, Plt count, fibrinogen, and D-dimer test.

D-Dimer Test: The D-dimer test is a variant of the FSP test. However, the antibodies employed are directed against the D-dimer (a crosslinked dimer released by plasmin from factor XIII crosslinked fibrin). Peptides split from fibrinogen do not react. Hence with this assay one is able to distinguish between fibrinolysis and fibrinogenolysis. See: D-DIMER TEST.

The conditions associated with increased fibrin split products are those associated with disseminated intravascular coagulopathy (DIC); see DISSEMINATED INTRAVASCULAR COAGULOPATHY(DIC) PROFILE.

FIBRINOGEN

SPECIMEN: Blue (citrate) top tube with liquid anticoagulant, fill tube to capacity; separate plasma; freeze plasma if determination is not done immediately.
REFERENCE RANGE: 200-400mg/dL. To convert conventional units in mg/dL to international units in g/liter, multiply conventional units by 0.01. A patient with dysfibrinogenemia may have normal levels of fibrinogen antigen by immunological assay, but not by clottable activity.
METHOD: May assay by function (clottable fibrinogen) or immunologically (fibrinogen antigen). Light scattering; gel diffusion; antisera; functional ability to form a clot. The most common assay is a modification of the thrombin time. Heparin and FSP may interfere. During thrombolytic therapy a fibrinolytic inhibitor may have to be added to the collection tube.
INTERPRETATION: Fibrinogen is decreased in the plasma in the following conditions:

Causes of Decreased Plasma Fibrinogen
Acquired Deficiencies:
Liver Disease (Dysfibrogenemia)
Disseminated Intravascular Coagulopathy(DIC)
Hemorrhage
Congenital Hypofibrinogenemias:
Afibrinogenemia
Hypofibrinogenemia
Dysfibrinogenemia

In hypofibrinogenemia, both PT and PTT are prolonged; the bleeding time is normal. Hypofibrinogenemia can be distinguished from dysfibrinogenemia by a comparison of clottable and antigenic fibrinogen. Levels of fibrinogen below 50-100mg/dL during thrombolysis places the patient at increased risk of bleeding.

Test results useful in diagnosing dysfibrinogenemia are listed in the next Table:

Results - Diagnosis of Dysfibrinogenemia
Prolonged PT, PTT, Thrombin Time
Failure of Prolonged PT, PTT, Thrombin Time to Correct in Mixing Studies
Prolonged Reptilase Time
Difference between Antigenic and Clottable Fibrinogen

Fibrinogen is increased in the conditions listed in the next Table:

Causes of Increased Plasma Fibrinogen
Pregnancy
Oral Contraceptives
Acute Inflammation and Tissue Damage
Malignancy

FLUORESCENT TREPONEMAL ANTIBODY-ABSORBED TEST (FTA-ABS)

SPECIMEN: Red top tube, separate serum
REFERENCE RANGE: Negative
METHOD: Antigen derived from T. pallidum is used.
INTERPRETATION: The FTA-ABS test is a test for both IgG and IgM antibodies in the serum of patients with syphilis. This test is most commonly used to determine whether the results of a nontreponemal test are due to syphilis or due to a condition causing a false-positive. Although the FTA-ABS test is also subject to false-positives, sequential use of nontreponemal and treponemal tests greatly improves the accuracy of serologic diagnosis. This test may also be used to detect syphilis in patients with negative nontreponemal test results but with clinical evidence of late syphilis. In addition, the FTA-ABS test becomes positive before nontreponemal tests (VDRL, RPR) and may be useful for the early diagnosis of primary syphilis. The use of FTA-ABS on CSF for diagnosis of neurosyphilis is controversial. VDRL is the standard CSF test. FTA-ABS has more false positives but is highly specific when negative. This test is strong evidence against neurosyphilis (Centers for Disease Control. MMWR. 1989; 38,S8:5-15). Uses of the FTA-ABS test are given in the following Table:

Uses of the FTA-ABS
Confirmatory testing following positive nontreponemal (VDRL,RPR) test.
Detect syphilis in patients with negative nontreponemal test and probable late syphilis.
Early diagnosis of primary syphilis (before nontreponemal tests become positive).
Neurosyphilis-negative test eliminates diagnosis

The reactivity of nontreponemal and treponemal tests in untreated syphilis is given in the next Figure (Henry JB [ed.] Todd-Sanford-Davidsohn. Clinical Diagnosis by Laboratory Methods. 16 ed., W.B. Saunders Co. Phila. PA. 1979; II:1890):

Reactivity of Nontreponemal and Treponemal Tests in Untreated Syphilis

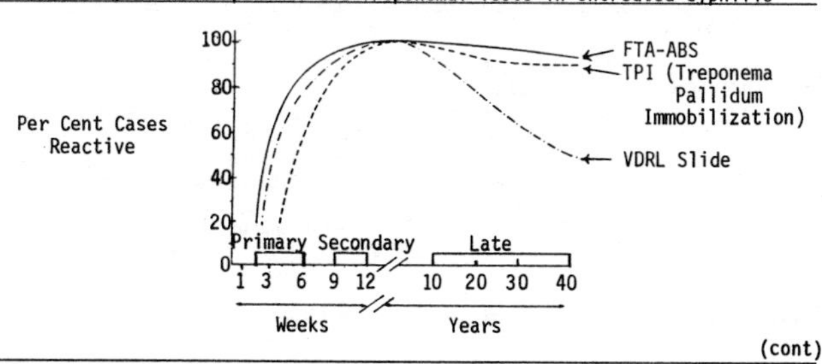

(cont)

Fluorescent Treponemal Antibody Test (Cont)

Effect of Therapy: The FTA-ABS test results remain positive following therapy in most patients. In one study, 24 percent of patients seroconverted to negative after 3 years; all seroconversions occurred in persons with primary syphilis. With successful therapy, the titer of VDRL will tend to fall and will become negative in at least two-thirds of patients if treatment is given in the primary or secondary stages (Romanowski B, et al. Ann Intern Med. 1991; 114:1005-1009).

Congenital Syphilis: The incidence of congenital syphilis has increased rapidly in recent years. The serologic diagnosis of syphilis in infants is problematic due to passive transfer of maternal antibody. Evaluation of possible congenital syphilis and indications for treatment are given in the following Table (Centers for Disease Control. MMWR. 1989; 38(Suppl.S-8):5-15):

Evaluation of Possible Congenital Syphilis and Indications for Treatment
Evaluation of Infant with Possible Congenital Syphilis:
Physical examination
Nontreponemal Antibody Titer
CSF analysis for cells, protein, and VDRL
Long bone x-rays
FTA-ABS-19S-IgM antibody
Indications for Treatment:
Evidence of active disease (physical exam, x-ray)
Reactive CSF-VDRL
Abnormal CSF (WBC >5/mm^3 or protein >50mg/dL) regardless of CSF serology
Quantitative nontreponemal serology fourfold higher than mother's
Positive FTA-ABS-19S-IgM antibody

The FTA-ABS-19S-IgM antibody test is available from the CDC and reference laboratories (Lewis LL. Infect Dis Clin N Amer. 1992; 6:31-39; Ikeda MK, Jenson HB. J Pediatr. 1990; 117:843-852).

Syphilis and HIV: Patients with HIV are at increased risk for syphilis and vice versa. Patients with HIV are more likely to have false-positive nontreponemal tests. In patients with late or advanced HIV, serologic testing for syphilis may be delayed or absent. In general, however, serologic testing provides accurate information for patients with HIV (Hook EW, Marra CM. N Engl J Med. 1992; 326:1060-1069; Haas JS, et al. J Infect Dis. 1990; 162:862-866; Johnson PDR, et al. AIDS. 1991; 5:419-423).

False positive reactions with the FTA-ABS test occur in the conditions listed in the next Table (Hook EW, Marra CM. N Engl J Med. 1992; 326:1060-1069; Hart G. Ann Intern Med. 1986; 104:368-376):

False Positive Reactions with the FTA-ABS Test
Lyme Disease
Leprosy
Malaria
Infections Mononucleosis
Relapsing Fever
Leptospirosis
Genital Herpes
Narcotic Addiction
Systemic Lupus Erythematosus

See also SYPHILIS ANTIBODIES.

FOLATE (FOLIC ACID), RED BLOOD CELL

SPECIMEN: Assay usually includes serum folate. Two blood specimens are required: red top tube and a lavender (purple)(EDTA) top tube for red cell folate and hematocrit. Red top tube, separate serum and freeze serum. Lavender top tube, mix by inverting at least six times; add 0.5 mL of EDTA blood to a vial containing 4.5 mL of freshly prepared 1% ascorbic acid solution; then freeze hemolysate.

REFERENCE RANGE: 160-700 ng/mL. Borderline 140-160 ng/mL. To convert conventional units in ng/mL to international units in nmol/liter, multiply conventional units by 2.266.

METHOD: RIA

INTERPRETATION: RBC folate in conjunction with serum folate is a good index for diagnosing folate deficiency.

Serum folate values fluctuate significantly with diet; measurement of red cell folate more closely reflects tissue folate stores. Following dietary deprivation or poor absorption of folate, serum folate levels will be decreased within 3 weeks while the RBC folate, representing the storage form, will remain normal for 3-4 weeks. Sequential changes and time course for the development of megaloblastic anemia due to folate deficiency are given in the following Table (Herbert V. Am J Hematol. 1987; 26:199-207):

Sequential Development and Time Course of Folate Deficient Megaloblastic Anemia

Sequential Changes	Time (weeks)
Low Serum Folate (<3ng/mL)	3
Hypersegmentation of Neutrophils in Bone Marrow	5
Hypersegmentation of Neutrophils in Peripheral Blood; Abnormal Mitoses; Basophilic Intermediate Megaloblasts in Bone Marrow	7
Large Metamyelocytes and Polychromatophilic Intermediate Megaloblasts in Bone Marrow	10
Orthochromatic Intermediate Megaloblasts in Bone Marrow	14
Low RBC Folate	17
Macroovalocytosis; Many Large Metamyelocytes in Bone Marrow	18
Overtly Megaloblastic Bone Marrow	19
Anemia	20

See FOLATE (FOLIC ACID), SERUM for causes of folate deficiency.

FOLATE (FOLIC ACID), SERUM

SPECIMEN: Red top tube, separate serum. The test should not be ordered for patients who have recently received a radioisotope, methotrexate, or other folic acid antagonist. Strongly consider ordering vitamin B_{12} determination.
REFERENCE RANGE: 3-14ng/mL. To convert conventional units in ng/mL to international units in nmol/liter, multiply conventional units by 2.266.
METHOD: RIA
INTERPRETATION: The causes of folate deficiency are given in the next Table:

Causes of Folate Deficiency

Inadequate Intake:
- Alcoholism
- Nutritional Deficiencies

Relative Inadequate Intake (Increased Requirements):
- Pregnancy
- Hemolytic or Sideroblastic Anemia
- Chronic Hemodialysis or Peritoneal Dialysis (Folate Dialyzable)
- Chronic Myelofibrosis
- Hyperthyroidism
- Exfoliative Dermatitis

Inadequate Absorption or Increased Excretion:
- Malabsorption Syndromes:
  - Tropical Sprue
  - Gluten-Sensitive Enteropathy (Non-Tropical Sprue)
  - Crohn's Disease
  - Lymphoma or Amyloidosis of Small Bowel
- Diabetic Enteropathy
- Intestinal Resections or Diversions

Interference with Folic Acid Metabolism
- Drugs Blocking Action of Dihydrofolate Reductase: Methotrexate, Aminopterin, Trimethoprim, Pyrimethamine, Triamterene
- Drugs Interfering by Unknown Means: Anticonvulsants (Phenobarbital, Primidone, Phenytoin), Ethanol, Cycloserine, Oral Contraceptives (Ethinylestradiol/Norethisterone Acetate)

Folate deficiency is most commonly encountered in pregnancy and alcoholism; the requirement during pregnancy is about 4 times that of the normal requirement. Decreased red blood cell folate and normal serum folate occasionally occur with pure vitamin B_{12} deficiency (folate-trap hypothesis).

Folate deficiency is associated with megaloblastic anemia.(cont)

Folate (Folic Acid) Serum (Cont)

References: Kones R. South Med J. 1990; 83:1454-1458; Bailey LB. J Nutr. 1990; 120:1508-1511; Marcus DL, Freedman ML. J Amer Geriatr Soc. 1985; 33:552-558; Lambie DG, Johnson RH. Drugs. 1985; 30:145-155.

Folate and Neural Tube Defects: Multiple studies have documented a decreased risk of neural tube defects in children born to mothers who received folate supplementation early in pregnancy. The current recommendation is that women of childbearing age who are capable of becoming pregnant should consume 0.4mg of folic acid daily (not more than 1mg daily) for the purpose of reducing the risk of neural tube defects (spina bifida and anencephaly) in infants born to these mothers. Relevant studies have been reviewed and recommendations were developed by the Centers for Disease Control, Food and Drug Administration, Health Resources and Services Administration, and National Institutes of Health (CDC, MMWR. 1992; 41(RR-14):1-7).

FOLLICLE STIMULATING HORMONE(FSH)

SPECIMEN: Red top tube, separate serum from cells as soon as possible; if serum is not analyzed immediately, store at -20°C. Repeated freezing and thawing should be avoided. There are rapid cyclical changes in concentration; thus, optimally, three blood specimens should be drawn at 60 minute intervals; combine the serum of the three specimens.

REFERENCE RANGE: Units(IU/liter or mIU/mL); FSH levels are influenced by age, sex, degree of sexual development, drugs, and in females, menstrual status and pregnancy. Prepubertal: Up to 5 IU/liter. In the female, FSH concentrations start to increase at about age 10 or 11 and plateau at the time of menarche; in the male, FSH concentrations start to increase about 2 years later. Female: Changes in FSH during the menstrual cycle are given in the next Figure:

Changes in FSH During Menstrual Cycle

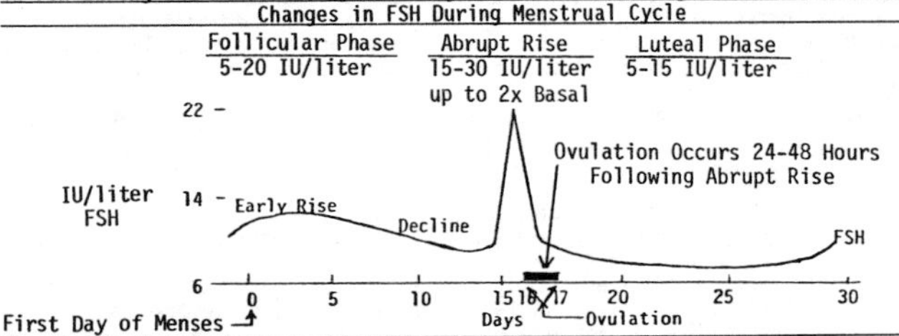

Postmenopausal Female: Up to 200 IU/liter. Male: Adult: 4-20 IU/liter; Castrate: Up to 200 IU/liter.

METHOD: RIA, others.

INTERPRETATION: In females, FSH stimulates the development of ovarian follicles and, in conjunction with LH, estrogen secretion and ovulation. In males, FSH stimulates spermatogenesis.

FSH determinations are indicated in the work-up of patients with disorders of puberty (delayed or precocious), subfertility (gonadal failure, polycystic ovarian disease, disorders of hypothalamic-pituitary-gonadal interaction) and pituitary disorders (tumors, hypopituitarism)(Beastall GH, et al. Ann Clin Biochem. 1987; 24:246-262). In infertility, FSH levels are done to determine whether there is primary failure of ovaries or testicles, or failure secondary to pituitary hypofunction. The test results, expected in pituitary, ovarian or testicular and end-organ failure, are given in the next Table:

FSH Levels in Pituitary, Ovarian or Testicular and End-Organ Failure

Failure	Serum FSH
Pituitary (Consider Prolactinoma)	Decreased
Ovarian (Consider Polycystic Ovarian Disease)	Increased
End-Organ	Normal

Persistent elevation of FSH levels (>40 IU/liter) is indicative of ovarian failure or "resistant ovary" syndrome (Blacker CM. Endocrinol Metab Clin N Amer. 1992; 21:57-84).

The gonadotropin (FSH or LH) levels for various conditions is given in the following Table (Beastall GH, et al. Ann Clin Biochem. 1987; 24:246-262):

FSH, LH, and Sex Steroid Levels in Various Conditions

Condition	Sex Steroids	FSH	LH
Delayed Puberty			
Gonadal Dysgenesis/Failure	Decr	Incr	Incr
Hypothalamic/Pituitary Disorders	Decr	Decr	Decr
Precocious Puberty			
True Precocious Puberty	Incr	Incr	Incr
Pseudo Precocious Puberty	Incr	Decr	Decr
Subfertility			
Gonadal Failure	Decr	Incr	Incr
Disorders of Hypothalamic-Pituitary-Gonadal Interaction	NL-Decr	NL-Decr	NL-Decr
Polycystic Ovarian Disease	NL	NL-Decr	Incr
Pituitary Disorders			
Primary or Secondary Gonadotroph Tumors	NL-Incr	Incr	Incr
Hypopituitarism	Decr	Decr	Decr

In isolated pituitary gonadotropin (FSH and LH) deficiency, there are no clinical signs before the age of puberty. Secondary sexual development fails to occur during adolescence and puberty is delayed. If isolated pituitary gonadotropin deficiency occurs in the female adult, amenorrhea develops.

Increased levels of FSH are found in conditions listed in the next Table:

Increased Serum Levels of FSH

Females	Males
Primary Ovarian Failure:	Primary Testicular Failure:
Menopause, Premature Menopause	Undescended Testis; Radiation
Ovariectomy	Castration
Ovarian Agenesis, Dysgenesis	Anorchia (Testicular Agenesis)
Genetic Disorders: Turner's Syndrome	Genetic Disorders: Klinefelter's Syndrome
	Sertoli-Cell-Only Syndrome

Decreased levels of FSH are found in conditions listed in the next Table:

Decreased Serum Levels of FSH

Females	Males
Medications: Estrogens, Testosterone, Oral Contraceptives	Medications: Testosterone or Estrogen
Hypopituitarism; Hypophysectomy	Hypopituitarism
Adrenal or Ovarian Neoplasm	Hypophysectomy
Polycystic Ovarian Disease	Adrenal Neoplasm
Various Menstrual Disorders	
Sheehan's Syndrome (Postpartum Pituitary Necrosis)	

See also Simoni M, Neischlag E. J Endocrinol Invest. 1991; 14:983-997.

FRACTIONAL EXCRETION OF FILTERED SODIUM
(see EXCRETION FRACTION OF FILTERED SODIUM)

FREE ERYTHROCYTE PROTOPORPHYRIN(FEP)
(see ZINC PROTOPORPHYRIN(ZPP), BLOOD)

FRUCTOSAMINE

Synonyms: Glycated/Glycosylated Serum Proteins
SPECIMEN: Red top tube, separate serum.
METHOD: Determination of glycated serum proteins by nitroblue tetrazolium (NBT) colorimetry. Second generation assays incorporate uricase, nonionic surfactants, and a glycated polylysive standard.
NORMAL RANGE: 205-285micromol/L.
INTERPRETATION: The serum fructosamine assay is used to monitor short term clinical glycemic control. Glycated serum proteins with short half life (mean 20 days) will reflect glycemic control over the previous 2-4 weeks in contrast to glycated hemoglobin which reflects control over a 2-3 month period. Fructosamine does not refer to a single type of molecule, but rather a group of serum proteins which have reacted with glucose. Assay of these various glycated proteins depends on their ability to reduce tetrazolium dyes.

The fructosamine assay has been shown to discriminate between normals, diabetics with good/moderate glycemic control, and those with poor control (Cefalu WT, et al. Clin Chem. 1991; 37:1252-1256). Deterioration of glycemic control appears to result in a rapid increase in fructosamine, while fructosamine levels appear to decrease slowly during improvement (Baker JR, et al. BMJ. 1984; 288:1484-1486). This may be a reflection of the heterogeneity of the glycated proteins. Absolute correlation between fructosamine and HgbAlc (glycohemoglobin) should not be expected because these assays depend on the half-lives and chemical reactivities of different proteins. The 1st generation NBT assays may be affected by uric acid, bilirubin, lipids and level of albumin concentration.

The fructosamine assay appears to be a faster, more convenient method of accessing glycemic control than the more labor-intensive HgbAlc and glycosylated hemoglobin (affinity chromatography) techniques (Gebhart SSP, et al. Arch Intern Med. 1991; 151:1133-1137). However, the fructosamine assay is reported to be unreliable following hemodialysis. The glycosylated hemoglobin by affinity chromatography is said to be superior in this condition and unaffected by the cyanated hemoglobin produced during uremia (Nunoi K, et al. Metabolism. 1991; 40:986-989). See also HEMOGLOBIN Alc.

FUNGAL PANEL, CEREBROSPINAL FLUID(CSF)

The fungi found in the CSF and their distinguishing features are given in the next Table (Bigner SH. Am Soc Clin Path Check Sample, Cytopathology II, No. 5, 1983):

Fungi in CSF and Their Distinguishing Features

Organism	Shape and Size	Capsule	Bud Type	Predisposition
Cryptococcus Neoformans	Round Yeast, 2 to 15 Microns	Thick; Muco-polysaccharide; Stains with Alcian Blue and Mucicarmine	Thin-necked	AIDS, Lymphoma or Steroids; occ. Healthy Individuals
Blastomyces Dermatitidis	Round Yeast, Usually 1 to 5 Microns	None	Thick-necked	Healthy Individuals
Histoplasma Capsulatum	Round Yeast, 1 to 5 Microns	None	Thin-necked	AIDS, Healthy Individuals
Candida Albicans	Round, Oval or Elongated, 2 to 5 Microns	None	Thin-necked	Lymphoma, Steroids or Cytotoxic Agents
Coccidioides Immitis	Round or Oval Endospores, up to 60 Microns	None	Rarely seen	Healthy Individuals, AIDS
Aspergillus	Septate Filaments 3 to 4 Microns	-	-	Antibiotics or Steroids

Cryptococcus neoformans is the most common fungus involving the human central nervous system.

Differentiation of common fungi based on morphology is given in the following Table (Kaufman L. CID. 1992; 14(Suppl 1):S23-S29):

Morphologic Identification of Fungi

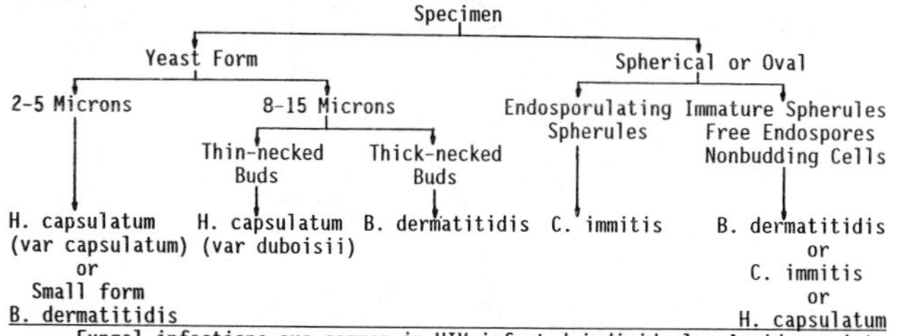

Fungal infections are common in HIV infected individuals. A wide variety of infections can occur; the most common cause of meningoencephalitis in AIDS patients is Cryptococcus neoformans. Review of fungal infections in AIDS, see Macher AM, et al. "AIDS and the Mycoses." Infect Dis Clin N Amer. 1988; 2:827-839.

Fungal Serological Panel

FUNGAL SEROLOGICAL PANEL

The common invasive fungi are listed in the next Table (Penn RL, et al. Arch Intern Med. 1983; 143:1215-1220):

Common Invasive Fungi

Pathogenic Fungi	Opportunistic Fungi
Blastomyces Dermatitidis	Aspergillus Species
Coccidioides Immitis	Candida Species
Cryptococcus Neoformans	
Histoplasma Capsulatum	

The pathogenic fungi can infect normal individuals; the opportunistic fungi usually cause infections in immunocompromised hosts or in patients with indwelling catheters.

The serologic tests that are most commonly done for diagnosis of invasive fungi are given in the next Table (Kaufman L. CID. 1992; 14(Suppl 1):S23-S29):

Serologic Tests for Invasive Fungi

Fungi	Serologic Tests
Blastomyces Dermatitidis	Sensitivity 40-88%; Specificity 92-100%; Direct exam and culture are primary diagnostic tests.
Coccidioides Immitis	Tube Precipitin(TP) Test for IgM Antibodies Complement Fixation(CF) Test for IgG Antibodies
Cryptococcus Neoformans	Latex Agglutination(LA) Test for Cryptococcal Antigen is more than 90% sensitive. Antibody testing is not useful.
Histoplasma Capsulatum	Complement Fixation(CF) Test for Antibodies is frequently used for diagnosis. Antigen testing may be useful.
Aspergillus Species	Immunodiffusion(ID), CF, etc. for Antibodies to Aspergillus necessary for diagnosis of allergic bronchopulmonary aspergillosis(ABPA).
Candida Species	Antigen and Antibody Testing are generally unreliable for diagnosis of invasive candidal infection.

Testing for fungal pathogens are also discussed as individual topics.

GAMMA GLUTAMYL TRANSPEPTIDASE (GGTP)

SPECIMEN: Red top tube, separate serum.
REFERENCE RANGE: Male: 11-51IU/liter; Female: 7-33IU/liter
METHOD: Kinetic
INTERPRETATION: Assay of gamma glutamyl transpeptidase is most often used to differentiate the source of an elevated serum alkaline phosphatase, e.g., liver or bone; gamma glutamyl transpeptidase is not present in bone. Another common use is in the evaluation of alcohol abuse. Some of the diseases that are associated with increased serum GGTP are given in the next Table:

Conditions Associated with Elevated Serum GGTP
Alcoholism
Liver Disease
Obstructive Jaundice
Metastatic Disease to Liver
Congested Liver
Infectious Mononucleosis
Pancreatic Disease
Diabetes Mellitus (Boone et al. Am J Clin Pathol. 1974; 61:321-327).
Myocardial Infarction (Betro et al. Am J Clin Pathol. 1973; 60:679-683).
Congestive Heart Failure
Neurologic Disorders
Trauma
Nephrotic Syndrome
Chronic Renal Failure (occasionally)
Sepsis (Fang MH, et al. Gastroenterology. 1980; 78:592-597).
Drugs, such as Phenobarbital, other Barbiturates and Phenytoin (Dilantin), Antipyrine

Serum gamma glutamyl transpeptidase (GGTP) activity is elevated in all forms of liver disease and in intra-hepatic or extra-hepatic biliary obstruction. It is a sensitive method to determine whether bone or liver is the source of increased serum alkaline phosphatase and has gained wide popularity for this purpose, specifically in differentiating neoplasms metastatic to liver from those involving bone. Serum alkaline phosphatase is increased during the growth period because of increased osteoblastic activity associated with bone growth and during the last trimester of pregnancy because of the high content of alkaline phosphatase in the placenta. Serum GGTP is absent from the bone and the placenta. If alkaline phosphatase is high due to bone disease, serum GGTP is normal.

Serum GGTP is normal is the following conditions:

Normal Serum Gamma Glutamyl Transpeptidase (GGTP)
Children
Adolescents
Pregnancy
Bone Disease
Muscle Disease

Alcoholism: GGTP is the most sensitive enzyme used to detect liver damage from excessive alcohol intake. GGTP is situated on the smooth endoplasmic reticulum. Any substance, such as alcohol or barbiturates, which causes microsomal proliferation will cause an increase in serum GGTP. Drugs such as the antiepileptics, phenobarbital and phenytoin, or other drugs which stimulate the smooth endoplasmic reticulum, cause elevation of GGTP (Berg B, Tryding N. Lancet. 1981; 1162). Boone et al. (Boone DJ, et al. Ann Clin Lab Sci. 1977; 7:25-28) found that GGTP activity is useful in the assessment of alcohol-induced liver disease and for monitoring the progress of therapy as well as alleged abstention from alcohol in known alcoholics. Chronic alcoholics are more likely to have an elevation in GGTP than binge drinkers and those drinking less than 2 drinks per day.

Lamy et al (Lamy J, et al. Nov Presse Med. 1975; 4:487-490) studied the change of GGTP activity in the serum of "heavy drinkers" and alcoholics during detoxification; the decrease in enzyme activity is shown in the next Figure:

Decrease in Gamma-Glutamyl Transferase Activity in Serum of "Heavy Drinkers" and "Alcoholics" during Detoxification

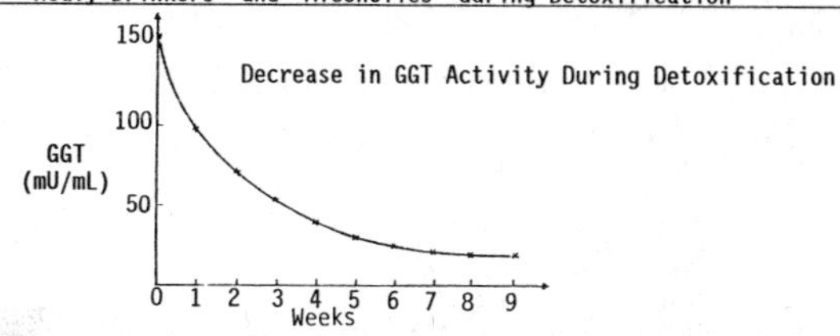

As seen in the above figure, after alcohol deprivation, the GGTP activity of alcoholics decreased in the first few days exponentially with a half-life of 5 to 17 days.

Elevated serum GGTP activity in the presence of normal results of other liver function tests and in the absence of jaundice and hepatomegaly is apparently a valuable indicator of significant alcohol consumption in adolescents who consumed six or more drinks per day (Westwood et al. Pediatr. 1978; 62:560-562). Patients with an isolated GGTP with other liver function tests normal, have only mild reversible change on liver biopsy. An isolated elevation of GGTP does not justify liver biopsy in alcoholic disease (Ireland A, et al. Br Med J. 1991; 302:388-389).

GASTRIN

SPECIMEN: Fasting 8-12 hours overnight; red top tube; pour serum into plastic tube; freeze immediately.
REFERENCE RANGE: Normal following overnight fast: 0 to 100pg/mL; borderline: 100-200pg/mL; elevated: 200pg/mL or more. To convert conventional units in pg/mL to international units in ng/Liter, multiply conventional units by 1.0.
METHOD: RIA
INTERPRETATION: Serum gastrin is elevated in the conditions listed in the next Table; (Essop AR, et al. N Engl J Med. 1982; 307:192):

Causes of Elevated Serum Gastrin
Zollinger-Ellison Syndrome
Pernicious Anemia
Duodenal Ulcers
Pyloric Stenosis
Chronic Renal Failure
Atrophic Gastritis
Intestinal Resection
Antral G-Cell Hyperplasia
Rheumatoid Arthritis
Ulcerative Colitis
Administration of Steroids and Calcium

A fasting gastrin level greater than 600pg/mL (normals and duodenal ulcer patients, less than 200pg/mL) with acid hypersecretion (greater than 15mmol/hour) is virtually diagnostic of Zollinger-Ellison syndrome.

The characteristic findings in Zollinger-Ellison syndrome are given in the following Table:

Characteristics of Zollinger-Ellison Syndrome
(1) Presence of non-beta cell tumor secreting gastrin (Gastrinoma)
(2) Marked secretion of acid from gastric parietal cells
(3) Single ulcers of gastric or duodenal mucosa in two-thirds of patients. Approximately one-quarter will have ulcers in unusual locations, e.g., distal duodenum or ileum.

GASTROINTESTINAL HEMORRHAGE PANEL

Gastrointestinal hemorrhage panel is given in the next Table:

Gastrointestinal Hemorrhage Panel
Initial Laboratory Tests:
Stool for Blood
Complete Blood Count(CBC)
Urinalysis
Blood Urea Nitrogen(BUN)
Electrolytes: Sodium, Potassium, Chloride, CO_2 Content
Coagulation Screen:
Prothrombin Time(PT)
Partial Thromboplastin Time(PTT)
Fibrinogen Level
Platelet Count
Liver Function Tests and Liver Enzymes:
Bilirubin, Total and Direct
Aspartate Aminotransferase (AST = SGOT)
Alanine Aminotransferase (ALT = SGPT)
Alkaline Phosphatase
Albumin, Serum
Amylase, Serum or Urine
Ammonia (Patients with Hepatic Dysfunction)
Gastric Aspirate for Blood
Routine X-Ray Films of the Chest and Abdomen
Electrocardiogram for Older Patients - Silent Infarct may Occur with Significant Gastrointestinal Bleeding
Specific Diagnosis:
Nasogastric Aspiration to Rule-out Upper Gastrointestinal Hemorrhage
Barium Enema for Diverticulosis
Fiberoptic Colonoscopy
Sigmoidoscopy
Selective Angiography
Tagged Erythrocyte Study
Scan for Meckel's Diverticulum
Upper Gastrointestinal Series with Small Bowel Follow-Through
Inspection of the Anus and Rectum

Ament ME. Pediatr Rev. 1990; 12:107-116; Colacchio TA. Am J Surg. 1982; 143:607-610.

Causes of gastrointestinal hemorrhage are given in the next Table:

Causes of Gastrointestinal Hemorrhage
Upper Gastrointestinal Hemorrhage: Peptic Ulcer Disease; Esophageal Varices; Gastritis; "Stress" Ulcerations; Mallory-Weiss Syndrome; Cancer; Gastroesophageal Reflux with Esophagitis; Drugs (Steroids, Aspirin, Iron, Alcohol)
Lower Gastrointestinal Hemorrhage: Diverticular Bleeding; Vascular and Elastic Tissue Disorders; Colorectal Carcinoma; Infectious Enterocolitis; Colonic Polyps; Intussusception; Volvulus; Inflammatory Bowel Diseases
Gastrointestinal Bleeding in Infants and Children:
Neonates:
Anorectal Fissure; Hemorrhagic Disease of the Newborn; Infectious Diarrhea; Volvulus; Vascular Malformations
Infants up to Age Two: Infectious Diarrhea; Anal Fissure; Intussusception; Meckel's Diverticulum; Peptic Ulcer; Volvulus
Children, 2 to 12 Years: Infectious Diarrhea; Juvenile Polyp; Peptic Ulcer; Esophageal Varices; Inflammatory Bowel Disease; Meckel's Diverticulum; Intestinal Duplication; Intussusception; Vascular Malformation
Adolescents: Peptic Ulcer; Inflammatory Bowel Disease; Esophageal Varices

References: Geier DL, Cooke AR. J Crit Illness. 1992; 7:1676-1695; Mann NS, Mann SK. Hosp Phys. March 1989; 47-54; Ament ME, already cited.

GENTAMICIN

SPECIMEN: Red top tube, separate serum and freeze. Heparinized tubes should not be used to collect specimens. Obtain serum specimens as follows:
(1) 24-48 hours after starting therapy if loading dose is not given.
(2) 5 to 30 minutes before I.V. gentamicin (trough).
(3) 30 minutes after a 30 minute I.V. infusion of gentamicin (peak).
I.M. administration: 30 minutes to one hour.

The following times should be recorded on the laboratory requisition form and on the patient's chart:

Trough Specimen Drawn _________(Time)
Gentamicin Started _________(Time)
Gentamicin Completed _________(Time)
Peak Specimen Drawn _________(Time)

High concentrations of beta-lactam antibiotics inactive aminoglycosides (gentamicin, streptomycin, amikacin, tobramycin and kanamycin); to reduce this interaction, specimens containing both classes of antibiotics should either be assayed immediately using a rapid method or stored frozen.

REFERENCE RANGE: Therapeutic: 4-10mcg/mL; Toxic: >12mcg/mL; Peak Values: 4-10mcg/mL; Trough: <2mcg/mL. Time to Steady State: Adults (<30 years), 2.5-15 hours; Adults (>30 years), 7.5-75 hours; Children, 2.5-12.5 hours; Neonates, 10-45 hours. Time to steady state and time to peak concentration may be significantly prolonged in patients with renal dysfunction.

METHOD: RIA, EMIT

INTERPRETATION: Gentamicin is an aminoglycoside antibiotic which is used frequently in hospitals to treat patients who have serious aerobic gram-negative bacterial infections (e.g., a number of the Enterobacter species and P. aeruginosa), especially septicemia and staphylococcal infections, untreatable with penicillins.

The mechanism of action of gentamicin is like other aminoglycosides in that it binds to the bacterial 30S ribosomal subunit to produce a non-functional 70S initiation complex that results in the inhibition of bacterial cell protein synthesis and misreading of the genetic code.

The three main toxic side effects of gentamicin are ototoxicity, nephrotoxicity and neuromuscular blockage (Smith CR, et al. N Engl J Med. 1980; 302:1106); it is important to control the dose given by monitoring peak and trough levels of the drug, particularly in patients with any degree of renal failure. As renal function declines, drug half-life increases to up to 50 hours.

To minimize risk of toxicity, it has been recommended that peak levels not exceed 10mcg/mL and that trough levels should fall between 1 and 2 mcg/mL.

Gentamicin is eliminated exclusively by renal excretion; excessive serum concentrations may occur and lead to further renal impairment. The renal damage is to the renal proximal tubules and is usually reversible if discovered early.

Ototoxicity is usually due to vestibular damage and is often not reversible.

Electrolyte disturbances may also occur, as given in the following Table:

Electrolyte Disturbances with Aminoglycoside Therapy
Hypomagnesemia
Hypocalcemia
Hypokalemia

Hypomagnesemia (<1.6mg/dL) is observed in almost 40% of patients; hypocalcemia and hypokalemia may also occur. It may be advisable to monitor serum magnesium and use replacement therapy (Zaloga GP, et al. Surg Gynecol Obstet. 1984; 158:561-565).

GIARDIA TESTING

SPECIMEN: Stool; Fluid from duodeno-jejunal junction; Small intestine biopsy. Antibody Detection: Serum.

REFERENCE RANGE: Negative

METHODS: Direct Wet Saline Preparation: Microscopic examination of stool or fluid from the small intestine is examined at x40, x100, or x450. Fluid from the duodeno-jejunal junction can be obtained by endoscopy or by the String test. In this test, the patient swallows a pill attached to a string. One end of the string is taped to the patient's cheek. The other end of the string attached to a weight which is liberated when the pill dissolves, and this end extends into the duodenum. After four hours, the string is withdrawn. The fluid from the bile-stained portion of the string is squeezed out and examined microscopically.

Giardia Testing (Cont)

Antigen Detection: CIE or EIA. Cyst Detection: Indirect Immunofluorescence. Small Bowel Biopsy: Endoscopic biopsy specimens of the duodeno-jejunal junction are obtained and imprint smears are prepared for Giemsa or Masson's trichrome stain. Antibody Tests (IgG, IgM) - Not commercially available.

INTERPRETATION: Giardia lamblia is a common cause of clinically significant diarrhea. Typical situations in which Giardia should be considered as a likely cause of diarrhea are given in the following Table:

Giardia as a Likely Cause of Diarrhea
Travelers Diarrhea
Drinking Contaminated Water
Day Care Centers
Homosexual Males

Symptoms of Giardia infection are variable, and some individuals are asymptomatic. Typically, incubation period is 9 to 15 days; acute symptoms include nausea, anorexia, low grade fever, and chills. Subsequently, severe diarrhea may develop, with explosive, watery, malodorous diarrhea, lasting 3 or 4 days. Blood and mucous are not usually present. Following the acute infection, chronic infection may develop.

Most cases of giardiasis are diagnosed by identification of Giardia cysts and trophozoites in stool specimens. Stool specimens should be collected every other day until three specimens are obtained. These specimens may be examined microscopically as a wet preparation. Alternatively, the specimen may undergo rapid detection assays for antigen detection (CIE or EIA) or cyst detection (indirect immunofluorescence). Negative tests do not exclude infection. If there is clinical suspicion of Giardia infection with negative stool studies, direct sampling of duodeno-jejunal fluid or tissue by the string test or endoscopy with biopsy may be considered. Giardia antibody tests for IgG and IgM are not commercially available at this time.

Reference: See excellent review: Wolfe MS. Clin Microbiol Rev. 1992; 5:93-100.

GLIADIN ANTIBODIES

Synonym: Screening Test for Celiac Disease; Anti-Gliadin Antibody

SPECIMEN: Red top tube, separate serum.

REFERENCE RANGE: Negative, None detected.

METHOD: Fluorescent immunosorbent test; ELISA.

INTERPRETATION: The assay of gliadin antibody is useful as a screen for celiac disease in children; some state a negative test result essentially excludes the diagnosis of celiac disease. However, sensitivity and specificity vary between laboratories. The definitive diagnosis of celiac disease includes a mucosal biopsy before and after a gluten free diet and again after rechallenge. Assay for antibodies to gliadin has been examined in children with malabsorptive disorders in a prospective multicenter study. The test had a sensitivity of 100 percent and a specificity of 84 percent for the diagnosis of childhood celiac disease. Among the 16 percent of the patients with other malabsorptive disorders who had gliadin antibodies, 12 percent had a low titer and 3.5 percent showed moderate to high titers (Burgin-Wolff A, et al. J Pediatr. 1983; 102:655-660).

Anti-gliadin antibodies, usually IgA type, rise and fall in response to gluten challenge. These antibodies may also be found in 25% of patients with dermatitis herpetiformis (Kumar V, et al. Int Arch Allergy Appl Immunol. 1987; 83:155-159). Patients with IgA deficiency are at increased risk for celiac disease. In these cases, assay of IgG-anti-gliadin antibody assays may be of use. Individuals with celiac disease also may have anti-endomysial antibodies reactive against components of smooth muscle. Anti-reticulin antibodies are also highly specific for celiac disease (>98%) but have a relatively low sensitivity (Unsworth DJ, et al. Lancet. 1983; 1:874-875).

Other associated conditions and laboratory findings may include iron deficiency anemia, folate deficiency, malabsorption, low serum cholesterol, carotene, vitamin D, calcium, magnesium, phosphorus, potassium and albumin. There may be hyposplenism associated Howell-Jolly bodies, thrombocytosis, and deformed RBC (30-50% of untreated adults). Osteopenia and elevation of alkaline phosphatase is often present. The clinical and laboratory aspects of celiac disease are reviewed in: Trier JS. N Engl J Med. 1991; 325:1709-1719.

GLUCOCEREBROSIDASE (FOR GAUCHER'S DISEASE)

SPECIMEN: The preferred specimen is the fibroblast. Obtain a skin biopsy (4mm punch). Use culture media for transport; maintain specimen at 4°C.

White blood cells are used as an alternate specimen.

REFERENCE RANGE: Normal enzyme activity

METHOD: Culture; measure enzyme; report results in terms of units/gram cellular protein; the enzyme assay reaction is as follows:

$$\text{4-Methylumbelliferyl-Beta-D-Glucoside} + H_2O \xrightarrow{\text{Glucocerebrosidase}} \text{D-Glucose} + \text{4-Methylumbelliferone (Fluorescent)}$$

The activity of beta-glucosidase is measured by following the release of the fluorescent compound 4-methylumbelliferone.

INTERPRETATION: Gaucher's disease is the most common lysosomal storage disease; it is transmitted by an autosomal recessive gene. The incidence of Gaucher's disease is 1 in 450 births in Jews of eastern European origin (Ashkenazim). The incidence in the general population is 1 in 40,000 births. There is a deficiency of the enzyme, glucocerebrosidase, which catalyzes the reaction:

$$\text{Glucosylceramide} + H_2O \xrightarrow{\text{Glucocerebrosidase}} \text{D-Glucose} + \text{Ceramide}$$

The glucolipid, ceramide-glucose (glucocerebroside) accumulates in reticuloendothelial cells of the liver, spleen and bone marrow causing enlargement of the liver and spleen and erosion of the cortices of the long bones and pelvis.

There are three forms of the disease, Type I: Adult; Type II: Infantile; and Type III: Juvenile. 99 percent of cases are Type I Gaucher's Disease.

Type I; Adult: Most patients with adult-type Gaucher's disease are asymptomatic. The most common symptoms are progressive splenic and hepatic enlargement, leukopenia, thrombocytopenia, bleeding tendency, bone pain, fractures and anemia. Neurologic involvement does not occur. There is a moderate reduction in the enzyme beta-glucosidase (10-40%) of mean normal values. Increased serum acid phosphatase activity (non-prostatic) is a frequent finding in Type I.

The Gaucher cell may be seen on bone marrow; the macrophages have fibrillated cytoplasm, characteristic of Gaucher cell.

Type II, Infantile: The usual onset occurs acutely at 2-3 months; neuropathic abnormalities are severe and death usually occurs within two years.

Type III; Juvenile: The onset is subacute; neuropathic involvement occurs.

In Type II and Type III, there is a marked reduction of the enzyme, glucocerebrosidase, in tissues.

Enzyme replacement with alglucerase is very effective in reducing hepatosplenomegaly, improving hematologic studies, improving activity and height and weight gain in children; cost is $100,000-$300,000 annually.

References:

Wittington R, Goa KL. Drugs. 1992; 44:72-93. Beutler E. N Engl J Med. 1991; 325:1354-1360. Sidransky E, Ginns EI. JAMA. 1993; 269:1154-1157.

GLUCOSE

SPECIMEN: Red top tube, separate serum. Separate cells and run without delay to avoid cellular metabolism of glucose. If a delay in measurement is anticipated use plasma, grey top tube, containing oxalate and fluoride. Fluoride will inhibit glycolysis. If whole blood is analyzed, the glucose values for whole blood are approximately 10% less than those of serum, since the RBC's contain a slightly lower concentration of glucose than serum.

REFERENCE RANGE: Newborn: Premature: 20-80mg/dL; full-term: 20-90mg/dL; infants and children: 60-115mg/dL, adults: 65-120mg/dL. To convert conventional units in mg/dL to international units in mmol/liter, multiply conventional units by 0.05551.

METHOD: Glucose oxidase or hexokinase method.

INTERPRETATION: Glucose levels are employed in the diagnosis of hyperglycemia and hypoglycemia. Hypoglycemia is discussed under the topic GLUCOSE, HYPOGLYCEMIA.

(cont)

Glucose (Cont)

The causes of hyperglycemia are given in the next Table:

Causes of Hyperglycemia
Diabetes Mellitus
Acute Pancreatitis
Endocrine Hyperfunctions:
Cushing's Syndrome and Disease; Pheochromocytoma; Acromegaly; Hypothalamic Lesions; Carcinoid Syndrome; Thyrotoxicosis; Glucagonoma; Somatostatinoma
Hemochromatosis
Ataxia Telangiectasia
Drugs:
Anabolic Steroids; Epinephrine and Norepinephrine; Benzothiadiazine Diuretics; Phenytoin
Note: Increased with I.V. Glucose; Stress; Non-Fasting Specimen

Stress including trauma, shock, CVA, MI and burns will raise the blood glucose. A blood glucose >120 mg/dL during the early hours following the onset of ischemic stroke is associated with a poor neurologic outcome. The mechanism by which hyperglycemia contributes to morbidity in ischemic brain damage is unknown (Longstreth WT, et al. N Engl J Med. 1983; 308:1378; Pulsinelli WA, et al. Am J Med. 1983; 74:540). The prevalence of impaired glucose tolerance and diabetes is elevated in patients with cirrhosis compared to those with chronic active hepatitis and may represent a risk factor for development of glucose metabolic alterations (Cacciatore L, et al. Diabetes-Res. 1988; 7:185-188).

Diagnosis of Diabetes Mellitus

A metabolic disease resulting in a "state of chronic hyperglycemia" (WHO Expert Committee on Diabetes. Diabetes. 1979; 28:1039-1057). It may be primary or secondary to other causes. Diabetes may be asymptomatic and detected only on routine screening, or severe with the classical symptoms of polydipsia, polyuria, weight loss, and ketonuria. Complications include vascular disease, neuropathy, renal failure, and visual impairment.

Diabetes is divided into insulin-dependent diabetes mellitus-IDDM (juvenile diabetes, ketosis-prone, brittle diabetes) and non-insulin-dependent diabetes mellitus-NIDDM (maturity-onset diabetes, non-ketosis prone diabetes mellitus). The former typically develop diabetes before age 20 and have little or no endogenous insulin. Onset is rapid and accompanied by inflammation of the islets ("islitis"). Anti-islet antibodies are present. A subgroup of approximately 10% of juvenile diabetics have a clinical course more typical of the adult NIDDM-type diabetic.

NIDDM accounts for 90-95% of diabetics. These patients are often obese, over 40, often have a family history of diabetes, and have variable amounts of insulin. These individuals may develop hyperosmolar coma and may become ketotic during stress or infection. They do not have anti-islet antibodies. Inflammation of the islets is not seen, and some of them have deposits of amyloid in the islets. NIDDM has been divided into those diabetics in which obesity is present (60-90%), and into a non-obese group. Many endocrinologists believe NIDDM diabetics have insulin resistance as a result of either decreased insulin receptor number or decreased affinity for the hormone.

There are other types of diabetes and related abnormalities of glucose tolerance, as given in the following Table:

Types of Diabetes and Abnormalities of Glucose Tolerance
Types of Diabetes Mellitus
1. Insulin-dependent diabetes mellitus - IDDM
2. Non-insulin dependent diabetes mellitus - NIDDM
3. Gestational diabetes mellitus
4. Diabetes associated with other conditions (secondary diabetes)
Non-Diabetics with Previous or Partial Abnormalities of Glucose Tolerance
1. Impaired glucose tolerance (asymptomatic, chemical, subclinical borderline, latent diabetes)
2. Previous abnormal glucose tolerance (latent diabetes or prediabetes)

The "gold standard" for the diagnosis of diabetes mellitus is the demonstration of hyperglycemia. The diagnosis is made on the basis of the following criteria:

Criteria for the Diagnosis of Diabetes Mellitus
1. Unequivocal elevation of plasma glucose together with the classical symptoms of disease.
OR
2. Elevation of fasting plasma glucose on more than one occasion.
OR
3. Elevated plasma glucose after glucose challenge on more than one occasion

Diagnosis in children requires either (1) or (2) and (3). These criteria were established by the National Diabetes Data Group and are published in: Diabetes. 1979; 28:1039-1057. Detailed discussion of diagnostic protocols are discussed under the headings: GLUCOSE, DIAGNOSIS OF DIABETES MELLITUS. See also DIABETIC KETOACIDOSIS PANEL.

Other tests specifically of value in the diagnosis and management of diabetes include glycohemoglobin (hemoglobin Alc), fructosamine, urine protein for micro-albuminuria, ketones, and electrolytes.

GLUCOSE, DIAGNOSIS OF DIABETES MELLITUS IN ADULTS

SPECIMEN: See GLUCOSE for details.
REFERENCE RANGE: See below.
INTERPRETATION: The protocol for the diagnosis of diabetes mellitus in non-pregnant adults is given in the next Table (National Diabetes Data Group, Diabetes. 1979; 28:1039-1057; Keen H, et al. Diabetologia. 1979; 26:283-285):

Diabetes of Diabetes in Adults
1. Classical symptoms
PLUS
Random glucose ≥ 200mg/dL
OR
2. AM fasting glucose ≥ 140mg/dL on two occasions
OR
3. Oral glucose challenge (75g) Glucose values ≥ 200mg/dL at 2 hours <u>and</u> some time between 0 and 2 hours on more than one occasion

Glucose tolerance tests require specialized preparation and are discussed separately under the heading GLUCOSE TOLERANCE TESTS.

The World Health Organization has advocated a single 2-hour post glucose load level as a screening test for diabetes in epidemiological studies. This test has a within subject CV of 32.4% which emphasizes the point that individuals should not be classified as diabetic on the basis of a single test (Forrest RD, et al. Diabetic Med. 1988; 5:557-561).

Impaired Glucose Tolerance: Some individuals have fasting glucose levels less than that required for a diagnosis of diabetes, but have glucose tolerance test results intermediate between normal and diabetic. These individuals have a fasting value ≤ 140 mg/dL and a two hour value > 140 mg/dL but < 200 mg/dL. (The normal mean 2 hour value is 115 mg/dL). One of the values between 0 and 2 hours must be ≥ 200 mg/dL. While not diabetic, they are at higher risk than the general population for the development of diabetes (1-5% per year). A large proportion of these individuals revert to normal glucose tolerance, while others remain in this borderline category. Generally, this latter group does not develop the visual or renal complications of diabetes. However, some studies show an increased frequency of atherosclerotic disease.

Previous Abnormality of Glucose Tolerance: An important group of patients. Would include those previous NIDDM who have converted to normal glucose tolerance following weight reduction. Also would include those with previous gestational diabetes who have normal glucose tolerance following delivery.

GLUCOSE, DIAGNOSIS OF DIABETES MELLITUS IN CHILDREN

SPECIMEN: See GLUCOSE for details. Patient should be fasted overnight. For glucose tolerance tests, samples are drawn at 0, 1 and 2 hours following a glucose challenge.
REFERENCE RANGE: See below.
INTERPRETATION: Diabetes in children is typically insulin-dependent diabetes mellitus-IDDM (juvenile diabetes, ketosis-prone, brittle diabetes). Children typically develop diabetes before age 20 and have little or no endogenous insulin. Onset is rapid and accompanied by inflammation of the islets ("islitis"). Anti-islet antibodies are present. There is a minority of childhood diabetics (10%) who have a non-insulin dependent diabetes mellitus like pattern. Some of these patients have glucokinase mutations (Fraguel P, et al. N Engl J Med. 1993; 328:697-702).

The criteria for the diagnosis of diabetes in childhood were established by the National Diabetes Data Group: Diabetes. 1979; 28:1039-1057, and are given in the following table:

Diagnosis of Diabetes in Children
1. Classical symptoms*
PLUS
Random glucose ≥ 200mg/dL
OR
2. AM fasting glucose ≥ 140mg/dL
AND
Oral glucose challenge (1.75g/kg up to 75g maximum) Glucose values ≥ 200mg/dL at 2 hours and some time between 0 and 2 hours.

*The classical symptoms of diabetes are polydipsia, polyuria, weight loss, and ketonuria.

GLUCOSE, DIAGNOSIS OF DIABETES MELLITUS IN PREGNANCY

SPECIMEN: See GLUCOSE for details. Patient should be fasted 10-16 hours and on a 150g/day carbohydrate diet for 3 days. Samples are drawn at 0, 1, 2 and 3 hours following a 100g oral glucose challenge.
REFERENCE RANGE: See below. To convert conventional units in mg/dL to international units in mmol/L, multiply conventional units by 0.05551.
INTERPRETATION: Infants of mothers with gestational diabetes are at risk for congenital anomalies, hypoglycemia, hypocalcemia, polycythemia and hyperbilirubinemia. Gestational diabetes, carbohydrate intolerance with onset or first recognition during pregnancy, appears in 2-3% of all pregnant women in the U.S. It is felt to be a phenotypically and genotypically heterogeneous condition, which leads to an increase in both fetal and maternal morbidity and mortality. Diagnosis of glucose intolerance during pregnancy is important in view of the high incidence of fetal malformations associated with elevated blood glucose. Education and intensive management of glycemic control before and during early pregnancy in diabetic women has been shown to profoundly reduce the rate of congenital anomalies (10.9% reduced to 1.2%) in their children (Kitzmiller JL, et al. JAMA. 1991; 265:731-736).

There is some disagreement on the most appropriate diagnostic criteria resulting in understandable confusion (Ziporyn T. JAMA. 1985; 254:465-470). The diagnostic criteria below are that of the National Diabetes Data Group. Diabetes. 1979; 28:1039-1057.

Diagnosis of Gestational Diabetes: Oral Glucose Challenge (100g)		
Fasting	≥	105mg/dL
1 hour	≥	190mg/dL
2 hour	≥	165mg/dL
3 hour	≥	145mg/dL
Two or more of the above values elevated		

The diagnostic criteria given in a previous table are those arbitrarily set in 1964 as being values 2 SD above the mean in pregnant women. New diagnostic criteria for the diagnosis of diabetes during pregnancy have been proposed, but they have not been widely tested. Hence the National Diabetes Data Group decided to continue with the older, established criteria. The need for a standardized 3 day dietary preparation prior to the glucose tolerance test in obstetrical patients has been questioned (Harlass FE, et al. J Reprod Med. 1991; 36:147-150).

The definitive diagnosis of gestational diabetes is by a 100g glucose tolerance test with reference values as given in the table above. There is a need for simplified tests to determine which women should undergo the definite full tolerance test. One proposed test is performed at 24-28 weeks gestation and consists of a 50g oral glucose screening test followed by a plasma glucose measurement on a specimen taken 1 hour later. A value of 140ng/dL (7.8mmol/L) indicates a need for further testing (Berger W, Misteli F. Ther Unsch. 1990; 47:71-79; Harlass FE, et al. Am J Obstet Gynecol. 1991; 164:564-568; Cousins L, et al. Am J Obstet Gynecol. 1991; 165:493-496). There are also studies attempting to establish the conditions and reference values for diagnosing antecedent gestational diabetes in the postpartum (48 hour) period (Carpenter MW, et al. Am J Obstet Gynecol. 1988; 159:1128-1131).

In one large series employing the above criteria recommended by the National Diabetes Data Group, 3.2% of patients had gestational diabetes (Magee MS, et al. JAMA. 1993; 269:609-615). These authors suggested that a significantly greater number of patients with gestational diabetes (5% of pregnant patients) could be identified by employing the modified criteria of Carpenter and Coustan. In this approach a woman is considered to have gestational diabetes if any one of the following glucose values is exceeded during a 100g glucose tolerance test: 0 time, 95mg/dL; 1 hour, 180mg/dL; 2 hour, 155mg/dL; 3 hour, 140mg/dL (Carpenter MW, Coustan DR. Am J Obstet Gynecol. 1982; 144:768-773). Screening is also proposed based on fasting values. In another study, fasting plasma glucose levels exceeding 88mg/dL were 80% sensitive in detecting gestational diabetes (Sacks DA, et al. J Reprod Med. 1992; 37:907-909).

Measurement of glycosylated hemoglobin (HgAlc) appears insensitive, and the fructosamine assay is controversial in screening for gestational diabetes (Reece EA, et al. Obstet Gynecol Surv. 1991; 46:1-14).

Diabetes Testing Following Delivery: Following delivery these patients must be retested to determine if they are (1) still diabetic (by the usual adult criteria), (2) if they fall into the class of impaired glucose tolerance, or (3) the class of previous abnormality of glucose tolerance (glucose tolerance reverts to normal). Women with gestational diabetes are at increased risk for perinatal mortality, increased fetal loss, and increased fetal malformation. They are also at higher risk (30%) for developing diabetes 5-10 years following parturition.

Gestational diabetes, by definition, manifests itself during pregnancy and tends to disappear after parturition. Patients with gestational diabetes who convert to normal glucose tolerance following delivery are given the diagnosis, previous abnormality of glucose tolerance. There is a concern that perhaps 1/3 - 1/2 of the women with gestational diabetes will ultimately develop permanent diabetes (Ziporyn, already cited).

GLUCOSE, HYPOGLYCEMIA

SPECIMEN: See GLUCOSE for details.
REFERENCE RANGE: see below. To convert conventional units in mg/dL to international units in mmol/L, multiply conventional units by 0.05551.
INTERPRETATION: Hypoglycemia was previously defined as a plasma glucose level less than 40 mg/dL (50 mg/dL in the over 60 age group). It is more recently defined as serum glucose level less than 50 mg/dL in all age groups since many patients are significantly symptomatic in the 45-50 mg/dL range. Symptoms may be conveniently grouped into those resulting from adrenergic discharge, and those resulting from neuroglycopenia. Newborn and infants with the disorder present with irritability, feeding difficulties, lethargy, cyanosis, and tachypnea. A listing of the many signs and symptoms found in hypoglycemic patients is given below:

Symptoms of Hypoglycemia

Adrenergic	Neuroglycopenia
Anxiety	Headache
Nervousness	Blurred vision
Tremulousness	Paresthesias
Sweating	Weakness
Hunger	Tiredness
Palpitations	Confusion
Irritability	Dizziness
Pallor	Amnesia
Nausea	Incoordination
Flushing	Abnormal mentation
Angina	Behavioral change
	Feeling cold
	Difficulty waking in the morning
	Senile dementia
	Organic personality syndrome
	Transient hemiplegia
	Transient aphasia
	Seizures
	Coma

From: Field JB. Endocrin Metab Clin N Amer. 1990; 18:1.

Although "true" hypoglycemia is relatively uncommon, there are a large number of causes of this disorder. Hypoglycemic disorders may be conveniently grouped under: 1) Fasting hypoglycemia - which are triggered by fasting, and relived by feeding, and 2) Reactive/postprandial hypoglycemia - those which are triggered by a meal. A simplified classification system which lists the general causes of hypoglycemia and gives illustrative examples is given below:

Types and Causes of Hypoglycemia

Type of Hypoglycemia	Cause
Fasting	Insulinoma, liver disease
Reactive	Postgastrectomy
Factitious	Insulin abuse
Artifactual	Specimen preparation, leukocytosis
Idiopathic postprandial	NOT A TRUE HYPOGLYCEMIA STATE

A more complete list is given in the following Table:

Causes of Hypoglycemia
Postprandial (Reactive) Hypoglycemia: Alimentary Hyperinsulinism, e.g., Gastrectomy, Gastrojejunostomy, Pyloroplasty or Vagotomy; Hereditary Fructose Intolerance; Galactosemia; Leucine Sensitivity; and Idiopathic.
Falciparum Malaria
Reye's Syndrome
Liver Disease (advanced)
Malnutrition
Renal Glycosuria
Islet Beta Cell Tumor (Insulinoma)
Malignancy (large tumors)
Hypoglycin Ingestion (Ackee fruit)
Neonatal Hypoglycemia
Dormandy's Syndrome (Familial Fructose and Galactose Intolerance)
Endocrine Hypofunctions: Anterior Pituitary; Addison's Disease
Enzyme Deficiencies: Glycogen Storage Diseases
Factitious Hypoglycemia: Insulin; Oral Hypoglycemic Agents
Children of Diabetic Mothers
Erythroblastosis Fetalis
Beckwith Syndrome
High Dose Salicylates
Glycogenic Enzyme Deficiencies
Stimulatory Insulin-Receptor Antibodies
Autoimmune Insulin Syndrome
Artifactual: Polycythemia Vera, Bacterial Contamination of Specimen; Failure to Separate Clot from Serum Promptly

Reactive hypoglycemia (e.g. after gastrectomy) follows meals because of rapid gastric emptying with rapid absorption of glucose and excessive insulin release. Glucose concentrations decrease rapidly leading to hypoglycemia.

Hypoglycemia occurred in 8 percent of patients with falciparum malaria. The hypoglycemia is induced by quinine; quinine is the only available parenteral treatment for severe chloroquine-resistant falciparum malaria. Quinine stimulates insulin secretion which may partially account for the hypoglycemia. Other possible mechanisms include large glucose requirements of the malaria parasites (White NJ, et al. N Engl J Med. 1983; 309:61-66).

Hypoglycemia may also be classified into those cases associated with elevated insulin levels (insulinoma) and those in which the low glucose levels are associated with low insulin levels. In a normal individual, insulin secretion should be almost totally suppressed with glucose levels below 45mg/dL.

Non-insulinoma tumors causing hypoglycemia are usually large and usually are mesenchymal tumors or rarely hepatomas and adreno-cortical adenomas. These tumors do not secrete insulin. However, there is some evidence that they secrete a variant of IGF-II (Axelrod L, Rod D. N Engl J Med. 1988; 319:1477-1478).

Insulinoma and Fasting Hypoglycemia: Following a fast, women reach lower glucose levels than men. After a prolonged fast, a few women reach glucose levels in the hypoglycemic range, but remain asymptomatic. These women are usually lean subjects. The abnormally low glucose levels can be raised by exercise, presumably by gluconeogenesis from the exercise produced lactate.

The following are characteristic findings following a fast in normal individuals.

1) No subject exhibits glucose levels lower than 60 mg/dL in the first 24 hours of the fast.
2) Insulin levels fall in parallel with the falling glucose levels.
3) The ratio of immunoreactive insulin to glucose levels (IRI/G) remain constant throughout the fast.

By contrast, fasted individual with insulinoma exhibit the following characteristics:

Findings in Individuals with Insulinoma
1) Hypoglycemia within 24 hours of fast
2) Insulin levels fail to change
3) The immunoreactive insulin/glucose ratio increases as glucose levels fall

(cont)

Glucose, Hypoglycemia (Cont)

The fasted state can be documented in patients by the presence of a strongly positive ketonuria after about 24 hours. However, this does not occur in all patients with an insulinoma. Insulin levels may be increased in obese individuals, or those with insulin resistance, or anti-insulin antibodies. Heparin may either raise or lower insulin levels. One must also realize that some insulinomas produce primarily proinsulin, such that not all assays are equivalent. During the fasted state, clearance of C-peptide is less than that of insulin, so C-peptide assays may be of some use.

Two additional specialized tests may be of use in the diagnosis of insulinoma:

1) **C-peptide suppression**
   C-peptide levels in insulinoma patients will not be suppressed during hypoglycemia induced by the injection of exogenous insulin
2) **Tolbutamide tolerance test**
   Injection of tolbutamide into normal individuals will lower the glucose levels at 30 minutes with a return to normal after 2 hours. In the insulinoma patient, abnormal insulin levels will be reached within 2 minutes, there will be a profound hypoglycemia at 30 minutes, and the glucose will remain depressed after 2 hours.

Other Causes of Fasting Hypoglycemia: Non-insulinoma causes of fasting-type hypoglycemia are given below:

The Liver and Fasting Hypoglycemia

Type of Disease	Examples
Structural	Cirrhosis
Substrate Lack	Malnutrition
Hormonal Deficiency	Deficiency of Glucagon
Enzyme Defects	Glycogen Metabolism
Drugs	Alcohol
Autoimmune	Autoimmune Insulin Syndrome

A short discussion of the use of insulin levels in the workup of patients with hypoglycemia is given in: Polansky KS. N Engl J Med. 1992; 326:1020-1021.

Factitious Hypoglycemia: Insulin and C-peptide assays test are used to document hyperinsulinemic hypoglycemia and aid in the exclusion of factitial hypoglycemia. In fact, insulin levels should be either elevated or inappropriately normal in the face of a hypoglycemia. C-peptide should be secreted in parallel with insulin if it originates from the pancreas. This is diagrammed below:

Patterns of Insulin and C-Peptide Levels in Hypoglycemia

	Fasted Glucose	Insulin	C-peptide
Insulinoma	Low	High	High
Factitious (Insulin Injection)	Low	High	Low
Factitious (Oral Hypoglycemics)	Low	High	High

Notice that clever individuals who abuse oral hypoglycemics cannot be detected by C-peptide assays, since they stimulate release of native insulin by its normal pathway. These drugs may be detected in the urine. The only other case besides factitious insulin injection in which there is a similar disparity between measured insulin levels and C-peptide levels is in individuals with autoimmune insulin syndrome where insulin levels are artifactually increased relative to C-peptide (Polansky KS. N Engl J Med. 1992; 326:1020-1021).

Idiopathic Postprandial Syndrome: Some individuals have symptoms suggestive of postprandial hypoglycemia. However, when they are given a "mixed meal" in conjunction with timed glucose measurements, one finds that although adrenergic-type symptoms might develop, chemical hypoglycemia is not observed. There is no consistent relationship between glucose levels and the symptoms experienced. Because the cause of this condition is not known, the condition has been termed idiopathic postprandial syndrome. Because hypoglycemia has not been observed, previously used terms like "reactive" and "postprandial" hypoglycemia have been discarded (Charles MA, et al. Diabetes. 1981; 30:465-470; Hogan MJ, et al. Mayo Clin Proc. 1983; 58:491-496; Field JB, already cited).

GLUCOSE, SELF-MONITORING

SPECIMEN: Capillary blood obtained on finger-stick using automated, spring-operated lance; a commercially available lance is the Autolet.
REFERENCE RANGE: Infants and Children: 60-115mg/dL; adults: 65-120mg/dL. To convert conventional units in mg/dL to international units in mmol/liter, multiply conventional units by 0.05551.
METHOD: A reagent strip is impregnated with glucose oxidase, reduced chromogen and peroxidase; in the presence of glucose, the following reaction takes place:

$$\text{Glucose} + O_2 + H_2O \xrightarrow[\text{Oxidase}]{\text{Glucose}} H_2O_2 + \text{Gluconic acid}$$

$$H_2O_2 + \text{Reduced Chromogen} \xrightarrow{\text{Peroxidase}} \text{Oxidized Chromogen} + H_2O$$

Reagent strips may be interpreted visually by comparing the color generated to a color chart or reagent strips may be read with a reflectance meter that generates a numerical value for blood glucose. Visual interpretation is convenient; a range, rather than a precise number for blood glucose, is obtained. Diabetics treated with photocoagulation may have problems in properly interpreting strips.
INTERPRETATION: Maintenance of normal glucose levels in diabetics is important in retarding the development of complications of diabetes mellitus. The U.S. resident with diabetes makes an average of 2.7 visits/year for continuing care, with approximately 2 tests/year of blood glucose. Self-monitoring of blood glucose is performed by less than 10% of diabetic patients (Harris MI. Diabetes-Care. 1990; 13:419-426). Self-monitoring of glycemic control by type II diabetics by either blood or urine testing resulted in significant improvement in fasting plasma glucose. Self-monitoring of blood glucose was 8-12 times more expensive than urine testing, but no more effective (Allen BT, et al. Diabetes-Care. 1990; 13:1044-1050). Some diabetics prefer testing blood to urine and urine testing may be insufficiently sensitive (Holman RR, Turner RC. Lancet. 1977; 1:469-474). However, there are occasional discrepancies between home blood glucose measurements and laboratory values caused by physical or cognitive disabilities, or psychological stress (Campbell LV, et al. Br Med J. 1992; 305:1194-1196). Poorly controlled diabetes is associated with a higher incidence of major congenital anomalies in the offspring of pregnant women (Miller E, et al. N Engl J Med. 1981; 304:1331-1334; Frienkel N. N Engl J Med. 1981; 304:1357-1359). Close control of blood glucose in the pregnant diabetic has been shown to decrease infant mortality and maternal morbidity (Jovanovic L, Peterson CM. Diabetes Care. 1982; 5(Suppl 1):24). The development of diabetic neuropathy has been related to the degree of hyperglycemia (Porte D, et al. Am J Med. 1981; 70:195-200). However, no improvement was noted in any measurement of nerve function associated with improved blood glucose control (Service et al. Mayo Clin Proc. 1983; 58:283-289). The benefits of self-monitoring of blood glucose have been reviewed (Bell PM, Walshe K. Br Med J. 1983; 286:1230-1231) and arguments for randomized clinical trials are given (Barbosa J. Arch Intern Med. 1983; 43:1118-1119).

Capillary reflectance meter testing compared to standard laboratory glucose measurement demonstrates good correlation (r=0.90) and is cost effective, but there is controversy as to whether it should replace plasma glucose measurements in screening for and diagnosis of glucose intolerance and diabetes (Dacus JV, et al. J Reprod Med. 1990; 35:1150-1152; Yoo T, Chao J. J Fam Pract. 1989; 29:41-44). Fingerstick blood glucose measurements have been shown to be inaccurate when taken from patients in shock (Atkin S, et al. Ann Intern Med. 1991; 114:1020-1024).

GLUCOSE TOLERANCE TESTS

SPECIMEN: See GLUCOSE for details.
REFERENCE RANGE: Nonpregnant Adults: Normal values of glucose levels during glucose tolerance tests were highly variable previously because of a lack of standardized procedures, and a variability in patient preparation and the dose of glucose load administered. The following values are for venous plasma glucose during a glucose tolerance test performed with a standardized 75g oral glucose challenge (National Diabetes Data Group. Diabetes. 1979; 28:1039-1057).

Fasting Value: <115mg/dL (6.4 millimol/L)

Value at ½, 1, or 1½ hours: <200mg/dL (11.1 millimol/L)

Value at 2 hours: <140 mg/dL (7.8 millimol/L)

Children: (Mietes S. Ed. Pediatric Clinical Chemistry, Am Assoc Clin Chem. 1725 K. Street, N.W. Washington, D.C. 20006, pg. 286).

Blood Glucose Response to Oral Glucose Administration (1.75g/kg Body Weight) in Children

Time Min.	Serum Glucose (mg/dL)	Insulin (Micro USP units/mL)	Phosphorus (mg/dL)
Fasting	56-96	5-40	3.2-4.9
30	91-185	36-110	2.0-4.4
60	66-164	22-124	1.8-3.6
120	66-122	6-84	1.8-4.2
180	47-99	2-46	2.0-4.6
240	61-93	3-32	2.7-4.3

METHOD: Glucose measured by glucose oxidase or hexokinase method. Oral glucose challenge of 75g in adults, 100g in pregnant women, and 1.75g/kg up to 75g total in children. Timed specimens are taken. The oral glucose tolerance test should be performed in the morning after three days of unrestricted carbohydrate (>150g/day) and activity. The patient should be fasted 10-16 hours. Water is permitted but coffee and smoking are not. The patient should remain seated during the test. The presently advocated procedures are described in: National Diabetes Data Group Classification. Diabetes. 1979; 28:1039-1055; Nelson RL. Mayo Clin Proc. 1988; 63:263-269).
INTERPRETATION: Glucose tolerance tests are used most commonly in the diagnosis of diabetes. There has been an attempt to standardize glucose tolerance tests and diagnostic criteria in order to improve diagnostic accuracy and to facilitate epidemiological studies. Notice that the oral glucose challenge varies according to age and pregnancy. The oral test should be performed in the morning after 3 days of unrestricted diet.

Previously, glucose tolerance tests were generally performed in adults with a 50g glucose challenge, while those in Europe with a 100g challenge. The presently advocated 75g is a compromise between the two. Glucose load for pregnant women is increased (100g). Dose is based on the weight of the child in pediatric testing. The tolerance test is performed with blood samples taken at 0, 1/2, 1 and 2 hours. In pregnant women the 30 minute sample is not taken, and a 3 hour sample added.

Glucose tolerance tests are generally performed in order to establish the diagnosis of diabetes. There are a number of conditions in which abnormal appearing glucose tolerance tests are seen. These conditions and the observed findings are listed below:

Abnormal Glucose Tolerance Tests	
Diabetes Mellitus	Elevated glucose values; delay in euglycemia; values are in the range established as diagnostic for diabetes.
Impaired Glucose Tolerance	Elevated glucose values; Values are abnormal but not diagnostic for diabetes.
Malabsorption	Little change in glucose values. "Flat curve."
Gastrectomy	Markedly elevated glucose values at early time points, followed by low glucose values.
Early Diabetes	Elevated glucose values followed by low values at 4 to 5 hours post challenge.
Insulinoma	Response may vary. Low glucose values especially 2 to 3 hours post challenge. An exaggerated response. Diagnosis is usually made by observation of fasting hypoglycemia.
Idiopathic Postprandial Syndrome	No change in glucose tolerance.

Impaired Glucose Tolerance: Some adult individuals have fasting glucose levels less than that required for a diagnosis of diabetes, but have glucose tolerance test results intermediate between normal and diabetic. These individuals have a fasting value ≤ 140 mg/dL and a two hour value > 140 mg/dL but < 200 mg/dL (one of the values between 0 and 2 hours must be ≥ 200 mg/dL for diagnosis of diabetes). While not diabetic, they are at higher risk than the general population for the development of diabetes (1-5% per year). A large proportion of these individuals revert to normal glucose tolerance, while others remain in this borderline category. Generally, this latter group does not develop the visual or renal complications of diabetes. However, some studies show an increased frequency of atherosclerotic disease.
Previous Abnormality of Glucose Tolerance: An important group of patients. Would include those previous NIDDM who have converted to normal glucose tolerance following weight reduction. Also would include those with previous gestational diabetes who have normal glucose tolerance following delivery.
Invalid Glucose Tolerance Tests: These can result from administration of the test in the afternoon, after less than 10 hours or more than 16 hours fasting, and after physical inactivity or restricted carbohydrate diet (less than 150g) in the days prior to testing. In some patients with bowel disease and decreased oral absorption, the glucose challenge can be administered i.v.

GLYCOHEMOGLOBIN (see HEMOGLOBIN A_{1C})

GONOCOCCAL ANTIGEN ASSAY

SPECIMEN: This assay is not used for specimens from the throat or anus. Swabs and transport media are provided by the manufacturer.
Cervix: Specimens from the cervix are obtained as follows: Use a sterile bivalve speculum, moisten the vaginal speculum with warm water before introduction. Do not use lubricating jelly because it may be lethal for gonococci. Wipe the cervix with sterile cotton swabs to remove vaginal secretions. Gently compress the cervix between the blades of the speculum to help to produce endocervical exudate. Rotate the swab in the cervix from 10 to 30 seconds to ensure adequate sampling and absorption by the swab. Avoid contamination by not allowing the swab to contact the vaginal walls.
Urethra in Males: Use a sterile wire swab to obtain specimen from anterior urethra by gently scraping the mucosa.
REFERENCE RANGE: Negative
METHOD: Enzyme immunoassay(EIA) by Abbott Diagnostics Division, North Chicago, Illinois. The test procedure is rapid, taking less than one hour. The swab specimen is incubated with beads; N. gonorrhoeae adheres to the bead. The bead is incubated with anti-gonococcal antibodies which react with gonococci on the bead. Next, the bead is incubated with horseradish peroxidase-labelled sheep anti-rabbit globulins, which reacted with the antigen-antibody complex on the bead. After another incubation with an appropriate substrate the samples are read with a spectrophotometer to determine the relative quantity of N. gonorrhoeae antigens absorbed to the bead.
INTERPRETATION: The Gonozyme test is highly sensitive and specific for men with gonococcal urethritis, but less sensitive than culture for detecting gonococcal infections in women. This test does not have the accuracy when used as a screening test in a low prevalence population. Gonozyme testing may be useful for evaluation of high risk populations in which microbiology laboratory access is limited (Dallabetta G, Hook EW. Infect Dis Clin N Amer. 1987; 1:25-54; Ehret JM, Knapp JS. Clin Lab Med. 1989; 9:445-480; Zenilman JM. Hosp Practice. 1993; 29-50).

GONORRHEA CULTURE

SPECIMEN: Specimens may be obtained from cervix, vagina, urethral specimens from males, throat, or joint fluid. Specimens are obtained using a swab. Do not allow specimen to dry and avoid refrigeration. Highest yield of positive cultures is obtained with immediate plating onto selective media and incubation at 35-37°C under 5 to 8 percent carbon dioxide. Nutritive and non-nutritive transport systems are available if immediate plating is not possible.
Cervix: Specimens from the cervix are obtained as follows: Use a sterile bivalve speculum; moisten the vaginal speculum with warm water before introduction. Do not use lubricating jelly because it may be lethal for gonococci. Wipe the cervix with sterile cotton swabs to remove vaginal secretions. Use a swab to obtain specimens for culture. The swab should be rotated for 20 to 30 seconds within the endocervical canal and removed without touching the vaginal walls. Gram Stain is insensitive (30-70%) but is very specific (95%); this test has been recommended for all women presenting to an STD clinic, or for high risk patients (pelvic inflammatory disease, endocervical mucopurulent discharge, or history of gonorrhea contact)(Zenilman JM. Hosp Practice. 1993; 29-50; Judson FN. Med Clin N Amer. 1990; 74:1353-1366). Definitive diagnosis of gonococcal infection is based on culture.
Rectal Cultures: Rectal cultures should be done on all women at the same time cervical cultures are done because the yield is increased, and should be done on all patients that practice rectal sexual activity. Anogenital contact is not necessary in women, most anal infections result from spread through endocervical exudate. Both rectal and cervical cultures may be plated on a single plate, spaced separately. Gram stain of rectal exudate with gram-negative diplococci within PMN's is highly predictive of a positive culture for gonorrhea (Dallabetta G, Hook EW. Infect Dis Clin N Amer. 1987; 1:25-54).
Urethra in Males: Specimens from the urethra in males are obtained as follows: Urethral specimens should not be collected until at least one hour after urination. Collect urethral discharge directly or from discharge obtained by "milking" the urethra. If no discharge is available, insert an unmoistened thin swab into the distal urethra for approximately 2cm and gently rotate it. Gram stain should be done; a positive smear, i.e., typical Gram negative diplococci within neutrophils is sufficient for office diagnosis; confirmation should be obtained by culture.

Throat Culture: Depress tongue and expose pharynx. Swab posterior pharynx, tonsils, and tonsillar fossae vigorously. Gram stains are insensitive and nonspecific and should not be done.

Specimens should be inoculated on selective medium in an enhanced CO_2 atmosphere. Specimen must be inoculated on Martin-Lewis or Thayer-Martin or New York City media. Neisseria gonorrhoeae is a very fragile bacteria and will lose viability if allowed to dry. Plates for inoculation must be at room temperature before use. Specimen must not be refrigerated. Plates are placed in a CO_2 incubator.

REFERENCE RANGE: No N. gonorrheae organisms isolated.

METHOD: Neisseria gonorrhoeae is an aerobic oxidase-positive, Gram negative diplococcus. Specimens are almost always contaminated and thus must be inoculated on selective media that will allow Neisseria to grow (chocolate agar or other appropriate media) but containing antibiotics and antifungal agents that inhibit the growth of contaminants. For instance, the modified Thayer-Martin medium also contains vancomycin, colistin, nystatin and trimethoprim; the Martin-Lewis medium also contains vancomycin, colistin, trimethoprim and anisomycin (an antifungal agent). Bring the media to room temperature before inoculation. The swab is used to make a "Z" streak onto the surface of the medium; then a needle is used to streak the inoculum. The plates are then incubated in a CO_2 incubator or a candle jar at 35-37°C; moisture is maintained in the jar to prevent the medium from drying out.

The oxidase test is used to confirm N. Gonorrheae and is done on isolated colonies; morphology (gram-negative diplococci) should be confirmed by Gram stain. The organisms, Moraxella and N. meningitidis, are part of the normal pharyngeal flora and morphologically resemble N. Gonorrhoeae; these organisms are also oxidase-positive. Therefore, diagnosis must be confirmed by carbohydrate fermentation pattern or by serologic methods.

INTERPRETATION: The sensitivity and specificity of gram stain for the diagnosis of gonorrhea is given in the following Table (Dallabetta G, Hook EW, already cited):

Sensitivity and Specificity of Gram Stain for the Diagnosis of Gonorrhea

Site	Gram Stain Sensitivity	Gram Stain Specificity
Urethra, symptomatic males	95	95
Urethra, asymptomatic males	60	95
Rectum, symptomatic males	30-65	95
Cervix, symptomatic females	40-70	95
Pharynx	Not Recommended	

The sensitivity of gonorrhea culture based on site is given in the following Table (Ehret JM, Knapp JS. Clin Lab Med. 1989; 9:445-480):

Sensitivity of Gonorrhea Culture

Site	Sensitivity
Males (Heterosexual)	
Urethra-Symptomatic	94-98
Urethra-Asymptomatic	84
Males (Homosexual)	
Urethra	60-98
Rectum	40-85
Pharynx	50-70
Females (Nonhysterectomized)	
Cervix	86-96
Vagina	55-90
Urethra	60-86
Rectum	70-85
Pharynx	50-70
Females (Hysterectomized)	
Urethra	89
Vagina	56
Rectum	41

GRAM STAIN

SPECIMEN: Two specimens should be obtained; one for gram stain and the other for culture. Gram stained smears of exudates, abscesses or infected body fluids should be obtained; Gram stained smear of exudates should be obtained from all patients with acute urethritis, pelvic inflammatory disease, pneumonia (where there is productive cough), selected patients with infections of the skin or urinary tract.

For specimen collection, use sterile container, culturette; transport specimen to the laboratory as soon as possible and refrigerate.

REFERENCE RANGE: Depends on site of origin of specimen.

METHOD: Gram stain technique is done as follows (Gulick P, et al. Medical Clinics of North America. 1983; 67:39-55):

Gram Stain Technique

1. Make a thin smear of the material onto a clean slide and allow the slide to air dry.
2. Fix the material onto the slide. This can be done by placing the slide on the slide warmer set at about 70°C or by passing the slide, right side up, through a Bunsen burner flame 3 or 4 times.
3. Overlay the smear with crystal violet solution for 1 minute.
4. Wash slide thoroughly with water.
5. Overlay the smear with Gram iodine for 1 minute.
6. Wash slide thoroughly with water.
7. Flood the surface of the slide with the decolorizer, acetone-alcohol, until no violet color washes off the slide. This usually requires about 10 seconds or less.
8. Wash slide thoroughly with water.
9. Overlay the smear with safranin counterstain for 1 minute.
10. Wash the slide thoroughly with water and allow the excess water to run off the slide, then allow the slide to air dry or blot dry with bibulous paper.
11. Examine the stained smear under low power to observe the nature of the specimen, then examine under 100 x (oil immersion) to observe the bacteria present. Gram-positive bacteria stain dark blue to purple and gram-negative bacteria stain pink-red; the nuclei of polymorphonuclear leukocytes should stain pink-red.

INTERPRETATION: Gram stains of some clinically important bacteria are given in the following Table:

Gram Stain of Clinically Important Bacteria

Gram-Positive Cocci
Staphylococcus
  Aureus
  Epidermidis
Streptococcus
  Pyogenes (Grp A,C,G)
  Group B
  Viridans
  Bovis
  Enterococcus
  Pneumoniae (Pneumococcus)
  Anaerobes (eg, Peptostreptococcus)

Gram-Negative Cocci
Neisseria
  Gonorrhoeae (Gonococcus)
  Meningitidis (Meningococcus)
Moraxella (Branhamella) Catarrhalis
Anaerobes (Veillonella)

Gram-Positive Bacilli
Bacillus Anthracis (Anthrax)
Clostridium
  Perfringens
  Tetani
  Difficile
Corynebacterium
  Diphtheriae
Listeria Monocytogenes

Gram-Negative Bacilli
Haemophilus
  Influenzae
  Ducreyi (Chancroid)
Pseudomonas
  Aeruginosa
  Actinobacillus
  Pseudomallei
Bordetella Pertussis
Brucella (Brucellosis)
Gardnerella Vaginalis
Legionella
Enteric Bacilli
  Proteus Mirabilis
  E. coli
  Klebsiella Pneumoniae
  Campylobacter
  Enterobacter
  Salmonella Typhi
  Shigella
  Bacteroides Fragilis
  Providencia
  Serratia
Yersinia Pestis (Plaque)
Vibrio Cholerae (Cholera)
Acinetobacter
Pasteurella

GROWTH HORMONE

SPECIMEN: Red top tube, separate serum. A.M. levels are obtained after an overnight fast. Urine
REFERENCE RANGE: Children and adult, baseline, resting after overnight fast; children: <0.7-6ng/mL (mean = 2.4); Adults <0.7-6ng/mL (mean = 1.8). Following stimulation a normal response is a rise up to 10ng/mL; a value intermediate between 7-10 ng/mL may mean partial deficiency and requires further evaluation. To convert conventional units in ng/mL to international units in mcg/liter, multiply conventional units by 1.00.
METHOD: RIA; Immunoradiometric Assay (IRMA)
INTERPRETATION: Growth hormone assays are done to evaluate: (1) Deficiency of growth hormone in small children or (2) Excess of growth hormone such as occurs in gigantism or acromegaly.
Growth Hormone Deficiency is usually idiopathic; the deficiency may be isolated or it may be accompanied by a deficiency of other pituitary hormones. Growth hormone is low in normal subjects; secretion is pulsatile and variable. Diagnostic usefulness of spontaneous growth hormone levels is inferior to stimulation tests when used to diagnose growth hormone deficiency in prepubescent children (Rose SR, et al. N Engl J Med. 1988; 319:201-207). A variety of stimuli are used to evaluate impaired growth hormone secretion as given in the following Table (Martinez NJ. Resident and Staff Physician. July 1991; S15-S19; Frasier SD. Pediatr. 1974; 53:929):

Stimuli for Growth Hormone
Insulin
Arginine
L-Dopa
Glucagon
Propranolol
Clonidine
15 Minutes following Vigorous Exercise

Arginine-Insulin Stimulation Test: Arginine and insulin are often given in a piggy-back fashion in order to approach 100 percent sensitivity for growth-hormone stimulation test. In this test, arginine (0.5 gm/kg) is infused over 30 minutes; blood specimens are obtained at 15 minute intervals. At 60 minutes, give 0.05-0.1 U/kg regular insulin I.V. and collect blood specimens at 15, 30, 45 and 60 minutes. Observe the patient continuously for hypoglycemia; give glucose if the patient develops clinical signs and symptoms of hypoglycemia. As collected, each sample should be centrifuged and separated. A decrease in serum glucose of 50 percent or more is required for the insulin infusion test to be valid. Follow the schedule for injection of arginine and insulin and drawing blood specimens as given in the next Figure:

Schedule for Arginine and Insulin Infusion and Time to Obtain Serum Specimens for Growth Hormone Stimulation Tests

X = Specimens in Red Top Tubes

Start I.V. with Normal Saline to Keep Vein Open	X-30		X-0	X+15	X+30	X+45	X+60	X+75	X+90	X+105	X+120
	-30	-15	0	+15	+30	+45	+60	+75	+90	+105	+120min.
	↑ Infuse Arginine (0.5 Gms per kg body weight)		↑ Stop Arginine; Switch to Normal Saline to Keep Vein Open				↑ Infuse Insulin (1 unit= 1 mL;0.05 to 0.1 unit/kg)	↑ Vital Signs	↑ Vital Signs	↑ Vital Signs	↑ Vital Signs

The blood specimens in red top tubes collected at -30, 0, +15, +30 and at +45 minutes are for growth hormone assay; the blood specimens in red top tubes collected at +60, +75, +90, +105 and +120 minutes are for growth hormone, glucose and cortisol assays.

Results of arginine-insulin stimulation test are shown in the next Figure:

Arginine-Insulin Stimulation Test for GH Deficiency

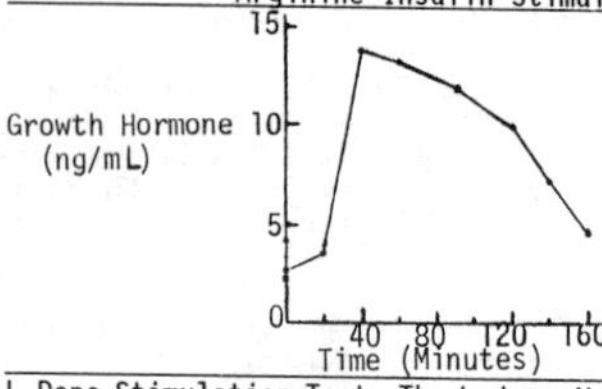

The results in this Figure show stimulation of growth hormone, indicating a normal response; subjects with growth deficiency show no stimulation of growth hormone during the arginine-insulin stimulation test.

L-Dopa Stimulation Test: The L-dopa (L-dihydroxyphenylalanine, levodopa) test is done as follows: Give L-dopa orally (less than 30 pounds, 125 mg; 30 to 70 pounds, 250 mg; >70 pounds, 500 mg). Obtain serum specimens as follows: -15, 0, 30, 60, 90 and 120 minutes for growth hormone assay. Serum growth hormone normally peaks between 30 and 120 minutes.

Other Growth Hormone Stimulation Tests: Following administration of the secretagogue, serum specimens for growth hormone assay are collected at baseline and every 30 minutes for two hours (Moore KC, et al. J Pediatr. 1993; 122:687-692).

Urinary Growth Hormone Assay: Urinary growth hormone assay may be useful as a diagnostic test for growth hormone deficiency and excess (Pholséna M, et al. Acta Endocrinol. 1993; 128:9-14; Walker JM, et al. Arch Dis Child. 1990; 65:89-92). See also Linder B, Cassorla F. "Short Stature-Etiology, Diagnosis and Treatment." JAMA. 1988; 260:3171-3175).

Growth Hormone Excess: Usually growth hormone assay per se is sufficient to indicate the diagnosis of acromegaly; growth hormone is almost always increased in acromegaly.

HAPTOGLOBIN

SPECIMEN: Red top tube; separate serum

REFERENCE RANGE: 27-140mg/dL; haptoglobin is produced in premature infants as early as 28 weeks gestation. At term, cord blood contains less than 10mg/dL. Haptoglobin reaches adult levels within 3 to 12 months.

To convert conventional units in mg/dL to international units in g/liter, multiply conventional units by 0.01.

METHOD: Radial immunodiffusion; Nephelometry

INTERPRETATION: Haptoglobin is a serum protein that has the capacity to bind hemoglobin in vivo and in vitro. This test is generally used to help in the diagnosis of anemia. The changes of serum haptoglobin in disease are shown in the next Table:

Changes of Serum Haptoglobin in Disease

Decreased	Increased
Hemolytic Anemia	Tissue Destruction:
Megaloblastic Anemia	Infections; Malignant Neoplasms;
Chronic Hepatocellular Disease	Collagen Diseases
Infectious Mononucleosis	Obstructive Jaundice
Toxoplasmosis	Inflammatory Reactions and
Newborns	Tissue Proliferation
Ineffective Erythropoiesis	Steroid Therapy
Hemolysis of any Cause	
Congenital Defect of Synthesis	

The clinical significance of hypohaptoglobinemia is limited. Haptoglobin may be used to differentiate hemoglobinuria from myoglobinuria. Hemoglobinuria cannot occur if the plasma contains free haptoglobin (plasma free hemoglobin must be in excess of haptoglobin for renal excretion to occur).

Elevated haptoglobin levels occur as a result of increased synthesis. Haptoglobin is an acute phase reactant.

Haptoglobin migrates with alpha-2 globulin band on protein electrophoresis.

Reference: Javid J. Curr Topics in Hematol. 1978; 1:151-192.

HEAVY METAL SCREEN, URINE

SPECIMEN: Urine; 24 hour specimen preferred. Refrigerate at 4°C.
REFERENCE RANGE: Negative
METHOD: Atomic Absorption
INTERPRETATION: A large number of heavy metals may be detected. Among these are antimony, arsenic, bismuth, boron, cadmium, cobalt, copper, lead, mercury, selenium, tellurium, thallium and zinc. See ARSENIC, COPPER, LEAD, MERCURY, and ZINC.

HEINZ BODY STAIN

SPECIMEN: Lavender (EDTA) top tube
REFERENCE RANGE: No Heinz bodies identified
METHOD: Oxidative denaturation of hemoglobin to Heinz bodies using acetylphenylhydrazine; staining of Heinz bodies using crystal violet.
INTERPRETATION: The three main conditions associated with the presence of Heinz bodies are given in the next Table:

Conditions Associated with Heinz Bodies
Exposure to Certain Chemicals or Drugs
Deficiency of one of the Reducing Systems of the Blood such as Glucose-6-Phosphate Dehydrogenase Deficiency
Presence of an Unstable Hemoglobin

Heinz body formation is also observed in some patients with red cell membrane abnormalities such as hereditary red cell membrane high phosphatidyl choline hemolytic anemia, and other conditions such as congenital hemolytic anemia of unknown etiology, acquired hyperlipidemia, and paroxysmal nocturnal hemoglobinuria (Mannoji M, et al. Scand J Haematol. 1985; 35:257-263).

Heinz bodies are insoluble inclusions of hemoglobin precipitated within the red blood cell.

HEMATOCRIT

SPECIMEN: Lavender(EDTA) top tube or microhematocrit (lavender) capillary tube containing EDTA. Specimens may be stored for 8 hours at room temperature or 24 hours in the refrigerator. Although EDTA anticoagulated blood is preferred, specimens in other anticoagulants may be used.
REFERENCE RANGE: Microhematocrit results average 3 percent higher than a hematocrit obtained by calculation, that is, Hct = MCV x RBC's; trapping of leukocytes, platelets and plasma occurs with "spun" hematocrits.

Values of "spun" hematocrit are given in the next Table:

Normal Values of "Spun" Hematocrit	
Age	Percent
Birth	44-64
14-90 days	35-49
6 months-1 year	30-40
4-10 years	31-43
Adult, Male	42-52
Adult, Female	37-47

METHOD: Microhematocrit using capillary tube; calculated from determination of MCV and RBC's using Coulter counter or other automated counter from the relationship Hct = MCV x RBC's.
INTERPRETATION: The hematocrit is determined to access red cell mass as part of routine testing or in the evaluation of blood loss, anemia, state of hydration, and of various polycythemic states. The hematocrit is decreased in the conditions listed in the next Table:

Decrease in Hematocrit
Anemias (Irrespective of Cause)
Recovery Stage After Blood Loss
Hemodilution
Pregnancy
Sample drawn above i.v. line
Edematous states
Recumbency

(cont)

Hematocrit (Cont)

The hematocrit is increased in the conditions listed in the next Table:

Increase in Hematocrit
Polycythemia
Erythrocytosis of Dehydration
Hemoconcentration as in Shock associated with Trauma, Surgery and Burns
High Altitude

HEMOGLOBIN

SPECIMEN: Lavender(EDTA) top tube or microtube (lavender) which contains EDTA. The specimen is stable at room temperature for up to 8 hours and in the refrigerator for up to 24 hours. The anticoagulant, heparin (green top tube) may also be used.

REFERENCE RANGE: The reference range, with age, is given in the next Table:

Reference Range with Age

Age	g/dL
Cord	14-20
Birth	15.0-24.0
1 Week	13.0-20.0
1 Month	11.0-17.0
6 Months	10.5-14.5
1 Year	11.0-15.0
10 Years	11.0-16.0
15 Years (Male)	14-18
(Female)	12-16

At birth, hemoglobin (Hgb) values are very high (15-24 g/dL); this high Hgb concentration is attributable to the relatively low levels of oxygen in-utero. It decreases markedly and at two to three months of age, it reaches a value of about 10-14 g/dL. Then, Hgb slowly increases until it reaches the adult value at about age 15.

Fetal Hgb constitutes about 60 percent of the Hgb at term; by 6 months of age, fetal Hgb constitutes 5 to 6 percent of the total and 1 to 2 percent by adulthood.

METHOD: The most popular method is the cyanmethemoglobin method.

INTERPRETATION: Hgb levels are performed in order to determine the oxygen-carrying capacity of blood, and to assess anemia, polycythemia, and their response to therapy. Decreased Hgb is caused by anemia of all types (see ANEMIA PANEL). Hgb concentration is also decreased with blood loss and fluid reconstitution, edematous states, pregnancy, and if the specimen is drawn upstream of an i.v. line. Hgb values fall after an individual assumes a recumbent position.

The causes of increased Hgb are polycythemia vera, secondary polycythemia, vigorous exercise and high altitude. Hgb levels rise with hemoconcentration as may be seen in dehydration and in untreated shock associated with fluid loss.

HEMOGLOBIN A1c (GLYCOHEMOGLOBIN)

SPECIMEN: The patient need not be fasting when the blood is drawn. Anticoagulated blood is collected in a grey top tube (oxalate, fluoride); or lavender (EDTA) or green (heparin) top tubes. Blood is centrifuged; the red blood cells are separated and hemolyzed. A lavender top tube (EDTA) is used for the glycohemoglobin measurement.
REFERENCE RANGE: Varies by method. Normal: 4.0-8.2% for Hgb A1c. Higher for glycohemoglobin by affinity chromatography.
METHOD: Hemoglobin A1c is measured by electrophoresis or cation-exchange column chromatography, such as Bio-Rad columns. Hemoglobin F has been reported to co-chromatograph with HgbA1c; thus, conditions associated with elevated levels of hemoglobin F (thalassemia and certain hemoglobinopathies) are associated with "false" elevations of hemoglobin A1c (Goldstein DE, et al. Diabetes. 1982; 31(Suppl 3):70-78). When Hgb reacts with glucose it's charge may be changed, and the modified hemoglobins may be identified by electrophoresis or cation exchange chromatography. HgbA1c is the predominant form of glycosylated hemoglobin which can be identified by it's change in charge. Unfortunately, Hgb variants associated with various hemoglobinopathies may comigrate or interfere. Alternatively, glycosylated Hgb may be measured by affinity chromatography. Affinity chromatography will measure many different glycosylated Hgb in addition to HgbA1c. Variant Hgbs do not interfere in the affinity chromatography methods.

When glycohemoglobin is assayed by affinity column chromatography, hemoglobin A1c plus the other forms of glycosylated hemoglobins are measured. Hemoglobins F, S and C do not interfere with the measurement of glycohemoglobin by affinity chromatography. Other techniques such as HPLC and electrophoretic techniques have also been introduced.
INTERPRETATION: Glycosylated hemoglobin is a measure of chronic blood sugar control in the diabetic. Free amino groups of hemoglobin form reversible shiff base intermediates with the aldehyde group of glucose. This is followed by irreversible rearrangement. The extent of glycosylation depends upon glucose levels and the length of time the protein is in circulation. In Hgb A1c there is glycosylation of the amino terminal of the ß-chain. Glycosylated hemoglobins consist of 10% HbA1a+b, 52% HbA1c, and 38% HbAo-like components (Abraham EC, et al. J Lab Clin Med. 1983; 1:187-197). Assay of hemoglobin A1c is useful as a means of monitoring carbohydrate control in diabetic patients. It is of greatest use in diabetics receiving insulin under poor control whose glucose levels fluctuate greatly. It is not as useful in well controlled stable diabetics (Nathan DM. N Engl J Med. 1990; 323:1062-1064).

When diabetic patients are carefully and optimally regulated, the levels of glycosylated hemoglobin begin to drop toward normal in from three to five weeks (Koenig et al. N Engl J Med. 1976; 295:417-420). Hemoglobin A1c can be measured at infrequent intervals in non-fasted patients to determine whether the patient's diabetes is well controlled; this monitoring will allow a more objective assessment of therapeutic efficacy (Koenig RJ, Cerami A. Ann Rev Med. 1980; 11:29-34; Tegos C, Beutler E. Blood. 1980; 56:571-572; Nathan DM, et al. N Engl J Med. 1984; 310:341-346; Goldstein DE, N Engl J Med. 1984; 310:384-385).

The mean blood glucose concentration(MBG) can be calculated from the value of the HgbA1c by the following equation:

$$MBG = 33.3(HgbA1c) - 86$$

However, an error of 1 percent in the measurement of glycosylated hemoglobin leads to an error of approximately 35 mg per deciliter in the blood glucose level. Measurement of hemoglobin A1c has been proposed as a means of detecting diabetes. Many claim it lacks sensitivity for screening (Mulkerrin EC, et al. Age Aging. 1992; 21:175-177). It is not as economical as a blood glucose and it is not one of the criteria used to establish the diagnosis of diabetes. Glucose tolerance tests were found to be superior to repeated glucose determinations and measurement of hemoglobin A1c (Gerken KL; Van Lente F. Arch Pathol Lab Med. 1990; 114:201-203).(cont)

Hemoglobin A1c (Cont)

Hemoglobin Alc is artifactually increased in patients with elevated hemoglobin F and decreased in patients with hemoglobins S and C. In these instances a measurement of glycohemoglobin by affinity chromatography is preferred. Glycosylated hemoglobin cannot be used to monitor control in diabetic patients with chronic renal failure. Glycosylated hemoglobin is significantly lower in patients with chronic renal failure than in normal controls. This is due to shortened erythrocyte survival secondary to chronic renal failure and not due to hemodialysis (Freedman D, et al. J Clin Path. 1982; 35:737-739). HgbAlc levels do not correlate with glucose tolerance in cirrhotic patients (Cacciatore L, et al. Diabetes Res. 1988; 7:185-188).

There is a significantly higher incidence of major congenital anomalies in the offspring of women with elevated HgbAlc values in early pregnancy as compared to women with normal levels of HgbAlc (Miller E, et al. N Engl J Med. 1981; 304:1331-1334; Freinkel N. N Engl J Med. 1981; 304:1357-1359).

The mechanism that has been proposed for nonenzymatic glycosylation of proteins is shown in the next Figure (Guthrow et al. Proc Natl Acad Sci. 1979; 76:4258-4261; Brownlea M, et al. N Engl J Med. 1988; 318:1315-1321).

Nonenzymatic Glycosylation of Proteins

```
                                     H             H H
H-C=O                           H-C=N-Protein   H-C-N-Protein                     H
H-C-OH                          H-C-OH            C=O           H    ---O CH2-N-Protein
HO-C-H  + H2N Protein  <=>   HO-C-H   Amadori->  OH-C-H    <=>  H/H          \
H-C-OH                          H-C-OH          H-C-OH          HO\H      OH/ OH
H-C-OH                          H-C-OH          H-C-OH             OH
  CH2OH                           CH2OH           CH2OH
Glucose                       Schiff Base     Ketoamine         Glycosylated
                                              Structure           Protein
```

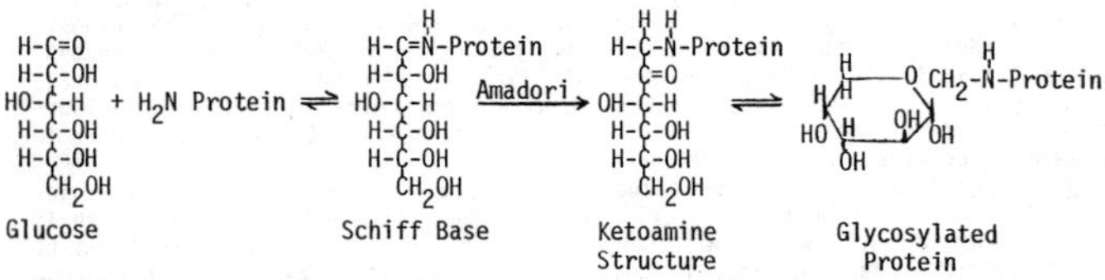

Advanced glycosylation end products form spontaneously from glucose-derived Amadori products and are postulated to play a role in the pathology of aging and diabetes (Makita Z, et al. Science. 1992; 258:651-653; Browlee M, et al. N Engl J Med. 1988; 318:1315-1321; Markita Z, et al. N Engl J Med. 1991; 325:836-842).

Albumin is glycosylated in a similar manner. Normal levels of glycosylated albumin are 6-15%. Poorly controlled diabetics may have levels as high as 30%. Glycosylated albumin may serve as an index of short term glucose control (1-2 weeks). The fructosamine test is also proposed as a similar indicator of short term glucose control. See FRUCTOSAMINE. The fructosamine test measures all forms of glycosylated proteins in the serum including albumin. It does not measure the glycosylated hemoglobins since the test is performed on serum.

HEMOGLOBIN BART'S (ALPHA-THALASSEMIA SCREEN)

SPECIMEN: Lavender(Purple)(EDTA) top tube.
REFERENCE RANGE: <1 percent; hemoglobin Bart's may be found in the blood of normal infants; it disappears at about 3 months of age.
METHOD: Hemoglobin Bart's may be detected or quantitated by electrophoretic or column techniques. Hemoglobin Bart's is quantitated by applying a freshly prepared hemolysate to a prepacked ready to use column (Isolab, Akron, Ohio). Hemoglobin Bart's is eluted in the first fraction; remaining hemoglobins are eluted with a second buffer. The absorbancies at 415nm of each fraction are used to calculate the percent hemoglobin Bart's.

The migration of Bart's on electrophoresis at pH 8.6 on cellulose acetate is shown in the next Figure:

Electrophoresis of Hemoglobins on Cellulose Acetate at pH 8.6

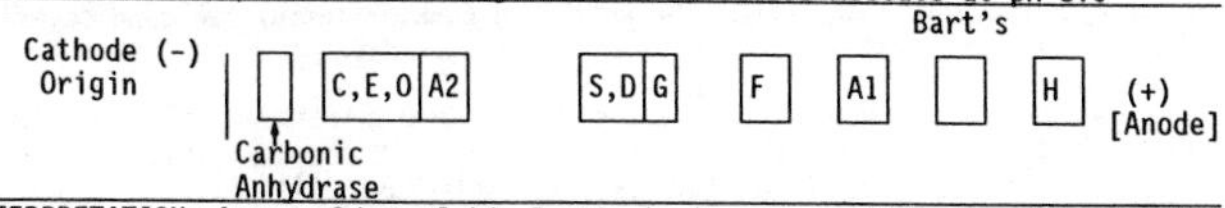

Cathode (-) Origin | [ ] | C,E,O A2 | S,D G | F | A1 | Bart's [ ] | H | (+) [Anode]

Carbonic Anhydrase

INTERPRETATION: Assay of hemoglobin Bart's is done on cord blood or blood of the newborn suspected of having alpha-thalassemia. Alpha-thalassemia is a condition characterized by reduced synthesis of alpha chains. The normal hemoglobins that contain alpha chains are hemoglobin A (two alpha chains and two beta chains), hemoglobin A_2 (two alpha chains and two delta chains) and hemoglobin F (two alpha chains and two gamma chains). Hemoglobin Bart's (four gamma chains) is formed when extra gamma chains of hemoglobin F combine in tetramers. Hemoglobin H has four beta chains.

The analysis of thalassemia is complicated by the fact that each individual inherits four copies of the allele coding for the alpha-globulin chain (2 copies on each of the paired chromosomes). This allows for 6 patterns of inheritance. The characteristics of alpha-thalassemias are given in the next Table:

Characteristics of Alpha Thalassemias

Condition	Genotype and Risk	% Bart's in Cord Blood	Hemoglobin Pattern
Normal		<1%	
Heterozygous (Silent Carrier) (Alpha-Thal-2 Trait)	(Silent Carrier x Normal)(Risk = 1/2) One of 4 Alpha-Globulin Genes Deleted	0.8 to 3	Normal
Heterozygous (Alpha-Thal-1 Trait)	(Alpha-Thal Trait x Normal)(Risk = 1/2) Two of 4 Alpha-Globulin Genes Deleted	3 to 10	In Adults, Hemoglobin H Inclusion Bodies are Found in About 1 in 10^5 Red Cells. Diff. Dx.: Iron Deficiency Anemia, Beta-Thal. or Decr. Alpha-Chain/Beta Chain Ratio (approx. 0.6)
Homozygous (Alpha-Thal-2 Trait)	Alpha-Thal Trait x Normal) Two of 4 Alpha-Globulin Genes Deleted	3 to 10	
Hemoglobin H Disease	(Alpha-Thal Trait x Silent Carrier) (Alpha-Thal Trait x Hemoglobin CS Hetero-zygote)(Risk = 1/4) Three of 4 Alpha-Globulin Genes Deleted	20 to 30; Level in Adults Variable	4-30% Hemoglobin H
Homozygous	(Risk = 1/4) All Alpha-Globulin Genes Deleted	>80	No Hemoglobin A or F; Hemoglobin H Present

(cont)

Hemoglobin Bart's (Alpha-Thalassemia Screen) (Cont)

Clinical Severity:

Heterozygous (Silent Carrier)(Alpha-Thal-2 Trait): Benign condition; no hemolytic abnormalities; diagnosis not made reliably in adults.

Heterozygous (Alpha-Thal-1 Trait): Mild; very mild in African Americans; clinical picture similar to beta-thal traits with very mild anemia and microcytosis.

Hemoglobin H Disease: Variable; usually chronic anemia; MCV and MCH are decreased; hypochromia; reticulocytes = 4 to 5 percent.

Homozygous: Lethal; hydrops fetalis with hemoglobin Bart's.

Hemoglobin Constant Spring: Hemoglobin Constant Spring is protein product of a prematurely truncated α-globin gene. Constant Spring can be inherited together with α-thalassemia trait to give rise to clinical disease. Hemoglobin Constant Spring is relatively common among Orientals. A combination of two gene deletion α-thalassemia and hemoglobin Constant Spring can give rise to hemoglobin H disease.

The levels of Hemoglobin Bart's found in variants of alpha-globulin deficiency states (alpha-thalassemia) are given in the next Table:

Levels of Hgb Bart's in Various Conditions

Condition	% in Cord Blood
Normal	<1%
Silent Carrier (α-Thal-2 Trait)	0.8-3%
Heterozygous (α-Thal-1 and -2 Trait)	3-10%
Hgb H Disease	20-30%
Homozygous (α-Thal)	>80%

Disorders of hemoglobulin structure and synthesis produce a selective advantage to the carrier state in malaria infection. As a result these disorders are common in people who live in or who have migrated from the areas with malaria. Alpha-thalassemia, a disorder of alpha-globulin chain synthesis is the most common genetic trait in the world and usually is due to deletions of the α-globulin gene. In some African countries, 40% of the people are estimated to have this trait. In some Asian countries the estimate is 15% (Kan YW. JAMA. 1993; 267:1532-1536). The severe form of α-thalassemia is most prevalent in Southeast Asia, the African form is milder.

Alpha-thalassemia-2 trait occurs in approximately 28% of African Americans and homozygous alpha-thalassemia-2 occurs in 2-3% of African Americans. Dried blood smears may be stained in the brilliant cresyl blue dye test. Erythrocytes from patients with hemoglobin H disease will be stained in a distinctive "raspberry" pattern. Hemoglobin H inclusions can also sometimes be detected in alpha-thalassemia-1 trait. However, the most reliable way to identify hemoglobin H is by electrophoresis and can be confirmed by an isopropanol denaturation test (unstable hemoglobins are denatured and precipitated following incubation in isopropanol solutions).

DNA analysis is helping to clarify and aid in the diagnosis of the thalassemias (Kan YW. JAMA. 1992; 267:1532-1536).

HEMOGLOBIN ELECTROPHORESIS

SPECIMEN: Lavender (EDTA) top tube
REFERENCE RANGE:

Normal Values of Hemoglobin

Normal Values:	Adult	Newborn
$HgbA_1$	96-98.5%	20%
$HgbA_2$	1.5-4%	2%
HgbF	0-2.0%	80%

METHOD: Electrophoresis at alkaline pH(8.6) and acid pH using citrate buffer. Electrophoresis at both pH's may be necessary to distinguish between some of these abnormal hemoglobin variants. The electrophoretic migration of the hemoglobins on cellulose acetate at pH 8.6 is shown in the next Figure:

Electrophoresis of Hemoglobins on Cellulose Acetate at pH 8.6

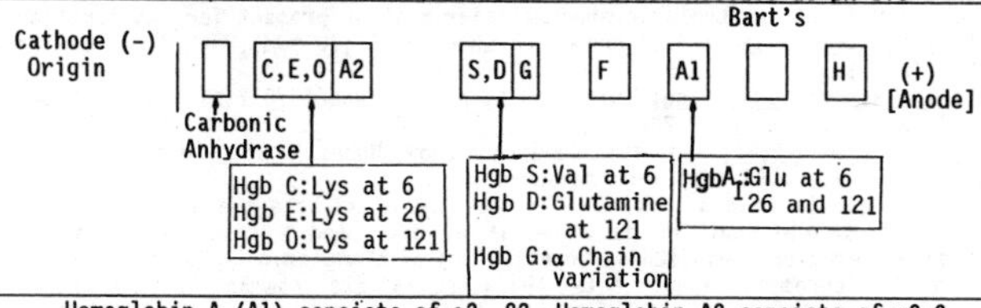

Hemoglobin A (A1) consists of α2, β2. Hemoglobin A2 consists of α2Δ2.

The electrophoretic migration of the hemoglobins on citrate agar at pH 6.2 is shown in the next Figure:

Electrophoresis of Hemoglobins on Citrate Agar of pH 6.2

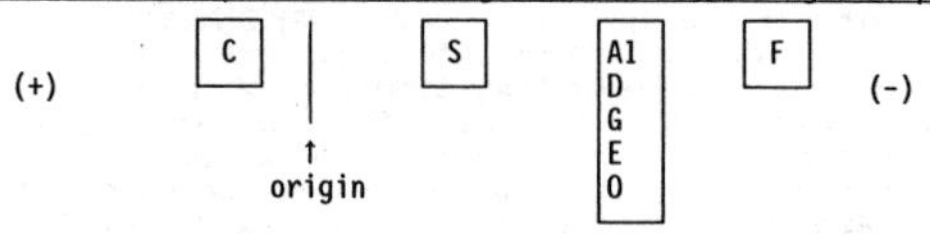

INTERPRETATION: Hemoglobin electrophoresis is performed in the evaluation of suspected hemoglobinopathies. This is done following an abnormal screening test, in family studies, and in the work-up of patients with hemolytic anemias possibly due to abnormal hemoglobins.

Characteristic findings in some common hemoglobinopathies are given in the next Table:

Characteristic Findings in Hemoglobinopathies

Hemoglobinopathy	Hgb A1	Hgb S	Hgb C	Hg A2	Hgb F	Other
Normal	96-98.5%	--	--	1.5-4%	0-2.0%	
Sickle Cell Trait	60-80%	20-40%	--	--	N	
Sickle Cell Anemia	--	60-99%	--	--	1-40%	
C Trait	70%	--	30%	--	--	
C Disease	--	--	95%	--	5%	
Sickle C Disease	--	45-55%	45-55%	--	--	
Beta Thal. Minor (↓ Beta Chains)	100-A2-F	--	--	N to ↑	N to 7	
Beta Thal. Major (↓ Beta Chains)	--	--	--	Low or ↑	7 to 100	
Alpha Thal. Minor (↓ Alpha Chains)	--	--	--	N to ↓	N to ↓	γ_4

A hemoglobin composed of four beta chains (B4) is known as HgbH. Hemoglobin Bart's is composed of four gamma chains (γ4).

Sickle Trait: This is a heterozygous state showing HgbA1, and HgbS with a normal amount of HgbA2. The HgbS concentration may vary from 20-50%.

Sickle Cell Anemia: This is a homozygous state showing almost exclusive HgbS. A small amount of HgbF (>15%) may also be present. However, in patients from Saudi Arabia and India with sickle cell anemia, HgbF levels average 25%, and these patients have mild or no clinical disease. In this disorder there is a Val at position 6 of the beta chain substituted for the normally occurring Glu.(cont)

Hemoglobin Electrophoresis (Cont)

C Trait: This is a heterozygous state showing HgbAl, and HgbC with a normal amount of HgbA2. The HgbC concentration is usually about 30%.

C Disease: This is a homozygous state showing almost exclusively HgbC plus up to 5% HgbF. In this disorder there is a Lys at position 6 of the beta chain substituting for the normally occurring Glu.

Sickle C Disease (SC): This is a heterozygous state showing HgbS (45-55%) and HgbC (45-55%).

E Hemoglobin: Hemoglobin E is perhaps the most common abnormal hemoglobin worldwide and is common in Oriental patients.

Beta Thalassemia Minor: This state shows HgbAl and HgbA2. The HgbA2 band is elevated to twice normal levels with a concentration of 5.8% in approximately 60% of the cases. HgbA2 values between 4-5% probably represent thalassemia minor, but repeated red cell morphologic abnormalities must be present for confirmation.

Beta Thalassemia Major: This condition shows HgbF (70-90%), HgbAl, (2-20%) and HgbA2 (2-8%).

Beta Thalassemia-S Disease: This condition shows HgbAl (0-25%), HgbS (50-90%), HgbA2 (3-8%) and HgbF (2-20%).

Beta Thalassemia-C Disease: This condition shows HgbAl (0-25%), HgbF (2-20%) and HgbC (60-90%), HgbA2 cannot be measured.

Hemoglobin A2 ($\alpha 2 \Delta 2$) is variably elevated in the beta-thalassemias and some ß chain hemoglobinopathies as a result of decreased ß chain production and decreased hemoglobin Al ($\alpha 2 \beta 2$). Quantitation of hemoglobin A2 is best performed by column chromatography. Beta thalassemias are common in the Eastern Mediterranean, Asia and parts of Africa. They may be found in patients of Italian and Greek descent as well as African Americans and Asians. It is a complex heterogeneous group of disorders, and beta-globin gene expression may be either partially or completely suppressed. There are over 100 point mutations which have been associated with ß-thalassemia (Kan YW. JAMA. 1992; 267:1532-1536).

Red cell sickling can be confirmed with a test such as the sickle cell solubility test (Sickle Cell Preparation, Sickledex). This is based on the insolubility of hemoglobin S in the presence of dithionate (reducing agent) and concentrated phosphate buffer. Infants under 6 months may not give a reliable result due to increased Hemoglobin F. Some unusual hemoglobins may also give positive results. As a result, many have stated that hemoglobin electrophoresis is the preferred method for primary screening of hemoglobinopathies including sickle trait and SS disease. Approximately half the patients with severe sickling disease will have HgbS/C or HgbS/Beta-thalassemia, both of which would be misclassified by a sickle solubility test alone.

DNA analysis is now being used to clarify and to aid in the diagnosis of the thalassemias and hemoglobinopathies (Kan YW. JAMA. 1992; 267:1532-1536).

HEMOGLOBIN F

SPECIMEN: Lavender(Purple)(EDTA) top tube

REFERENCE RANGE: The level of fetal hemoglobin F (HgbF) with age is shown in the next Table:

Fetal Hemoglobin with Age

Age	Hemoglobin F(Percent)
Newborn(Cord Blood)	50-85
1 Month	50-75
2 Months	25-60
3 Months	10-35
6 Months	8
1 Year	<2
Normal Adult	<2

Most adults <1%, some blacks <2%.

METHOD: Hgb F resists denaturation under alkaline conditions; most other Hgbs are denatured at basic pH and are then precipitated with ammonium sulfate. Fetal Hgb remains soluble and is quantitated spectrophotometrically (NCCLS Proposed Standard: H13-P, Guidelines for the quantitative measurement of HgbF by the alkali denaturation method). HgbF may also be assayed desitometrically on electrophoresis, column chromatography, or by immunological techniques. Electrophoresis tends to overestimate HgbF concentration in the lower range, the alkaline denaturation assay tends to underestimate the levels in the upper range.

INTERPRETATION: Hemoglobin F measurements are performed to evaluate suspected thalassemias, hemoglobinopathies, hemolytic anemias, patients with possible hereditary persistence of fetal hemoglobin, and a variety of other hematological conditions.

Hgb F consists of two alpha chains and two gamma chains (α2 2). Conditions associated with an increase in HgbF are given in the next Table:

Conditions Associated with Increased Fetal Hemoglobin

Condition	Hemoglobin F(Percent)	
Thalassemia Major	10-90	
Thalassemia Minor	2-12	Normal in more than 50 percent of patients; may be as high as 12 percent
HgbS-Thalassemia	5-20	
HgbSS	1-40	
HgbC	<7	
Hereditary Persistence of Fetal Hemoglobin:		
Homozygote	100	
Heterozygote	20-30	

HgbF is elevated in certain hemoglobinopathies and thalassemia. HgbF is most commonly elevated in the beta-thalassemias (2-90%) depending upon the variant. But it may be also elevated in the hemoglobin H variety of alpha thalassemia. It may be elevated in acquired conditions, e.g., megaloblastic anemia, myelofibrosis, aplastic anemia, leukemias, erythroleukemia, refractory anemias, pregnancy and paroxysmal nocturnal hemoglobinuria. Patients with hereditary persistence of HgbF apparently suffer no ill effects from the lifelong production of this hemoglobin.

HgbF is elevated in sickle cell anemia patients from Saudi Arabia and India. In these patients, HgbF levels are typically 25% and they have mild or no clinical disease. Levels of HgbF in African Americans, which are relatively low, can be raised with butyrate which stimulates the fetal -chain gene (Perrine SP, et al. N Engl J Med. 1993; 328:81-86).

HEMOLYTIC DISEASE OF THE NEWBORN(HDN) PANEL

Specimens and proposed tests for the work-up of infants with HDN are given in the next Table (Komarmy L. Pathologist. 1983; 266-267):

Specimens and Proposed Tests for Infants with HDN

	Test Group	Test	Specimen
Infant	Blood Bank	Blood Type	Red Top Tube
		Direct Coombs Test	Lavender(EDTA) Top Tube
		Crossmatch	
	Routine Hematology	Hemoglobin, Hematocrit,	Lavender(EDTA) Top Tube
		White Cell Count	Lavender(EDTA) Top Tube
		Wright Stained Peripheral Blood Smear	Finger-Stick
	Arterial Blood Gases	pH, pO_2, pCO_2	
Mother	Blood Bank	Blood Type	Red Top Tube
		Antibody Screen	Red Top Tube
	Routine Hematology	CBC with Indices	Lavender(EDTA) Top Tube
Father	Routine Hematology	CBC with Indices	Lavender(EDTA) Top Tube

Hemolytic disease of the newborn is caused by maternal-fetal blood group incompatibility. Clinically significant disease may develop due to various blood group antigens including RhD, ABO, and other Rh, Kell, Duffy, Kidd and MNSs groups. The extreme consequence of severe hemolytic disease of the newborn is hydrops fetalis. The pathophysiology of this process is as follows: IgG receptors on monocytes, macrophages and killer lymphocytes recognize antibody-coated red cells and hemolytic anemia results. Patients develop severe anemia and extramedullary hematopoiesis in the liver and spleen. Extramedullary hematopoiesis causes hepatosplenomegaly which interferes with organ function; the result is portal hypertension, hypoalbuminemia, edema and ascites. See also BILIRUBIN, AMNIOTIC FLUID.

Review: Gibble JW, Ness PM. "Maternal Immunity to Red Cell Antigens and Fetal Transfusion." Clin Lab Med. 1992; 12:553-576.

HEMOSIDERIN, URINE

SPECIMEN: 30 mL of random urine specimen; no preservative. The specimen must reach the laboratory within one hour of collection; refrigerate if determination not done immediately. Forward to outside laboratory frozen in plastic vial on dry ice.
REFERENCE RANGE: Negative
METHOD: Centrifuge specimen. Examine specimen microscopically for coarse brown granules in epithelial cells; if granules are seen, suspend the rest of the sediment in a fresh mixture of 5 mL of 2 percent potassium ferrocyanide and 5 mL of 1 percent HCl and allow to stand for 10 minutes. If iron is present, the ferric ion reacts with potassium ferrocyanide, to form ferric ferrocyanide, (Prussian blue). Centrifuge and examine the sediment microscopically. Coarse granules of hemosiderin appear blue.
INTERPRETATION: Urine hemosiderin may be detected in cases of extensive intravascular hemolysis. When excessive hemolysis occurs within the vascular space, hemoglobin is released. Up to 150 mg/dL may be bound to haptoglobin. Amounts in excess of this level are filtered through the renal tubules. Hemosiderin is formed in the tubular cells from hemoglobin; these cells slough into the urine. Urine hemosiderin is only found with intravascular hemolysis. The causes of intravascular and extravascular hemolysis are given in the next Table:

Causes of Intravascular and Extravascular Hemolysis

Intravascular	Extravascular
Hemolytic Transfusion Reactions	Spherocytosis
Autoimmune Hemolytic Anemia	Sickle Cell Trait
Drug-Induced Hemolysis	Thalassemia Minor
Massive Burns	Other Hemoglobinopathies
Sickle Cell Anemia(Occ.)	
Thalassemia Major(Occ.)	
Paroxysmal Nocturnal Hemoglobinuria	
Intravascular Water (e.g., Prostatic Resection)	
Hemochromatosis	
Microangiopathic Hemolytic Anemia	

HEPATIC PANEL

The following tests may be useful in the screening of liver disease.

Hepatic Panel
Aspartate Aminotransferase(AST)(SGOT)
Alanine Aminotransferase(ALT)(SGPT)
Gamma Glutamyl Transpeptidase(GGTP)
Alkaline Phosphatase (Alk Phos)
Bilirubin, Total and Direct
Urine Bilirubin
Urine Urobilinogen
Prothrombin Time(PT)
Partial Thromboplastin Time(PTT)
Serum Total Protein, Albumin, Protein Electrophoresis
Percent Transferrin Saturation

Elevation of the AST and ALT are consistent with hepatocellular damage. The GGTP and Alk Phos are elevated in space occupying lesions and biliary tract disease. The GGTP has also been used as a monitor of alcohol abuse. In view of the widespread occurrence of hemochromatosis in the American population, some articles advocate screening for this disorder. Percent transferrin saturation is the preferred test; serum ferritin may be suitable, but not as good. Further details can be found under the respective test headings.

HEPATITIS A ANTIBODIES, IgM AND IgG

SPECIMEN: Red top tube, separate serum
REFERENCE RANGE: Negative
METHOD: Competitive binding of antibody in patient's serum or plasma with the binding of a known amount of I-125 labeled or enzyme labelled antibody to hepatitis A virus (HAV) coated on to a solid phase.
INTERPRETATION: IgM antibodies reflect recent acute infection with HAV, and IgG antibodies reflect infection which occurred months to years before. The changes in relative concentration of hepatitis A antibodies, IgG and IgM, following exposure to hepatitis A, are shown in the next Figure:

Changes in Relative Concentration of Hepatitis A Antibodies, IgG and IgM

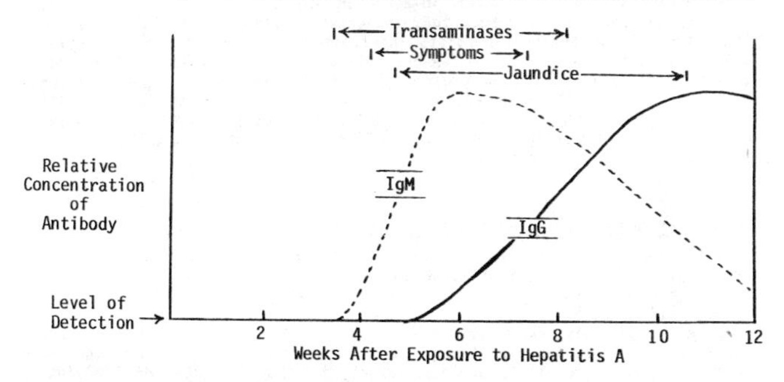

IgM antibodies are detected in the serum at the same time as symptoms develop and stay elevated for several months; IgG antibodies are detected for years after the acute illness. Onset of illness (15-45 days) is earlier than for hepatitis B (30-180 days).

Most people with hepatitis A, perhaps 90%, do not develop overt disease. When it is overt, illness ranges from mild malaise, to prostration with nausea, vomiting and diarrhea, fever, headaches and myalgia. Hepatomegaly is common but not splenomegaly. Clinical disease resolves in 0.5 to 1.5 months. Hepatitis A virus does not appear to cause chronic hepatitis or cirrhosis. However, approximately 1% of patients can develop a potentially fatal fulminant hepatitis.

When hepatitis A occurs in epidemics, it is usually referable to a common source such as contaminated water or the ingestion of shellfish containing the virus. The majority of U.S. Nationals returning from travel abroad with acute viral hepatitis probably have type A hepatitis. There is a high incidence of HAV disease in military personnel and Peace Corps volunteers living abroad. Travelers to countries with endemic HAV are recommended to receive prophylaxis with immune globulin. However because this is prepared from plasma collected primarily from donors in the USA, it is unlikely to prevent hepatitis E virus (HEV) infection, another common cause of fecal-oral transmitted hepatitis in many parts of the world (CDC Editorial, MMWR. JAMA. 1993; 269:845-846).

HEPATITIS B CORE ANTIBODIES, IgM AND IgG

Synonyms: Anti-HBc; HBcAb; Antibody to Hepatitis B Core Antigen; Core Antibody
SPECIMEN: Red top tube, separate serum
REFERENCE RANGE: Negative
METHOD: RIA, ELISA, or EIA
INTERPRETATION: Anti-HBc, IgM, is most useful as a marker for hepatitis B infection in the "window" between the time that the surface antigen (HBsAg) disappears and the time that the antibody to the surface antigen (Anti-HBs) appears. Hence in some patients anti-HBc may be the only marker of HBV infection at a particular stage in acute infection. The presence of anti-HBc, IgM, indicates current hepatitis B infection. Donated blood is presently being screened for anti-HBc antibodies in addition to HBsAg.

The presence of anti-HBc, IgG, indicates previous hepatitis B infection. The anti-HBc, IgG antibody persists indefinitely, in contrast to the anti-HBs which may be lost with time. A few patients will show a persistent isolated anti-HBc antibody without the presence of anti-HBs antibody. Response of these patients to hepatitis B immunization is able to distinguish false positive results from post-hepatitis infection (McIntyre A, et al. Aust NZ J Med. 1992; 22:19-22). Some reference laboratories offer measurement of total, IgM, or IgG anti-hepatitis B core antibodies.

HEPATITIS B SURFACE ANTIBODY (Anti-HBs)

Synonyms: Antibody to hepatitis B surface antigen; HBsAb; HBsAgAb.
SPECIMEN: Red top tube, separate serum
REFERENCE RANGE: Negative
METHOD: RIA or ELISA
INTERPRETATION: This test is run to diagnose previous exposure to HBV, to access the immune status of individuals exposed to blood or needlestick injury (possible need for immune globulin), and to evaluate the immune status of individuals following immunization. The presence of hepatitis B antibody indicates previous hepatitis B infection, exposure or vaccination; it usually appears about 5 months after exposure to hepatitis. Presence of anti-HBs is taken as a sign of recovery. anti-HBs levels may decline and even disappear with time.

Individuals with anti-HBs are considered protected from HBV infection. The procedure to follow after exposure to possibly infected blood or fluids is as follows: 1) If exposed individual is anti-HBs positive, they are considered immune. No further action. 2) If exposed individual is anti-HBs negative and blood or fluid is from an individual anti-HBs negative or felt at low risk for HBV, immunize exposed individual with HBV vaccine. 3) If exposed individual is anti-HBs negative and patient blood or fluid is HBsAg positive, or status of individual is unknown, immunize individual with HBV vaccine and treat with anti-HBV immune globulin. If individual is treated with both vaccine and immune globulin, these are given in different arms. Guidelines to follow after HBV exposure are given in: Kuhl T, Cherry J. Infect Control. 1987; 8:211-213; MMWR. 1990; 39(No. RR-1).

Not all individuals respond to immunization. Approximately 10-20% of immunized individuals do not develop anti-HBs after an initial course of immunizations; approximately half of these will respond to a second course.

HEPATITIS B SURFACE ANTIGEN (HBsAg)

Synonyms: Australia Antigen, Hepatitis Associated Antigen, HAA
SPECIMEN: Red top tube, separate serum
REFERENCE RANGE: Negative
METHOD: RIA, ELISA or EIA. New ELISA methods (Auszyme II) are at least as sensitive as RIA methods.
INTERPRETATION: Testing for HBsAg is performed to diagnose hepatitis B, diagnose carriers of HBV, and to access the progression and prognosis of the disease process. It is also used to screen blood donors.

HBsAg is the surface antigen of the hepatitis B virus. The HBsAg is detected between one and 4 months following exposure to the hepatitis B virus (rarely one to two weeks post exposure). The presence of HBsAg in blood indicates infection with the B virus. HBsAg persist in the blood in 3% to 5% of subjects; these patients often develop chronic liver disease. If the HBsAg persists for longer than 6 months, these patients are considered to have chronic hepatitis. Some patients may develop chronic liver disease and be HBsAg positive but have no history of acute hepatitis. HBsAg may be present in some people who have no history of hepatitis or signs of chronic liver disease; these people are referred to as "carriers." The changes in the different antigens and antibodies in typical serum hepatitis are shown in the next Figure:

Changes in Antigens and Antibodies in Typical Serum Hepatitis
Parenterally Transmitted

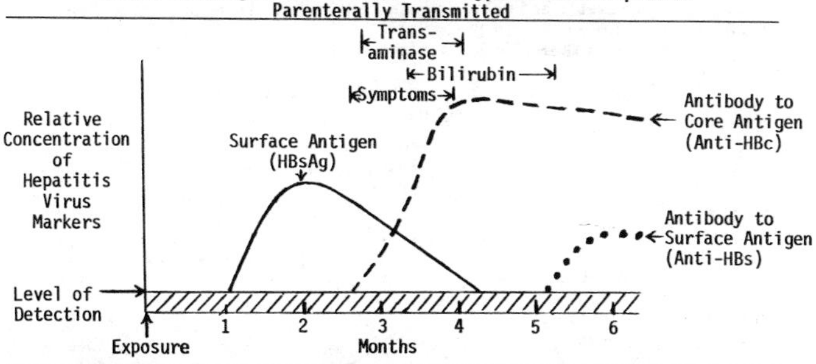

Notice that there is a fairly long interval between exposure, presence of HBsAg, and the finding of anti-HBs in the serum. Patients with chronic hepatitis have a persistence of HBsAg >6 months and no detectible anti-HBs antibody. Notice also that antibodies to the core antigen (anti-HBc antibody) appear before antibodies to the surface antigen, anti-HBs antibody. Hence it may be necessary to test for anti-HBc antibody in this "window" period between disappearance of HBsAg and appearance of anti-HBs.

HEPATITIS Be ANTIGEN (HBeAg)

SPECIMEN: Red top tube, separate serum
REFERENCE RANGE: Negative
METHOD: RIA; EIA
INTERPRETATION: Measurement of HbeAg is used to assess the infectiousness of a patient. HbeAg appears at the same time as the viral DNA polymerase and is associated with viral replication and infectivity. The presence of the "e" antigen indicates the blood is particularly infectious. HbeAg typically lasts for 2 to 6 weeks. The persistence of HBe from 8 to 10 weeks following acute infection indicates the development of the chronic carrier state. Seroconversion of HBeAg to anti-HBe in the acute stages of infection indicates reduced viral replication and pending resolution of infection.

The "e" antigen is not found in healthy carriers but is often present in patients with chronic liver disease particularly those with chronic aggressive hepatitis. Seroconversion from HBeAg to anti-HBe in the chronic carrier state is prognostic for an improvement in the patient's liver disease status (Lofgren B, Nordenfelt E. J Med Virology. 1980; 5:323-330). Carriers with HBeAg are particularly infectious. Approximately 30% of patients with chronic hepatitis will have chronic active hepatitis with both HBsAg and HBeAg present. Approximately 60% of chronic cases will have chronic persistent hepatitis with HBsAg and anti-HBe present. This later group is still potentially infectious, but is usually asymptomatic and well. In both groups anti-HBc is present and both may be subsequently infected with HDV.

The changes in the different antigens and antibodies in serum hepatitis are shown in the next Figure:

Changes in Antigens and Antibodies in Serum Hepatitis
Parenterally Transmitted

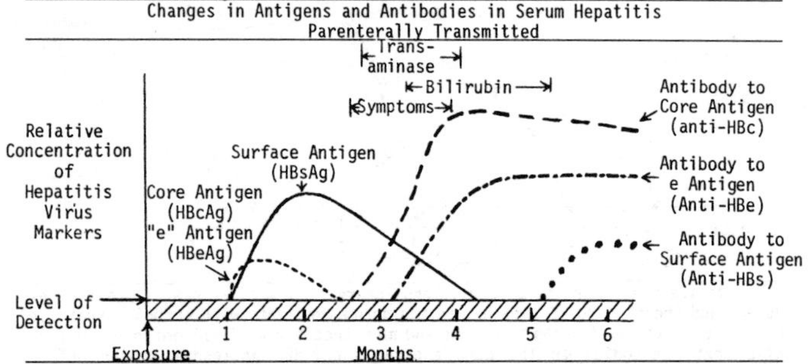

HBeAg positive mothers will infect over 90 percent of their newborns, most of whom will become chronic carriers; these carriers tend to develop serious chronic sequelae of hepatitis B infection. In general, these infants recover well from the acute episode, however, the majority become chronic carriers. Hepatitis B immunoglobulins should be administered to all newborns of HBeAg positive mothers and a course of immunization begun. Hepatitis immunization is recommended for all children, beginning immediately after delivery.

HEPATITIS C ANTIBODIES

Synonym: Antibodies to hepatitis C virus; Anti-HCV
SPECIMEN: Red top tube, separate serum
REFERENCE RANGE: Negative
METHOD: ELISA; Recombinant Immunoblotting Assay (RIBA).
INTERPRETATION: Hepatitis C antibodies (anti-HCV) are measured in the diagnosis of liver disease and in the screen of blood donors. Testing of blood for HCV was initiated in April 1990. The epidemiology and natural history is still to be fully defined. HCV was first described in 1974 (Price AM, et al. Lancet. 1974; 2:241-246; Feinstone SM, et al. N Engl J Med. 1975; 292:767-770). At present, it is estimated that approximately 80-90% of post-transfusion hepatitis is due to HCV (Aach RD, et al. N Engl J Med. 1991; 325:1325-1329). Recent studies suggest that sexual transmission of HCV occurs infrequently (Weinslock HS, et al. JAMA. 1993; 269:392-394; Osmond DH, et al. JAMA. 1993; 269:361-365). Presently blood is tested for syphilis,HBsAg, HIV-1, Anti-HBc, HTLV-1, Anti-HCV and the surrogate marker ALT (McCullough J. JAMA. 1993; 269:2239-2245). In patients who became HCV infected after blood transfusion, approximately 50% were asymptomatic, and only 20% were jaundiced in the acute stage. However, there was biochemical evidence of evolution toward chronic hepatitis occurring in 77% (Lee S, et al. J Infect Dis. 1991; 163:1354-1357). In a series of patients with HCV infection, 62% developed chronic hepatitis. Of 106 patients with HCV infection, 93 were positive for anti-HCV (second generation assay) and 13 were positive only for HCV RNA or HCVAg (Alter MJ, et al. N Engl J Med. 1992; 327:1899-1905). Anti-HCV became undetectable in very few of these patients, with an antibody loss of 0.6 per 100 person years. Some individuals with HCV hepatitis experience multiple episodes of acute hepatitis. Evidence suggests that HCV does not elicit protective immunity against reinfection (Farci P, et al. Science. 1992; 258:135-140). The value of ordinary gamma globulin in protecting against hepatitis C in individuals exposed to HCV is unproven.

The first generation ELISA test detected antibody to recombinant HCV antigen from the nonstructural region (c100). Second-generation ELISA tests detect antibodies to antigen from three regions (c100, c200, and c22) from both the nonstructural and core regions. There are also confirmatory immunoblot assays developed. A comparison of the results obtained with these assays in blood donors and organ recipients has been performed (Kleinman S, et al. Transfusion. 1992; 32:805-813; Pereira BJG, et al. N Engl J Med. 1992; 327:910-915). The currently available tests that detect anti-HCV significantly underestimate the incidence of transmission of HCV. In at least 10% of patients tested by second generation tests, no antibodies to the virus was detected (Alter MJ, et al. N Engl J Med. 1992; 327:1899-1905). There are also a fair number of false positive anti-HCV tests. Most blood donors who test positive for HCV by the screening ELISA test are falsely positive (Alberti A, et al. Ann Intern Med. 1991; 114:1010-1012). The confirmatory RIBA test was much more specific than the screening test. False-positive tests for HCV have been described following influenza vaccination (MacKenzie WR, et al. JAMA. 1992; 268:1015-1017).

Patients with HCV antibody have a fourfold increase in hepatocellular carcinoma risk compared to a sevenfold increased risk in patients with HBsAg (Tsukuma H, et al. N Engl J Med. 1993; 328:1797-1801). HCV infection has now been associated with cryoglobulinemia, polyarteritis, and glomerulonephritis (Appel GB. N Engl J Med. 1993; 328:505-506; Agnello V, et al. N Engl J Med. 1992; 327:1490-1495; Johnson RJ, et al. N Engl J Med. 1993; 328:465-470).

HEPATITIS, CHRONIC, PANEL

The conditions to consider and the tests to perform in patients with chronic hepatitis are given in the next Table:

Chronic Hepatitis

Condition	Tests
Hepatitis B	Hepatitis B Surface Antigen (HBsAg) Hepatitis B Surface Antibody (Anti-HBs) Core Antibody (Anti-HBc)
Hepatitis C (non-A, non-B)	Hepatitis C Virus Antibody
Cytomegalovirus	Cytomegalovirus Antibody Titer
Infectious Mononucleosis	Heterophil Antibody
"Autoimmune" Hepatitis	Anti-Smooth Muscle(ASM) Antibody
Primary Biliary Cirrhosis	Anti-Mitochondrial Antibody
Metabolic Causes:	
Wilson's Disease	Ceruloplasmin; Urinary Copper
Alpha-1-Antitrypsin Deficiency	Serum Electrophoresis; Quant. of Alpha-1-Antitrypsin
Hemochromatosis	Serum Iron and Iron Binding Capacity, Serum Ferritin, % Transferrin Saturation
Infantile Cirrhosis in India	Ceruloplasmin; Serum and Urine Copper

Liver biopsy is often the most definitive approach. Hepatitis A virus does not lead to chronic hepatitis or cirrhosis. Infantile cirrhosis in India has recently been shown to result from ingestion of excess copper leached from cooking and storage vessels. The % transferrin saturation is the best screening test for hemochromatosis. Testing for HAV is not performed in chronic hepatitis since this virus does not produce chronic disease.

HEPATITIS DELTA ANTIBODY

Synonym: Anti-Delta antibody; Delta Antigen Antibody; Hepatitis D Antibody
SPECIMEN: Red top tube, separate serum
REFERENCE RANGE: Negative
METHOD: RIA; EIA
INTERPRETATION: This test is performed to diagnose coinfection with the delta agent of patients with HBV infection. The delta agent discovered in 1977 (Rizzetto M, et al. Gut. 1977; 18:997-1003) is a virus-like particle consisting of delta antigen and a ribonucleic acid core; this defective viral agent requires hepatitis B virus coinfection for replication. The presence of the delta agent is associated with a higher incidence of fulminant hepatitis and chronic hepatitis.

Delta agent has been found in groups at high risk for hepatitis B; these include persons from southern Italy, parts of Africa, drug addicts and hemophiliacs in western Europe and the United States. Infection by the delta agent can occur as a coinfection with hepatitis B, which usually causes acute hepatitis that resolves. When delta agent occurs as a superinfection of a hepatitis B carrier, an episode of acute hepatitis and the establishment of persistent delta infection may lead to chronic active hepatitis and cirrhosis (Rizzetto M, et al. Ann Intern Med. 1983; 98:437-441).

Approximately 30-50% of fulminant hepatitis B is associated with coinfection or superinfection with delta agent (Smedile A, et al. Lancet. 1982; 2:945-947). Delta agent may cause epidemic hepatitis in areas where hepatitis B infection is endemic (Hadler SC, et al. Ann Intern Med. 1984; 100:339-344).

HEPATITIS PANEL

SPECIMEN: Red top tube, separate serum
REFERENCE RANGE: Negative
METHOD: RIA or ELISA
INTERPRETATION: This profile is used for the assigning of a viral etiology for patients with suspected infectious hepatitis. The interpretation of results is given in the next Table:

Interpretation of Results of Tests for Hepatitis

HBsAg	Anti-HBs	Anti-HBc IgM	Anti-HBc IgG	Anti-HAV IgM	Anti-HAV IgG	Anti-HCV	Interpretation
+	-	+	-	-	-	-	Early Acute Hepatitis B
-	-	+	-	-	-	-	Early Acute Hepatitis B in "Window" Period
-	+	-	+	-	-	-	Convalescent Hepatitis B
-	-	-	+	-	-	-	Late Convalescent Hepatitis B
-	+	-	-	-	-	-	Vaccinated for HBV
-	-	-	-	+	-	-	Acute Hepatitis A
-	-	-	-	-	+	-	Hepatitis A in the Past
-	-	-	-	-	-	+	Compatible with Hepatitis C
-	-	-	-	-	-	-	Compatible with a Non-Viral Etiology or False Negative HCV

Some patients with fulminant hepatitis B may have coinfection with the delta agent (HDV) as evidenced by the presence of anti-HDV antibodies. Antibodies to the hepatitis C virus (anti-HCV) are detected in a variable number of cases of non-A, non-B hepatitis (approximately 60% of cases). There is a 1-3 month delay between clinical onset and anti-HCV development. Sometimes it may be up to a year with presently used tests. There are a significant number of false positives and false negatives with this test. Some patients with prior history of HBV infection may lose the anti-HBs titers but remain anti-HBc positive.

HEPATITIS, VIRAL, TYPES, NEONATE

Types of viral hepatitis and the tests to detect these different types are given in the next Table (Zucker GM, Clayman CB. JAMA. 1982; 247:2011):

Viral Hepatitis

Virus	Assay
Hepatitis A	Hepatitis A, Antibodies IgM: Early Acute Hepatitis A IgG: Hepatitis A in the Past
Hepatitis B	Hepatitis B, Antibodies and Antigens HBsAg; anti-HBcAg-IgM, HBeAg: Acute Infection Anti-HBs alone: Infection in past or immune
Hepatitis C (non-A, non-B)	Anti-HCV: Many false(+) and (-). Serial Test.
Hepatitis D (delta)	Excluded if Hepatitis B infection is excluded
Hepatitis E	Not generally tested for; all infections have been contracted outside USA. Has been reported in Mexico
Cytomegalovirus	Cytomegalovirus(CMV) Antibodies IgM: Current Infection IgG: Infection in the Past
Rubella (Congenital or Acquired)	Rubella (German Measles) Antibodies IgM: Current Infection IgG: Infection in the Past
Herpes Simplex (Type 2 more than Type 1)	Herpes Simplex Antibodies IgM: Current Infection IgG: Infection in the Past These Tests Usually Do Not Distinguish Between Type 1 and 2

In the neonate, hepatitis associated with viruses other than A, B or C may occur; these are listed in the previous Table. Other stigmata of these viral diseases are usually present. A test for HCV surface antigen is now available.

Hepatitis, Viral, Types, Neonate (Cont)
Serial test and look for rising titers. This is to exclude the many false (+) being observed. There are also a lot of false (-) tests as a result of the long period of time it takes some patients to develop these antibodies (up to 12 months). In some instances polymerase chain reaction (PCR) techniques for the virus are necessary. When viral isolation and serologic findings are negative, and metabolic and hereditary causes are excluded, there remains up to 75 percent of cases in which the etiology remains unknown. See specific topics for further details.

HEROIN

SPECIMEN: 50mL random urine without preservatives; store at -20°C.
REFERENCE RANGE: None detected.
METHODS: Thin-layer chromatography; radioimmunoassay; gas liquid chromatography; latex agglutination-inhibition; hemagglutination-inhibition; spectrophotofluoremetry; free radical assay.
INTERPRETATION: Assessment of heroin abuse is by qualitative urine screening for morphine, its major metabolite. Heroin or morphine levels are rarely quantitated. Heroin is diacetylmorphine and, in the body, is rapidly deacetylated to morphine; urine is the major route of excretion and morphine is excreted either free (5 to 20 percent) or conjugated, primarily to glucuronide. A finding of morphine in the urine does not necessarily indicate heroin use since other opioids and opiates are also metabolized to morphine. Following injection, 50 percent of the total dose is excreted in the urine within the first 8 hours and about 90 percent during the first 24 hours. Morphine may appear in the urine within six minutes of an intravenous injection of heroin. Measurable urinary free morphine persists for 48 hours or longer. The time since the last dose at which morphine is detected in the urine varies with the method of detection as follows: hemagglutination-inhibition, >4 days; RIA, >4 days; thin-layer chromatography, 4 days (Catlin DH. Am J Clin Path. 1973; 60;719-728). The sensitivity of thin-layer chromatography is about 1000ng/mL of urine. Immunological techniques are more sensitive than TLC. A positive qualitative screen should be confirmed by gas chromatography/mass spectrometry. Ingestion of poppy seed containing foodstuffs can result in a positive opiate screen (Hayes LW, et al. Clin Chem. 1987; 33:806-808; Abelson JL. JAMA. 1991; 266:3130-3131).

Quinine is a common diluent of heroin; it may be detected in urine for several days after detectable morphine has disappeared. Street heroin is generally 5-10% actual heroin; the usual euphoric dose taken by abusers is equivalent to 10-20 mg of morphine. The classical triad of narcotic (heroin, morphine and codeine) overdose is coma, respiratory depression, and miosis (pinpoint pupils).

In the treatment of narcotic overdose, a specific antidote exists; naloxone (Narcan) can reverse the pharmacological effects of opioid agents. It reverses CNS depression caused by narcotics but has no effect on CNS depression of non-narcotic etiology. The most common complication of narcotic overdose is pulmonary edema (Cuddy PG. Crit Care Quart. 1982; 4:65-74).

HERPES SIMPLEX CULTURE

SPECIMEN: Specimens may be obtained from vesicular skin lesions and other body sites, specifically the nasopharynx, oropharynx, conjunctivae, stool, urine, leukocytes and cerebrospinal fluid. Use viral transport media. Mucous membranes or other lesions should be swabbed; the swab is rinsed vigorously in the transport media and squeezed-out against the glass and discarded. Calcium alginate swabs should not be used since viral growth may be inhibited. Specimens should be brought to the laboratory as soon as possible; if delay is unavoidable, specimens must be stored at 4°C.
REFERENCE RANGE: Negative
METHOD: Herpes simplex virus grows readily on tissue culture cells; cell strains and lines of human origin (human embryo, Hela, etc) support the growth of herpes simplex virus. Growth is rapid and the characteristic cytopathogenic effects, with ballooning and rounding of cells, are visible within 24 to 48 hours. With HSV-2 strains, syncytial formation usually occurs. Definitive typing of the virus, type 1, type 2, may be necessary.

INTERPRETATION: Virus isolation is the best method to confirm infection with herpes and should be used wherever possible in place of serology. Herpes simplex virus infection is probably the most common viral disease in man; the vast majority are not clinically apparent.

Two main strains of the virus exist; type 1 causes most nongenital infections and type 2 causes genital infections. Type 1 usually produces infections above the waist while type 2 primarily produces venereal lesions in males and females.

Serologic tests that differentiate between HSV-1 and HSV-2 are not commonly available. Serologic testing of single urban adults demonstrated that two-thirds of this population have HSV-1 antibodies and one-third have HSV-2 antibodies. Since many patients have unrecognized or subclinical infections, serologic testing may be useful to identify such patients (Siegel D, et al. JAMA. 1992; 268:1702-1708; Koutsky LA, et al. N Engl J Med. 1992; 326:1533-1539).

Genital Herpes: About 700,000 cases of genital herpes are diagnosed each year in the United States. Following sexual contact, the incubation period ranges from 1 to 26 days (average 7 days). In the male, the glans penis, the skin, the mucosal surfaces of the prepuce and the frenal area are commonly involved. In the female, lesions are common on the mucosal surfaces of the labia minora, clitoral hood, skin of the labia majora, buttocks, and thighs in severe cases; lesions heal in about 3 weeks.

Recurrent herpes lesions are less severe and are not due to reinfection; the mean time interval between initial and recurrent infection is about 120 days (25-360 days).

Acyclovir is recommended for treatment of first episodes, treatment of recurrences, and suppression of recurrences. Acyclovir is usually given orally, but may be given IV for more severe cases. Review of genital herpes, See Wang LL, Liu C. Infect in Med. November 1991; 42-47.

Neonatal Herpes: The incidence of neonatal herpes virus infection is estimated to be 400 to 1000 cases per year with a rate of infant infection of 1 per 3500 to 8000 deliveries (Stone KM, et al. Sex Transm Dis. 1989; 16:152-156). Most cases of neonatal HSV infection occur when mothers with subclinical HSV infection have asymptomatic shedding of the virus at the time of delivery since the current standard of care in the United States is to perform cesarian sections on all women with herpetic lesions present during labor (Gibbs RS, et al. Obstet Gynecol. 1988; 71:779-780). This practice has recently been questioned (Randolph AG, et al. JAMA. 1993; 270:77-82). Of asymptomatic women who shed HSV in early labor, women who have recently acquired genital HSV are ten times more likely to have an infant with neonatal infection than women with HSV reactivation (Brown ZA, et al. N Engl J Med. 1991; 324:1247-1252).

There are three patterns of neonatal infection: localized (confined to skin, eyes, or mouth), encephalitis, or disseminated infection. In one study, there were no deaths among the neonates with localized infection, but 6 percent of patients had neurologic impairment. In patients with encephalitis, the mortality rate was 15 percent; mortality was associated with semicomatose or comatose state or prematurity, seizures were associated with morbidity but not mortality. In patients with disseminated disease, mortality was associated with disseminated intravascular coagulopathy or pneumonia; semicomatose or comatose state was associated with both morbidity and mortality. In all of the patients, HSV-1 was associated with increased mortality whereas HSV-2 was associated with increased morbidity (Whitley R, et al. N Engl J Med. 1991; 324:450-454). Review: Jenkins M, Kohl S. Infect Dis Clin N Amer. 1992; 6:57-74.

HETEROPHIL ANTIBODIES (see MONOSPOT SCREEN)

HIGH DENSITY LIPOPROTEINS(HDL) (see LIPOPROTEINS)

HISTOPLASMOSIS TESTING

SPECIMEN: Culture: Sputum, Urine, Lung, Lymph Nodes, Bone Marrow, Liver, CSF, Blood; Antibody Tests: Serum, CSF; Antigen Tests: Serum, CSF, Urine.
REFERENCE RANGE: Negative; a positive histoplasmin skin test induces elevated histoplasmin CF (antibody) titers in 12 to 27 percent of persons (Buechner NA, et al. Chest. 1973; 63:259-270).
METHODS: Culture: Special media to reduce contamination by bacteria or other fungi; Antibody Tests: Complement fixation, Immunodiffusion; Latex Agglutination; Enzyme Immunoassay; Antigen Tests: Enzyme Immunoassay.
INTERPRETATION:

Culture: Cultures of involved tissues are frequently positive in disseminated or chronic pulmonary histoplasmosis, but are rarely positive in acute self-limiting illness. Culture takes 2-4 weeks for isolation and identification of histoplasma capsulatum. Exoantigen or DNA probe tests may be used for more rapid identification of histoplasma in mature culture.

Antibody Tests: Antibody testing of serum or CSF is used as the basis for diagnosis in most cases. Complement fixation tests detect yeast(Y) antibody or histoplasmin(M) antibody; sensitivities are 90% and 80% respectively. Immunodiffusion tests detect histoplasmin H and M antigens. Presence of both M and H precipitins suggests active histoplasmosis. The M band may be present after histoplasmin skin testing, but is also present in early infection, active infection, and past infection. M and H precipitins in the CSF occurs in meningeal histoplasmosis. Latex agglutination test for histoplasmin is useful for detection of early acute primary infection but may be negative in patients with chronic histoplasmosis. Radioimmunoassay or enzyme immunoassay are used to detect IgG, IgM, or IgA antibodies to histoplasma. There are inherent limitations to all antibody tests. The test may be falsely negative in early infection or immunocompromised patients. False positive tests may occur with other fungal diseases or tuberculosis. Positive tests may also indicate past infection rather than acute infection.

Antigen Tests: Histoplasma antigen may be detected in urine, serum, or CSF. Antigenuria detection allows rapid early diagnosis of histoplasmosis, but false positive tests may occur with other fungal diseases (Wheat LJ, et al. Am J Med. 1989; 87:396-400).

Skin Tests: Skin testing in histoplasmosis is not helpful and may increase antibody titers making interpretation of serology more difficult.

Reviews of histoplasmosis testing: Kaufman L. Laboratory Methods for the Diagnosis and Confirmation of Systemic Mycoses." Clin Infect Dis. 1992; 14(Suppl 1):S23-S29; Wheat LJ. "Histoplasmosis." Infect Dis Clin N Amer. 1988; 2:841-859).

HUMAN CHORIONIC GONADOTROPIN (HCG), SERUM

Synonym: Serum pregnancy test; beta-subunit human chorionic gonadotrophin
SPECIMEN: Red top tube, separate serum
REFERENCE RANGE: Negative in non-pregnant women. See below for range in intrauterine versus ectopic pregnancy. Great confusion is possible because various assay techniques have been standardized against different reference HCG preparations and the reference ranges may vary by as much as 50%. Some laboratory assays are standardized against the 2nd International Standard of HCG (1964) which was the recognized standard for 20 years and the standard against which most of the clinical literature is based. The 1st International Reference Preparation (1974) provided a purer, more homogeneous standard and methods standardized against this preparation yield HCG values approximately double those obtained with the older method. The newer intact-molecule HCG assays are almost all standardized against the new 1st International Reference Preparation. Knowing the normal reference range appropriate to the method employed in your laboratory is important. It is particularly essential in the interpretation of pregnancy viability and diagnosis of ectopic pregnancy (Painter PC. Diagnostic Clin Testing. 1989; 27:20-24).
METHOD: RIA; different assays have and are being developed. Some methods employ antibody specific for the beta-subunit portion of the hormone. Other methods employ an intact molecule-specific technique.
INTERPRETATION: Human chorionic gonadotrophin (HCG) levels are used to diagnose pregnancy, diagnose ectopic pregnancy, and to diagnose and monitor therapy of gestational trophoblastic tumors and other HCG producing neoplasms. Serum HCG is increased in the conditions listed in the next Table:

Increase of Serum HCG in Different Conditions
Normal Pregnancy
Ectopic Pregnancy
Abortion
Gestational Trophoblastic Tumors:
Hydatiform Mole, Invasive Mole, and Choriocarcinoma
Gonadal Tumors - Germinal Cell Origin:
Choriocarcinoma; Other Gonadal Neoplasms with Syncytiotrophoblastic Giant Cells
Other Tumors: Some Gastric Carcinomas; Some Hepatomas; Some Pancreatic Carcinomas and Other Tumors

Normal Pregnancy: The change of human chorionic gonadotropin (HCG) level in maternal blood during pregnancy is shown in the next Figure (Krieg AF. in Clinical Diagnosis and Management by Laboratory Methods. Henry JB (ed). W.B. Saunders, Co., Phila. 1979, pg. 685):

Human Chorionic Gonadotropin (HCG) in Pregnancy

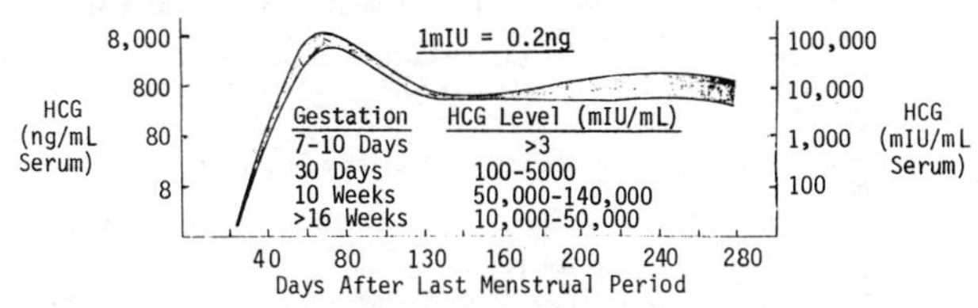

With new HCG assays serum HCG becomes detectable within 24-48 hours after implantation (5IU/liter), increases progressively and peaks about 10 weeks after the last menstrual period. By contrast office "slide tests" are usually positive approximately 14 days after implantation. The Hybriteck ICON™ test kit for serum or urine will detect HCG levels at approximately 20 IU/liter or about 7 days after implantation. Elevated levels may be found in multiple pregnancies, polyhydramnios, eclampsia and erythroblastosis fetalis.

False negative results or low levels may be seen in patients with ectopic pregnancy, non-viable pregnancy, or threatened abortion. False positive (pregnancy) results are seen in patients with trophoblastic disease or other HCG-producing neoplasms. With the very sensitive assays, low level physiological fluctuations in HCG may sometimes be detected.

Ectopic Pregnancy: Ectopic pregnancy in the U.S. has increased to 16.8 ectopic pregnancies per 1000 pregnancies. Improved diagnosis of ectopic pregnancy has resulted from development of highly specific HCG assays, high-resolution ultrasonography, and frequent use of laparoscopy (reviewed by Ory SJ. JAMA. 1992; 267:534-537).

Urinary qualitative pregnancy tests used 20 years ago were positive in only 50% of patients with ectopic pregnancy. By contrast currently available HCG-RIA assays are more than 99% sensitive in detecting ectopic pregnancy. Abnormal pregnancies (ectopic or threatened abortion) are associated with impaired HCG production. In 1981 it was reported that HCG levels should double every 1.98 days in normal early intra-uterine pregnancy (Kadar N, et al. Obstet Gynecol. 1981; 58:162-166). These authors described a method for screening for ectopic pregnancy based on this observation. They claim that an increase in HCG less than 66% over a 48 hour period is predictive of ectopic pregnancy or threatened abortion. The concept of measuring doubling time increases in ectopic pregnancy is a useful concept, but should be used in conjunction with ultrasonography and other modalities. A prospective study reported a sensitivity of 36% and specificity of 63-71% for the HCG doubling time test (Sheperd RW, et al. Obstet Gynecol. 1990; 74:417-420). (cont)

Human Chorionic Gonadotrophin (HCG) (Cont)

The absence of an intrauterine gestational sac on ultrasonography after a certain time period may be used to exclude intrauterine pregnancy. Using transabdominal ultrasonography, Kadar et al. proposed a discriminatory zone at a HCG level of 6000-6500 IU/liter (1st International Reference Preparation) above which a gestational sac could be visualized in 94% of cases (Kadar N, et al. Obstet Gynecol. 1981; 58:156). Many centers now report of HCG range of 1200-1500 IU/liter (1st International Reference Preparation) as a discriminatory zone for vaginal transducers (Fossum G, et al. Fertil Steril. 1988; 49:788-791).

Presence of an intrauterine gestation sac does not necessarily exclude ectopic pregnancy. Recent estimates of combined intrauterine and tubal pregnancy are 1:4000 in the general population, and 1:100 in the in-vitro pregnancy population.

Levels of HCG may be detectible for 4-6 weeks following removal of the ectopic pregnancy. Some patients are now being treated conservatively. Those undergoing conservative surgery should be followed with serial weekly HCG levels. If the HCG levels fail to decline, it indicates persistent trophoblastic tissue and the need for additional treatment.

Chemotherapy is considered for patients with ectopic pregnancies less than 3.5 cm in diameter and HCG levels <3500 IU/L. There have been anecdotal reports of tubal rupture in patients with low and declining HCG levels (Ory SJ, already cited).

Abortion: Following a complete first trimester abortion, HCG disappears in about 40 days.

Gestational Trophoblastic Tumors and other HCG-Producing Neoplasms: Gestational trophoblastic tumors, hydatiform mole and choriocarcinoma, often present with markedly elevated levels of HCG. Following complete removal of hydatiform mole, serum HCG declines steadily and disappears in 100 days, on average (Franke HR, et al. Obstet Gynecol. 1983; 62:467-473).

HCG is present in the serum in patients with tumors that may have syncytiotrophoblastic cells such as choriocarcinoma, embryonal carcinoma, seminoma, or dysgerminoma. In a relatively high percent of patients with malignant insulinoma, there is increased plasma levels of alpha-HCG (57%), beta-HCG (21%) or immunoreactive HCG (25%). Plasma levels of HCG were not elevated in any of 41 benign insulinomas (Kahn et al. N Engl J Med. 1977; 297:565-569).

HUMAN CHORIONIC GONADOTROPIN(HCG) STIMULATION TEST

SPECIMEN: Blood specimens for testosterone, androstenedione and 17-hydroxyprogesterone are collected in red top tubes at 0900 hours on day 1 and days 4 and 5; separate serum. In some protocols only testosterone may be measured.

REFERENCE RANGE: Normal response is an increase in steroids to above the upper limit of the normal range in at least one of the test samples. The degree of stimulation will vary depending on the age of the patient. Prepubertal males approximately 70-fold; puberty approximately 6-fold; adult male approximately 2-4 fold.

METHOD: Various protocols are described. The following is one. On day 1, collect blood specimen at 0900; then inject intramuscularly 3000 units of HCG per square meter body surface area. Repeat intramuscular injection on days 2, 3 and 4. On day 4, collect blood specimen prior to injection of HCG. On day 5, collect blood specimen.

INTERPRETATION: This procedure is used to assess hypogonadism in males and to distinguish primary from secondary disease. Human chorionic gonadotropin(HCG) stimulates interstitial cells and is used as a test of the ability of interstitial cells to secrete testosterone. A normal response is an increase in plasma testosterone to above the upper limit of the normal range in at least one of the test samples. In primary testicular disease, there is a reduced or absent response; in secondary hypogonadism (hypogonadotrophism), a normal response may be obtained.

HUMAN IMMUNODEFICIENCY VIRUS (HIV) TESTING

The incidence of HIV infection in the United States is estimated to be 1 million as of late 1990; worldwide, approximately 10 million people are infected. As of January 1992, more than 200,000 AIDS cases have been reported to the Center for Disease Control (CDC)(CDC. MMWR. 1991; 40:357-369; CDC. MMWR; 1992; 41:28-29; CDC. MMWR. 1992; 41[RR-18]:1-29).

Retroviruses that cause human infection include at least HIV-1, HIV-2, HTLVI and HTLVII. The predominant cause of clinical acquired immunodeficiency syndrome (AIDS) in the United States is HIV-1, which will be abbreviated "HIV."

HIV virus is transmitted by exposure to contaminated blood or blood products predominantly by sexual contact, perinatal exposure, parenteral (IV drugs) or mucous membrane exposure. The clinical stages, duration, and corresponding CD4 lymphocyte count is given in the following Table (Kessler HA, et al. Disease-a-Month. 1992; 38:635-690):

Clinical Stages, Duration, and CD4 Lymphocyte Counts in HIV Infection

Clinical Stage	Duration	CD4 Lymphocyte Count (per mm^3)
Incubation	2 to 10 weeks	
Acute HIV Infection	1 to 2 weeks	$\geq$750
Asymptomatic	2 to >10 years	200 to 750
Symptomatic	0 to 5+ years	100 to 500
Advanced HIV Disease	0 to 3+ years	<200
End-Stage HIV Disease	0 to 2+ years	<50

Structure of the Virus: The structure of the HIV virus, illustrating various antigens, is given in the next Figure (Kessler HA, et al, already cited):

Human Immunodeficiency Virus(HIV)

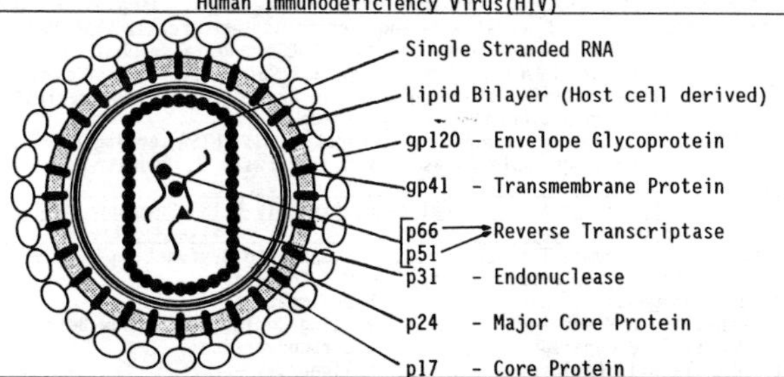

Life Cycle: HIV attaches to cells by interaction of envelope glycoprotein (gp120) and cell surface receptor, CD4 molecule (Attachment). Transmembrane protein gp41 and gp120 are both involved in fusion of HIV to the host cell; the virus core is then incorporated into the cell. The virus core is partially uncoated (Uncoating) and the viral single stranded RNA is converted to double stranded DNA (Reverse Transcription) which circularizes (Circularization) before migration into the host cell nucleus. In the nucleus, proviral DNA becomes incorporated into the host cell genome (Integration) using a viral endonuclease, integrase. Once integration has occurred, HIV virus can be produced by the host cell. Virion RNA is transcribed from DNA (Transcription) and virus proteins are translated from RNA (Translation). Components of the virus core are assembled (Core Particle Assembly). The virus core migrates to an area of the cell membrane containing HIV-specific gp41 and gp120 where the virus buds off the host cell (Final Assembly/Budding). The steps in the life cycle of HIV virus are summarized in the following Table (Kessler HA, et al, already cited; Bryant ML, Ratner L. Pediatr Infect Dis J. 1992; 11:390-400):

Life Cycle of HIV Virus
Attachment
Uncoating
Reverse Transcription
Circularization
Integration
Transcription
Translation
Core Particle Assembly
Final Assembly/Budding

HIV Testing (Cont)

Diagnosis: Various laboratory tests are used to diagnose HIV infection, as given in the following Table (Kessler HA, et al, already cited):

Tests Used to Detect HIV Infection
Anti-HIV Antibody Tests
Enzyme Immunoassays (EIA or ELISA)
Western Blot
Immunofluorescence Assays
Radioimmunoprecipitation
Latex Agglutination
Whole Blood Immunoassay
HIV Antigen Tests
Enzyme Immunoassays
HIV Nucleic Acid Tests
Polymerase Chain Reaction
In Situ Hybridization
HIV Culture
Coculture of Peripheral Blood Mononuclear Cells
Plasma Culture
Detection of Immunodeficiency
Lymphocyte Subset Analysis
Lymphocyte Functional Assays (In Vitro)

Antibody Tests: The standard method used to diagnose HIV infection is measurement of antibody by enzyme linked immunosorbent assay (ELISA) as a screening test followed by Western blot for confirmation of ELISA positive samples. ELISA is an extremely accurate test, sensitivity and specificity are greater than 99 percent. The frequency of false-positive tests is 0.0007% and false-negative tests is 0.0003% (Burke DS, et al. N Engl J Med. 1988; 319:961-966). The most common cause of false-negative tests is obtaining a sample during the "window period" - the time after infection but before seroconversion.

Causes of false-positive and false-negative HIV ELISA serologic testing are given in the following Table (Krasinski K, Borkowsky W. Pediatr Clin N Amer. 1991; 38:17-35):

False-Positive and False-Negative HIV ELISA Serology	
False-Positive	False-Negative
Antibody Against Human Tissues (eg, smooth muscle, mitochondria, T-cell antigens)	B-Lymphocyte Dysfunction with Severe HIV Infection
Anti-Hepatitis A Virus IgM	Hypogammaglobulinemia
Anti-Hepatitis B Core IgM	Testing During the Window Before Seroconversion
Anti-HLA Class II Antigens (eg, HLA-DR4, HLA-DQw3)	Malignancy
Severe Liver Disease	Immunosuppressive Therapy
Syphilis Serology Positive	Exchange Transfusion
Malignancy	Bone Marrow Transplant
Cross-Reacting Viral Infections (eg, HIV-2, HTLVI/II)	Heat Inactivation
Renal Failure	
Passively Acquired HIV-1 Antibody (Transplacental, Hepatitis B Immune Globulin)	

Western blot is used as a confirmatory test for a positive ELISA test. The Western blot detects antibodies to specific components of the HIV virus. HIV proteins and glycoproteins are electrophoretically separated and transferred to nitrocellulose paper. The antigen-impregnated paper is reacted with patient serum; antibodies bind specific HIV antigens. HIV antigen-antibody bands are visualized by using antihuman immunoglobulin conjugated with biotin or an enzyme.

Criteria for a positive Western blot test are given in the following Table (Proffitt MR, Yen-Lieberman B. Infect Dis Clin N Amer. 1993; 7:203-219):

Criteria for a Positive Western Blot	
Organization	Criteria
Assoc of State and Territorial Public Health Laboratory Directors/CDC	Any two of: p24, gp41, gp120/gp160
FDA Licensed DuPont Test	p24, p31, and gp41 or gp120/gp160
American Red Cross	≥3 bands (one from each gene-producing group) gag (p55, p17, p24, p15) and pol (p31, p66/p51) and env (gp120/gp160, gp41)
Consortium for Retrovirus Standards	≥2 bands: p24 or p31 and gp41 or gp120/gp160

Immunofluorescence assays use HIV infected cells to detect the presence of HIV antibody in serum or plasma. A slide preparation of inactivated HIV infected cells and control cells is reacted with the patient's serum or plasma. The antigen-antibody reaction is detected by fluoresceinated anti-human immunoglobulin. The reaction is detected by fluorescence microscopy.

Other antibody detection tests such as radioimmunoprecipitation, latex agglutination, and whole blood immunoassay have been developed but are less frequently used.

Antigen Tests: The most common test used to detect HIV antigen is the enzyme immunoassay (EIA) which detects the p24 core associated viral protein. This test is very sensitive; antigen may be detected to about 10 picograms/mL. HIV antigen EIA is most useful in the diagnosis of acute HIV infection during the "window period" prior to host antibody response. This test has also been used to monitor the clinical stage of HIV infection and to evaluate and monitor the effects of antiviral therapy. False-negative test results may occur due to formation of immune complexes; HIV-specific antibodies can render the antigen undetectable. Immune complex dissociation increases the sensitivity of this test. EIA testing for p24 antigen may be negative during latent infection or when viral replication is taking place slowly (Proffitt MR, Yen-Lieberman B, already cited; Kessler HA, et al, already cited; Krasinski K, Borkowsky W, already cited).

Nucleic Acid Tests: The most sensitive test for detection of HIV infection is the polymerase chain reaction (PCR); this test can detect one molecule of proviral DNA in 10 microliters of blood. This test amplifies the nucleic acid "signal" to a more easily detectable level. Test cells are lysed and the double-stranded DNA is denatured to single stranded DNA by heating. Two short DNA segments that are complementary to the HIV proviral DNA are added to the sample. Bacterial DNA polymerase (Taq polymerase) is added and the proviral DNA replicates. Each replication cycle doubles the amount of DNA; after 30 cycles, there is a millionfold increase in the number of target sequences. The DNA sequences are detected by use of radioisotopes, biotinylated probes, or by oligomer restriction testing. The PCR has been used to quantitate the amount of HIV virus in infected individuals. The PCR has also been used to identify infants with true HIV infection. Infants may have passively acquired maternal HIV antibodies for up to 15 months; identification of HIV nucleic acids definitively identifies infected babies (Sloand EM, et al. JAMA. 1991; 266:2861-2866). PCR may also be useful in patients with indeterminate Western blot, seronegative high-risk individuals, sexual partners of known HIV infected individuals, and for individuals exposed to HIV virus in other setting (eg, health care workers)(Kessler HA, et al, already cited). One of the limitations of PCR is that automation is problematic because PCR is extremely sensitive to cross-contamination by small quantities of nucleic acids from previous samples.

In situ hybridization directly identifies HIV proviral or extrachromosomal DNA in peripheral blood mononuclear cells. This test is less sensitive than the PCR technique described above.

HIV Testing (Cont)

HIV Culture: The most specific way to diagnose HIV infection is by virus isolation. Specimens for virus isolation are obtained from peripheral mononuclear cells or from body fluids. Plasma can also be cultured for the presence of free HIV. Samples are inoculated into peripheral blood mononuclear cells from HIV negative individuals which have been stimulated with interleukin-2 and phytohemagglutinin. Cultures are maintained for up to 1 month; supernatants from the cultures are sampled every 3 or 4 days for the presence of p24 antigen or retroviral reverse transcriptase. HIV culture is expensive, time consuming, and labor intensive. Success of culture depends on the disease stage and viral burden. False-negative tests may also result from the presence of CD8-bearing T lymphocytes in the culture. Thus, a negative test does not exclude HIV infection.

Detection of Immunodeficiency: Abnormalities of immune function associated with HIV infection are given in the following Table (Noel GJ. Pediatr Clin N Amer. 1991; 38:37-43):

Immunologic Abnormalities Associated with HIV Infection

Cell Type	Proposed Mechanism	Laboratory and Clinical Findings
Lymphocytes		
T-cells	Lysis of CD4 positive cells	Lymphopenia Decreased CD4/CD8 ratio
	Impaired cell-cell interaction	Decreased response to mitogens and T lymphocyte-dependent antigens Anergy
	Decreased cytokine production	Impaired interleukin-2-mediated responses Decreased interferon-gamma-dependent macrophage activation Impaired granuloma formation
B-cells	Polyclonal activation	Hypergammaglobulinemia Decreased antibody production to T lymphocyte-dependent and independent antigens
Natural Killer Cells	Maturation arrest	Decreased cytotoxic responses
Mononuclear Phagocytes		
Monocytes	Lysis Cytoskeletal dysfunction	Decreased circulating monocytes Impaired chemotaxis and phagocytosis
Monocytes/ Macrophages	Decreased production of cytokines	Impaired interleukin-1-mediated responses Impaired granuloma formation
	Impaired expression of Fc and complement receptors	Decreased Fc- and complement-mediated clearance by the reticuloendothelial system Increased susceptibility to systemic bacterial infection
	Impaired expression of HLA-DR surface antigen	Decreased response to MHC class II antigens
Polymorphonuclear Phagocytes		
	Decreased granulocytopoiesis	Neutropenia
	Impaired receptor-mediated functions	Decreased chemotaxis and phagocytosis

Once a patient has been diagnosed as HIV infected close monitoring for infectious complications is necessary.

Suggested initial laboratory evaluation for patients with newly diagnosed HIV infection is given in the following Table (Gold JWM. Med Clin N Amer. 1992; 76:1-18):

Initial Laboratory Evaluation of Newly Diagnosed HIV Infection
Complete Blood Count, Differential, Platelets
Biochemistry Screening Profile
Urinalysis
Tuberculin Test with Anergy Panel
Serologic Test for Syphilis
Toxoplasma Serology
T Lymphocyte Subsets
Hepatitis B Serology
Chest X-ray

Tests for Other Retroviruses: Laboratory tests for retroviruses related to HIV-1 have been developed. HIV-2 is associated with AIDS in Western Africa; ELISA antibody tests using synthetic antigens based on transmembrane glycoproteins have been developed. Combination tests which detect both HIV-1 and HIV-2 are also available. Through 1991, 17 cases of HIV-2 infection have been reported in the United States (O'Brien TR, et al. JAMA. 1992; 267:2775-2779). Tests for antibodies to HTLV-I and HTLV-II are also available.

HUMAN LEUKOCYTE ANTIGEN(HLA)

SPECIMEN: Three green (heparin) top tubes. Do not refrigerate or freeze. Specimens must arrive in laboratory within 24 hours of drawing.
REFERENCE RANGE: Antigens present are reported.
METHOD: In the lymphocytotoxicity test, purified lymphocytes are mixed with typing sera (known antibody) and incubated. Complement is added, and if typing antibody is bound to the indicator cell, the cell is killed. The three HLA loci, HLA-A,B and C are determined in this manner. The HLA-D locus (DR or D-related) is determined by mixed lymphocyte culture. Other techniques are being developed including gene probes.
INTERPRETATION: HLA typing is performed for transplantation candidate matching, blood product matching, paternity testing, and correlation with various disease states. Assay for HLA-B27 is useful for diagnosis of ankylosing spondylitis. HLA typing is performed for selection of HLA-matched platelets in patients who have not responded well to previous platelet transfusions.

HLA are antigens located on the surface of leukocytes as well as on the surface of all nucleated cells. They are not present on the surfaces of erythrocytes. In HLA testing, the HLA antigens on lymphocytes are determined; lymphocytes have a relatively high concentration of HLA antigens and lack ABO antigens. HLA are glycoproteins and most can be demonstrated by serologic tests.

Genetic coding for HLA is found on the short arm of chromosome 6. There are four loci of the HLA region; these loci are called A, B, C, D and DR (for D-related). D-antigens are tested by a different method, that is, mixed lymphocyte culture, because their immune function differs from that of other antigens.

Some of the HLA antigens are associated with disease; these are listed in the next Table:

HLA and Disease

Disease	HLA
Ankylosing Spondylitis Reiter's Syndrome Yersinia Enterocolitica Arthritis	B-27
Multiple Sclerosis	DW-2
Chronic Active Hepatitis Gluten-Sensitive Enteropathy	B-8
Systemic Lupus Erythematosus	B-15
Hemochromatosis	A-3
Diabetes Mellitus	Multiple Loci

Human Leukocyte Antigen (HLA) (Cont)

Juvenile- or early-onset insulin-dependent diabetes mellitus(IDDM) in Caucasians is associated with several HLA antigens, HLA-B8, HLA-B15, HLA-DR3 and HLA-DR4 (Ginsberg-Fellner F, et al. Diabetes. 1982; 31:292-298; Pittman WB, et al. Diabetes. 1982; 31:122-125; Patel R, et al. Metabolism. 1977; 26:487-492). In adult- or late-onset IDDM, there is significant increase in the frequency of HLA-DRW and a decreased frequency of HLA-DR2 (Pittman WB, et al. Diabetes. 1982; 31:122-125).

Antibodies to HLA-A, B and C are found in the following:

Antibodies to HLA Antigens
Polytransfused Patient Who Has Received Whole Blood
Multiparous Women
Recipients of Allotransplants
Kidney Dialysis Patients
Volunteers Immunized with Cells

Most of the antisera contain multispecific antibodies.

HUMAN TUMOR STEM-CELL ASSAY

Synonyms: Human tumor clonogenic assay; clonogenic assay. Related techniques: antimetabolic assay of chemosensitivity; MTT assay; inhibition of succinate dehydrogenase assay; fluorescent cytoprint assay, FCA; fast green assay, FGA.
SPECIMEN: Surgical specimen: Tumor Tissue; remove connective and adipose tissues; weigh specimen; mince into pieces less than 2mm in diameter in the presence of appropriate media.
REFERENCE RANGE: Growth or inhibition of growth in presence of a particular drug.
METHOD: The cells of the tumor are enzymatically dissociated using enzymes such as DNase and collagenase. The tumor fragments are stirred and the free cells are decanted through gauze, centrifuged and the free single cells are resuspended; drugs are added to the tumor cell suspension. The cells are grown on soft agar placed in wells. The plates are examined microscopically about twice a week; the number of colonies reach maximum in two to three weeks. There are a large number of assay variants (see below) that differ in the preparation of the specimen - single cells or tissue fragments, the time of incubation, and the technique of accessing tumor growth or death.
INTERPRETATION: The human tumor stem-cell assay has been used as an in-vitro test for determining the sensitivity or resistance of an individual patient's tumor to an anticancer drug. The objective is to allow the selection of drugs to which patients are more likely to respond. The concept is similar to in-vitro testing of antibiotics to isolated bacteria. The human tumor stem-cell assay is commercially available and many tumor specimens have been tested.

There are data that have shown that in-vitro responses of individual tumors to selected drugs correspond well with the clinical responses of the patient to those drugs. Typically, using the results of the human tumor stem-cell assay, the accuracy rate for prediction of drug sensitivity is correct in 40 to 90 percent of cases and the accuracy rate for prediction of drug resistance is correct in 90 to 95 percent of patients. The data indicate that this stem-cell assay may allow selection of patients who are more likely than others to respond to specific drug therapy.

There are problems with the use and interpretation of these tests; usually, less than 50 percent of the tumors cultured have sufficient in-vitro growth for drug testing. Other problems include insufficient tumor cells isolated for growth studies, presence of excessive non-neoplastic cells, and bacterial contamination of cultures. Criticisms to this assay are discussed (Selby et al. N Engl J Med. 1983; 308:129-134; Von Hoff DD. N Engl J Med. 1983; 308:154-155). However, other studies indicate highly significant associations between in-vitro chemosensitivities and the clinical course of the patient (Kern DH, et al. Ann Clin Lab Science. 1983; 13:10-15). This technique has been reviewed (Hanauske AR, et al. Curr Prob Cancer. 1985; 9:1-50). (cont)

Variants of the clonogenic assay and other assays designed to assess tumor drug sensitivities include: the fluorescent cytoprint assay employing drug screening on micro fragments of tumor and increase in fluorescence as an indicator of growth (Meitner PA. Oncology. 1991; 5:75-81; Leone LA, et al. Cancer Invest. 1991; 9:491-503); subrenal capsular tumor growth in drug treated animals (Drago JR, Smith JJ. In Vitro. 1987; 1:57-59) ; in-vitro drug resistance accessed by dye exclusion assay, e.g., fast green (fast green assay, FGA)(Weisenthal LM, et al. Cancer Res. 1983; 43:258-264); inhibition of tumor cell succinate dehydrogenase employing tetrazolium dye (MTT) as electron acceptor with purified cells (Yamane H, et al. Eur J Cancer. 1991; 27:1258-1263); growth inhibition in a collagen matrix (Furukawa T, et al. J Surg Oncol. 1992; 49:86-92); and inhibition of nucleotide incorporation into DNA and RNA - the antimetabolic assay of chemosensitivity (AAC)(Franchi F, et al. Oncology. 1991; 48:510-516).

Hydrogen Breath Analysis

HYDROGEN BREATH ANALYSIS

SPECIMEN: Breath specimens for hydrogen analysis following oral administration of carbohydrate
REFERENCE RANGE: Compare to normal curves; 2 to 20ppm.
METHOD: Gas chromatography using thermal-conductivity detector.
INTERPRETATION: This test is performed in patients with suspected carbohydrate malabsorption, most often due to absence of digestive enzymes of intestinal origin. The most common indication for use of breath hydrogen analysis is in the patient who is suspected of having lactose malabsorption. Other major clinical applications of the hydrogen breath test are in the work-up of patients with possible sucrose malabsorption and D-xylose malabsorption.

Hydrogen gas originates solely from the metabolism of carbohydrate from the bacteria in the gastrointestinal tract. Ninety percent of hydrogen production originates in the colon. In order for the hydrogen to be generated in the colon, the carbohydrate must pass unabsorbed by the small intestine to the colon. Thus, hydrogen in the breath reflects bacterial metabolism of carbohydrate in the colon. The hydrogen gas is absorbed into the venous blood; in the lung it diffuses into the alveoli and is finally expired.

Malabsorption of a carbohydrate is detected by an increase in hydrogen in the breath; values in excess of 20ppm are considered abnormal.

Major clinical applications of the hydrogen breath test in gastrointestinal disease are given in the left hand column of the next Figure with test results in the right hand column (Solomons NW. Comprehensive Therapy. 1981; 7:7-15; Brugge WR. Resident and Staff Physician. 1983; 29PC-37PC):

Major Clinical Applications of the Hydrogen Breath Test

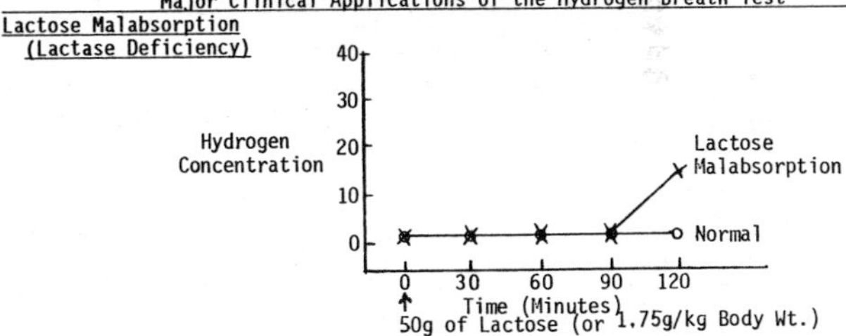

Patients with lactase deficiency do not absorb lactose in the small bowel; lactose passes unchanged into the colon where it is broken down to hydrogen and other products. The hydrogen is detected in the breath. Defects in digestion and absorption of other carbohydrates, e.g., sucrose, are performed in a similar manner.

Hydrogen, Breath Analysis (Cont)

Major Clinical Applications of the Hydrogen Breath Test (Cont.)

Bacterial Overgrowth, Small Bowel

e.g. Fistulae
- Blind Loop
- Jejunal Diverticulosis
- Diabetic Gastroparesis

Hydrogen Concentration
40
30
20
10
0
Bacterial Overgrowth
Normal
0 30 60 90 120 150
Time (Minutes)
10g of Lactulose with 400mL of Water

In patients with bacterial growth in the small bowel, the bacteria will metabolize lactulose to hydrogen and hydrogen will appear in the breath at an earlier time as compared to normal.

Intestinal Transit

Decreased Transit Time
- Irritable Bowel Syndrome
- Some patients with Vagotomy
- Some patients with Gastro-jejunostomy or Fistula

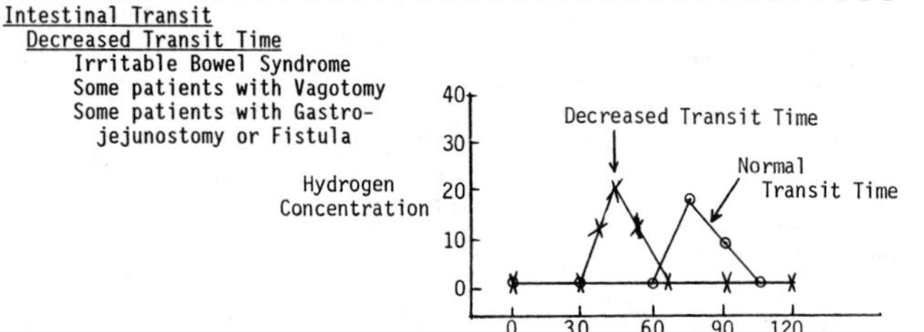

Factors that influence the interpretation of hydrogen breath tests are given in the next Table (Solomons NW. Comprehensive Therapy. 1981; 7:7-15):

Influences on Hydrogen Breath Tests
Idiopathic Absence of the Appropriate Flora
Iatrogenically Induced Absence of the Appropriate Flora
Active Diarrhea
Acidic Colonic Milieu
Delayed Gastric Emptying
Uninterrupted Sleep During Test
Tobacco Smoking
Inclusion of Fiber or Oligosaccharides in Test Meals

17-HYDROXYCORTICOSTEROIDS (17-OHCS) IN URINE

SPECIMEN: 24 hour urine. Place 10gm boric acid in container prior to collection. Instruct the patient to void at 8:00 A.M. and discard the specimen. Then collect all urine including the 8:00 A.M. specimen at the end of the 24 hour collection period. Refrigerate urine as each specimen is collected. The specimen must be refrigerated or frozen after collection since 17-OHCS are destroyed at room temperature. The patient should be off all drugs with the exception of aspirin and barbiturates. Record the 24 hour volume. 50mL aliquot is required for the assay. Any test requiring a 24 hour urine collection may also be run on this specimen, e.g., 17-keto-steroids.
REFERENCE RANGE: Male: 6-16mg/24 hours; Female: 4-8mg/24 hours; Children: 2-10mg/24 hours; 0-2 years: 2-4mg/24 hours; 2-6 years: 6-10mg/24 hours; 6-10 years: 6-8mg/24 hours; 10-14 years: 8-10mg/24 hours.
METHOD: Three steroids are assayed; these are deoxycortisol (Cpd S), cortisol (Cpd F) and Cpd E; all of these compounds have the dihydroxy acetone side chain configuration in ring structure D. In the color-reaction 17-hydroxy-steroid reacts with phenylhydrazine to yield the phenylhydrazone which absorbs at 410nm.
INTERPRETATION: Urine 17-OHCS are measured in the work-up of suspected disorders of adrenal steroid metabolism. Causes of increased or decreased urinary 17-OHCS are listed in the next Table:

Causes of Increased Urinary 17-OHCS
Cushing's Syndrome
Medical or Surgical Stress
Hyperthyroidism
Obesity (occasionally)
ACTH, Cortisone or Cortisol Therapy
11-Hydroxylase Deficiency

Causes of Decreased Urinary 17-OHCS
Addison's Disease
Pituitary Deficiency of ACTH Secretion
Administration of Potent Synthetic Corticosteroids
21-Hydroxylase Deficiency; 17-Hydroxylase Deficiency
Inanition States, i.e., Anorexia Nervosa
Liver Disease
Hypothyroidism
Newborn Period (Synthesis Normal, Glucuronidation Decreased)

The steroids that are assayed as 17-hydroxycorticosteroids in the pathway for the synthesis of steroids by the adrenal are given in the next Figure:

Steroids Assayed as 17-Hydroxycorticosteroids

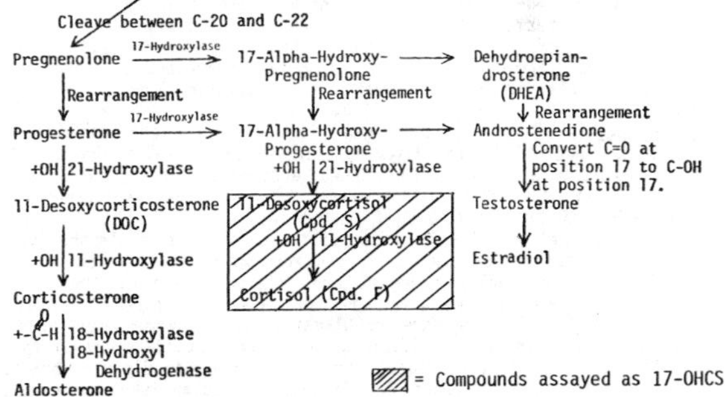

A variety of chromogenic substances can interfere with 17-OHCS measurements. Measurement of 17-OHCS has been supplanted in many instances by measurement of plasma cortisol, urinary free cortisol, or other intermediates of steroid metabolism.

5-HYDROXYINDOLE ACETIC ACID (5-HIAA)

SPECIMEN: Random urine for screening test. Quantitative Test: Collect 24 hour urine in bottle containing 12g of boric acid, acetic acid or 25mL 6N HCl; remove a 100mL aliquot for analysis.
REFERENCE RANGE: Screening test positive if 5-HIAA present in large amount. Quantitative: 1.8-6.0mg/24 hour. Abnormal >10mg/24 hours, diagnostic >30mg/24 hours. To convert conventional units in mg/24 hour to international units in micromol/day, multiply conventional units by 5.230.
METHOD: Spectrophotometric following extraction.
INTERPRETATION: 5-Hydroxyindole acetic acid is measured in the workup of possible carcinoid tumors. Carcinoid tumors (argentaffin cells), especially mid-gut carcinoids, synthesize 5-hydroxytryptamine (serotonin) which is metabolized to 5-HIAA and excreted in the urine. In one study, a single urinary 5-HIAA measurement had a sensitivity of 73% and a specificity of 100% for the diagnosis of carcinoid (Feldman J, O'Doriso TM. Am J Med. 1986; 81:Suppl.6B:41-48). The causes of an elevation of urine 5-HIAA are given in the next Table:

Elevation of Urine 5-HIAA
Carcinoid Tumor
Food Products: Banana; Pineapple; Tomato; Avocado; Plums; Eggplant; Walnut
Whipple's Disease
Celiac Disease
Drugs: Glyceryl Guaiacolate, Acetaminophen, Caffeine, Phenacetin

A more complete list of foods and of drugs which may interfere with 5-HIAA assays (elevation or decrease) is given in a review (Roberts LJ. Endocrin Met Clin N Amer. 1988; 17:415-436).

Carcinoid tumors are found in the respiratory tract and throughout the gastrointestinal tract from the stomach to the rectum. They are also found in the biliary tree, pancreatic duct or gonads. They occur most commonly in the terminal ileum and appendix; when they occur in the appendix, they usually cause appendicitis or are incidental findings at operation or autopsy.

The carcinoid syndrome is usually caused by wide-spread hepatic metastasis of a carcinoid tumor. The hallmark of the carcinoid syndrome is the occurrence of attacks of flushing. These attacks do not occur or are minor in one-third of the patients. Other hallmarks of the carcinoid syndrome are abdominal pain, cramps, diarrhea, weight loss, facial or nasal cyanosis with telangiectasia; hypotension may occur. The right side of the heart and pulmonic valve may become involved.

17-HYDROXYPROGESTERONE (17-OH PROG.), SERUM

SPECIMEN: Serum, red top tube; freeze and keep frozen in dry ice.
REFERENCE RANGE: Varies by method, sex, menstrual history. Male: 0.4-4.0ng/mL; Female: 0.1-3.3ng/mL; Pregnant Women: 2.3-7.6ng/mL; Postmenopausal: 0.3-0.9ng/mL; Children: up to 0.5ng/mL. See below. To convert conventional units in ng/mL to international units in nmol/liter, multiply conventional units by 3.026.
METHOD: RIA
INTERPRETATION: 17-hydroxyprogesterone measurements are utilized in the evaluation of congenital adrenal hyperplasia (adrenogenital syndromes). Serum 17-OH progesterone is elevated in the adrenal enzyme deficiency states, 21- and 11-hydroxylase deficiency; it is decreased in 17-hydroxylase deficiency. See: ADRENAL PANEL, CONGENITAL ADRENAL HYPERPLASIA.

The principal use of 17OH progesterone measurements is in the diagnosis of 21-hydroxylase deficiency. In the severe form with virilization of the female infant levels are usually >10ng/mL and interpretation of the data is not difficult. Diagnosis of the attenuated or late-onset form of homozygous 21-hydroxylase deficiency is more problematic. These women may present with hirsutism and be clinically indistinguishable from Stein-Leventhal (polycystic ovary) syndrome. The true prevalence of this disorder is disputed but estimates vary from 1-12% of hirsute women. The diagnosis is established by measurement of 17OH-progesterone levels following ACTH (cosyntropin) stimulation (30-60 minutes). The expected values in women with the disorder compared to reference groups before and after cosyntropin stimulation are given in the next Table:

Screening for 21-Hydroxylase Deficiency	
Basal	Levels of 17OH-PROG.
Week One Post Menses	1-2 ng/mL
Luteal Phase	<2.5ng/mL
Polycystic Ovary Syndrome	2-10ng/mL(<3.5 usually)
Heterozygous Attenuated 21OH Deficiency	2-10ng/mL
Severe 21OH Deficiency	>10 ng/mL
Cosyntropin Stimulated	
Normal	<3.5ng/mL
Polycystic Ovary Syndrome	<3.5ng/mL
Attenuated 21OH Deficiency	>10 ng/mL
Severe 21OH Deficiency	Not Needed

Modified from: Burch Jr. WM. NC Med J. 1992; 53:521.

HYDROXYPROLINE, URINE

SPECIMEN: Add 20mL of toluene as preservative to urine container. Some laboratories prefer 6N HCl. The patient discards the first voided AM specimen; then the patient collects all voided urine specimens until the same time on the following day, 24 hours later. Mix before obtaining 20mL sample. Indicate patient's age and the 24 hour urine volume on request form. Patient should be on collagen free diet for 24 hours.
REFERENCE RANGE: Adults: 14-45mg/24 hour; children: no normals (Mayo Medical Laboratories). Reference range, two hour urines: Male: 0.4-0.5mg/2 hour; Female: 0.4-2.9mg/2 hour. To convert conventional units in mg/24 hours to international units in mcmol/day, multiply conventional units by 7.626.
METHOD: Amino acid analysis, quantitated by ninhydrin reaction following hydrolysis; HPLC.
INTERPRETATION: Urinary hydroxyproline is a measure of resorption of bone collagen. The total urine hydroxyproline represents hydroxyproline found free and in small hydroxyproline containing peptides. Hydroxyproline is an amino acid found only in collagen; during periods of increased resorption of collagen, especially collagen of bone, hydroxyproline increases in the urine. The causes of increased urinary hydroproline are listed in the next Table:

Causes of Increased Urinary Hydroxyproline
Growth Period
Bone Diseases:
Primary and Secondary Hyperparathyroidism
Paget's Disease
Myeloma
Osteoporosis
Osteomalacia/Rickets
Congenital Hypophosphatasia
Acromegaly
Rheumatoid Arthritis
Metastatic Neoplasms
Marfan's Syndrome
Scleroderma
Hyperthyroidism

A 2 hour urinary hydroxyproline test is sometimes performed in lieu of the 24 hour test. The patient should be on a 24 hour collagen free diet. The patient should fast from 6 P.M. the previous day. Urine is collected from 8 to 10 A.M., provided the patient is adequately hydrated.

A free hydroxyproline assay is offered by some reference laboratories. Usually only 10% of hydroxyproline is found as the free amino acid, the rest is present in small peptides. This measurement reflects the level of free hydroxyproline in blood and is elevated in the exceedingly rare condition, hyper-hydroxyprolinemia. Usually, free hydroxyproline should be undetectable or only in trace amounts in urine (2mg/24 hours).

25-HYDROXYVITAMIN D (see VITAMIN D)

HYPERTENSION PANEL

The causes of hypertension, as reflected by referrals to a community-based hypertension clinic, are given in the next Table (Ferguson RK. Ann Intern Med. 1975; 82:761):

Causes of Hypertension - Patients with Hypertension Referred to a Community-Based Referral Hypertension Clinic

Diagnosis	Percent	Suggestive of Diagnosis
Essential Hypertension	66	
Borderline Hypertension	15	
No Diagnostic Hypertension	10	
Secondary Hypertension		
(1) Oral Contraceptive-Induced	3.3 (11/331 Patients)	History
(2) Renal Hypertension	3.9 (13/331 Patients)	
Renal Arterial Hypertension	2.1 (7/331 Patients)	
Chronic Renal Disease	0.9 (3/331 Patients)	Urinalysis
Polycystic Kidney Disease	0.9 (3/331 Patients)	Urinalysis
(3) Endocrine	0.6 (2/331 Patients)	
Primary Aldosteronism	0.3 (1/331 Patients)	Decreased Potassium, Increased Sodium
17-Hydroxylase Deficiency	0.3 (1/331 Patients)	Decreased Potassium, Increased Sodium

Diagnosis of Chronic Renal Disease: All 3 patients who had chronic renal disease had at least one abnormal urinalysis containing cells, casts or protein.
Diagnosis of Polycystic Kidneys: These 3 patients with polycystic kidneys had microscopic hematuria or casts or both in urinalysis. The diagnosis was confirmed radiologically.
Endocrine Hypertension: Two patients had probable endocrine hypertension, one due to primary aldosteronism and the other was a suspected 17-hydroxylase deficiency. Diagnostic clues were provided by serum potassium which was low to borderline low. Additional support for the diagnosis of primary aldosteronism was obtained by decreased plasma renin and increased plasma aldosterone. Most of the secondary causes of hypertension are suggested from history, urinalysis and electrolyte analysis.

Curable causes of hypertension are given in the next Table:

Curable Forms of Hypertension
Renovascular Disease (The Major Cause of Secondary Hypertension)
Primary Aldosteronism
Pheochromocytoma
Coarctation of the Aorta
Unilateral Renal Parenchymal Disease
Cushing's Syndrome
Adrenogenital Syndromes (17- and 11-Hydroxylase Deficiency)
Oral Contraceptives
Licorice Induced Hypertension
Renin-Secreting Neoplasms

The most common causes of hypertension are essential hypertension and malignant hypertension. As seen from the list in the Table, most of the potentially curable forms of hypertension are secondary to renal or adrenal diseases.

Screening tests for secondary hypertension are given in the next Table:

Screening Tests for Secondary Hypertension

Suspected Condition	Test
Renovascular Hypertension	Intravenous Pyelogram Suppressed Plasma Renin Activity(PRA) or Stimulated Plasma Renin Activity(PRA)
Primary Aldosterism	Serum and Urine Potassium Stimulated Plasma Renin Activity(PRA) PRA/Aldosterone Ratio
Pheochromocytoma	Urinary Catecholamines, Metanephrine, VMA
Cushing's Syndrome	Dexamethasone Suppression Test, Serum Cortisol, Urinary Free Cortisol
Coarctation	Chest X-Ray, BP Upper/Lower Extremities

Tests performed in the laboratory evaluation of patients with hypertension to evaluate possible complications associated with hypertension are given in the next Table (Townsend RR, DiPette DJ. Clin Lab Med. 1993; 13:287-302):

Laboratory Evaluation of Hypertension

Test	Possible Diagnostic Information
Blood:	
Complete Blood Count	Anemia or Polycythemia
Electrolytes	Hypokalemia - Mineralocorticoid Excess
	Alkalemia - Mineralocorticoid Excess
	Acidosis - Renal Insufficiency
Creatinine	Renal Insufficiency
Glucose	Diabetes Mellitus
Calcium	Parathyroid Disease
Cholesterol/Triglycerides	Detects Another Cardiovascular Risk Factor
Uric Acid	Detects Gout; Detects Another Cardiovascular Risk Factor
T4, T3 Uptake, TSH	Thyroid Disease
Urine:	
Urinalysis	Glucose Positive in Diabetes
	Blood or Protein Positive in Renal Disease

IMIPRAMINE (TOFRANIL)(see TRICYCLIC ANTIDEPRESSANTS)

IMMUNE COMPLEX PROFILE

SPECIMEN: Red top tube; separate serum and freeze immediately. Ship frozen on dry ice. See COMPLEMENT, Clq BINDING ASSAY and RAJI CELL ASSAY
REFERENCE RANGE: None detected
METHOD: Clq Binding Assay; Raji Cell Assay; PEG-ELISA Precipitation Assay. See COMPLEMENT Clq BINDING ASSAY and RAJI CELL ASSAY; results are method dependent since different methods detect complexes of different compositions.
INTERPRETATION: Immune complex assays are run on serum from patients with diseases associated with circulating immune complexes. The Raji cell assay detects immune complexes containing C3. Some diseases associated with immune complexes are given in the next Table (Michael M. Frank, NIH, 1980):

Diseases Associated with Immune Complexes

Rheumatologic Diseases:
Systemic Lupus Erythematosus, Rheumatoid Arthritis, Sjögrens Syndrome, Mixed Connective Tissue Disease, Relapsing Polychondritis, Mixed Cryoglobulinemia, Necrotizing Vasculitis

Glomerulonephritis:
Rapidly Progressive Glomerulonephritis, Systemic Lupus Erythematosus, Acute Glomerulonephritis

Infectious Diseases:
Bacterial Endocarditis, Disseminated Gonorrhea, Acute and Chronic Schistosomiasis, Malaria, Leprosy, Viral Hepatitis, Dengue Fever

Neoplastic Diseases:
Hodgkin's Disease, Leukemias, Solid Tumors, Burkitt's Lymphoma, Malignant Melanoma

Miscellaneous:
Primary Biliary Cirrhosis, Chronic Active Hepatitis, Idiopathic Pulmonary Fibrosis

Immune complex formation is illustrated in the next Figure:

Immune Complex Formation

Antigen + Antibody → Antigen-Antibody Complex

Antigen-Antibody Complex + Complement → Antigen-Antibody-Complement

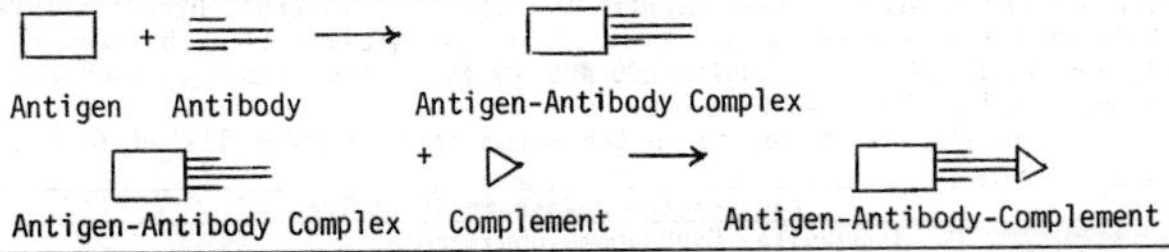

Antigen-antibody soluble complexes may develop in the circulation and vasculitis and glomerulonephritis may occur as a result. Continued elevation on multiple sampling is more significant than an isolated elevation on a single measurement. Rheumatoid factor and cryoglobulins may lead to a false positive result.

IMMUNOGLOBULIN E (IgE)

SPECIMEN: Red top tube, separate serum

REFERENCE RANGE:

Age	Mean ± 2SD IgE (U/mL)	Age	Mean ± 2SD IgE (U/mL)
Newborn	(0.14-2)	8 years	(1.3 -270)
1-11 months	(0.11-56)	9 years	(0.6 -464)
1 year	(0.1 -83)	10 years	(1.9 -421)
2 years	(0.3 -133)	11-14 years	(1.6 -456)
3 years	(1.4 -101)	15-19 years	(1.5 -384)
4 years	(0.4 -144)	20-30 years	(0.86-239)
5 years	(3.00-148)	31-50 years	(1.2 -324)
6 years	(0.44-573)	51-80 years	(0.70-197)
7 years	(0.35-552)		

Reference: Pharmacia Diagnostics

To convert conventional units in U/mL to international units in mcg/liter, multiply conventional units by 2.4.

METHOD: PRIST (paper-radioimmunosorbent test) Pharmacia Fine Chemicals, Piscataway, N.J.; methods include both RIA and EIA (enzyme-immunoassay).

INTERPRETATION: Measurement of IgE levels may be useful in the assessment of allergic disorders. The causes of elevation of serum IgE are given in the next Table (Ali M, et al. Diagnostic Medicine. 1982; CPC. Am J Med. 1983; 24:887-897):

Causes of Elevation of Serum IgE

Parasitic Infestations:
- Ascariasis; Visceral Larva Migrans, (Toxocara);
- Capillariasis; Echinococcosis;
- Hookworm (Necator); Amebiasis

Allergic Disorder:
- Asthma
- Allergic Rhinitis
- Hayfever
- Inhalant Allergy
- Atopic Rhinitis and Sinusitis
- Atopic Dermatitis and Urticaria
- Allergic Bronchopulmonary Aspergillosis(ABPA)
- Hypersensitivity Pneumonitis
- Drug and Food Allergies

Immunological Disorders of Uncertain Pathogenesis:
- Hyper-IgE and Recurrent Pyoderma (Job-Buckley Syndrome)
- Thymic Dysplasias and Deficiencies
- Wiskott-Aldrich Syndrome
- Pemphigoid
- Polyarteritis Nodosa
- Hypereosinophilic Syndrome

Neoplasms
- IgE Myeloma

The most common causes of elevation of serum IgE are parasitic infestations (ascariasis, Toxocara canis, intestinal capillariasis, bilharziasis, hookworm, Echinococcus, trichinosis) and atopic disorders. Atopic allergy implies a familial tendency to manifest alone, or in combination, clinical conditions as bronchial asthma, rhinitis, urticaria and eczematous dermatitis (atopic dermatitis).

IgE is frequently elevated in asthma, hay fever and other allergic disorders and tends to correlate with the degree of allergic hypersensitivity. Levels of IgE associated with high risk of allergic disease: children (1 year old) >60 IU (100% risk); adults 200-400 IU (35% risk)(Asser S, Hamburger RN. Schumpert Med Q. 1986; 4:272-280).

IgE levels are decreased in the serum in conditions listed in the next Table:

Causes of Decrease in Serum IgE
- Congenital Hypogammaglobulinemia
- Acquired Hypogammaglobulinemia
- Sex-linked Hypogamaglobulinemia
- Ataxia-telangiectasia
- IgE Deficiency

IMMUNOGLOBULIN PROFILE (QUANTITATIVE SERUM IMMUNOGLOBULINS G, A, M)

SPECIMEN: Red top tube, separate serum and store at -20°C.
REFERENCE RANGE:

Normal Levels for IgG, IgM and IgA by Age

Age	IgG (mg/dL)	IgM (mg/dL)	IgA (mg/dL)
Newborn	1,031 ± 200	11 ± 5	2 ± 3
1-3 mo	430 ± 119	30 ± 11	21 ± 13
4-6 mo	427 ± 186	43 ± 17	28 ± 18
7-12 mo	661 ± 219	54 ± 23	37 ± 18
13-24 mo	762 ± 209	54 ± 23	50 ± 24
25-36 mo	892 ± 183	61 ± 19	71 ± 37
3-5 yr	929 ± 228	56 ± 18	93 ± 27
6-8 yr	923 ± 256	65 ± 25	124 ± 45
9-11 yr	1,124 ± 235	79 ± 33	131 ± 60
12-16 yr	946 ± 124	59 ± 20	148 ± 63
Adults	1,158 ± 305	99 ± 27	200 ± 61

Ref.: Lou K, Shanbrom E. JAMA. 1967; 200:323.

Reference ranges vary around the world, probably as a result of variable immune stimulation. IgD levels appear to be of little clinical use. IgE is discussed under IMMUNOGLOBULIN E (IgE).

IgG, IgM, IgA: To convert conventional units in mg/dL to international units in g/liter, multiply conventional units by 0.01.

METHOD: Single radial diffusion; turbidity

INTERPRETATION: Immunoglobulin levels are quantitated in the work-up of suspected immunodeficiency states, to monitor replacement therapy in patients with immunodeficiencies, to evaluate humoral immunity, and to document cases of suspected polyclonal and monoclonal immunoglobulinemia. The basic structure and properties of immunoglobulins are shown in the next Figure:

Properties of Immunoglobulins

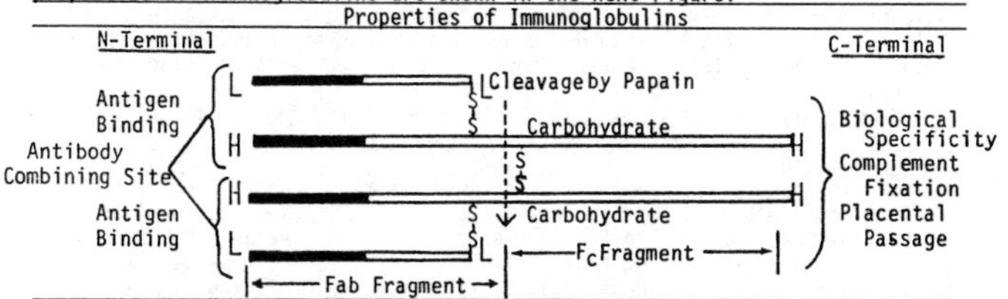

Immunoglobulin molecules consist of heavy (H) and light (L) chains. The classification of immunoglobulins is made on the basis of differences in the amino acids of the constant regions of the heavy (H) chains. Assay for immunoglobulins are useful for determination of specific changes that are associated with diseases and for following the course of therapy. The diseases that are reflected in the alteration of the concentration of these components are shown in the next Table:

Decreased Values of Immunoglobulins

IgG	IgA	IgM
Selective Hypo-IgG	Selective Hypo-IgA	Selective Hypo-IgM
Protein-losing Enteropathies	Protein-losing Enteropathies	Protein-losing Enteropathies
	Hereditary Ataxia Telangiectasia	
Nephrotic Syndrome	Nephrotic Syndrome	

In addition, suppression of normal immunoglobulin production is seen in multiple myeloma. The concentration of the class of immunoglobulins to which the myeloma protein belongs would, of course, be elevated.

A common clinical indication for immunoglobulin assay is in patients, particularly children, who have repeated severe infections. The infections may be related to impaired synthesis of one or more of the immunoglobulins. The most frequent site of infection is the respiratory tract. The age of onset of illness in hypoimmunoglobulinemia depends upon whether the deficiency is congenital or acquired later in life. (cont)

Immunoglobulin Profile (Cont)

The genetics and serum levels of immunoglobulins in various hereditary conditions are given in the next Table:

Genetics and Serum Levels of Immunoglobulins

DISEASE	GENETICS	Serum Levels of Immunoglobulins IgG	IgA	IgM
IgA Deficiency	Unknown	N	↓	N
Hereditary Agammaglobulinemia (Bruton)	Sex-linked	↓ to A	↓ to A	↓ to A
Combined Immunodeficiency	Autosomal Recessive or Sex-linked		Same as above	
Primary Acquired	Unknown	↓	↓	↓
Immunoglobulin M Deficiency	Sex-linked	N	N	↓
Hereditary Ataxia Telangiectasia	Autosomal Recessive	N	Def. in 60%	N
Wiskott-Aldrich Syndrome	Sex-linked	N	↑	N to ↓
Thymic Hypoplasia (DiGeorge)	Unknown	N	N	N

↓ = Decrease; ↑ = Increase; A = Absent

As seen in the above table, there may be a deficiency in one or more of the immunoglobulins in these diseases.

IgA deficiency is the most frequently recognized selective hypogammaglobulinemia; the incidence is estimated at 1 in 500 to 3500 persons. Both serum and secretory IgA are decreased or absent; the secretory piece is produced. Patients commonly have an associated autoimmune disease. Patients with IgA deficiency who receive blood transfusions are at risk for transfusion reactions during subsequent transfusions. Many of these patients develop anti-IgA antibodies.

Gamma globulins may be elevated in the conditions listed in the next Table:

Elevated Levels of Serum Immunoglobulins

IgG	IgA	IgM
Chronic Infections	Infections of Respiratory, GI, Renal Tract	Acute Viral Infections
Autoimmune Diseases	Portal Cirrhosis	Recent Immunization
Multiple Myeloma	Multiple Myeloma	Parasitic Disorders: Trypanosomiasis Toxoplasmosis
	Hepatitis (Late)	Nephrotic Syndrome
	IgA Nephropathy (Berger's)	Waldenström's Macroglobulinemia
	Henoch-Schönlein Purpura	Hepatitis (Early)

In multiple myeloma associated with monoclonal bands, there is suppression of normal immunoglobulin production.

Secretory immunoglobulin A is also present in circulating blood and concentrations are high in patients with carcinomas and chronic infectious diseases; a column enzyme immunoassay has been devised (Yamamoto R, et al. Clin Chem. 1983; 29:151-153).

Waldenström's Macroglobulinemia: Laboratory abnormalities in patients with macroglobulinemia are as follows: anemia (88%), positive Sia test (76%), increased serum viscosity (41%), cryoglobulinemia (37%), Bence-Jones proteinuria (25%), thrombocytopenia (6%), leukocytosis (4%)(Deuel TF. Arch Intern Med. 1983; 143:986-988).

IMMUNOREACTIVE TRYPSIN (see TRYPSIN/TRYPSINOGEN)

INAPPROPRIATE ANTI-DIURETIC HORMONE (see SYNDROME INAPPROPRIATE ADH)

INBORN ERRORS OF METABOLISM PANEL

Evaluation of patients for inborn errors of metabolism includes tests on urine and blood. There are certain clues which raise the suspicion of such disorders:

Urine: Odor, Ketones, Reducing Substances

Blood: Anemia, Leukopenia, Thrombocytopenia, Acidosis, Hypoglycemia, Ketonemia

Laboratory tests for evaluation of patients suspected of having an inborn error of metabolism are given in the next Table (Cederbaum SD in Genetic Disease, Diagnosis and Treatment, Dietz AA ed., Am Assoc Clin Chem, 1725 K. Street, N.W., Washington, DC. pg. 149, 1983):

Laboratory Evaluation of An Inborn Error of Metabolism

Blood/Plasma	Urine
Blood Sugar	Urine Metabolic Screening Tests
Blood Electrolytes, pH, and Bicarbonate	Urine Amino Acids
Blood Lactate and Pyruvate	Urine Organic Acids
Plasma Ketones	Urine Ketones
Plasma Ammonia	
Plasma Amino Acids	
Plasma Organic Acids	

INBORN ERRORS OF METABOLISM SCREEN, URINE

SPECIMEN: Obtain about 20mL of urine specimen, refrigerate

REFERENCE RANGE: Negative

METHOD: See below

INTERPRETATION: Urine is used to screen for many metabolic diseases. Reactions of the screening reagents, $FeCl3$, DNP-hydrazine, cyanide-nitroprusside and Benedict's with urine are given in the next Table:

Reactions of Metabolic Screening Tests in Urine

Disease	FeCl3	DNPH	Nitroprusside	Reducing Subst
Phenylketonuria	Blue-Green	+++	-	-
Maple Syrup Urine	Blue	+++	-	-
Tyrosinosis (emia)	Green (Fades)	+++	-	±
Tyrosyluria	Green	+++	-	-
Histidinemia	Green	++	-	-
Hyperglycemia	-	+++	-	-
Hartnup's Disease	-	-	-	-
Fructose Intolerance	-	-	-	+
Galactosemia	-	-	-	+
Homocystinuria	-	-	+	-
Cystinuria	-	-	+	-
Wilson's Disease	-	-	-	-
Cystinosis	-	-	-	±
Lowe's Syndrome	-	-	-	±
Mucopolysaccharidoses	-	-	-	-
Salicylates	Purple	-	-	-
Phenothiazines	Purple	-	-	-
Acetone		+++		

The sensitivity of these screening tests for the amino acidurias is usually not as great as the quantitative column methods for amino acid analysis.

Ferric Chloride Test: This is positive for several amino acids and other metabolic abnormalities. See FERRIC CHLORIDE TEST.

DNP-Hydrazine Test (2,4-Dinitrophenyl Hydrazine): A yellow precipitate forms in the presence of ketones, aldehydes, and keto acids. A positive test occurs in the following conditions:

Positive 2,4-Dinitrophenyl Hydrazine Test

Condition	Metabolite in Urine
Maple Syrup Urine Disease	Keto Acids
Phenylketonuria (PKU)	Phenylpyruvic Acid
Histidinemia	Imidazole Pyruvic Acid
Oasthouse Syndrome (Methionine Malabsorption)	Alpha Ketopyruvic Acid
Hyperglycinemia	Acetone
Glycogen Storage Diseases Types 1,3,5 and 6	Acetone
Acetonemia (Ketonuria) from any cause	

(cont)

Inborn Error of Metabolism Screen, Urine (Cont)

Cyanide-Nitroprusside Test: This is a test for cystine and homocystine (cystine is reduced by cyanide to cysteine and the sulhydryl groups then react with nitroprusside to produce a red-purple color).
Urine Reducing Substances (Benedict's Test): (cupric sulfate (blue) ⟶ cuprous oxide (red)) for sugars and other reducing substances. The reactions with Benedict's test of substances found in the urine are given in the next Table:

Positive Benedict's Test on Urine
Glucose
Other Sugars: Fructose, Galactose, Lactose, Maltose, Pentose, Sucrose
Urine Constituents: Homogentisic Acid (Alkaptonuria)
Creatinine, Uric Acid, and many other reducing agents may cause false positives

The glucose oxidase test may be used to confirm or exclude glucose as the reason for a positive Benedict's test. False positive urine glucose oxidase tests may result from urine contamination by H_2O_2 or hypochlorite. False negative tests may result from large amounts of ascorbic acid (vitamin C). See REDUCING SUBSTANCES.

INDIRECT COOMBS (see COOMBS, INDIRECT)

INFANT, IRRITABLE, PANEL

Infants presenting with acute, unexplained, excessive crying may have a wide variety of diagnoses; final diagnoses for 56 infants is given in the following Table (Poole SR. Pediatr. 1991; 88:450-455):

Diagnoses in 56 Infants with Unexplained Excessive Crying*
Idiopathic (10)
Colic (6)
Infections: Otitis Media (10); Viral Illness (2); Urinary Tract Infection (1); Prodrome of Gastroenteritis (1); Herpangina (1); Herpes Stomatitis (1)
Trauma: Corneal Abrasion (3); Foreign Body in Eye (1); Foreign Body in Oropharynx (1); Tibial Fracture (1); Clavicular Fracture (1); Brown Recluse Spider Bite (1); Hair Tourniquet Syndrome - Toe (1)
Gastrointestinal: Constipation (3); Intussusception (1); Gastroesophageal Reflux with Esophagitis (1)
Central Nervous System: Subdural Hematoma (1); Encephalitis (1); Pseudotumor Cerebri (1)
Drug Reaction/Overdose: DTP Reaction (1); Inadvertent Pseudoephedrine Overdose (1)
Behavioral: Night Terrors (1); Overstimulation (1)
Cardiovascular: Supraventricular Tachycardia (2)
Metabolic: Glutaric Aciduria Type I (1)

*Number of patients with a specific diagnosis are given in parentheses.

In this group of patients, the diagnosis was made by physical examination (41%), follow-up (39%), skeletal x-ray or computed tomography (11%), laboratory tests (5%), or electrocardiography (4%). Studies that may be helpful in the evaluation of infants with irritability are given in the following Table:

Laboratory Studies for Irritable Infants
Complete Blood Count(CBC)
Urinalysis
Chemistries(Serum):
Glucose
Electrolytes (Na^+, K, Cl^-, CO_2 Content)
BUN or Creatinine
Calcium
Specific Drug Assays if Indicated
Lumbar Puncture
Amino and Organic Acid Studies
Radiologic Studies:
Skeletal Roentgenography
Computed Tomography of the Head
Barium Enema
Esophagram
Abdominal Films
Electrocardiography
Fluorescein Stain in the Eyes (Corneal Abrasion)

See Review: Poole SR. "The Infant with Acute, Unexplained, Excessive Crying." Pediatr. 1991; 88:450-455.

INFECTIOUS DISEASE PANELS

The frequency of infectious diseases in ambulatory patients is shown in the next Table (McHenry MC, Weinstein AJ. Med Clin of N Amer. 1983; 67:3-16):

Frequency of Infectious Diseases in Ambulatory Patients

Ambulatory Patients	Frequency (Percent)
Family Practice	78
Pediatric Practice	73
Emergency Room	28

Infectious diseases are the most common illnesses in ambulatory patients, accounting for 87 percent of illnesses (Dingle JH, et al. Illness in the Home, A Study of a Group of Cleveland Families, Cleveland, Ohio, Case Western University Press. 1964; 19-32); 78 percent of illnesses in family practice and 28 percent of visits to the emergency room of a large general hospital (Moffet HL. Ann Intern Med. 1978; 89:264-277).

The frequency of various infections in ambulatory patients is shown in the next Table (Moffet HL, already cited):

Frequency of Various Infections in Ambulatory Patients

Infections	Frequency (Percent)
Upper and Lower Respiratory Infections and Pararespiratory Infections, such as, Pharyngitis, Otitis Media and Sinusitis	70
Gastrointestinal Infections	8
Sexually Transmitted Diseases	5
Dermatologic Infections	5
Urinary Tract Infections	4
Miscellaneous Infections	8

Examples of infections amenable to antimicrobial chemotherapy in ambulatory patients are given in the next Table (McHenry MC, Weinstein AJ. Already cited):

Examples of Infections Amenable to Antimicrobial Chemotherapy in Ambulatory Patients

Infection	Usual Causative Organism(s)
Streptococcal Pharyngitis and Tonsillitis	Group A Streptococcus
Acute Exacerbation of Chronic Bronchitis in Patients with Chronic Obstructive Pulmonary Disease (COPD)	Streptococcus Pneumoniae, Hemophilus Influenzae
Uncomplicated Community-Acquired Pneumonia	Mycoplasma Pneumoniae, Streptococcus Pneumoniae
Dental Abscess or Cellulitis	Penicillin-Susceptible Anaerobes
Otitis Media (Patients Over 2 Months of Age)	Streptococcus Pneumoniae, Hemophilus Influenzae Moraxella Catarrhalis
Sinusitis, Acute	Streptococcus Pneumoniae, Hemophilus Influenzae Moraxella Catarrhalis
Sinusitis, Chronic	Penicillin-Susceptible Anaerobes
Urethritis	Chlamydia Trachomatis, Neisseria Gonorrhoeae, Ureaplasma Urealyticum
Salpingitis (PID)	Neisseria Gonorrhoeae

PHARYNX: Bacteria associated with pharyngitis are listed in the next Table:

Bacteria Associated with Pharyngitis
Group A Streptococci (Streptococcus Pyogenes)
Neisseria Gonorrhoeae
Corynebacterium Diphtheriae

Infectious Diseases Panel (Cont)

The normal flora (bacteria) of the oropharynx are listed in the next Table:

Normal Flora of the Oropharynx
Alpha-Hemolytic Streptococci
Non-Hemolytic Streptococci
Streptococcus Pneumoniae
Neisseria Species
Staphylococcus Epidermidis
Diphtheroids

Throat cultures most often reveal acute streptococcal infections as the cause of acute pharyngitis and tonsillitis. Most acute streptococcal diseases result from Group A organisms; sequelae of Group A streptococcal infections include acute rheumatic fever and acute glomerulonephritis. Group D enterococci may cause a variety of diseases and group B may cause puerperal and neonatal infections. The differentiation of Group A from non-group A beta hemolytic streptococci is of clinical importance primarily in terms of the preventing of rheumatic fever and acute glomerulonephritis.

Methods of diagnosis of agents that cause exudative pharyngitis are given in the next Table (Levy ML, et al. Med Clin of N Amer. 1983; 67:153-171):

Methods of Diagnosis of Agents that Cause Exudative Pharyngitis

Organism or Condition	Method of Diagnosis: Readily Available	Method of Diagnosis: Generally Available
Group A beta hemolytic streptococcus	Throat Culture Rapid Antigen Testing	Serologic tests: ASO, AHT*
Epstein-Barr virus	Complete blood cell count with differential Heterophile	Specific Epstein-Barr antibodies titer
C. diphtheriae	Culture	Toxin assay
F. tularensis	Serologic test (agglutination)	Culture on appropriate media**
Vincent's angina	Gram's stain of mouth lesion	
Viruses (adenovirus, coxsackie-viruses, herpes simplex)	Usually none	Viral culture with confirmatory serologic test

*ASO = antistreptolysin-O; AHT = antihemolysin
**Poses risk of creating infectious aerosol in the laboratory.

PNEUMONIA: Causes of overwhelming pneumonia are given in the next Table (Bradsher RW. Med Clin of N Amer. 1983; 67:1233-1250):

Causes of Overwhelming Pneumonia

	Normal Host	Abnormal Host
Usual Organisms	Pneumococcus Mycoplasma Pneumoniae Hemophilus Influenzae Influenza Viruses Fungi	Pneumococcus Gram-negative Bacilli Anaerobic Bacteria Staphylococcus Aureus Fungi
Unusual Organisms	Legionella Chlamydia Franciscella Tularensis Yersinia Pestis Coxiella Burnetii Meningococcus Group A Streptococcus Actinomyces Israelii Mycobacterium Hantavirus	Aspergillus Mucor, Absidia, Rhizopus Candida Nocardia Varicella-Zoster Virus Cytomegalovirus Pneumocystis Carinii Strongyloides Stercoralis Noninfectious

INFECTIOUS DIARRHEA: Clues that are associated with specific infectious agents are given in the next Table (Satterwhite TK, DuPont HL. Med Clin of N Amer. 1983; 67:203-220):

Infectious Diarrhea - Clues Associated with Specific Infectious Agents

Clue	Suspect Agent
Recent administration of antibiotics	Clostridium Difficile
Contact with children attending day care center	Shigella, Campylobacter, Rotavirus or Giardia
Foodborne illness without fever	Staphylococcus Aureus, Clostridium Perfringens, Bacillus Cereus, or Enterotoxigenic Escherichia Coli
Foodborne illness with fever	Salmonella, Shigella, Vibrio Parahaemolyticus, Campylobacter, or Yersinia
Specific food item incriminated	
poultry	Salmonella
eggs	Salmonella or Campylobacter
seafood	V. parahaemolyticus
raw milk	Salmonella, Campylobacter, or Yersinia
fried rice or bean sprouts	B. cereus
Travel in a developing region	Enterotoxigenic E. Coli in one half of cases; other agents are Shigella (15%), Salmonella (10%), Giardia (4%), and Campylobacter (3%) (Hoffman T. Hosp Pract. 1984; 111-112)
Homosexual activity	Gonococcal and Chlamydial Proctitis, Amoeba, Shigella, or Campylobacter

Laboratory identification of pathogens is given in the next Table (Satterwhite TK, DuPont HL. Med Clin of N Am. 1983; 67:203-220):

Laboratory Identification of Pathogens Involved in Infectious Diarrhea

Agents detected by routine laboratory testing:
- Salmonella
- Shigella
- Campylobacter
- Entamoeba
- Giardia

Agents that can be detected by most laboratories by special request:
- Vibro parahaemolyticus
- Rotavirus
- Yersinia

Agents that are identified only by special laboratories:
- Enterotoxigenic Escherichia coli
- Invasive Escherichia coli
- Clostridium perfringens
- Clostridium difficile
- Norwalk agent

Agents that require quantitative food microbiology or toxin assay:
- Staphylococcus aureus
- Clostridium perfringens
- Clostridium difficile
- Bacillus cereus

Ref.: Wolf JL. Patient Care. May 15, 1983; 79-125.

Infectious Diseases Panel (Cont)

VAGINITIS: The most frequent causes and methods of diagnosis of vaginitis are shown in the next Table:

Causes and Methods of Diagnosis of Vaginitis

Causes	Methods of Diagnosis
Candida Albicans ("Yeast Vaginitis")	Potassium Hydroxide (Preferred Method) Gram Smears Saline Wet Mounts Routine Culture for Fungi Unnecessary
Trichomonas Vaginalis	Saline Wet Mount Culture (Sensitive but Expensive)
Gardnerella Vaginalis (Hemophilus vaginalis)	"Clue Cells" on Methylene Blue or Gram Stain; "Clue Cells" are Epithelial Cells whose surface is covered with Coccobacilli

Vaginitis is commonly due to yeast and Trichomonas; herpes can also cause symptoms. "Non-specific" vaginitis is related to the bacterium Gardnerella vaginitis (Amsel R, et al. Am J Med. 1983; 74:14).

CERVICITIS: The most frequent causes of cervicitis are N. Gonorrhoeae; Chlamydia Trachomatis; Herpes Simplex.

SEXUALLY TRANSMITTED DISEASES: These are syphilis, gonorrhea, lymphogranuloma venereum, granuloma inguinale, chancroid and chlamydia trachomatis. Sexual promiscuity is a risk factor in the transmission of herpes, hepatitis B, and HIV infection.

The sequelae of sexually transmitted diseases are given in the next Table (Rudbach JA. Infectious Disease Forum, Abbott Laboratories, Dec. 1983):

Sexually Transmitted Disease

Disease	Sequelae
Gonorrhea	Females: Pelvic Inflammatory Disease (PID) including Fallopian Tube Damage; Pelvic Adhesions; Ectopic(Tubal) Pregnancy; Sterility; Disseminated Infection Males: Urethritis, Stricture, Epididymitis which can lead to Sterility; Disseminated Infection
Syphilis	Brain and Heart Damage Birth Defects
Chlamydia Trachomatis	Pelvic Inflammatory Disease (PID) Epididymitis
Genital Herpes	Transmission to the Newborn; may cause death or severe retardation
HIV	Bacterial, viral, and fungal infections; eventually leads to death

Chlamydia trachomatis is one of the most common sexually transmitted pathogens.

PERINATAL INFECTIONS (Bolande R. Personal Communication): The causes of perinatal infections are given in the next Table:

Perinatal Infections

Transplacental Infections of Fetus
- Congenital Syphilis
- Toxoplasmosis
- Rubella Syndrome
- Cytomegalovirus Infection
- Coxsackie B. Encephalo-myocarditis

Amniotic Infection Syndrome
- Coliform organisms
- Group B Streptococci
- Listeriosis
- Mycoplasma

Intrapartum Infections
- Herpes Simplex
- Chlamydia

INSULIN ANTIBODIES

SPECIMEN: Red top tube, separate serum
REFERENCE RANGE: Negative
METHOD: RIA
INTERPRETATION: Anti-insulin antibodies are measured in the diagnosis of exogenous (factious) insulin administration, or in conjunction with insulin assays. Almost all patients (90%) receiving exogenous (porcine or bovine) insulin develop serum antibodies after several months. The concentration of antibodies is usually moderate. The similarity in structure to human insulin is responsible for the usual low antigenicity of the commercial preparations. There are two types of antibodies, high affinity and low affinity. There is no apparent clinical significance except in the special case of the development of insulin resistance.

The presence of insulin antibodies make it impossible to obtain accurate values for serum insulin; in the presence of insulin antibodies, C-peptide is the test of choice to reflect insulin secretion. Approximately 50% of the patients who regularly inject themselves with recombinant human insulin will develop detectible antibodies (Saudek CD. JAMA. 1990; 264:2791-2794).

Some individuals with immunological disorders (SLE, rheumatoid arthritis, Grave's disease, multiple myeloma, and benign monoclonal gammopathy) will develop hypoglycemia associated with anti-insulin antibodies. This "autoimmune insulin syndrome" is characterized by postprandial hyperglycemia and fasting hypoglycemia (Polonsky K. N Engl J Med. 1992; 326:1020-1021). In most of these cases there is an artifactual increase in measured insulin concentration and normal or decreased C-peptide levels.

INSULIN WITH GLUCOSE ASSAY

SPECIMEN: For fasting insulin levels patient should be fasting 7 hours. Red top tube, separate serum and freeze immediately; hemolysis destroys insulin; heparin gives spuriously high values; if the patient has been treated with pork or beef insulin, antibodies usually develop and cause invalid test results.
REFERENCE RANGE: Values may vary between laboratories and assays. Cord Blood <2-84microunits/mL; Infants and Prepubertal Children <2-13microunits/mL (fasted); Pubertal Children and Adults <2-17microunits/mL (fasted). Mean fasting insulin values are lower in infants and children than in adults. To convert conventional units in microunits/mL to international units in pmol/liter, multiply conventional units by 7.175.
METHOD: RIA using antibodies derived from guinea pigs.
INTERPRETATION: This test is used in the diagnosis of insulin producing neoplasms. The laboratory features of insulinoma are as follows:

(a) Hypoglycemia (glucose less than 60mg/dL) with simultaneous hyperinsulinemia (serum insulin >6 microunits/mL).
(b) Absence of insulin antibodies: Absence of insulin antibodies helps to rule out surreptitious self-administration of insulin.
(c) "C"-peptide level; both insulin and "C" peptide are secreted in equimolar amounts in insulinoma.
(d) If proinsulin measurements are available: Insulinomas are characterized by decreased proinsulin to insulin conversion. There are increased proinsulin levels and an increased proinsulin: insulin ratio.

Insulinoma: Insulinoma is a rare tumor. Eighty percent of patients have a single benign tumor, usually less than 2 cm in diameter, located about equally in head, body or tail of the pancreas; about 10 percent have multiple adenomas often associated with multiple endocrine neoplasia, Type I syndrome. The remaining 10 percent of patients have metastatic malignant insulinoma.
Whipple's Triad: (1) Symptoms precipitated by fasting or exercise; (2) Association of symptoms of hypoglycemia with a low circulating glucose concentration; and (3) Relief of symptoms after administration of glucose.

Insulinomas usually present with fasting hypoglycemia. Occasionally they may present with postprandial hypoglycemia (Sandek CD. JAMA. 1990; 264:2791-2794). Isolated measurements of insulin levels and "C" peptide may not be elevated. If the ratio of glucose (mg/dL) to immunoreactive insulin (microunits/mL) is less than 3:1 during hypoglycemia, the reason is probably insulinoma (Sandek, already cited). Difficult diagnostic cases may have to undergo a supervised 72 hour fasting study. Over 98% of cases may be diagnosed in this manner (Service FJ, et al. Mayo Clin Proc. 1976; 51:417-429). (cont)

Insulin With Glucose Assay (Cont)

Tests for Malignant Insulinoma: In a relatively high percent of patients with malignant insulinoma, there are increased plasma levels of alpha-HCG (57%), beta-HCG (21%) or immunoreactive HCG (25%). Plasma levels of HCG were not elevated in any of 41 benign insulinomas (Kahn et al. N Engl J Med. 1977; 297:565-569).

Conditions in which plasma insulin secretion is increased or decreased in response to a glucose load are given in the next Table:

Conditions Associated with Increase or Decrease in Plasma Insulin
Decrease in Plasma Insulin:
Type I Diabetes - IDDM
Post Pancreatectomy
Increase in Plasma Insulin:
Type II Diabetes - NIDDM
Hormonogenic Diabetes
Malaria (Quinine Therapy)

The inadequate production and/or secretion of insulin is the primary cause of insulin-dependent diabetes (Type I). In insulin-dependent diabetes, there tends to be a decrease in islet cells; this could cause decrease of plasma insulin. Patients often have anti-islet cell antibodies. In non-insulin-dependent diabetes (Type II), there may be increased plasma insulin; it may be that in this disease, the action of insulin is blocked. Insulin is increased and glucose is decreased in about 8 percent of patients with falciparum malaria; quinine may induce insulin secretion (White NJ, et al. N Engl J Med. 1983; 309:61-66). A high fasting and low 2-hour insulin concentration following glucose load is said to be predictive of conversion to NIDDM from impaired glucose tolerance (Charles ME, et al. Diabetes. 1991; 40:796-799).

IRON, SERUM

SPECIMEN: Red top tube; separate serum from cells as soon as possible. The serum specimen is stable up to 4 days at room temperature and one week in the refrigerator at 4°C. Interferences: Gross hemolysis. The anticoagulants citrate, EDTA and fluoride-oxalate cause significant depression of serum iron values.

REFERENCE RANGE: Changes of serum iron with age are given in the next Table:

Changes of Serum Iron with Age

Age	Total Serum Iron (mcg/dL)
Newborn	100-250
Infant	40-100
Child	50-120
Adult:	
Male	50-160
Female	40-150

Iron-Binding Capacity: Infant: 100-400mcg/dL; Thereafter: 255-450mcg/dL.

To convert conventional units in mcg/dL to international units in mcmol/liter, multiply conventional units by 0.1791.

METHOD: Iron (Fe^{+3}) is liberated from transferrin at pH 4.2; then reduced. The iron is then reacted with a reagent to form a colored complex.

INTERPRETATION: Serum iron is usually measured in conjunction with the iron-binding capacity (total iron-binding capacity; TIBC) in the work-up of anemias, iron overload, hemochromatosis, and thalassemia. The causes of a decrease in serum iron and the laboratory findings are given in the next Table (Cook JD. Seminars in Hematology. 1982; 19:6-18):

Causes and Laboratory Findings in Conditions Associated with Decreased Serum Iron

Condition	Serum Iron	Iron Binding Capacity	Serum Ferritin
Iron Deficiency Anemia	Decreased	Increased	Decreased
Nutritional			
Chronic Blood Loss			
Achlorhydria, Gastrectomy			
Small Bowel Disease			
Increased Demand			
Anemia of Chronic Disease	Decreased	Low or Normal	Normal or Increased
Chronic Renal Failure	Decreased	Low or Normal	<50mcg/liter

Nutritional Iron Deficiency: This is by far the most common cause of anemia in infants, children and premenopausal women. Iron deficiency anemia occurs in infants (milk anemia) who are fed only milk for the first six to nine months of life. Iron supplemented formula usually prevents this problem. Iron deficiency in childhood and infancy has been reviewed: Oski FA. N Engl J Med. 1993; 329:190-193. It is associated with anemia: Hgb <11g/dL and iron deficiency: transferrin saturation <10%. Other laboratory findings are tabulated in Oski, already cited.
Chronic Blood Loss: Chronic blood loss may occur from gastrointestinal bleeding lesions of any type. Excessive blood loss may occur during menstrual periods.
Achlorhydria: In the stomach, acid causes the conversion of ferric iron to the absorbable ferrous iron. In the absence of acid, achlorhydria, or gastrectomy, there is an inadequate amount of ferrous iron available for absorption.
Defective Absorption: Iron is absorbed in the duodenum and jejunum; conditions that alter the mucosa, such as sprue or steatorrhea, may cause a decrease in iron absorption.
Increased Demand: During pregnancy, the iron requirements increase significantly for two basic reasons; increased iron requirements are necessary to meet the needs of the fetus and for the enlarging red cell mass.
Anemia of Chronic Disease (ACD): ACD is the most common cause of anemia in a hospital population. This is thought to be the result of IL-1 production in inflammatory states. Serum ferritin may be used to reliably distinguish iron deficiency anemia from ACD; in iron deficiency anemia, serum ferritin is low; in ACD, serum ferritin is normal or elevated.
Chronic Renal Disease: In patients on chronic hemodialysis, the typical picture is anemia of chronic renal failure and superimposed iron deficiency from blood loss associated with maintenance dialysis. A serum ferritin below 50mcg/liter is highly suggestive if not diagnostic of iron deficiency in patients with chronic renal failure; serum ferritin measurements should be performed at 4-6 week intervals from the onset of maintenance hemodialysis (Cook JD. Seminars in Hematology. 1982; 19:6-16).

The causes of an increase in serum iron are given in the next Table (Halliday JW, Powell LW. Seminars in Hematology. 1982; 19:42-53):

Increase in Serum Iron
Idiopathic (primary hereditary) Hemochromatosis
Secondary Hemochromatosis
(1) Anemia and Ineffective Erythropoiesis
(a) Thalassemia Major
(b) Sideroblastic Anemia
(c) Hemolytic Anemias
(2) Liver Disease
(a) Alcoholic Cirrhosis
(b) After Portacaval Anastomosis
(3) Excessively High Oral Intake

The most common causes of hemochromatosis are idiopathic hemochromatosis and hemochromatosis secondary to ineffective erythropoiesis (e.g., thalassemia and sideroblastic anemias). The laboratory tests that are done in idiopathic hemochromatosis and secondary hemochromatosis are shown in the next Table:

Typical Test Results in Primary or Secondary Hemochromatosis

Test	Hemochromatosis
Serum Iron	>175mcg/dL
Saturation of Transferrin	>50% female, >60% male
Serum Ferritin	>250ng/mL

Idiopathic Hemochromatosis: Liver biopsy is the definitive test for confirmation of increased iron stores. Since this disease is genetically determined (autosomal recessive inheritance; full expression in homozygote), it is possible to identify within families homozygous and heterozygous relatives by comparison of their HLA antigens with those of the affected who is presumed homozygote (Bassett ML, et al. Hepatology. 1981; 1:120-126). The preferred test to screen for hemochromatosis is the % saturation of iron binding capacity.
Secondary Hemochromatosis: In hemochromatosis secondary to anemia, the diagnosis of the underlying anemia is usually apparent after routine hematologic studies. The diagnosis of the iron overload is made in the same manner as for primary hemochromatosis. In iron loading anemias, there are two groups: (a) hypoplastic bone marrow (e.g., aplastic anemia) whose major source of excess iron is blood transfusion and (b) patients with hyperplastic bone marrow but ineffective erythropoiesis. In this latter group, the excess iron results from increased iron absorption secondary to the ineffective erythropoiesis and to blood transfusions.

Iron (Cont)

Alcoholic Liver Disease: (Halliday JW, Powell LW. Seminars in Hematology. 1982; 19:42-53).

Oral Intake: Iron is an important cause of accidental poisoning death in children. The average lethal dose is about 200 to 250mg/kg body weight. Iron causes a necrotizing gastroenteritis with bleeding and accumulation of blood in the lumen and bloody diarrhea. After 16-24 hours, the patient may develop metabolic acidosis. Specific treatment is with the chelating agent for iron, deferoxamine (Desferal).

IRON-BINDING CAPACITY, TOTAL (% SATURATION)

Synonym: Total iron binding capacity; TIBC; (% Transferrin Saturation)

SPECIMEN: Red top tube; separate serum from cells as soon as possible. The serum specimen is stable up to 4 days at room temperature and one week in the refrigerator at 4°C. Interferences: gross hemolysis. The anticoagulants, citrate, EDTA and fluoride-oxalate cause significant depression of iron-binding capacity.

REFERENCE RANGE: Iron-Binding Capacity: Infant: 100-400mcg/dL; Thereafter: 255-450mcg/dL. To convert conventional units in mcg/dL to international units in mcmol/liter, multiply conventional units by 0.1791. % Saturation: 20-50%.

METHOD: Add ferric ions to serum; absorb excess iron with magnesium carbonate; assay for iron.

INTERPRETATION: Iron-binding capacity is usually run in conjunction with serum iron levels in the evaluation and diagnosis of anemia, especially iron deficiency anemia, anemia of chronic disease, and thalassemia. It may also be of use in the diagnosis of hemochromatosis and iron toxicity.

Decreased Levels: The causes of decreased serum total iron-binding capacity are given in the next Table:

Decreased Serum Total Iron-Binding Capacity(TIBC)
Anemia of Chronic Disorders
Hemochromatosis
Sideroblastic Anemia
Protein Deficiency

Increased Levels: The causes of increased serum total iron-binding capacity (TIBC) are given in the next Table:

Increased Serum Total Iron-Binding Capacity (TIBC)
Iron Deficiency Anemia
Pregnancy
Oral Contraceptives

Percent Saturation: The percent saturation is calculated from the serum iron (mcg/dL) and the total iron-binding capacity (TIBC) (mcg/dL) as follows:

$$\frac{\text{Serum Iron (mcg/dL)}}{\text{Total Iron-Binding Capacity (mcg/dL)}} \times 100 = \text{\% Saturation}$$

The normal value for percent saturation is 33%. The alterations in iron metabolism in several disorders are given in the next Table (Cartwright GE. Diagnostic Laboratory Hematology, Grune and Stratton, New York, 4th Ed., 1968; 215):

Alterations in Iron Metabolism in Several Disorders

Disorder	Serum Iron (mcg/dL)	Total Iron Binding Capacity (mcg/dL)	Sat.
Iron Deficiency	D	I	D
Pregnancy	D	I	D
Anemia of Chronic Disorders	D	D	D
Hemochromatosis	I	D	I
Sideroblastic Anemia	I	D	I
Protein Deficiency	D	D	N

In iron deficiency % saturation is typically <15% (low serum iron and increased TIBC). In anemia of chronic disease the % saturation may also be <15% and be confused with iron deficiency. In such instances a serum ferritin may be helpful. In iron deficiency the serum ferritin will be low, while in chronic disease it may be normal or even elevated. Serum transferrin levels may be calculated from serum total iron-binding capacity(TIBC) utilizing the conversion formula:

$$\text{Serum Transferrin} = 0.8 \times \text{TIBC} - 43$$

In hemochromatosis, the transferrin or iron-binding capacity are typically decreased, but the % saturation is increased.

Iron-Binding Capacity and Screening for Hemochromatosis:

Hereditary (idiopathic) hemochromatosis is a common disorder and it is estimated that perhaps 0.5-1% of the white population is affected. However, clinically recognized disease is about one tenth of this (reviewed in: Rovault TA. "Hereditary Hemochromatosis." JAMA. 1993; 269:3152-3154 and Edwards CQ, Kushner JP. N Engl J Med. 1993; 238:1616-1620). There is tight linkage to HLA-A3. The most common presentations include: weakness, abdominal pain, diabetes mellitus, arthralgias, decreased libido, amenorrhea, dyspnea.

The single best screening test for hemochromatosis is serum transferrin saturation (% TIBC saturation). Suggested screening cutoffs are >60% in males and >50% in females. If levels are borderline (40-60%), it is suggested the test be repeated under fasting conditions. The serum ferritin has been used in some screening studies. A normal ferritin level does not exclude disease, it means that there is not yet substantial hepatic iron overload. Serum ferritin can be used to gauge the optimal frequency of phlebotomy. A suggested protocol for hemochromatosis screening and treatment is given in the following Table (Edwards CQ, Kushner JP. N Engl J Med. 1993; 328:1616-1620):

Protocol for Hemochromatosis Screening and Treatment

Measure Transferrin Saturation (Saturation of TIBC)
If >60% in men or >50% in women

↓

Measure Serum Ferritin

Elevated for Age and Sex	Normal
↓	↓
Liver Biopsy	Measure transferrin saturation and ferritin every 2 years
↓	Initiate phlebotomy if ferritin is elevated
Initiate Phlebotomy - Initially, rapid sequence	OR
Measure ferritin yearly	Initiate phlebotomy(3-4 units/year)
Adjust frequency of phlebotomy to maintain normal ferritin level	Measure ferritin every 2 years
	Adjust frequency of phlebotomy to maintain normal ferritin level.

Laboratory testing for hemochromatosis may be difficult in patients with chronic hepatitis as discussed in DiBisceglie AM, et al. Gastroenterology. 1992; 102:2108-2113. In patients who have neither cirrhosis or diabetes, iron depletion results in a normal life expectancy (Niederan C, et al. N Engl J Med. 1985; 313:1256-1262).

Iron overload is sub-Sahara Africa is thought to be due to interaction between increased dietary intake and a gene distinct from any HLA-linked gene (Gordeuk V, et al. N Engl J Med. 1992; 326:95-100). The nature of this postulated hereditary defect is still to be determined.

Increases in % saturation may also be seen in cases of iron overload including secondary hemochromatosis, extensive transfusions, and hemolytic anemia. % saturation is also increased in cirrhosis and thalassemia and any other anemia associated with ineffective erythropoiesis.

ISOPROPYL ALCOHOL

SPECIMEN: Red top tube
REFERENCE RANGE: Negative. To convert conventional units in mg/dL to international units in mmol/liter, multiply conventional units by 0.1664.
METHOD: Gas-liquid chromatography
INTERPRETATION: Isopropyl alcohol is measured in patients suspected of ingestion, toxic inhalation, or excessive dermal absorption. Rubbing alcohol contains 70 percent isopropyl alcohol (isopropanol) and is sometimes ingested by accident or in desperation by alcoholics. Individuals sometimes mix ethanol and isopropanol so a positive blood alcohol does not exclude exposure to other intoxicants. In early toxicity, an isopropyl alcohol value may be elevated. In late toxicity, only acetone may be elevated. In some cases, comparison of measured serum osmolality with calculated serum osmolality may provide evidence for the presence of an unsuspected alcohol intoxication (isopropanol, methanol, ethylene glycol). In the case of isopropyl alcohol, the osmolality will be increased, acetone increased, and no metabolic acidosis present. The expected osmolality may be calculated from the formula:

$$\text{Osmolality} = 2 \times \text{Na} + \frac{\text{glucose(mg/dL)}}{18} + \frac{\text{BUN(mg/dL)}}{2.8}$$

The clinical signs and symptoms of isopropyl alcohol ingestion are hypothermia, deep coma, areflexia, and hypotension. The lethal dose of isopropyl alcohol is about 8 ounces or 240 mL. Isopropanol is metabolized to acetone; blood and urine acetone are very high. A patient survived with an isopropanol blood level 440mg/dL (King LH, et al. JAMA. 1970; 211:1855), and another patient survived with a blood acetone level of 1,878mg/dL (Dua SL. JAMA. 1974; 230:35); these patients were treated with hemodialysis and peritoneal dialysis respectively.

JOINT FLUID ANALYSIS (see ARTHRITIC JOINT PANEL)

17-KETOGENIC STEROIDS (17-KGS) URINE

SPECIMEN: 24 hour urine. Place 10gm boric acid in container prior to collection. Instruct the patient to void at 8:00 A.M. and discard the specimen. Then collect all urine including the 8:00 A.M. specimen at the end of the 24 hour collection period. Refrigerate urine as each specimen is collected. The specimen must be refrigerated or frozen after collection since 17-OHCS (included in the 17-KGS) are destroyed at room temperature. The patient should be off all drugs with the exception of aspirin and barbiturates. Record the 24 hour volume. 50mL aliquot is required for the assay. Any test requiring a 24 hour urine collection may also be run on this specimen, e.g., 17-keto-steroids.
REFERENCE RANGE: Female: 3-15mg/24 hours; Male: 5-22mg/24 hours.
METHOD: The 17-KGS include the 17-OHCS (Cpd S and Cpd F) plus 17-OH precursors of these compounds. The 17-KGS are steroids which can be oxidized to 17-ketosteroids. The 17-KGS value is the difference between 17-ketosteroids before and after oxidation. The assay for 17-KGS entails a series of reduction and oxidation steps whereby the 17-KGS are reacted with bismuthate or periodate to yield 17-keto compounds. They then undergo the Zimmerman reaction to give a colored reaction.
INTERPRETATION: The 17-KGS may be useful in the work-up of a patient with congenital adrenal hyperplasia due to a 21-hydroxylase or 11-hydroxylase deficiency. In such situations the 17-KGS are elevated. 17-OHCS (Cpd S and Cpd F) plus 17-OHCS precursors are assayed; 17-KS are not assayed by this method.

The 17-KGS may also be assayed as part of the metyrapone test (used to discriminate causes of Cushing's syndrome). In a normal person and in most persons with a pituitary ACTH producing adenoma the 17-KGS should be elevated following a metyrapone test (17-KGS and 11-deoxycortisol levels usually increase). In persons with functional adrenal cortical adenomas the 17-KGS and 11-deoxycortisol do not increase.

This assay has largely been replaced by direct measurement of 11-deoxycortisol in the metyrapone test, and by direct measurement of 17-OH progesterone in the workup of hydroxylase deficiencies.

KETONES, SERUM (see ACETONE, SERUM)

17-KETOSTEROIDS (17-KS), URINE

SPECIMEN: 24 hour urine. Place 10gm boric acid in container prior to collection. Instruct the patient to void at 8:00 A.M. and discard the specimen. Then collect all urine including the 8:00 A.M. specimen at the end of the 24 hour collection period. Refrigerate urine as each specimen is collected. The specimen must be refrigerated or frozen after collection since 17-KS are destroyed at room temperature. The patient should be off all drugs with the exception of aspirin and barbiturates. Record the 24 hour volume. 50mL aliquot is required for the assay. Any test requiring a 24 hour urine collection may also be run on this specimen, e.g., 17-hydroxycorticosteroids (17-OHCS).

REFERENCE RANGE: Male: 9-22mg/24 hour; Female: 4-14mg/24 hour; Neonate: up to 2mg/24 hour; 1 month-5 year: $\leq$ 0.5mg/24 hour; 6-8 year: 1-2mg/24 hour.

METHOD: Steroids having a ketone group at position 17 in the ring structure D are assayed; testosterone has a hydroxyl group at position D and hence does not react. In the color reaction the 17-keto-group reacts with meta-dinitrobenzene to yield a purple compound that absorbs at 520nm.

INTERPRETATION: 17-KS are measured in the evaluation of adrenal cortical disorders. The causes of increased urinary 17-KS are listed in the next Table:

Causes of Increased Urinary 17-KS
Cushing's Syndrome
11-Hydroxylase and 21-Hydroxylase Deficiency
Stressful Illness, e.g., burns, etc.
ACTH, Cortisone or Androgens (except Testosterone)
Androgen Producing Gonadal Tumors

The causes of decreased urinary 17-KS are listed in the next Table:

Causes of Decreased Urinary 17-KS
Addison's Disease
Anorexia Nervosa
Panhypopituitarism

Measurement of 17-KS has been largely replaced by other more specific assays (plasma cortisol, urine free cortisol, 17-OH progesterone, 11-deoxycortisol, DHEA-S) in the evaluation of patients with Cushing's syndrome, congenital adrenal hyperplasia (21 and 11-hydroxylase deficiency), and hirsutism. In patients with adrenal cortical neoplasms, elevation of 17-KS may favor diagnosis of adrenal carcinoma over benign adenoma.

The steroids that are assayed as 17-ketosteroids are given in the next Table:

Steroids Assayed as 17-Ketosteroids

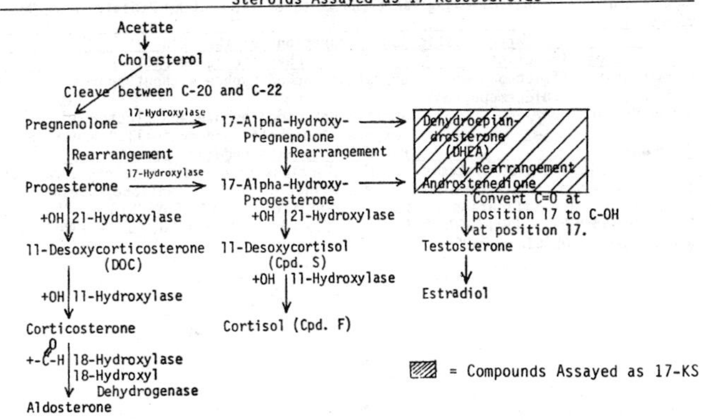

KIDNEY STONE ANALYSIS (see RENAL STONE ANALYSIS)

KLEIHAUER-BETKE STAIN (see FETAL HEMOGLOBIN STAIN)

KOH EXAMINATION

SPECIMEN: Obtain fluid, such as sputum, rather than a swab of fluid.
REFERENCE RANGE: Negative
METHOD: Suspend a drop of exudate such as vaginal discharge in a drop of 10 or 20 percent KOH on a clean glass slide. Cover the drop with a clean glass cover slip and press gently to make a thin mount. Allow specimen to sit at room temperature for about 20 to 30 min. Scan under low power with reduced lighting; change to high power to inspect for presence of suspected fungi.
INTERPRETATION: The polysaccharides of fungal cell walls are relatively resistant to alkali that dissolves mammalian cells or renders them translucent.

Fungal diseases, known as mycoses, are classified as systemic (deep), subcutaneous or superficial (cutaneous) mycoses. Fungi capable of causing mycoses are given in the next Table:

Fungi Causing Mycoses		
Systemic	Subcutaneous	Cutaneous
Cryptococcus Neoformans	Sporotrichum Schenckii	Trichophyton
Coccidioides Immitis	Candida Albicans	Epidermophyton
Histoplasma Capsulatum		Keratinomyces
Blastomyces Dermatitides		Candida Albicans
Candida Albicans		
Aspergillus Fumigatus or Flavus		
Mucormycosis*		

*Mucormycosis refers to a group of diseases caused by fungi of the order Mucorales. Species include Absidia, Mucor, Rhizomucor, and Rhizopus.

The KOH preparation is used to examine vaginal discharge for Candida albicans ("yeast vaginitis"); typical yeast cells and pseudomycelia may be observed.

The KOH preparation may also be a useful method for diagnosing infections of the skin, nails, or hair. Characteristic patterns of KOH preparations of the skin are given in the following Table (Guzzo C, et al. Contemp Pediatr. May 1986; 53-78):

Characteristic KOH Preparations of the Skin	
Fungal Infection	Characteristic Pattern
Dermatophytes (Trichophyton, Microsporum, Epidermophyton)	Septate branched hyphae without spores
Tinea Versicolor	Short blunt-ended hyphae and cluster of spores "spaghetti and meatballs" appearance
Candida	Pseudohyphae with budding spores

The diagnosis of systemic fungal infections frequently requires a variety of laboratory studies including direct examination culture, antibody stains, and exoantigen identification (see ASPERGILLUS, BLASTOMYCOSIS, CANDIDA, COCCIDIOIDOMYCOSIS, CRYPTOCOCCOSIS, and HISTOPLASMOSIS TESTING, FUNGAL PANEL, CSF and FUNGAL SEROLOGICAL PANEL).

LACTASE (LACTOSE TOLERANCE TEST)

SPECIMEN: Red top tube (serum); specimens obtained at time 0, 30 min., 1 hour and 2 hours following oral dose of lactose 50g/m^2 body surface area. Different protocols employ different doses. Some give 50g lactose others call for 100g. In a patient with severe lactase deficiency the lower dose would be better tolerated. Doses for children: Infants: 0.6-1.3g/kg body weight; older children to 12 years: 1.7g/kg body weight up to a maximum of 50g. Patient should be fasted.

REFERENCE RANGE: Rise of serum glucose greater than 20mg/dL between 30 minutes and 1 hour following lactose load.

METHOD: Glucose assay.

INTERPRETATION: The lactose tolerance test is employed in the diagnosis of lactase deficiency, malabsorption, unexplained diarrhea and intestinal gas and cramps. A significant number of patients with the irritable bowel syndrome (gas and bloating, abdominal pain and diarrhea) have lactase deficiency and relief may be obtained by restricting the intake of milk. High prevalence of lactase deficiency occurs in African Americans, Orientals, Jews and Native Americans.

Lactase deficiency is normally defined for subjects who have a blood glucose rise of less than 20mg/dL after 50g/m^2 of a oral lactose dose; a tolerance test with 25g of glucose and 25g of galactose should be normal. The results of an oral lactose dose in normal subjects and in subjects with lactase deficiency are shown in the next Figure:

Oral Lactose Test

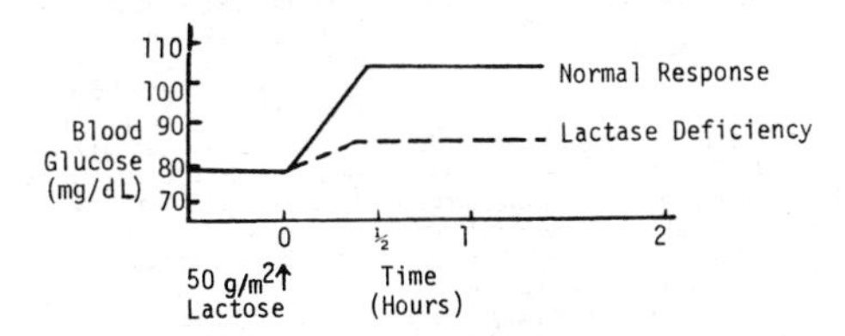

Abnormal results should be confirmed in conjunction with a glucose tolerance test. Hypolactasia is a common cause of gastrointestinal symptoms. In patients with lactase deficiency, lactose remains within the intestinal lumen; it produces an osmotic diarrhea and may be associated with secondary malabsorption of other nutrients. Lactase deficiency may be acquired and seen in conjunction with other bowel disease. Abnormal lactase tolerance is seen with small bowel resection, small bowel disease - Whipple's, Crohn's, Giardiasis and Cystic Fibrosis.

Lactase deficiency has been infrequently found to be responsible for symptoms among patient with the irritable bowel syndrome who are white, of north-western European extraction, and without a milk intolerance (Newcomer AD, McGill DB. Mayo Clin Proc. 1983; 58:339-341).

Lactase deficiency may also be detected by analysis of respiratory hydrogen. The test is based on the principle that lactose (or other carbohydrate) that is not absorbed in the small intestine is metabolized by colonic bacteria to various products, including H_2. The presence of H_2 in the breath is indicative of bacterial degradation. The diagnosis of lactase deficiency is determined by measurement of respiratory H_2 excretion before and 1, 2, 3 and 4 hours after a lactose load of 50g dissolved in 500mL of water. Respiratory H_2 values of more than 0.30mL/minute above the fasting value are considered indicative of lactase deficiency. See also HYDROGEN BREATH ANALYSIS.

A test on urine, using a simple strip impregnated with galactose oxidase, has been developed for possible use in mass screening to detect subjects with lactose malabsorption (Arola H, et al. The Lancet. 1982; 524-525). Individuals who cannot tolerate lactose in milk may be able to tolerate the same sugar in yogurt (Kolars JC, et al. N Engl J Med. 1984; 310:1-3).

LACTIC ACID

SPECIMEN: Arterial blood preferred because contraction of muscles can cause increase in lactate in venous blood. Grey (sodium fluoride) top tube; keep tube on ice until delivered to laboratory; separate plasma and refrigerate plasma.
REFERENCE RANGE: Arterial blood: 0.5-1.66 mmol/liter (4.5-14.4mg/dL). Venous blood: 0.5-2.2 mmol/liter (4.5-19.8mg/dL). To convert conventional units in mg/dL to international units in mmol/liter, multiply conventional units by 0.1110.

METHOD: $\text{Lactate} + \text{NAD}^+ \xrightarrow{\text{LDH}} \text{Pyruvate} + \text{NADH} + \text{H}^+$ (340nm)

INTERPRETATION: Lactic acidosis may be divided into two types, in one type, there is poor tissue oxygenation and in another type, there is no clear evidence of tissue hypoxia. The two types and the conditions associated with the types are given in the next Table (Cohen RD, Woods HF. Clinical and Biochemical Aspects of Lactic Acidosis. Boston, Blackwell Scientific Publications, 1976; Mizock BA. Disease a Month. 1989; 35:237-300):

Types of Lactic Acidosis

Poor Tissue Oxygenation (Type A)	Non-Hypoxic Conditions (Type B)
Shock (cardiogenic, septic, hypovolemic)	Common Disorders:
Regional Hypoperfusion	Diabetes Mellitus
Severe Hypoxemia	Liver Disease
Severe Anemia	Sepsis
Carbon Monoxide Poisoning	Malignancies
	Thiamine Deficiency
	Pheochromocytoma
	Drugs/Toxins
	Hereditary Enzyme Defects
	Miscellaneous (e.g. Hypoglycemia)

Lactic acid is produced from pyruvic acid via anaerobic glycolysis with a resultant ATP production of 2mol ATP per mole of glucose compared to 38mol ATP per mole of glucose produced during aerobic glycolysis via the citric acid cycle.

The normal arterial concentration of lactate is less than 1.6 mmol/liter (approximately 15mg/dL). A concentration greater than 2 mmol/liter is abnormal. When lactic acidosis is present, the arterial lactate concentration is usually higher than 7 mmol/liter. The normal arterial concentrate of pyruvate is less than 0.15 mmol/liter. The arterial lactate-pyruvate ratio is normally about 10:1 (1.5 mmol/liter lactate: 0.15 mmol/liter pyruvate). When lactic acidosis is present, the ratio may rise to 60/1 (Narins RG, et al. Hosp Pract. 1980; 91-98). Lactic acidosis has been successfully treated with dichloroacetate (Stacpoole PW, et al. N Engl J Med. 1983; 309:390-396; Stacpoole PW, et al. Ann Intern Med. 1988; 108:58-63). Dichloroacetate has several modes of action: it reduces circulating lactate concentrating by stimulating the activity of pyruvate dehydrogenase, the enzyme that catalyzes pyruvate to acetyl CoA; it has a positive inotropic effect by increasing myocardial ATP levels; it induced peripheral vasodilatation which may enhance myocardial performance (Mizock BA, Falk JL. Crit Care Med. 1992; 20:80-93).

Anion gap has become a less sensitive test as a screen to detect hyperlactatemia. The most likely reason for this decrease in sensitivity is an upward shift in chloride normal values caused by wide application of chloride specific electrodes (e.g., ASTRA analyzers) which shifts the reference range for anion gap downward (to 3-11 mmol/liter)(Winter SD, et al. Arch Intern Med. 1990; 150:311-313). 60% of critically ill surgical patients with elevated lactic acid levels did not have an anion gap > 16 mmol/liter (Ibert TJ, et al. Crit Care Med. 1990; 18:275-277).

LACTIC DEHYDROGENASE (LDH)

SPECIMEN: Special care must be taken in obtaining samples since minimal hemolysis will cause elevated LDH. Do not freeze or refrigerate specimens. LDH may be measured on a variety of body fluids also.
REFERENCE RANGE: Adult Male: 118-273IU/Liter; Adult Female: 122-220IU/Liter.
METHOD: Kinetic
INTERPRETATION: LDH is principally measured to diagnose conditions in which there is tissue damage. Serum lactate dehydrogenase is elevated in a wide variety of conditions reflecting its widespread tissue distribution; it is elevated in the conditions listed in the next Table:

Causes of Increased Lactic Dehydrogenase
Myocardial Infarction (Acute)
Myocarditis - from any source
Liver Disease
Congestive Heart Failure
Infectious Mononucleosis
Malignancy
Hemolytic Anemia
Reabsorbing Extravasated Blood
Megaloblastic Anemia
Skeletal Muscle Disease
Trauma
Tissue Necrosis
Shock
Acute Pancreatitis (Occ.)
Multisystem Diseases
Collagen-Vascular Diseases
Cerebrovascular Accidents
Acute Renal Infarction
Pulmonary Infarction
Pulmonary Disease (Active)
Pulmonary Alveolar Proteinosis
Hypothyroidism
Transfusions
Hemolysis
Failure to Separate Clot
Heparin Therapy
Pulmonary Toxoplasma

This enzyme is elevated in acute myocardial infarct, liver necrosis, acute pulmonary infarct, primary muscular dystrophy, pernicious anemia, hemolytic anemia, cerebral infarcts and malignancy. These diseases may be differentiated by LDH isoenzyme patterns. The typical uncomplicated myocardial infarct exhibits a rise in LDH levels (LDH1>LDH2) beginning at 8-12 hours, peaks at 24-72 hours, and remains elevated for 7 to 12 days.

Lactic dehydrogenase has been recommended as a prognostic factor in following patients with advanced colorectal carcinoma; patients with an initially normal level of LDH versus those with an abnormal level of LDH had median survivals of 16 and 7.0 months respectively (Kemeny N, Braun DW. Am J Med. 1983; 74:786-794).

Sixty percent of hypertensive patients with pheochromocytoma had elevated serum LDH (O'Connor DT, Gochman N. JAMA. 1983; 249:383-385). Lactic dehydrogenase (LDH) was elevated in patients with ovarian dysgerminoma (Awais GM. Obstet Gynecol. 1983; 61:99-101). LDH is elevated in about one-third of patients on heparin therapy; LDH-5, the liver band, is the fraction elevated (Dukes GE, et al. Ann Intern Med. 1984; 100:645-650). LDH is measured in body fluids and compared with the level found in serum. If the LDH level in the fluid is approximately equal to that of the serum, this is taken as evidence of an inflammatory process. Serum LDH levels have been proposed to distinguish pulmonary toxoplasmosis from P. carinii pneumonia (Pugin J, et al. N Engl J Med 1992; 326:1226).

LACTIC DEHYDROGENASE ISOENZYMES(LDH ISOENZYMES)

SPECIMEN: Red top tube; separate serum; do not freeze because freezing destroys LDH-4 and LDH-5; many laboratories store at room temperature; do not use hemolyzed specimen - hemolysis causes elevation of LDH-1 and LDH-2. Isoenzymes should not be done if total LDH is less than 130IU/L.

REFERENCE RANGE:

Adult Values:

Isoenzyme	%	IU/L
LDH-1 (Heart)	22 to 37	22 to 85
LDH-2	30 to 45	30 to 100
LDH-3	15 to 30	15 to 65
LDH-4	5 to 11	5 to 25
LDH-5	2 to 11	2 to 25
	Total:	100 to 225

Neonates: As compared to adults, the percentage of LDH-5 is increased and the percentage of LDH-1 decreased in neonates.

METHOD: Electrophoresis; LDH isoenzymes should be reported in both percent and international units.

INTERPRETATION: The principal use of LDH isoenzymes is in the diagnosis of acute myocardial infarct. It is principally employed in the patient who has delayed seeking medical attention a few days following a cardiac event and in those patients with equivocal or difficult to interpret findings. LDH isoenzyme patterns are, in general, not as useful or as specific as CPK isoenzyme studies. In many hospital laboratories, serial CPK isoenzyme studies are performed, and LDH isoenzyme studies are optional. The normal LDH isoenzyme pattern of serum is given in the next Figure. The tissues having high concentration of LDH-1, LDH-3, and LDH-5 are given just below the corresponding isoenzyme. Those that have clinical significance are underlined.

Serum Isoenzyme Pattern and Tissue Distribution of LDH Isoenzymes

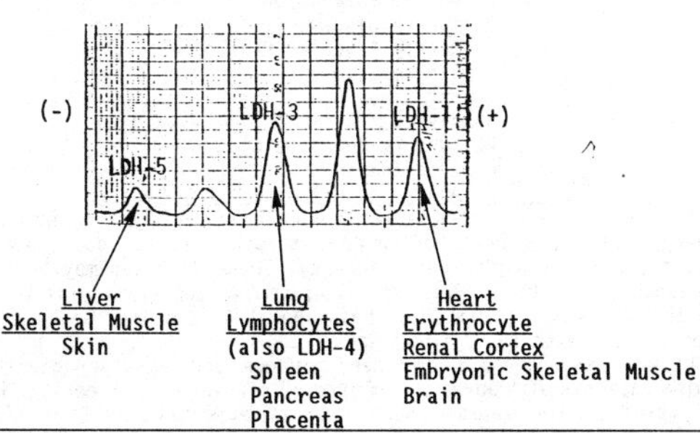

Serum LDH isoenzyme patterns found in various diseases are given in the next six Figures:

Elevation of LDH-1:

Elevation of LDH-1

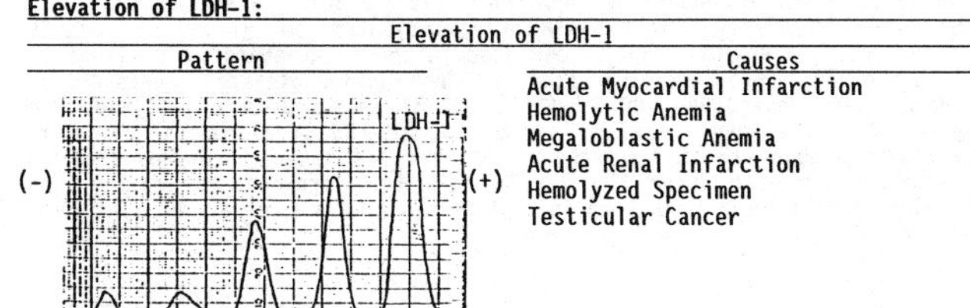

Causes

Acute Myocardial Infarction
Hemolytic Anemia
Megaloblastic Anemia
Acute Renal Infarction
Hemolyzed Specimen
Testicular Cancer

Elevation of LDH-1: As expected from the tissue distribution (see previous Figure), elevation of serum LDH-1 is seen in acute myocardial infarction; hemolytic anemia; megaloblastic anemia (both folic acid deficiency and vitamin B_{12} deficiency) and acute renal infarction. LDH-1 occurs in the brain but does not normally pass the blood-brain barrier. The LDH-1, which may be elevated in patients with strokes, probably originates from the red blood cells in the thrombus, or in the area of hemorrhage in the brain. Elevated LDH-1 may also be seen in the muscular dystrophies, myositis and other conditions in which muscle is responding to chronic injury. Hemolyzed serum specimen will cause an elevated LDH-1. Serum LDH-1 elevation may occur in patients with testicular carcinoma (Liu F, et al. Am J Clin Path. 1983; 78:178-183), and in patients with ovarian carcinoma. LDH1/LDH2 ratios are increased in hemolytic anemia principally in patients with reticulocytosis (Kazmierczak SC, et al. Clin Chem. 1990; 36:1638-1641).

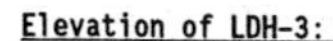

Elevation of LDH-3:

Elevation of LDH-3

Pattern	Causes
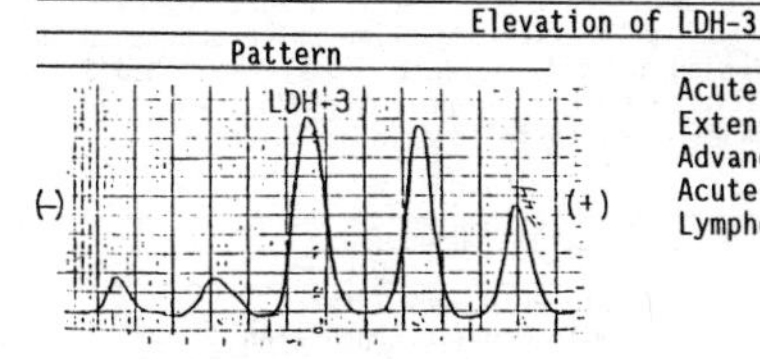	Acute Pulmonary Infarction Extensive Pneumonia Advanced Cancer Acute Pancreatitis Lymphocytosis

Elevation of LDH-3 occurs occasionally in acute pulmonary infarction and extensive pneumonia. Elevation of LDH-3 may occur in patients with advanced cancer and may be useful in following the effectiveness of therapy for cancer.

Elevation of LDH-5:

Elevation of LDH-5

Pattern	Causes
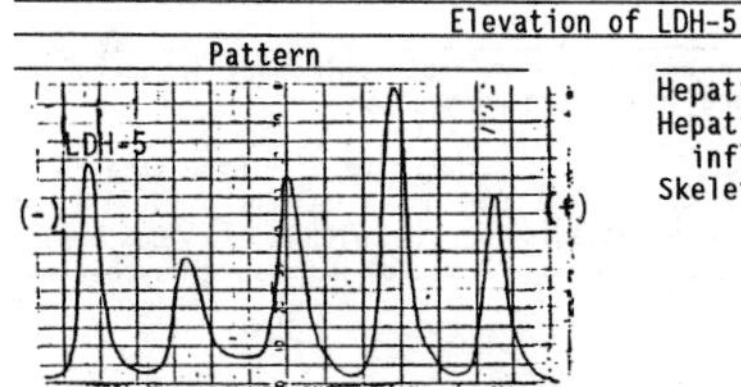	Hepatic Congestion Hepatitis or other liver injury or inflammation Skeletal Muscle Injury

Elevation of LDH-5 may occur in congestive heart failure, and in acute, subacute, and chronic hepatitis, including liver involvement due to infectious mononucleosis (Rutenberg Z, et al. Clin Chem. 1991; 37:116-117); it is not elevated in intra- or extrahepatic bile duct obstruction. LDH-5 is elevated in skeletal muscle injury whether caused by trauma or by surgery.

Elevation of All LDH Isoenzymes:

Elevation of All Isoenzymes

Pattern	Causes
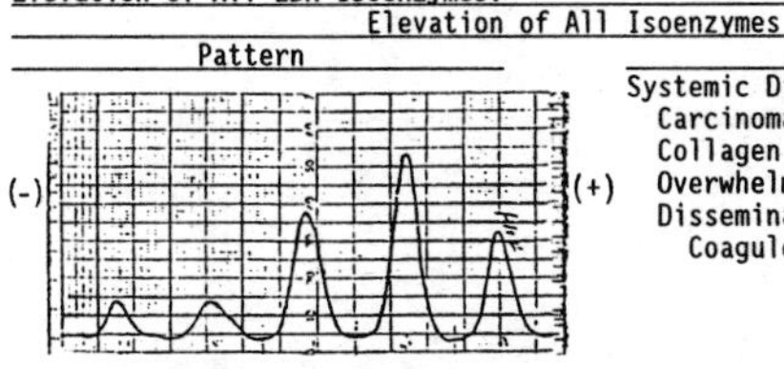	Systemic Diseases: Carcinomatosis Collagen Vascular Disease Overwhelming Sepsis Disseminated Intravascular Coagulopathy

Elevation of all isoenzymes of LDH occur in systemic diseases, such as carcinomatosis, collagen vascular disease, overwhelming sepsis, disseminated intravascular coagulopathy.

LDH Isoenzymes (Cont)

Elevation of Both LDH-1 and LDH-5:

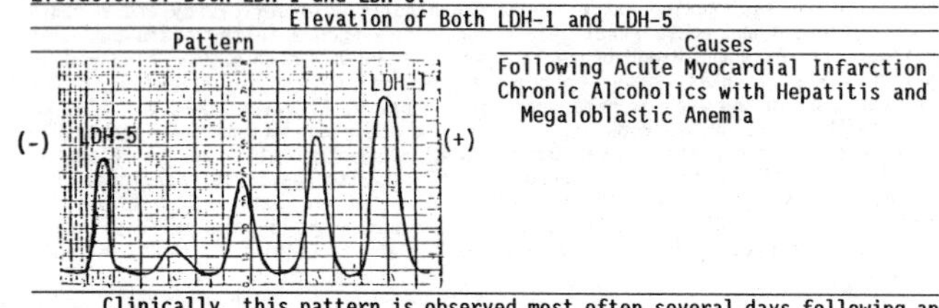

Elevation of Both LDH-1 and LDH-5

Pattern	Causes
(-) LDH-5 ... LDH-1 (+)	Following Acute Myocardial Infarction Chronic Alcoholics with Hepatitis and Megaloblastic Anemia

Clinically, this pattern is observed most often several days following an acute myocardial infarction; elevation of LDH-1 is due to the acute myocardial infarction; elevation of LDH-5 is due to hepatic congestion which occurs in 30 percent of patients 1.5 days following infarction (West, et al. Am J Med Sci. 1961; 241:350). This pattern may also be seen in chronic alcoholics who develop both hepatitis and megaloblastic anemia due to folic acid deficiency.

Elevation of LDH-3 and LDH-5:

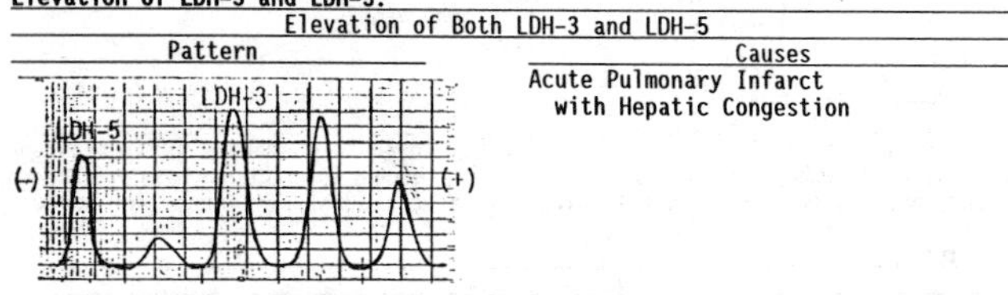

Elevation of Both LDH-3 and LDH-5

Pattern	Causes
(-) LDH-5 ... LDH-3 ... (+)	Acute Pulmonary Infarct with Hepatic Congestion

This pattern may be seen in patients with acute pulmonary infarcts and right-sided heart failure resulting in hepatic congestion.

LDH-Isoenzymes in Other Fluids: LDH isoenzyme studies have also been studied in other fluids. A LDH1/LDH2 ratio <1 in the CSF is a strong indicator of carcinomatous meningitis or cerebral metastasis compared with primary CNS benign or malignant tumors - ratio >1 (Lampl Y, et al. J Neurol Neurosurg Psychiatry. 1990; 53:697-699).

LATEX AGGLUTINATION(LA)**(see BACTERIAL MENINGITIS ANTIGENS)**

LEAD, BLOOD

SPECIMEN: Lavender (EDTA) top tube; or green (heparin) top tube.
REFERENCE RANGE: Acceptable levels continue to change. The permissible level for children has recently been revised (1991). Children: <10mcg/dL; adults: <40mcg/dL; industrial exposure: <60mcg/dL. The WHO Regional Office for Europe recommends the adult population should have a blood lead level below 20mcg/dL (WHO Regional Publications, European Series. No. 23. Copenhagen: WHO. 1987; 242-261). To convert conventional units in mcg/dL to international units in mcmol/liter, multiply conventional units by 0.04826.
METHOD: Atomic absorption; anodic stripping voltammetry (ESA Trace Metals Analyzer).
INTERPRETATION: **The CDC now calls for lead screening to be performed by direct measurement of blood levels** (Center for Disease Control. Preventing lead poisoning in young children. Atlanta: Department of Health and Human Services, 1991). Finger stick specimens are often contaminated and not recommended. The concentration of lead in whole blood is about 75 times that of serum or plasma, hence measurements are done on whole blood. The laboratory tests that have been previously used in evaluating lead exposure are listed in the next Table:

Laboratory Tests in Evaluating Lead Poisoning
Increased lead levels in blood, urine and/or hair (not recommended)
Increased delta aminolevulinic acid(DALA) and coproporphyrin III in urine
Decreased delta aminolevulinic acid(ALA) dehydrase
Increased erythrocyte zinc protoporphyrin(ZPP)

The inhibition of the activity of delta ALA dehydrase is a sensitive indicator of acute and chronic lead poisoning. The activity of this enzyme decreases with increasing lead levels. Measurement of erythrocyte zinc protoporphyrin was a popular and easy way to screen for lead exposure in children. However, it is now found to be too insensitive, and the CDC recommends direct measurement of blood lead. Results on fingerstick specimens should be confirmed on venous blood samples to rule out contamination.

Toxic effects of lead are observed in the systems listed in the next Table:

Systems Involved in Toxic Effects of Lead
Hematopoietic System
Nervous System
Gastrointestinal System
Kidneys

Hematopoietic System: The hematologic manifestations of lead poisoning are listed in the next Table:

Hematologic Manifestations of Lead Poisoning
Basophilic stippling of red blood cells and precursors
Anemia
Reticulocytosis
Erythroid hyperplasia with dyserythropoiesis
Autofluorescence of red blood cells and precursors

Lead interferes at multiple steps with the biosynthesis of heme. The location of metabolic blocks in the biosynthesis of heme is shown in the next Figure:

Pathway for Synthesis of Heme and Enzymes Inhibited by Lead

```
"Activated   +   Succinyl
 Glycine"          CoA
    |_______________|
            ↓ ALA Synthetase (↑Activity)
Delta-Aminolevulinic Acid
            ‡ ALA Dehydrase
     Porphobilinogen
            | Uroporphyrinogen Synthase
            ↓ Uroporphyrinogen Isomerase
    Uroporphyrinogen III
            ↓ Uroporphyrinogen Decarboxylase
   Coproporphyrinogen III
            ‡ Coproporphyrinogen Oxidase
   Protoporphyrinogen III
            ↓          +Zn++
     Protoporphyrin III ------> Zinc Protoporphyrin
       +Fe+2  ‡ Ferrochelatase
            ↓
           Heme
```

(cont)

Lead, Blood (Cont)

Lead inhibits the activity of three enzymes in the synthesis of heme. These are indicated in the previous Figure (Cambell BC, et al. Clinical Science and Molecular Medicine. 1971; 53:335-340).

Lead is an electropositive metal and has a high affinity for the negatively charged sulfhydryl group; ALA dehydrase, coproporphyrinogen oxidase and ferrochelatase are sulfhydryl-dependent enzymes. There is increased activity of the enzyme, ALA synthetase, probably secondary to decreased production of heme. ALA-dehydrase is inhibited at blood lead levels 10-15mcg/dL; hemoglobin is reduced at blood lead levels >40mcg/dL.

Nervous System: Both the central and peripheral nervous system may be affected. The characteristic neurologic abnormalities are motor paralysis (wrist drop) and acute encephalopathy. There may be nonspecific symptoms such as headache, dizziness, sleep disturbances, memory deficit and increased irritability, and in severe cases, profound central nervous system disturbances with convulsions. Low-level lead exposure in early life is associated with decreased intellectual and neuropsychological development (Boghurst PA, et al. N Engl J Med. 1992; 327:1279-1284).

Gastrointestinal System: Gastrointestinal symptoms may be related to disturbed innervation of the intestines; there may be loss of appetite, weight loss, abdominal discomfort and crampy abdominal pains, e.g., "lead colic."

Kidney: Lead exposure may cause interstitial nephritis. Children with lead intoxication have aminoaciduria, glycosuria, and may develop Fanconi's syndrome. In the general adult population there was an impairment of renal function with increased blood lead levels (Staessen JA, et al. N Engl J Med. 1992; 327:151-156). Uric acid excretion and erythropoietin production may be decreased.

LEAD EXPOSURE PANEL

The laboratory tests for evaluation of lead exposure are listed in the next Table:

Laboratory Tests for Evaluation of Lead Exposure

Abnormal Test Result	Blood Lead Level
ALA-Dehydrase Inhibition	10-15mcg/dL
Elevated Erythrocyte Zinc Protoporphyrin	15-20mcg/dL
Blood Lead Level	Definitive Test

The inhibition of Delta-ALA dehydrase activity is a sensitive indicator of acute and chronic lead poisoning; this enzyme's activity decreases with increasing lead levels. Erythrocyte zinc protoporphyrin(ZPP) was considered an excellent screening test to detect lead exposure (see ZINC PROTOPORPHYRIN). However, the CDC has recently lowered the acceptable lead blood level in children to 10mcg/dL due to findings of neurobehavioral/developmental deficits in children at blood lead levels of 10-15mcg/dL. **Unfortunately the ZPP test (the test recommended in 1985 guidelines) is not abnormal until blood levels reach 15-20mcg/mL and not considered reliable for blood lead levels <25mcg/dL** (Brewer RD, et al. NCMJ. 1992; 53:149-152; Turk DS, et al. N Engl J Med. 1992; 326:137-138).

The blood lead level is a definitive indicator of the extent of current or recent lead absorption. **The CDC now calls for screening for lead by direct measurement of blood levels** (Centers for Disease Control. Preventing lead poisoning in young children. Atlanta: Department of Health and Human Services, 1991). Following exposure to the metal, the whole-blood concentration of lead (which localizes primarily in the erythrocytes) is about 75 times that in the serum or plasma. The reference levels for lead in the blood are <10mcg/dL for children and <40 mcg/dL for adults; the industrial exposure level is <60 mcg/dL. Toxicity is probable when the blood lead level exceeds these limits. Fingerstick results should be confirmed with venous blood samples to avoid spurious results due to contamination. Lead can be quantitated with atomic absorption(AA) spectrometry or anodic striping voltammetry(ASV).

LEAD MOBILIZATION TEST (see LEAD, URINE)

LEAD, SERUM

SPECIMEN: Red top tube, separate serum.
REFERENCE RANGE: 0.80-2.46ng/mL.
METHOD: Atomic absorption; anodic stripping voltammetry.
INTERPRETATION: The concentration of lead in whole blood is about 75 times that of serum or plasma. Thus, whole blood is the preferred sample for determination of toxic lead levels. See LEAD, BLOOD. This test should not be ordered for lead screening.

LEAD, URINE

SPECIMEN: Collect 24 hour urine in acid washed-containers. Use plastic containers (polyethylene or polypropylene); add 10% HCl solution or a nitric acid solution to the container and allow to "soak" for 10 minutes; rinse with five volumes of tap water and then five volumes of deionized or distilled water. The patient should urinate at 8:00 A.M. and the urine is discarded. Then, urine is collected for 24 hours including the next day 8:00 A.M. specimen. Indicate 24 hour volume. A 50mL aliquot is used for analysis.
REFERENCE RANGE: <50mcg/24 hours, abnormal >125mcg/24 hours. To convert conventional units in mcg/24 hours to international units in mcmol/day, multiply conventional units by 0.004826.
METHOD: Atomic absorption; anodic stripping voltammetry.
INTERPRETATION: Urine lead levels are useful in monitoring the rate of lead excretion, especially during chelation therapy. See LEAD, BLOOD. Urine lead levels may not correlate with clinical symptoms of lead toxicity or blood lead levels; "normal" urinary lead excretions have been observed in the presence of significantly elevated blood lead levels and clinical evidence of lead poisoning (Berman E. Laboratory Medicine. 1981; 12:677-684).

The EDTA lead mobilization test has been used to identify lead nephropathy. This test is done as follows: Urine is collected for 24 hours for lead analysis as control. Then, 1 gram of $CaNa_2EDTA$ is given I.V. in 250mL of 5% dextrose/water over 1 hour. Urine is collected for lead assay for 3 days. Normal subjects excrete less than 600mcg during the 3 days following parenteral EDTA; subjects with lead nephropathy excrete more than 600mcg (Wedeen RP. Clin Exper Dialysis and Apheresis. 1982; 6:113-146; Wedeen RP, et al. Am J Med. 1975; 59:630; Emmerson BT. Kidney Int. 1973; 4:1). The EDTA lead-mobilization test may also be performed by intramuscular injection of a total of 2 g of EDTA with 1.0 mL of 2 percent xylocaine, given in two divided, one gram, doses 8 to 12 hours apart. Twenty-four hour urinary excretion of lead is measured over three consecutive days, starting with the first injection. Subjects without unusual lead exposure excrete less than 600mcg (2.9 micromol) of lead per three days during this test (Batuman V, et al. N Engl J Med. 1983; 309:17-21).

LECITHIN/SPHINGOMYELIN (L/S) RATIO AND PHOSPHATIDYLGLYCEROL (PG)

SPECIMEN: Amniotic fluid. Store at 4° and run within 24 hours. Protect from light. Interpretation may be difficult if contaminated with blood or meconium. Centrifuge to remove debris, but not long enough to pellet lamellar bodies (e.g. 500g for 3 minutes). Transfer to new tube.
REFERENCE RANGE: An L/S ratio >2:1 usually indicates fetal lung maturity; a ratio <1.5:1 suggests immaturity. False predications of maturity sometimes occur, especially in gestational diabetes.
METHOD: Thin-layer chromatography (Kulovich MV, et al. Am J Obstet Gynecol. 1979; 135:57-63; Kulovich MV, Gluck L. Am J Obstet Gynecol. 1979; 135:64-70). Also may use agglutination, fluorescence polarization and "shake test" to assess lecithin content.
INTERPRETATION: The L/S ratio is determined on amnionic fluid to access fetal lung maturity in order to determine the ability of the infant to breath without danger of respiratory distress syndrome.

The surface-active lipid, lecithin, originates from Type II alveolar epithelial cells of the fetal lung and during gestation finds its way into amniotic fluid. It's presence and chemical character (palmitic acid content) reflect fetal pulmonary maturation. The changes in amniotic fluid lecithin and sphingomyelin during gestation are shown in the next Figure:

Lecithin/Sphingomyelin (L/S) Ratio (Cont)

Amniotic Fluid Lecithin/Sphingomyelin Ratio

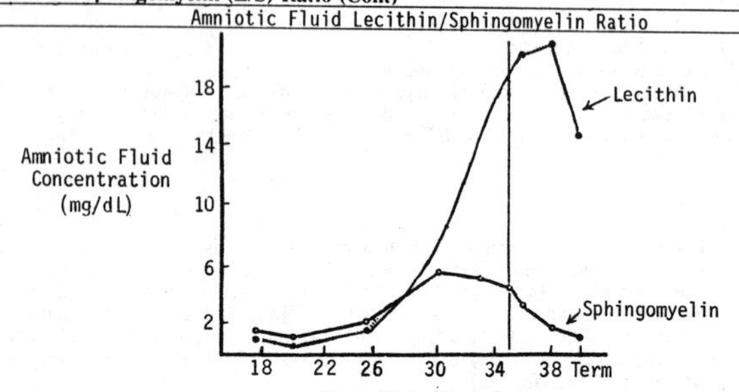

The phospholipid, lecithin, increases with gestational age; another phospholipid, sphingomyelin, remains at a relatively low and constant concentration throughout gestation and can be used as a baseline for comparison. The L/S ratio is close to 1:1 until the 35th week of gestation. At that time, the lecithin content increases dramatically, attaining a ratio of 2:1, indicating pulmonary maturity.

A L/S ratio >2:1 generally indicates that the infant will not develop RDS. However, the infant delivered with a ratio <2:1 will not invariably develop RDS.

Phosphatidylglycerol (PG) in Amniotic Fluid: Phosphatidyl glycerol (PG) may also be measured in amnionic fluid as an indicator of fetal lung maturity.

The content of PG in amniotic fluid during gestation is shown in the next Figure (Saunders BS, et al. Clin Perinatology. 1978; 5:231-242):

Phosphatidylglycerol in Amniotic Fluid During Gestation

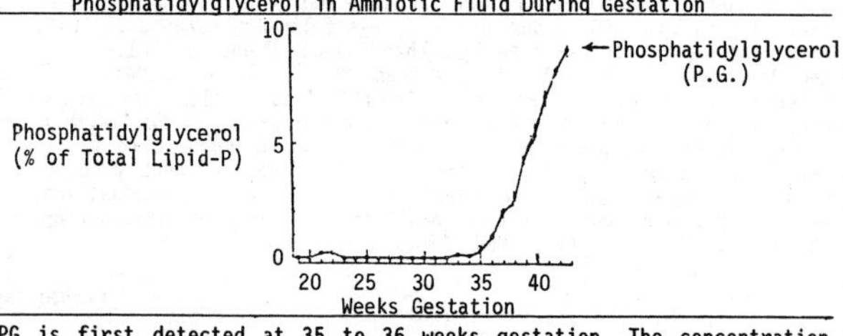

PG is first detected at 35 to 36 weeks gestation. The concentration increases and it is the second most common phospholipid at term. The presence of PG in amniotic fluid is indicative of pulmonary maturity.

Other Assays: Among other assays for fetal maturity is the surfactant-albumin ratio. In this assay a fluorescent dye is partitioned between the lipid and albumin in amnionic fluid. Results are reported in mg surfactant per gram albumin (mg/g). Values <50mg/g indicate immaturity; 50-69mg/g are borderline; values >70mg/g are mature. This test has been reviewed and considered a rapid reliable test for lung maturity (Steinfeld JD, et al. Obstet Gynecol. 1992; 79:460-464).

Problems in Interpretation: In diabetes mellitus, a L/S ratio of 2.0 or greater does not insure fetal lung maturity. However, the presence of 3% or more PG in surfactant in amniotic fluid is used as the guideline for delivery in diabetics; RDS has not been observed when PG is 3% or greater. In mild hypertension, the development of surfactant, as reflected by L/S ratio and PG, is similar to the normal. In severe hypertension, there may be accelerated maturation, sometimes with L/S ratios >2.0 and early appearance of PG. In premature rupture of membranes (PROM), there is acceleration of fetal lung maturation with early elevation of L/S ratios and early appearance of PG. One of the dangers of PROM is chorioamnionitis. In one study, assay of amnionic fluid glucose levels were instructive. Fluid glucoses <5mg/dL were 90% predictive of infection, while glucoses >20mg/dL were 98% predictive of no infection (Kilty RJ, et al. Obstet Gynecol. 1991; 78:619-622).

LEE-WHITE CLOTTING TIME (see CLOTTING TIME)

LEGIONNAIRE'S DISEASE, ANTIBODY SERUM

SPECIMEN: Red top tube, separate and refrigerate serum. Obtain two blood specimens: an acute phase specimen plus a convalescent phase specimen. The convalescent phase specimen is obtained 21 or more days after onset of fever.
REFERENCE RANGE: Acute infection is indicated by a fourfold rise in titer between acute and convalescent samples. A single titer greater or equal to 1:256 with compatible clinical symptoms allows a presumptive diagnosis of Legionella pneumonia.
METHOD: Indirect immunofluorescence
INTERPRETATION: In Legionella infections, antibody titers tend to be low during the first week, rise during the second and third weeks and reach maximum levels at about five weeks. After recovery, titers drop slowly and tend to remain elevated for many years. 10-20 percent of patients do not seroconvert. This test is not helpful for the initial diagnosis of acute pneumonia. Serologic diagnosis may be problematic due to cross reactivity between Legionella pneumophilia and non-pneumophilia legionella species and other gram-negative organisms; false positive tests may occur with tuberculosis (Ching WTW, Meyer RD. Infect Dis Clin N Amer. 1987; 1:595-614). Optimum sensitivity requires collecting specimens during the acute phase and 6-9 weeks later. Seroconversion is 75% sensitive and 95-99% specific for diagnosis of legionnaire's disease (Edelstein PH. CID. 1993; 16:741-749).

LEGIONNAIRE'S DISEASE, CULTURE

SPECIMEN: Fresh lung tissue, such as needle biopsy, pleural fluid, bronchial washings or brushings, transtracheal aspirates, blood, pus, sputum.

Collect specimen in sterile container. Transport to laboratory immediately. Specimen may be forwarded to reference laboratory in sterile, screw-capped container on ice. Do not freeze.
REFERENCE RANGE: Negative. Positive culture in 2-7 days (average: 3 days).
METHOD: Selective media. Maximum yield from sputum requires multiple selective media containing antibiotics and dyes; pretreatment of sputum samples with acid wash improves sensitivity by 10-20 percent (Nguyen MH, et al. Infect Dis Clin N Amer. 1991; 5:561-584).
INTERPRETATION: The etiologic agent of Legionnaire's disease is Legionella pneumophilia and related species; it is a gram-negative, non-acid fast bacillus. The pathologic findings are largely limited to the lungs, in which pneumonia and acute diffuse alveolar damage are seen. Extrapulmonary manifestations include sinusitis, wound infection, perirectal abscess, pericarditis, endocarditis, pyelonephritis, peritonitis, and pancreatitis.

The incubation period is two to ten days; patients with underlying pulmonary disease and smokers are particularly susceptible. Patients develop fever, headache, malaise and pneumonia. Nonspecific laboratory findings include leukocytosis with immature cells, elevated erythrocyte sedimentation rate, elevated aspartate aminotransferase, alanine aminotransferase, lactic dehydrogenase and alkaline phosphatase. Hyponatremia (probably due to SIADH) and hypophosphatemia may occur.

Epidemics are associated with public buildings, such as hotels and hospitals; it has been found in the water supply.

Legionella causes two types of illness: a severe form associated with pneumonia (Legionnaire's Disease) and a mild self-limiting form characterized by general symptoms of fever and aching (Pontiac fever). Erythromycin is the drug of choice; rifampin is used if there is not a satisfactory response to erythromycin alone (Nguyen MH, et al. already cited; Ching WTW, Meyers RD. Infect Dis Clin N Amer. 1987; 1:595-614).

The sensitivity and specificity of culture in the diagnosis of legionnaire's disease is given in the following Table (Edelstein RH. CID. 1993; 16:741-749):

Culture for Diagnosis of Legionnaires' Disease

Culture Specimen	Sensitivity(%)	Specificity(%)
Sputum; Bronchoalveolar Lavage	80-90	100
Lung Biopsy Specimen	90-99	100
Blood	10-30	100

LEGIONNAIRE'S DISEASE, DIRECT IMMUNOFLUORESCENCE

SPECIMEN: Lung tissue, respiratory tract fluids, sputum, pleural fluid, pus
REFERENCE RANGE: Negative
METHOD: Direct immunofluorescence
INTERPRETATION: This test detects bacteria in various fluids and can be performed rapidly (1-3 hours). Sensitivity varies from 25-80% (Nguyen MH, et al. Infect Dis Clin N Amer. 1991; 5:561-584). This test is highly specific (95-99%)(Edelstein PH. CID. 1993; 16:741-749).

LEGIONNAIRE'S DISEASE, URINE ANTIGEN

SPECIMEN: Urine
REFERENCE RANGE: Negative
METHOD: Enzyme Immunoassay
INTERPRETATION: Legionella pneumophilia antigen occurs in both serum and urine but is 30 to 100-fold more concentrated in urine. Excretion of antigen occurs within days and persists for several weeks. This test may be performed rapidly and will remain positive after antibiotic therapy has been initiated (Grimont PAD. Isr J Med Sci. 1986; 22:697-702). Urinary antigen detection is 80-99% sensitive and 99% specific for diagnosis of legionnaire's disease (Edelstein RH. CID. 1993; 16:741-749).

LEUKEMIA, ACUTE, PANEL

Acute leukemia had been divided into two divisions: acute lymphocytic leukemia (ALL) and acute nonlymphocytic leukemia (ANLL). However, beginning in 1976 and modified subsequently, French, American and British hematologists established the FAB classification for acute leukemia. This subdivides the acute leukemias into further groupings. The FAB classification is given in the following Table:

FAB Classification of Acute Leukemia		
Acute Lymphocytic Leukemia:		
L1		Acute lymphoblastic leukemia (most common childhood ALL)
L2		Acute lymphoblastic leukemia (most common adult ALL)
L3		Acute lymphoblastic leukemia (Burkitt's type)
Acute Nonlymphocytic Leukemia:		
M1		Acute myeloblastic leukemia (without maturation)
M2		Acute myeloblastic leukemia (with maturation)
M3		Acute promyelocytic leukemia (usual type)
	M3V	Microgranular variant
M4		Acute myelomonocytic leukemia (usual type)
	M4Eo	With abnormal eosinophils
M5		Acute monocytic leukemia
	M5a	Poorly differentiated
	M5b	Differentiated
M6		Acute erythroleukemia
M7		Acute megakaryocytic leukemia

There are occasional difficulties in classifying some acute leukemias, and the above classification scheme is subject to further revision. Diagnosis of acute leukemia requires the presence of 30% or greater leukemic blasts in the bone marrow (normal up to 3%). Subclassification of acute leukemias in the FAB system is based on morphology and cytochemistry. Immunophenotyping is not required with the exception of M7 (megakaryocytic leukemia) in which immunophenotyping has been accepted as necessary for accurate classification.
Cytochemical Stains: Accurate classification of acute leukemias requires cytochemical and immunochemical examination of the abnormal cells with a battery of stains. A list of the usually employed stains is given below:

Cytochemical Stains and Acute Leukemia
Myeloperoxidase
Sudan Black B
Terminal Deoxynucleotidyl Transferase (TdT)
Chloroacetate Esterase (Specific Esterase)
Butyrate or Acetate Esterase (Non-specific Esterase)
Acid Phosphatase
PAS
Iron

These stains are not entirely specific for one type of acute leukemia. Myeloperoxidase is the most important of the above stains. Myeloperoxidase will in general separate the ALLs (L1-3) from the ANLLs (M1-7). Sometimes the myeloperoxidase stain is not evident at the light microscopic level, but is visualized with electron microscopy. Not all M5 and M6 leukemias stain for myeloperoxidase and M7 is myeloperoxidase negative.

Immunophenotyping, Cell Surface Markers: Immunophenotyping is not strictly necessary for classification of acute leukemia in the FAB system, with the exception of M7. However, it is being widely employed, and has been of use in difficult cases where there is uncertainty or a discordance between morphology and cytochemical staining. It has been estimated that by morphology alone, 60-70% of acute leukemias may be accurately diagnosed; approximately 90% by combined morphology and cytochemistry; and close to 99% when immunophenotyping is added (Duque RE, et al. Clin Immunol Newsletter. 1990; 10:43-62). For further details on immunophenotyping see LYMPHOCYTE DIFFERENTIAL.

Other studies now available for further classification of acute leukemic cell populations include analysis of rearrangement of IgH and IgL genes (to indicate a B-cell lineage) and rearrangement of the βTCR and γTCR genes (to indicate a T-cell lineage). Rarely a B-cell line will exhibit a TCR gene rearrangement. The presence of a clone of B or T-cells (5% or more of the population), will be detected as a distinctive DNA banding pattern when hybridized to the appropriate probe.

Favorable Prognosis in Acute Nonlymphocytic Leukemia: In general, there is little difference in prognosis between FAB subgroups, with the exception of FAB subgroup M3 (acute promyelocytic leukemia) which tends to do slightly better. This form of leukemia usually presents with a bleeding diathesis which is exacerbated by chemotherapy. Retinoic acid derivatives have been recently shown to be effective in inducing differentiation of acute promyelocytic leukemia cells and reducing complications during induction therapy (Warrell RP, et al. N Engl J Med. 1993; 329:177-189). Acute promyelocytic leukemia is associated with a (15;17) translocation which fuses the PML gene (myl gene, older terminology) with the retinoic acid receptor-alpha (RAR-α) gene (Larson RA, et al. Am J Med. 1984; 76:827-841. Chomienne C, et al. Leukemia. 1990; 4:802-807).

There have been attempts to ascertain other criteria by which patients and their leukemias may be grouped into those with a good prognosis compared to those with a poor prognosis. In general, patients who are <60 years old,have an absence of previous hematological disease, have a normal peripheral blood count, and have an absence of other concurrent disease have a better prognosis compared to patients without these characteristics.

Laboratory tests may also be run on the leukemic cells in order to assess prognosis. These tests are listed in the next Table:

Tumor Characteristics Associated with a Favorable or Unfavorable Prognosis

Characteristic	Favorable Prognosis	Unfavorable Prognosis
Cytogenetics:	Inversion (16) Translocation (15;17) Translocation (8;21)	Trisomy 8 Deletion (-5/-7)
Immunophenotypes:	CD2; CD19	CD7, CD34
Cell Cycle Kinetics:	Low Rate	High Rate
Other:	Presence of Auer Rods	In Vitro Proliferation Multiple Drug Resistance

The (8;21) translocation is usually associated with the FAB M2 subgroup; the (16) inversion is associated with FAB M4 with eosinophilia.

Approximately 15% of acute myelogenous leukemia patients will exhibit the above favorable prognostic features.

LEUKEMIA, CHRONIC, PANEL

CHRONIC LYMPHOCYTIC LEUKEMIA

Chronic lymphocytic leukemia (CLL) constitutes a spectrum of disorders, subclassified with the aid of immunological and cytochemical techniques. Subtypes of CLL are given below:

Subtypes of Chronic Lymphocytic Leukemia

Subtype	Incidence
B-CLL	90-95%
T-CLL	1-5%
null-CLL	1-2%
Hairy Cell Leukemia	

In addition to the above, there are examples of prolymphocytic leukemia (PLL), both B and T cell subtypes being recognized.

Subclassification of CLL may include the various tests listed in the next Table:

Laboratory Tests to Subclassify Chronic Lymphocytic Leukemia

Test	Remarks
Sheep RBC Rosettes	T-cells
Mouse RBC Rosettes	B-cells
SIg	B-cells
CIg	B-cells
Ia (HLA-DR)	B-cells
Acid Phosphatase	T>B-cells
ß-Glucuronidase	T>B-cells
α-Naphthyl Acetate Esterase	T>B-cells
Dipeptidyl Amino Peptidase	T>B-cells
Immunophenotyping:	
OKT 3,4,5,8,11, Leu 1,4,5 (CD 3,4,8,2,5)	T-cells
B1,2, Leu 12,14 (CD 20,21)	B-cells
OKT 4, Leu 3 (CD4)	T Helper-cells
OKT 8, Leu 2 (CD8)	T Suppressor-cells
TAC, HC-1,2, Leu 14, M5, B1 (CD25)	Hairy-cells
Immunoglobulin Gene Rearrangement	B-cells
ßTCR and γTCR Gene Rearrangement	T-cells (rarely B-cells)

The T-helper phenotype CLL is more aggressive than the T-suppressor CLL. Approximately 5-10% of CLL evolves into an acute phase. This may represent: 1) a prolymphoblastic leukemia; 2) diffuse lymphoma (Richter's syndrome), or 3) acute blastic phase.

Null-Cell Chronic Lymphocytic Leukemia: Some lymphocytic leukemias were previously classified as non-T, non-B or "null" cell because of failure to rosette with SRBC, lack of SIg and lack of T-cell antigens. Many null cell leukemias will rosette with mouse RBC and are HLA-DR positive and hence are probably B-cells with weak SIg. These are now recognized as pre B-cells in view of the finding of rearranged IgH genes. In these leukemias, expression of CD19 and Ia precedes IgLg rearrangement, CD10 (CALLA) and CD20 (Korsmeyer SJ, et al. J Clin Invest. 1983; 71:301-313).

Hairy Cell Leukemia: Hairy cell leukemia is represented by cells with prominent microvilli found in blood and bone marrow. These cells usually infiltrate the spleen and often infiltrate the liver. The cell of origin was unknown for a long time, but is now thought to represent an activated B-cell lineage, but lacking plasmacytoid differentiation. Cell markers are listed below:

Cell Markers in Hairy Cell Leukemia

Acid Phosphatase (tartrate resistant)	Strongly Positive
SIg (light chain)	Positive
CIg	Positive
Immunoglobulin Gene Rearrangement	Positive
Immunophenotyping: CD19, 20	Positive

Hairy cell leukemic cells are strongly and diffusely positive for tartrate resistant acid phosphatase isoenzyme 5. Sezary cells and activated lymphocytes may also be positive, but the activity is weak and focal.

CHRONIC MYELOCYTIC LEUKEMIA (CML)

Philadelphia Chromosome and abl-bcr: The most important laboratory marker in CML is the Philadelphia chromosome (Ph[1]) which represents an abnormally small chromosome 22 resulting from a translocation between chromosomes 9 and 22 - t ($9q^+$; $22q^-$). The majority of CML is Ph[1] positive (90-95%) and typically has an insidious onset, progressive splenomegaly, anemia, WBC approximately 200,000/mm^3 in the chronic phase, with a hypercellular marrow, increased eosinophils, basophils, and platelets. Ph[1] negative CML is less characteristic.

It has recently been determined that there is a reciprocal translocation of the cellular analogue of the transforming sequence of the Abelson murine leukemia virus (c-abl) on chromosome 9. This is transferred to a breakpoint cluster region (bcr) on chromosome 22 to give a modified gene coding for a hybrid 210KD phosphoprotein gene product (p210 bcr-abl). New techniques are being developed to detect the new mutant gene, the hybrid m-RNA produced, or the fusion gene protein product to replace karyotyping. This translocation is detected in approximate 95% of patients with CML, 5-10% with ALL and 1-2% with AML.

Philadelphia Chromosome Negative Chronic Myelocytic Leukemia: Approximately 5-10% of CML cases are Ph[1] negative. Many of these have the translocation masked, but do have the abl-bcr rearrangement as detected by restriction fragment analysis (bcr gene rearrangement assay). These behave the same as classical Ph[1] positive CML. The other Ph[1] negative patients do poorly. Some of these have been misdiagnosed as CML, while in others the genetic cause of their CML is still to be elucidated.

Approximately 30-36 months after initial diagnosis, the disease enters an accelerated phase or a blast crisis. The blast phase is usually represented by myeloblasts, but in approximately 20-30% of the cases, lymphoblasts are found. A lymphoid versus a myeloid lineage of these blasts can be determined by the presence of terminal deoxynucleotide transferase, TdT, in the lymphoblasts.

LEUKOCYTE ALKALINE PHOSPHATASE (LAP) STAIN

Synonym: LAP smear; LAP score.

SPECIMEN: Six fresh finger stick smears; air dry.

REFERENCE RANGE: Score of 50-150; each laboratory should determine its own normal range.

METHOD: Positive and negative controls must be run with each unknown; a positive control is obtained from women in their third trimester of pregnancy or within a few days following delivery. Negative controls are prepared by immersing an air dried fixed smear into boiling water for 1 minute. Slides are fixed in formalin:methanol (1:9) for 30 sec. at 0-5°C; wash in running water and air dry; incubate with substrate, naphthol phosphate. The liberated naphthol is then conjugated with a diazo compound.

Alkaline phosphatase activity is reflected by colored granules (usually red, blue or purple depending on diazo cpd) in the cytoplasm of mature leukocytes; cells other than reticulum cells, osteoblasts and endothelial cells show no activity. A kit for determining LAP activity is available from Sigma Corp. St. Louis, Mo.

Using oil immersion, 100 consecutive segmented and band form neutrophils are rated 0 to 4+ on the basis of granule size, staining and background of cytoplasm. The score is the sum of these grades; the maximum score is 400.

INTERPRETATION: LAP activity is assessed on blood smears in order to discriminate among a number of hematological conditions. The principal uses of the LAP stain are in distinguishing CML from other causes of neutrophilia and in distinguishing polycythemia vera from secondary polycythemia. LAP may be useful in differentiating the conditions listed in the next Table: (cont)

Leukocyte Alkaline Phosphatase Stain (Cont)

Differentiation Using Leukocyte Alkaline Phosphatase (LAP)

Low LAP		Elevated LAP
Chronic Myelocytic Leukemia	versus	Polycythemia Vera Granulocytic Leukemoid Reactions Myelosclerosis Myeloid Metaplasia
Normal or Low LAP		**Elevated LAP**
Acute Myelocytic Leukemia	versus	Acute Lymphocytic Leukemia Acute Myelomonocytic Leukemia
Normal LAP		**High LAP**
Secondary Polycythemia	versus	Polycythemia Vera

LAP has no relationship to serum alkaline phosphatase. The causes of increased LAP are given in the next Table:

Causes of Increased Leukocyte Alkaline Phosphatase
Neutrophilia Secondary to Infections
Stress, Trauma and Burns, Immediate Postoperative Period, Tissue Necrosis
Lymphomas: Acute Lymphocytic Leukemia; Hodgkin's Disease
Myeloproliferative Disorders: Polycythemia Vera; Myelofibrosis; Agnogenic Myeloid Metaplasia
Pregnancy and During Lactation
Oral Contraceptives
Newborns to 2 weeks of age

The causes of decreased LAP are given in the next Table:

Causes of Decreased Leukocyte Alkaline Phosphatase (LAP)
Acute (50%) and Chronic Myelocytic Leukemia
Erythroleukemia
Infectious Mononucleosis (Early)
Collagen Diseases
Paroxysmal Nocturnal Hemoglobinuria
Hypophosphatasia
ITP

LAP is normal in secondary polycythemia, chronic lymphocytic leukemia and viral infections.

LIDOCAINE (XYLOCAINE)

SPECIMEN: Red top tube, separate serum and refrigerate; or green (heparin) top tube. Spuriously low plasma lidocaine concentrations may be obtained when blood is collected in certain commercial rubber-stoppered collection tubes. Use all-glass tubes or test any commercial collecting tube thoroughly before use.
REFERENCE RANGE: Therapeutic: 2-5mcg/mL. Time to Steady State (Sampling Time): 12 hours after beginning lidocaine. Half-Life: 74-140 minutes. Cardiac Failure: 115 minutes. Uremia: 77 minutes. Cirrhosis: 296 minutes. Toxic: >5mcg/mL.
METHOD: EMIT; GLC; HPLC; Fluorescence Polarization (Abbott).
INTERPRETATION: Lidocaine is a class IB antiarrhythmic drug (lidocaine, mexiletine, phenytoin, tocainide); lidocaine has local anesthetic action (sodium channel blockade) and shortens action potential duration. Lidocaine is used for the prevention and treatment of ventricular arrhythmias. It had been used as a local anesthetic and for treatment of seizures. Current recommendations for use of lidocaine in acute myocardial infarction are listed in the following Table (Gunnar RM, et al. JACC. 1990; 16:249-292):

Lidocaine Use in Acute Myocardial Infarction
Frequent Ventricular Premature Beats (VPB)(>6/min)
Closely Coupled VPB's (R on T Phenomenon)
Multiform VPB's
Short Bursts of VPB's (Three or More)
Ventricular Tachycardia or Fibrillation Resistant to Defibrillation

Routine lidocaine prophylaxis is not recommended following acute myocardial infarction due to the lack of clear benefit and high incidence of adverse effects (Zehender M, et al. Clin Cardiol. 1990; 13:534-539; MacMahon S, et al. JAMA. 1988; 260:1910-1916; Gunnar RM, et al. already cited).

Guidelines for therapy and serum specimen monitoring are given in the next Table:

Guidelines for Therapy and Monitoring

Therapy	Monitoring
Loading Dose: 1mg/kg (max 100mg) Repeat 0.5mg/kg every 8-10 min to max total dose 4mg/kg if needed Maintenance: 20-50mcg/kg/min (1.4-3.5mg/min for 70kg patient)	12 hours after beginning lidocaine infusion; at least every 12 hours in patients with signs or symptoms of cardiac or hepatic insufficiency. Obtain specimen whenever toxicity is suspected. Monitor whenever ventricular arrhythmias occur despite lidocaine administration.

The majority of lidocaine's adverse effects involve the central nervous or cardiovascular systems. Milder adverse reactions, such as drowsiness, dizziness, transient paresthesias, muscle twitching, and nausea, are common after loading doses or can occur at plasma levels greater than 5mcg/mL. Severe adverse reactions, such as central nervous system, cardiovascular and respiratory depression, convulsions, coma, may appear at concentrations greater than 9mcg/mL. The incidence of adverse reactions to lidocaine ranges from 4 to 85 percent (average, 35 percent)(Review: Tisdale JE. Henry Ford Hosp Med J. 1991; 39:217-225).

Plasma levels and signs or symptoms of toxicity need to be closely monitored in patients with altered drug elimination such as cardiac failure, acute myocardial infarction, liver disease and prolonged intravenous infusion. Drug Interaction: Propranolol and cimetidine reduce the systemic clearance of lidocaine by reducing hepatic blood flow and hepatic metabolism of lidocaine; thus, serum concentration of lidocaine may increase.

When lidocaine is administered with other antiarrhythmic drugs such as procainamide, quinidine, phenytoin or propranolol, the cardiac effects may be additive or antagonistic and toxic effects may be additive.

Concurrent use of tocainide can cause seizures, possibly from additive neurotoxicity.

LIPASE

SPECIMEN: Red top tube, separate serum; store refrigerated or frozen.
REFERENCE RANGE: <130U/liter
METHOD: Turbidimetric method; thin-film (Ektachem) method; Titrimetry.
INTERPRETATION: Lipase activity is present in significant quantities only in the pancreas. Serum lipase is elevated in patients with acute pancreatitis. The change in serum lipase, serum amylase and urine amylase in acute pancreatitis is illustrated in the next Figure:

Serum Amylase, Serum Lipase and Urine Amylase Following Acute Pancreatitis

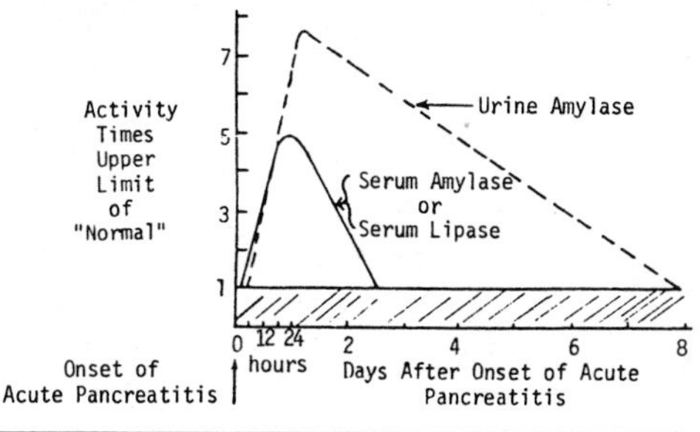

Serum lipase changes in a manner similar to that of serum amylase following onset of acute pancreatitis. It begins to rise in 2 to 6 hours, reaches a maximum in 12 to 30 hours and remains elevated for 2 to 4 days.

Lipase (Cont)

Serum lipase has a clinical sensitivity greater than 80% and a specificity of 60% for acute pancreatitis, which is superior to total amylase (Lott JA, et al. Clin Chem. 1986; 32(7):1290-1302). Using a level of three times normal, serum lipase has a very high sensitivity (100%) and specificity (100%) in the detection of acute alcoholic pancreatitis (Gumaste V, et al. Am J Med. 1992; 92:239-242).

LIPOPROTEIN (a), Lp(a)

SPECIMEN: Lavender (EDTA) top tube, separate plasma.
REFERENCE RANGE: Less than 30mg/dL.
METHOD: Immunological
INTERPRETATION: Lp(a) is a macromolecular complex of LDL which has been modified by covalent addition of a molecule of apolipoprotein (a). One molecule of apolipoprotein (a) is linked to one molecule of apolipoprotein (a) is linked to one molecule of apo B via a disulfide bond. Lipoprotein (a) is important because it is an independent risk factor for coronary atherosclerosis and also an excellent predictor of restenosis of vein grafts and following percutaneous transluminal angioplasty (Rath M, et al. Arteriosclerosis. 1989; 9:579-592; Scanu AM. JAMA. 1992; 267:3326-3329).

Lp(a) is highly polymorphic with a MW varying from 280,000 to 500,000. There is an inverse relationship between the size of the apo (a) isoprotein and Lp(a) concentration. The smallest isoforms are associated with the highest levels (Unterman G. Science. 1989; 246:904-910). The biological function of Lp(a) is unknown. However, its gene has recently been shown to be highly homologous to the gene for plasminogen (McLean JW, et al. Nature. 1987; 330:132-137). It is suggested that Lp(a) may be prothrombogenic by competing with plasmin and thereby inhibiting clot dissolution.

Concentrations of Lp(a) greater than 30mg/dL are associated with a two-fold greater risk of developing coronary atherosclerotic disease (Rader DJ, Brewer HB. JAMA. 1992; 267:1109-1112). The distribution of plasma Lp(a) varies greatly between ethnic groups (Unterman, already cited). It is higher in African Americans than whites and is probably a major risk factor in this group (Guyton JR, et al. Arteriosclerosis. 1985; 5:265-272). Causes of an elevated Lp(a) are given below (Rader and Brewer, already cited):

Causes of an Elevated Lp(a)
Genetics (Certain Apo(a) Phenotypes)
Familial Hypercholesterolemia
Renal Disease
Nephrotic Syndrome
Diabetes Mellitus with Microalbuminuria

LIPOPROTEIN PROFILE/LIPOPROTEIN ELECTROPHORESIS, CHOLESTEROL, HIGH DENSITY LIPOPROTEIN (HDL), LOW DENSITY LIPOPROTEIN (LDL)

SPECIMEN: Fasting, 12-14 hours; red top tube, separate serum.
REFERENCE RANGE: The normal distributions for plasma lipids and lipoprotein cholesterol are given in the next Table; (Manual of Laboratory Operations, Vol. I, N.I.H., Bethesda, Maryland 20014, DHEW Publication No. (NIH. 75-628, pg. 73); Consensus Conference. JAMA. 1985; 253:2080-2086).

Normal Distributions for Plasma Lipids and Lipoprotein Cholesterol

Age	Total Plasma Cholesterol mg/dL	Plasma Triglyceride mg/dL	VLDL Cholesterol mg/dL	LDL Cholesterol mg/dL	HDL Cholesterol mg/dL	
					Male	Female
0-19	<180	10-140	5-25	50-170	30-65	30-70
20-29	<200	10-140	5-25	60-170	35-70	35-75
30-39	<220	10-150	5-35	70-190	30-65	35-80
40-49	<240	10-160	5-35	80-190	30-65	40-95
50-59	<240	10-190	10-40	80-210	30-65	35-85

Cholesterol: To convert conventional units in mg/dL to international units in mmol/liter, multiply conventional units by 0.02586.

Triglycerides: To convert conventional units in mg/dL to international units in mmol/liter, multiply conventional units by 0.01129.

More detailed tables containing lipid values for the American population may be found in the appendix of: Expert Panel, National Cholesterol Education Program. Arch Intern Med. 1988; 148:36-69.

METHOD: Cholesterol is assayed by a cholesterol oxidase method; triglycerides are assayed by an enzymatic method following hydrolysis and release of glycerol. HDL is measured as cholesterol in the supernatant of serum after precipitation of other lipoproteins with a polyanion-divalent cation combination such as phosphotungstate-Mg^{++} or heparin-Mn^{++}. In most instances, LDL is measured indirectly and calculated from total cholesterol minus the HDL and approximately one-fifth the triglyceride. There are new methods being developed to measure LDL directly including affinity purified antibodies and others which depend on circular diachromism. The lipoproteins are separated electrophoretically; the chylomicrons are found at the line of application; presence of chylomicrons can be ascertained by a lipid layer following overnight refrigeration; LDL migrates to the beta-globulin region; VLDL migrates to the pre-beta region and HDL migrates to the alpha region. Individual apolipoproteins are determined by immunological procedures.

INTERPRETATION: Lipoprotein profiles are primarily determined to assess cardiac risk and to aid in the diagnosis of certain disorders of lipoprotein metabolism.

Analysis of a lipoprotein profile and its various components (VLDL, LDL, HDL and triglycerides (TG) requires a 12-14 hour fasted sample. There is a fair amount of within-individual variability observed when measuring these components: cholesterol CV=6%, TG CV=25%, HDL CV=8% (Cooper GR, et al. JAMA. 1992; 267:1652-1660).

Summary of Lipoprotein Characteristics: Lipoproteins may be classified according to their density into chylomicrons, very low density lipoproteins (VLDL), low density lipoproteins (LDL), and high density lipoproteins (HDL). This is a tedious ultracentrifugation procedure which is not used in laboratory diagnosis, but which resulted in the terminology presently used.

Lipoproteins may also be separated by electrophoresis into the beta (LDL), prebeta (VLDL), and alpha (HDL) lipoproteins. Electrophoresis was used in the original Fredrickson classification of primary hyperlipidemias. Electrophoresis rarely has to be performed in present clinical practice. The type of hyperlipidemia can usually be determined by consideration of cholesterol and TG levels. The relationship between lipoprotein density, electrophoretic mobility, and lipid content is given in the next Table:

Characteristics of Lipoproteins

Lipoprotein Fraction	Electrophoretic Mobility	Composition
Chylomicrons	-	TG rich (80%)
VLDL	Pre-beta	TG rich (55%)
LDL	Beta	Cholesterol rich (45%)
HDL	Alpha	Protein rich (45%)

The Fredrickson Classification System: Fredrickson classified primary hyperlipoproteinemias into 5, and later 6, groups based on electrophoretic patterns and clinical presentation (Fredrickson DS, Less RS. Circulation. 1965; 31:321-327; Benmont JL, et al. Bull WHO. 1970; 43:891-915).

Lipoprteins (Cont)

The laboratory findings with regard to electrophoretic banding and cholesterol and TG levels is given in the next Table:

Findings in Fredrickson Classification

Type of Hyperlipidemia	Lipoprotein Band: Chylomicrons	Beta	Pre-Beta	Lipid: Cholesterol	TG
I	↑				↑
IIa		↑		↑	
IIb		↑	↑	↑	↑
III		↑*		↑	↑
IV			↑		↑
V	↑		↑		↑

* In type III the abnormal lipoprotein is composed of VLDL remnants. These have the density of the VLDL but migrate in the beta region.

Classification of hyperlipidemias may also be done on the basis of fasting serum levels of cholesterol and triglycerides alone. The lipoprotein pattern is normal if plasma cholesterol and triglycerides are both normal. If fasting plasma cholesterol is elevated and plasma triglycerides are normal, then the patient has a Type IIa hyperlipoproteinemia. If the fasting plasma triglycerides are elevated and plasma cholesterol is normal, then VLDL or chylomicrons are elevated; fasting elevated triglyceride is usually due to elevated VLDL and Type IV hyperlipoproteinemia; if chylomicrons are also elevated, then the patient may have Type V; if only chylomicrons are elevated, then the patient has the rare Type I hyperlipidemia. The alterations of lipids and possible interpretation are summarized in the next Table:

Alterations of Plasma Lipids and Fredrickson's Classification

Lipid	Interpretation
Cholesterol Increased; Triglyceride Normal	Type IIa only
Triglyceride Increased; Cholesterol Normal	Type IV, V, I
Both Cholesterol and Triglyceride Increased	Type IIb, III, IV

Differentiation of Type IIb from Type IV: It is often necessary to differentiate whether a patient with elevated triglyceride with accompanying elevated cholesterol be classified as Type IIb or Type IV. Is hypercholesterolemia due to abnormally high concentration of VLDL or of LDL? The following formula may be used if the triglyceride concentration is less than 400mg/dl.

$$\text{LDL} = \text{total cholesterol} - \left(\frac{\text{TG}}{5} + \text{HDL}\right)$$

In type IV hyperlipidemia, VLDL is elevated and TG levels are in the range of 250-750mg/dL. If the TG level is >750mg/dL, this is probably a type V hyperlipidemia, or rarely type I. Another form of hyperlipidemia is termed Familial Combined Hyperlipidemia. This is found in families with no specific genetic defect but overproduction of lipoproteins. First degree relatives may exhibit multiple phenotypes including IIa, IIb, IV and V. It is estimated that approximately 15% of the people with coronary heart disease under age 60 have this condition.

A number of inherited abnormalities of enzymes, apolipoproteins, and lipoprotein receptors are now known to result in the phenotypic presentations subclassified by the Fredrickson system (Manley RW, et al. JAMA. 1991; 265:78-83) as given in the next table:

Fredrickson Hyperlipidemias and Possible Responsible Genetic Defects

Type Hyperlipidemia	Genetic Defect	Principal Laboratory Finding
I	Lipoprotein Lipase Deficiency C-II apolipoprotein deficiency	↑ Chylomicron (↑TG)
IIa	Defective LDL (apolipoprotein B) receptors	↑ LDL (↑ cholesterol)
III	Mutations of apolipoprotein E	↑ Chylomicron and VLDL remnants (↑ cholesterol, ↑ TG)
V	Heparin induced (activated) lipoprotein lipase deficiency	↑ Chylomicrons and VLDL (↑ TG)

Classification of Primary Hyperlipoproteinemia (Goldstein): A simpler way of classifying primary hyperlipoproteinemia is that of Goldstein, et al. J Clin Invest. 1973; 52:1544-1568 as seen in the next Table:

Classification of Primary Hyperlipoproteinemia (Goldstein)
Familial Hypercholesterolemia
Familial Triglyceridemia
Combined Hypercholesterolemia and Triglyceridemia

Secondary Hyperlipoproteinemia: It is not necessary to perform lipoprotein electrophoresis on the secondary hyperlipoproteinemias. The causes of secondary hyperlipoproteinemia and the change in plasma cholesterol and triglycerides are given in the next Table:

Changes in Plasma Lipids in Secondary Hyperlipoproteinemias

Condition	Cholesterol	Triglycerides
Pregnancy	↑	
Hypothyroidism	↑	
Cholestasis; Intra or Extrahepatic	↑	
Acute Intermittent Porphyria	↑	
- - - - - - - -		
Nephrotic Syndrome	↑	↑
Chronic Renal Failure	↑	↑
Steroid Immunosuppressive Therapy	↑	↑
- - - - - - - -		
Oral Contraceptives		↑
Diabetes Mellitus		↑
Acute Alcoholism		↑
Acute Pancreatitis		↑
Gout		↑
Gram-Negative Septicemia		↑
Type I Glycogen Storage Disease		↑

The conditions above the first dashed line cause elevation of cholesterol; those between the dashed lines cause elevation of both cholesterol and triglycerides while those below the second dashed line cause elevation of triglycerides.

High Density Lipoprotein (HDL): It has been found that decreased levels of high density lipoprotein (HDL) in plasma lead to an increased risk in males of coronary heart disease. Among the various lipid risk factors, HDL cholesterol appears to have the strongest relationship to coronary heart disease. It may also be that raised levels exert a protective effect in that premenopausal women have HDL concentrations 30 to 60 percent higher than their male counterparts and subjects with familial hyper-alphalipoproteinemia have an above-average life expectancy. In a survey of patients with very high values of HDL (>100mg/dL) the majority 43 out of 46 were women. Many of these were treated with estrogen, used H2-blockers or alcohol (Wetzman JB, Vladutin AO. Arch Pathol Lab Med. 1992; 116:831-836). Many had evidence of coronary atherosclerosis. Other causes of HDL elevation include familial hyperalphalipoproteinemia and hypobetalipoproteinemia.

The incidence of coronary heart disease by HDL cholesterol level is shown in the next Table (Gordon et al. Am J Med. 1977; 62:707-714):

Incidence of Coronary Heart Disease by HDL Cholesterol Level

HDL Cholesterol Level (mg/dL)	Incidence of Coronary Heart Disease	
	Men	Women
Patient Population	8%	4%
<25	18%	-
25-34	10%	17%
35-44	10%	5%
45-54	5%	5%
55-64	6%	4%
65-74	2.5%	1.4%
>75	0%	2%

Lipoprteins (Cont)

HDL levels can be reported as absolute values or as the total cholesterol/HDL cholesterol ratio. This ratio does not require a fasted specimen or a triglyceride measurement.

Low levels of HDL (<35mg/dL - 0.9 mmol/L) are associated with increased myocardial risk. Low levels of HDL are also associated with increased risk of coronary angioplasty restenosis (Shah PK, Amin J. Circulation. 1992; 85:1279-1285). The major causes of reduced serum HDL are listed below:

Major Causes of Reduced Serum HDL
Cigarette Smoking
Obesity
Lack of Exercise
Androgenic and Related Steroids
B-adrenergic Blocking Agents
Hypertriglyceridemia
Genetic Factors (eg, Primary Hypoalphalipoproteinemia)

HDL can be fractionated into two measurable subfractions, HDL_2 and HDL_3. Alcohol intake appears to elevate HLD_3, apo A-I and apo A-II (Taskinen M-R, et al. Am Heart J. 1987; 113:458-463). Diet may elevate HDL by increasing the cholesterol content of the HDL while exercise may increase the number of HDL particles (Schwartz R. Metabolism. 1987; 36:165-171). Although HDL measurements are important in predicting myocardial risk, detailed laboratory measurement of HDL_2, HDL_3, or the associated apolipoproteins A-I or A-II do not add to the diagnostic accuracy (Stamper MJ, et al. N Engl J Med. 1991; 325:373-381).

Low Density Lipoprotein (LDL): A LDL greater than 170 mg/dL is associated with increased myocardial risk, even though the total cholesterol is in the desirable or borderline range. The LDL level is determined on a fasting sample and calculated from the values for total cholesterol, HDL and TG as follows:

$$LDL = \text{total cholesterol} - \left(\frac{TG}{5} + HDL\right)$$

Some prefer to use "6.25" in the denominator in place of "5".

Low LDL is seen in abetalipoproteinemia and in familial hypoproteinemia. In the former there is an absence of LDL, low total cholesterol (20-50mg/dL), a defect in vitamin E delivery, and symptoms of anemia (acanthocytes), progressive pigmented retinopathy and neurological symptoms. This is probably due to lack of a microsomal transfer protein. However, most patients with low LDL have the latter disorder, familial hypobetalipoproteinemia. These patients have a truncated apo-B protein. A third disorder with a low LDL is Anderson's disease, a disorder of unknown etiology. These disorders are discussed and contrasted in: Rader DJ, Brewer HB. JAMA. 1993; 270:865-869.

The role of triglyceride as an independent risk factor in risk of coronary atherosclerotic disease is controversial.

Other topics: See under respective headings: APOLIPOPROTEINS, LIPOPROTEIN(a), CARDIAC RISK ASSESSMENT, TRIGLYCERIDE.

LITHIUM

SPECIMEN: Red top tube, separate serum; refrigerate at 4°C; stable for 24 hours at room temperature. Draw specimen for trough concentration 8-12 hours after last oral dose. Reject specimen if collected in lithium heparin.

REFERENCE RANGE: Serum therapeutic range: 0.8mmol/liter to 1.3mmol/liter for the treatment of acute mania; 0.5 to 1.0mmol/liter for sustained prophylactic use. The levels in saliva are two times greater than those in serum. For the average adult with normal creatinine clearance, these serum levels can be achieved with a dose of lithium carbonate of approximately 300mg to 600mg orally three times daily. However, the dosage must be individualized because there is variation in the rate at which lithium is eliminated by the kidney.

METHOD: Flame emission photometry; atomic absorption.

INTERPRETATION: Lithium (more precisely lithium ion) is used for the treatment of patients with acute mania and for prophylaxis against recurrent manic depressive illness. It is also used for treatment of migraine and cluster headache. Lithium may prevent depression but is not effective in lifting a patient out of the depressive phase. The serum half-life varies from about 18 hours in the young to about 36 hours in the elderly.

Lithium is rapidly absorbed following oral doses with peak plasma levels occurring in one to three hours. Therapeutic and toxic drug levels are obtained on specimens drawn 8 to 12 hours after administration of the last oral dose.

Side Effects of Lithium: The common side-effects of lithium therapy are given in the next Table:

Common Side Effects of Lithium Therapy

Thyroid:
- Transitory Biochemical Changes (Elevated Thyroid Stimulating Hormone (TSH); Decreased Free Thyroxine (T-4) and Triiodothyronine (T-3) Clinical Change)
- Goiter with Euthyroidism (4%)
- Hypothyroidism (4% to 15%)
  - (a) Thyroiditis with Goiter; High Thyroid Antibody Titer; Low RAIU
  - (b) No Goiter; No Antibodies
- Hyperthyroidism (rare)

Renal Function:
- Polyuria (40% of patients) Plasma Osmolality, Normal; Urine Osmolality, Decreased

Granulocytes:
- Benign, reversible leukocytosis (14,000-24,000/cu mm) or total count normal, increase in granulocytes and platelets

Thyroid: Patients who have a personal or family history of thyroid disorder may be especially prone to develop hypothyroidism; the incidence in women is about ten times that in men. Hypothyroidism may occur any time during the course of therapy. The usual course of events is that thyroxine levels decrease initially and the thyrotropin (TSH) levels become elevated in the first few weeks and finally stabilize. Thyrotropin (TSH) levels have been found to be abnormal in up to 30% of lithium treated patients at some time during the course of therapy. They return to normal in about half of these patients over a period of time (Emerson CH, et al. Clin Endocrinol Metab. 1973; 36:338-346; Calabrese JR, Gulledge AD. Cleve Clin. 1983; 50:32-33). Withdrawal of lithium usually results in the return of normal thyroid function.

Renal Function: Polyuria has been reported to occur in over 40% of patients on lithium; in some patients the urinary output has exceeded 11 liters/day. Lithium is excreted almost entirely by the kidneys; in states of sodium depletion, lithium will be reabsorbed by the tubules, leading to increase in serum and tissue lithium levels and development of clinical lithium toxicity. Conditions associated with sodium loss, such as diuretics, renal failure, gastrointestinal losses, and excessive perspiration may be associated with lithium toxicity.

The thiazide diuretics cause a significant rise in the serum lithium. These diuretics cause a net sodium loss, which may lead to a compensatory increase in proximal sodium and lithium reabsorption. Furosemide, which acts on the cells of the ascending limb of the loop of Henle proximal to the thiazide diuretics, does not cause a significant increase in serum lithium levels (Jefferson TW, Kalin NH. JAMA. 1979; 241:1134-1136).

Lithium (Cont)

Toxic renal effects including tubular lesions, interstitial fibrosis and decreased creatinine clearance have been reported in patients treated with lithium; these are uncommon (Hansen HE. Drugs. 1981; 22:461; Ramsey TA, Cox M. Am J Psychiatry. 1982; 139:443).

Granulocytes: A benign, reversible leukocytosis is frequently seen during lithium therapy (14,000-24,000/cu mm) or the total white count may remain normal with an absolute increase in mature granulocytes and platelets. The leukocytosis begins within a week of the start of administration, persists during therapy, and resolves within a week of discontinuance of the drug. This lithium effect is of no clinical significance other than to avoid confusion with an occult infection or blood dyscrasia (Editorial, The Lancet. 1980; 626-627).

The side effects of lithium, that is, hypothyroidism, polyuria and leukocytosis have been utilized in treatment of patients with hyperthyroidism, SIADH, and leukopenia, respectively.

Laboratory Studies Performed Before Lithium Administration: Before administration of lithium the laboratory studies listed in the next Table should be performed:

Laboratory Studies Performed Before Lithium Administration

Test	Possible Change During Lithium Therapy
Urinalysis	Lithium retention in renal disease; for baseline specific gravity.
Serum Creatinine or BUN	Lithium retention in renal disease.
Serum Sodium	Low sodium associated with retention; high sodium associated with elimination.
Thyroxine (T-4) TSH Thyroid Microsomal Antibody Titers	Hypothyroidism - High incidence (4% to 15%)
White Cell Count	Leukocytosis
Electrocardiogram	Flattened or Inverted Wave - Benign and Reversible Arrhythmias, Ventricular Tachycardia, Sinoatrial and Atrioventricular Block (Jaffe CM. Am J Psy. 1977; 134:88-89). Frequent ECG monitoring recommended

Toxic Manifestations of Lithium: Toxic manifestations of lithium generally do not appear unless plasma levels exceed 1.5mmol/L. The toxic manifestations are listed in the next Table (Coyle JT. Med Clin N Am. 1977; 61(4):891-905):

Toxic Manifestations of Lithium Therapy

Plasma Level (mmol/L)	Symptoms
<1.5 (Therapeutic)	Nausea (Especially on Initiation of Treatment) Fine Tremor: Mild Polyuria
>1.5 <2.5	Diarrhea and Vomiting: Polyuria; Coarse Tremor, Ataxia; Muscle Weakness and Fasciculation; Sedation, Langour
>2.5 <4.0	Muscle Hypertonia; Choreiform Movements; Increased Deep Tendon Reflexes; Impairment of Consciousness with Somnolence, Confusion, Stupor; Transient Focal Neurologic; Signs; Seizures
>4.0	Coma: Death

Most of the effects are due to the action of lithium on the central nervous system. Neurological side effects are many and vary from tremor, usually seen at the start of treatment and occurring at therapeutic drug levels, to marked alterations in consciousness, neuromuscular irritability, seizures, incoordination, delirium, irreversible brain damage and death.

The degree of toxic symptoms does not always correlate well with serum lithium levels and in fact, individuals have been seen with documented lithium toxicity in spite of "normal" lithium serum values (Strayhorn JM, Nash JL. Diseases of the Nervous System. 1977; 38:17-111).

LIVER ANTIBODY PANEL

Measurement of autoantibodies in a patient with liver disease may be useful in select cases. This is especially true of those disorders in which an immunological component is suspected: primary biliary cirrhosis, chronic active hepatitis, and drug induced hepatitis. Incidence of elevated smooth muscle antibodies, anti-mitochondrial antibodies and antinuclear antibodies in hepatic conditions are given in the next Table modified from Mayo Medical Laboratories Test Catalogue. 1983; 319.

Elevated Antibodies in Various Hepatic Conditions (%)

Condition	Anti-Mitochondrial	Anti-Smooth Muscle	Anti-Nuclear
Biliary Cirrhosis, Primary	90-95	0-50	25-50
Chronic Active Hepatitis (Autoimmune Type; Lupoid Hepatitis)	0-30	50-97*	20-30
Extra-Hepatic Biliary Obstruction	0-5	0	0
Cryptogenic Cirrhosis	0-25	0-1	0-1
Viral (Infectious) Hepatitis	0	1-2	
Drug-Induced Hepatitis	50-80**	---	---

*Values are variable. Anti-smooth muscle antibodies (anti-actin) have been reported in high titer in up to 97% of patients with autoimmune chronic active hepatitis (Lidman K, et al. Clin Exp Immunol. 1976; 24:266-272).
**Especially iproniazid and halothane-induced.

LIVER DISEASES: CONGENITAL, GENETIC AND CHILDHOOD DISEASE PANEL

The conditions and tests in childhood liver diseases are given in the next Table:

Congenital, Genetic and Childhood Liver Diseases

Condition	Tests
Defects in Bilirubin Metabolism: Gilbert's Syndrome	Isolated Elevation of Serum Bilirubin Increased Indirect Bilirubin; Bile Acids are Normal; Caloric Restriction: 200 Calorie Diet for 24-48 hours results in increased serum bilirubin by two-fold or more.
Crigler-Najjar Type I	Indirect Bilirubin, >20 mg/dL Direct Bilirubin, Absent Response to Phenobarbital: None
Type II	Indirect Bilirubin, <20mg/dL Direct Bilirubin, Absent Response to Phenobarbital: Lowers Bilirubin Concentration
Dubin-Johnson Syndrome	Increased Direct Bilirubin
Hereditary Hemochromatosis	Serum Iron Increased; Decreased Iron Binding Capacity, % Saturation Increased: >60% (male) >50% (female) Increased Serum Ferritin Biopsy of Liver: Increased Iron
Hepatolenticular Degeneration (Wilson's Disease)	Decreased Ceruloplasmin Serum Decreased Total Copper in Serum: Decreased Indirect Reacting Copper Increased Direct Reacting Copper Increased Copper in Urine Increased Copper Deposited in Liver
Alpha-1-Antitrypsin Deficiency	Serum Protein Electrophoresis; Carefully examine Alpha-1-Band (low or absent) Assay Alpha-1-Antitrypsin when Alpha-1-Band is Decreased or Absent Perform Phenotyping if Indicated

(cont)

Liver Diseases: Congenital, Genetic, and Childhood Disease Panel (Cont)

Congenital, Genetic, and Childhood Diseases (Cont)	
Neonatal Hepatitis	Liver Function Tests
Biliary Atresia	Liver Function Tests
Infectious Hepatitis	Liver Function Tests. Appropriate antibodies to suspected viral agents.
Infantile Cirrhosis in India	Recently shown to be due to excessive copper intake from cooking and storage vessels. Increased copper stores in liver. Test ceruloplasmin; serum and urine copper.

Reference for hyperbilirubinemia: Berk PD, Javitt NB. Am J Med. 1978; 64:311-326.

LOW DENSITY LIPOPROTEIN (see LIPOPROTEIN PROFILE)

LUPUS ANTICOAGULANT

Synonym: Circulating anticoagulant
Related Tests: Anti-cardiolipin Antibody/Anti-phospholipid Antibody
REFERENCE RANGE: None or <1% of the population.
SPECIMEN: Platelet free (<15,000/mm^3). Many labs perform double centrifugation, others micropore filters or plasma separators.
METHOD: Usually detected in screening as a prolongation of the PTT using a lupus anticoagulant sensitive PTT reagent. Further documentation is made with mixing studies and more specialized coagulation tests. Guidelines for testing and revised diagnostic criteria may be found in: Exner T, et al. Thromb Haem. 1991; 65:320.
INTERPRETATION: Testing for the lupus anticoagulant is usually performed in patients with recurrent or unexplained thrombosis and recurrent fetal loss (typically second trimester or later). Clinical conditions and complications associated with the lupus anticoagulant are tabulated below (Sammaritana LR, et al. Semin Arch Rheum. 1990; 20:81-96):

Clinical Conditions and Complications Associated with the Lupus Anticoagulant

Vascular
- Arterial and Venous Thrombosis
- Associated Conditions (Stroke, etc.)

Obstetrical
- Fetal Growth Retardation
- Pre-eclampsia
- Intrauterine Death and Abortion (Late)
- Diminished Amnionic Fluid

Hematological
- Thrombocytopenia
- Hemolytic Anemia (Coombs Positive)

Other
- Valvular Heart Disease
- Neurological Disease

Dermatological
- Livedo Reticularis
- Cutaneous Necrosis

In those patients with thrombotic episodes, 30% are arterial, and approximately 70% involve the lower extremity.

The lupus anticoagulant is probably a subpopulation of anti-phospholipid/anti-cardiolipin antibodies which in vivo cause thrombosis, but which in vitro result in a paradoxical prolongation of the phospholipid dependent coagulation tests. The lupus anticoagulant is measured in coagulation testing, while the closely related anti-cardiolipin/anti-phospholipid antibodies are measured in immunological assays (see ANTI-CARDIOLIPIN ANTIBODY). Not all individuals with anti-cardiolipin antibodies will have the lupus anticoagulant, and vice versa (Lockskin MD. JAMA. 1992; 268:1451-1453). In general, the anti-cardiolipin assay is easier and more sensitive, while the lupus anticoagulant tests are more specific. Estimated prevalences of the lupus anticoagulant and anti-phospholipid antibody in various conditions is given in the next Table:

Estimated Prevalence of Anti-phospholipid Antibody (APL) and Lupus Anticoagulant (LA)

Type Patient	% + APL	% + LA
Normal	<2	<1
Normal Pregnant	<2	<2
SLE	30-40	10-20
Recurrent Fetal Death (non-SLE)	10-20	10-20
RA	10*	<5
HIV	60-90*	20-40
Stroke, MI <40	10-20	

* Low Titer. Sammaritana LR, et al. (1990).

The lupus anticoagulant is classically associated with SLE. However, there are many other situations in which lupus anticoagulants have been described as listed below. In some of these conditions, the anticoagulant is found in low titer, and in many of these conditions is thought not to be clinically significant.

Conditions Associated with Lupus Anticoagulant or Anti-phospholipid Antibody
Autoimmune Disease: SLE, RA
Drugs: Chlorpromazine, Procainamide, Hydralazine, Phenytoin, Quinine, Certain Antibiotics
Infections: Bacterial, Protozoal and Viral (AIDS)
Malignancy: Hematopoietic and Lymphatic
Familial

Patients with lupus anticoagulants related to infections do not appear to have the same thrombotic risk as those occurring in autoimmune type responses. The most common cause of a prolonged PTT and lupus anticoagulant in children is probably infectious episodes (Currimbhoy Z. Am J Pediatr Hematol Oncol. 1984; 6:210-212). In such children a careful bleeding history should be taken to exclude factor VIII or IX deficiency.

The lupus anticoagulant is first manifest in the laboratory as a prolongation of one of the phospholipid dependent coagulation tests, usually the PTT. In order to document this prolongation, it is essential that the specimen be platelet free, and that the PTT reagent is lupus anticoagulant sensitive. Some laboratories employ a diluted PTT reagent for these studies (limiting the quantity of phospholipid in these assays improves sensitivity). Presence of the coagulation inhibitor as the cause of the prolonged PTT can be confirmed by mixing studies. Mixing with normal plasma will not correct the PTT if an inhibitor is present, whereas it will be corrected if due to the deficiency of a coagulation factor. Distinguishing a heparin effect from lupus anticoagulant may be accomplished by a Thrombin Test (abnormal with heparin; normal with lupus anticoagulant). Further confirmation requires more detailed and specialized testing. Presence of anti-cardiolipin/anti-phospholipid antibodies may be confirmed by immunological tests.

Detection and definition of the lupus anticoagulant may be difficult because of the different assays used, and different reagents employed. The PTT is the most commonly employed test and is available in most laboratories. However, the kaolin clotting time and Russell viper venom time tests are said to be more sensitive. A dilute Russell viper venom test is confirmed for the presence of lupus anticoagulant by addition of excess phospholipid (Thiagarajun P, et al. Blood. 1986; 68:869). Discussion of the various tests used in demonstration of the lupus anticoagulant may be found in: (Exner T. Thrombosis and Haem. 1985; 53:15-18. Triplett DA, et al. Am J Clin Path. 1983; 79:678-682).

LUTEINIZING HORMONE (LH)

SPECIMEN: Hormone levels undergo rapid and large oscillations. It is recommended to obtain three equally-spaced samples at about 15 to 20 minute intervals and pool the three specimens. Red top tube, separate serum.
REFERENCE RANGE: Comparison of reference values between various assays may be difficult because of standardization against different reference preparations. Representative values are as follows: Prepubertal children: less than 5 IU/L; LH values are high in the first several months of life. Adult female: follicular: 3-14 IU/L; luteal: 2-14 IU/L. Post Menopausal or Castrate: 15-70 IU/L. Adult Male: 3-12 IU/L. Primary hypogonadism, male and female: LH > 25 IU/L.
METHOD: RIA
INTERPRETATION: LH determinations are indicated in the work-up of patients with disorders of puberty (delayed or precocious), subfertility (gonadal failure; polycystic ovarian disease, disorders of hypothalamic-pituitary-gonadal interaction) and pituitary disorders (tumors, hypopituitarism). The gonadotropin (FSH or LH) levels for various conditions is given in the following Table (Beastall GH, et al. Ann Clin Biochem. 1987; 24:246-262):

FSH, LH, and Sex Steroid Levels in Various Conditions

Condition	Sex Steroids	FSH	LH
Delayed Puberty			
Gonadal Dysgenesis/Failure	Decr	Incr	Incr
Hypothalamic/Pituitary Disorders	Decr	Decr	Decr
Precocious Puberty			
True Precocious Puberty	Incr	Incr	Incr
Pseudo Precocious Puberty	Incr	Decr	Decr
Subfertility			
Gonadal Failure	Decr	Incr	Incr
Disorders of Hypothalamic-Pituitary-Gonadal Interaction	NL-Decr	NL-Decr	NL-Decr
Polycystic Ovary Disease	NL	NL-Decr	Incr
Pituitary Disorders			
Primary or Secondary Gonadotroph Tumors	NL-Incr	Incr	Incr
Hypopituitarism	Decr	Decr	Decr

(1) Male: The gonadotropins, LH and FSH, are under the control of a single hypothalamic hormone, the decapeptide gonadotropin releasing hormone (GnRH) also known as luteinizing hormone releasing hormone (LH-RH). Pulsatile release of GnRH begins early in adolescence; the gonadotropin ratios (FSH, LH) shift from the prepubertal pattern of FSH dominance to the postpubertal pattern of LH dominance. Without GnRH pulses, there is no puberty. FSH stimulates the seminiferous tubules and spermatogenesis; LH stimulates the Leydig cells which produce testosterone (Spark RF. Ann Intern Med. 1983; 98:103-105).

Knowledge of the characteristics of the pathway from the hypothalamus to the Leydig cells and secretion of testosterone is utilized in the successful treatment of patients with advanced prostatic cancer; these patients were treated with repeated administration of a long acting analogue of GnRH; suppression of the gonadotropins and testosterone and objective and subjective signs of regression of disease occurred in 75% of the patients (Waxman JH, et al. Brit Med. 1983; 286:1309-1312).

(a) Testicular Failure (Hypergonadotropic Hypogonadism): In these conditions there is testicular failure: LH and FSH are elevated and testosterone is decreased; parenteral testosterone is the preferred treatment. Conditions associated with hypergonadotropic hypogonadism are shown in the next Table:

Conditions Associated with Testicular Failure (Hypergonadotropic Hypogonadism)

Condition	Prepubertal	Postpubertal
Damaged seminiferous tubules with variable Leydig cell failure	Klinefelter's syndrome Germinal aplasia Reifenstein syndrome (and other related syndromes)	Cryptorchidism Mumps and other orchitis Irradiation Alkylating agents Myotonia dystrophica Malnutrition Uremia Idiopathic
Total gonadal failure	Functional prepubertal castrate (anorchia, etc.)	Adult castrate (trauma, surgery)

Klinefelter's syndrome presents as an adolescent boy with gynecomastia, a married male with infertility, or an adult male with a clinical history of testosterone deficiency. Reifenstein Syndrome is a hereditary condition characterized by hypospadia, postpubertal atrophy of the seminiferous tubules, and varying degrees of eunuchoidism and gynecomastia.

(b) Pituitary Failure (Hypogonadotropic Hypogonadism): When gonadotropin levels are low, or low-normal, hypothalamic pituitary disease must be suspected; further evaluation must include determination of serum prolactin values for hyperprolactinemia (Carter JN, et al. N Engl J Med. 1978; 299:847-852). Conditions associated with pituitary failure are shown in the next Table:

Conditions Associated with Pituitary Failure (Hypogonadotropic Hypogonadism)

Condition	Prepubertal	Postpubertal
LH failure alone	"Fertile eunuch"	
FSH and LH failure	Delayed puberty Hypogonadotrophic eunuchoidism: (a) with anosmia (Kallman Syndrome) (b) without anosmia (c) Lawrence-Moon-Biedl	Adult isolated gonadotropin failure (rare)
Failure of all pituitary hormones	Prepubertal panhypopituitarism	Adult panhypopituitarism

In Kallman's syndrome, hypogonadism is associated with anosmia or hyposmia and often with other neurologic defects (anosmia - loss or impairment of the sense of smell). The presumed cause of Kallman's syndrome is lack of pulsatile release of GnRH; sexual function can be restored by administering GnRH via an infusion pump programmed to mimic normal pulsatile secretion. Lawrence-Moon-Biedl Syndrome is characterized by retinitis pigmentosa, obesity, mental deficiency, polydactylia and hypogonadotropic hypogonadism.

(c) Precocious Puberty: Premature gonadotropin releasing hormone (GnRH) pulses cause precocious puberty (Crowley WF, et al. J Clin Endocrinol Metab. 1981; 52:370-372).

(2) Female: Serum LH levels are low in pituitary insufficiency and sustained elevations are seen in ovarian failure or postmenopausally. LH levels are decreased in anorexia nervosa, amenorrhea-galactorrhea syndrome (prolactin increased), and in women taking oral contraceptives. LH levels are usually increased in Stein Leventhal syndrome. The tests that may be used to differentiate pituitary, ovarian and end-organ failure, as a cause of amenorrhea, are given in the next Table:

Evaluation of Amenorrhea

Failure	Plasma FSH and LH	Urinary Estrogens	Urinary Estrogens Following HCG
Pituitary	↓	↓	↑
Ovarian	↑	↓	No Change
End-Organ	N	N	↑

In isolated pituitary gonadotropin (FSH and LH) deficiency, there are no clinical signs before the age of puberty. Secondary sexual development fails to occur during adolescence. If isolated pituitary gonadotropin deficiency occurs in the female adult, amenorrhea develops.

In amenorrhea associated with galactorrhea, there is a strong probability of prolactin elevation due to pituitary adenoma or dysfunction.

The elevated levels of LH in primary hypogonadism may be confused with the ovulatory peak in normal women. In the latter case, FSH levels will be normal range, and in the former, sex hormones are low. The combination of an elevated LH and normal FSH may suggest polycystic ovary disease. Many of these patients have a LH:FSH ratio > 2.5.

LH in Urine: LH is filtered at a rate of 10-20% of glomerular filtration rate. Urinary LH has been used to evaluate children with precocious puberty and to detect ovulation in the care of infertile couples.

See also Moghissi KS. Endocrinol Metab Clin N Amer. 1992; 21:39-55; Blacker CM. Endocrinol Metab Clin N Amer. 1992; 21:57-84.

LUTEINIZING HORMONE-RELEASING HORMONE TEST (LH-RH TEST)

Synonym: Gonadotropin Releasing Hormone Stimulation Test
SPECIMEN: Red top tubes for each blood specimen collected at -20 minutes and at 0, 20, 40, 60, 90, 120 and 180 minutes for LH and FSH assay; allow blood to clot in tubes, centrifuge and separate serum. In some protocols, FSH is not measured since its response is less.
REFERENCE RANGE: Normal responses for prepubertal subjects, adult males, and adult females, early follicular, preovulatory and luteal phase are given in the next Table:

Gonadotropin-Releasing Hormone(LH-RH) Test Results

Subject	Results Following LH-RH	
	LH	FSH
Prepubertal	Little Change	Increase: 1/2 to 2-Fold
Adult Male	Increase: 4 to 10-Fold	Increase: 1/2 to 2-Fold
Adult Female, Early Follicular Phase	Increase: 3 to 4-Fold	Increase: 1/2 to 2-Fold
Adult Female Preovulatory	Increase: 3 to 5-Fold	Increase: 1/2 to 2-Fold
Adult Female, Luteal	Increase: 8 to 10-Fold	Increase: 1/2 to 2-Fold

METHOD: NPO from evening snack or midnight. In A.M., start slow I.V. drip of normal saline solution. Obtain blood specimen at -20 and at 0 minutes for LH and FSH assay. After 0 specimen is drawn, give luteinizing releasing hormone(LH-RH), dose 100 mcg in children, 100 to 150 mcg in adults, I.V., push. Then, obtain blood specimens.
INTERPRETATION: This test is a measure of the functional integrity of the hypothalamic-pituitary-gonadal axis. The gonadotropin-releasing hormone or luteinizing hormone-releasing hormone(LH-RH) stimulates LH and, to a lesser degree, FSH secretion by the pituitary. If the gonadotropin values(LH and FSH) are unequivocally elevated, there is no need to perform a LH-RH test. The drug, clomiphene may also be used to stimulate LH and FSH secretion.

Females: The LH-RH test is most often used to evaluate females presenting with menstrual disturbances or infertility. The conditions in which the LH-RH test may be useful are given in the next Table (Wills MR, Harvard B. "Laboratory Investigation of Endocrine Disorders." Butterworths, London, 2nd edition, pgs. 20-23, 1983):

LH-RH Test in Females
Hypothalamic Lesions
Isolated Hypogonadotrophic Hypogonadism
Post-Operative Assessment Following Hypophysectomy
Polycystic Ovaries (Stein-Leventhal Syndrome)
Anorexia Nervosa
Primary Amenorrhea
Secondary Amenorrhea
Without Galactorrhea
With Galactorrhea

Amenorrhea with galactorrhea is commonly associated with prolactin producing pituitary adenomas.

Males: The LH-RH test is used in the investigation of males with hypogonadotropic hypogonadism in conditions listed in the next Table:

LH-RH Test in Males
Primary (Hypergonadotropic) Hypogonadism:
Adult
Pubertal
Cryptorchism
Secondary (Hypogonadotropic) Hypogonadism:
Systemic Diseases
Traumatic
Tumors
Infections
Psychogenic

Disease State Responses: LH and FSH are low in panhypopituitarism and response to stimulation is absent or blunted. A blunted response also occurs in the majority of patients with hypothalamic lesions. An exaggerated response with high basal values and a delay in return to baseline levels (>3 hours) indicates primary gonadal failure.

LYME DISEASE TESTING

SPECIMEN: Serology: Red top tube, separate serum. CSF and synovial fluid may also be tested. Culture: skin, blood, CSF, synovial fluid.
REFERENCE RANGE: Depends on specific test and laboratory. Mayo Medical Laboratories reports ELISA Nonreactive: less than 249 antibody response units (ABR); Weakly Reactive: 250-999 ABR; Reactive >1000 ABR. WESTERN BLOT: Normal less than or equal to 5 bands. Culture: Negative.
METHODS: Antibody Detection: ELISA, Immunofluorescence, Western Blot. Other: Culture, Direct Visualization, T-Cell Proliferation Assay, Urinary Spirochetal Antigen, Polymerase Chain Reaction.
INTERPRETATION: Lyme disease is caused by the spirochete Borrelia burgdorferi which infects humans via a tick vector. Lyme disease occurs throughout the world; in the United States, this disease is particularly prevalent in certain geographic areas including the northeastern states (Massachusetts to Maryland), upper midwestern states (Minnesota and Wisconsin), and certain western states (California, Oregon, Nevada, and Utah). The characteristic feature of early Lyme disease is the skin rash, erythema chronicum migrans. The clinical stages of Lyme disease are given in the following Table (Duffy J, et al. Mayo Clin Proc. 1988; 63:1116-1121):

Clinical Stages of Lyme Disease

Stage	Timing	Findings
1	Median of 4 Weeks	Erythema Migrans Influenza-Like Illness Severe Fatigue Musculoskeletal Pain Headache Stiff Neck
2	Days to Months	CNS Disease (Meningitis, Encephalitis, Bell's Palsy) Peripheral Nervous System Disease (Radiculopathy or Neuropathy) Cardiac Involvement (Heart Block, Myopericarditis, Congestive Heart Failure) Ophthalmitis
3	Months to Years	Arthritis: Asymmetric, Pauciarticular. Often intermittent, chronic in 10% CNS Disease: Severe, Chronic (Encephalitis, Demyelination, Psychiatric Disorders)

The time course and clinical manifestations of Lyme disease are illustrated in the following Figure (Reproduced with permission from Borenstein D. Lyme Disease Drug Therapy. July, 1990; 86-90):

Time Course and Clinical Manifestations of Lyme Disease

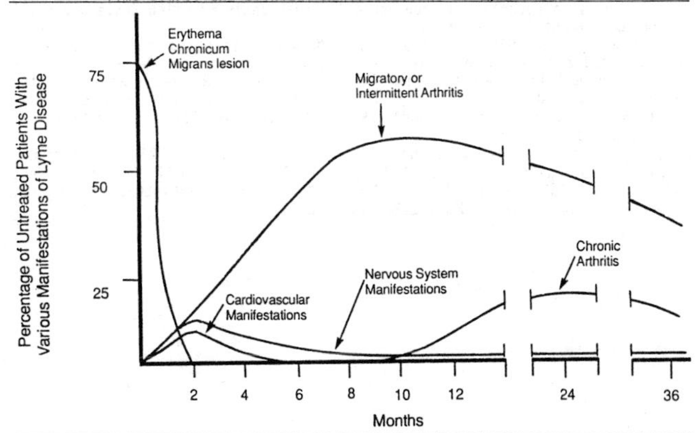

Laboratory Testing: Lyme disease is a clinical diagnosis supported by serologic testing. Antibodies (IgG or IgM) to Borrelia burgdorferi are most commonly detected by enzyme linked immunoassay (ELISA). In untreated patients, IgM antibody response occurs initially, peaking 3 to 6 weeks after infection and waning afterwards. IgG antibody response occurs later, during stage 2 or 3. Numerous studies have documented variable and often poor sensitivities, false positives, and significant intralaboratory and interlaboratory variability partly due to a lack of standardization of antigen preparation and quality control (Bakken LL, et al. JAMA. 1992; 268:891-895; Corpuz M, et al. Arch Intern Med. 1991; 151:1837-1840; Luger SW, Krauss E. Arch Intern Med. 1990; 150:761-763; Schwartz BS, et al. JAMA. 1989; 262:3431-3434; others).

Indirect immunofluorescence was developed prior to ELISA; this method is time consuming, and has largely been replaced by the ELISA technique. Western blot is more sensitive but less specific than ELISA, and is used to confirm ELISA results. Problems associated with serologic testing for Lyme disease are given in the following Table (Ostruv BE, Athreya BU. Pediatr Clin N Amer. 1991; 38:535-553):

Problems Associated with Serologic Testing for Lyme Disease
Assay Insensitivity and Nonspecificity
Interlaboratory Variability
Asymptomatic Seroconversion
False-Positive
Laboratory error
Other spirochetal infections (e.g. syphilis) or normal flora
Polyclonal B cell activation (rheumatoid arthritis, systemic lupus erythematosus, scleroderma, infectious mononucleosis, others)
Interference with assay by high titer ANA or RF
False-Negative
Laboratory error
Measurement too early in disease
Delayed immune response with late seroconversion
High cutoff for positive result (some positive results called negative)
Antibody sequestration in immune complexes
Aborted antibody response due to early antibiotic treatment
Pregnancy

(For individual references, see Ostruv and Athreya).

Culture and direct visualization using monoclonal antibodies are used occasionally. These methods of testing are insensitive; the poor recovery rate is presumably due to the short duration of spirochetemia and the low density of organisms in tissues (Duffy J, et al, already cited).

T-cell proliferation assay may be useful in a subgroup of patients with late Lyme disease and a negative or indeterminate antibody response by ELISA (Dressler F, et al. Ann Intern Med. 1991; 115:533-539).

Urinary spirochetal antigen and polymerase chain reaction tests are still in the experimental stage of development.

Lyme disease has been overdiagnosed and overtreated for a number of reasons. Media coverage has alerted the general public to the risk of this disease. The multisystem nature and slow evolution of Lyme disease as well as the insensitivity and nonspecificity of laboratory tests make this diagnosis problematic (Steere AC, et al. JAMA. 1993; 269:1812-1816; Ostrov BE, Athreya BH, already cited).

References: See also Steere AC. Hosp Practice. April 15, 1993; 37-44; Gerber MA, Shapiro ED. J Pediatr. 1992; 121:157-162; Rahn DW, Malawista SE. Ann Intern Med. 1991; 114:472-481; Barbour AG. Ann Intern Med. 1989; 110:501-502; Hedberg CW, Osterholm MT. Arch Intern Med. 1990; 150:732-733).

LYMPHOCYTE DIFFERENTIAL

Synonyms: T and B lymphocyte enumeration; Lymphocyte surface marker analysis; lymphocyte immunophenotyping; lymphocyte subset analysis.

Specimen: For marker studies on peripheral blood the minimum specimen is one 7mL green top tube (heparin) along with a lavender top tube for WBC and differential. Specimens may be held up to 24 hours at room temperature. For lymphoid tissues, the minimum is 5gm of tissue. Thin sections are placed in 10x tissue volume of RPMI/5% FCS and held up to 24 hours at 4°C. Bone marrow is usually aspirated into a heparinized syringe or placed in an equal volume of RPMI/5% FCS as above. If storage >24 hours is necessary, specimens are cryopreserved in liquid nitrogen. Body fluids typically require >2mL of fluid for analysis. In some laboratories, greater volumes of blood and other materials may be necessary for a complete analysis.

REFERENCE RANGE: Varies somewhat depending on methodology and reagents used. Values as a function of age are tabulated below (Modified from: Hannet I, et al. Immunology Today. 1992; 13:215).

	Cord Blood	Day 1-11 Months	1-6 Years	7-17 Years	18-80 Years
WBC					
Absolute Count	12(10-15)	9(6.4-11)	7.8(6.8-10)	6.0(4.7-7.3)	5.9(4.6-7.1)
Lymphocytes					
%	41(35-47)	47(39-59)	46(38-53)	40(36-43)	32(28-39)
Absolute Count	5.4(4.2-6.9)	4.1(2.7-5.4)	3.6(2.9-5.1)	2.4(2.0-2.7)	2.1(1.6-2.4)
T-Cells					
%	55(49-62)	64(58-67)	64(62-69)	70(66-76)	72(67-76)
Absolute Count	3.1(2.4-3.7)	2.5(1.7-3.6)	2.5(1.8-3.0)	1.8(1.4-2.0)	1.4(1.1-1.7)
B-Cells					
%	20(14-23)	23(19-31)	24(21-28)	16(12-22)	13(11-16)
Absolute Count	1.0(0.7-1.5)	0.9(0.9-1.5)	0.9(0.3-1.3)	0.4(0.3-0.5)	0.3(0.2-0.4)
$CD4^+$ T-Cells					
%	35(28-42)	41(38-50)	37(30-40)	37(33-41)	42(38-46)
Absolute Count	1.9(1.5-2.4)	2.2(1.7-2.8)	1.6(1.0-1.8)	0.8(0.7-1.1)	0.8(0.7-1.1)
$CD8^+$ T-Cells					
%	29(26-33)	21(18-25)	29(25-32)	30(27-35)	35(31-40)
Absolute Count	1.5(1.2-2.0)	0.9(0.8-1.2)	0.9(0.8-1.5)	0.8(0.6-0.9)	0.7(0.5-0.9)
CD4:CD8 Ratio	1.2(0.8-1.8)	1.9(1.5-2.9)	1.3(1.1-1.4)	1.3(1.1-1.4)	1.2(1.0-1.5)

Absolute counts are in 10^3 cells per cumm. Data are expressed as median together with percentile P25 and P75 values in parentheses. T-cells were derived from CD3 analysis, B-cells from CD19 or CD20 analysis. Normal values for children under 5 are also given in: Denny T, et al. JAMA. 1992; 267:1484-1488.

METHOD: Enumeration of T and B cells was originally performed by E rosette techniques with sheep RBC (T-cells) and mouse RBC (some B-cells). Detailed subtyping of lymphocytes, including T-helper and T-suppressor subpopulations, now commonly employs fluorescently labelled monoclonal antibodies to specific lymphocyte subset antigens. The labelled lymphocytes are then enumerated by flow cytometry or fluorescence microscopy. Problems related to standardization of T-cell subset enumeration are given in: Cunningham-Rundles S. JAMA. 1989; 262:1526.

INTERPRETATION: Classification and enumeration of lymphocyte populations has become an increasingly important laboratory test. Classification and enumeration of lymphocyte subpopulations is performed to: establish the identity of abnormal cell populations (e.g. leukemic cells), determine absolute T4-cell counts and T4:T8-cell ratios in AIDS patients, diagnose hereditary and acquired immunodeficiency syndromes, evaluate lymphocyte subpopulation numbers in autoimmunity disorders, monitor transplant patients.

Lymphocyte Differential (Cont)

Classification of Lymphocyte Populations: Lymphocytes are classified by immunophenotyping employing a variety of monoclonal antibodies directed against specific determinates on the surface of lymphocyte subpopulations. These determinants are sometimes found on more than one type of lymphocyte, and in some instances on blood cells other than lymphocytes. They are often found on a particular cell population, only at a particular stage of that cell's development. Classification of abnormal cell populations is often useful in the classification of certain leukemias/lymphomas (CLL in particular).

Monoclonal antibodies to particular determinants (antigens), have been produced by a number of manufacturers, each with their own designation. This has led to a great deal of confusion. A detailed review of the various antibodies marketed and their specificities is given in: Foon KA, Todd RF. Blood. 1986; 68:1-31. This has been partially resolved by referring to these cellular antigens by the CD (cluster of differentiation, cluster designation) nomenclature.

The CDs (Clusters of Differentiation, Cluster Designations) can be roughly divided into subsets as follows:

Overview of CD Specificities

Cell Specificity	CD
T-Cells	CD 1-8
Pre-B-Cells	CD 9-10
Myeloid Cells	CD 11-18
B-Cells/Pre-B-Cells	CD 19-24
Miscellaneous	CD 25-45

(From: Hansel TT. Lancet. 1987; II:1382-1383).

The above is a gross simplification. Multiple CDs are found on a given cell type and a given CD may be found on multiple cells. Expression of a CD may depend on the developmental state of a cell, activation, etc. Abnormal (leukemic) cells may not always express these antigens in the same manner as normal cells.

A list of the commonly tested CD antigens and their associated cell populations is given in the next table:

Lymphocyte Populations and CD Phenotype

Cell Type	Marker	Cell Type
T-Cell	CD 1	Cortical Thymocytes
	CD 2	T-Cells (SRBC Receptor)
	CD 3	T-Cells (TcR)
	CD 4	T-Helper Subset/Monocyte Subset
	CD 5	T-Cells/B-Cell Subset/Thymocyte
	CD 7	T and NK Cells
	CD 8	T-Suppressor Subset
	CD 25	Interleukin-2 Receptor
B-Cell	CD 10	Granulocyte/CALLA
	CD 19	All B-Cells/Pre-B ALL
	CD 20	All B-Cells
	CD 21	Resting B-Cells (EBV Receptor)
	CD 22	Resting B-Cells/Hairy Cell Leukemia
Myeloid	CD 13	Monocytes/Granulocytes
	CD 14	Monocytes/Few Granulocytes
	CD 33	Early Myeloid Progenitors/Myeloid/AML
Other	CD 11b	Neutrophils, Monocytes, LGL
	CD 15	(Leu-M1) Reed Sternberg Cells
	CD 16/56	(Dual Marker) NK Cells
	CD 24	B-Cells; Granulocytes
	CD 30	Activated T and B cells; Reed Sternberg Cells
	CD 34	Myeloid Progenitors; Stem Cells
	CD 38	Plasma Cells
	CD 45	(Leukocyte Common Antigen) All Leukocytes
	HLA-DR	Activated B-Cells

Classification of abnormal cell populations may also require cytochemical stains, assay for cytoplasmic and surface immunoglobulin (CIg and SIg), terminal deoxynucleotide transferase (TdT), HLA-DR, glycophorin, and gene rearrangement studies (see LEUKEMIA PANELS).

Lymphocyte Subpopulations and AIDS: Enumeration of the percent and absolute cell counts of the T-4 helper and T-8 suppressor subsets, and the T4:T8 cell ratio are important in the management of HIV infected pediatric and adult patients. At present, about 750 U.S. laboratories conduct T-lymphocyte phenotyping. In general, absolute WBC and lymphocyte counts are highest at birth and decline with time. By contrast, the percent of T-4 helper (CD4) and T-8 suppressor (CD8) cells progressively increase with time.

Guidelines for performance of CD4 T-cell determinations in persons with HIV infection have been published in the May 8, 1992 CDC "Morbidity and Mortality Weekly Report." The CDC has expanded the AIDS surveillance case definition to include all HIV infected persons with <200 CD4 positive T-lymphocytes/microliter or CD4 positive T lymphocyte to total lymphocyte percentage <14. A new classification system is also established based on three clinical categories (A,B,C) and three CD4 categories (1,2,3). The ranges for the three CD4 positive T-lymphocyte categories are: category 1: >500 CD4 cells/cumm; category 2: 200-499 CD4 cells/cumm; category 3: <200 CD4 cells/cumm (MMWR, CDC. JAMA. 1993; 269:729-730). The PHS has recommended CD4 counts be monitored every 3-6 months in HIV infected persons. There are studies which relate total lymphocyte count to the absolute CD4 count in HIV infected individuals (Blatt SP, et al. JAMA. 1993; 269:622-626).

Lymphocyte Subpopulations and Primary Immunodeficiencies: Enumeration of lymphocyte subpopulations is important in establishing the diagnosis of the rare patient with primary immunodeficiency. Subpopulation changes found in the various immunodeficiency states is given in the next table:

B- and T-Cell Immunodeficiency Disorders

Immunodeficiency Disorder	Blood Lymphocytes	Major Immune Defect
Bruton-Type X-Linked Infantile Agammaglobulinemia	Absent B-Cells	B-Cell Defect
Severe Combined Immuno-Deficiency Disorders(SCID)	Decreased B-Cells Decreased T-Cells	B- and T-Cell Defects
Thymic Dysplasia (DiGeorge Syndrome)	Decreased T-Cells Increased B-Cells	T-Cell Defect
Ataxia Telangiectasia	Decreased T-Cells Increased B-Cells	B- and T-Cell Defects
Wiskott-Aldrich Syndrome	Decreased T-Cells Dec. to Inc. B-Cells	B- and T-Cell Defects

References: Gupta S, Good RA. Seminars in Hematology. 1980; 17:1-29; Foucar K, Goeken JA. Laboratory Medicine. 1982; 13:403-414; deSaint Basile G, Fischer A. Immunol Today. 1991; 12:456-458.

Subset Analysis for Immunodeficiency and Transplant Patients: A simplified panel of subset analyses for immunodeficiency and transplant patients is given below:

Immunodeficiency and Transplant Panel

CD3, CD2	T Cells
CD4	T Helper Cells
CD8	T Suppressor Cells
CD19	B Cells
CD56	NK Cells

LYSERGIC ACID DIETHYLAMIDE (LSD)

SPECIMEN: Urine or plasma
REFERENCE RANGE: Negative
METHOD: RIA; fluorometric
INTERPRETATION: Lysergic acid diethylamide (LSD) is a potent hallucinogen derived from ergot, the fungus that spoils rye grain. It may cause a delirium, accompanied by hallucinations and delusions. LSD has a very high safety margin, but fatal ingestions may occur with massive overdosage. LSD is not detected by comprehensive drugs screens due to the potency of the drug and low concentration. Specific urine testing for LSD may remain positive for 24 hours after ingestion of 200-400 micrograms of the drug; typical dose is 100-250 micrograms (range 25-500 micrograms). Reviews: Kulig K. "LSD." Emerg Med Clin N Amer. 1990; 8:551-558; Brown RT, Braden NJ. "Hallucinogens." Pediatr Clin N Amer. 1987; 34:341-347.

LYSOZYME, (MURAMIDASE) SERUM

SPECIMEN: Red top tube, separate serum and freeze
REFERENCE RANGE: 3.0-12.8 mg/liter
METHOD: Calorimetric
INTERPRETATION: Lysozyme is a lysosomal enzyme which is useful in diagnosing and following the course of certain leukemias and granulomatous diseases, in particular, sarcoidosis. Lysozyme is found in a variety of tissues, but relatively high concentrations occur in monocytes and polymorphonuclear leukocytes. Conditions associated with an increase in serum lysozyme are given in the next Table:

Conditions Associated with Increased Lysozyme
Leukemias:
Acute Monocytic Leukemia
Acute Myelocytic
Acute Myelomonocytic Leukemia
Chronic Myelocytic Leukemia (Philadelphia Chromosome Positive)
Histiocytic Medullary Reticulosis
Sarcoidosis
Crohn's Disease
Others:
Renal Disease; Acute Bacterial Infections; Tuberculosis;
Leukemoid Reactions; Megaloblastic Anemias;
Neutropenic Disorders Associated with Increased Granulocytic Turnover

Increased lysozyme in serum or urine has been used to verify a monocytic or myelomonocytic leukemia. However, its use in this situation is being replaced by esterase or immunophenotyping of the abnormal cells (Perillie, Finch. Med Clin N Amer. 1973; 57:395). In the acute leukemias, marked lysozymuria is virtually pathognomonic of acute myelomonocytic leukemia or acute monocytic leukemia. Serial measurements of serum lysozyme in those acute leukemias associated with marked elevation may be useful in following the effect of therapy. Serum enzyme measurements are not diagnostically useful in chronic leukemia.

Histiocytic Medullary Reticulosis has also been followed with lysozyme measurements (Duffy et al. N Engl J Med. 1976; 294:167).

Falchuk et al (N Engl J Med. 1975; 292:395) found serum lysozyme levels to be useful in the diagnosis and follow-up of patients with Crohn's disease. Mean lysozyme concentrations are greater in patients with active Crohn's disease than in those with uncomplicated ulcerative colitis; elevated lysozyme levels are also found in patients with complicated ulcerative colitis (Falchuk and Perrott. Gastroenterology. 1975; 68:890).

Mononuclear phagocytes contain high lysozyme activity. The serum lysozyme tends to be elevated in granulomatous diseases such as sarcoidosis and tuberculosis which are associated with increased monocytes (Pascual et al. N Engl J Med. 1973; 289:1074).

LYSOZYME, (MURAMIDASE) URINE

SPECIMEN: 50mL random urine, freeze
REFERENCE RANGE: 0-2.9 mg/liter.
METHOD: Colorimetric
INTERPRETATION: Urine lysozyme is used to differentiate glomerular disease from renal tubular disease in that lysozyme is excreted in renal tubular disease and minimally in glomerular disease (Daniels, et al. Rep Biol Med. 1972; 30:1): Lysozyme has a molecular weight between 14,000 and 15,000; the glomerular membrane is about 0.8 as permeable to lysozyme as it is to creatinine. The proximal tubule is the site of catabolism and reabsorption of lysozyme (Ottonsen and Naunsback, Kidney International. 1973; 3:315-326). Lysozyme is found in the urine in a wide variety of renal diseases including uremia.

Urine lysozyme measurements have also been used in the diagnosis and management of certain leukemias. Specifically, elevated lysozyme levels have been associated with monocytic and myelomonocytic leukemias. See LYSOZYME, SERUM.

MAGNESIUM, SERUM

SPECIMEN: Red top tube, separate serum. Hemolysis may yield elevated results; the red cell magnesium level concentration is almost three times that of serum magnesium level.
REFERENCE RANGE:

Change of Serum Magnesium with Age

Age	mg/dL	mEq/liter	mmol/liter(SI Units)
Newborn	1.8-2.8	1.5-2.3	0.75-1.15
Children	1.7-2.3	1.4-1.9	0.7-0.95
Adults	1.7-2.4	1.4-2.0	0.7-1.0

Convert mg/dL to mmol/liter (SI units): multiply by 0.4114.
METHOD: Fluorometric, 8-hydroxyquinoline; colorimetric, titan yellow; others.
INTERPRETATION: The causes of hypomagnesemia are given in the next Table (Salem M, et al. "Hypomagnesemia in Critical Illness," Crit Care Clin. 1991; 7:225-252; Massry SG. "Magnesium Homeostasis and Its Clinical Pathophysiology," Resident and Staff Physician. 1981; 105-109; Rude BK, Singer FR. Annu Rev Med. 1981; 32:245-259):

Causes of Hypomagnesemia

Excessive Urinary Losses	Increased Intestinal Losses
Alcoholism	Malabsorption syndromes
Renal Tubular Acidosis	Massive surgical resection of small intestine
Glomerulonephritis	Prolonged nasogastric suction
Interstitial Nephritis	Excessive use of laxatives
Diuretic phase of acute renal failure	Intestinal and biliary fistulas
Chronic renal failure with renal magnesium wasting	Severe diarrhea, as in ulcerative colitis and infantile gastroenteritis
Idiopathic renal magnesium wasting	Rarely, prolonged lactation
Hypercalcemic states: malignancy, hyperparathyroidism and vitamin D excess	Primary hypomagnesemia (Hennekam RCM, Donckerwolcke RA. Lancet. 1983; 1:927)
Hypophosphatemia	Pancreatitis
Acidemia	**Alterations in Distribution**
SIADH	Cardiopulmonary bypass surgery
Hyperaldosteronism	Acute administration of glucose, insulin, or amino acids
Hyperglycemia	Sepsis syndrome
Diabetic Ketoacidosis	Hungry bone syndrome
Hyperthyroidism	Multiple transfusions or exchange transfusions with citrated blood
Hypoparathyroidism	Pancreatitis
Drug-Induced (see below)	Alkalemia
Decreased Intake	Thermal injury
Protein-calorie malnutrition	Catecholamines
Starvation	
Prolonged intravenous therapy	
Chronic ethanol abuse	
Pregnancy	

Magnesium Serum (Cont)

Probably the most common cause of hypomagnesemia is acute or chronic alcoholism; the excretion of magnesium by the kidney is increased during the time when there is a significant concentration of alcohol in the blood.

Drugs associated with hypomagnesemia include: furosemide, ethacrynic acid, bumetanide, thiazides, mannitol, urea, digoxin, amphotericin B, aminoglycosides, carbenicillin, cis-platinum, methotrexate, cyclosporine, ethanol, calcium, citrate, terbutaline, and pentamidine.

In the presence of magnesium deficiency, hypocalcemia and hypokalemia cannot be corrected without magnesium replacement.

The clinical and laboratory findings in magnesium depletion are given in the next Table (Elin RJ. Disease-a-Month. 1988; 34:163-218):

Clinical and Laboratory Findings in Magnesium Deficiency
Neuromuscular: Weakness, Tremors, Muscle fasciculation, Positive Chvostek's sign, Positive Trousseau's sign, Dysphagia
Cardiac: Arrhythmias, ECG Changes
CNS: Depression, Agitation, Psychosis, Nystagmus, Seizures
Laboratory: Hypomagnesemia, Hypocalcemia, Hypokalemia, Hypophosphatemia (occasionally, hyperphosphatemia), Low urinary Mg and calcium

The conditions associated with hypermagnesemia are listed in the next Table (Van Hook JW. Crit Care Clin. 1991; 7:215-223):

Conditions Associated with Hypermagnesemia
Acute renal failure
Chronic renal failure
Infants of mothers treated with Mg for eclampsia
Adrenal insufficiency
Administration of pharmacologic doses of Mg and use of oral purgatives or rectal enemas containing Mg, especially in patients with impaired renal function
Other: Hyperparathyroidism; Hypocalciuric Hypercalcemia; Lithium therapy

The most common cause of hypermagnesemia is acute or chronic renal failure; the elevation of serum magnesium tends to occur when creatinine clearance is less than 30 mL/min.

The clinical findings associated with hypermagnesemia are given in the following Table (Van Hook JW, already cited):

Clinical Findings Associated with Hypermagnesemia
Cardiovascular: Delayed intraventricular conduction; Prolonged Q-T interval; Atrioventricular block; Cardiac arrest
Neuromuscular: Hyperreflexia; Respiratory depression

Magnesium and Cardiovascular Disease: Magnesium infusion in acute myocardial infarction has been evaluated in a randomized, double blind, placebo controlled study involving 2316 patients (LIMIT-2). Intravenous magnesium sulphate (8mmol over 5 min followed by 65mmol over 24 h) raised serum magnesium levels to about twice normal. Mortality at 28 days was 10.3% in the placebo group and 7.8% in the magnesium group; incidence of left ventricular failure was 14.9% in the placebo group and 11.2% in the magnesium group. These findings were statistically significant. Magnesium was thought to provide a direct protective action on the myocardium (Woods KL, et al. Lancet. 1992; 339:1553-1558).

MALABSORPTION PANEL **(see FAT MALABSORPTION PANEL)**

MALIGNANT HYPERTHERMIA PANEL

Malignant hyperthermia is a group of inherited disorders which predispose patients to a fulminant hypermetabolic crisis triggered by anesthetic agents. Drugs that may precipitate malignant hyperthermia include succinylcholine, halothane, methoxyflurane, cyclopropane and ketamine.

In malignant hyperthermia, patients develop fever (39 to 44°C), generalized skeletal muscle rigidity, tachycardia and hypotension. Early metabolic manifestations include respiratory and metabolic acidosis, hyperkalemia, hypermagnesemia, hyperphosphatemia, hypercalcemia, elevated serum myoglobin, myoglobinuria and creatine phosphokinase. Patients may develop disseminated intravascular coagulation and renal failure. Laboratory findings in malignant hyperthermia are listed in the following Table:

Laboratory Tests in Malignant Hyperthermia

Early:	Respiratory and Metabolic Acidosis
	Electrolyte Disorders (Elevated Potassium, Magnesium, Phosphorus, Calcium)
	Elevated Lactate and Pyruvate
	Elevated Creatine Phosphokinase
Late:	Rhabdomyolysis (Serum and Urine Myoglobin)
	Renal Failure
	Disseminated Intravascular Coagulation (DIC)

Treatment of malignant hyperthermia includes hyperventilation with 100% oxygen, dantrolene 3mg/kg, repeat doses up to 10mg/kg. Treat hyperkalemia (bicarbonate, insulin, etc). Actively cool the patient and maintain urine output with furosemide (Lasix) and mannitol.

MARIJUANA

SPECIMEN: Random urine
REFERENCE RANGE: None; the cutoff level for cannabinoid detection in the workplace is 50-100 nanograms per mL.
METHOD: EMIT (Limit 20-100ng/mL); Thin-layer chromatography (Limit 20ng/mL); Gas Chromatography/Mass Spectroscopy (Limit 2ng/mL).
INTERPRETATION: The pharmacokinetics of marijuana are given in the following Table (Selden BS, et al. Emerg Med Clin N Amer. 1990; 8:527-539):

Pharmacokinetics of Marijuana

Route	Absorption	Peak Level	Peak Effect
Inhalation	10-50%	7-8 min	20-30 min
Oral	1-10%	45 min	2-3 hrs

The duration of positive tests after abstinence from marijuana varies with the dose of drug and level of sensitivity of the test. Moderate users will often have a positive test 3 to 5 days after last use. Chronic heavy users tested positive for a mean time of 27 days; one person tested positive for 77 days (Ellis GM, et al. Clin Pharmacol Ther. 1985; 38:572-578).

Various adulterants may lead to false-negative tests. Addition of salt, vinegar, hypochlorite bleach, liquid detergent, soap, or blood, precipitation of normal phosphates and oxalates due to refrigeration of samples, diuretic use, ingestion of large amounts of fluid, or dilution or replacement of the sample with tap water all may lead to false-negative results (Schwartz RH, et al. AJDC. 1985; 139:1093-1096).

There are great variations in absorption and distribution of the drug, manner of ingestion and urinary volume of THC; thus, assay is useful only as an indicator of recent use and not as a measure of intoxication. Psychological effects do not correlate with urinary metabolite levels.

Marijuana is an Asiatic herb; the principal active ingredient is tetrahydrocannabinol (THC). The parts with the highest THC content are the flowering tops of the plant. Hashish (hash) is the dark brown resin that is obtained from the tops of the plants. Marijuana or hashish is generally smoked in self-rolled cigarettes called "joints" or in pipes; they may be added to food or drinks.

The effects of marijuana vary widely; it may act as a stimulant or a depressant, or as a hallucinogen with sedative properties. Chronic heavy marijuana use has been associated with depression of plasma testosterone levels, gynecomastia, damage to the bronchial tract and lungs, and neural and possibly immunologic effects.

(See also references: Schwartz RH. Arch Intern Med. 1988; 148:2407-2412; Schwartz RH. Pediatr Clin N Amer. 1987; 34:305-317; Schwartz RH, Hawks RL. JAMA. 1985; 254:788-792).

MEAN CORPUSCULAR HEMOGLOBIN (MCH)
(see RED CELL INDICES)

MEAN CORPUSCULAR HEMOGLOBIN CONCENTRATION (MCHC)
(see RED CELL INDICES)

MEAN CORPUSCULAR VOLUME (MCV) (see RED CELL INDICES)

MELANIN, URINE

SPECIMEN: Freshly voided random urine
REFERENCE RANGE: Negative
METHOD: Thormahlen's Test: Nitroprusside $\xrightarrow{\text{Melanin}}$ Prussian Blue; Ferric chloride test.
INTERPRETATION: This test is used to evaluate urine melanin in patients with known or suspected metastatic melanoma. Excreted colorless melanogen may slowly convert to dark-black melanin in the urine of these patients. Dark urine, red-brown to black, may also be formed from homogentisic acid in patients with alkaptonuria. In the ferric chloride test, melanogens give a brown to black color, homogentisic acid gives a blue-green color.

MERCURY, URINE

SPECIMEN: Urine: Collect 24 hour urine in acid washed-containers. Use plastic containers (borosilicate, polyethylene or polypropylene); add 10% HCl solution to the container and allow to "soak" for 10 minutes; rinse with five volumes of tap water and then five volumes of deionized or distilled water. The patient should urinate at 8:00 A.M. and the urine is discarded. Then, urine is collected for 24 hours including the next day 8:00 A.M. specimen. Indicate 24 hour volume. A 50mL aliquot is used for analysis. Blood: Lavender (EDTA) top tube, special metal-free tube is used.
REFERENCE RANGE: Urine: 20mcg/liter; symptoms: >200mcg/liter; remove from occupational exposure: >300mcg/liter; toxicity in chronic exposure can be seen at levels: 50-100mcg/liter. Permissible levels of exposure to heavy metals are being frequently changed downward. Blood: less than 0.005mcg/mL; toxic: >0.05mcg/mL. To convert conventional units in mcg to international units in nmol, multiply conventional units by 4.985.
METHOD: Qualitative: Reinsch test sensitive to 1000mcg/liter; Quantitative: atomic absorption.
INTERPRETATION: Mercury measurements are performed in order to access possible acute poisoning, or the degree of chronic occupational exposure. Urine measurements appear to be more reliable in chronic exposure than blood measurements. Employees with potential exposure should receive a baseline pre-employment level followed by monitoring urine levels taken at 3-6 month intervals. In acute toxic encounters, blood levels are preferred to urine measurements. Gastric contents may be used in oral overdoses. The organ systems involved in acute mercury poisoning (acrodynia) are gastrointestinal tract, the kidneys and the central nervous system; in chronic mercury poisoning, the same organ systems are involved. The three cardinal signs of mercury poisoning are dysarthria, ataxia and constricted visual fields. In addition, pulmonary injury occurs when mercury vapor is inhaled.

Both inorganic (elemental) mercury and organic mercury poisoning occur. Ingestion of metallic mercury is generally harmless. Spills of metallic mercury creates the potential for release of mercury vapor since vacuumed mercury may be hazardous (Zelman M, et al. Clin Pediatr. 1991; 30:121-123). Organic mercury compounds were responsible for Minamata Disease - an epidemic of mercury poisoning in Minamata Bay Japan in the 1950's and 60's in which inorganic mercury industrial contaminants were converted to organic mercury, ingested by fish which were subsequently ingested by humans. Chronic elemental mercury poisoning is insidious in onset, and may take years to develop. It is an occupational disease of miners, mirror makers, gilders, hatters ("mad as a hatter," Lewis Carrol's "Alice-in-Wonderland"), factory and laboratory workers. Organic mercury poisoning is usually a more serious disease and may develop precipitously.

One death and several instances of elevated serum mercury levels have been reported from the practice of instilling merthiolate into external ears (FDA Drug Bulletin. April 1983; 13:No.1.

Reference: Mack RB. Contemp Pediatr. Aug 1989; 139-148.

METABOLIC SCREEN (see INBORN ERRORS OF METABOLISM SCREEN)

METANEPHRINES, TOTAL URINE

SPECIMEN: 24 hour urine. A quantitative 24 hour urine is the "gold standard." Some labs will perform a "screen" on a first morning random urine. The specimen may also be used for assay of catecholamines and for vanillylmandelic acid (VMA). Instruct the patient to void at 8:00 A.M. and discard the specimen. Add 25mL of 6N HCl to container prior to collection. Be aware that skin burns may result if care is not taken. Then, collect all urine including the 8:00 A.M. specimen at the end of the 24 hour collection period. Refrigerate container as each specimen is collected. Following collection, add 6N HCl to pH 1 to 2 (Do not use boric acid). Metanephrines are unstable in urine at alkaline pH. Record 24 hour urine volume. In urine collected at a pH of less than 3 and stored at 4°C, the metanephrines are stable for at least one week. Forward a 100mL aliquot of 24 hour urine for analysis. A urine pH >3 is cause for rejection. Many drugs affect metanephrine levels and some of the assay procedures. Lists of these drugs may be found in: Spilker B, et al. Ann Clin Lab Sci. 1983; 13:16-19 and Stein PP, Black HR. Medicine. 1991; 70:46-66. In order to ensure reliability of test results, discontinue all drugs if possible for at least 3 days before urine is collected.

REFERENCE RANGE: Adult: <1.0mg/24 hours. The reference range may also be reported as the ratio mcg metanephrine/mg creatinine. The reference range for urinary metanephrines is given in the next Table (Gitlow SE, et al. J Lab Clin Med. 1968; 72:612-620).

Reference Range for Metanephrines

Age (years)	Upper Limit (mcg/mg Creatinine)
<1	4.3
1-2	3.9
2-10	2.8
10-15	1.6
15-Adults	1.0

There are a number of assay procedures available. In general, the more specific the method, the lower the reference range will be.

METHOD: Conjugates of metanephrines are hydrolyzed and the metanephrines are absorbed onto an ion-exchange resin; after elution metanephrines are oxidized to vanillin which is measured spectrophotometrically. Initial methods were spectrophotometric. Extraction procedures were added to remove interfering substances such as from the diet or drugs. HPLC or radioimmunoassays are now the most specific techniques. Reference range will vary according to method.

INTERPRETATION: Urinary metanephrines are often measured with urinary catecholamines and vanillylmandelic acid (VMA) in the work-up of patients with possible pheochromocytoma, neuroblastoma, ganglioneuroma and ganglioneuroblastoma. Urinary metanephrines are elevated in 95% of patients with pheochromocytoma (Stein and Black, already cited). Also see PHEOCHROMOCYTOMA SCREEN, CATECHOLAMINES and VMA for further discussion.

METHANOL (METHYL ALCOHOL)

SPECIMEN: Red top tube or random urine or gastric contents; store at 4°C.
REFERENCE RANGE: Normally not present; toxicity: >250mg/liter (25mg/dL) in serum. To convert conventional units in mg/dL to international units in mmol/liter, multiply conventional units by 0.3121.
METHOD: Gas-liquid Chromatography; Sunshine, Methodology for Analytical Toxicology, CRC Press, 1975.
INTERPRETATION: Most laboratories do not offer gas-liquid chromatography on a "stat" basis. However, there are two tests that may be done to screen for methanol ingestion; these are: osmolality and acid-base imbalance with increased anion gap metabolic acidosis. The increase in serum osmolality for each 100mg/dL of methanol is 31 mosm/liter. Only slightly elevated or decreased serum osmolality suggests other causes of coma such as trauma or drugs other than methanol. Most patients with methanol toxicity will have elevated serum and urine amylase.

The minimum lethal dose of methanol is approximately 30g. Methanol itself is non-toxic; its metabolic products, formaldehyde and formic acid, are toxic. Following ingestion of methanol, it takes 12 to 24 hours for the toxic products to form and take effect. The mortality rate is related to both duration and severity of the metabolic acidosis which develops in 8 to 12 hours (Pappas SC, Silverman M. Can Med Assoc J. 1982; 126:1391-1394). The clinical effects of methanol are as follows: drunkenness without alcohol on the breath, convulsions, coma, retinal edema, blindness due to optic nerve damage and atrophy

Treatment is directed toward correction of metabolic acidosis (usually bicarbonate) and removal of methanol and toxic metabolites (hemodialysis). Dialysis is indicated when a patient has metabolic acidosis, mental, visual, or fundoscopic changes, blood methanol greater than 50mg/dL or has ingested more than 30g of methanol. Ethanol serves as an antidote for methanol by competitive inhibition since both are metabolized by alcohol dehydrogenase.
Ref. Editorial, The Lancet. April 23, 1983; 1:910-912.

METHAQUALONE (QUAALUDES)

SPECIMEN: Quantitation: Red top tube, separate serum. Qualitative: 50mL random urine.
REFERENCE RANGE: Toxic: 1.0-3.2mg/dL; Lethal: >3.0mg/dL. Half-life: 10-42 hours.
METHOD: Gas-liquid chromatography; mass spectroscopy; RIA
INTERPRETATION: Methaqualone is a prescription sedative-hypnotic drug which is used as an illegal "street drug" obtained by production in illegal laboratories or by illegal importation. When used recreationally, methaqualone produces a feeling of both sedation and mild euphoria.

Methaqualone and alcohol act synergistically; most methaqualone-related deaths result from overdose plus ethanol.

The clinical symptoms include hypertonia, vomiting, salivation, delirium, muscle twitching, tonic-clonic seizures and coma.

METHEMOGLOBIN (FERRI-HEMOGLOBIN)

SPECIMEN: Green (heparin) or lavender (EDTA) or blue (citrate) top tube. Transport specimen to laboratory on ice immediately. The specimen is rejected if specimen takes too long in transit (2-8 hours depending on laboratory) or specimen clotted.
REFERENCE RANGE: 0.5-1.5% of total hemoglobin.
METHOD: Spectrophotometric before and after cyanide addition.
INTERPRETATION: Methemoglobin measurements are performed in the evaluation of cyanosis, toxic encounters, polycythemia, certain suspected RBC enzyme defects, and certain hemoglobinopathies. Methemoglobin is hemoglobin with iron in the ferric form; methemoglobin is not capable of carrying oxygen. Assay of methemoglobin is done to evaluate cyanosis; concentration of methemoglobin greater than 10% of the total hemoglobin or 1.5 to 2.5g/dL of methemoglobin may cause visible cyanosis.

The diagnosis can be suspected by the characteristic chocolate brown color of a freshly obtained blood sample. The blood sample does not become bright red when oxygen is bubbled through it. A screening test is as follows: Place one drop of patient's blood and one drop of your own blood or other appropriate blood specimen as control side by side, on a piece of filter paper. Chocolate color indicates presence of methemoglobin.

The causes of methemoglobinemia are given in the next Table (Cartwright GE. Diagnostic Laboratory Hematology, Grune and Stratton, New York, 4th Edition, 1968, pg. 234):

Causes of Methemoglobinemia

I. Acquired:
   - Nitrites, Nitrates, Chlorates, Chloroquine, Quinones, Aminobenzenes, Nitrobenzenes, Nitrotoluenes, Ferrous Sulfate (large doses), Some Sulfonamides.
   - Aniline Dye Derivatives, Phenacetin, Acetanilid Pyridium, Benzocaine

II. Hereditary:
   - A. Enzymatic or metabolic
     - NADH-Methemoglobin reductase (diaphorase) deficiency (recessive)
     - NADPH-Methemoglobin reductase deficiency (recessive)
     - Glutathione deficiency (dominant)
   - B. Hemoglobin M (dominant)

The causes of methemoglobinemia are acquired or hereditary; most causes are acquired and are due to drugs and chemicals especially those containing nitro- and amino-groups such as aniline and derivatives, nitrites and some sulfonamide, large doses of ferrous sulfate and some bacteria.

Patients with methemoglobinemia may develop polycythemia as a compensatory mechanism and may have dyspnea and headache. Acute methemoglobinemia may be treated with the reducing agent methylene blue; treatment is not necessary unless 35% or more of the hemoglobin is in the methemoglobin form. The definitive treatment of significant methemoglobinemia includes intravenous 1% methylene blue; the dose is 1-2mg/kg given slowly over a 5-minute period; give oxygen. The lethal concentration of methemoglobin is probably 70% and above.
References: Chilcote R, et al. Pediatrics. 1977; 59:280-282; Mack RB, North Carolina Medical Journal. 1982; 43:292-293).

METHOTREXATE

SPECIMEN: Red top tube, separate serum as soon as possible. Stable at 4°C for at least 24 hours and at -20°C for one month. Keep specimen in the dark.
REFERENCE RANGE: More than 90 percent of methotrexate is excreted by the kidney and renal toxicity is common; determine serum creatinine and creatinine clearance before starting therapy.

High dose methotrexate is usually accompanied by leucovorin rescue depending on the concentration of methotrexate at 24, 48 and 72 hours.
Sampling Times: Sampling times are usually 24, 48 and 72 hours after starting methotrexate infusion, and ideally everyday during leucovorin rescue.
Toxicity: Without leucovorin rescue: $>8 \times 10^{-8}$M for > 42 hours. With leucovorin rescue: 10^{-7}M when leucovorin is discontinued; 10^{-5}M at 24 hours; 10^{-6}M at 48 hours; 10^{-7}M at 72 hours.
Half-Life: After intravenous infusion, methotrexate is distributed within the total body water and the initial half-life is about one hour. After this initial distribution, the half-life is about 3 hours.
METHOD: EMIT; RIA
INTERPRETATION: Methotrexate in high-dose is used to treat cancer. Methotrexate in low doses has been used to treat juvenile rheumatoid arthritis with good results and low toxicity (Giannini EH, et al. N Engl J Med. 1992; 326:1043-1049; Graham LD, et al. J Pediatr. 1992; 120:468-473; Rose CD, et al. J Pediatr. 1990; 117:653-659). Low dose methotrexate has been used in other rheumatic diseases including rheumatoid arthritis, psoriatic arthritis, polymyositis and Reiter's Syndrome (Review: Healey LA. Bull on the Rheumatic Dis. 1986; 36:1-10). Methotrexate has been used to treat severe asthma; some studies show benefit (Guss S, Portnoy J. Pediatr. 1992; 89:635-639; Mullarkey MF, et al. Ann Intern Med. 1990; 112:577-581; Mullarkey MF, et al. N Engl J Med. 1988; 318:603-607); another study has failed to show benefit (Erzurum SC, et al. Ann Intern Med. 1991; 114:353-360). (cont)

Methotrexate (Cont)

Massive doses of methotrexate must be coupled with leucovorin because of the toxicity induced by methotrexate. Methotrexate blocks the action of the enzyme, dihydrofolate reductase, as illustrated in the next Figure:

Methotrexate and Folate Metabolism

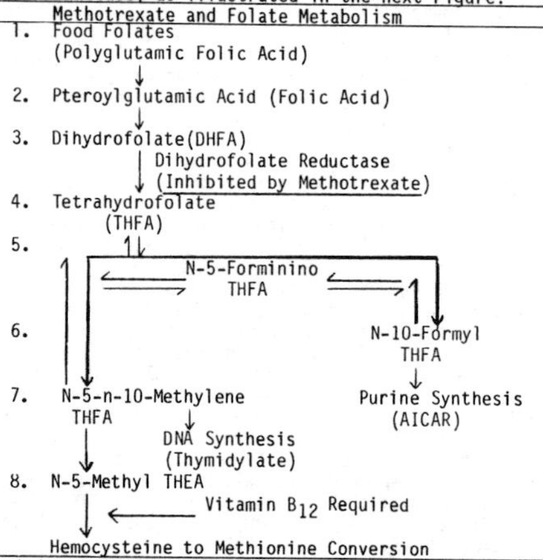

As illustrated in this Figure, methotrexate inhibits the synthesis of thymidine, which is required for DNA synthesis and cell replication.

Toxicity: With high-dose methotrexate, nephrotoxicity is a common problem. More than 90 percent of methotrexate is excreted by the kidney within 24 to 48 hours. At acid pH, methotrexate may deposit in the renal cells; one of the major metabolites, 7-OH methotrexate, is relatively soluble at acid pH. Nephrotoxicity is reversible. Toxicities, other than nephrotoxicity that occurs with high-dose methotrexate include B-lymphocyte dysfunction, acute dermatitis, bone marrow depression, leukopenia, thrombocytopenia, anemia, stomatitis, vomiting and diarrhea. Chronic methotrexate therapy may lead to progressive fibrosis of the liver leading to cirrhosis and methotrexate pneumonitis.

METYRAPONE (METOPIRONE) TEST

Specimen and Procedure: The metyrapone test may be performed in a variety of ways. Traditionally it consisted of baseline measurements of steroid production followed by administration of 3 gm metyrapone per day in divided doses over a 2 to 3 day period. On the third day, a 24 hour urine was assayed for 17-ketogenic steroids. Some now use an overnight metyrapone test (Spiger M, et al. Arch Intern Med. 1975; 698-700). Three grams metyrapone is administered between 11 and 12 P.M. and then plasma cortisol and 11-deoxycortisol is measured in the A.M. and compared to baseline values. The cortisol should fall if the dose of medication is adequate. The metyrapone test may not be performed within 2 to 3 days after a high dose dexamethasone suppression test.

A positive test in Cushing's disease (ACTH producing pituitary adenoma) should show a five-fold increase in 11-deoxycortisol with a fall in cortisol to less than 8mcg/dL (Carpenter PC. Endocrin Met Clin N Amer. 1988; 17:445-472).

INTERPRETATION: Metyrapone is a drug that is used to: 1) differentiate adrenal hyperplasia from adrenal tumor and 2) measure pituitary reserve. Metyrapone selectively inhibits the synthesis of cortisol from 11-deoxycortisol. In Cushing's disease there is a fall in cortisol production with a resultant increase in ACTH and 11-deoxycortisol, but there is no change in steroid synthesis in Cushing's syndrome due to an adrenal adenoma. The metabolic pathway, which is blocked by metyrapone, is shown in the next Figure:

Inhibition of Pathway for Synthesis of Cortisol by Metyrapone

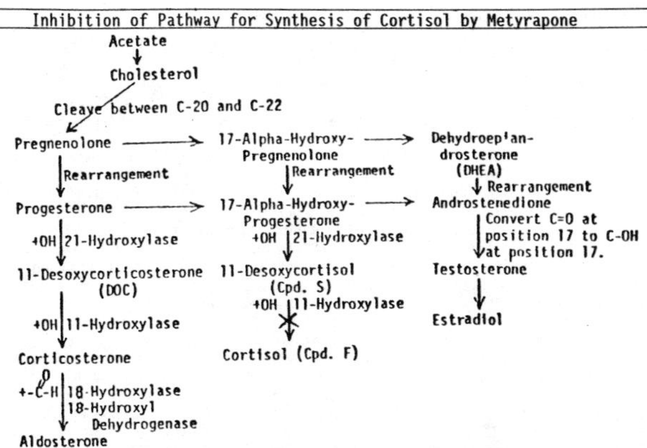

In normal patients, the resultant fall in serum cortisol causes an increase in ACTH, which causes increased steroid synthesis to the point of inhibition (11-deoxycortisol). This is illustrated in the next Figure:

Secretion of ACTH in Subjects Treated with Metyrapone

Normal and ACTH Producing Pituitary Adenoma (Cushing's Disease): Plasma cortisol production is inhibited after administration of metyrapone. Urinary 17-OHCS and 17-KGS, and plasma ACTH and 11-deoxycortisol increase.

Abnormal: Little or no increase in urinary 17-OHCS, 17-KGS, or plasma 11-deoxycortisol, is seen with ACTH deficiency, hypothalamic tumors, adrenal cortical adenomas, and adrenal suppression secondary to exogenous pharmacologic doses of steroids.

The metyrapone test is useful to differentiate the various causes of Cushing's syndrome. The metyrapone test may be used to differentiate adrenal hyperplasia from adrenal adenoma in that there is an increase in urinary steroids and 11-deoxycortisol in 90% of adrenal hyperplasia but these levels are not changed in patients with adrenal adenoma. This test was commonly used for this purpose before reliable ACTH assays were available.

MINIMUM INHIBITORY CONCENTRATION (MIC)

SPECIMEN: Isolated culture of organisms to be tested, prepared by the microbiology laboratory. The test cannot be done if bacterium isolated from patient is not available or fails to grow. The physician must request MIC testing within 48 hours of submission of the specimen for initial culture so that the laboratory will save the patient's isolate. If the isolate has not been saved, the test cannot be performed.

REFERENCE RANGE: Reference range is the least concentration (minimum inhibitory concentration) of antibiotic which will inhibit growth of that organism.

METHOD: Various methods for MIC testing are available including antibiotic dilution tests, agar diffusion tests and antibiotic disk susceptibility tests. The microdilution susceptibility test is performed using a molded plastic plate with numerous wells. The wells contain antimicrobic agents as multiple dilutions, one or more singles at breakpoint concentrations plus sterility and growth controls. Different drugs are presented in panels for testing gram-positive or gram-negative organisms. The wells are inoculated, incubated for 15-18 hours, and read for the presence of visible turbidity. The lowest concentration of each antimicrobic which demonstrates no visible turbidity (no visible growth) is recorded as the minimum inhibitory concentration for the antimicrobic in mcg/mL against the test organism (Barry AL. Clin Lab Med. 1989; 9:203-219).

INTERPRETATION: MIC is done in order to determine the susceptibility of a given organism to an antimicrobic. Information on susceptibility is important in order to select intelligently the choice of antibiotic to be used in therapy. Antimicrobic panels must be developed based on recommendations of the National Committee for Clinical Laboratory Standards (NCCLS) and resistance patterns of endemic and epidemic strains within an institution. Recommendations for susceptibility testing for various organisms is given in the following Table (NCCLS, Antimicrobial Susceptibility Testing. 215 525-2435):

Antimicrobial Susceptibility Testing for Selected Organisms

Organism	Group A: Primary Test and Report	Group B: Primary Test Report Selectively
Enterobacteriaceae	Ampicillin Cefazolin Cephalothin Gentamicin	Mezlocillin *or* Piperacillin Ticarcillin Amoxicillin/Clavulanic Acid *or* Ampicillin/Sulbactam Ticarcillin/Clavulanic Acid Cefmetazole Cefoperazone Cefotetan Cefoxitin Cefamandole *or* Cefonicid *or* Cefuroxime Cefotaxime *or* Ceftizoxime *or* Ceftriaxone Imipenem Amikacin Ciprofloxacin Trimethoprim/Sulfamethoxazole
Pseudomonas Aerugenosa Acinetobacter	Mezlocillin Ticarcillin Gentamicin	Azlocillin *or* Piperacillin Cefoperazone Ceftazidime Aztreonam Imipenam Amikacin Tobramycin
Staphylococci	Penicillin G Oxacillin *or* Methacillin Cefazolin *or* Cephalothin	Amoxicillin/Clavulanic Acid *or* Ampicillin/Sulbactam Vancomycin Clindamycin Azithromycin *or* Erythromycin Trimethoprim/Sulfamethoxazole
Enterococci	Penicillin G *or* Ampicillin	Vancomycin

Antimicrobial Susceptibility Testing (Cont.)		
Streptococci (Not Enterococci)	Penicillin G	Cephalothin Vancomycin Chloramphenicol Clindamycin Azithromycin or Erythromycin
Haemophilus	Ampicillin Trimethoprim/Sulfamethoxazole	Amoxicillin/Clavulanic Acid or Ampicillin/Sulbactam Azithromycin Cefaclor or Cefprozil or Loracarbef Cefixime Cefuroxime Cefotaxime or Ceftazidime or Ceftizoxime or Ceftriaxone Chloramphenicol Tetracycline

Antibiotics in Group A are the primary antibiotics tested and reported. Antibiotics in Group B may warrant primary testing, but should be reported selectively when the organism is resistant to agents in the same family, or for particular specimen sources, or for patients with allergy to particular antibiotics. Additional lists of antibiotics appropriate for testing resistant organisms or specifically for testing urine are available from NCCLS.

MITOCHONDRIAL ANTIBODIES

SPECIMEN: Red top tube, separate serum.
REFERENCE RANGE: Negative; positive titer at 1:10
METHOD: Indirect immunofluorescence. The usual substrate is rat kidney and a granular pattern is seen in epithelial cells.
INTERPRETATION: These antibodies are rarely found in extrahepatic biliary tract obstruction or normal patients; thus, this test is very valuable in the differential diagnosis of primary biliary cirrhosis versus extrahepatic biliary tract obstruction. Titers greater than 1:40 suggest primary biliary cirrhosis. Antibody levels roughly parallel disease progression. Less than 1% of healthy persons have anti-mitochondrial antibodies. Antibodies against mitochondria are found in the sera of 93 percent of patients with primary biliary cirrhosis (Sherlock S, Scheuer PJ. N Engl J Med. 1973; 298:674-678; Yeamon SJ, et al. Lancet. 1988; 1:1067-1070).

Mitochondrial antibodies may also be found in the serum of patients with other liver diseases: chronic hepatitis (low titer) and cryptogenic (obscure origin) cirrhosis - 25%; viral hepatitis - less than 2%, and drug hypersensitivity, especially iproniazid and halothane jaundice. A raised titer of mitochondrial antibodies may be found in all types of non-hepatic autoimmune disease.

Primary Biliary Cirrhosis: Primary biliary cirrhosis is a disease of unknown etiology characterized by severe derangement of hepatic excretory function and progressive cirrhosis. Clinical features of the disease are listed in the next Table:

Clinical Features of Primary Biliary Cirrhosis
Middle-aged Females
Insidious Onset
Pruritis, Skin Pigmentation, Jaundice
Skin Xanthomata, Xanthelasmata
Hepatomegaly, Splenomegaly
Steatorrhea (Fat in Stool)
Signs and Symptoms of Portal Hypertension
Associated Diseases (Sicca Syndrome, Thyroiditis, etc.)
Death in Liver Failure

(cont)

Mitochondrial Antibodies (Cont)

About 90 percent of the patients are women. The disease starts insidiously. Pruritis may be the presenting complaint, and usually precedes jaundice. Pruritis is usually attributed to raised bile acid concentrations. Hepatosplenomegaly is not usually a presenting finding. Steatorrhea is caused by defective bile acid excretion and thus poor fat and fat-soluble vitamin absorption leading to vitamin K deficiency and vitamin D deficiency. As the disease progresses, portal hypertension may develop. Primary biliary cirrhosis often occurs in association with other autoimmune conditions such as CREST or scleroderma (3%), Hashimoto's thyroiditis or Sjogrens syndrome, myocarditis and syphilis.

Diagnosis of Primary Biliary Cirrhosis: Laboratory findings in primary biliary cirrhosis are those of biliary obstruction, e.g., serum alkaline phosphatase increased, serum bilirubin increased, serum cholesterol increased, serum IgM values usually increased and positive serum mitochondrial antibody test. Periportal granulomas are seen on liver biopsy as well as Mallory hyaline in some cases.

MONOSPOT SCREEN

Synonyms: Heterophil Antibodies, Qualitative; Infectious Mononucleosis Antibodies.

SPECIMEN: Red top tube, separate serum and refrigerate.

REFERENCE RANGE: Negative

METHOD: Hemagglutination of horse RBC. The test is performed on a slide as illustrated in the next Figure:

Monospot Screen

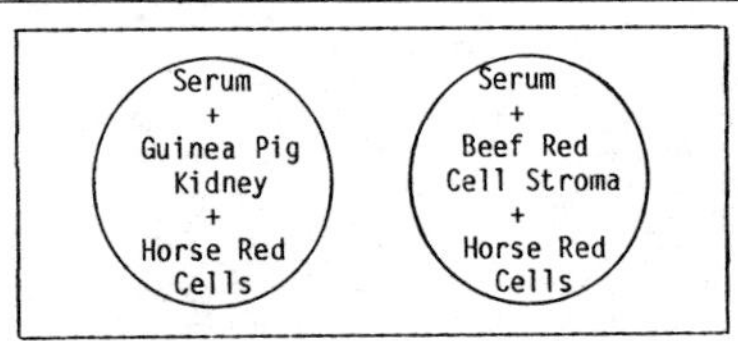

Absence of Antibody to Infectious Mononucleosis:	No Agglutination	No Agglutination
Presence of Antibody to Infectious Mononucleosis:	Agglutination	No Agglutination

Antibodies for horse red cells may occur in normal sera and in conditions other than infectious mononucleosis such as serum sickness and other infections; these nonspecific antibodies are absorbed by guinea pig kidney but not by beef red cells. By contrast, the specific antibodies developed in infectious mononucleosis are removed by beef red cells but are not absorbed by guinea pig kidney.

INTERPRETATION: This test is used as an aid in the diagnosis of infectious mononucleosis. Infectious mononucleosis is caused by the Epstein-Barr virus. The virus preferentially infects B lymphocytes; the immune response includes the activation of T cells which, when present in the blood, appear as atypical lymphocytes (altered T cells).

The laboratory hallmark of infectious mononucleosis is the presence of heterophil antibodies, IgM immunoglobulins. These antibodies are present in over 50 percent of patients during the first two weeks of illness and up to 95 percent at the end of one month; these disappear within four to six months. The test results are typically reliable in adults, but are often negative (unreliable) in younger children.

Positive results may be obtained in patients who have high titers of serum sickness antibodies and rarely in patients with leukemia, cytomegalovirus, Burkitt's lymphoma, rheumatoid arthritis and viral hepatitis. Also see EPSTEIN-BARR VIRUS (EBV) ANTIBODIES, SERUM.

MONOSPOT SCREEN PLUS HETEROPHILE ABSORPTION

Synonyms: Heterophil Antibodies, Quantitative; Davidsohn Differential
SPECIMEN: Red top tube, separate serum.
REFERENCE RANGE: Monospot: Negative. Heterophile Absorption: Negative titer <1:56.
METHOD: Absorption by tissues followed by sheep RBC agglutination. When the monospot test is positive and antibody titers are desired then the heterophil absorption test may be requested. In this test, titration is done using sheep cells (classical), guinea pig kidney antigen and beef red cell stroma.

There are three possible sources of heterophile antibody; these are infectious mononucleosis, serum sickness and Forssman antibodies. The differentiation of these heterophile antibodies by the Davidsohn differential test is given in the next Table:

Differentiation of Antibodies Formed in Infectious Mononucleosis, Serum Sickness and Normal Forssman Antibodies

	Observation after Absorption with Guinea Pig Kidney	Beef RBC
Disease:		
Infectious Mononucleosis	Antibody Remains	Antibody Absorbed
Serum Sickness	Antibody Absorbed	Antibody Absorbed
No Disease:		
Forssman Antibodies	Antibody Absorbed	Antibody Remains

INTERPRETATION: The heterophile absorption test is done only when the monospot test is positive. The sensitivity and specificity of the heterophile agglutination test are very high; the sensitivity of the heterophile test is 80% to 95% in adults after one month. Test results are not as reliable in younger children. This test was classically performed using sheep RBC. Use of horse RBC has improved the sensitivity of the test (Monospot). See also EPSTEIN-BARR VIRUS(EBV) ANTIBODIES, SERUM.

MORPHINE

SPECIMEN: 50mL random urine without preservatives, store at -20°C
REFERENCE RANGE: None detected
METHOD: Thin-layer chromatography; radioimmunoassay; gas-liquid chromatography/ mass spectrometry; latex agglutination-inhibition; hemagglutination-inhibition, spectrophotofluorometry; free radical assay.
INTERPRETATION: Morphine is measured as an indicator of opiate (morphine, codeine) or opioid (e.g. heroin, fentanyl, meperidine, methadone, propoxyphene) use. Heroin is diacetylmorphine and, in the body, is rapidly deactelylated to morphine; urine is the major route of excretion and morphine is excreted either free (5 to 20 percent) or conjugated, primarily to glucuronide. Since most of the semisynthetic and synthetic opioids are metabolized to morphine, measurement of this substance will indicated medical use or abuse of this class of drugs. The immunological methods are more sensitive than TLC. Morphine levels are usually not quantitated. Because of social and legal considerations, a positive qualitative screen should be confirmed by a gas chromatography/mass spectrometry method.

Opiate and most opioid action can be quickly reversed by naloxone administration (Goldfrank LR. Emerg Med. 1984; 16:105).

MUCIN COAGULATION, JOINT FLUID (ROPE'S TEST)

SPECIMEN: 2mL of joint fluid; no anticoagulant
REFERENCE RANGE: Normal: A tight, ropy clump or soft mass with a clear or slightly cloudy solution. Abnormal: A small, friable mass in a cloudy solution or a few flecks in a cloudy solution.
METHOD: A few drops of synovial fluid are added to 20mL of 5% acetic acid.
INTERPRETATION: The mucin coagulation test provides a estimate of joint fluid quality. The mucin coagulation test reflects polymerization of hyaluronate; the normal concentration of hyaluronate in synovial fluid is 0.3 to 0.4g/L. Normally, a clot forms within 1 minute; a "poor" clot indicates inflammation and degregation. Mucin clot test results in different conditions are shown in the next Table:

Abnormal Mucin Clot Results in Different Joint Conditions
Good to Fair Mucin Clot Formation: Osteoarthritis, Traumatic Arthritis, Systemic Lupus Erythematosus, Rheumatic Fever
Fair to Poor Mucin Clot Formation: Rheumatoid Arthritis, Gout, Pseudogout
Poor Mucin Clot Formation: Acute Bacterial Arthritis, Tuberculous Arthritis

Both viscosity and clot formation reflect polymerization of hyaluronate; synovial fluids with poor viscosity also form poor clots.

MUCOPOLYSACCHARIDES (MPS), QUALITATIVE, URINE

SPECIMEN: Random urine; no preservative. Stable for 1 week at 4°C. Stable indefinitely at -20°C.
REFERENCE RANGE: Negative
METHOD: Acid turbidity test: The acid albumin turbidity test is based on the reaction of acid MPS with albumin, at acid pH, to form a precipitate. A spot test may be done by spotting urine on filter paper and then adding toluidine blue or alcian blue. Another spot test is based on the reaction of cetyltrimethylammonium bromide with mucopolysaccharide. Laboratory tests employed in the diagnosis of the mucopolysaccharidoses are given in (Dorman A, Matalon R. in "The Metabolic Basis of Inherited Disease," 3:1246-1256. Stanbury JB, Wyngaarden JB, Fredrickson DS, eds., N.Y., McGraw-Hill Book Co., 1972).
INTERPRETATION: Testing for MPS in urine is employed in the diagnosis of the mucopolysaccharidoses. These screening tests are applicable to all these conditions except Morquio's Syndrome.

The mucopolysaccharidoses (MPS) are a group of heritable diseases characterized by intracellular storage and abnormal excretion of urinary MPS secondary to enzyme deficiency. All are inherited as autosomal recessives, except Hunter's Syndrome in which both an autosomal recessive and x-linked recessive are known. The mucopolysaccharidoses (MPS) are listed in the next Table:

Mucopolysaccharidoses

Disease	Excessive Urinary MPS	Enzyme Deficiency
Hurler's Syndrome	Dermatan Sulfate Heparin Sulfate	Alpha-L-Iduronidase
Scheie Syndrome	Dermatan Sulfate	Alpha-L-Iduronidase
Hunter Syndrome	Dermatan Sulfate Heparin Sulfate	Sulfoiduronate Sulfatase
Sanfilippo Syndrome A	Heparin Sulfate	Heparin Sulfate Sulfatase
Sanfilippo Syndrome B	Heparin Sulfate	N-Acetyl-Alpha-D Glucosaminidase
Morquio's Syndrome	Keratin Sulfate	Chondroitin Sulfate N-Acetylhexosamine Sulfate Sulfatase
Maroteaux-Lamy Syndrome	Dermatan Sulfate	Arylsulfatase B
Beta-Glucuronidase Deficiency	Dermatan Sulfate	Beta-Glucuronidase

There are additional variants of MPS not tabulated above. Heparin sulfate and/or dermatan sulfate are excreted in all of the MPS except for Morquio's syndrome in which keratin sulfate is excreted. Specific diagnosis is done by demonstrating deficiency of a specific enzyme; this is usually done on cultured fibroblasts obtained following skin biopsy.

Quantitative determinations and identifications of excreted products are performed on 24 hour urines (see MUCOPOLYSACCHARIDES (MPS) QUANTITATIVE, URINE).

MUCOPOLYSACCHARIDES (MPS), QUANTITATIVE, URINE

SPECIMEN: 24 hour urine collection. The patient voids at 8:00 A.M. and discards the specimen. Then, all urine is collected over the next 24 hours including the 8:00 A.M. specimen collected the next morning. The urine should be kept cold during the collection. The specimen is stable for one week at room temperature.
REFERENCE RANGE: Method dependent; age-dependent. See Dorfman A, Matalon R. in "The Metabolic Basis of Inherited Diseases," 3:1248-1249. ed., Stanbury JB, Wyngaarden JB, Fredrickson DS., eds. N.Y., McGraw-Hill Book Co., 1972. Rapid identification of MPS excreted in urine may be made by electrophoresis. In another method, concentrated urine is placed on a Sephadex G-25 column; the MPS fraction is collected. The acid MPS are precipitated with cetylpyridinium chloride; the polysaccharide-cetylpyridinium complex is dissolved and uronic acid is determined by the carbazole method. The acid MPS keratin sulfate, which is excreted in Morquio's syndrome, is not detected using this method.
INTERPRETATION: See MUCOPOLYSACCHARIDES, QUALITATIVE, URINE. There has been some speculation that levels of MPS excretion correlate with severity of disease. Mental retardation is postulated to be linked with heparin sulfate excretion, while corneal opacification and aortic disease are linked to dermatin sulfate excretion.

MULTIPLE SCLEROSIS, PANEL

A multiple sclerosis panel is given in the next Table:

Multiple Sclerosis - Panel
Cerebrospinal Fluid:
CSF Protein Electrophoresis
Myelin Basic Protein

Typical CSF findings in multiple sclerosis are given in the next Table:

Typical CSF Findings in Multiple Sclerosis

Condition	Pressure mm CSF	Gross Appearance	Cells (cu. mm.)	Protein (mg/dL)	Glucose (mg/dL)
Multiple Sclerosis	Normal	Clear, Colorless	Normal or 10-50 Lymphs	Normal or 45-100	Normal

Typical findings in a patient with active multiple sclerosis include:

1) Elevated CSF gamma globulin (75%)
2) Oligoclonal banding on CSF-protein electrophoresis (IgG-region)
3) Elevated CSF-myelin basic protein levels

CSF Protein: In multiple sclerosis, CSF gamma globulin is found to be elevated in about 75% of patients with an established clinical diagnosis of multiple sclerosis. Other causes of increased CSF gamma globulin include encephalitis, meningitis, neurosyphilis, arachnoiditis, and some cases of benign and malignant intracranial neoplasms.

Oligoclonal Banding: Normal CSF immunoglobulin migrates as a diffuse band; abnormal immunoglobulins migrate as discrete sharp bands or oligoclonal bands. This is the pattern observed in multiple sclerosis; this pattern of discrete bands within the gamma globulins supports the hypothesis that these globulins arise from a few clones of cells. It is most helpful when there is a comparison of oligoclonal banding in a CSF specimen compared to the pattern in a serum specimen taken at the same time. If oligoclonal banding is found in the CSF, but not in the serum, this is taken as evidence of an intrathecal immune response. The majority of such findings are seen in multiple sclerosis (Zeman A, et al. J Neurol Neurosurg Psychiatry. 1993; 56:32-35). By contrast, identical oligoclonal banding found in both CSF and serum is due to passive movement of immunoglobulin from serum into the CSF. However, oligoclonal bands are not unique to multiple sclerosis and are observed in the conditions listed in the next Table:

Oligoclonal Bands on Electrophoresis
Multiple Sclerosis
Inflammatory Polyneuropathy
Neurosyphilis
Cryptococcal Meningitis
Chronic Rubella Panencephalitis
Subacute Sclerosing Panencephalitis
HIV Infection

Oligoclonal bands are observed using agarose electrophoresis; 3mL of CSF is concentrated to 0.05mL. A specimen of serum from the patient is electrophoresed at the same time. The results of electrophoresis are illustrated in the next Figure (Johnson KP, Hosein Z. Lab Man. 1981; 36-40):

Agarose Electrophoresis of Serum and CSF Specimens

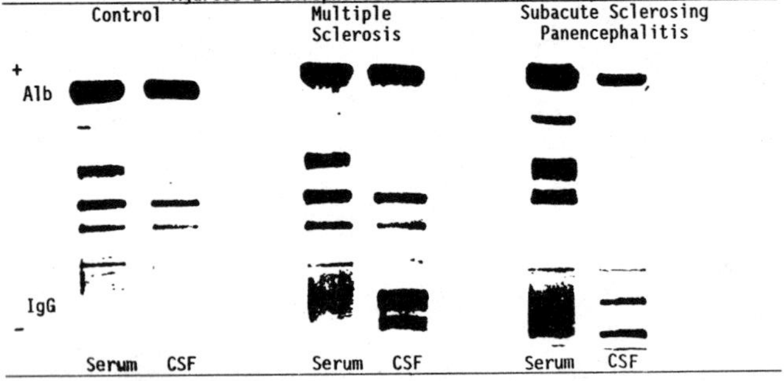

Control
Multiple Sclerosis
Subacute Sclerosing Panencephalitis
+
Alb
IgG
-
Serum CSF
Serum CSF
Serum CSF

An abnormal result is the finding of two or more bands in the CSF that are not present in the serum specimen. The sensitivity of CSF oligoclonal bands in the diagnosis of multiple sclerosis in 79-94% (Markowitz H, Kokmen E. Mayo Clin Proc. 1983; 58:273-274). A scan of cerebrospinal fluid, obtained from a patient with multiple sclerosis, using high resolution, is shown in the next Figure:

High Resolution Scan of Cerebrospinal Fluid

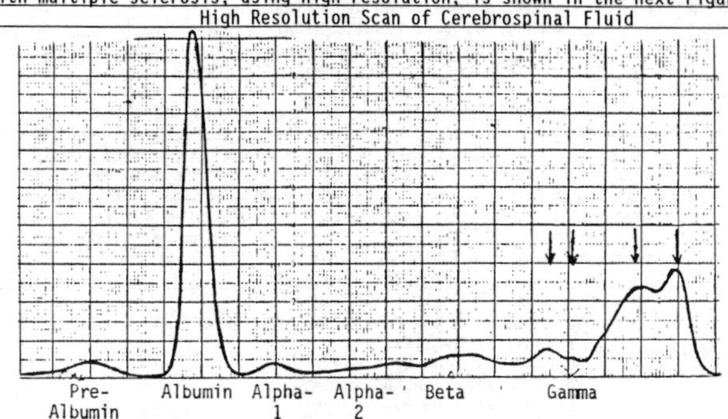

Myelin Basic Protein: The conditions associated with elevated levels of myelin basic protein in cerebrospinal fluid (CSF) are given in the next Table:

Conditions Associated with Elevated Levels of Myelin Basic Protein in Cerebrospinal Fluid (CSF)
Multiple Sclerosis
Myelinopathies Other Than Multiple Sclerosis
Transverse Myelitis
Hereditary Leukodystrophy
Metachromatic Leukodystrophy
Central Pontine Myelinolysis
Peripheral Neuropathy
Neurologic Disease:
Encephalitis
Post Neurosurgery
CNS-SLE
Strokes Near the Surface of the Brain
Brain Damage After Head Injury
Subarachnoid Hemorrhage

Patients with acute multiple sclerosis have very high levels of myelin basic protein in their cerebrospinal fluid (CSF); those patients whose disease is in remission and clinically inactive have no detectable myelin basic protein. The levels of myelin basic protein are absent or low in those patients whose disease is chronically active (Cohen SR, et al. N Engl J Med. 1976; 295:1455-1457).

Myelin basic protein may also be detected in the serum of patients with brain damage after head injury; in these patients, levels of myelin basic protein correlate well with the type and severity of brain damage and also with final outcome (Thomas et al. Lancet. 1978; 1:113-115; Palfreyman et al. Clin Chim Acta. 1979; 92:403-409).

MURAMIDASE, SERUM (see LYSOZYME, SERUM)

MURAMIDASE, URINE (see LYSOZYME, URINE)

MYELIN BASIC PROTEIN

SPECIMEN: Cerebrospinal fluid (CSF)

REFERENCE RANGE: MetPath: Reference Range of Myelin Basic Protein

Result	Interpretation
Less than 4ng/mL	Negative result: No demyelination
4-8ng/mL	Weakly positive result: Slowly progressive form of demyelination or the recovery phase from acute exacerbation
Greater than 9ng/mL	Positive result: Consistent with active demyelination

METHOD: RIA

INTERPRETATION: Myelin basic protein may be measured in CSF to aid in the diagnosis and evaluation of a number of demyelinating neurological conditions. It is elevated during active multiple sclerosis (MS), other myelinopathies, demyelination during chemotherapy or radiation therapy of the CNS, and following brain injury. Serial determinations may be performed on MS patients to monitor response to therapy. The conditions associated with elevated levels of myelin basic protein in cerebrospinal fluid (CSF) are given in the next Table:

Conditions Associated with Elevated Levels of Myelin Basic Protein in Cerebrospinal Fluid (CSF)

- Multiple Sclerosis
- Myelinopathies Other Than Multiple Sclerosis
  - Transverse Myelitis
  - Hereditary Leukodystrophy
  - Metachromatic Leukodystrophy
  - Central Pontine Myelinolysis
  - Peripheral Neuropathy
- Neurological Disease:
  - Encephalitis
  - Post-Neurosurgery
  - CNS-SLE
  - Strokes Near the Surface of the Brain
  - Brain Damage After Head Injury
  - Subarachnoid Hemorrhage

Patients with acute multiple sclerosis have very high levels of myelin basic protein in their cerebrospinal fluid (CSF); those patients whose disease is in remission and clinically inactive have no detectable myelin basic protein. The levels of myelin basic protein are intermediate to absent in those patients whose disease is chronically active (Cohen SR, et al. N Engl J Med. 1976; 295:1455-1457). See MULTIPLE SCLEROSIS, PANEL.

Myelin basic protein may also be detected in the serum of patients with brain damage after head injury; in these patients, levels of myelin basic protein correlate well with the type and severity of brain damage and also with final outcome (Thomas et al. Lancet. 1978; 1:113-115; Palfreyman, et al. Clin Chim Acta. 1979; 92:403-409).

MYOCARDIAL INFARCTION PANEL

(see **ACUTE MYOCARDIAL INFARCTION PANEL**)

MYOGLOBIN, SERUM

SPECIMEN: Red top tube, separate serum

REFERENCE RANGE: 30-90ng/mL (or mcg/liter)

METHOD: RIA, Latex Agglutination

INTERPRETATION: Myoglobin is the main protein of striated muscle (skeletal and cardiac). It resembles hemoglobin; however, it is unable to release oxygen except at extremely low oxygen tension. Myoglobin is increased in the serum in the conditions listed in the next Table:

Increase in Serum Myoglobin

- Acute Myocardial Infarction
- Skeletal Muscle Conditions: Trauma (Crush Injury); Surgical Procedures; Intramuscular Injections; Acute Alcoholism with Delirium Tremens; Any Condition Associated with Rhabdomyolysis (crush injury to a limb, alcoholism, drugs, overuse of skeletal muscle, seizures, burns, sepsis, viral infections, myopathies, heat, hypokalemia, neuroleptic-malignant syndrome, others)

Acute Myocardial Infarction: Serum myoglobin is elevated following onset of acute myocardial infarction, as follows: 4 hours, 89% of patients; 6 hours, 92% of patients. It reaches a peak within 12 hours.

Myoglobin concentration may be determined rapidly by latex agglutination. The overall cumulative release of myoglobin correlates with infarct size. Myoglobin determination may be useful in patients with suspected myocardial infarction who do not have ST elevation on ECG - in this setting, sensitivity is 82%, specificity is 84% (Ohman EM, et al. Brit Heart J. 1990; 63:335-338). Early diagnosis of acute myocardial infarction by serum myoglobin determination can complement serial enzyme determinations and may be of value in the early confirmation of the diagnosis of acute myocardial infarction and early initiation of thrombolytic therapy.

Rhabdomyolysis: Rhabdomyolysis is associated with many conditions including crush injury to a limb, alcoholism, drugs (crack cocaine, lovastatin-gemfibrozil combination, others), overuse of skeletal muscle, seizures, burns, sepsis, viral infections, myopathies, heat, hypokalemia, neuroleptic malignant syndrome and others (Odeh M. N Engl J Med. 1991; 324:1417-1422; Ward MM. Arch Intern Med. 1988; 148:1553-1557).

Acute renal failure is reported to occur in 16.5% of patients with rhabdomyolysis (Ward MM, already cited). Although serum myoglobin is often elevated in rhabdomyolysis, demonstration of a five-fold increase in serum creatinine kinase (CPK) in the absence of cardiac or brain injury is the single most sensitive diagnostic test for rhabdomyolysis (Gabow PA, et al. Medicine. 1983; 61:141-152).

MYOGLOBIN, URINE

SPECIMEN: Random urine, refrigerate
REFERENCE RANGE: Qualitative: Negative; Quantitative: 0-2mcg/mL
METHOD: Qualitative: Dipstick (Ames or BMC) will detect both hemoglobin and myoglobin; if test is positive use ammonium sulfate to precipitate hemoglobin, filter and test supernatant for myoglobin with dipstick. Quantitative: Antigen-antibody reaction measured by nephelometry or RIA.
INTERPRETATION: This test may be of use in establishing muscle damage, skeletal or myocardial. Myoglobin is increased in the urine in the conditions listed in the next Table:

Increase in Urine Myoglobin
Acute Myocardial Infarction
Viral Infection (Cunningham E, et al. JAMA. 1979; 24: 2428-2429)
Skeletal Muscle Conditions
Trauma (Crush Injury)(Review: Better OS, Stein JH. N Engl J Med. 1990; 322:825-829)
Surgical Procedures
Polymyositis
Acute Alcoholism with Delirium Tremens
Exertional Myoglobinuria in Untrained Individuals
Metabolic Conditions Affecting Striated Muscle
Any Condition Associated with Rhabdomyolysis (see MYOGLOBIN, SERUM)

Following acute myocardial infarction, myoglobin appears in the urine within 48 hours.

Myoglobinuria depends on the total amount of myoglobin released from muscle, the serum myoglobin concentration, the glomerular filtration rate, and the urine output. Myoglobinuria correlates poorly with myoglobinemia.

In one study, 26% of patients with rhabdomyolysis documented by CPK elevation had negative urine dipstick for myoglobin within 24 to 48 hours (Gabow PA, et al. Medicine. 1983; 61:141-152). Myoglobin is nephrotoxic, particularly with associated oliguria, aciduria, and uricosuria. Patients with traumatic rhabdomyolysis should have forced diuresis while maintaining urinary pH above 6.5 until myoglobinuria resolves (Better OS, Stein JH. N Engl J Med. 1990; 322:825-829).

MYSOLINE (see PRIMIDONE)

NEURON-SPECIFIC ENOLASE(NSE)

SPECIMEN: Red top tube, separate serum; freeze serum.
REFERENCE RANGE: Normal: 5.4-12.9ng/mL; Children with Metastatic Neuroblastoma: 10-1240ng/mL (Zeltzer PM, et al. Lancet. 1983; ii:361-363).
METHOD: Double-antibody RIA, EIA.
INTERPRETATION: The serum neuron-specific enolase(NSE) level is a very sensitive marker and a prognostic indicator in children with metastatic neuroblastoma. The NSE has also been used as a serial marker to study the response to therapy in patients with undifferentiated small cell carcinoma of lung. Use of the ratio of serum neuron-specific enolase to serum nonneuronal enolase (NSE/NNE ratio) improves the specificity of the test for small cell carcinoma of the lung compared to NSE alone (Viallard JL, et al. Chest. 1988; 93:1225-1233).

Neuron-specific enolase(NSE) is an isoenzyme of the glycolytic enzyme, 2-phospho-D-glycerate-hydrolase; the enzyme is found in neuronal and neuroendocrine cells of the central and peripheral nervous systems.

Serum NSE levels has prognostic value for patients with neuroblastoma. NSE levels below 100ng/mL correlate with longer survival; this relation is highly significant in the subgroup of infants less than 1 year old at diagnosis (Zeltzer, already cited).

CSF NSE levels may be a nonspecific marker for neuronal damage in comatose children with a variety of CNS pathologies (encephalitis, encephalopathy, Reye's syndrome)(Nara T, et al. AJDC. 1988; 142:173-174).

NITROBLUE TETRAZOLIUM (NBT) TEST

SPECIMEN: Variable. Green or yellow top tube. Must be transported and tested rapidly.
REFERENCE RANGE: Normal Neutrophil Function: NBT is reduced to blue formazan. Chronic Granulomatous Disease: NBT is not reduced; no blue color is identified.
METHOD: Oxidized NBT is colorless. During phagocytosis, NBT is reduced and precipitates in the cytoplasm as blue formazan. The NBT test involves mixing the patient's neutrophils with NBT dye in the presence of a stimulant such as phorbol myristate acetate (PMA). Test results may be evaluated qualitatively by performing the test on a slide and evaluating the slide under a microscope. Quantitative tests may be performed by extraction of reduced dye and subsequent spectrophotometric measurement. Other methods may also be used to quantitate the NBT test (Lopez M, et al. JAMA. 1992; 268:2970-2990).
INTERPRETATION: The nitroblue tetrazolium (NBT) dye reduction test is used to diagnose chronic granulomatous disease (CGD). CGD is a group of disorders characterized by recurrent microbial infections with organisms such as Staphylococcus aureus, Klebsiella species, Serratia marcescens, Escherichia coli, Aspergillus species, Pseudomonas cepacia, Candida albicans, Salmonella species, other enteric bacteria, Nocardia species, Mycobacteria, and Pneumocystis carinii. Clinically, patients present with pneumonia, lymphadenitis, or abscesses involving the lungs, liver, lymph nodes, or skin.

Neutrophils in CGD have normal phagocytic function, but have defective intracellular killing. The defect is in NADPH oxidase, the membrane-bound enzyme responsible for the respiratory burst. When bacteria are ingested, CGD neutrophils do not undergo a respiratory burst; therefore, oxidizing radicals are not produced and certain microorganisms, particularly catalase positive organisms that break down hydrogen peroxide, are not killed. The inheritance of CGD is usually X-linked recessive, but autosomal recessive forms also exist (Hopkins PJ, et al. Clin Lab Med. 1992; 12:277-304; Smith RM, Curnutte JT.
J Amer Soc Hematol. 1991; 77:673-686; Shyur S, Hill HR. Pediatr Infect Dis J. 1991; 10:595-611; Gallin JI, Malech HL. JAMA. 1990; 263:1533-1537; Baehner RL, Nathan DG. N Engl J Med. 1968; 278:971-976).

NITROGEN BALANCE

SPECIMEN: Twenty-four hour urine urea nitrogen determination is required. Collect urine as follows: Refrigerate specimen as it is collected. Instruct the patient to void at 8:00 A.M. and discard the specimen. Then collect all urine including the final specimen voided at the end of the 24 hour collection; i.e., 8:00 A.M. the next morning. Transport to lab and refrigerate specimen at 4°C.
REFERENCE RANGE: Normally, the 24 hour urinary output of nitrogen is 12 to 24 grams. The range of values for nitrogen balance in adults is positive 4 to negative 20 grams of nitrogen per day.
METHOD: Urine urea nitrogen, see Urea Nitrogen.
INTERPRETATION: Nitrogen balance is used as a measure of the effectiveness of nutritional therapy; it is calculated from the formula:

$$\text{Nitrogen Balance} = \frac{\text{24 Hr Protein Intake(g)}}{6.25} - [\text{24 Hr Urine Urea Nitrogen(g)} + 4]$$

With this equation, the net gains (positive nitrogen balance) or losses (negative nitrogen balance) by the body are assessed. A gram of nitrogen excreted represents the degradation of 6.25 grams of protein. The 4 in the above equation is an estimate of the non-urinary nitrogen loss, i.e., skin and feces and non-urea urinary losses; this approximation is reasonable in the absence of severe diarrhea, hemorrhage or fistula losses.

Normally, the 24 hour urinary output of urea nitrogen is 12 to 20 grams. The range of values for nitrogen balance seen clinically in adults is from positive 4 to negative 20 grams of nitrogen per day. One gram of nitrogen represents the protein content of one ounce of lean tissue. A high-protein diet will not necessarily result in a positive nitrogen balance because ingestion of large amounts of protein without an adequate intake of calories results in utilization of much of the protein as energy. Endogenous protein breakdown can be avoided by adequate glucose, fat, or branched-chain amino acid administration (protein-sparing nutrition). Nitrogen balance is increased by increasing nitrogen or energy intake or both (Berger R, Adams L. Chest. 1989; 96:139-150; D'Attellis NP, et al. J Crit Illness. 1988; 3:49-68).

NORPACE (see DISOPYRAMIDE)

NORTRIPTYLINE (see TRICYCLIC ANTIDEPRESSANTS)

NUTRITION, PARENTERAL, PEDIATRIC MONITORING PANEL

Some of the indications for parenteral nutrition in infants are given in the next Table (Silber G. Pediatric Review. 1990; 3:3-7):

Indications for Parenteral Nutrition in Infants and Children
Surgical Gastrointestinal Disorders (Gastroschisis, Omphalocele, Hirschprung's Disease, Diaphragmatic Hernia, Small or Large Bowel Atresia, Tracheoesophageal Fistula, Obstruction, Imperforate Anus, Malrotation)
Intractable Diarrhea of Infancy
Short Bowel Syndrome
Necrotizing Enterocolitis
Pancreatitis
Inflammatory Bowel Disease
Chronic Idiopathic Intestinal Pseudo-obstruction
Gastrointestinal Fistulas
Supportive Care for Patients with Malignancy
Severe Burns and Trauma
Malabsorption Syndromes

(cont)

Nutrition, Parenteral, Pediatric Monitoring Panel (Cont)

A suggested schedule for monitoring parenteral nutrition in pediatric patients is given in the next Table (Committee on Parenteral Nutrition. Pediatrics. 1983; 71:547-552; Heird WC in Pediatric Nutrition Handbook; Evanston, IL. Am Acad Pediatr. 1979; 392-408):

Suggested Monitoring Schedule During Total Parenteral Nutrition

Variable Monitored	Suggested Frequency: Initial Period	Later Period
Serum Electrolytes	Daily	2-3 times/wk
Serum Urea Nitrogen	3 times/wk	2 times/wk
Serum Calcium, Magnesium, Phosphorus	3 times/wk	2 times/wk
Serum Glucose	See Below	See Below
Serum Acid-Base Status	3-4 times/wk	2-3 times/wk
Serum Protein, Albumin	Weekly	Weekly
AST, ALT	Weekly	Weekly
Hemoglobin	2 times/wk	2 times/wk
Urine Glucose	Daily	Daily
Clinical Observations (Activity, Temperature, etc.)	Daily	Daily
WBC Count and Differential Count	As Indicated	As Indicated
Cultures	As Indicated	As Indicated
Serum Triglyceride	As Indicated	As Indicated

Initial period is period before glucose intake is achieved, or any period of metabolic instability.
Later period is period during which patient is in a metabolic steady state.
Blood glucose should be monitored closely during period of glucosuria (to determine degree of hyperglycemia) and for two to three days after cessation of parenteral nutrition (to detect hypoglycemia). In latter instance, frequent Dextrostix determination constituents adequate screening. Total parenteral intravenous nutrition is given to those patients that are unable to ingest, digest or absorb sufficient nutrients. Total parenteral nutrition (TPN) can preserve body weight in almost all groups of patients.

NUTRITIONAL PROFILE: PROTEIN-CALORIE MALNUTRITION

The objective evaluation of protein-calorie malnutrition states is based on clinical laboratory testing, evaluation of immunocompetence and anthropometric measurements. Malnutrition is classified into three general categories; these are called marasmus, kwashiorkor-like syndrome, and combinations of these two states.
Marasmus: (derived from the Greek marasmus meaning "a dying away") is a variant of protein-calorie malnutrition, characterized by a depression of the anthropometric measurements, weight and creatinine-height index in the presence of normal serum albumin. There is a loss of lean body mass and fat stores and in the more advanced cases, depressed immune function. This condition is brought about by prolonged inadequate intake of both protein and calories (prolonged negative nitrogen balance). This condition is often readily recognized clinically; edema is absent.
Kwashiorkor-like syndrome: (Ghana for "red boy") occurs as a result of inadequate protein intake with a low or normal caloric intake; this syndrome is characterized by depletion of visceral protein stores, i.e. decreased serum albumin and serum transferrin with normal anthropometric measurements. In addition, there is decreased immunological resistance reflected by low total lymphocyte count and delayed hypersensitivity as measured by skin tests. Clinically, patients have pitting edema, ascites, growth failure, liver enlargement and diarrhea.

Combined marasmus and kwashiorkor: These conditions commonly coexist in malnourished patients.

Characteristics of marasmus and kwashiorkor are shown in the next Table (Blackburn GL, et al. J Parenter Nutr. 1977; 1:11-22):

Classification of Malnutrition

Marasmus:	Percent Ideal Weight	Creatinine-Height Index (Percent)	Immune Competence Skin Tests (mm)
Moderate	60-80	60-80	
Severe	<60	<60	<5

Kwashiorkor-Like:	Serum Albumin (g/dL)	Serum Transferrin (mg/dL)	Total Lymphocyte Count	Skin Tests (mm)
Moderate	2.1-3.0	100-150	800-1200	<5
Severe	<2.1	<100	<800	<5

Patients who are at particularly high risk for complications of malnutrition are surgical patients and alcoholics.

Clinical Laboratory Testing: The clinical laboratory assays used to evaluate malnutrition are given in the next Table:

Nutritional Assessment Profile

- Measurement of Serum Proteins (Hepatic Secretory Proteins)
  - Albumin
  - Transferrin or Total Iron-Binding Capacity(TIBC)
    - (Transferrin = (0.8 x TIBC)-43)
  - Thyroxine-Binding Prealbumin(TBPA)
  - Retinol-Binding Protein(RBP)
- Immunocompetence:
  - Total Lymphocyte Count
  - Delayed Cutaneous Hypersensitivity

Protein Malnutrition: Measuring the proteins, serum albumin and serum transferrin, has been the mainstay of nutritional screening in clinical practice. In a comparative study of four serum proteins: albumin, transferrin, thyroxine-binding prealbumin (TBPA) and retinol binding protein (RBP), it has been shown that albumin has low sensitivity and transferrin has intermediate sensitivity while TBPA and RBP have the greatest sensitivity to an alteration in the nutritional state (Ingenbleek Y, et al. Clin Chim Acta. 1975; 63:61-67).

Serum albumin is not sensitive to short-term low protein diet because it has a relatively long half-life (15 to 19 days), large total albumin mass, capacity to maintain hepatic synthesis of albumin and reduced albumin catabolism. Intravascular albumin mass is also maintained by mobilization of albumin from the extravascular pool on dietary restriction (Hoffenberg R, et al. J Clin Invest. 1966; 45:143-151). Thus, given all these factors and others, it is concluded that albumin is a poor index of the short-term adequacy of protein and energy intake (Shetty PS, et al. Lancet. 1979; 230-232).

Serum transferrin is reduced in severe protein-energy malnutrition; however, serum transferrin is often not reliable in mild cases of malnutrition and the response to treatment is unpredictable (Ingenbleek Y, et al, already cited). Serum transferrin is not sensitive to short-term restriction of protein or energy: transferrin has a relatively long half-life (8 days); it is increased in iron deficiency; it is an "acute phase" reactant and is increased in patients with infections and as a response to stress.

The serum proteins, thyroxine-binding prealbumin (TBPA) and retinol-binding protein (RBP) are more sensitive indices of protein-energy malnutrition than either serum albumin or serum transferrin. Both TBPA and RBP are synthesized in the liver; the half-life of TBPA is 2 days while that of RBP is only 12 hours, TBPA and RBP could be used to detect subclinical malnutrition and monitor the effectiveness of dietary treatment (Shetty PS, et al, already cited).

Evaluation of Immunocompetence: The lymphocyte count and skin sensitivity reactions are used to reflect immunocompetence; immune response is modified by inherited and environmental factors; dietary intake is the most important environmental factor and malnutrition is the most frequent cause of secondary immunodeficiency. Lymphoid tissue, such as the thymus and tonsils, are small in subjects with malnutrition; this is reflected in a decreased number of lymphocytes in the peripheral blood.

Nutritional Profile (Cont)

Intradermal skin tests are used employing recall antigens, e.g., mumps skin test antigen, diphtheria toxoid, streptokinase/streptodornase, Candida albicans skin test antigen, Trichophytin, and tuberculin purified protein derivative (PPD). The diameter of induration is measured and recorded at 24 and 48 hours. The skin test is regarded as positive or reactive if the induration at either reading is 5mm or greater for any of the three antigens (Mullen JL, et al. Arch Surg. 1979; 114:121-125). The utility of skin testing in nutritional assessment has been critically reviewed (Twomey P, et al. J Parenter Nutr. 1982; 6:50-58).
Anthropometric Measurements: Anthropometric measurements, used to evaluate the malnutrition found in marasmus, are given in the next Table:

Anthropometric Measurements
Height
Calculate Creatinine-Height Index
Weight
Calculate Percent Ideal Weight
Arm Measurements:
Triceps Skinfold Thickness(mm)
Arm Circumference(cm)
Calculate Arm Muscle Circumference(cm)
Midarm Muscle Area (mm^2)

Creatinine-Height Index (CHI): The creatinine-height index (CHI) is a measure of lean muscle mass; it is given by the equation:

$$\text{Creatinine Height Index (CHI)} = \frac{\text{Actual Urinary Creatinine}}{\text{Ideal Urinary Creatinine}} \times 100$$

Two measurements are required, height(cm) and 24 hour urinary creatinine. The ideal urinary creatinine is obtained from standard Tables (Blackburn GL, et al, already cited).
Percent Ideal Weight: Percent ideal weight is calculated from the formula:

$$\text{Percent Ideal Body Weight} = \frac{\text{Actual Weight}}{\text{Ideal Body Weight}} \times 100$$

The ideal body weight is obtained from standard Tables (Blackburn GL, et al, already cited).
Arm Measurements: Arm measurements are made to estimate muscle mass and fat stores; arm muscle circumference is calculated from the equation:

$$\text{Arm Muscle Circumference(cm)} = \text{Arm Circumference (cm)} - [0.314 \times \text{Triceps Skinfold(mm)}]$$

Triceps Skinfold Thickness: There is a good correlation between triceps skinfold and body fat (Forse RA, et al. Surgery. 1980; 88:17-24). The triceps skinfold is measured on the nondominant arm with a caliper; tables of average values for triceps skinfold are given in the article by Butterworth and Blackburn (Butterworth CE, Blackburn GL. Nutrition Today. March/April 1975; 8-18). The percentage of body fat may be predicted from the skinfold thickness measurements of the triceps, subscapular and suprailiac folds (Shephard RJ. Diagnosis. 1984; 157-172).
Mid-Upper Arm Circumference: The mid-arm circumference reflects both caloric adequacy and muscle mass. It is measured using a tape measure.
Midarm Muscle Area: Upper arm area, upper arm muscle area and upper arm fat area are derived from measurements of upper arm circumference and triceps skinfold. Calculations and normals are available (Frisancho AR. Am J Clin Nutr. 1981; 34:2540-2545).
Therapy: During the course of nutritional therapy, the following measurements are obtained (Blackburn GL, et al. J Parenter Nutr. 1977; 1:11-22):

Protocol for Monitoring Efficacy of Nutritional Therapy

Variable Monitored	Suggested Frequency
Body Weight	Daily
Nitrogen Balance	Twice Weekly
Total Lymphocyte Count	Weekly
Anthropometrics	Every Three Weeks
Serum Transferrin	
Skin Tests	

OSMOLALITY, SERUM

SPECIMEN: Red top tube, separate serum
REFERENCE RANGE: 285-295milliosmols/liter
METHOD: Freezing point lowering or vapor pressure; Vapor pressure method less precise and will not detect volatiles.
INTERPRETATION: Serum osmolality measurements are done for two reasons: Determine whether serum water content deviates widely from normal; or screen for the presence of foreign low molecular-weight substances in the blood.

The osmolality of the serum is approximately 285 to 295; all but 25 mOsm are contributed by sodium ion and the anions, chloride and bicarbonate. An osmolality above 350mOsm per liter or below 250mOsm per liter is usually necessary to produce clinical signs and symptoms. However, symptoms may occur at values closer to the normal range if the change in osmolality has developed rapidly.

There is little evidence that hyperosmolality per se is harmful; hyperosmolality may reflect severe fluid shifts. Fluid shifts depend upon whether or not the substance has free access to intracellular water; substances with ready access, such as ethanol and urea, cause little fluid shift; substances with limited access, such as glucose and sodium ion, cause significant fluid shift.

The serum osmolality is increased in the conditions listed in the next Table:

Causes of Increased Serum Osmolality
Increase in Normal Constituents of Serum:
Increased Sodium Ion (Hypernatremia)
Increased Glucose (Hyperglycemia)
Increased BUN (Uremia)
Increased Toxic Substances:
Ethanol; Methanol; Ethylene Glycol; Others
Decrease in Water:
Dehydration
Diabetes Insipidus

Serum osmolality is decreased in the conditions listed in the next Table:

Causes of Decreased Serum Osmolality
Overhydration
Syndrome Inappropriate Antidiuretic Hormone (SIADH)
Loss of sodium ion (hyponatremia)

The steps in the evaluation of serum osmolality are given in the next Table (Gennari FJ. N Engl J Med. 1984; 310:102-105):

Steps in the Evaluation of Serum Osmolality

1. Measure Serum Osmolality, Sodium, Glucose and Urea.
2. Calculate Osmolality:

$$mOsm/kg\ H_2O = 2Na^+ + \frac{Glu}{18} + \frac{BUN}{2.8}$$

3. If Measured minus Calculated Osmolality >10 mOsm/kg H_2O, then consider the following Differential:
   - Decreased Serum Water Content
     - Hyperlipidemia
     - Hyperproteinemia (If Total Protein > 10 g/dL)
   - Low-Molecular-Weight Substances in Serum: Ethanol; Methanol; Ethylene Glycol; Mannitol; Acetone, Isopropanol, Paraldehyde, etc.
   - Laboratory or Calculation Error
4. If Measured minus Calculated Osmolality < 10 mOsm/kg H_2O, then the Above Possibilities are Ruled Out.

The contribution to the serum osmolality of the following substances is given in the next Table:

Contribution to Serum Osmolality

Substance	Contribution
Glucose	5.5 (Each 100mg/dL)
BUN	7.1 (Each 20mg/dL)
Ethanol	21.7 (Each 100mg/dL)
Methanol	31.0 (Each 100mg/dL)
Ethylene Glycol	16.3 (Each 100mg/dL)

OSMOTIC FRAGILITY

SPECIMEN: Green (heparin) top tube; fresh blood must be used. Fresh defibrinated blood may also be used.

REFERENCE RANGE: Normal Values for Osmotic Fragility

Sodium Chloride (Conc. in %)	Percent Hemolysis	
	Prior to Incubation	Following Incubation
0.85	0	0
0.75	0	0-2
0.65	0	0-20
0.60	0	10-40
0.55	0	15-70
0.50	0-5	40-85
0.45	5-45	55-95
0.40	50-90	65-100
0.35	90-99	75-100
0.30	97-100	80-100
0.20	100	91-100
0.10	100	100

Normal red blood cells will begin lysis at 0.5% and lysis is complete at about 0.3%. Following incubation, the normal red blood cells are more sensitive and will begin lysis at a higher concentration of saline.

METHOD: Red blood cells are placed in decreasing concentrations of NaCl phosphate buffered solutions (1.0 to 0.1%). The suspensions are mixed and allowed to stand at room temperature for 30 minutes. Then, the suspensions are remixed, centrifuged for 5 minutes at 2000RPM and the supernatant is pipetted off. Measure the absorbance of the supernatant in each tube at 540nm.

The values for percent hemolysis are recorded and plotted on ordinary graph paper with percent hemolysis on the ordinate and the concentrations of sodium chloride on the abscissa.

Incubation: Blood is defibrinated by shaking with glass beads; duplicate samples of the defibrinated blood are incubated at 37°C for 24 hours in sterile screw-capped containers; then, osmotic fragility is determined as described.

INTERPRETATION: This test is used in the evaluation of hemolytic anemias, especially when hereditary spherocytosis is suspected. Historically this test was also used to screen for thalassemia. Red cells obtained from patients with hereditary spherocytosis are especially susceptible to rupture when placed in hypotonic solution. These cells are relatively more susceptible to rupture than normal cells because of their shape; the spherocyte has a smaller surface area to cell size as compared to the normal cell which has the shape of a biconcave disc. Incubation increases the sensitivity of the test; mild cases of hereditary spherocytosis can be more readily detected following incubation. The causes of increased and decreased susceptibility to osmotic fragility are given in the next Table:

Osmotic Fragility	
Increased Susceptibility to Osmotic Fragility	Decreased Susceptibility to Osmotic Fragility
Hereditary Spherocytosis	Glucose-6-Phosphate Dehydrogenase (G6PD)
Acquired Immune Hemolytic Anemias	Iron Deficiency
ABO Hemolytic Disease	Liver Diseases (Obstructive Jaundice, Cirrhosis, Gilbert's Disease)
Antihuman Globulin-Positive Anemias	Hereditary Anemias: Sickle Cell Anemia, Hgb C Disease, Hgb E Disease, Thalassemia Major
Zieve's Syndrome	Plumbism
Pregnancy (32 percent in last trimester)	

Increased osmotic fragility is seen in any condition in which there is a spherocytosis.

Hereditary Spherocytosis: Hereditary spherocytosis is inherited as an autosomal dominant. The circulating red cells are rounder (spherocytes) and sometimes smaller than normal. The life span of the spherocyte is shorter than normal. The disease usually manifests itself in children but may occur at any age. The triad in the young is usually anemia, jaundice and splenomegaly. Treatment is splenectomy.

OSTEOMALACIA PANEL

The tests for osteomalacia are given in the next Table:

Screening Tests for Osteomalacia

Test	Expected Value in Osteomalacia
Screening:	
Serum Calcium (Corrected for Serum Albumin)	Decreased
Serum Phosphate	Decreased
Serum Alkaline Phosphatase	Increased
25-Hydroxyvitamin D	Decreased
Definitive:	
Bone Biopsy	Histology

If the corrected serum calcium or serum phosphate level is below or the serum alkaline phosphatase level is above the normal range, then bone biopsy may be considered.

Serum alkaline phosphatase is the best single routine biochemical screening test for osteomalacia; however, the false-negative rate is 10% and the false-positive rate is 32% (Peach H, et al. J Clin Pathol. 1982; 35:625-630).

New assays for bone turnover based on detection of urinary pyridinoline (a breakdown product of crosslinked collagen) are now available (Robins SP, et al. Eur J Clin Invest. 1991; 21:310-315; Delmas PD, et al. J Bone Miner Res. 191; 6:639-644). These may prove useful in the future study of metabolic bone disease.

Bone Biopsy: Osteomalacia is characterized histologically by defective bone mineralization; there is an increase in osteoid volume and seam thickness. Transiliac biopsy specimens are obtained. The antibiotic, tetracycline, taken orally, will deposit in sites of bone formation and subsequently can be studied in uncalcified sections by fluorescence microscopy (Frost HM. Calc Tiss Res. 1969; 3:211-237).

OXYGEN DISSOCIATION CURVE (P-50)

SPECIMEN: Green (heparin) top tube; also submit a normal control.
REFERENCE RANGE: See curve below.
METHOD: Sample of blood is exposed to varying oxygen partial pressure (PO_2) and the fraction of oxyhemoglobin ($HgbO_2$) is measured. The PO_2 is measured by an oxygen electrode. The $HgbO_2$ is measured by dual wavelength spectrometry at 560 and 576nm. The data are plotted (PO_2 versus $HgbO_2$ fraction); the P-50, which is defined as the partial pressure of oxygen at which the given hemoglobin sample is 50 percent saturated, is read from the curve.
INTERPRETATION: A wide variety of conditions will cause a shift of the dissociation curve of hemoglobin. These conditions may cause a shift of the dissociation curve of oxyhemoglobin to the left or to the right.

The principal significance of the shift in the oxyhemoglobin dissociation curve in different conditions lies in the delivery of oxygen to the tissue. When the curve is shifted to the left, less oxygen is delivered to the tissue for a given percent saturation of hemoglobin. When the curve is shifted to the right, more oxygen is delivered to the tissue for a given percent saturation of hemoglobin. This is illustrated in the next Figure:

Conditions Influencing Oxyhemoglobin Dissociation Curve

Shift to the Left	Shift to the Right
High Affinity of Hgb for Oxygen	Low Affinity of Hgb for Oxygen
Delivers Less Oxygen to Tissue	Delivers More Oxygen to Tissue
Alkalosis	Acidosis
Decreased PCO_2; 2,3-DPG; ATP; and Temperature	Increased PCO_2, ATP and Temperature
Increased CO Hgb and Met Hgb	Increased 2,3-DPG
Certain Abnormal Hemoglobinopathies	Living at High Altitude
Defects in Glycolytic Pathway before Formation of 2,3-DPG e.g. Hexokinase Def.	Cyanotic Heart Disease
Newborn (Fetal Hgb)	Females
	Chronic Anemia
	Chronic Respiratory Disease
	Pyruvate Kinase Deficiency
	Certain Abnormal Hemoglobinopathies

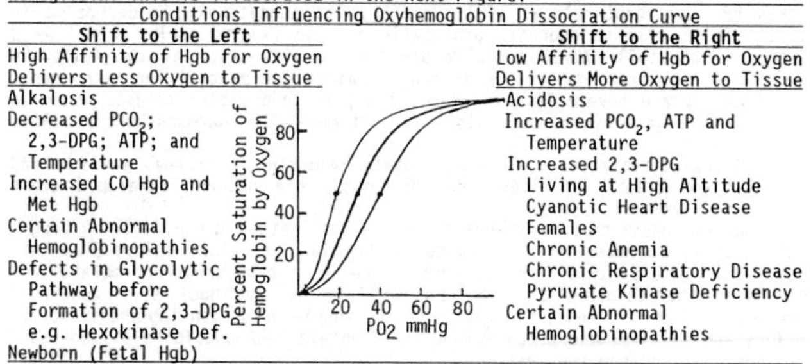

In the example graphed above, at 50 percent saturation of hemoglobin by oxygen, PO_2 at the tissue level is 20mmHg instead of 27mmHg when the curve is shifted to the left and PO_2 is 40mmHg instead of 27mmHg when the curve is shifted to the right.

PANCREATIC INSUFFICIENCY PANEL

Tests for pancreatic insufficiency are listed in the next Table:

Tests for Pancreatic Insufficiency
Tests for Steatorrhea: (see FAT MALABSORPTION PANEL)
Fecal Fat (72 Hours)
Microscopic Examination of the Stool Fat
Tests for Pancreatic Insufficiency:
Bentiromide
Trypsin, Immunoreactive Serum
Abdominal Radiograph (Pancreatic Calcification)
Schilling Test
[^{14}C] Triolein Breath Test
Stimulating Tests
Secretin Test
Secretin-Cholecystokinin Test
Lundh Test
Fatty Meal Test

The secretin and secretin-cholecystokinin tests require instrumentation of the patient with aspiration of the duodenal contents. Administration of secretin promotes pancreatic secretion of fluids and electrolytes. Cholecystokinin promotes secretion of pancreatic enzymes. In a secretin test the volume and electrolyte content of the duodenal fluids are measured; in a combined secretin-cholecystokinin test, these plus enzyme content are measured. The Lundh test is simpler and requires intubation of the duodenum, ingestion of a test meal, and collection of duodenal contents (Lundh G. Gasteroenterology. 1962; 42:275). The fatty meal test has been designed to be a less invasive test of pancreatic function. A fatty meal is ingested followed by analysis of blood lipids, in particular the triglycerides and chylomicrons. This test assumes normal bowel absorptive function. This test avoids stool analysis and intubation (Goldstein R, et al. Am J Clin Nutr. 1983; 38:763-767).

The bentiromide test and the assay of immunoreactive trypsin in serum are two new simple outpatient tests for diagnosis of pancreatic exocrine insufficiency. See TRYPSIN, IMMUNOREACTIVE, SERUM.

Bentiromide Test (Tripeptide Hydrolysis Test): (The Medical Letter on Drugs and Therapeutics, 1984; 26:50) Bentiromide (Chymex-Adria) has been approved by the Food and Drug Administration for clinical use as a diagnostic procedure for evaluation of pancreatic insufficiency in both children and adults. This test may be done on outpatients. The patient fasts overnight; the patient should be observed for about 30 minutes after taking bentiromide as a precautionary measure in case of a rare adverse reaction.

Bentiromide is taken orally; it is cleaved by pancreatic chymotrypsin to yield Para-aminobenzoic acid (PABA). PABA is absorbed in the intestine, conjugated in the liver and excreted in the urine; urinary PABA is measured in a 6 hour urine sample.

The sensitivity (positivity in disease) of the bentiromide test for pancreatic insufficiency is 86-100% (Lankisch PG, et al. Dig Dis. 1983; 28:490; Toskes PP. Gastroenterology. 1983; 85:565). The sensitivity is decreased in patients with mild to moderate pancreatic disease (Ventrucci M, et al. Am J Gastroenterol. 1983; 78:806). False positive results occur in the following conditions: vomiting, gastric retention, impaired gut mucosal function, renal insufficiency and severe liver disease. Thus, false-positive results occur in patients with gastrointestinal diseases (Mitchell CJ. Pharmacotherapy. 1984; 4:79).

Adverse effects are as follows: nausea, vomiting, diarrhea, headache and transient elevations of liver function tests; one patient developed acute respiratory symptoms.

Interferences that occur with the analytical method are as follows; Drugs: acetaminophen, phenacetin, benzocaine, lidocaine, procaine, chloramphenicol, procainamide, sulfanamides and thiazide diuretics. These drugs are metabolized to products that absorb at the same wavelength as PABA and should be discontinued three days before the test. Pancreatic enzyme supplements, tanning and sunscreen lotions and multivitamin preparations that contain PABA should be discontinued for seven days before the test.

PANCREATITIS, ACUTE, PANEL

SPECIMEN: Red top tube, separate serum for serum amylase, serum lipase, serum trypsin, serum calcium; random urine specimen for amylase.
INTERPRETATION: Changes in enzymes in acute pancreatitis are summarized in the next Table:

Enzyme Changes in Acute Pancreatitis

Enzyme	Beginning of Increase (hrs.)	Maximum (hrs.)	Return to Normal (Days)
Serum Amylase	2-6	12-30	2-4
Serum Lipase	2-6	12-30	2-4
Urine Amylase	4-8	18-36	7-10

It is important to note that urine amylase is elevated when serum amylase is normal. This occurs because renal glomerular filtration for amylase is increased in acute pancreatitis; thus, amylase appears in the urine several days after the serum amylase returns to normal. The changes in serum amylase, serum lipase and urine amylase in acute pancreatitis are illustrated in the next Figure:

Serum Amylase, Serum Lipase and Urine Amylase Following Acute Pancreatitis

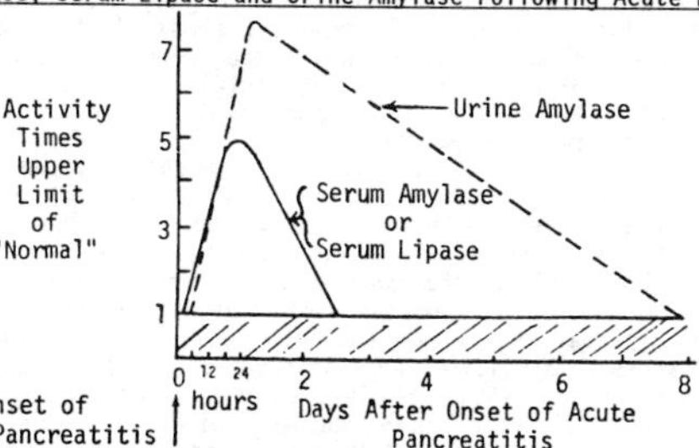

Following onset of acute pancreatitis, serum amylase begins to rise in 2 to 6 hours, reaches a maximum in 12 to 30 hours and remains elevated for 2 to 4 days. Urine amylase begins to rise 4 to 8 hours following onset of acute pancreatitis; this is several hours after the initial increase of serum amylase. Urine amylase remains elevated for 7 to 10 days, about five days after the serum amylase returns to normal. There is an increased renal clearance of amylase. Serum lipase changes in a manner similar to that of serum amylase following onset of acute pancreatitis. Apparently, serum lipase is elevated as often as serum amylase in patients with acute pancreatitis. Serum trypsinogen (immunoreactive trypsin) is considered to be the most sensitive and specific test for acute pancreatitis. Serum trypsin levels increase 5-10 times normal and remain elevated for 4-5 days following the acute injury. However, the immunoreactive trypsin test is not available in all laboratories, and certainly not on a 24-hour basis.

Serum calcium is often depressed during the first 24 hours following onset of acute pancreatitis; fatty acids released from triglycerides, combine with calcium to form insoluble calcium salts.

There are two types of acute pancreatitis, edematous pancreatitis and hemorrhagic pancreatitis. Hemorrhagic pancreatitis has a high mortality. If methemalbumin is detected in the serum of a patient with acute pancreatitis, then the diagnosis of hemorrhagic pancreatitis should be strongly considered (Geokas MC, et al. Ann Intern Med. 1972; 76:105-117).

Increased amylase in pleural fluid is practically pathognomonic of acute pancreatitis. Pleural effusion occurs in approximately 10 percent of patients with acute pancreatitis.

Prolonged elevation of serum and urine amylase and serum lipase suggests either continued inflammation, pseudocyst or renal disease. However, in renal diseases, serum amylase activity rarely exceeds twice the upper limit of normal. Prolonged elevation of amylase may also be due to macroamylasemia. In macroamylasemia, amylase is combined in a macromolecular complex, too large to be excreted in the urine; this condition may be diagnosed by finding a very low amylase/creatinine clearance.

Pancreatitis, Acute, Panel (Cont)

There are no diagnostic chemical tests with high sensitivity and specificity for chronic pancreatitis. Pancreatic function, including enzyme output, may be assessed in duodenal aspirates following stimulation with secretion - cholescytokinin. A newer test of pancreatic function measures the hydrolysis in the bowel of orally administered para-aminobenzoic acid conjugated peptides (Bentiromide test; tripeptide hydrolysis test)(see PANCREATIC INSUFFICIENCY PANEL). The released para-aminobenzoic acid can be measured in serum or urine. Calcium precipitates may be detected by X-ray. If there is extensive destruction of acinar tissue, malabsorption will occur. If there is extensive destruction of islets of Langerhans, which tend to be preserved, diabetes mellitus will occur.

PARATHYROID HORMONE (PTH)

SPECIMEN: Red top tube, separate serum and freeze. Fasting morning specimen preferred.
REFERENCE RANGE: As new reagents and methods are introduced to assay for PTH, the reference range continues to change. There are also multiple forms of PTH present in circulation, each with a different reference range. It is also important sometimes to relate plasma PTH levels to the total plasma calcium (or ionized calcium). Range for intact PTH is 10-65 pg/mL. Adult range for C-terminal/mid-region is <10-80pg/mL (with normal calcium).
METHOD: RIA. Assays continue to be developed. Immunological heterogeneity of circulating PTH and its low concentration has hindered the development of assays. It is now possible to measure intact, active PTH employing two antibodies directed at two distinct regions of the PTH molecule. The measurement of intact or active PTH will probably be the assay of choice in the near future.
INTERPRETATION: PTH measurements are performed primarily in the investigation of the patient with hypercalcemia, or the patient suspected of having disease of the parathyroids (hyperparathyroidism, hypoparathyroidism). In most patients with primary hyperparathyroidism (90%), PTH levels are elevated. In the other 10%, PTH levels are inappropriately elevated for the level of blood calcium. The majority of cases of primary hyperparathyroidism are due to a solitary adenoma (80%). Hyperplasia is responsible for 15% and parathyroid carcinoma accounts for 1-5%. The incidence of primary hyperparathyroidism is 5 per 10,000 per year.

PTH is a hormone of 84 amino acids. This is subsequently split into an amino terminal fragment, a mid-region fragment, and a carboxy terminal fragment. Hence there are four possible forms of PTH in the circulation as tabulated below:

Forms of Circulating PTH

	Activity	Circulation Time
Intact PTH	Yes	Short
Amino Terminal	Yes	Short
Mid Region	No	Intermediate
Carboxy Terminal	No	Long

The active forms of the hormone circulate briefly and are found in low concentration. Conversely, the inactive forms circulate for a long time and are found at higher concentrations. The inactive carboxy terminal fragment is excreted by the kidney. As a result, in renal failure, some assays will yield a high PTH level although the amount of active hormone is low.

The previous years have seen a lot of controversy concerning the measurement of which fragment of PTH is most diagnostic. It would appear that many centers are now promoting the measurement of the active intact PTH molecule.

The parathyroid hormone is increased in two conditions: primary and secondary hyperparathyroidism; it is decreased in hypoparathyroidism. Conditions associated with increased parathyroid hormone are listed in the next Table:

Conditions Associated with Increased Parathyroid Hormone

- Primary Hyperparathyroidism
- Secondary Hyperparathyroidism (Associated with Hypocalcemia)
  - Chronic Renal Failure
  - Pseudohypoparathyroidism, Types I and II
  - Pseudoidiopathic Hypoparathyroidism
  - Magnesium Deficiency (Some Patients)
  - Other Causes of Hypocalcemia
    - Drug-Induced: Anticonvulsants
    - Rickets
    - Osteomalacia

PTH is increased in primary hyperparathyroidism because the parathyroid gland secretes PTH autonomously since feedback is not intact. In most conditions associated with hypocalcemia they are associated with increased PTH. In hypocalcemia associated with secondary hyperparathyroidism, suboptimal levels of serum calcium stimulate PTH. In secondary hyperparathyroidism due to renal failure, erythropoietin requirements appear to be directly correlated with the degree of PTH elevation (Rao DS, et al. N Engl J Med. 1993; 328:171-175). Many of the effects of uremia are now thought to be mediated by elevated PTH via the toxic effects of increased cellular calcium (Massry SG, Fadda GZ. Am J Kidney Dis. 1993; 21:81-86).

There are a number of causes of a decreased PTH. These are given in the next Table. Conditions associated with decreased PTH are listed in the next Table:

Conditions Associated with Decreased PTH
Hypercalcemic states other than primary hyperparathyroidism
Primary Hypoparathyroidism
DiGeorge's Syndrome
Multiple Inherited Conditions
Postsurgery
Parathyroidectomy
Thyroid Surgery (Transient)
Wilson's Disease (Carpenter TO, et al. N Engl Med. 1983; 309:873-877)

The most common cause of a decreased PTH is due to hypercalcemia (non-primary hyperparathyroidism hypercalcemia). There are at least three forms of inherited hypoparathyroidism, the most common is autosomal dominant and associated with sensorineural deafness and renal dysplasia (Bilous RW, et al. N Engl J Med. 1992; 327:1069-1074). Approximately 22% of patients experience low PTH and hypocalcemia after thyroidectomy and this is correlated with the extent of resection (Demeester-Mirkine N, et al. Arch Surg. 1992; 127:854-858).

Using older PTH assays, it was often necessary to relate PTH levels to the calcium level. Levels of plasma PTH and plasma calcium in different conditions are given in the next Table (MetPath Lab Report):

Levels of Plasma PTH and Plasma Calcium in Different Conditions

Calcium (mg/dL)	PTH (pg/mL)	Result Consistent with	Most Common Causes
8.8-10.8	100-600	Normal Parathyroid Function	
11.0-16.2	>310	Primary Hyperparathyroidism	Benign Adenoma Hyperplasia Carcinoma
11.2-21.0	<280	Nonparathyroid Hypercalcemia	See list; Causes of hypercalcemia
<9.4	>800	Secondary Hyperparathyroidism	Chronic Renal Disease
<7.5	<300	Hypoparathyroidism	
<7.0	350-700	Pseudohypoparathyroidism	

With pseudohypoparathyroidism, there is a lack of PTH receptor response, hence serum calciums are low, and PTH production is increased in response to the low calcium.

Additional findings which may help in the diagnosis of primary hyperparathyroidism are hypercalcemia, hypercalcuria, hypophosphatemia, hyperphosphaturia, elevated serum chloride, a tendency to a metabolic acidosis, and increased urine cAMP. A chloride: phosphate ratio >33 is said to consistent with primary hyperparathyroidism (Palmer FJ, et al. Ann Intern Med. 1974; 80:200-204). Levels of 1,25-dihydroxy vitamin D3 also tend to be elevated.

At one time it was believed that some hypercalcemia of malignancy was due to ectopic PTH production. Now there is evidence that a PTH-related protein is produced by some malignant tissues. This PTH-related protein is produced by some malignancies, normal skin, placenta, pregnant uterus and lactating mammary gland. Elevations of PTH-related protein have been reported in cases of "humoral" hypercalcemia of malignancy, and during pregnancy and lactation (Lepe F, et al. N Engl J Med. 1993; 328:666-667). In such situations there is hypercalcemia, increased PTH-related protein, and decreased intact PTH.

PARAXYSMAL NOCTURNAL HEMOGLOBINURIA (PNH) SCREEN

Synonyms: Acidified serum lysis test (Ham Test); Sugar Water Test, Sucrose Hemolysis Test

SPECIMEN: Purple(EDTA) top tube, refrigerate. Any anticoagulant may be used if the assay procedure uses washed cells. If unwashed cells will be used (some sugar water tests) use citrated or oxalated blood. EDTA and heparin will inhibit complement activity.

REFERENCE RANGE: Negative. No hemolysis observed.

METHOD: Red cells are incubated at 37°C with complement containing serum at pH 6.5 to 7.0 (Ham acid hemolysis test) or placed in a medium of low ionic strength (sugar or sugar water hemolysis). Only PNH cells are lysed.

INTERPRETATION: These tests are used to diagnose paroxysmal nocturnal hemoglobinuria(PNH). The pathognomonic feature of PNH is that a population of the red cells is markedly sensitive to fixation and lysis by normal complement due to a membrane defect found on RBC, WBC and platelets.

Paroxysmal nocturnal hemoglobinuria is a rare acquired chronic hemolytic anemia associated with hemosiderinuria and pancytopenia; episodic urinary excretion of hemoglobin after a night's sleep occurs in 25 percent of cases. In some cases there is an associated iron deficiency anemia. Half of PNH cases develop between ages 20 and 40; other conditions that may be associated with PNH are aplastic anemia, leukemia, and myelofibrosis. PNH cells have deficient acetylcholinesterase activity.

The sucrose hemolysis test (sugar water test) is typically used as a screening test for PNH, to be followed by an acidified serum lysis test for confirmation. In the sucrose hemolysis test, red blood cells in the presence of complement are suspended in isotonic sucrose. The low ionic strength medium promotes complement fixation and hemolysis ensues. False positive tests are sometimes seen with RBC's from cases of megaloblastic anemia and autoimmune hemolytic anemia. An assay for direct antiglobulins (direct Coombs) should exclude the later possibility. False negative tests may be the result of inactivation of complement.

The acidified serum lysis test (Ham test) is usually performed as a confirmatory test. It is considered more specific than the sucrose hemolysis test. However, it is a low volume test that requires careful performance and attention to detail. Complement is activated at pH 6.8 and then added to washed RBC. In PNH there will be complement dependent hemolysis. False positive hemolysis can be seen in other disorders (hereditary spherocytosis, old transfused blood, and autoimmune hemolytic anemia). However, in these conditions the hemolysis does not depend on the addition of acidified serum.

PARTIAL THROMBOPLASTIN TIME, ACTIVATED (PTT)

Synonym: Partial Thromboplastin Time (PTT)
SPECIMEN: Blue (citrate) top tube filled to capacity (9:1 ratio of blood to anticoagulant); or pediatric blue top tube; avoid contamination with tissue thromboplastin as follows: If multiple tests are being drawn, draw coagulation studies last; if only an PTT is being drawn, draw 1 to 2mL into another vacutainer, discard, and then collect blood for PTT. If the test cannot be assayed immediately, centrifuge, separate plasma from cells and freeze plasma.

The specimen is rejected if the tube is not full, specimen hemolyzed, specimen clotted or specimen received more than 2 hours after collection.
REFERENCE RANGE: 20 to 34 seconds. Varies between laboratories. May be different for bedside whole blood PTT monitors.
METHOD: Reagents: Phospholipid from brain or soybean as a substitute for platelet factor 3, i.e., partial thromboplastin; calcium chloride; activator is celite kaolin. Platelet-poor plasma is incubated with the phospholipid (platelet substitute) and surface activator. After this activation step, calcium is added, and the time to clot formation determined.
INTERPRETATION: This test is used most often to monitor heparin therapy. It is also prolonged with deficiencies of clotting factors of the intrinsic system, and the common pathway (Factors X, V, thrombin and fibrinogen). Presence of anti-factor antibodies, and other inhibitors (lupus anticoagulant) may also be detected with this test (Triplett DA. Anticoagulant Therapy: Monitoring Techniques, Laboratory Management. 1982; 20:31-42). The PTT test is a measure of all of the blood factors except factor VII. The activated partial thromboplastin time (APPT) is a modified form of the PTT (activated by addition of a surface activator). Although not precisely correct, the terminology for APTT and PTT are used interchangeably by most clinicians; in this write-up, PTT will be used to denote activated partial thromboplastin time (APTT). The causes of prolonged PTT are given in the next Table:

Causes of Prolonged Activated Partial Thromboplastin Time
Heparin Therapy
Vitamin K Deficiency
Liver Disease
Hypofibrinogenemia
Disseminated Intravascular Coagulation(DIC)
Factor Deficiency:
Factor VIII (Classical Hemophilia; Hemophilia A)
Factor IX (Christmas Disease; Hemophilia B)
Factor XII (Hageman Factor)
Factor XI (Partial Thromboplastin Antecedent)
Factor X (Stuart-Prower Factor)
Factor V (Labile Factor)
Factor II (Prothrombin)
Factor I (Fibrinogen)
Anti-Factor Antibodies
Lupus Anticoagulant
Nephrotic Syndrome
Deficiencies of Prekallikrein and Kininogens
Dysproteinemias, i.e., Waldenström's
Some Dysfibrinogenemias
Gaucher's Disease

Blood coagulation may be initiated by (1) contact activation triggered in-vivo by skin and collagen (intrinsic system) or (2) release of tissue thromboplastin (tissue activated extrinsic system). The following factors are involved in the intrinsic system: XII, XI, IX, VIII, X, V, II, I; the intrinsic system is measured by the PTT.

Partial Thromboplastin Time (PTT) (Cont)

This pathway does not require tissue thromboplastin and Factor VII. The coagulation cascade is shown in the next Figure:

Coagulation Cascade Table

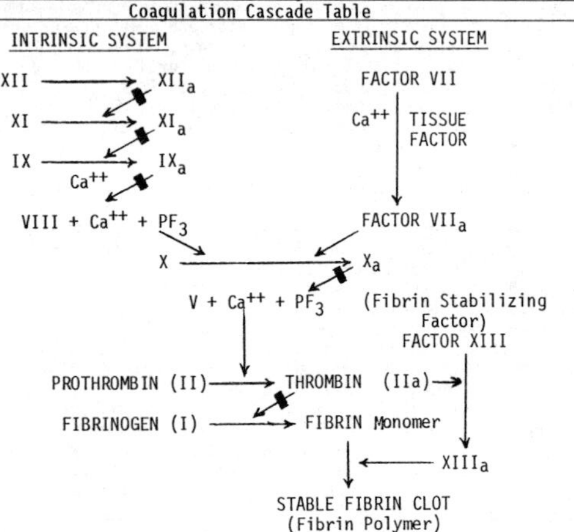

■ = Block by Heparin $\quad$ PF_3 = Platelet Factor

The actual in vivo sequence of blood coagulation is being questioned. It is presently felt that activation of factor VII is the initiating event in vivo and that certain components of the intrinsic pathway (factor XI) are then activated by the generated thrombin (Gailani D, Broze GJ. Science. 1991; 909-912). Deficiency of factor XII does not lead to clinical bleeding. However, in the in vitro PTT test, factor IX is activated by contact activated XII. In addition, factor IX may be activated by factor VII - tissue factor complex (reviewed in Furie B, Furie BC. N Engl J Med. 1992; 326:800-806).

Heparin mediates its action by first interacting with anti-thrombin III, converting anti-thrombin to its active form. This complex then inhibits the active site of the activated blood coagulation factors XIIa, XIa, IXa, Xa, and thrombin. Heparin also inhibits plasmin, the active serine protease of the fibrinolytic system.

The laboratory control of oral anticoagulation (warfarin) is usually done using the prothrombin-time (PT); Factor VII is the procoagulant first depressed with coumadin therapy and the PTT does not measure Factor VII.

Heparin Therapy: Heparin is used in the prophylaxis and management of thromboembolic disease on the venous side of the circulation, that is, venous thrombosis and pulmonary embolism. In heparin therapy of thromboembolism, the PTT is usually maintained at from 1.5 to 2.5 times the upper limits of normal (50 to 65 seconds) corresponding to a heparin level of 0.2-0.4 units/mL by protamine titration (Hirsh J. N Engl J Med. 1991; 324:1565-1574). In another approach, the PTT is maintained at from 1.5 to 2.5 times the patient's admission PTT. For preventive therapy, it is held at the upper normal range (Hirsh, already cited).

The activated clotting time (ACT) is used for monitoring the large amounts of heparin given during cardiopulmonary bypass since the PTT is not clottable with these large amounts of heparin. Many labs then switch to the PTT following heparin reversal to determine a residual heparin effect.

Anticoagulation after Percutaneous Transluminal Coronary Angioplasty (PTCA):

It is suggested that the PTT should be maintained at 3 times the control value by heparinization for 18-24 hours after elective PTCA to avoid abrupt coronary artery disease, and ischemic events (McGarry TF, et al. Am Heart J. 1992; 123:1445-1451).

Thrombocytopenia may occur in association with heparin therapy as a result of anti-platelet antibody formation. A platelet count every two to three days allows this complication to be detected early (Stein PD, Willis PW. Arch Intern Med. 1983; 143:991-994). See PLATELET COUNT.

<u>Liver Disease</u>: In liver disease, levels of factors XII, XI, IX, X, VII, V, II and I may be deficient. Factor VIII activity remains unchanged. Deficiency of clotting factors may be due to defective synthesis such as occurs in parenchymal liver disease; or deficiency may be due to lack of vitamin K as in biliary obstruction; vitamin K is required for the synthesis of factors IX, X, VII, II, protein C and protein S.

<u>Disseminated Intravascular Coagulation (DIC)</u>: The PTT is usually prolonged (see DISSEMINATED INTRAVASCULAR COAGULOPATHY (DIC) PANEL).

<u>Factor Deficiency</u>: Deficiency of factors in the intrinsic pathway may be suspected by the finding of a prolonged PTT, and normal PT. The most common hereditary factor deficiencies are factor VIII and factor IX deficiencies. The presence of a factor deficiency can be substantiated by mixing studies. Mixing of patient factor deficient plasma with normal plasma will lead to correction of the clotting time.

<u>Factor Inhibitors</u>: If the prolonged PTT is the result of an antifactor antibody, the mixing studies will not correct. The most common antibodies are antifactor VIII found in patients with severe factor VIII deficiency, and the lupus anticoagulant (anti-lipid antibody) found in some patients with SLE and some patients who experience recurrent late abortion.

Deficiency of individual factors are measured by specific assays; the most common disorders are classic hemophilia, A or B, von Willebrand's disease and anticoagulants used in therapy. Factors VIII and IX deficiencies are sex-linked; women with congenital defects of coagulation usually have either factor XI deficiency or von Willebrand's disease.

<u>von Willebrand's Disease</u>: In von Willebrand's disease, factor VIII levels are usually low (pseudohemophilia), and there is a prolongation of bleeding time (vascular hemophilia). The best single test for von Willebrand's disease is inhibition of platelet aggregation in the presence of the antibiotic, ristocetin. See VON WILLEBRAND DISEASE PANEL.

<u>Nephrotic Syndrome</u>: About 10% of patients with the nephrotic syndrome in relapse have a factor IX deficiency secondary to loss of factor IX in the urine. Factor IX deficiency is accompanied by a prolonged PTT and normal PT. However, most patients with nephrotic syndrome clinically have hypercoagulability syndrome secondary to loss of anti-thrombin-III (Roberta Gray, Personal Communication).

Pericardial Effusion Analysis

PERICARDIAL EFFUSION ANALYSIS

<u>SPECIMEN</u>: Collect 4 tubes; purple (EDTA) for cell count; special vials for microbiological studies, aerobic, and anaerobic; red or green (heparin) top tube for chemistries.

<u>REFERENCE RANGE</u>: Normal volume, 10 to 50 mL, clear and pale yellow.

<u>METHOD</u>: See individual methods.

<u>INTERPRETATION</u>: An algorithm for differentiating causes of pericardial effusion is given in the next Figure (Kindig JR, Goodman MR. Am J Med. 1983; 75:1077-1079):

Algorithm for Pericardial Effusion

WBC's>15,000 per cubic mm		
Glucose<55 mg/dL	Glucose, 45-130 mg/dL	Glucose, 95-120 mg/dL
pH <7.1 ↓	pH 7.2-7.4 ↓	pH >7.4 ↓
Connective Tissue Disorder Bacterial Effusion	Neoplasm Idiopathic Tuberculosis Uremia	Hypothyroidism Post-Pericardiotomy

Additional data should be obtained in order to confirm the results of Kindig and Goodman.

Peritoneal Fluid Analysis

PERITONEAL FLUID ANALYSIS (see ASCITIC FLUID ANALYSIS)

PERTUSSIS, FLUORESCENT ANTIBODY

SPECIMEN: Nasopharyngeal secretions are obtained (see PERTUSSIS, NASOPHARYNGEAL CULTURE).
REFERENCE RANGE: Presence of Bordetella pertussis by immunofluorescence staining.
METHOD: Nasopharyngeal secretions are smeared onto slides and heat fixed. Slides are stained with fluorescein-isothiocyanate-conjugated anti-B. pertussis antibody. Positive and negative controls should be included. The test is positive if three or more organisms are identified with bright fluorescence and typical morphology (Halperin SA, et al. J Clin Microbiol. 1989; 27:752-757).
INTERPRETATION: Pertussis fluorescent antibody testing is more rapid than culture and does not require viable organisms, but does require trained personnel and special equipment. Fluorescent antibody testing is subject to some of the same limitations as pertussis culture; sensitivity is decreased by antibiotic use, previous immunization, and increasing time since onset of symptoms. This test has a high rate of false-positive and false-negative results (Onorato IM, Wassilak SGF. Pediatr Infect Dis. 1987; 6:145-151; Halperin SA, et al. Already cited).

PERTUSSIS, NASOPHARYNGEAL CULTURE

SPECIMEN: A nasopharyngeal (not throat) swab is used. Cotton swabs inhibit the growth of Bordetella pertussis. Dacron or calcium alginate swabs are used. The organism is more difficult to culture from patients treated with antibiotics (erythromycin, tetracycline, or trimethaprim-sulfamethoxazole) or patients previously immunized (Onorato IM, Wassilak SGF. Pediatr Infect Dis J. 1987; 6:145-151).

With the patient's head immobilized, pass the swab gently into the nostril until it reaches the posterior nares and is left in place for 15 to 30 seconds. If resistance is encountered during insertion of the swab, the other nostril should be tried.

Direct plating on special media, such as modified Stainer-Scholte, Regan Lowe, or Bordet-Gengou agar, ideally should be done at the bedside. Therefore, notify the laboratory beforehand so that the media can be warmed.
REFERENCE RANGE: No B. pertussis or B. parapertussis.
METHOD: The genus Bordetella consists of aerobic, gram-negative, coccoid bacilli; Bordetella pertussis is strictly a human parasite. Growth of B. pertussis requires the presence of blood; charcoal or ion-exchange resin are incorporated in the medium to absorb substances, such as peroxides, fatty acids, etc., that inhibit the growth of B. pertussis. The media should contain cephalexin to prevent the overgrowth of indigenous organisms. The organism grows on charcoal agar after 3 to 7 days of incubation at 35° with added CO_2. Microscopically, the Bordetella are non-spore-forming, encapsulated, bipolar, pale-obtaining, small, gram-negative bacilli.
INTERPRETATION: Bordetella pertussis is the causative agent of pertussis. B. parapertussis causes a pertussis-like illness. The organism multiplies in the nasopharynx, releases toxins, causing inflammation.

In the classic pertussis syndrome, there is an incubation period of 7 to 10 days, followed by coughing which progresses over 1 to 2 weeks. The "whoop" represents a forced inspiration over a partially closed glottis. Lymphocytosis occurs during this period. Infants less than 6 months account for the most severe cases; clinical presentation in this age group is severe coughing paroxysms with apnea and cyanosis. Adults have less severe symptoms than children. Typical symptoms include paroxysmal cough (worse at night), shortness of breath, and a tingling sensation in the throat. Lymphocytosis is not seen. Adults were the infection source for infants in 16% of cases (Aoyama T, et al. AJDC. 1992; 146:163-166).

Although culture is considered the "gold standard" for diagnosis of pertussis, this test is far from ideal. Sensitivity is decreased by previous immunization or antibiotic use. Percentage of positive cultures drops to 25%, 14%, and 0% by the third, fourth, and fifth weeks respectively (Steketee RW, et al. J Infect Dis. 1988; 157:441-449). The organism is fastidious and slow growing, requiring special media, skills, and up to 7 days to culture (Onorato IM, Wassilak SGF, already cited).

The only other test typically used to diagnose pertussis is the fluorescent antibody test (see PERTUSSIS, FLUORESCENT ANTIBODY). Serologic tests (ELISA, Agglutination reaction, Neutralizing antibody) lack sensitivity and specificity. Secretory antibody detection (IgA) may be positive late in the disease process. Detection of the organism or its products (Gene probes, Adenylate cyclase, CIE, ELISA with monoclonal antibodies) may be useful in the future (Review: Onorato IM, Wassilak SGF. "Laboratory Diagnosis of Pertussis: the State of the Art." Pediatr Infect Dis J. 1987; 6:145-151).

PHENCYCLIDINE

Synonyms: "PCP", "Angel Dust", "Hog"
SPECIMEN: 50mL random urine; serum or gastric contents may be tested.
REFERENCE RANGE: None detected
METHOD: Thin-layer chromatography; mass spectrometry; Enzyme Immunoassay
INTERPRETATION: Phencyclidine (PCP) was used as an intravenous anaesthetic in the late 1950s; however, 10 to 20 percent of the patients developed postanesthetic delirium with unmanageable behavior and, thus, human use was discontinued. Ketamine is a phencyclidine derivative that is still used in humans.

PCP is a commonly used animal tranquilizer and popular in the drug culture. PCP is taken either orally or inhaled smoking ("angel dust"), mixed with marijuana or other smokable substances.

The plasma half-life of PCP ranges from 11 hours to 4 days. There is a poor correlation between degree of toxicity and urine levels. PCP is rapidly metabolized to hydroxylated derivatives; there is little physiological activity of PCP metabolites.

The clinical presentation of PCP intoxication is extremely variable, including CNS stimulation or depression, cholinergic, anticholinergic or adrenergic effects. The most common effects are nystagmus (horizontal, vertical, or rotatory) and hypertension. Alterations in mental status includes confusion, disorientation, hallucinations, bizarre or violent behavior. Patients may have seizures, hyperthermia or rhabdomyolysis. Deaths can occur due to violent behavior (self-directed or other-directed), seizures and cerebral hemorrhage. Some PCP intoxicated patients will have negative urine thin layer chromatography (TLC) screen. Immunoassay methods may be used to detect PCP (Baldridge EB, Bessen HA. Emerg Med Clin N Amer. 1990; 8:541-550; Schwartz RH. Arch Intern Med. 1988; 148:2407-2412; Brown RT, Braden NJ. Pediatr Clin N Amer. 1987; 34:341-347).

PHENOBARBITAL

SPECIMEN: Red top tube, separate serum; or green (heparin) top tube, separate plasma.

REFERENCE RANGE: Therapeutic: 15-40mcg/mL. Time to Obtain Specimens (Steady State): 11-25 days (Adults); 8-15 days (Children); Half-Life: 50-120 hours (Adults); 40-70 hours (Children).

METHOD: EMIT; RIA; GLC; HPLC; TLC (qualitative); Nephelometry(ICS); Fluorescence Polarization(Abbott)

INTERPRETATION: Phenobarbital is given IV or orally and is used as an anticonvulsant; it is effective in all seizure disorders including status epilepticus except absence (petit mal) seizures (Medical Letter. 1989; 31:1-4).

Dosage guidelines and blood specimen monitoring schedule are given in the next Table:

The Dosage Guidelines and the Blood Specimen Monitoring for Phenobarbital

Dosage Guidelines	Blood Monitoring
Newborns: Loading Dose: 20mg/kg I.V. Repeat 10mg/kg if necessary Maintenance Dose: 5-6mg/kg/day	Frequent because the liver of the neonate metabolizes phenobarbital slowly during the first 4 weeks of life.
Children: Loading Dose: 16-20mg/kg Maintenance Dose: Infant: 5-8mg/kg/day Children: 3-5mg/kg/day	Monitor in 10-15 days after starting therapy; then monitor every 3 to 4 months
Adults: Loading Dose: 10-20mg/kg Maintenance Dose: 1.5-3.0mg/kg/day	Monitor in 15-25 days

Therapeutic monitoring may be done at random. Monitoring should be for the following patients: poorly controlled, toxic symptoms, change of medication or dosage allowing 2 to 3 weeks for new steady state and when drugs, such as primidone are given. Primidone is metabolized to phenobarbital.

May cause respiratory depression, especially with rapid IV administration or with concomitant administration of other anticonvulsants.

Sedation and behavior disturbances including hyperactivity, loss of concentration, and depression are the principal adverse effects (Brent DA, et al. Pediatrics. 1987; 80:909). Paradoxical excitement and hyperactivity are more common in children and the elderly. Occasionally, skin rashes, disturbances in motor function (ataxia) and megaloblastic anemia may occur.

With acute toxicity, CNS depression, respiratory depression, shock syndrome, congestive heart failure and cardiac arrhythmias may develop.

Rarely, Stevens-Johnson syndrome, exfoliative dermatitis, photosensitivity, hepatitis and jaundice have been reported.

Drug Interactions: Drug interactions are given in the next Table:

Drug Interactions of Phenobarbital

Drug	Comment
Phenytoin	Phenobarbital may either increase or decrease the serum concentration of phenytoin; phenytoin's effect on phenobarbital is variable
Valproic Acid	Valproic acid increases the serum conc. of phenobarbital
Folic Acid Carbamazepine	These drugs decrease the serum conc. of phenobarbital
Oral Anticoagulants, Oral Contraceptives, Griseofulvin, Quinidine, Chloramphenicol, Rifampin, Tetracycline	Phenobarbital induces cytochrome P-450 enzymes
Other Sedative Drugs	Additive Effect

PHENOTHIAZINES, URINE

SPECIMEN: Random urine
REFERENCE RANGE: None detected
METHOD: Ferric chloride (Phenistix-Ames) urine test, TLC
INTERPRETATION: Phenothiazines are assayed in urine in cases of suspected toxicity. A rapid simple ferric chloride screening test is available as is a more complex TLC test. Therapeutic drug monitoring is usually not practical or performed. There are a large number of phenothiazines used, their therapeutic and toxic serum levels vary greatly, and many are rapidly metabolized. Phenothiazines may be excreted for months after ingestion. Treatment is based on signs and symptoms and not on drug levels.

Diagnosis of phenothiazine toxicity can be aided by history of ingestion, suspicious physical findings, ferric chloride (Phenistix) urine test and a plain abdominal film - phenothiazines are radio-opaque. Blood levels do not correlate well with symptoms or prognosis (Mack RB. N Carolina Med J. 1982; 43:222-223).

The ferric chloride (Phenistix) test is non-specific; a list of selected substances in urine that react with ferric chloride are given in the next Table (Bradley M, Schumann GB in Clinical Diagnosis and Management, 17 Ed., Editor, Henry JB. W.B. Saunders Co., Phila., 1984; 451):

Ferric Chloride Test in Urine

Substance	Color Change
Acetoacetic Acid	Red or Red-Brown
Pyruvic Acid	Deep Gold-Yellow or Green
Drugs:	
Aminosalicylic Acid	Red-Brown
Antipyrines and Acetophenetidines	Red
Cyanates	Red
Phenol Derivatives	Violet
Phenothiazine Derivatives	Purple-Pink
Salicylates	Stable Purple

A more complete list of substances that react with ferric chloride is given under the FERRIC CHLORIDE TEST.

Phenothiazines are used for the following conditions: antipsychotic, antinausea, antiemetic and antihistaminic and are used to potentiate analgesics, sedatives and general anesthetics. Toxic reactions associated with phenothiazines include: peculiar posturing, weakness and muscular fatigue, motor restlessness and jerking movements. Other toxic reactions include hypotension, miosis, hypothermia and myocardial depression.

PHENYLALANINE BLOOD OR PLASMA

SPECIMEN: Collect specimen immediately prior to discharge from newborn nursery. If an infant has been tested prior to 24 hours of age, rescreen; collect a second specimen no sooner than 96 hours and no later than three weeks of age (American Academy of Pediatrics, Committee on Genetics. Pediatrics. 1982; 69:104).
Blood: Special filter paper is provided by the laboratory as illustrated in the next Figure:

Filter Paper for Blood Specimens for Phenylalanine Testing in the Newborn

All three circles are soaked with blood; be sure blood soaks through. Use several drops of blood applied to each circle. These specimens on filter paper may be used to test for phenylalanine, thyroid stimulating hormone(TSH), galactose and cystine.
Plasma: The filter paper method is used for neonatal screening; the method is semi-quantitative only; quantitative values are obtained from plasma specimens. Green (heparin) top tube, separate plasma promptly from cells from fasting patient; a fast of 4 hours or more in infants. Send specimen frozen in plastic vial on dry ice.
REFERENCE RANGE: <2 mg/dL for samples collected between 12 and 48 hours of age; <4 mg/dL on the third day of life or later, (McCabe ERB, et al. Pediatrics. 1983; 72:390-398). Premature infants have mildly elevated values.

Phenylalanine (Cont)
METHOD: Fluorometric; ion-exchange, bacterial inhibition assay (Guthrie Test) or paper chromatography.
INTERPRETATION: The filter paper specimen for phenylalanine testing is employed in newborn screening for phenylketonuria. The quantitative plasma determination is used to evaluate subjects with suspected phenylketonuria(PKU), to monitor therapy of PKU patients who are on a phenylalanine restricted diet or to test siblings of patients known to have PKU.

Phenylketonuria(PKU) is an autosomal recessive disease; each sibling of an identified patient has a 25 percent chance of having PKU. The carrier rate in the United States is 2%. Phenylketonuria(PKU) is caused by a deficiency of the enzyme phenylalanine hydroxylase and is characterized by an increased level of the amino acid phenylalanine in the blood; the enzyme, phenylalanine hydroxylase, catalyzes the conversion of phenylalanine to tyrosine.

Clinically, the main signs and symptoms relate to neurologic findings and to the finding of mental retardation due to the accumulation of the amino acid phenylalanine. Prevention of mental retardation associated with the disease depends on a diet low in the amino acid phenylalanine. The goal is to maintain phenylalanine levels <15 mg/dL. Pregnant women with this disorder should be on strict dietary control to avoid fetal damage.
Diagnosis of PKU: The diagnosis of phenylketonuria(PKU) is made by testing the level of phenylalanine in the plasma of the newborn 24 hours following feeding of protein; misdiagnosis is associated with early discharge from the hospital, e.g., before milk feeding has been established. In patients with PKU, tested within the first two or three days following delivery, the plasma level of phenylalanine is usually above normal (2 mg/dL), but not in the classic PKU range, e.g., above 20 mg/dL. The plasma phenylalanine level tends to rise markedly following adequate protein (milk) intake; this tends to occur after the third day of life.

PKU has been subdivided into three forms: 1) Classical PKU (little or no phenylalanine hydroxylase present); 2) variant PKU (low amounts of phenylalanine hydroxylase, 5-10% present); 3) Benign PKU (significant amounts of phenylalanine hydroxylase present).

The following abnormalities must be present for the diagnosis of classical PKU: 1) serum phenylalanine >20 mg/dL on two occasions > 24 hours apart; 2) serum tyrosine <5 mg/dL; 3) metabolic products of phenylalanine present in urine (Williamson ML, et al. Pediatrics. 1977; 60:815-821).

In the variant form of PKU, phenylalanine levels are 10-20mg/dL, tyrosine levels are within normal limits, and abnormal phenylalanine metabolites may or may not be present in urine. In benign PKU, phenylalanine levels are 4-10mg/dL, tyrosine levels are normal, and no abnormal metabolites are found in urine. Classical and variant forms are sometimes distinguished by an oral phenylalanine challenge.
Detection of Carriers of Phenylalanine(PKU): Heterozygotes for PKU are clinically normal and can be detected by biochemical tests, the most widely used being loading tests with phenylalanine given either orally or intravenously. Loading tests indirectly assess the subject's ability to convert phenylalanine to tyrosine. Compared with controls, carriers have higher phenylalanine levels and lower tyrosine levels.

The human gene for the enzyme, phenylalanine hydroxylase has been cloned. Using the techniques developed, prenatal diagnosis is possible by analysis of fetal DNA, obtained through amniocentesis, for matings between 2 heterozygotes or a homozygote with a heterozygote (Woo SL, et al. Nature. 1983; 306:151).

PHENYTOIN (DILANTIN)

SPECIMEN: Specimens for monitoring therapeutic response should be drawn just before next dose. Specimens for suspected toxicity are obtained at least six hours after last dose. Blood: Red top tube, separate serum and freeze; or green (heparin) top tube. Urine: 50mL random urine.

Specimens should be obtained as follows:

Oral: Draw trough concentrations one week after initiating therapy and then again 3-5 weeks later.

I.V.: Obtain specimen 2-4 hours after I.V. loading dose or just prior to initiating maintenance dose.

REFERENCE RANGE: Adults, Children, Neonates, Pre-term to 12 Weeks: 10-20 mcg/mL; free phenytoin levels: 1-2mcg/mL.

METHOD: RIA; EMIT; GLC; HPLC: SLFIA(Ames); Nephelometry(ICS); Fluorescence Polarization(Abbott).

INTERPRETATION: Phenytoin is given by mouth or intravenously; it should not be given intramuscularly. Phenytoin is one of the most widely used anticonvulsants; it is effective against both grand mal and the generalized spread of seizures with focal origin. It is also used as an antiarrhythmic, in particular for the treatment of digoxin toxicity. The characteristics of Phenytoin are as follows: half-life, adults: 18-30 hours; children, 12-22 hours; neonates 30-60 hours; time to steady state, adults, 4-6 days; children, 2-5 days; time to peak plasma level, 4-8 hours. Phenytoin exhibits nonlinear pharmacokinetics, resulting in highly variable time to steady state. The effectiveness of phenytoin is increased at higher plasma levels within the "therapeutic range."

Therapeutic monitoring is done for the reasons listed in the next Table (Baer M. "Interpretation of Drug Concentration." Am Soc Clin Path. 1981):

Monitoring Dilantin Levels for Patients
Seizures Poorly Controlled
Toxic Symptoms
Change of Medication or Dosage; Allow One Week to Reach Steady State
Children 10 to 13 years old every Three to Four Months until Dose-Serum Relationship has Stabilized

Phenytoin may be given I.V. for acute treatment of seizures at a rate of 40mg/min or less using an infusion pump (Pediatrics: <1mg/kg/min) (Ernest MP, et al. JAMA. 1983; 249:762-765). IV preparation of phenytoin has a pH of 12 and is extremely caustic to veins. Extravasation may cause ischemia or gangrene. Phenytoin should be given in a large vein and well diluted.

Toxicity: Correlation of serum level of phenytoin and symptoms are given in the next Table:

Serum Level (mcg/mL)	Symptoms
20mcg/mL	Nystagmus
30mcg/mL	Ataxia
40mcg/mL	Disorientation and Somnolence

The most common side effect of phenytoin therapy is gingival hyperplasia (overgrowth of the gums over the teeth); this occurs in about 20 percent of all patients receiving the drug.

Patients with low serum albumin have a higher percentage of free to bound drug ratio (protein binding, 90 percent) and thus therapeutic or toxic reactions occur at a lower serum level. Hypoalbuminemia in liver disease and renal disease is associated with increased percentage of free phenytoin. Critically ill children and adults may have significantly elevated free phenytoin levels with total phenytoin levels in the normal range (Driscoll DF, et al. Crit Care Med. 1988; 16:1248-1249; Griebel ML, et al. Crit Care Med. 1990; 18:385-391; Others).

In addition, elevated metabolic waste products, bilirubin and urea, can displace phenytoin from albumin binding sites. Accumulation of phenytoin metabolites in patients with end-stage renal disease is associated with artifactually elevated serum concentrations of phenytoin; phenytoin concentrations in uremic serum were 20% (fluorescence polarization immunoassay); 60% (enzyme immunoassay) and 80% (rate nephelometric inhibition immunoassay) higher than corresponding values determined by high performance liquid chromatography (Haughey DB, et al. J Anal Toxicol. 1984; 8:106-111).

Serum free testosterone is decreased in patients on phenytoin although total testosterone is elevated; these patients have decreased libido. Phenytoin induces the synthesis of sex hormone binding protein; the increase in the circulating level of sex hormone binding globulin elevates total serum testosterone but depresses free testosterone (Toone BK, et al. J Neurol Neurosurg and Psych. 1983; 46:824-826).

Phenytoin's metabolite, HPPH (see metabolism below), which inhibits the binding of phenytoin to albumin, may accumulate in renal failure. Since the free phenytoin increases in these conditions, optimal anticonvulsive activity is attained at lower total serum phenytoin concentration. Thus, direct measurement of free phenytoin would be the best index of anticonvulsive activity (Finn AL, Olanow CW. in Individualized Drug Therapy, Vol. 2, Taylor WJ, Finn A, eds. Gross, Townsend, Frank, Inc., N.Y., N.Y. 1981; 64-85). Serum phenytoin is unaltered by dialysis.

Phenytoin may cause hepatotoxicity, non-dose related, in susceptible individuals; this reaction is characterized by fever, skin rash, lymphadenopathy, eosinophilia, leukocytosis and hemolytic anemia (Spielberg SP, et al. N Engl J Med. 1981; 305:722-727).

One of the less common complications of this drug is the induction of the movement disorder, choreoathetosis. This disorder is associated with underlying central nervous system disorders and toxic phenytoin levels (Filloux F, Thompson JA. J Pediatr. 1987; 110:639-641).

Hypotension may occur even with slow phenytoin infusion in septic patients - presumably due to negative inotropic and vasodilator effects (Isenstein D, Nasraway SA. Crit Care Med. 1990; 18:1036-1038).

Fetal phenytoin exposure may lead to fetal hydantoin syndrome (craniofacial anomalies, prenatal and postnatal growth deficiency, mental retardation and limb defects). This syndrome may be due to the accumulation of toxic intermediary metabolites in the fetus; it may be possible in the future to identify fetuses at risk by amniocentesis (Buehler BA, et al. N Engl J Med. 1990; 322:1567-1572).

<u>Metabolism</u>: Phenytoin is oxidized in the liver by cytochrome P-450; about 60 to 70 percent of phenytoin is metabolized to an inactive, hydroxylated derivative (5-p-hydroxyphenyl-5-phenylhydantoin, HPPH) which is conjugated and excreted in the urine. Phenytoin potentiates the metabolism of other drugs, e.g., phenobarbital, by this same system (P-450). Five percent of phenytoin appears in the urine unchanged.

The enzymatically mediated reaction utilized for the metabolism of phenytoin is nearly saturated at therapeutic plasma concentrations of the drug. The metabolism of phenytoin is thus said to be saturable, or capacity-limited. When the steady-state serum phenytoin level is within the therapeutic range of 10 to 20 mcg/mL, a very small increase in dose may result in clinical evidence of toxicity and serum levels well above the 20 mcg/mL (Raebel MA. N Engl J Med. 1983; 309:925).

PHEOCHROMOCYTOMA SCREEN

Although only 0.1-0.3% of hypertension appears due to pheochromocytoma, testing for this disorder is an important aspect of laboratory medicine. The majority (90%) of patients with pheochromocytoma are hypertensive and 50% exhibit orthostatic hypotension. Most present with the classical triad of: 1) headache, 2) diaphoresis, and 3) palpitations. Pheochromocytomas are 10% bilateral, 10% extra-adrenal, and 10% malignant.

Most cases can be diagnosed by a combination of: 1) the clinical symptoms, 2) abnormal laboratory tests of catecholamines and their metabolites, and 3) imaging studies. Diagnosis may be missed or delayed in the cases of atypical presentation or labile hypertension. Pheochromocytomas secreting epinephrine alone are reported not to result in hypertension (Bachmann AW, et al. Clin Exp Pharmacol-Physiol. 1989; 16:275-279). Pheochromocytoma in children is characterized by a higher incidence of familial association, bilaterality, and extra-adrenal location (Caty MG, et al. Arch Surg. 1990; 125:978-981). In the familial MEN-2 syndrome, only 41% of gene carriers present with the syndrome by age 70. By contrast laboratory screening may detect biochemical abnormalities in 93% by age 31 (Easton DF, et al. Am J Hum Genet. 1989; 44:208-215). Self administration of epinephrine to produce factitious pheochromocytoma has been reported (Keiser HR. JAMA. 1991; 266:1553-1555).

The major laboratory tests available for use in the diagnosis of pheochromocytoma are listed in the table below (Townsend RR, DiPette DJ. Clin Lab Med. 1993; 13:287-302):

Laboratory Tests Used in the Diagnosis of Pheochromocytoma

Test	Normal Range
Blood:	
Norepinephrine (NE)	<750pg/mL
Epinephrine (E)	<110pg/mL
NE 2-3 h after 0.3mg Clonidine	<500pg/mL and >40% decr from baseline
NE 2-3 min after 2mg Glucagon IV	<750pg/mL; values often >2000pg/mL with 2-6x incr from baseline in pheochromocytoma
3,4-Dihydroxyphenylglycol (DHPG)	5.75pmol/mL (mean value); Clinically, ratio of DHPG/NE is used; Normal ratio is >2:1
Urine:	
Norepinephrine (NE)	<85-100mcg/24 h
Epinephrine (E)	<25-30mcg/24 h
Total Catecholamines (NE + E)	<120mcg/24 h
Vanillylmandelic acid (VMA)	<9mg/24 h
Metanephrines	<1.3mg/24 h
Normetanephrine	<0.9mg/24 h
Metanephrine	<0.4mg/24 h
3,4-Dihydroxyphenylglycol (DHPG)	<308nmol/24 h

Assays of catecholamines and their metabolites were initially calorimetric or spectrophotometric. Extraction steps were subsequently added to make than more specific and less prone to diet and drug interference. Fluorometric assays were developed for substances at low concentration. Presently highly specific HPLC and radioimmuno or immunoenzyme assays are being developed or employed. As these more specific assays are introduced, the reference ranges continue to change. Drug interference with assay procedures are not as problematic as before, but drugs may have profound effects on catecholamine metabolism, either raising, lowering or having no effect on the various assays. Lists of these drugs and their effects may be found in: Spilker, et al. Ann Clin Lab Sci. 1983; 13:16-19 and Stein PP, Black HR. Medicine. 1991; 70:46-66.

Diagnosis of pheochromocytoma relies on the concept of combined testing - i.e. diagnosis of pheochromocytoma may be missed in a minority of patients if only one laboratory test is performed, but will be almost certainly made if two different tests of catecholamine or catecholamine metabolites are performed. A summary of the sensitivity of 24 hour urine tests in the diagnosis of pheochromocytoma is given below (Stein and Black, already cited):

Pheochromocytoma Screen (Cont)

Results of 24 hour Urine Measurements in Patients with Pheochromocytoma	Normal Level	Elevated Level
VMA (n=384)	11%	89%
MN (n=271)	5%	95%
UFC (n=319)	4%	96%

MN = Metanephrine; UFC = Urinary Free Catecholamines.

Other studies relating to the sensitivity and specificity of urinary tests are published (Smythe GA, et al. Clin Chem. 1992; 38:486-492). Some claim that a VMA test by itself is not adequate as a screen since it may miss up to 50% of patients with a pheochromocytoma.

Plasma catecholamines (PC) may be measured but may be only episodically elevated in patients with labile hypertension and require careful patient preparation. Urine tests (24 hour) are less susceptible to fluctuations in catecholamine levels. Spot and overnight urine tests are being developed. However, PC measurements are finding use in provocative stimulation or suppression testing. Urinary free catecholamine measurements appear more useful in detection of small neoplasms, while large tumors produce more of the metabolites (MN or VMA).

Epinephrine (E) can be resolved from norepinephrine (NE) and dopamine by HPLC. The E to NE ratio in UFC is usually less than 20%. An increase in E or in this ratio points to an adrenal or organ of Zuckekandel origin of the pheochromocytoma. The ratio is also increased in patients with neoplasms associated with MEN II. Only 50% of patients with pheochromocytoma have elevations in dopamine. Dopamine production is more often associated with the malignant pheochromocytomas, or pheochromocytomas not associated with hypertension.

Both CT scan and ultrasound are able to localize pheochromocytomas. CT scan is said to be 86% sensitive. With new generation scanners and MRI, the sensitivity approaches 100%. Pheochromocytoma may be discriminated from cortical adenomas; however, there is difficulty in distinguishing pheochromocytoma from malignant adrenal disease (Kier R, McCarthy S. Radiol. 1989; 171:671-674). MIBG scanning is positive in about 86% of patients with pheochromocytoma and appears especially useful in localization of extra-adrenal and small lesions. With the increased availability of new imaging techniques, there are a large number of incidental adrenal masses being detected (0.6-1.3%). Most of these are hormonally inactive. The laboratory workup of a patient with an incidental adrenal mass is discussed: Ross NS, Aron DC. N Engl J Med. 1990; 323:1401-1405.

These tests are not entirely specific for pheochromocytoma. Approximately 12% of non-stressed hypertensive patients will have elevations in UFC, MN or VMA. However, in no instance were these values 2x greater than the upper limit of normal (Stein and Black, already cited). Increased plasma E levels can be associated with emotional or physical stress, illness, hypoglycemia, excessive caffeine or tobacco use, adrenal medullary hyperplasia in family members of relatives with MEN II syndrome, and in patients with adrenal cysts (Streeter DH, et al. Arch Intern Med. 1990; 150:1528-1533).

<u>Provocative Testing: Glucagon Stimulation; Clonidine Suppression</u>:

Histamine, tyramine and glucagon will stimulate catecholamine release. Glucagon administration to patients with pheochromocytoma results in 80% responding with an increase in blood pressure and elevated PC. A positive test is indicated by a 3-fold rise in PC or a PC level >2000 pg/mL (11.83 nmol/L)(Grossman E, et al. Hypertension. 1991; 17:733-741). The stimulation test may provoke dangerous increases in blood pressure in patients with pheochromocytoma. Safety is said to be enhanced in a modified version in which the blood pressure response, but not the PC response, is attenuated by alpha-receptor blockade (Elliott WJ, et al. Arch Intern Med. 1989; 149:214-216).

Release of PC may be measured 3 hours following administration of clonidine. This agent should lower PC in normal and hypertensive patients (Bravo E, et al. N Engl J Med. 1981; 305:623-626). A PC value >500 pg/mL (2.96 nmol/L), or a less than 50% decrease from baseline is considered diagnostic. The test is claimed to be 97% sensitive with a specificity of 67% (Grossman et al., 1991). This test appears most useful in patients in whom a pheochromocytoma is suspected, but whose baseline PC (norepinephrine) values are <2000pg/mL (Sjoberg RJ, et al. Arch Intern Med. 1992; 152:1193-1197). Hypotension may result from the test. Diuretics and beta-blockers interfere with the test.

Further details are entered under the headings: PLASMA CATECHOLAMINES, URINE FREE CATECHOLAMINES, METANEPHRINES, and VANILLYMANDELIC ACID (VMA).

PHOSPHORUS, SERUM

SPECIMEN: Red top tube, separate serum; refrigerate at 4°C.
REFERENCE RANGE: Adults: 2.3-4.3mg/dL; Cord: 3.7-8.1mg/dL; Premature: 5.4-10.9mg/dL; Newborn: 3.5-8.6mg/dL; Infant: 4.5-6.7mg/dL; Child: 4.5-5.5mg/dL. To convert conventional units in mg/dL to international units in mmol/liter, multiply conventional units by 0.3229.
METHOD: Phosphomolydate
INTERPRETATION: The causes of increased serum phosphorus are listed in the next Table (Peppers MP, et al. Crit Care Clin. 1991; 7:201-214; Others):

Causes of Increased Serum Phosphorus
Renal Disease
Malignancy Involving Bone
Immobilization
Magnesium Deficiency
Drug Induced:
Ca^{++} Containing Antacids plus Milk (Milk-Alkali Syndrome)
Vitamin D Intoxication
Etidronate Disodium
Phosphate Enemias or Laxatives
Rhabdomyolysis
Tumor Lysis Syndrome
Sarcoidosis
Hypoparathyroidism
Pseudohypoparathyroidism
Cushing's Disease
Acromegaly (growth hormone excess)
Transfusions
Intravenous or Oral Phosphate Administration
Hemolysis
Failure to Separate Clot from Serum Promptly

Elevation of serum phosphorus in patients with ischemic bowel disease indicates extensive bowel injury, acute renal insufficiency, and acidosis; mortality is significantly increased in these patients (May LD, Berenson MM. Am J Surg. 1983; 146:266-268).

The clinical consequences of hyperphosphatemia are primarily due to the resulting hypocalcemia. An in-vitro study has suggested that an increase of 2.0mg/dL (0.65mmol/L) of serum phosphate may decrease serum ionized calcium by 1.0mg/dL (0.25mmol/L)(Lehmann M, Mimouni F. AJDC. 1989; 143:1340-1341).

The cause of hypophosphatemia are given in the next Table (Zaloga GP. J Crit Illness. 1992; 7:364-375; Peppers MP, et al. Crit Care Clin. 1991; 7:201-214):

Causes of Hypophosphatemia
Intracellular Phosphorus Shift
Alkalosis (Especially Respiratory)
Recovery from Acidosis
Carbohydrate Administration
Recovery from Malnutrition (Refeeding Syndrome)
Beta-2-Adrenergic Agents
Gastrointestinal Phosphorus Losses
Diarrhea
Malabsorption
Nasogastric Suctioning
Aluminum-Containing (Phosphorus-Binding) Antacids
Renal Phosphorus Losses
Hypomagnesemia
Hypokalemia
Renal Tubular Defects
Diuretic Therapy (Thiazides, Loop Diuretics, Acetazolamide, Mannitol)
Corticosteroids
Xanthine Derivatives

In hospitalized patients, medications alone or in combination were the cause of severe hypophosphatemia in more than 80% of patients. Intravenous glucose was the most common cause (Halvey J, Bulvick S. Arch Intern Med. 1988; 148:153).

Certain diseases are commonly associated with hypophosphatemia; often, the causes are multifactorial. For example, alcoholics have multiple factors contributing to hypophosphatemia. Alcoholic gastritis may lead to chronic antacid ingestion. Poor nutrition may lead to inadequate nutritional intake. Alcoholics may have respiratory alkalosis leading to intracellular phosphate shift. Glucose administration may lead to exacerbation of this situation (Peppers MP, et al. Already cited). Diseases commonly associated with hypophosphatemia are given in the following Table (Peppers MP, et al. and Zaloga GP. already cited):

Diseases Associated with Hypophosphatemia
Diabetic Ketoacidosis
Chronic Obstructive Pulmonary Disease (COPD)
Asthma
Alcoholism
Hyperparathyroidism
Diuretic Therapy
Trauma
Burns
Gram Negative Sepsis
Familial Hypophosphatasia

The clinical consequences of hypophosphatemia involve multiple organ systems. The most serious consequences are myocardial dysfunction and respiratory failure. Other manifestations include neurologic symptoms (encephalopathy, seizures, coma, paresthesias, peripheral neuropathy), musculoskeletal symptoms (weakness, rhabdomyolysis), hematologic symptoms (hemolysis, thrombocytopenia, left shift of oxyhemoglobin dissociation curve), renal symptoms (glycosuria, hypercalciuria, hypermagnesuria), metabolic symptoms (glucose intolerance, insulin resistance) and gastrointestinal symptoms (anorexia, nausea, vomiting)(Besunder JB, Smith PG. Crit Care Clin. 1991; 7:659-693).

PITUITARY PANEL

ANTERIOR PITUITARY: The tests on serum that are performed in the work-up of patients with possible pituitary tumors are listed in the next Table (Tucker HG, et al. Ann Intern Med. 1981; 94:302-307):

Serum Tests in Work-Up of Patients with Possible Pituitary Tumors
Prolactin(PRL)
Growth Hormone(GH)
Adrenocorticotropin(ACTH)
Follicle Stimulating Hormone(FSH)
Luteinizing Hormone(LH)
Thyroid Stimulating Hormone(TSH)
Total Thyroxine(T-4), T-3 Resin Uptake, and FTI
Cortisol
Estradiol

Pituitary reserve: Pituitary reserve may be determined in a number of ways. These would be performed in a patient with suspected pituitary insufficiency. 17-hydroxycorticosteroid (17-OHCS) or serum 11-deoxycortisol can be determined before and after administration of oral metyrapone, 750mg every 4 hours for six doses. (See METYRAPONE TEST)

ACTH secretion can be stimulated by hypoglycemic stress. Intravenous injection of insulin is given and blood is drawn every 30 minutes for 2 hours and analyzed for glucose, prolactin, growth hormone and cortisol. The test is considered invalid if the blood sugar does not fall below 45mg/dL. Alternatively, ACTH release may be stimulated by ovine corticotrophin releasing hormone (CRH) administration.

TSH producing capability may be assessed in a TRH stimulation test. Intravenous thyrotropin releasing hormone(TRH) is given; serum specimens for prolactin, growth hormone and TSH are obtained before and every 15 minutes for 1 hour.

These and other stimulation tests (response to insulin, response to thyrotropin-releasing hormone, and response to luteinizing hormone-releasing hormone) may be performed simultaneously on a single day (Lufkin ED, et al. Am J Med. 1983; 75:471-475). This approach saves a patient's time and money.

Pituitary adenoma: The relative incidence of pituitary adenoma cell types in patients with pituitary tumors is given in the next table (Kovacs et al. Pathology Annual. 1977; 12:341):

Adenoma Type in Pituitary Tumors

Pituitary Tumor	Incidence (%)
Prolactin Cell Adenoma	32
Undifferentiated Cell Adenoma	23
Growth Hormone Cell Adenoma	21
Corticotroph Cell Adenoma	13
Mixed Growth Hormone and Prolactin Cell Adenoma	6
Acidophil Stem Cell Adenoma	3.5
Gonadotroph Cell Adenoma	1
Thyrotroph Cell Adenoma	0.5

Of the secreting neoplasms, prolactin-secreting tumors are the most common. In the study by Randall et al (Mayo Clin Proc. 1983; 58:108-121) prolactin-secreting pituitary adenomas constituted 40 percent of the patients with pituitary adenomas. In women, this is usually manifest by amenorrhea, with or without galactorrhea; in men, this is usually manifest by infertility, decreased libido, and sexual potency.

Jordan et al (Jordan RM, et al. Ann Intern Med. 1976; 85:49-55) measured ACTH, GH, TSH, prolactin, LH and FSH in CSF in patients with pituitary tumors; their data suggest that an elevated CSF pituitary hormone is a sensitive indicator of suprasellar extension of a pituitary tumor, and post-treatment measurements are useful in determining efficacy of treatment.

POSTERIOR PITUITARY: The principal disorders of the posterior pituitary are diabetes insipidus(DI), and syndrome of inappropriate ADH secretion(SIADH).

Diabetes Insipidus: The causes of diabetes insipidus are given in the next Table (Coggins CH. N Engl J Med. 1983; 309:420):

Causes of Centeral Diabetes Insipidus

- Idiopathic - 30 Percent
  - Familial - Rare
- Traumatic - 30 Percent
  - Head Injury; Neurosurgical Operation
- Neoplastic - 30 Percent
  - Primary Brain Tumor
    - Pituitary Adenoma; Craniopharyngioma; Meningioma; Optic Glioma
  - Metastatic Tumor
    - Lung; Breast; Colon and Others
- Vascular
  - Hemorrhage; Aneurysm; Hemangioma; Postpartum Necrosis; Postanoxic
- Infection
  - Bacterial or Fungal Abscess; Meningitis/Encephalitis; Tuberculoma
- Systemic
  - Sarcoidosis; Langerhans-Cell Granulomatosis; Wegener's Granulomatosis

Diabetes insipidus may be either central - due to a pituitary problem, or nephrogenic-renal in origin. The clinical findings in diabetes insipidus are thirst and polyuria. The laboratory findings include hypernatremia, increased plasma osmolality, decreased urine sodium concentration, and increased urine volume. The diagnosis can be made directly by measurement of ADH or indirectly by osmolality measurements over a 3 hour period during water restriction (Zerbe and Robertson, N Engl J Med. 1981; 305:1539). The response to administered ADH may be useful in distinguishing pituitary (central) from nephrogenic diabetes insipidus.

Syndrome Inappropriate Antidiuretic Hormone (SIADH): SIADH results in serum dilution manifest by hyponatremia, reduced osmolality, renal Na loss, low BUN, and low uric acid. SIADH can result from increased native ADH secretion or ectopic ADH production. It is usually of CNS or pulmonary origin. Some of these conditions are listed below:

Conditions Associated with SIADH

- Small cell undifferentiated carcinoma
- Tumors of thymus and pancreas
- TB of lung
- Meningitis
- Brain abscess
- Post CNS surgery
- Stroke
- CNS tumors
- Vincristine therapy

Pituitary Panel (Cont)

The pathological effects of SIADH results from extracellular hyponatremia, over-hydration of the CNS, leading to increased intracranial pressure. Anorexia, nausea and vomiting is associated with serum sodiums of 120 mmol/L or less; CNS symptoms are seen at sodium levels below 120 mmol/L. Symptoms depend not only on the absolute level of serum sodium, but also the rate at which the low level is reached. Patients with a rapid drop in serum sodium may be symptomatic at higher levels.

The diagnosis of SIADH is based on the fact that these patients experience serum dilution with increased renal Na loss and urine concentration. There is decreased serum osmolality with increased urine osmolality. Serum sodium by itself is useless in substantiating the diagnosis since there are many causes of hyponatremia. Diagnosis is made by a comparison of serum and urine osmolalities, or by measurement of ADH. Serum uric acid also appears to be of use in the diagnosis of SIADH. In a series of patients with hyponatremia, serum uric acid was markedly decreased in those with SIADH (Beck. N Engl J Med. 1979; 301:528).

At times it may be important to determine the cause of the SIADH, i.e. ectopic tumor ADH versus CNS ADH. An alcohol stimulation test has been developed based on the premise that tumor ADH is not under physiological control. The patient is given a standard dose of a large amount of alcohol. If there is physiological control and the ADH is produced by the hypothalamus, polyuria, decreased urine concentration, and decreased urine osmolality develops. A negative test results is supportive of a tumor etiology of the SIADH.
(See SYNDROME OF INAPPROPRIATE ANTIDIURETIC HORMONE PANEL)

PLASMA EXCHANGE (see THERAPEUTIC PHERESIS)

PLASMINOGEN

SPECIMEN: Blue (citrate) top tube, separate plasma. Do not use plasma collected in the presence of fluoride, EDTA or heparin.
REFERENCE RANGE: 73-122% (DuPont ACA Methodology).
METHOD: The plasminogen method is based on the reaction of streptokinase, which has been added in excess, to form an enzymatically active complex with plasminogen in the sample. The active complex then hydrolyses a synthetic substrate; the product reacts with a chromogen to form a complex that absorbs at 405 nm. The increase in absorbance at 405 nm is directly proportional to the amount of functional plasminogen. Inhibitors of plasminogen activation are added to specimens drawn from patients on fibrinolytic therapy. An immunological test for plasminogen antigen may also be available.
INTERPRETATION: Plasminogen levels are measured in patients with a thrombotic tendency (possible absence of anti-thrombin III, proteins C or S, or deficiency of plasminogen) or in the patient on thrombolytic therapy (streptokinase, urokinase, tPA). In interpreting results, the plasminogen level of a pool of normal plasmas is taken as 100%. Levels of plasminogen decrease following fibrinolytic therapy.

Plasminogen is proteolytically activated to plasmin; plasmin attacks fibrinogen and fibrin to produce fibrin split products(FSP).

Depressed values are seen in active fibrinolysis such as fibrinolytic therapy or disseminated intravascular coagulopathy(DIC), liver disease, and in hereditary disorders of plasminogen production.

PLATELET AGGREGATION

SPECIMEN: Blue (citrate) top tube; do not refrigerate. Patient should not receive aspirin, phenylbutazone, phenothiazines or antihistamines for 10 days prior to the test. Platelet count should be less than 100,000/cu mm.
REFERENCE RANGE: Interpreted by laboratory
METHOD: Platelet aggregometer; this test is based on an increase in the transmission of light as platelets aggregate in response to an aggregating agent.
INTERPRETATION: Platelet aggregation studies are done to evaluate platelet function. This is a specialized test and would normally be performed in patients with an abnormal bleeding time or some other indicator of a qualitative platelet disorder.

Qualitative platelet disorders may be subdivided into those with defective:
1) Adhesion (e.g. von Willebrand disease, Bernard-Soulier syndrome)
2) Secretion/release (e.g. storage pool defects, release defects, e.g. aspirin)
3) Aggregation (e.g. Glanzmann's)

The commonly used agents in platelet aggregation studies are ristocetin, ADP, collagen and epinephrine. Ristocetin causes von Willebrand factor binding and will cause platelet aggregation independent of the storage/release mechanism. ADP and epinephrine will induce a "two wave" aggregation phenomenon, the second wave is dependent upon the release mechanism.

The following conditions cause alteration of platelet aggregation:

Alterations of Platelet Aggregation
Glanzmann's Thrombasthenia
Abnormality of Platelet Release Mechanism:
Aspirin, Uremia, Myeloproliferative Disorders, Severe Liver Disease, Dysproteinemia
Impaired Aggregation to Ristocetin with Normal Aggregation to ADP, Epinephrine and Collagen:
von Willebrand's disease
Other Platelet Abnormalities

The typical findings in various platelet disorders are summarized in the following table:

Aggregation Studies in Platelet Defects			
Disorder	Ristocetin	ADP/Epinephrine	Mechanism
von Willebrand Disease	(-)	Nl	Absence of vWF
Bernard-Soulier Syndrome	(-)	Nl	Lack of vWF receptor (GpIb)
Glanzmann's Thrombasthenia*	Nl	(-)	Lack of fibrinogen receptor (GpIIbIIIa)
Multiple Conditions	Nl	Defective Second Wave	Lack of Storage Granules/ Defect Granule Release

* Afibrinogenemia will also give a similar result.

von Willebrand's disease is the most common of the congenital platelet defects. There are three major types of von Willebrand disease presently described and multiple subtypes. In all three forms the bleeding time is increased and the PT normal. The PTT, Factor VIII level, antigenic vW Factor results are variable. Detailed analysis of von Willebrand disease patients is desirable because of differing responses to therapy among these various subgroups. See VON WILLEBRAND DISEASE PROFILE.

PLATELET ANTIBODIES

SPECIMEN: Red top tube, separate serum
REFERENCE RANGE: Negative or <1000 molecules IgG/platelet
METHOD: Indirect immunofluorescence; flow cytometry; immunoassays.
INTERPRETATION: Platelet antibodies are determined in the investigation of patients with unexplained thrombocytopenia and patients refractory to platelet transfusions. The conditions that are associated with platelet antibodies are given in the next Table (Mayo Medical Laboratories Communique. 1983; 8:12):

Conditions Associated with Platelet Antibodies
Idiopathic Thrombocytopenia Purpura (ITP)
Post-Transfusion Purpura
Platelet Refractoriness
Neonatal Isoimmune Purpura
Drug-Induced Thrombocytopenia (Quinidine, Quinine, Furosemide, Sulfonamides)

Antibodies that develop to platelet antigens are of two types, autoantibodies and alloantibodies. Autoantibodies are antibodies that develop in response to ones own platelets; these antibodies are found in idiopathic thrombocytopenia purpura(ITP). Alloantibodies develop following exposure to antigens from foreign platelets.

Idiopathic Thrombocytopenia Purpura(ITP): About 90 percent of children with acute ITP have autoantibodies in their serum against platelets. In children, ITP is usually an acute process while in adults, it is usually chronic.

Post-Transfusion Purpura (PTP): Following whole blood transfusion, antibody (anti-PLA1) against a platelet specific antigen found in 98% of the population may develop. Profound thrombocytopenia, manifested by purpura and mucosal bleeding, occurs about one week after the transfusion; the antibody tends to disappear in about six weeks. It may recur with subsequent transfusions. Women may be sensitized to platelet transfusions by a previous pregnancy.

Platelet Refractoriness: On repeated infusions of platelet concentrates, the increment in platelet counts after each infusion may become steadily smaller; this is due to development of antibodies in the recipient to HLA antigens on the infused platelets. HLA-matched platelets will usually increase the platelet count.

Neonatal Isoimmune Purpura (Neonatal Alloimmune Thrombocytopenic Purpura [NAITP]): The phenomena observed in neonatal isoimmune purpura is similar to that of hemolytic disease of the newborn(HDN)(see COOMBS, DIRECT) in that, the mother, on exposure to fetal platelets antigens of the father, develops antibodies to these antigens. These antibodies pass from maternal blood into the fetus. In the fetus the antibodies combine with the antigens on the fetal platelets. Subsequent pregnancies may also be affected. If a pregnant mother has ITP, antibodies can cross the placenta and cause a similar reaction.

Drug-Induced Thrombocytopenia: Drugs, such as quinidine, quinine, furosemide and sulfonamides, may be associated with thrombocytopenia. The fall in platelets may occur acutely following administration of the drug or it may develop weeks after the drug is discontinued. The thrombocytopenia may be due to antiplatelet antibodies or to progressive bone marrow damage.

Heparin-Induced Thrombocytopenia: In some patients, heparin therapy results in thrombocytopenia with paradoxical thrombosis. In this condition, the antibody binds to platelets only in the presence of heparin and results in platelet aggregation and a paradoxical initiation of thrombosis. The platelet count should be periodically measured during heparin therapy.

HIV-Associated Thrombocytopenia: Thrombocytopenia in HIV infected patients is associated with increased platelet associated antibody (IgG, IgM and C3). However, there appears to be no specificity for platelet specific antigens and there is speculation that the observed thrombocytopenia is due to direct HIV-megakaryocyte precursor infection and decreased platelet production (Ballem PJ, et al. N Engl J Med. 1992; 327:1779-1784).

PLATELET COUNT

SPECIMEN: Lavender (EDTA) top tube; specimen rejected if clotted.
REFERENCE RANGE: Adults: 150,000-400,000/mm^3; cord: 100,000-300,000/mm^3, premature: 100,000-300,000/mm^3; newborn: 140,000-300,000/mm^3; neonate: 150,000-300,000/mm^3; infant: 200,000-475,000/mm^3; child: 150,000-450,000/mm^3
METHOD: Automated methodology; rapid estimation of the platelet count can be made from the blood smear; there should be about one platelet for every 20 red cells; an estimation may be obtained by multiplying the number of platelets per oil immersion field by 20,000.
INTERPRETATION: Platelet counts are determined in patients with suspected bleeding disorders, patients with purpura or petechia, those with a prolonged bleeding time, those with leukemia/lymphoma, DIC, and various platelet disorders, patients on chemotherapy, and to determine the response to patients receiving platelet transfusions.
Thrombocytopenia: The causes of decreased platelet count are given in the next Table (modified from Bauer JD in Gradwohl's Clinical Laboratory Methods and Diagnosis, Sonnenwirth AC, Jarett L, eds., 8th Ed., CV Mosby Co., St. Louis, 1980; 990):

Causes of Decreased Platelet Count

Decreased Production:
- Marrow Depression: Aplastic Anemia, Radiation, Chemotherapy, Drugs
- Marrow Infiltration: Acute Leukemia, Carcinoma
- Myelofibrosis, Multiple Myeloma
- Megaloblastic Anemia
- Congenital: Wiskott-Aldrich Syndrome, Fanconi Syndrome, Immune Deficiency States, Bernard-Sonlier Syndrome, Thrombocytopenia with Absent Radius, May-Hegglin Anomaly, Hereditary Thrombocytopenia Resembling ITP, Gaucher's Disease

Increased Destruction:
- Immunologic
  - Isoimmune: Post Transfusion Purpura, Neonatal Thrombocytopenia
  - Autoimmune: Idiopathic Thrombocytopenic Purpura (ITP), Evans' Syndrome, Infectious Mononucleosis
  - Antigen-Antibody Complexes: Systemic Lupus Erythematosus, Lymphoma, Chronic Lymphocytic Leukemia
- Others
  - Drugs (Cimo PL. Arch Intern Med. 1983; 143:1117-1118)
  - Dilution: Exchange Transfusion
  - Coagulopathies: Disseminated Intravascular Coagulation, Septicemia, Hemolytic-Uremic Syndrome, Thrombotic Thrombocytopenic Purpura, Large or Multiple Hemangiomas, Heart Valve, Eclampsia
  - Severe Hemorrhage
  - Hypersplenism
  - Heparin Induced
  - Spurious: Platelet Aggregation, Large Platelets

There is little tendency to bleed until the platelet count falls below 20,000-50,000/mm^3. Bleeding due to low platelet counts typically present as petechiae, epistaxis and gingival bleeding. For surgery, platelet counts above 50,000/mm^3 are desired.

Approximately 50% of women in one study admitted with severe preeclampsia had thrombocytopenia (counts <150,000 mm^3). It was found that the degree of thrombocytopenia was the best indicator of maternal and fetal mortality and morbidity. Abnormalities of PT, PTT and fibrinogen did not occur in the absence of thrombocytopenia (Leduc L, et al. Obstet Gynecol. 1992; 79:14-18).

Drug-induced thrombocytopenia is rare; the number of drugs that have been invoked as possible cause of thrombocytopenia is extensive.

Platelet Count (Cont)

Complications of Heparin Therapy: Heparin therapy may induce thrombocytopenia and is usually an asymptomatic complication. It is more common with heparin from bovine lung than from porcine gut (15.6% versus 5.8% of patients respectively; King DJ, Kelton JG. Ann Intern Med. 1984; 100:535-540). In it's mild form, reduced counts range 100,000-150,000/mm^3, develops 2-4 days after initiation of therapy and disappears in 1-5 days even on therapy (Cola C. Amer Heart J. 1990; 119:368-374). The more serious form is characterized by platelet counts <100,000/mm^3 and may induce "white-clot syndrome." Heparin-induced "white clot syndrome" was diagnosed in one study on 0.48% of 2,500 patients on heparin (Abu Rahma AF, et al. Am J Surg. 1991; 162:175-179). Diagnosis was based on: 1) development of plt counts <100,000/mm^3 or 50% decrease from admission value, 2) normalization of count following heparin discontinuation, 3) exclusion of other causes of thrombocytopenia, 4) presence of thrombotic complications, 5) positive test for heparin-induced platelet aggregation or documentation of white clots in surgical or autopsy material. Thrombocytopenia occurred 1-9 days (mean 5.5) after starting heparin therapy and the mean nadir of plt counts approximately 27,000/mm^3. This syndrome is due to a heparin dependant IgG antibody, and cessation of heparin often leads to a rebound thrombocytosis. Monitoring of platelet counts every 2-3 days in patients receiving heparin is suggested.

Thrombocytosis: The causes of increased platelet count are given in the next Table:

Causes of Increased Platelet Count
Reactive Thrombocytosis: Infection, Acute Blood Loss, Disseminated Carcinoma, Splenectomy (e.g. in Hereditary Spherocytosis), Tissue Damage, Chronic Inflammation, Surgery (Stress)
Thrombocythemia:
Myeloproliferative Disorders, Polycythemia Vera, Chronic Granulocytic Leukemia, Hemolytic Anemia and Myelosclerosis
Essential Thrombocythemia

The relative incidence of these various causes in a teaching hospital have been given in: Santhosh-Kumar CR, et al. J Intern Med. 1991; 229:493-495.

PLEURAL FLUID ANALYSIS

SPECIMEN: Collect 3 tubes; purple (EDTA) for cell count; special Bactec-vials for microbiological studies, blue (aerobic), yellow (anaerobic); red or green (heparin) top tube for chemistries. Culture, gram stain and Ziehl-Nielsen staining should be done on a centrifuged specimen.

Red top tube, separate serum (for serum protein, LDH, bilirubin and cholesterol analysis). Obtain within 30 minutes of pleural specimen.

REFERENCE RANGE: Appearance: Clear and colorless to pale yellow; less than 1000 WBC/cu mm; less than 25% polys; 0 RBC's; glucose level approximates serum glucose level.

METHOD: The tests are given in the next Table:

Tests of Pleural Fluid
Specific Gravity
Total Protein (Serum and Pleural Fluid)
Glucose
Lactate Dehydrogenase (LDH) (Serum and Pleural Fluid)
Cholesterol (Serum and Pleural Fluid)
Bilirubin (Serum and Pleural Fluid)
White Blood Cell Count and Differential
Red Cell Count
Culture, Gram Stain, Ziehl-Nielsen stain for Mycobacteria
Cytologic Examination
CEA for Tumor
Amylase for Diagnosis of Acute Pancreatitis

INTERPRETATION: The causes of pleural fluid effusions are given in the next Table; (Krieg AF in Henry JB. Clinical Diagnosis and Management. W.B. Saunders, Philadelphia, 17 Edition, pg. 484, 1984):

Pleural Fluid Effusions	
Transudates	**Exudates**
Congestive Heart Failure Hepatic Cirrhosis Hypoproteinemia (e.g. Nephrotic Syndrome)	Neoplasms Bronchogenic Carcinoma Metastatic Carcinoma Lymphoma Mesothelioma (Increased Hyaluronate in Effusion) Infections Tuberculosis Bacterial Pneumonia Viral or Mycoplasma Pneumonia Trauma Pulmonary Infarction Rheumatoid Disease(Usually Low Glucose) Systemic Lupus (Occ. L.E. Cells) Pancreatitis (Elev. Amylase) Ruptured Esophagus (Elevated Amylase, Low pH) Chylous Effusion: Damage or obstructed Thoracic Duct

Differentiation of transudate from exudate is given in the next Table:

Differentiation of Transudate from Exudate

Characteristic	Transudate	Exudate	Reference
Specific Gravity	<1.016	>1.016	
Protein	<3 gms/dL	>3 gms/dL	
LDH	<200 IU/L	>200 IU/L	Light RW, et al. Ann Intern Med. 1972; 77:507-513.
Pleural Fluid Protein / Serum Protein =	<0.5	>0.5	
Pleural Fluid LDH / Serum LDH =	<0.6	>0.6	
Serum Albumin Minus Pleural Fluid Albumin	>1.2g/dL	<1.2g/dL	Roth BJ, et al. Chest. 1990; 98:546-549.
Pleural Fluid Bilirubin / Serum Bilirubin	<0.6	>0.6	Meisel S, et al. Chest. 1990; 98:141-144.
Cholesterol	<55-60mg/dL	>55-60mg/dL	Hamm H, et al. Chest. 1987; 92:296-302.
Pleural Fluid Cholesterol / Serum Cholesterol	<0.3	>0.3	Valdes L, et al. Chest. 1991; 99: 1097-1102.

Although newer tests appear to be appropriate methods to differentiate between transudates and exudates, the criteria developed by Light remain valid. All of these methods have limitations.

Treatment of congestive heart failure with transudate may convert the effusion into a "pseudoexudate" with pleural effusion protein elevation (2.2 to 3.2g/dL), LDH elevation (116 to 183 units/L) protein ratio elevation (0.34 to 0.47) and LDH ratio elevation (0.39 to 0.64) with aggressive diuretic therapy (Chakko SC, et al. Chest. 1989; 95:798-802).

Pleural effusions have been reported to occur in 3 to 18 percent of patients with acute pancreatitis or pseudocyst. The effusions are more commonly left-sided. A level of amylase in pleural fluid, which is significantly raised above normal and higher than simultaneous serum values, is practically pathognomonic of pancreatitis. Increased amylase in pleural fluid may occur by one of the following mechanisms: transdiaphragmatic from peritoneal cavity to pleural cavity by lymphatic transfer; intrapleural rupture of mediastinal extensions of pseudocysts; or diaphragmatic perforation. Increased amylase in pleural fluid may also be caused by esophageal perforation and malignancy.

PNEUMOCYSTIS CARINII TESTING

SPECIMEN: Sputum, Bronchoalveolar lavage, Lung tissue (Direct or indirect staining techniques); Serum (Antibody or antigen detection); Sputum, Bronchoalveolar lavage, Blood (DNA amplification)
REFERENCE RANGE: Negative
METHOD: Direct Stains: Grocott-Gomori methenamine silver nitrate, Toluidine blue O, Giemsa, others; Indirect Stains: Immunofluorescence with monoclonal antibodies (Kovacs JA, et al. N Engl J Med. 1988; 318:589-593); DNA amplification with polymerase chain reaction (Lipschik GY, et al. Lancet. 1992; 340:203-206); Serology: serum antibody or antigen detection (ELISA for IgG; latex agglutination for antigenemia) is not reliable for definitive diagnosis of pneumocystis carinii infection (Pifer LLW, et al. AJDC. 1988; 142:36-39).
INTERPRETATION: Pneumocystis carinii is an opportunistic parasite which can cause interstitial pneumonia in the immunocompromised patient.

Pneumocystis carinii pneumonia(PCP) is the most common life threatening opportunistic infection in AIDS patients. About fifty percent of patients with acquired immunodeficiency syndrome(AIDS) present with pneumocystic carinii pneumonia(PCP). The survival rate for AIDS patients with PCP and acute respiratory failure was very low during the initial stages of the AIDS epidemic. More recent data shows a much better long-term survival with immediate hospital survival rate of 47% and one year survival rate of 37% (Friedman Y, et al. JAMA. 1991; 266:89-92).

Pneumocystic carinii prophylaxis in patients with HIV infection includes trimethoprim-sulfamethoxazole (TMP-SMX), trimethoprim-dapsone or aerosolized pentamidine. Treatment of PCP depends on severity of symptoms and response to therapy. Parenteral TMP-SMX is the therapy of choice; alternatively, parenteral pentamidine may be used. Trimetrexate is available on a compassionate basis for patients who fail conventional therapy. Corticosteroids are effective adjunctive therapy to improve survival and decrease the occurrence of respiratory failure in patients with AIDS and severe PCP (Gagnon S, et al. N Engl J Med. 1990; 323:1444-1450). Other therapies are under investigation (see review: Sattler FR, Feinberg J. "New Developments in the Treatment of Pneumocystis Carinii Pneumonia." Chest. 1992; 101:451-457).

PORPHOBILINOGEN, QUALITATIVE AND QUANTITATIVE, URINE

SPECIMEN: Qualitative: 50mL of fresh random urine; keep specimen protected from light during transit to lab by wrapping in foil; freeze until ready for assay. Quantitative: The specimen should be protected from light at all times. Wrap container in foil. The patient should discard the urine specimen on arising in the morning; then collect urine specimens including the specimen obtained just prior to completion of 24 hour collection. Record the total urine volume and freeze a 100mL aliquot. Specimens are best taken during an attack of suspected porphyria.
REFERENCE RANGE: Qualitative, negative; quantitative: <2mg/24 hours. To convert conventional units in mg/day to international units in micromol/day, multiply conventional units by 4.420.
METHOD: Column chromatography followed by reaction with Ehrlich's reagent (dimethylaminobenzaldehyde).
INTERPRETATION: Porphobilinogen is measured in the workup of patients with a suspected defect in heme metabolism, the porphyrias, and in the screening of asymptomatic carriers. Porphobilinogen is increased in the conditions listed in the next Table:

Increase in Porphobilinogen
Hereditary Hepatic Porphyrias:
Acute Intermittent Porphyria (AIP)
Variegate Porphyria (VP)
Hereditary Coproporphyrin (HCP)

The two most common forms of porphyria are porphyria cutanea tarda and acute intermittent porphyria (AIP). AIP results in elevated levels of delta-aminolevulinic acid and porphobilinogen (Kusher JP. N Engl J Med. 1991; 324:1432-1434). Attacks of AIP may be triggered by stress, surgery, infection and drugs (e.g., barbiturates, anti-convulsants, steroids, O.C.). Porphobilinogen is not increased in the urine in porphyria cutanea tarda (PCT) nor lead poisoning. The major metabolites elevated in lead poisoning are aminolevalinic acid and coproporphyrinogen III. The metabolic defect in AIP is shown in the next Figure:

Metabolic Defect in Acute Intermittent Porphyria (AIP)

```
"Activated   +   Succinyl
 Glycine"          CoA
    |________________|
            ↓  ALA Synthetase (↑ Activity)
2 Delta-Aminolevulinic Acid
            ↓  ALA Dehydrase
      Porphobilinogen
            ┼  Uroporphyrinogen I Synthase
            ↓  Uroporphyrinogen Isomerase
    Uroporphyrinogen III
            ↓  Uroporphyrinogen Decarboxylase
   Coproporphyrinogen III
            ↓  Coproporphyrinogen Oxidase
   Protoporphyrinogen III
            ↓
     Protoporphyrin III
      +Fe+2 ↓  Ferrochelatase
           Heme
```

The enzyme, uroporphyrinogen I synthase, is also known as porphobilinogen deaminase. Following an acute attack, delta aminolevulinic acid and porphobilinogen levels fall, but rarely into the normal range. Most mutated porphobilinogen deaminase carriers do not exhibit the clinical phenotype. These silent carriers do not excrete excess porphobilinogen or delta-aminolevulinic acid. The reasons for these observations are unknown. During an acute neurological attack the clinical and laboratory manifestations of AIP and VP may be identical. However, during neurologically asymptomatic periods, the diagnosis of VP may be made by the increase of porphyrins in the bile or stool (sometimes a difficult procedure).

POTASSIUM, SERUM

SPECIMEN: Red top tube or green top (heparin) tube; the serum or plasma must be separated from the red cells promptly; otherwise spurious elevation of serum potassium occurs.
REFERENCE RANGE: Adult: 3.5-5.3 mmol/liter; premature (cord): 5.0-10.2 mmol/liter; premature (48 hours): 3.0-6.0 mmol/liter; newborn (cord): 5.6-12.0 mmol/liter; newborn: 3.7-5.0 mmol/liter; infant: 4.1-5.3 mmol/liter; child: 3.4-4.7 mmol/liter. Conventional units (mEq/L) equal international units (mmol/L). The following are life-threatening values and, after confirmation, preferably by repeat determination on a different specimen, should be telephoned to the responsible nursing staff or to the responsible physician so that corrective therapy may be immediately undertaken: serum potassium: <2.5 mmol/liters; serum potassium: >6.5 mmol/liter.
METHOD: Ion-selective electrode or flame emission.
INTERPRETATION: The concentration of intracellular K^+ is 130 mmol/liter; the serum concentration of K^+ is 3.5 to 5.0 mmol/liter. Thus, only two to three percent of the K^+ is extracellular. Generally, serum K^+ reflects K^+ stores.

There are generally three mechanisms to consider in hypokalemia (1) urinary loss, (2) gastrointestinal loss and (3) movement of potassium from the extracellular to the intracellular fluid. These mechanisms are listed in the next Table (Freedman BI, Burkart JM. Crit Care Clin. 1991; 7:143-153):

Causes of Hypokalemia

G.I. Loss	Urine Loss
Vomiting	Diuretics: Thiazides; Loop diuretics (furosemide, ethacrynic acid)
Nasogastric suction	Magnesium depletion
Pyloric obstruction	Antibiotics: Carbenicillin, Amphotericin B
Diarrhea	Increased Mineralocorticoid: Florinef; Carbenoxolone; Licorice; Chewing Tobacco
Malabsorption	Renal Tubular Acidosis Type I (distal) or Type II (proximal)
Villous adenoma	Bartter's Syndrome
Enema and laxative abuse	Hyperaldosteronism
Biliary drainage	Cushing's Syndrome or Disease
Enteric fistula	Congenital Adrenal Hyperplasia (11 or 17 hydroxylase def)
Redistribution into the Intracellular Space	High Renin States
Alkalosis	**Other**
Na-K-ATPase Stimulation: Insulin; β_2 agonists; Catecholamines	Decreased K^+ intake
Familial Hypokalemic Periodic Paralysis	Sweating
Barium Intoxication	
Hypothermia	
Frozen deglycerolyzed RBC transfusion	
Acute Myeloid Leukemia	

Diuretic therapy and gastrointestinal loss are major causes of hypokalemia.

Drug-induced hypokalemia is usually due to diuretics, adrenal corticosteroids, or carbenoxolone; carbenoxolone has a mineralocorticoid-like action on the distal tubule.

Small potassium supplements do not reliably prevent hypokalemia in patients taking thiazides or loop diuretics. Elderly individuals and patients taking both thiazides or loop diuretics plus digoxin should receive potassium supplements sufficient to maintain a normal plasma potassium concentration. The blood potassium concentration should be checked at least once every 3 months.

Generally, serum K^+ reflects K^+ stores; two important exceptions are alkalosis and insulin hypersecretion which promote entry of K^+ into cells resulting in hypokalemia despite adequate or increased intracellular K^+. Excessive cell breakdown, e.g., hemolysis, results in a relative excess of extracellular K^+ despite normal stores.

More than 90 percent of potassium is excreted in the urine; it is filtered and totally reabsorbed proximally and is excreted by the distal tubules. The factors that determine the degree of distal tubular secretion are dietary K^+ intake, mineralocorticoid secretion, distal tubular flow and the anion accompanying Na^+ to the distal tubular site. Both increased dietary K^+ intake and aldosterone promote movement of K^+ into cells and favor secretion into the tubular lumen and subsequent excretion.

Normally, less than 10 percent of potassium that is excreted appears in the stool; however, gastrointestinal abnormalities can cause loss of K^+.

The increase of serum potassium in different conditions is shown in the next Table:

Increase of Serum Potassium in Different Conditions

(1) Redistribution:
- Metabolic or Respiratory Acidosis
- Drugs: Insulin; Beta adrenergic blockade; Arginine infusion; Succinylcholine; Digitalis Toxicity
- Hyperkalemic Periodic Paralysis

(2) Renal Failure: Acute or Chronic with Oliguria

(3) Aldosterone Antagonists: Spironolactone; Triamterene; Amiloride

(4) Adrenogenital Syndrome: 21-Hydroxylase Deficiency

(5) Adrenal Insufficiency, e.g., Patient on Long Term Corticosteroids and Sudden Discontinuance

(6) Hypoaldosteronism: Addison's Disease

(7) Potassium Load:
- Massive Muscle Necrosis
- I.V. Therapy, i.e., Especially with K^+ supplements or to patients with renal disease
- Blood Transfusions of Aged Blood

(8) Significant Thrombocytosis or Leukocytosis

(9) Artefactual Increase:
- Hemolyzed Serum
- Repeated Fist Clenching during Venipuncture
- Delayed Separation of Serum from Cells

Acute acidemia usually results in hyperkalemia. Hyperkalemia occurs most often during the course of acute renal failure. Hyperkalemia may be drug-induced, such as, distal tubular diuretic, e.g., aldosterone antagonists, to a patient with renal failure; potassium supplement together with an aldosterone antagonist as diuretic. Acute hyperkalemia may occur in recipients of massive blood transfusions; it can be avoided by use of blood less than one week old.

The causes of hyperkalemia in a hospital population are given in the next Table (Paice B, et al. Brit Med J. 1983; 286:1189-1192):

Hyperkalemia in a Hospital Population

- Renal Disease, Acute or Chronic
- Drug Treatment:
  - Oral Potassium Supplements
  - I.V. Potassium (Overenthusiastic)
  - Aldosterone Antagonists
- Redistribution of Potassium
  - Catabolic States
  - Severe Acidosis, e.g., Diabetic Ketoacidosis

Review of hyperkalemia: Williams ME. Crit Care Clin. 1991; 7:155-174.

ECG changes that occur in hypokalemia and hyperkalemia are given in the following Table:

ECG Changes in Hypokalemia or Hyperkalemia

HYPOKALEMIA	Flattened T-Wave Prominent U-Wave
NORMAL	
MILD HYPERKALEMIA (<6.5mmol/L)	Peaked T-Wave
MODERATE HYPERKALEMIA (6.8-8.0mmol/L)	Flattened P-Wave Prolonged PR Interval Widened QRS Complex Deep S-Wave
SEVERE HYPERKALEMIA (>8.0mmol/L)	Sine Wave Pattern

PREGNANCY TEST, SERUM (see HUMAN CHORIONIC GONADOTROPIN, HCG)

PREGNANCY TEST, URINE

Synonym: Human Chorionic Gonadotropin, Urine
SPECIMEN: First morning voided urine; refrigerate or freeze.
REFERENCE RANGE: Negative; positive in normal pregnant female 0-3 weeks after missed period. Sensitivity will vary by technique.
METHOD: Newer agglutination methods use antibody to beta chain of human chorionic gonadotropin; Immunoenzymatic methods. Sensitivity will vary by technique.
INTERPRETATION: Used in the diagnosis of pregnancy. These tests should not be used in evaluation of ectopic pregnancy, problem pregnancy, or trophoblastic disease. Intra-uterine pregnancy can be confirmed by ultrasonography. The most sensitive urine tests are positive at the first missed period and most kits will yield positive results two or three weeks after the first missed period. False positive results are obtained with patients who have hematuria, proteinuria, or opiates.

The serum pregnancy test (HUMAN CHORIONIC GONADOTROPIN, SERUM), instead of the urine pregnancy test, should be used for early detection of pregnancy, detection of ectopic pregnancy and when a false-positive or negative test result is suspected.

PRENATAL SCREENING PANEL

Various recommendations have been made regarding laboratory tests for prenatal screening. A prenatal laboratory screening panel is given in the next Table (Rosen MG, et al. "Caring for our Future: A Report by the Expert Panel on the Content of Prenatal Care." Obstet Gynecol. 1991; 77:782-787):

Laboratory Prenatal Screening Panel

Pre-Conception Care*:	
Recommended for All	Offer to Some or All
Hemoglobin/Hematocrit	Cytomegalovirus
Rh Factor	Herpes Simplex
Rubella Titer	Toxoplasmosis
Urine (Protein/Sugar)	Varicella
Papanicolaou Smear	Recommended for Some
Gonococcal Culture	Tuberculosis Screen
Syphilis Test	Chlamydia Culture or Rapid Screen
Hepatitis B	Hemoglobinopathies
Offer to All	Tay-Sachs Screen
Human Immunodeficiency Virus (HIV)	Parental Karyotype
Toxic Drug Screen	
First Pregnancy Visit:	
Recommended for All	Recommended for Some
Hemoglobin/Hematocrit	Rh Screen
Urine Culture	Syphilis Test
Blood Group**	Blood Glucose Level
Atypical Antibody Screen**	Gonococcal Culture
Subsequent Pregnancy Visit:	
Recommended for All	Recommended for Some
Hemoglobin/Hematocrit (24-28 wks)	Rh Screen (26-28 wks)
Diabetic Screen (26-28 wks)	Repeat Syphilis
Serum Alpha-Fetoprotein (14-16 wks)	Repeat Gonococcal Culture
	Repeat HIV Testing

*Preconception care is recommended, but often does not occur. In such cases, appropriate tests may be incorporated into the first pregnancy visit.
**Blood Group and Atypical Antibody Screen: Although not specifically suggested by these authors, blood group and atypical antibody screen are usually done. In one study of prenatal screening for antibodies, 3 percent of Rh-negative patients and 1.8 percent of Rh-positive patients had irregular antibodies. However, of the Rh-positive patients who had atypical antibodies, only 6 percent had irregular antibodies that were considered potentially clinically significant (anti-M, anti-Kell and anti-E)(Solola A, et al. Obstet Gynecol. 1983; 61:25-29).

Human Immunodeficiency Virus (HIV) Testing: Routine prenatal HIV testing has important clinical and public health implications. Pre-test and post-test counseling of seronegative women may educate them about risk factors for HIV infection. Seropositive women may elect to terminate their pregnancy. Babies born to seropositive mothers may be followed for signs and symptoms of HIV infection (Barbacci M, et al. Lancet. 1991; 337:709-711).

PREOPERATIVE PANEL

"Routine" testing of patients prior to surgery has a relatively low yield. The purpose of presurgical testing is to detect unsuspected medical conditions, which may alter assessment of surgical risk or lead to interventions which lead to lower risk. In addition, baseline laboratory studies may be helpful in decision making during or after surgery. Routine preoperative testing of all patients prior to elective surgery is not indicated. Typical preoperative tests are listed in the following Table with recommendations for testing prior to elective surgery (Macpherson DS. Med Clin N Amer. 1993; 77:289-308):

Preoperative Tests and Recommendations Prior to Elective Surgery

Test	Recommendations
Urinalysis	Selective Use
Complete Blood Count(CBC)	
Hemoglobin	Prior to Major Surgery
White Blood Cell Count	Selective Use
Platelet Count	Selective Use
Prothrombin Time(PT), Partial Thromboplastin Time(PTT)	Selective Use
Bleeding Time	Selective Use
Chemistry Tests:	Selective Use
Electrolytes (Na^+, K^-, Cl^-, CO_2 Content)	
Glucose	
BUN or Creatinine	
AST	
Lactate Dehydrogenase(LDH)	
Bilirubin, Total and Direct	
Chest X-Ray	All Patients Over 60 Years
Electrocardiogram(ECG)	All Patients Over 40 Years
Blood Type and Screen	Prior to Major Surgery

Patients should be tested if abnormalities are suggested by the history and physical. A proposal for preoperative testing, generated to help eliminate unnecessary tests, is given in the next Table (Blery C, et al. Effective Health Care. 1983; 1:111-114):

Pre-Operative Testing - A Proposal

	CBC	Type & Screen	ECG	Chest X-Ray	Blood Glucose	Electrolytes, Creat.& BUN	PT PTT	Platelets, Bleeding Time
Surgical Procedures:								
Minor								
Major	X	X						
Age:								
<40								
40-70			X					
>70			X			X		
Associated Condition:								
Cardiovascular			X	X				
Pulmonary			X	X				
Malignant							X	
Hepato-Biliary							X	
Renal						X		
Bleeding							X	X
Diabetes					X	X		
Medications:								
Diuretics						X		
Digitalis						X		
Corticoids					X	X		
Anticoagulants							X	

For review of preoperative laboratory testing, see Macpherson DS. Med Clin N Amer. 1993; 77:289-308.

PRIMIDONE (MYSOLINE)

SPECIMEN: Red top tube, separate serum and refrigerate; green (heparin) top tube, separate plasma. Reject if specimen hemolyzed.
REFERENCE RANGE: Therapeutic(Primidone): 5-12 mcg/mL; Therapeutic(Phenobarbital): 15-40 mcg/mL; Time to Obtain Blood Specimen (Steady State): 4 to 7 days; Half-Life: Primidone: 10-12 hours; Phenobarbital: 50 to 120 hours (adults), 40 to 70 hours (children); Peak Time: 0.5-0.9 hours (variable).
METHOD: EMIT; GLC; HPLC; SLFIA(Ames); Nephelometry(ICS); Fluorescence Polarization(Abbott)
INTERPRETATION: Primidone is used as an anticonvulsant generally as alternative therapy for grand-mal and partial (complex) seizures (Medical Letter. 1989; 31:1-4). Primidone is metabolized in the liver to phenobarbital and phenylethylmalonamide (PEMA). Both have anticonvulsant activity. Dosage guidelines and time to obtain blood specimens are given in the next Table:

Dosage Guidelines and Blood Specimens

Dosage Guidelines	Blood Specimens
Adults: 500 to 1000mg (5 to 10mg/kg/day)	After 4 to 7 days and just before next dose
Pediatrics: 10 to 25mg/kg/day	

Toxicity: If toxicity is suspected, monitor at least six hours after last dose; primidone serum levels above 12mcg/mL are likely to be associated with serious toxicity. Severe reactions are rare. Mild adverse effects are drowsiness, ataxia, vertigo, anorexia, nausea and vomiting; ataxia and vertigo tend to disappear with continued therapy. Primidone can also cause disturbances in behavior, difficulty in concentration, and loss of libido.

Initial neurotoxicity is worse than for phenobarbital which may make primidone intolerable (Mattson RH, et al. N Engl J Med. 1985; 313:145-151).
Phenobarbital: Within 5 to 7 days, phenobarbital, from the metabolism of primidone, is detectable. Monitor blood phenobarbital levels; the optimal range of phenobarbital is 15-30mcg/mL; when both primidone and phenobarbital levels are measured together, the phenobarbital level is two to three times the primidone level.

PRIST (TOTAL IgE)**(see IgE)**

PROCAINAMIDE (PRONESTYL)

SPECIMEN: Red top tube, separate serum and refrigerate; reject if serum left standing on cells for several days or serum frozen on cells.
REFERENCE RANGE: Therapeutic: 4-8mcg/mL for procainamide; 10-30mcg/mL for procainamide plus N-acetyl procainamide (NAPA). Toxic: >30mcg/mL for both procainamide plus N-acetyl procainamide (NAPA); up to 30mcg/mL may be tolerated by some individuals. Half-Life: Procainamide: 3 hours in normal subjects; 9 hours with renal impairment. NAPA: 7 hours in normal subjects; 10 hours to 40 hours with renal impairment. Determine creatinine and BUN before starting therapy. Steady State (Time to obtain serum specimens): Oral: 48 hours; Intravenous: 24 hours after loading dose.
METHOD: EMIT (Two separate assays, one for procainamide, other for NAPA); HPLC and GLC (Differentiates procainamide and NAPA).
INTERPRETATION: Procainamide is a class IA antiarrhythmic agent (quinidine, procainamide, disopyramide); procainamide has a local anesthetic (sodium channel blockade) effect and prolongs action potential duration. Procainamide depresses the automaticity and excitability of cardiac muscle, slows conduction and prolongs refractory period. Procainamide is used to treat atrial and ventricular arrhythmias.

It is frequently used to treat life-threatening ventricular arrhythmias resistant to lidocaine and as prophylaxis against ventricular arrhythmias following acute myocardial infarction.

The major metabolite of procainamide is N-acetylprocainamide (NAPA) which also possesses anti-arrhythmic activity. The serum concentration of NAPA is genetically determined by the activity of the hepatic enzyme, N-acetyltransferase; different individuals acetylate procainamide at different rates. The marked variations in absorption, metabolism and elimination are reasons to monitor patients receiving procainamide.

Therapy and suggested monitoring schedule for procainamide are given in the following Table:

Procainamide Therapy and Monitoring Schedule

Intravenous Therapy	Blood Specimens
Loading Dose: 17mg/kg in 100mL D5W given over one hour or 275mcg/kg/min for 25 min	First blood specimen obtained at end of Loading Dose
Maintenance Dose: 2.8mg/kg/hour (Add total daily dose to 500mL D5W and infuse at 20mL/hour for 24 hours or 20-60 mcg/kg/min	Second: Two hours after maintenance infusion begins. Note: Loading infusion level should be higher than maintenance infusion level; neither sample should exceed 15mcg/mL. Third: Six to 12 hours after maintenance infusion starts. Fourth: At 24 hours (Steady State)
Oral Therapy	**Blood Specimens**
Dosing interval should not exceed 4 hours; a sustained release preparations is given every 6-8 hours. Oral Dose = maintenance infusion rate x dosing interval. Example: 2.8mg/kg/h x 4 h = 11.2mg/kg given every 4 h	Blood specimens should be obtained as follows: First Blood Specimen: 48 hours after starting oral therapy and when dose is administered. Second Blood Specimen: Middle of same dosing interval. Third Blood Specimen: End of same dosing interval. Obtain second and third blood specimens at equally spaced time intervals.

Renal Disease, Pulmonary Edema and Cardiogenic Shock: About half of the available dose of procainamide appears in the urine. In patients with renal impairment, the dose of procainamide must be adjusted downward.

Dose as follows in patients with renal failure:

Dosage of Procainamide in Renal Failure

Intravenous:
  Loading Dose (Normal): 17mg/kg
    Moderate Renal Failure (BUN 25 to 40mg/dL): 17mg/kg
    Severe Renal Failure (BUN >40mg/dL) or pulmonary edema or cardiogenic shock: 14mg/kg
  Maintenance Dose: Normal: 2.8mg/kg/hour
    Moderate Renal Failure (BUN 25 to 40mg/dL): 2mg/kg/hour
    Severe Renal Failure (BUN >40mg/dL) or pulmonary edema or cardiogenic shock: 1mg/kg/hour
Oral: Normal: 11.2mg/kg/4 hours
  Moderate Renal Failure: Dosing interval, 4 to 6 hours.
  Severe Renal Failure: Dosing interval, 6 to 12 hours.

Adverse Effects: Adverse effects of procainamide are listed in the next Table:

Adverse Effects of Procainamide

Gastrointestinal Disturbances:
  Anorexia, Nausea and Vomiting
Cardiovascular:
  Slow Intraventricular Conduction, A-V Block, Bradycardia, Hypotension, Syncope
  Ventricular Arrhythmias, Increased Ventricular Response to Atrial Flutter or Fibrillation
  QT Prolongation and Widening of the QRS Complex; QRS widening >0.02 seconds suggests toxicity
  Contraindicated in and may precipitate Torsades de Pointes
Adverse Effects of Prolonged Procainamide Therapy:
  Reversible Lupus Erythematosus-Like Syndrome (30% of patients). Antinuclear Antibodies: (50% to 80%).
  Hypersensitivity with Fever, Rash, Urticaria, Neutropenia
  Mental Disturbances

Severe neutropenia (granulocytopenia) is associated with sustained-release procainamide; granulocytopenia generally occurs in the first 3 months and it usually develops rapidly. Therefore, leukocyte determinations should be done at 2-week intervals. Discontinuation of therapy has generally resulted in a prompt return to normal leukocyte counts (Ellrodt AG, et al. Ann Intern Med. 1984; 100:197-201; Gabrielson RM, ibid. 1984; 100:766).

In one study, positive antinuclear antibodies occurred in 83% of patients; drug-related lupus (arthritis, fever, malar rash, pleural and pericardial effusions) occurred in 30% of patients (Mongey A-B, et al. Arth and Rheum. 1992; 35:219-223).

PROGESTERONE

SPECIMEN: Red top tube, separate serum and freeze.
REFERENCE RANGE: The reference range for progesterone in various clinical conditions is given in the following Table:

Reference Range for Serum Progesterone

Sex	Condition	Reference Range (ng/dL)
Female	Follicular Phase	<150
	Luteal Phase	>300
	Midluteal	may exceed 2000
	Pregnancy: First Trimester	1500-5000
	Third Trimester	8000-20,000
	Postmenopausal	<50
Male		<50

To convert conventional units in ng/dL to international units in nmol/liter, multiply conventional units by 0.03180.
METHOD: RIA
INTERPRETATION: Assay of serum progesterone is most often done to answer the question, "Has the patient ovulated?" It is the best single laboratory test for detecting infertility in females. It is measured to document ovulation, confirm the impression of basal body temperature measurements, assess the function of the corpus luteum, and to assess placental function. The change in serum progesterone during the menstrual cycle is given in the next Figure:

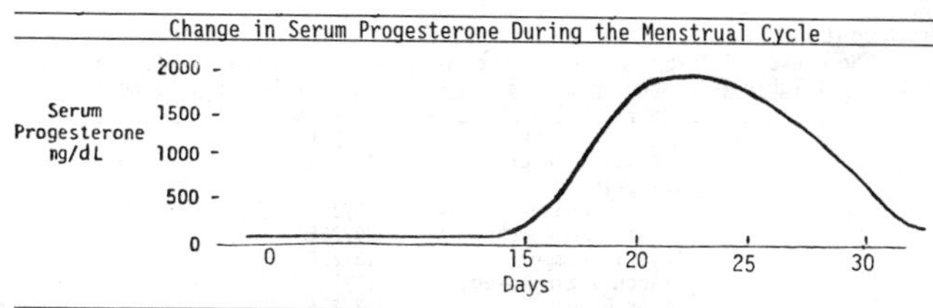

In the normal menstrual cycle, low levels of progesterone (of adrenal origin) are present in the blood during the follicular phase. Ovulation occurs as a result of the midcycle surge of luteinizing hormone (LH). Follicle cells undergo luteinization to form a corpus luteum which secretes progesterone; progesterone prepares the endometrium for implantation and development of the embryo and prepares the breast for lactation. Progesterone peaks in the mid-luteal phase.

To assess the formation and functional state of the corpus luteum, obtain blood specimens as follows: Mid-follicular phase: day 5 to 10; luteal phase: day 19 to 23 for normal menstrual cycle.

To pinpoint day of ovulation, obtain blood specimens on alternate days as follows: days 10, 12, 14, 16 of the menstrual cycle; for the most reliable results, all these samples should be assayed at the same time.

Progesterone assays are useful in evaluating the corpus luteum during the first trimester of pregnancy; some cases of spontaneous abortion have been associated with low progesterone reflecting luteal insufficiency (Hensleigh PA, Fainstat T. Fertility and Sterility. 1979; 32:396). Serial samples of progesterone should be obtained early in pregnancy and continuing through the twelfth week; at that time, the placenta produces progesterone.

PROGESTERONE RECEPTOR ASSAY
(see ESTROGEN/PROGESTERONE RECEPTORS)

PROLACTIN

SPECIMEN: Red top tube, separate serum and freeze; keep frozen during delivery to reference laboratory.
REFERENCE RANGE: <18.5ng/mL. Reference ranges continue to change as assays have become more specific. They also vary by sex, age and time of day. To convert conventional units in ng/mL to international units in mcg/liter, multiply conventional units by 1.00.
METHOD: RIA
INTERPRETATION: Prolactin levels are measured in the workup of galactorrhea, amenorrhea, infertility, hirsutism, impotence, and in cases of suspected pituitary neoplasm. An elevated prolactin classically presents with the syndrome of galactorrhea-amenorrhea in women, and the syndrome of infertility-impotence in men. Men with elevated prolactin typically have a low serum testosterone. However, testosterone replacement alone will not reverse the symptoms, the prolactin must also be reduced. (cont)

Prolactin (Cont)

The causes of inappropriate galactorrhea in one series of women are given in the next Table (Kleinberg DL, et al. N Engl J Med. 1977; 296:589-600):

Causes of Inappropriate Galactorrhea

Cause	Number
Pituitary Tumors	48/235
Idiopathic:	
With Menses	76/235
With Amenorrhea	20/235
Chiari-Frommel	18/235
Tranquilizing Drugs	13/235
Other Drugs	2/235
Post-Oral Contraceptives	12/235
Hypothyroidism	10/235
Empty Sella	5/235
Miscellaneous	30/235

Drugs that may cause galactorrhea include: chlorpromazine, phenothiazine, reserpine, cimetidine, oral contraceptives, and estrogens. Chiari-Frommel Syndrome is persistent galactorrhea and amenorrhea related to childbirth. Patients with empty sella syndrome have an enlarged sella turcica. The empty sella syndrome arises as a result of a defect in the dura which allows for prolapse of subarachnoid membranes into the sella fossa. Empty sella syndrome is not usually associated with hormonal loss.

Causes of an elevated prolactin are given below:

Causes of an Elevated Prolactin
Pituitary Adenoma
Compression of Pituitary Stalk (meningioma, CNS sarcoidosis, craniopharyngioma, eosinophilic granuloma)
Hypothyroidism
Drugs
Renal Failure
Idiopathic

A serum level of prolactin greater than 200ng/mL strongly suggests pituitary tumor; a serum level above 300ng/mL is practically pathognomonic of pituitary tumor (Edwards CRW, Fink CM. Brit Med J. 1981; 283:1561-1562).

In order to distinguish causes of galactorrhea, TRH stimulation may be done using thyrotropin-releasing hormone(TRH) with serum specimens for prolactin obtained at times 0, 15 and 30 minutes. The prolactin levels in patients with pituitary tumors show no increment, while prolactin levels in patients with other disorders increase 6- to 20-fold (Kleinberg DL, et al. N Engl J Med. 1977; 296:589-600). Pituitary adenomas are now routinely visualized by MRI.

The most common tumor of the pituitary is prolactin secreting adenoma; pituitary adenoma cell types in patients with pituitary tumors are given in the next Table (Kovacs, et al. Pathology Annual. 1977; 12:341):

Adenoma Type in Pituitary Tumors

Tumor	Incidence(%)
Prolactin Cell	32
Undifferentiated Cell	23
Growth Hormone Cell	21
Corticotroph Cell	13
Mixed Growth Hormone and Prolactin Cell	6
Acidophil Stem Cell	3.5
Gonadotroph Cell	1
Thyrotroph Cell	0.5

Many of the undifferentiated cell pituitary tumors express gonadotrophin m-RNA, but do not secrete these hormones. Treatment is by surgery (tumor), removal or reversal of the underlying condition (e.g., hypothyroidism), or administration of dopamine agonists (bromocriptin or perolide). Most of the prolactin secreting tumors respond to drugs and surgery is not necessary.

PROSTATE-SPECIFIC ANTIGEN (PSA)

SPECIMEN: Red top tube, serum.
METHOD: Immunoradiometric assay.
REFERENCE RANGE: May vary among assays. Generally values above 4mcg/L (4ng/mL) are considered abnormal. A value of 4mcg/L in a monoclonal assay approximates a level of 7mcg/L when assayed with polyclonal reagents. Normal PSA concentration does not exclude cancer. See Interpretation.
INTERPRETATION: Prostate-specific antigen (PSA) levels are used as an aid in the diagnosis of prostate cancer. Long term serial assays are used to follow response to therapy and to detect recurrent disease. The use of PSA as a screening test for prostatic cancer in asymptomatic men is also under investigation (Cantalona et al. N Engl J Med. 1991; 324:1156-1161). The use of the PSA has recently been reviewed by The Medical Letter. 1992; 34(Issue 880): 93-94. These consultants recommend a routine PSA in conjunction with an annual rectal exam for men older than 50.

PSA is a glycoprotein originally described in seminal fluids and later demonstrated to be specific for prostatic tissues. It is now probably the most useful tumor marker available for the diagnosis and management of prostatic carcinoma. PSA appears to be more sensitive than prostatic acid phosphatase for diagnosis of carcinoma of the prostate, but may be less specific since it appears to be more frequently elevated in benign prostatic conditions (hyperplasia and inflammation). Urinary retention, infection and biopsy can also elevate the PSA. Anti-PSA may be used as an immunohistochemical stain on tissue sections to affirm a prostatic origin of metastatic disease. Other techniques and assays used in the diagnosis and management of prostatic cancer include digital rectal examination, transrectal ultrasonography (hypoechoic areas in the periphery of the prostate), acid phosphatase measurements, needle biopsy studies, and alkaline phosphatase and bone scan studies for metastatic disease. Although the PSA is thought to be prostate specific, there are reports of women with renal cell carcinoma having positive results with polyclonal PSA assays (Pummer K, et al. J Urol. 1992; 148:21-23).

Values of PSA less than 4mcg/L are generally considered within the reference range. Approximately 22% of men with PSA levels between 4-9.9mcg/L had biopsy proven carcinoma, as did 67% of men with PSA levels higher than 10mcg/L (Catalona, et al. 1991). PSA levels are higher in patients with moderately differentiated tumors than in well or poorly differentiated tumors (Haapiainen RK, et al. Br J Urol. 1990; 66:635-638). Normal PSA levels do not appear to change for one hour following digital rectal examination, even in patients with prostatic hyperplasia (Oesterling JE. JAMA. 1992; 267:2236-2238; Crawford ED, et al. JAMA. 1992; 267:2227-2228). However, it is recommended that blood for the PSA be drawn within one hour or greater than 24 hours following digital rectal examination (Bruel J, et al. Eur Urol. 1992; 21:195-199; Yuan JJJ, et al. J Urol. 1992; 147:810-814). Values may increase following digital examination in those men with decidedly elevated levels (10-20mcg/L). Approximately 25% of men with benign prostatic hyperplasia, have an elevated PSA (Oesterling JE. J Urol. 1991; 145:907-923). Somewhere between 38-48% of men with organ-confined prostatic carcinoma, have a PSA level below 4mcg/L (Oesterling, already cited). In order to improve diagnostic sensitivity, attempts are being made to correlate presence of prostatic carcinoma with rate of change in PSA levels, PSA to prostate volume ratios, and PSA to testosterone ratios.

The average half-life of PSA levels following removal of prostatic carcinoma is approximately 46 hours (Pontes JE, et al. Urology. 1990; 36:415-419). Serial monitoring of PSA is useful in assessing response to therapy. PSA decrease within one month following endocrine therapy is associated with the best chance of survival, while persistent PSA elevation is associated with risk of disease progression (Arai Y, et al. J Urol. 1990; 144:1415-1419). Many would suggest that response to therapy be monitored by measurement of both PSA and prostatic acid phosphatase (Rainwater LM, et al. Mayo Clin Proc. 1990; 65:1118-1126). This is because some patients may have an elevated acid phosphatase but not PSA, and the fact that PSA expression requires presence of androgens. This is especially important in the patient treated with anti-androgens, a therapy which may mask the presence of tumor.

Large scale screening studies of asymptomatic men with PSA assays yield a detection rate of 2.2-2.6%, somewhat better than a yield of 1.3-1.5% given for digital rectal examination (Catalona WJ, et al. N Engl J Med. 1991; 324:1156-1161; Brawer MK, et al. J Urol. 1992; 147:841-845). PSA detected approximately 32% of the prostatic cancers missed by digital exam alone. However, normal levels of PSA were found in approximately 20% of the men with cancer detected by digital exam. Approximately 38-48% of men with cancer confined to the prostate have PSA values of <4mcg/L.

The ability of the PSA to detect occult prostate cancer can be improved by studying the PSA velocity - the annual change of PSA levels. In one study, the rate of change of the PSA in the control group was insignificant, while those patients with BPH had a gradual increase, and those with prostatic cancer an exponential increase. It was felt that an annual increased rate in the PSA of 0.75mcg/L/year would identify men who would benefit from further diagnostic studies (Carter HB, et al. JAMA. 1992; 267:2215-2220).

PROTEIN ELECTROPHORESIS, CSF

SPECIMEN: 2mL CSF

REFERENCE RANGE: Total CSF protein is 15 to 45mg/dL; the normal albumin/globulin ratio is 2:1. The IgG/albumin ratio should be less than 0.28. The IgG synthesis index calculated as the ratio of IgG(CSF)/albumin(CSF) to IgG(serum)/albumin(serum) should be less than 0.85. The mean value of normal IgG synthesis indices is around 0.5.

METHOD: Normal CSF contains only 0.5% to 1% of the protein concentration of serum (15mg/dL to 45mg/dL). Therefore, the CSF must be concentrated one hundred times to two hundred times prior to electrophoresis.

INTERPRETATION: The electrophoretic pattern of CSF resembles that of serum except for the pre-albumin fraction and lower proportion of gamma globulin. Interpretation of the pattern is largely dependent on the relative proportions of the protein fractions rather than their actual concentrations. Reduced levels of prealbumin, less than 2%, are associated with leakage across the blood brain barrier, as seen with some malignancies or with bacterial meningitis.

The most important clinical application of CSF protein electrophoresis is in the work-up of patients with possible multiple sclerosis. In multiple sclerosis, there is an increase in total protein concentration, primarily due to IgG; in addition, there is an increased cell count and unusual patterns on electrophoresis of the CSF proteins.

Increases in gamma globulin have been associated with disease. The CSF protein electrophoretic pattern of patients with active multiple sclerosis and other diseases associated with increased CSF gamma globulin fraction is shown in the next Figure:

Cerebrospinal Fluid Protein Electrophoretic Pattern in Active Multiple Sclerosis and other Conditions

Diseases Associated With Increased CSF Gamma Globulin

Active Multiple Sclerosis
Inflammatory Diseases of the CNS
- Encephalitis
- Meningitis (particularly T.B.)
- Neurosyphilis
- Arachnoiditis

Benign and Malignant Intracranial Tumors

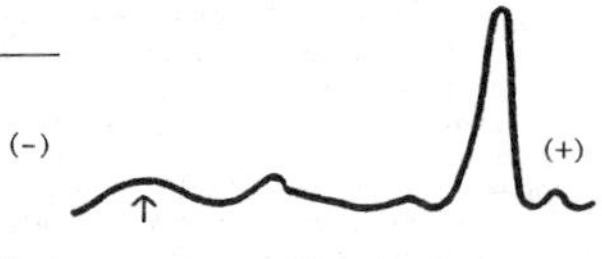

(-) (+)

In multiple sclerosis, CSF gamma globulin is found to be elevated in about 75% of patients with an established clinical diagnosis of multiple sclerosis. There is usually an elevated IgG/albumin ratio. Elevation of the IgG synthesis index implies local CNS synthesis of immunoglobulins.

Elevation of cerebrospinal fluid gamma globulin may also occur when the serum gamma globulin is elevated as in multiple myeloma, collagen disease, chronic liver disease and lymphomas. It is, therefore, recommended that a serum protein electrophoresis be done at about the same time as the cerebrospinal fluid protein.

Oligoclonal Banding: Normal CSF immunoglobulin migrates as a diffuse band; abnormal immunoglobulins migrate as discrete sharp bands or oligoclonal bands.

This is the pattern observed in multiple sclerosis; this pattern of discrete bands within the gamma globulins supports the hypothesis that these globulins arise from a few clones of cells. However, oligoclonal bands are not unique to multiple sclerosis and are observed in the conditions listed in the next Table:

Oligoclonal Bands on Electrophoresis
Multiple Sclerosis
Inflammatory Polyneuropathy
Neurosyphilis
Cryptococcal Meningitis
Chronic Rubella Panencephalitis
Subacute Sclerosing Panencephalitis (Persistent Rubella Viral Infection)
Burkitt's Lymphoma
HIV Infection

Oligoclonal CSF IgG was observed in 12 of 13 patients with Burkitt's lymphoma (Wallen WC, et al. Arch Neurol. 1983; 40:11-13). About 65% of HIV infected individuals exhibit oligoclonal bands in the CSF and 60% have an elevated IgG synthesis index. These findings do not appear to correlate with the stage of disease.

Representative agarose electrophoresis results from a control serum and CSF are compared with those from patients with multiple sclerosis and subacute sclerosing panencephalitis are shown below:

Agarose Electrophoresis of Serum and CSF Specimens

Control | Multiple Sclerosis | Subacute Sclerosing Panencephalitis

+ Alb

IgG -

Serum CSF | Serum CSF | Serum CSF

An abnormal result is the finding of two or more bands in the CSF that are not present in the serum specimen. The sensitivity of CSF oligoclonal bands in the diagnosis of multiple sclerosis is 79-94% (Markowitz H, Kokmen E. Mayo Clin Proc. 1983; 58:273-274). A scan of cerebrospinal fluid obtained from a patient with a multiple sclerosis, using high resolution electrophoresis, is shown in the next Figure:

High Resolution Scan of Cerebrospinal Fluid

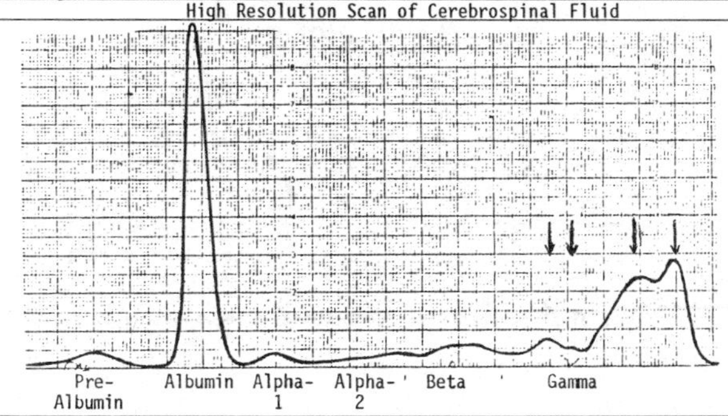

PROTEIN ELECTROPHORESIS, SERUM

SPECIMEN: Red top tube; separate serum; plasma should not be used because fibrinogen separates as an extra band; hemolyzed specimens should not be used since hemoglobin separates as an extra band. Fibrinogen can be removed from specimens from patients with DIC or on anti-coagulation by addition of thrombin.

REFERENCE RANGE:

	gm/dL
Total Protein	6.2 to 8.3
Albumin	3.6 to 5.2
Alpha-1-Globulin	0.15 to 0.40
Alpha-2-Globulin	0.50 to 1.00
Beta Globulin	0.60 to 1.20
Gamma Globulin	0.60 to 1.60

METHOD: Electrophoresis at pH 8.6. High resolution electrophoresis is performed with agarose gels. High resolution patterns allow better quantitation of α-1 antitrypsin and monoclonal "spikes."

INTERPRETATION: Although in many instances the analysis of albumin and total protein and the calculation of the globulins is sufficient, it is sometimes important to have a quantitative analysis of the globulins. This is done by protein electrophoresis followed by densitometric scanning of the stained gels. The patterns are diagnostic of some diseases and suggestive of others. Specific alterations in patterns that are diagnostic include analbuminemia and bisalbuminemia. The diseases that cause depression of albumin levels have been given (see ALBUMIN). Protein electrophoresis is helpful in the work-up of patients for α-1 antitrypsin deficiency, biliary tract disease, liver disease, myeloma, macroglobulinemia. Electrophoresis separates globulins into alpha-1, alpha-2, beta and gamma components. Certain disease processes cause change in the concentration of components as shown in the next Table:

Elevated Values of Globulins

alpha-1	alpha-2	beta	gamma
Acute and chronic infections	Biliary cirrhosis	Biliary cirrhosis	Chronic infections
Febrile reactions	Obstructive jaundice	Obstructive jaundice	Hepatic diseases
	Nephrosis	Multiple myeloma (occasionally)	Autoimmune diseases
	Multiple myeloma (rarely)		Collagen diseases
	Ulcerative colitis		Multiple myeloma
			Waldenstrom's macroglobulin
			Leukemias and some cancers

Depressed Values of Globulins

alpha-1	alpha-2	beta	gamma
Nephrosis	Acute Hemolytic Anemia	Nephrosis	Agammaglobulinemia
Alpha-1 Anti-Trypsin Deficiency			Hypogammaglobulinemia
			Nephrotic Syndrome

The major components of the alpha-1, alpha-2 and beta globulins and their relative positions are given in the next Figure:

Protein Electrophoretic Pattern-Major Components of Bands

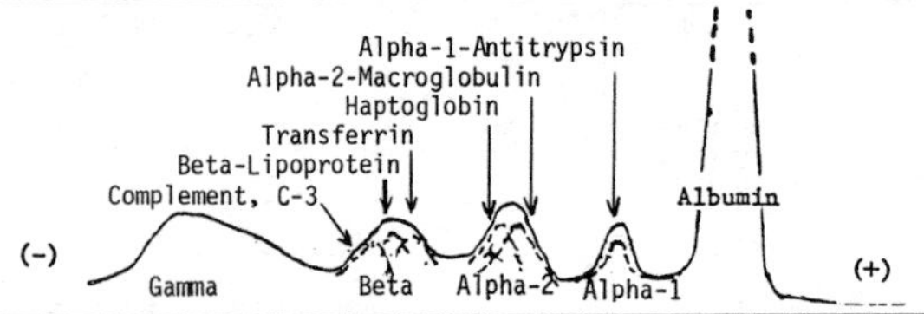

Note that the most positively charged proteins (IgG) migrate to the negative electrode (the cathode) and the most negatively charged (albumin) migrates towards the positive electrode (the anode). A split band in the beta region occurs with very fresh serum specimens, high resolution electrophoresis, excessively long electrophoresis runs, increase in beta lipoproteins or the presence of a paraprotein.

A summary of protein electrophoretic patterns is shown in the next Figure:

Protein Electrophoretic Patterns

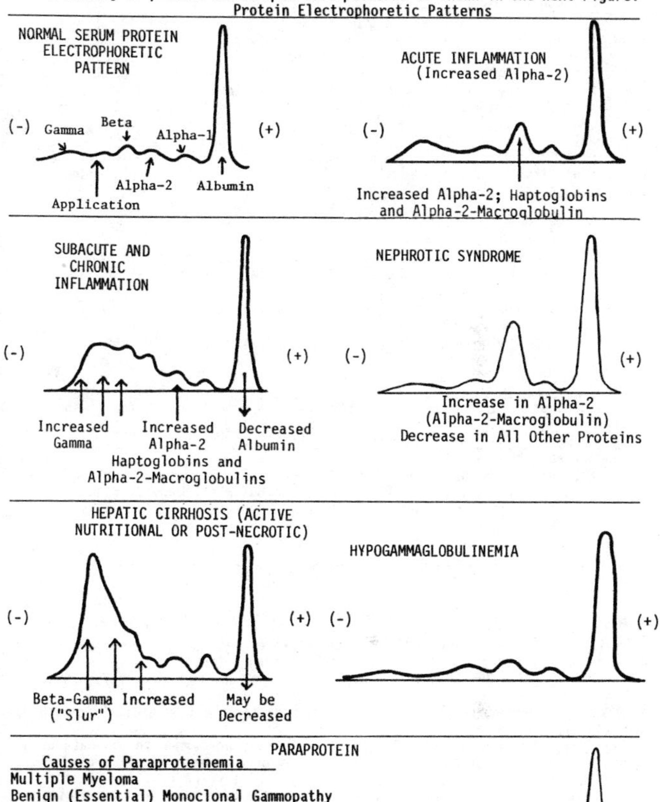

PARAPROTEIN

Causes of Paraproteinemia

Multiple Myeloma
Benign (Essential) Monoclonal Gammopathy
Macroglobulinemia of Waldenstrom
Other Malignant Lymphoproliferative Diseases
Franklin's Disease (Heavy Chain Disease)
Cryoglobulinemia

(-) (+)

Paraprotein Albumin

PROTEIN, TOTAL, SERUM

SPECIMEN: Red top tube, separate serum; refrigerate at 4°C.
REFERENCE RANGE: Adult: 6.0-8.5g/dL; Premature: 4.3-7.6g/dL; Newborn: 4.6-7.6g/dL; Child: 6.2-8.0g/dL. To convert conventional units in g/dL to international units in g/liter, multiply conventional units by 10.0.
METHOD: Biuret: Complex of cupric ions with protein or peptide bond. Reaction takes place at alkaline pH; Coomassie Blue; refractometry.
INTERPRETATION: The total serum protein is the sum of the circulating proteins in the serum. The significance of the level of total protein is difficult to interpret without knowledge of the level of the individual fractions, e.g., albumin and globulins. The quantitative distribution of these is obtained by protein electrophoresis. Total protein is increased in the serum in conditions listed in the next Table:

Causes of Increased Total Serum Protein
Chronic Infection
Liver Diseases
Collagen - Autoimmune Diseases; e.g., Vascular-Systemic Lupus Erythematosus; Scleroderma; Rheumatoid Arthritis
Hypersensitivity States
Sarcoidosis
Cryoglobulinemia
Monoclonal Gammopathies: Multiple Myeloma; Macroglobinemia
Franklin's Disease
Dehydration
Hemolysis

Total protein is decreased in the serum in conditions listed in the next Table:

Causes of Decreased Total Serum Protein
Decreased Albumin
Subacute and Chronic Debilitating Diseases; Liver Disease; Renal Disease; Nephrotic Syndrome; Malabsorption and Malnutrition Gastrointestinal Loss; Third Degree Burns; Exfoliative Dermatitis; Dilution by I.V. Fluids
Hypo- and Agammaglobulinemia
Pregnancy

PROTEIN, TOTAL, URINE

SPECIMEN: Twenty-four hour urine or random urine: this single specimen is used to obtain an estimate of the quantitative protein value.
REFERENCE RANGE: 25-150 mg/24 hours. <0.2 mg urinary protein/mg urinary creatinine.
METHOD: Trichloroacetic acid precipitation; Biuret method; turbidimetric method; nephelometric method.
INTERPRETATION: Quantitation of urinary protein is performed to evaluate urine protein loss over a 24 hour period and to provide a measure of all protein lost in the urine. Dipstick testing of urine will show a trace positive reaction at a protein concentration of 100-200mg/L. These strips are most sensitive to albumin, with less sensitivity to the globulins, and sometimes no reaction at all with Bence-Jones proteins.

The usual specimen for quantitating urinary protein is to obtain a 24 hour urine; however, one of the most difficult tasks, especially in a hospital setting, is accurate collection of a 24 hour urine.

The causes of increased protein in urine are given in the next Table:

Causes of Increased Protein in Urine

Level of Protein	Causes
Heavy (> 4g/day)	Nephrotic Syndrome Acute and Chronic Glomerulonephritis Lupus Nephritis Amyloid Disease Severe Venous Congestion of Kidney
Moderate (0.5 to 4g/day)	Diseases Listed Above Plus Nephrosclerosis Pyelonephritis with Hypertension Multiple Myeloma Diabetic Nephropathy Pre-Eclampsia of Pregnancy Toxic Nephropathies Including Radiation
Minimal (< 0.5g/day)	Diseases Listed Above Plus Chronic Pyelonephritis Polycystic Kidney Disease Renal Tubular Diseases "Benign" Postural Proteinurias

Differentiation of Glomerular Proteinuria versus Renal Tubular Proteinuria: The proteinuria of patients with renal tubular disease is qualitatively different from the proteinuria of glomerular disease. Both lysozyme (muramidase) and beta-2-microglobulin are increased in the urine in renal tubular disease and in renal parenchymal disease (Sherman RL, et al. Arch Intern Med. 1983; 143:1183-1185). Measurement of beta-2-microglobulin is a more sensitive marker for renal tubular disease than lysozyme (Barratt M. Brit Med J. 1983; 287:1489-1490). Lysozyme is increased 100-fold in the Fanconi syndrome (Barratt TM, Crawford R. Clin Sci. 1970; 39:457-465).

Trace Proteinuria: Trace proteinuria is associated with significant increases in both cardiovascular and overall mortality; trace proteinuria is much more common in persons with diabetes and those with hypertension (Framingham Population Study, Hosp Pract. 1982; 17(2):32). Microalbuminuria in patients with maturity-onset (Type II) diabetes is predictive of clinical proteinuria and increased mortality (Mogensen CE. N Engl J Med. 1984; 310:356-360).

PROTHROMBIN TIME (PT)

SPECIMEN: Blue (citrate) top tube filled to 4.5mL with blood; pediatric blue top tube filled to 2.7mL with blood; tubes must be filled to capacity (9:1 ratio blood to anticoagulant). Avoid contamination with tissue thromboplastin as follows: If multiple tests are being drawn, draw coagulation studies last; if only a PT is being drawn, draw 1 to 2mL of blood into another vacutainer, discard and then collect blood for PT. If test cannot be assayed immediately, centrifuge and separate plasma from cells and freeze plasma. The specimen is rejected if the tube is not full, specimen hemolyzed, specimen clotted or specimen received more than 4 hours after collection.

REFERENCE RANGE: 10 to 13 seconds. Patients results should be compared to a reference range determined by the reporting laboratory. Many labs will report an international normalized ratio (INR) for orally anticoagulated patients. See PT-INTERNATIONAL NORMALIZED RATIO (PT-INR).

METHOD: Tissue thromboplastin (complete thromboplastin) from brain; calcium chloride. Platelet-poor plasma is mixed with thromboplastin and calcium and the time to clot formation determined. There is a variation in thromboplastin concentrations and source between commercial suppliers.

INTERPRETATION: This test is often used to monitor warfarin (coumadin) effect. It may also be used to screen for hemostatic dysfunction involving the extrinsic system as a result of liver disease, vitamin K deficiency, factor deficiency, or DIC (Triplett DA. Anticoagulant Therapy: Monitoring Techniques, Laboratory Management. 1982; 20:31-42). The PT test is a measure of the extrinsic coagulation system and measures factors X, VII, V, II and I. The causes of prolonged PT are given in the next Table:

Prothrombin Time (PT) (Cont)

Causes of Prolonged Prothrombin(PT)
Warfarin (Coumadin) Therapy
Heparin Therapy (PTT > PT)
Vitamin K Deficiency
Liver Disease
Decreased Fibrinogen
Dysfibrinogenemia
Disseminated Intravascular Coagulation (DIC)
Factor Deficiency:
Factor VII (Proconvertin)
Factor X (Stuart Factor)
Factor V (Proaccelerin)
Factor II (Prothrombin)
Factor I (Fibrinogen)

The PT is usually not prolonged until the factors are decreased or less than 50% of normal, or the fibrinogen is decreased to less than 80 to 100mg/dL.
Warfarin: The optimum PT range when warfarin is used in North America has been recommended to be 1.3-1.5 times normal for deep-vein thrombosis and 1.5-2.0 for mechanical heart valves, based on a thromboplastin sensitivity of 2.2-2.6. The problems with standardizing anticoagulant therapy are discussed in Bussey HI, et al. Arch Intern Med. 1992; 152:278-282. Many labs will report an international normalized ratio (INR) for orally anticoagulated patients. See PT-INR. The standard PT-INR for most indications is 2.0-3.0; patients with mechanical heart valves are anticoagulated to a PT-INR of 2.5-3.5 (Hirsh J, et al. Chest. 1992; 102:3125-3265).

Warfarin acts by decreasing the synthesis of the vitamin K-dependent clotting factors: prothrombin, factor VII, factor IX, and factor X. After administration of an oral anticoagulant, the level of factor VII is the first to decrease followed by the level of factor IX, followed by factor X and then factor II. The following conditions cause an increased sensitivity to warfarin: diarrhea, abnormalities of the small intestine such as sprue, various drugs.

Therapy with warfarin is begun with 10 to 15mg given once daily for 2 consecutive days, with a prothrombin time determination made before the third daily dose. Daily determinations of the PT are then made until the PT is less than two-times normal, and the maintenance dose is determined. The dose of warfarin given for daily maintenance is generally in the range of 2 to 15mg. Drug interactions with oral anticoagulants are given in the Medical Letter. 1984; 26:11-14).

Heparin Therapy: see PARTIAL THROMBOPLASTIN TIME.

Abnormal Prothrombin: There is an abnormal prothrombin which circulates in the blood of some patients with primary hepatocellular carcinoma; this abnormal prothrombin may be useful in the laboratory diagnosis of primary hepatocellular carcinoma (Liebman HA, et al. N Engl J Med. 1984; 310:1427-1434).

PROTHROMBIN TIME - INTERNATIONAL NORMALIZED RATIO

SPECIMEN AND METHOD: Exactly as in the regular PT
REFERENCE RANGE: For orally anticoagulated patients only. Standard anticoagulation: INR 2.0-3.0; Patients with mechanical heart valves: INR 2.5-3.5.
INTERPRETATION: The International Normalized Ratio (INR) relates the sensitivity of PT reagents to an international standard so that the effectiveness of oral anticoagulant therapy can be monitored independent of changes in reagents. The INR can be simplistically thought of as a "corrected prothrombin test." A PT test is drawn and performed in the usual manner. The difference between the PT and a PT-INR is in the reporting of the test result.

The standard therapeutic range of INR for most patients is 2.0-3.0. Suggested target for patients with mechanical heart valves 2.5-3.5 (Hirsh J, et al. Chest. 1992; 102(suppl.):3125-3265). Values may fall out of this range due to: initial phase of therapy; non-standard therapy protocol; insufficient or excessive warfarin (coumadin) dosing; interaction with other drugs; patient non-compliance.

In theory, reporting values with the INR should normalize PTs performed by different methods and using different reagents. The INR value is calculated by the formula:

$$INR = (PT\ patient/PT\ normal)^{ISI}$$

where: PT patient is the patient PT
PT normal is the mean normal PT for the laboratory system
ISI is the International Sensitivity Index assigned to the system

The INR should be used only for orally anticoagulated patients, not for patients with other coagulation defects. At this time (1993) approximately 50% of U.S. hospitals will calculate and report a PT-INR.

References dealing with the problems of standardizing oral anticoagulant therapy include:

Hirsh J, et al. "Optimal Therapeutic Range for Oral Anticoagulation." Chest. 1992; 102(Suppl.):3125-3265.

Eckman MH, et al. "Effect of Laboratory Variation in the Prothrombin-Time Ratio on the Results of Oral Anticoagulant Therapy." N Engl J Med. 1993; 329:696-702.

PROTOPORPHYRIN, FREE (see ZINC PROTOPORPHYRIN)

PSEUDOCHOLINESTERASE (see CHOLINESTERASE)

PSITTACOSIS ANTIBODIES (see CHLAMYDIA ANTIBODIES)

PULMONARY CAPILLARY WEDGE PRESSURE (SWAN-GANZ CATHETER)

REFERENCE RANGE: <12 mm Hg

METHOD: A Swan-Ganz catheter is introduced usually in a vein in the antecubital area and then the tip is guided into a sufficiently large vein, e.g., the subclavian vein. Then, a balloon is inflated in the large vein; blood flow propels the balloon and catheter into the right atrium, right ventricle, pulmonary artery and pulmonary artery branch. Once the balloon is in position, the balloon is deflated and subsequently reinflated when pulmonary capillary wedge pressure is to be recorded. The catheter has three or four lumens: a lumen that terminates at the catheter tip and is used to measure pulmonary artery and pulmonary capillary wedge pressure and to sample blood; a lumen terminates 30 cm from the tip of the catheter and is used to measure right atrial or central venous pressure. Cardiac output can also be reliable determined by the thermodilution method. For more details about the procedure see McIntyre KM, Lewis AJ, eds. Textbook of Advanced Cardiac Life Support, American Heart Association, 1983; 183-194.

INTERPRETATION: Pulmonary capillary wedge pressure is a useful and reliable indicator of left ventricular dynamics and pulmonary congestion. It is especially important to assess left ventricular dynamics following acute myocardial infarction. Pulmonary capillary wedge pressure, chest X-ray findings and pulmonary congestion are closely correlated, while the older method of measuring central venous pressure is not well correlated with pulmonary congestion.

Hemodynamics in acute myocardial infarction are given in the next Table (Forrester JS, et al. N Engl J Med. 1976; 295:1356):

Pulmonary Capillary Wedge Pressure in Acute Myocardial Infarction

Clinical Classification	Mean PC Wedge Pressure (mmHg)	Cardiac Index	Mortality
No Pulmonary Congestion or Peripheral Hypoperfusion	≤18	>2.2	3%
Pulmonary Congestion without Hypoperfusion	>18	>2.2	9%
Peripheral Hypoperfusion without Pulmonary Congestion	≤18	≤2.2	23%
Both Peripheral Hypoperfusion and Pulmonary Congestion	>18	≤2.2	51%

Cardiac index is the cardiac output per square meter of body surface. Under basal conditions, it is equal to 2.5-4.2 liter/min/meter^2.

In the majority of patients who recover from acute myocardial infarction, the initially elevated pulmonary capillary wedge pressure returned toward normal over the first few days after the infarction.

The types of pulmonary edema are discriminated on the basis of pulmonary capillary wedge pressure as illustrated in the next Figure (McIntyre KM, Lewis AJ, eds. Textbook of Advanced Cardiac Life Support, American Heart Association, 1983; 237):

Pulmonary Edema and Pulmonary Capillary Wedge Pressure

Pulmonary Edema
↓
Pulmonary Capillary Wedge Pressure

> 21 mm Hg — Cardiogenic
< 12 mm Hg — Noncardiogenic → <4.0 mm Hg Hypo-Osmotic; >4.0 mm Hg ↑Capillary Permeability

> 21 mm Hg Cardiogenic	<4.0 mm Hg Hypo-Osmotic	>4.0 mm Hg ↑Capillary Permeability
Left Ventricular Failure:	(Effect 1° Due to ↓Albumin)	Aspiration
Pump Failure	Failure to make Albumin:	HCl, H_2O
Mitral Regurgitation	Starvation	Noxious Gases
Ventricular Septal Defect	Liver Disease	(SO_2, N_2O, NH_3, Cl_2)
Fluid Overload	Losses of Albumin:	Thermal Injuries
Mitral Stenosis	Renal	Oxygen Toxicity
Severe Hypertension	G.I.	Pulmonary Fat Emboli
	Third Space	Shock Lung Syndrome
		Gram-Negative Sepsis

Cardiogenic pulmonary edema can occur with pulmonary capillary wedge pressure <21 mmHg if the colloid osmotic wedge pressure is less than 25 mmHg. Osmotic pressure is reduced when plasma proteins, particularly albumin levels are reduced.

Complications associated with pulmonary artery catheterization are given in the following Table (Ermakov S, Hoyt JW. Crit Care Clin. 1992; 8:773-806):

Complications with Pulmonary Artery Catheterization

Complication	Incidence (%)
Associated with Central Venous Cannulation	
Local Thrombosis	66
Arterial Puncture	2-16
Pneumothorax	2-4
Hydrothorax	2
Local Hemorrhage	2
Air Embolism; Brachial Plexus Injury; Phrenic or Recurrent Laryngeal Nerve Injury	each <1
Cardiac Complications with Catheter Advancement	
Atrial and Ventricular Premature Beats	13-87
Ventricular Tachycardia	11-63
Right Bundle Branch Block	2.6-5.9
Ventricular Fibrillation	1.5-2.5
Complete Heart Block; Cardiac Perforation and Tamponade; Catheter Knotting	each <1
Complications with Catheter Maintenance	
Catheter-Related Endocardial Lesions	33-91
Mural Thrombus	14-91
Infection (Positive Catheter Tip Culture)	4.9-45
Infection (Sepsis)	0.3-0.5
Pulmonary Infarction	1-7
Pulmonary Artery Rupture/Hemorrhage	0.06-2

Date represents numerous studies as described in the review: Ermakov S, Hoyt JW. "Pulmonary Artery Catheterization." 1992; 8:773-806.

PULMONARY EMBOLI PANEL

Laboratory tests in patients with pulmonary emboli are given in the next Table (Hayes SP, Bone RC. Med Clin N Am. 1983; 67:1179-1191; Bone RC. JAMA. 1990; 263:2794-2795):

Laboratory Tests in Patients with Pulmonary Emboli
Definitive Diagnosis: Pulmonary Angiograms; Lung Scan (not truly definitive)
Required: Arterial Blood Gases; Chest X-Ray; Electrocardiogram; Noninvasive Leg Study
Rarely Helpful: Serum Bilirubin; Lactate Dehydrogenase(LDH); Aspartate Aminotransferase (AST); Leukocyte Count (Leukocytosis); Fibrin Split Products (Elevated)

Definitive Diagnosis:

Pulmonary Angiography: A bolus of radiopaque contrast medium is injected into a peripheral vessel. Serial films are obtained in an attempt to detect a filling defect. This is the "gold standard"!

Lung Scan: Technetium-99 albumin microspheres or technetium-99 macroaggregated albumin are injected intravenously. Particles become trapped in those areas of the pulmonary bed that are adequately perfused; regions of obstructed blood flow due to emboli are seen as nonradioactive defects. False positive results may occur, even when ventilation scans are added. Ventilation scans are performed by inhalation of xenon 133. A negative lung scan is reliable in excluding embolism in 96 percent of cases. The sensitivity and specificity of lung scans compared to angiography is given in the following Table (PIOPED Investigators, Value of the Ventilation/Perfusion Scan in Acute Pulmonary Embolism: Results of the Prospective Investigation of Pulmonary Embolism Diagnosis (PIOPED). JAMA. 1990; 263:2753-2759):

Lung Scan Diagnosis of Pulmonary Embolism Compared to Angiography

Scan Category	Sensitivity(%)	Specificity(%)
High Probability	41	97
High or Intermed Probability	82	52
High, Intermed or Low Probability	98	10

Pulmonary Emboli Panel (Cont)

Required Studies:

Arterial Blood Gases: PO_2 almost always below 80 mm Hg; patients usually develop a respiratory alkalosis.

Chest X-Ray: May be normal even with massive emboli; pleural effusions, etc.

Electrocardiogram: Abnormal in 80 percent of patients with massive emboli.

Noninvasive Leg Study: Since deep venous thrombi usually precede pulmonary emboli, noninvasive leg studies (impedance or duplex ultrasound) provide useful information. An algorithm for the diagnostic evaluation of pulmonary embolism is given in the following Table (Bone RC. JAMA. 1990; 263:2794-2795):

Diagnostic Evaluation of Pulmonary Embolism

Clinical Suspicion of Pulmonary Embolism

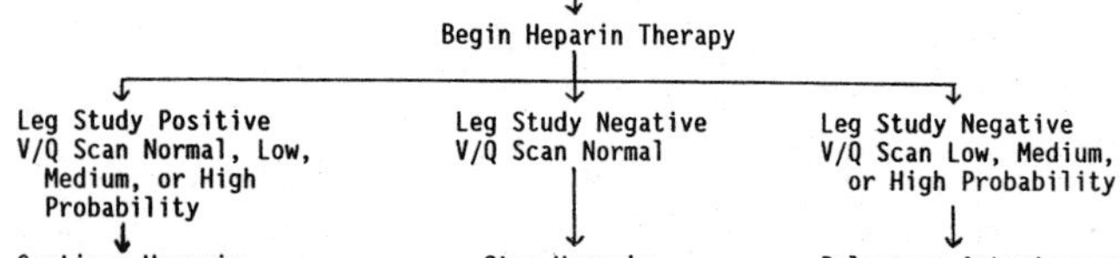

Treatment of Pulmonary Emboli: Medical therapy for pulmonary emboli consists of the anticoagulants, heparin and warfarin, and thrombolytic agents, streptokinase or urokinase. A heparin infusion is used during the acute phase and warfarin compounds for as long as 4 to 6 months.

See also Goldhaber SZ, et al. "Diagnosis, Treatment, and Prevention of Pulmonary Embolism-Report of the WHO/International Society and Federation of Cardiology Task Force," JAMA. 1992; 268:1727-1733; Carson JL, et al. "The Clinical Course of Pulmonary Embolism," N Engl J Med. 1992; 326:1240-1245; Dalen JE. "Clinical Diagnosis of Acute Pulmonary Embolism-When Should a V/Q Scan Be Ordered," Chest. 1991; 100:1185-1186; Stein PD, et al. "Clinical Laboratory, Roentgenographic, and Electrocardiographic Findings in Patients with Acute Pulmonary Embolism and No Pre-Existing Cardiac or Pulmonary Disease," Chest. 1991; 100:598-603; Kelley MA, et al. "Diagnosis Pulmonary Embolism: New Facts and Strategies," Ann Intern Med. 1991; 114:300-306.

PULMONARY FUNCTION TESTS

Tests that may be used to evaluate pulmonary function airway resistance using a simple office spirometer are given in the next Table:

Evaluation of Pulmonary Function
Forced Expiratory Volume in 1 Second(FEV-1)
Total Forced Expiratory Volume (Vital Capacity)(FVC)
Ratio: (FEV-1) x 100/FVC

Lung Volume: Lung volume subdivisions are illustrated in the next Figure:

Lung Volume Subdivisions

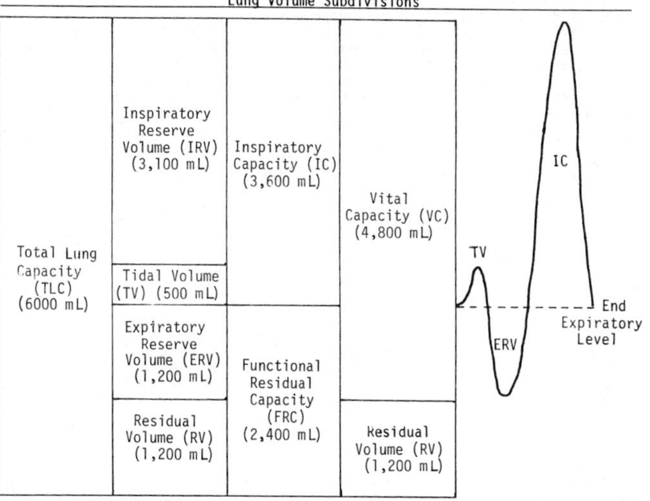

Spirometry: Hand-held peak flow meters, convenient for home or office use, are commonly available.

Total Lung Capacity(TLC): The maximum amount of air the lungs can hold; it is divided into four primary volumes:

(1) Tidal Volume(TV): The amount of air moved into or out of the lungs; resting tidal volume is the normal volume of air inhaled or exhaled in one breath.

(2) Inspiratory Reserve Volume(IRV): The amount of air that can be forcefully inspired after a normal tidal volume inhalation.

(3) Expiratory Reserve Volume(ERV): The amount of air that can be forcefully exhaled after a normal tidal volume exhalation.

(4) Residual Volume(RV): The amount of air left in the lungs after a forced exhalation.

Pediatric Lung Volumes: Normal pulmonary function values for newborns and adults are given in the following Table:

Normal Pulmonary Functions in Infants and Adults

Pulmonary Function Test	Newborn(mL/kg)	Adult(mL/kg)
Tidal Volume	6	6
Total Lung Capacity	63	86
Functional Residual Capacity	30	34
Vital Capacity	35	70
Residual Volume	23	16

FORCED EXPIRATORY VOLUME IN 1 SECOND(FEV-1)
SPECIMEN: After maximum inspiration, FEV-1 is the volume of a single, maximally fast expiration in the first second.
REFERENCE RANGE: Normally between 3 and 5 liters. FEV-1 is expressed in one of two ways: (1) as a percentage of the expected FEV-1 for a normal individual of the same age, height and sex; or (2) as a percentage of the individual's own total forced expiratory volume (vital capacity); that is, (FEV-1) x 100/FVC; this value is usually greater than 75 percent.
METHOD: Expiration into a spirometer.
INTERPRETATION: FEV-1 is the most important clinical tool for assessing the severity of airway obstructive disease. The FEV-1 is a measure of flow early in expiration and is somewhat dependent on effort. Forced expiratory volume in 1 second(FEV-1) is decreased in the presence of obstructive disease and restrictive disease. Typical patterns are shown in the next Figure (West JB. Respiratory Physiology - The Essentials. 2nd Edition. 1979. Williams and Wilkins. Baltimore, p. 144):

Patterns of FEV-1 in Normal, Obstructive and Restrictive Disease

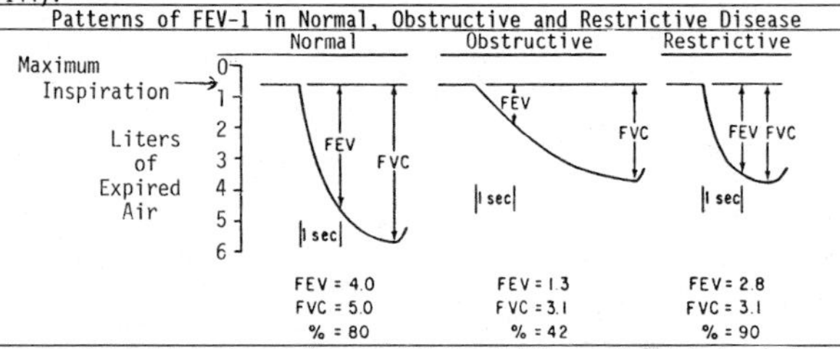

In obstructive airway disease, FEV-1 and FVC are diminished and the ratio of FEV-1 to FVC is diminished. In restrictive airway disease, FEV-1 is somewhat diminished, FVC is diminished but the ratio of FEV-1 to FVC is normal or supranormal.

RATIO: FORCED EXPIRATORY VOLUME IN 1 SECOND(FEV-1): TOTAL FORCED VITAL CAPACITY(FVC)
SPECIMEN: The patient takes the deepest inspiration possible; this is done with the respirometer attached. The inhaled air will not be measured. At the peak of inspiration, the patient is told to exhale as fast and completely as possible.
REFERENCE RANGE: Normally, FEV-1 is between 3 and 5 liters and FVC is between 4 and 6 liters; the ratio (FEV-1) x 100/FVC is usually greater than 75 percent.
METHOD: Expiration into a spirometer.
INTERPRETATION: Forced vital capacity reflects ventilatory reserve.

The degree of risk in obstructive lung disease, as calculated from the ratio, (FEV-1) x 100/FVC, is given in the next Table:

Degree of Risk in Obstructive Lung Disease

Degree of Risk	(FEV-1) x 100/FVC
Normal	>75
Mild	60-75
Moderate	45-60
Severe	35-45
Extreme	<35

Syndromes of Obstructive Lung Disease: Syndromes of obstructive lung disease include asthma, chronic bronchitis, bronchiectasis, and emphysema.
Asthma: Asthma is characterized by recurrent episodes of generalized airway obstruction that abate spontaneously or with treatment. The hallmark is reversible airway obstruction; symptoms result from the impedance of the movement of air into or out of the lungs. The obstruction is due to smooth muscle constriction, mucosal edema, and the secretion and/or lack of clearance of mucus.

Syndromes of Restrictive Lung Disease: Syndromes of restrictive lung disease are listed in the next Table:

Syndromes of Restrictive Lung Disease	
Chest Wall or Space Occupying Processes:	Interstitial and Infiltrative Disorders:
Kyphoscoliosis	Sarcoidosis
Post-Surgery	Environmental Diseases, e.g., Asbestosis, Silicosis, Coal, Graphite, Berylliosis, etc.
Paralysis	Interstitial Pneumonias
Obesity	Connective Tissue Disorders, e.g., Scleroderma, Lupus, Rheumatoid Lung
Ascites	Pulmonary Vascular Disease, e.g., Multiple Emboli and Idiopathic Hypertension
Pleuritis	
Pleural Effusion	

See reviews of pulmonary function testing in adults (Schwartzstein RM, Weinberger SE. Resident and Staff Physician. 1986; 32:43-50) and children (Pattishall EN. Pediatrics. 1990; 85:768-773; Eigen H. Pediatrics in Review. 1986; 7:235-245).

QUINIDINE

SPECIMEN: Red top tube, serum; or lavender (EDTA) top tube, plasma.
REFERENCE RANGE: Therapeutic: 2.0-5.0mcg/mL; Toxic: >5.0mcg/mL; Half Life: 6 hours; Steady State: 24 to 42 hours (4 to 7 half-lives)
METHOD: EMIT, RIA, Fluorometric, GLC
INTERPRETATION: Quinidine is a class IA antiarrhythmic agent (quinidine, procainamide, disopyramide); quinidine has a local anesthetic (sodium channel blockade) effect and prolongs action potential duration. Quinidine depresses the excitability of cardiac muscle, slows the heart rate and is used for prophylaxis and treatment of atrial and ventricular arrhythmias. Use is investigational for treatment of severe malaria (e.g., cerebral malaria caused by Plasmodium falciparum).
Specimens: The first serum specimen should be obtained 24 to 42 hours following initial dose of quinidine and just prior to the next dose (trough). Trough concentration should be above the minimum effective concentration to maintain the desired therapeutic effect (2.0mcg/mL). If it is necessary to determine the half-life of quinidine, collect at least 3 timed specimens 30-60 minutes apart, and plot log concentration versus time. Conditions which alter the half-life of quinidine are listed in the next Table:

Conditions Which Alter Half-Life of Quinidine
Decrease: Phenobarbital, Phenytoin, Nifedipine, Primidone, Rifampin
Increase: End Stage Liver Disease, Severe Heart Failure

The toxicity of numerous drugs is potentiated by quinidine including beta-blockers, tricyclic antidepressants, neuroleptics and antiarrhythmic agents (Caporaso NE, Shaw GL. Arch Intern Med. 1991; 151:1985-1992). Digoxin concentration increases 2-3 fold when quinidine therapy is initiated. Therefore, digoxin dose must be reduced (e.g. by half) and digoxin levels must be monitored carefully. Warfarin is displaced from plasma proteins which results in increased prothrombin time. Neuromuscular blockers are potentiated.

The effects of quinidine can be increased by concurrent use of antacids, amiodarone, carbonic anhydrase inhibitors, and cimetidine. Hypotension can occur with IV verapamil.
Toxicity: Toxicity may occur if serum levels exceed 5mcg/mL. The common toxic effects of quinidine are listed in the next Table:

Common Toxic Effects of Quinidine
Gastrointestinal Disturbances: Nausea, Vomiting, Anorexia, Diarrhea
Cardiovascular: Slow Intraventricular Conduction, Bradycardia, Hypotension, "Quinidine Syncope"
Cinchonism: "Tinnitus, Light-Headedness, Giddiness" etc.; Eight Cranial Nerve Damage

Adverse cardiovascular effects, e.g., vasodilatation and hypotension, are more common with IV than with oral administration. Serious ventricular tacharrhythmias associated with excessive QT prolongation (e.g., Torsade de Pointes) are the likely cause of "quinidine syncope," which is potentiated by hypokalemia. Reversible hepatitis and a lupus-like syndrome have been reported (Cohen MG, et al. Ann Intern Med. 1988; 108:369-371; Knobler H, et al. Arch Intern Med. 1986; 146:526; Lavie CJ, et al. Arch Intern Med. 1985; 145:446). Hypoglycemia and thrombocytopenia may occur.

Quinidine (Cont)

The major route of elimination is hepatic; renal failure does not generally lead to toxicity on the usual doses; however, active metabolites may accumulate and dosage adjustment may be required. Quinidine can cause cinchonism, headache, visual and auditory disturbances.

Quinidine has numerous adverse effects and drug interactions, not all of which are listed here. Physicians should be familiar with these effects and interactions prior to administering this drug.

Quinidine has a high rate of toxicity; in one study in an elderly population (age 62-100 years), 48 percent of patients developed adverse effects requiring discontinuation of therapy (Aronow WS, et al. Am J Cardiol. 1990; 66:423-428). In another study, 50 percent of quinidine treated patients reported side effects (Juul-Möller S, et al. Circulation. 1990; 82:1932-1939).

Meta-Analyses of previous studies have demonstrated an increased mortality with quinidine treatment for maintenance of sinus rhythm after cardioversion (Coplen SE, et al. Circulation. 1990; 82:1106-1116) and for ventricular arrhythmias (Morganroth J, Goin JE. Circulation. 1991; 84:1977-1983). The proarrhythmic effects of quinidine may cause increased mortality in these patients; the risk of long-term treatment with quinidine or other class I antiarrhythmic agents must be weighed against potential benefit (Salerno DM. Circulation. 1991; 84:2196-2198; Feld GK. Circulation. 1990; 82:2248-2250).

RAJI CELL ASSAY

SPECIMEN: Red top vacutainer; allow the blood to clot at 4°C; separate the serum and freeze it immediately. Ship the frozen specimen in dry ice.

REFERENCE RANGE: Varies between laboratories.

Normal = <12 microgram Aggregated Human Gammaglobulin Equivalents/mL
Borderline = 12 to 25 microgram Aggregated Human Gammaglobulin Equivalents/mL
Abnormal = >25 microgram Aggregated Human Gammaglobulin Equivalents/mL

METHOD: The Raji cell is a human lymphoblastoid derived cell line; it has B cell characteristics but no surface immunoglobulins. Raji cells have C-3 receptors and can bind immune complexes that have fixed C-3. This is illustrated in the next Figure:

Binding of Immune Complexes by Raji Cells

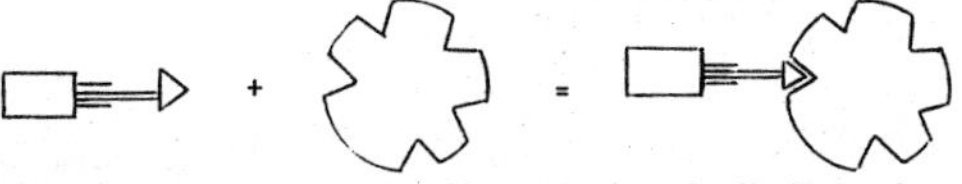

Antigen-Antibody-Complement   Raji Cell   Antigen-Antibody-Complement-Raji Cell

Patient serum is first reacted with Raji cells then I-125 anti-human IgG is added to label bound immune complexes.

INTERPRETATION: Some diseases reported to be associated with circulating immune complexes are given in the Table in IMMUNE COMPLEX PROFILE. Anti-lymphocyte antibodies found in some disease states will interfere with this assay causing false positives. Various assays for immune complexes are reviewed in: McDougal JS, et al. J Lab Clin Med. 1982; 100:705-719.

RAST (SPECIFIC IgE)

SPECIMEN: Red top tube, separate serum
REFERENCE RANGE: Negative
METHOD: RAST = Radioallergosorbent Test. Allergens conjugated to cellulose are used to bind specific IgE antibodies from the patient's sera; after incubation, radiolabeled anti-IgE antibody is added to detect the specific IgE-allergen complexes.
INTERPRETATION: Laboratory allergy tests (RAST) complement skin (or prick) tests in the work-up of patients with possible allergy. It can be particularly useful when skin conditions render skin testing unreliable. The allergens assayed by Dr. Donald Hoffman, Department of Pathology, East Carolina University, are given in the next Table:

Allergens Assayed by RAST Test (Dr. Donald Hoffman)		
Rye Grass	Bee Venom	Milk Casein
Timothy Grass	Yellow Jacket Venom	Milk Lactalbumin
Bermuda Grass	Yellow Hornet Venom	Milk Lactoglobulin
Johnson Grass	White Face Hornet Venom	Egg White
	Paper Wasp Venom	Codfish
	Fire Ant Venom	Shrimp
Alder or Birch Tree	Fire Ant or Honeybee WBE	Abalone
Box Elder Tree	Kissing Bug	Crab or Clam
Elm Tree		Lobster
Oak Tree	Alternaria	Oyster
Olive Tree	Aspergillus	Scallop
Poplar Tree	Chaetomium	Corn
Walnut Tree	Cladosporium	Oat
Mulberry Tree	Dreschlera	Rice
	Monilia	Wheat
	Penicillin	Kidney bean
English Plantain	Many Other Molds	Navy Bean
Pigweed	Available	Pea
Short Ragweed		Peanut
Russian Thistle	Penicilloyl	Soy
Sagebrush	KLH	Almond
Scale	Beef and Pork Insulin	Brazil Nut
Lambsquarter	Proquesterone BSA & BSA	Cashew Nut
		Hazel Nut
	Chocolate	Pecan or Walnut
House Dust and Mite	Coconut	Sesame or Orange
Cat, Dog, Horse Dander	Yeast or Buckwheat	Tomato or Cantaloupe

Specific antibodies are measured in patients with respiratory allergy (asthma, chronic bronchitis, persistent rhinitis, or sinusitis), food allergy, insect-venom allergy and drug hypersensitivity (Hoffman DR. Ann Allergy. 1979; 42:224-230; Hoffman DR, Haddad ZH. J Allergy Clin Immunol. 1974; 54:165-173; Hoffman DR, Haddad ZH. Pediatrics. 1974; 54:151-156).

When choice of allergens is a problem, then the common allergens are assayed. The most common environmental allergens are grass, pollens, house dust mite, cat or dog dander, and mold. In children, food allergens such as egg white or cow's milk are also important. House dust is the most common environmental allergen; it is not a single substance but is a group of substances, the most important of which is the house dust mite.

Total serum IgE levels should not be used as a screening test but should be used to complement specific IgE results obtained by RAST (Thompson and Bird. The Lancet. 1983; 169; Atkinson P, et al. The Lancet. 1983; 706-707).

In the skin test, allergen is applied to the skin by prick or scratch; a positive reaction occurs when allergen binds to specific IgE molecules which are bound to mast cells; this reaction causes release of vasoactive substances and histamine, from granules of mast cells (David J, Rocklin RE. Scientific American Medicine. 1983; 6:1-2).

RED BLOOD CELL COUNT

SPECIMEN: Lavender(EDTA) top tube or microtube containing EDTA. The specimen is stable at room temperature for up to 6 hours and in the refrigerator for up to 24 hours.

REFERENCE RANGE: The normal ranges at different ages are given in the next Table:

Red Blood Cell Count at Different Ages

Age	Red Blood Cell Count(millions/mm^3) Male	Female
Birth	5.0-6.3	5.0-6.3
1 Month	4.7-5.9	4.7-6.0
3 Months	3.8-5.2	3.8-5.2
6 Months	3.8-5.1	3.5-4.9
9 Months	3.7-5.2	3.5-4.9
1 Year	3.5-4.9	3.4-5.0
2 Years	3.5-4.9	3.5-5.0
4 Years	3.7-5.0	3.8-5.2
8 Years	4.0-5.1	3.9-5.1
14 Years	3.9-5.3	3.8-5.2
Adult	4.3-5.9	3.5-5.5

METHOD: Coulter Counter; Ortho; Technicon; and manual techniques.

Spurious Red-Cell Values with the Coulter Counter: Spurious red blood count, hematocrit and MCV are obtained using the Coulter counter when a patient has "cold" hemolytic anemia. Red cells are intermittently agglutinating; the particles being counted are larger than normal single red cells so the MCV's are elevated and the red cell count is decreased. The spurious lowering of the red-cell count is disproportionately greater than the false elevation of MCV, so the hematocrit is falsely depressed (Lawrence C, Zozicky O. N Engl J Med. 1983; 309:925-926).

INTERPRETATION: The causes of decreased red blood count are the anemias which are classified as follows: microcytic hypochromic anemia; normocytic, normochromic anemia, e.g., hemolytic anemia; and macrocytic anemia (see ANEMIA PANEL).

The causes of increased red cell count are polycythemia vera and secondary polycythemia, vigorous exercise and high altitude.

RED CELL INDICES (MCHC, MCV, MCH, RDW)

MCHC = Mean Cell Hemoglobin Concentration;
MCV = Mean Cell Volume;
MCH = Mean Cell Hemoglobin
RDW = RBC Distribution Width

SPECIMEN: Lavender (EDTA) top tube or EDTA microtainer.

REFERENCE RANGE: The value of the indices at different ages are given in the next Table (Guest GM, Brown EW. Am J Dis Child. 1957; 93:486):

Indices at Different Ages

Age	MCV	MCH	MCHC
Newborn	113 ± 0.8	36.9 ± 0.28	32.6 ± 0.18
1 day	110 ± 0.7	36.6 ± 0.25	33.5 ± 0.17
7 days	106 ± 0.6	36.2 ± 0.30	34.3 ± 0.18
20 days	100 ± 0.8	34.4 ± 0.32	34.5 ± 0.25
45 days	94 ± 0.8	32.8 ± 0.29	34.9 ± 0.16
76 days	88 ± 0.6	30.4 ± 0.21	34.6 ± 0.16
122 days	82 ± 0.7	27.7 ± 0.26	33.7 ± 0.17
6 months	78 ± 0.7	26.1 ± 0.28	33.5 ± 0.20
10 months	73 ± 1.0	23.7 ± 0.47	32.4 ± 0.28
12 months	73 ± 1.1	23.5 ± 0.51	32.3 ± 0.32
18 months	72 ± 1.2	23.2 ± 0.60	31.9 ± 0.39
2 years	75 ± 0.7	24.7 ± 0.35	32.6 ± 0.21
3 years	78 ± 0.7	26.6 ± 0.28	34.0 ± 0.20
5 years	80 ± 0.4	27.5 ± 0.15	34.2 ± 0.12
10 years	81 ± 0.6	27.6 ± 0.26	34.3 ± 0.14
14 years	81 ± 1.1	27.9 ± 0.48	34.3 ± 0.27
Adult	76-96	27-32	30-35

MCV decreases from birth to 18 months, and then increases slowly to adult level at ages 4 to 10. MCH changes in a similar fashion while MCHC remains relatively constant throughout life. The RDW is an expression of the size distribution curve obtained from automated instruments and may be given as the coefficient of variation or the standard deviation. Normal range 11.5-14.5; abnormal >15.

METHOD: Using Coulter, MCV, red blood count and hemoglobin are measured directly; other indices are calculated using formulas below; the hematocrit is calculated. Alternatively, these indices could be calculated from manually performed Hct, red blood count, and hemoglobin.

INTERPRETATION: The red cell indices are useful in the evaluation of anemias, polycythemia, and other hematological and nutritional disorders.

MCV (Mean Cell Volume): The MCV is very useful in the initial classification of anemias (macrocytic anemia, microcytic anemia).

The MCV is given by the formula:

$$MCV = \frac{\text{Hematocrit (\%)}}{\text{Red Cell Count (millions/mm}^3\text{)}} \text{ (femtoliters)}$$

The causes of increased MCV are given in the next Table:

Causes of Increased MCV
Megaloblastic Anemia
B-12 Deficiency
Folate Deficiency
Spurious Macrocytic Anemia
Aplasia
Myelofibrosis
Hyperglycemia
Cold Agglutinins
Reticulocytosis
Liver Disease
Drugs
AIDS (Zidovudine-Treated)

Megaloblastic anemias may result from either B-12 or folate deficiencies. In these conditions, the MCV may be abnormal before anemia develops. An increased MCV may be occasionally associated with liver disease. In one series of 146 children with an elevated MCV folate or B-12 deficiency was not found. The most common cause (35%) was anticonvulsant drugs. Other causes included heart disease, Down's syndrome, and miscellaneous hematological disorders (Pappo AS, et al. Pediatrics. 1992; 89:1063-1067). The MCV may be artifactually elevated by cold agglutinins and in hyperglycemia. Patients with AIDS and treated with zidovudine may develop an elevated MCV (Snower DP, Weill SC. Am J Clin Path. 1993; 99:57-60). This appears to be a common cause of an increased MCV.

The causes of decreased MCV are given in the next Table:

Causes of Decreased MCV
Iron Deficiency Anemia
Defects in Porphyrin Synthesis
Hereditary Sideroblastic Anemia
Acquired Sideroblastic Anemia
Pyridoxine-Responsive Anemia
Lead Poisoning
Hemolytic Anemia, especially Thalassemia Minor
Marked RBC Fragmentation
Hereditary Spherocytosis
Postsplenectomy

Thalassemia minor and iron deficiency are the major hematologic abnormalities that are associated with a marked decrease in the MCV. Patients with either iron deficiency or beta-thalassemia trait may be differentiated using serum ferritin and MCV (Hershko C. Acta Haematol. 1979; 62:236-239).

Red Cell Indices (Cont)

MCHC (Mean Cell Hemoglobin Concentration): The MCHC is of limited value in the evaluation of hematological disorders.

MCHC (Mean Cell Hemoglobin Concentration) is calculated from the formula:

$$\text{MCHC} = \frac{\text{Hemoglobin (g/dL)}}{\text{Hematocrit (\%)}} \text{ (grams/dL of RBCs)}$$

The causes of changes in the MCHC are given in the next Table:

Changes in the MCHC
Decreased:
Iron Deficiency Anemia
Defects in Porphyrin Synthesis
Hereditary Sideroblastic Anemia
Acquired Sideroblastic Anemia
Pyridoxine-Responsive Anemia
Lead Poisoning
Hemolytic Anemia, especially Thalassemia Minor
Increased:
Hereditary Spherocytosis

A low MCHC was associated with iron deficiency anemia when the RBC indices were performed manually. Using automated instruments a low MCHC is rarely seen except in severe iron deficiency. The originally derived low MCHC in iron deficiency anemia was caused by over estimation of the hematocrit due to increased plasma trapping during centrifugation (Fischer SL, et al. Arch Intern Med. 1983; 143:282-283).

MCH (Mean Cell Hemoglobin): The MCH is of little use clinically. Values of the MCH parallel those of the MCV.

The MCH is calculated from the formula:

$$\text{MCH} = \frac{\text{Hemoglobin (g/liter)}}{\text{Red Cell Count (millions/mm}^3\text{)}} \text{ (picograms)}$$

RDW (Red Blood Cell Distribution Width): This index is a measure of the variability of MCV distributions. It has been proposed to be used in the classification of anemias (Bessman JD, et al. Am J Clin Path. 1983; 80:322-326). The RDW is increased in conditions in which there is a large variation in RBC size. These are listed in the next Table:

Conditions Associated with an Increased RDW
Iron Deficiency Anemia
Folate Deficiency Anemia
B-12 Deficiency Anemia
Hemolytic Anemia
Sideroblastic Anemia
Transfusion
RBC Fragmentation
Hemoglobin H
Alcohol Abuse

The RDW appears to be the earliest hematologic manifestation of iron deficiency (Patton WN, et al. Clin Lab Haematol. 1991; 13:153-161). Heterozygous α or ß thalassemia trait or lead poisoning give normal RDW values (Oski FA. N Engl J Med. 1993; 329:190-193).

The RDW may be useful in distinguishing iron deficiency anemia from ß-thalassemia and anemia of chronic disease, the latter conditions associated with a normal RDW. Serum ferritin may also be useful in distinguishing between these conditions. Some cases of thalassemia may have only a slight increase in RDW. Alcohol abuse is associated with an increased RDW in approximately 40% of alcoholic men studied (Bessman JD, McClure S. JAMA. 1992; 267:1071). This appears to be independent of liver disease (Seppa K, Sillanankee P. JAMA. 1992; 268:1413).

RED CELL SURVIVAL

SPECIMEN: The first blood specimen is used to tag the hemoglobin of the red blood cells with ^{51}Cr. Then following incubation of the red blood cells with ^{51}Cr, the tagged red blood cells are injected and blood specimens are obtained at 24 hours and every two to three days for 30 days and counted. Specific instructions are available from the laboratory.
REFERENCE RANGE: 25-35 days. In this technique, cells are "randomly" labeled; thus the cells are not of a uniform age group but are erythrocytes of all ages representing the mixture present in the circulation. Thus, the red cell survival is an "average" survival and does not measure the actual longevity of the red cell.
METHOD: ^{51}Cr has a half-life of 27.8 days; the mode of decay is EC, gamma. The gamma ray energy is 322,000 electron volts. In comparison, the energies of other gamma emitters are as follows: ^{125}I, 35,000 e.v.; ^{131}I, 364,000 e.v.
INTERPRETATION: The red cell survival test is indicated when there is an obscure cause for anemia and in the evaluation of therapy for hemolytic anemia. It may be useful to document sequestration of RBC in the spleen prior to treatment by splenectomy in a number of conditions. The causes of shortened red cell survival are given in the next Table:

Causes of Shortened Red Cell Survival
Hemolysis
Sequestration of Red Cells by Spleen
Blood Loss

The causes of prolonged survival are prior splenectomy or an abnormality of red cell production.

The ^{51}Cr-labeled red blood cell survival test has been used in transfusion medicine to evaluate in-vivo survival of a small sample of donor red blood cells in patients with blood compatibility problems (Pineda AA, et al. Mayo Clin Proc. 1984; 59:25-30).

REDUCING SUBSTANCES

SPECIMEN: Urine; Stool specimen
REFERENCE RANGE: Negative
METHOD: Clinitest tablets (Ames) are used to detect reducing substances in urine or stool specimens. There are two methods used for urine testing - the 2-Drop and the 5-Drop methods. In the 2-Drop method, 2 drops of urine are placed in a test tube followed by 10 drops of water. One Clinitest tablet is added to the test tube; the sample will begin to boil. Wait 15 seconds, shake gently to mix, then compare the color of liquid to the color chart provided. Color range is blue to orange (negative, trace, ½, 1, 2, 3 or 5 percent). In the 5-Drop method, 5 drops of urine are used, but the rest of the technique is the same. Test result range is negative, ¼, ½, ¾, 1, or 2 percent. Stool is tested by using 5 drops of liquid stool or approximately the same volume of solid stool.

Clinitest tablets are a combination of cupric sulfate, citric acid, sodium carbonate, and anhydrous sodium hydroxide. Heat is produced by two exothermic reactions: addition of liquid to sodium hydroxide and reaction between sodium hydroxide and citric acid. Reducing substances in the urine reduce the cupric ions (blue color) to cuprous oxide (orange color).
INTERPRETATION: This test provides information about carbohydrate metabolism. Reducing sugars include glucose, lactose, fructose, galactose, and pentoses. Dipstick methods use the glucose oxidase reaction and are specific for glucose. If Clinitest is positive and dipstick is negative, the reducing substance is not glucose. Glucose may be present in the urine of diabetics, as well as in patients with renal disease. Non-glucose reducing sugars may be found in patients with hereditary fructose intolerance, galactosuria, and other conditions such as ingestion of large amounts of fruit, avocado, or cane sugar, and in lactating mothers. Numerous substances found in the urine also result in a positive test such as ascorbic acid, nalidixic acid, cephalosporins, and probenecid.

Stool may be positive for reducing substances when there is incomplete absorption of dietary sugars. A positive result must be interpreted in the context of the patient's underlying condition and diet.

Reference: Sonnenwirth AC, Jaret L. "Gradwohl's Clinical Laboratory Methods and Diagnosis," 8th Ed. St. Louis, CV Mosby Company, 1980.

RENAL FAILURE, ACUTE, PANEL

In acute renal failure, there is an abrupt decline in renal function with retention of creatinine and urea.

The tests that are useful to differentiate pre-renal, renal and post-renal are listed in the next Table:

Tests to Differentiate Pre-Renal, Renal and Post-Renal Oliguria
Examination of Urinary Sediment
Serum BUN/Serum Creatinine
Urine Specific Gravity
Urine Osmolality
Urine to Plasma Ratio for:
Osmolality
Creatinine
Urea
Urine Sodium
Excretion Fraction of Filtered Sodium

The test results that may be useful for differentiating of pre-renal, renal parenchymal and post-renal states are given in the next Table (Finn WF. Med Clin N Amer. 1990; 74:873-891; Nissenson AR. Hosp Med. Sept. 1979; 22-43; Schrier RW. Hosp Pract. March 1981; 93-112; Goldstein M. Med Clinics of N Amer. 1983; 67:1325-1344):

Differentiation of Pre-Renal, Renal Parenchymal and Post-Renal States

Test	Pre-Renal	Renal	Post-Renal
Examination of Urinary Sediment	Bland	Renal Tubular Cells; Granular, Hyaline and Cellular Casts; Occ. RBC's and RBC Casts	Bland
Serum BUN/Serum Creatinine	>10/1	10/1	5/1
Urine Specific Gravity	High	Low	Low
Urine Osmolality	>500 mOsm/kg	<350 mOsm/kg(ATN)	Non-Diagnostic
Urine/Plasma Osmolality	>1.1	<1.1	Non-Diagnostic
Urine Creatinine/ Plasma Creatinine	>40	<20(ATN) >40(GN)	<20
Urine Urea/ Plasma Urea	>8	<3(ATN) >8(GN)	<3
Urine Sodium	<20 mmol/L	>30-40 mmol/L	Non-Diagnostic
Excretion Fraction of Filtered Sodium	<1%	>2-3%(ATN) <1%(GN)	>2%

ATN = Acute Tubular Necrosis; GN = Glomerulonephritis

Examination of Urinary Sediment: The most important finding is the presence of casts. Erythrocyte casts are strongly suggestive of acute glomerulonephritis; granular casts indicate acute glomerulonephritis or acute tubular necrosis.

Pre-Renal: There is hypoperfusion of the kidney and avid reabsorption of sodium and water by the nephrons. The most important diagnostic criterion of pre-renal uremia is response to treatment (Editorial, Brit Med J. June 1980; 1333-1335). As body fluids deficits are made up, the flow of urine increases and the plasma concentration of urea will fall.

Ultrasound is useful to detect post-renal obstruction; if ultrasound is not available, intravenous pyelograph or retrograde pyelography may be considered. However, these procedures carry risk; intravenous pyelography-contrast induced acute tubular necrosis if serum creatinine exceeds 4 mg/dL; retrograde pyelography-general anaesthesia and should not be done in the presence of urinary tract sepsis.

See ACUTE GLOMERULONEPHRITIS PANEL for tests for glomerular diseases.

RENAL STONE ANALYSIS

SPECIMEN: Specimen should be washed free of tissue and blood and submitted in a clean, dry container.

METHOD: The stone is sectioned and examined under a dissecting microscope. Analysis is done on identifiable layers. Qualitative chemical analysis may be done for calcium, magnesium, oxalate, phosphate, urates and cystine; anions may be analyzed using spectroscopic procedures and cations by atomic absorption. Other methods of analysis include crystallographic analysis using a polarizing microscope, X-ray diffraction and infrared spectroscopy.

INTERPRETATION: Renal stones (calculi) are analyzed in order to gain an insight into the etiology of the stone formation and to guide subsequent therapy. It is important to differentiate the various stone compositions. There are four main types of stones; these are calcium, struvite, uric acid and cystine. Two-thirds of all renal stones are composed of either calcium oxalate or calcium oxalate mixed with calcium phosphate. A protein called matrix substance A has been found in all types of renal stones. Struvite stones are most likely caused by infection; the most common cause is urea-splitting bacteria, Proteus mirabilis. If the stone can be completely removed and infection stopped, the patient will have a very low rate of recurrence (Silverman DE, Stamey TA. Medicine. 1983; 62:44). In a pediatric population, stones of infection, e.g., calcium phosphate and magnesium ammonium phosphate, are the most common type (Walther PC, et al. Pediatrics. 1980; 65: 1068-1072).

The conditions that are associated with renal stone formation are listed in the next Table (The incidence of the various types are given in parenthesis):

Conditions Associated with Renal Stone Formation

Type	Causes
Calcium Containing Stones (70%)	Cause Unknown Hypercalciuric (see CALCIUM EXCRETION TEST) Idiopathic (Normocalcemic) Hypercalcemic (see SERUM CALCIUM)
Magnesium Ammonium Phosphate(20%)	Recurrent Urinary Tract Infections
Uric Acid (5%)	Hyperuricosuric States: Gout Secondary to Lymphoproliferative and Myeloproliferative Disorders Persistently Acid Urine: Chronic Diarrhea; Ileostomy
Cystine (3%)	Renal Tubular Defects of Cystine, Lysine Arginine and Ornithine
Oxalate (rare)	Hyperoxaluric States (Rare) Increased Ingestion Pyridoxine (Vitamin B-6) Deficiency Primary Hyperoxaluria Methoxyflurane Anaesthesia Secondary to Ileal Disease

Calcium Stones: Most calcium stone formers have normal serum calcium levels, (see CALCIUM EXCRETION TEST). Hypercalcemic conditions associated with recurrent stone formation are as follows: hyperparathyroidism, vitamin D intoxication, hyperthyroidism, sarcoidosis and distal hereditary renal tubule acidosis. In renal tubular acidosis, distal tubular type, there is an apparent defect in calcium transport allowing excessive calcium to leak into the urine.

Magnesium Ammonium Phosphate: There may be increased ammonia production in patients with chronic urinary tract infections, in particular, urea splitting organisms such as Proteus.

Uric Acid: Increased uric acid excretion occurs in gout and in malignancies associated with increased cell turnover such as occurs in lymphoproliferative and myeloproliferative disorders, and during chemotherapy.

Cystine: Excessive excretion of cystine may occur secondary to faulty renal tubular reabsorption. Cystinuria is the most common of the hereditary amino-acidurias.

Oxalate: Increased oxalate excretion occurs in the conditions listed in the previous Table. Patients with jejunoileal bypass, ileal disease and ileal resection have an enhanced absorption of dietary oxalate via the colon. Increased fatty acids secondary to malabsorption results in increased complex formation with calcium ions. Decreased free calcium ions within the lumen of the gut results in oxalate combining into more soluble sodium oxalate; concomitantly, there is the presence of unabsorbed bile salts leading to increased colonic permeability of oxalate and hyperoxaluria (Doffins JW, Binder HJ. N Engl J Med. 1977; 296:298-301).

The initial laboratory studies in the workup of patients with renal calculi are listed in the next Table:

Initial Laboratory Studies in Patients with Renal Calculi
Urinalysis - Look for Crystals
Urine Culture if there is Evidence of Infection on Urinalysis
Blood Studies: Serum Calcium, Serum Phosphate and Serum Uric Acid
Plain Film of the Abdomen

Urinalysis: Urinalysis is particularly important. There are three ways in which urinalysis may be of assistance. Alkaline urine suggests the syndrome of renal tubular acidosis or urinary tract infection with urea-splitting organisms; evidence of chronic infection in the urinary tract suggests triple phosphate stones; crystalluria may help identify the cause of stone disease, e.g., cystine or uric acid crystals. However, calcium oxalate crystals may be found in the urine of patients who have no history of stone formation. The appearance of crystals in the urine is illustrated in the next Figure:

Crystals in the Urine

"Envelope" Oxalate

"Coffin Lids" Magnesium Ammonium Phosphate

"Thorn Apple" Ammonium Biurate

"Hexagonal Plates" Cystine

"Flat-Notched Plates" Cholesterol

Urine Culture: See URINE CULTURE

As already noted, magnesium ammonium phosphate stones are associated with recurrent urinary tract infections with urea-splitting organisms, especially Proteus. A major complication of stone formation in the urinary tract is urinary tract infection.

Blood Studies: Serum calcium and phosphorus are done to detect hypercalcemia; serum uric acid is done to detect hyperuricemia as in gout.

Plain Film of the Abdomen: Diagnosis may be made by X-ray if the stones are radiopaque; radiopaque stones are composed of calcium and magnesium salts and comprise 80% to 95% of stones. The most common radiolucent stone is composed of uric acid.

Review: See DeVita MV, Zabetakis PM. Laboratory Investigation of Renal Stone Disease. Clin Lab Med. 1993; 13:225-234.

RENIN ACTIVITY, PLASMA

SPECIMEN: Record information on the status of the patient on the requisition form as follows: Patient fasting, or regular diet, or salt restricted diet, and/or diuretics; supine or upright, patient ambulated for ___ hours specimen obtained following catheterization of _______ vein.

The following schedule may be used in evaluating a patient in an office of outpatient setting. Discontinue all medication for one to four weeks before the test. Patient remains in an upright position for 3 hours prior to test. Blood specimens for renin and aldosterone are obtained. Draw blood into a chilled syringe; place blood into a chilled purple(EDTA) top tube; keep specimen on ice after collection; transport immediately to the laboratory. In the laboratory, keep specimen on ice or refrigerated until plasma is separated; centrifuge specimen and collect plasma. Freeze plasma immediately.

Plasma renin activity testing has not been fully standardized either from a clinical or laboratory standpoint. Factors contributing to the variability include the effects of salt intake, posture, drugs, and measurement technique (Davidson RA, Wilcox CS. JAMA. 1992; 268:3353-3358). Some authors do not recommend obtaining and processing samples in chilled tubes due to concern about cryoactivation of prerenin leading to artifactually elevated renin levels (Alderman MH, et al. N Engl J Med. 1991; 324:1098-1104; Sealey JE, et al. Endocr Rev. 1980; 1:365-391). The captopril test in which patients have plasma renin activity measured before and after oral captopril, may increase the sensitivity of screening of high-risk patients for renovascular hypertension (Davidson RA, Wilcox CS, already cited).

REFERENCE RANGE: Normal plasma renin levels, in different ages and conditions are shown in the next Table (Mayo Medical Laboratories):

Normal Plasma Renin Levels in ng/ml/hour

Age (years)	Conditions	Plasma Renin Activity (ng/mL/hour)
0-2		4.6
3-5		2.5
6-8		1.4
9-11		1.9
12-17		1.8
18-39	Sodium Depleted, Upright	10.8 (2.9-24.0)
	Sodium Repleted, Upright	1.9 (≤0.6-4.3)
≥40	Sodium Depleted, Upright	5.9 (2.9-10.8)
	Sodium Repleted, Upright	1.0 (<0.6-3.0)

To convert conventional units in ng/mL/hr to international units in ng/liter/sec., multiply conventional units by 0.2778.

Salt restricted diets and/or diuretics increase supine and upright renin levels.

METHOD: Plasma renin activity(PRA) is determined by measuring the rate of angiotensin I generated by plasma renin without addition of exogenous renin or angiotensin. The reaction is as follows:

$$\text{Angiotensinogen} \xrightarrow{\text{Renin}} \text{Angiotensin I}$$

Angiotensin I is assayed by radioimmunoassay(RIA).

This method for measuring PRA depends not only on the concentration of renin but also on the concentration of substrate in the specimen.

INTERPRETATION: Renin levels are determined in the differential diagnosis of hypertension. Renin is an enzyme which is synthesized in the juxta-glomerular cells lining the afferent arterioles of glomeruli of the kidney. Renin is released by change in pressure or change in stretch of the afferent arteriole. The action of renin is illustrated in the next Figure:

Action of Renin

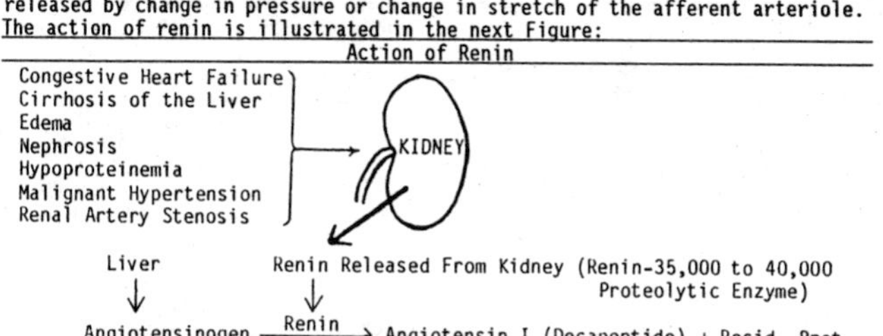

LUNG

$$\underset{\text{(Decapeptide)}}{\text{Angiotensin I}} \xrightarrow[\text{Lung}]{\text{Converting Enzyme}} \underset{\text{(Octapeptide)}}{\text{Angiotensin II}} + \text{2 Amino Acids}$$

Renin Activity, Plasma (Cont)

Angiotensin is a potent vasoconstrictor and stimulates the release of aldosterone from the adrenal cortex.

The relative levels in serum and urine of renin and aldosterone in patients with primary and secondary hyperaldosteronism are shown in the next Table:

Aldosterone and Renin Levels in Primary and Secondary Aldosteronism

Aldosterone	Aldosterone Level	Renin Level
Primary (60% Adenoma)	↑	↓
Secondary	↑	↑

The combination of elevated aldosterone and low plasma renin activity is practically pathognomonic of primary aldosteronism. See ALDOSTERONE, SERUM. Primary aldosteronism is usually associated with a serum potassium <3.5mmol/liter.

Plasma renin may be low, normal or elevated in essential hypertension. Plasma renin is increased with salt restricted diets, upright position, pregnancy, oral contraceptives, salt loss, Addison's disease, diuretics, and some renal diseases.

High plasma renin profile (high plasma renin and high urinary sodium) in patients with mild to moderate hypertension is associated with risk of myocardial infarction (Alderman MH, already cited; editorial: Dzau VJ. N Engl J Med. 1991; 324:1128-1130).

RESPIRATORY SYNCYTIAL VIRUS (RSV) TESTING

SPECIMEN: Epithelial cells from Nasopharynx, Bronchoalveolar Lavage Fluid, or Lung Tissue. Nasopharyngeal specimens may be obtained by swabbing or by washing; either method is used to obtain epithelial cells from which RSV is identified.
REFERENCE RANGE: Negative
METHOD: Antigen Detection: Direct Fluorescent Antibody (DFA) Staining or Enzyme-Linked Immunosorbent Assay (ELISA). Culture: Cell culture in HEp-2, monkey kidney, human neonatal kidney, human lung fibroblast, or other cell line.
INTERPRETATION: RSV is a major respiratory pathogen in infants; most cases of RSV are manifested as a common cold, but clinical presentation includes pneumonia, bronchiolitis, reactive airways disease, respiratory distress, cyanosis and apnea. RSV occurs as annual epidemics during winter and early spring. Significant morbidity and mortality may occur in normal infants, but risk is higher for infants with underlying conditions such as prematurity, congenital heart disease, underlying lung disease and immunocompromised patients (e.g. AIDS, chemotherapy patients, transplant immunosuppression). Rapid diagnosis of RSV is important because RSV is highly contagious; cohorting and contact isolation may be considered. Furthermore, antiviral therapy with ribavirin may be considered in some patients. Review of RSV infection: LaVia WV, et al. J Pediatr. 1992; 121:503-510.
Antigen Detection: Several methods are used to detect RSV antigens including direct and indirect fluorescent antibody techniques and enzyme immunoassays.

Direct and indirect fluorescent antibody techniques, in multiple studies, have a sensitivity of 73 to 95 percent and specificity of 69 to 99 percent; in general, sensitivity and specificity are greater than 85-90 percent.

Enzyme immunoassays consist of RSV antibodies attached to a solid phase which capture RSV antigens. An anti-RSV specific antibody attached to a detector system is used to quantitate the amount of RSV antigen. In multiple studies, sensitivity ranged from 37 to 100 percent and specificity ranged from 80 to 100 percent; in general, sensitivity is 75 to 95 percent and specificity is 85 to 100 percent (Woodin KA, Menegus MA. Pediatr Virol. 1991; 6(#1)1:5-8).

Antigen detection depends on many factors. An adequate sample should contain moderate to large numbers of exfoliated respiratory tract columnar epithelial cells. These tests perform best when RSV is abundant - early in the infection, in more severe infections, and in infants less than 6 months.
Culture: Culture is an imperfect "gold standard" for RSV infection. RSV is one of the most labile human viruses - it is very sensitive to pH, temperature, and repeated episodes of freeze-thawing. There is a 90% reduction in viral titer within 2 hours of collection. Specimen should be stored at 4°C for the first 3 to 5 hours and at -70°C thereafter. Sensitivity may be decreased by bacterial overgrowth, loss of cells, or by inhibition of the RSV-induced cytopathic effects by other viruses (e.g. enterovirus). The sensitivity of culture compared to rapid detection methods ranges from 57 to 92 percent. Culture does allow identification of other viruses with which the patient may be infected (Kellog JA. Arch Pathol Lab Med. 1991; 115:451-458).

RETICULOCYTE COUNT

SPECIMEN: Whole Blood: Lavender (EDTA) top tube.
REFERENCE RANGE: 0.5-1.5% in adults and children; 2-6% in full-term infants falling to adult level at 2 weeks. The normal range may be higher in menstruating women.
METHOD: Use mixture of blood plus reticulocyte stain such as New Methylene Blue N. The reticulocyte count is usually expressed as a percentage of the erythrocyte count. Fluorescent RNA dye and flow cytometry.
INTERPRETATION: The reticulocyte count is an index of erythropoiesis and is increased in the following conditions:

Increased Reticulocyte Count
Hemolytic Anemia and Hemolysis
Acute Blood Loss
Response of Anemia to Specific Therapy; e.g., Iron, Vitamin B_{12}, Folic Acid
Infiltrative Marrow Disorders

Following blood loss there is a doubling of the reticulocyte count in the first 24 hours due to premature release of "stress" reticulocytes. After about 7-10 days, reticulocyte counts should peak as newly produced reticulocytes enter the circulation. Levels should be about 3 times basal levels.

In infiltrative marrow disorders, immature red blood cells may "escape" into the circulation from the marrow and from sites of production in the sinusoids of the liver and spleen.

The reticulocyte count is decreased in the following conditions:

Decreased Reticulocyte Count
Marrow Aplasia
Replacement of Erythroid Precursors by Leukemic Cells
Infections
Toxin
Drugs: (more than 100 Drugs)
Renal Disease
Disorders of Maturation:
Iron Deficiency Anemia
Megaloblastic Anemia
Sideroblastic Anemia
Anemia of Chronic Disease

Correction of Reticulocyte Counts: Reticulocyte counts may be expressed as an absolute number (reference range 10-75,000/microliter). More commonly the % reticulocyte may be "corrected" for two effects: (1) the low Hct seen in many conditions; (2) the early release of large "young" or "stress" reticulocyte.

(1) The reticulocyte count, corrected for the patient's abnormal hematocrit only, is as follows:

$$\text{CORRECTED RETICULOCYTE COUNT} = \text{Reticulocytes (\%)} \times \frac{\text{Patients Hematocrit}}{45}$$

(2) In a further approximation, the reticulocyte count is "corrected" for the premature release of young reticulocytes. Under normal conditions, 1% of the reticulocytes are released daily into the peripheral blood; these cells remain in the circulation for one day before they lose their RNA reticulum and are converted into young erythrocytes. If red blood cell production in the bone marrow is increased (hypoxia or erythropoietin stimulation), premature delivery of large young reticulocytes occurs; these young reticulocytes survive for about two days, or twice as long as normal cells. A correction factor of 2 is applied to account for this doubling of survival time.

The Reticulocyte Production Index is used in order to correct the reticulocyte count, for both lower hematocrit and young "stress" reticulocytes. The following formula (Finch) is used:

$$\text{RETICULOCYTE PRODUCTION INDEX} = \frac{\text{Reticulocytes (\%)} \times \frac{\text{Patient's Hematocrit}}{45}}{\text{2 (Reticulocyte Maturation Time)}}$$

RETINOL-BINDING PROTEIN(RBP)

SPECIMEN: Red top tube, separate serum; plasma. Stress causes changes in RBP.
REFERENCE RANGE: Adult: 3-6mg/dL; Children: 1.5-3.0mg/dL; sharp increase in RBP at puberty.
METHOD: RIA; Nephelometry
INTERPRETATION: Retinol-binding protein measurements may be useful in the evaluation of nutritional states (vitamin A deficiency, malabsorption) and liver disease. Retinol-binding protein(RBP) is synthesized and stored in the liver; it circulates in the plasma complexed to thyroxine-binding pre-albumin. Vitamin A is transported by the RBP from its storage site in the liver to peripheral organs. The plasma concentration of RBP is regulated by vitamin A status; in vitamin A deficiency, RBP molecules are not secreted from the liver. The causes of decreased level of serum RBP are given in the next Table (Rask L, et al. Scand J Clin Lab Invest. 1980; 40:45-61):

Causes of Decreased Level of Serum RBP
Vitamin A Deficiency
Dietary Deficiency of Vitamin A
Fat Malabsorption Syndromes
(a) Deficiency of Pancreatic Digestive Enzymes
(b) Impairment of Intestinal Absorption
(c) Deficiency of Bile
Insufficient Synthesis of RBP
Chronic Liver Diseases
Cystic Fibrosis of the Pancreas
Protein-Energy Malnutrition
Hyperthyroidism
Infection

Vitamin A Deficiency: Vitamin A is a fat soluble vitamin and it decreases in dietary deficiency and the fat malabsorption syndromes. In vitamin A deficiency, RBP molecules are not secreted by the liver.
Insufficient Synthesis of RBP: Since RBP is synthesized in the liver, conditions involving the liver, such as chronic liver disease, cirrhosis, acute viral hepatitis, chronic active hepatitis cause decrease in this protein. RBP is increased in alcoholic fatty liver, alcoholic hepatitis, and chronic renal disease. Cystic fibrosis is associated with fat malabsorption and thus vitamin A deficiency and decrease in RBP occurs; in addition, cystic fibrosis is associated with insufficient synthesis of the carrier protein.

In protein-energy malnutrition, the liver may contain significant amounts of vitamin A but due to a diminished protein synthesis, an insufficient amount of RBP molecules are synthesized and symptoms of vitamin A deficiency occur in peripheral tissues. The concentrations of RBP and pre-albumin in serum are significantly reduced in protein-energy malnutrition (Rask L, et al. Scand J Clin Lab Invest. 1980; 40:45-61).

REYE'S SYNDROME

The abnormal laboratory values in Reye's Syndrome are listed in the next Table:

Abnormal Values in Reye's Syndrome
Elevated Transaminases (ALT, AST)
Hyperammonemia
Hypophosphatemia
Microvesicular lipid in hepatocytes

Hypoglycemia tends to be observed more often in patients less than 2 years of age. The definitive test for Reye's syndrome is liver biopsy; however, liver biopsy is seldom necessary because history of viral infection and laboratory values are sufficient for diagnosis. Reye's syndrome is associated with treatment of childhood fever, in particular viral illness with aspirin. The incidence of Reye's syndrome decreased dramatically following the decreased aspirin use in children.

RHABDOMYOLYSIS, PANEL

A screen for rhabdomyolysis is shown in the next Table (Liu ET, et al. Arch Intern Med. 1983; 143:154-157):

Screen for Rhabdomyolysis
Muscle Enzymes:
Creatine Phosphokinase (CPK)
Lactate Dehydrogenase (LDH)
Aspartate Aminotransferase (AST)
Hypocalcemia
Serum Myoglobin
Urinary Myoglobin

Creatine phosphokinase elevation to five times normal in the absence of cardiac or brain injury is the single most sensitive diagnostic test for rhabdomyolysis. Urinary myoglobin may be absent when serum myoglobin is markedly elevated.

Myoglobinuria depends on total amount of myoglobin released from muscle, the serum myoglobin concentration, the glomerular filtration rate, and the urine output.

In one study, 26% of patients with rhabdomyolysis documented by CPK elevation had a negative urine dipstick for myoglobin within 24 to 48 hours (Gabow PA, et al. Medicine. 1983; 61:141-152).

Rhabdomyolysis is associated with many conditions including crush injury to a limb, alcoholism, drugs (crack-cocaine, lovastatin-gemfibrozil combination, others) overuse of skeletal muscle, seizures, burns, sepsis, viral infections, myopathies, heat, hypokalemia, neuroleptic-malignant syndrome, and others. The management of traumatic rhabdomyolysis and the pathogenesis of crush syndrome have been recently reviewed (Better OS, Stein JH. N Engl J Med. 1990; 322:825-829; Odeh M. N Engl J Med. 1991; 324:1417-1422). Acute renal failure is reported to occur in 16.5% of patients (Ward MM. Arch Intern Med. 1988; 148:1553-1557).

RHEUMATOID FACTOR (RF, RA)

SPECIMEN: Red top tube, separate serum (No anticoagulant).
REFERENCE RANGE: Screen: Negative; Titer: Negative (< 1:20). Some normal aged individuals may have RF present.
METHOD: Latex agglutination of latex beads coated with heat denatured IgG
INTERPRETATION: Rheumatoid factor (RF) is measured in the diagnosis and evaluation of rheumatoid arthritis and other connective tissue diseases. If RF is present, the levels should be titered. Titers may be useful in assessing disease severity and the response to therapy. RF is an immunoglobulin (usually an IgM) present in the serum of patients with rheumatoid arthritis (75%) and some other conditions, especially connective tissue diseases. It reacts or binds to the Fc fragment of IgG.

Although RF may be present in a wide variety of conditions, the titer of RF is usually significantly higher in rheumatoid arthritis as compared to these other conditions. Conditions, other than rheumatoid arthritis, in which RF may be elevated, are listed in the next Table:

Conditions Associated with Elevated Rheumatoid Factor (RF)
"Collagen"-Vascular Diseases
Felty's Syndrome
Sjögren's Syndrome
Systemic Lupus Erythematosus
Scleroderma
Polyarteritis Nodosa
Infectious Conditions: Kala-azar; Leprosy; Tuberculosis; Syphilis; Bacterial Endocarditis; Viral Hepatitis; Chronic Hepatic Diseases
Myocardial Infarction
Renal Disease
Malignancy
Thyroid Diseases
Increased Age (Reference Range Increases)

(cont)

Rheumatoid Factor (Cont)

Typically, the rheumatoid factor is absent when the clinical signs and symptoms of rheumatoid arthritis first appear. Patients in which rheumatoid factor is demonstratable early in the course of rheumatoid arthritis have a greater risk of developing articular destruction and having sustained disabling disease. RF is typically absent in juvenile arthritis, ankylosing spondylitis, psoriatic arthritis, enteropathic arthritis, and Reiter's syndrome.

RF measurements have no value as a screening test. There is a high number of false positive tests, especially among the elderly (Shmerling RH, Delbanco TL. Arch Intern Med. 1992; 152:2417-2420; Am J Med. 1991; 91:528-534).

ROCKY MOUNTAIN SPOTTED FEVER(RMSF) ANTIBODIES

SPECIMEN: Red top tube, separate serum
REFERENCE RANGE: A fourfold increase in antibody titer between acute- and convalescent-phase serum specimens by indirect fluorescent antibody(IFA), indirect hemagglutination(IHA), latex agglutination, or microagglutination; or a single convalescent titer or 1:64 or higher(IFA) in a clinically compatible case. The sensitivity and specificity of tests used in the serologic diagnosis of RMSF are given in the following Table (Walter DH. Clin Microbiol Rev. 1989; 2:227-240):

	Serologic Diagnosis of RMSF			
Test	Diagnostic	Probable	Sensitivity	Specificity
IFA	4X rise or ≥ 64		94-100	100
IHA	4X rise	≥ 128	91-100	99
Latex Agglutination	4X rise	≥ 128	71-94	96-99
Complement Fixation*	4X rise or ≥ 16		0-63	100
Weil Felix				
P. Vulgaris OX-19		4X rise or ≥ 320	70	78
P. Vulgaris OX-2		4X rise or > 320	47	96

*Complement fixation testing is no longer available due to insufficient sensitivity. The CDC has ceased supplying the antigen to laboratories.
METHOD: Indirect fluorescent antibody(IFA); indirect hemagglutination; latex agglutination; or microagglutination. The Weil-Felix reaction, agglutination with Proteus vulgaris antigen, OX-2 or OX-19, give false negative and false positive reactions and is not reliable.
INTERPRETATION: Rocky Mountain Spotted Fever is caused by Rickettsia rickettsii (an intracellular parasite) and is transmitted by a tick bite; the parasite enters the blood stream, and then invades vascular endothelial cells, causing systemic vascular injury. The incidence of RMSF, by state, (cases per 100,000 population) in decreasing order, is as follows: Oklahoma (2.7), North Carolina (1.3), Kansas (1.2), Tennessee (1.2), South Carolina (1.1), Maryland (1.0); 39% of cases occur in the South Atlantic region, 20% from the West South Central region (Morbidity and Mortality Monthly Report. 1988; 37:388-389).

Note the high incidence in the eastern states which border the Appalachian mountains; it is rarely seen in the Rocky Mountains. The vast majority of cases occur between April 30 and September 30. About 50 percent of patients are under 20 years of age. Symptoms are as follows: fever (98%), headache (90%), rash on torso (90%) and rash on palms or soles of feet (70%).

Ticks can be readily removed by grasping as close to the skin as possible with curved forceps (alternatively, tweezers or protected finger) pull straight up with even pressure (Needham GR. Pediatr. 1985; 75:997-1002).

Rickettsial antibodies are first detectable within 7 to 10 days after onset of symptoms. However, there is a fatality rate of about 10 percent and treatment with tetracycline or chloramphenicol must be started before the results of these tests for antibodies are obtained. Tetracyclines are generally contraindicated in children prior to 9 years of age, but chloramphenicol has significant potential risks. As such, it has been suggested that physicians weigh benefits and risks of each drug as well as conditions in an individual case when choosing which antibiotic is most appropriate for a child with presumed RMSF (Abramson JS, Givner LB. Pediatr. 1990; 86:123-124).

Rapid diagnosis is done by identification of Rickettsia in biopsies of skin lesions. A full-thickness section of skin is obtained. Frozen tissue sections are cut and fixed, fluorescein-labeled antiserum to Rickettsia rickettsii is used for the direct identification of Rickettsia in the skin lesion (Walker DH, already cited).

LDH is invariably elevated in Rocky Mountain Spotted Fever; this is probably secondary to the diffuse vasculitis that occurs in this disease.

ROTAVIRUS ANTIGEN, FECES

SPECIMEN: Fecal specimen
REFERENCE RANGE: Negative
METHOD: EIA (enzyme immunoassay) or Latex Agglutination
INTERPRETATION: Rotavirus is the most frequent cause of viral gastroenteritis accounting for 60 to 80 percent of cases in infants and young children, most often in the 3 month to 2 year age group. In the United States, rotavirus accounts for 70,000 hospitalizations and 75-125 deaths yearly from severe diarrhea and dehydration in children less than 5 years of age. Worldwide, there are an estimated 873,000 deaths yearly (CDC, MMWR. 1990; 39RR-5:1-24; Glass RI, et al. J Pediatr. 1991; 118:27-33; Ho MS, et al. JAMA. 1988; 260:3281-3285). Infection occurs primarily in the winter. Epidemics have regional variation; in western states beginning in November and peaking in December-January, in eastern states beginning in January and peaking in February-March (CDC, MMWR. 1992; 41SS-3:47-51).

The incubation period is about 3 days. Profuse watery diarrhea lasts for 2 to 11 days and may be associated with vomiting (1 to 5 days) and fever (2 to 4 days); dehydration may require hospitalization. The laboratory findings include the following: Increase in urea nitrogen reflecting dehydration; decrease in serum sodium reflecting gastrointestinal loss; acidosis due to bicarbonate loss in the stool and/or lactate production. Treat patient by replacing lost fluids and correcting electrolyte abnormalities. Milk intolerance may develop after resolution of the gastroenteritis due to a deficiency in disaccharidases in regenerating gastrointestinal epithelium.

In neonates, the majority of rotavirus infections are mild or asymptomatic. Severe cases are rare, but severe diarrhea, necrotizing enterocolitis, bowel perforation, and death have been reported. The reason for the relative protective effect of young age is unknown, possibilities include:

(1) "nursery" strains of rotavirus may be less virulent,
(2) breast feeding may decrease severity of disease,
(3) immature gut may respond differently to rotavirus infection (Haffejee IE. Rev Infect Dis. 1991; 13:957-962).

Neonatal rotavirus infection does not confer immunity against reinfection but does protect against the development of clinical disease during reinfection (Bishop RF, et al. N Engl J Med. 1983; 309:72-76); serious infections have occurred in immunocompromised children and adults. The incidence of serious infection in adults is low and the infection is mild.

Rotavirus Testing: Rotavirus cannot be easily cultured. Virus can be identified by direct electron microscopy or immune electron microscopy. Rotavirus antigen can be identified by EIA or latex agglutination(LA); EIA has 95-100 percent sensitivity and specificity; LA is much faster, but is less sensitive and specific (Dennehy PH, et al. J Clin Microbiol. 1988; 26:1630-1634; Thorne GM. Infect Dis Clin N Amer. 1988; 2:768-769).

RPR (RAPID PLASMA REAGIN), TEST FOR SYPHILIS ANTIBODIES

Synonym: Serology Test for Syphilis (STS); Nontreponemal Test for Syphilis
SPECIMEN: Red top tube, separate serum; cord blood.
REFERENCE RANGE: Non-reactive
METHOD: The RPR uses an alcoholic extract of beef heart, cholesterol, lecithin and alcohol to give reproducible qualitative and quantitative agglutination reactions for the detection of antibody to treponemal infection.
INTERPRETATION: The RPR is a nontreponemal screening test for antibodies in the serum of patients with syphilis. It may also be used to follow therapy. The reactivity of nontreponemal and treponemal tests in untreated syphilis is given in the next Figure (Henry JB (ed.) Todd-Sanford-Davidsohn. Clinical Diagnosis by Laboratory Methods. 16 ed. 1979; 1890, Vol. II, W.B. Saunders Co., Phila., PA):

Reactivity of Nontreponemal and Treponemal Tests in Untreated Syphilis

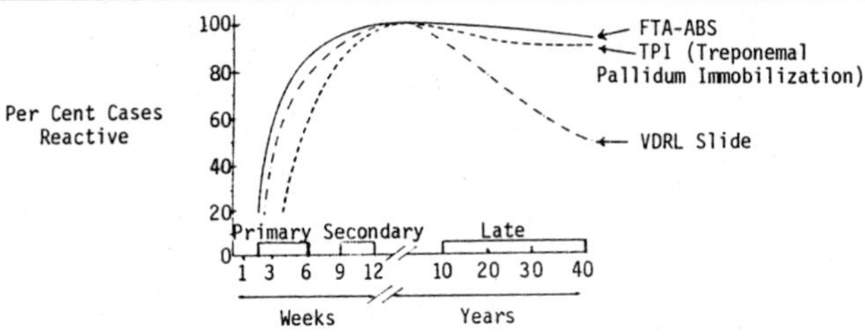

The percent positive for RPR in each stage of untreated syphilis is as follows: Primary Stage (Chancre Stage): 30% after one week; 90% after three weeks; Secondary Stage: 100%; Tertiary Stage: 90%; Late Syphilis: may become unreactive. In early and late syphilis, a nontreponemal test (FTA-ABS, MHA-TP) may be indicated.

This test is used to follow the effects of therapy; with successful therapy, the value of RPR will tend to fall and will become negative in the majority of patients if treatment is given in the primary or secondary stages (four-fold decrease within 3 months). However, treatment during late syphilis is infrequently associated with reversion of a reactive RPR to non-reactive. A discussion of the reversion rates of the RPR test in the various stages of syphilis may be found in: Romanowski B, et al. Ann Intern Med. 1991; 114:1005-1009.

Some hospitals routinely perform RPR testing on all cord blood; this is done as a screen for congenital syphilis. A positive test would have to be correlated with the maternal test. In asymptomatic infants, a positive test may represent passive transfer of maternal antibodies or early infection. For current recommendations for monitoring and treating infants exposed to maternal syphilis, see Committee on Infectious Diseases, Amer Acad Pediatr - 1991 Red Book, 22nd edition, 453-462. It is recommended that syphilis testing be performed in all newborns with fever, aseptic meningitis, hepatomegaly, and hematological abnormalities, even if previous serological studies were negative (Dorfman, DH, Glaser JH. N Engl J Med. 1990; 323:1299-1302).

False-positive reactions occur in a variety of acute and chronic conditions as listed in the next Table:

Conditions Causing False-Positive Using Nontreponemal Tests for Antibodies Produced in Response to Syphilis
Lupus Erythematosus
Drug Addiction
Infectious Diseases: Malaria; Infectious Mononucleosis; Infectious Hepatitis; Leprosy; Brucellosis; Atypical Pneumonia; Typhus; Related Treponemal Infections; Lyme Disease; HIV
Pregnancy

False-positive tests for syphilis, in patients with systemic lupus erythematosus, are caused by anti-DNA antibodies that cross-react with cardiolipin (Koike T, et al. Clin Exp Immunol. 1984; 56:193-199). Biologic false positive reactions (RPR test) were found in 3.8% of HIV-infected patients attending a sexually transmitted disease clinic compared to 0.8% of non-HIV infected patients (Rompalo AM, et al. J Infect Dis. 1992; 165:1124-1126).

RUBELLA (GERMAN MEASLES) ANTIBODIES, IgM AND IgG

SPECIMEN: Red top tube, separate serum. A single serum specimen is required for immune IgG status. For suspected infection, two serum specimens are required: the acute serum specimen should be obtained while the rash is present; convalescent serum is obtained 7 to 14 days later.

REFERENCE RANGE: The presence of IgM antibodies indicates current infection; a four-fold rise in IgG antibody titer is also indicative of an acute rubella infection; a stable or diminishing antibody titer is indicative of a past rubella infection.

METHOD: IgM: Enzyme immunoassay (EIA) or RIA (False-positive reactions may occur due to rheumatoid factor). IgG: EIA; Hemagglutination inhibition (HI); passive hemagglutination; complement fixation; latex agglutination; fluorescence immunoassay.

INTERPRETATION: Rubella or German measles is a mild acute viral illness that most commonly occurs in school-age children; however, maternal rubella infections acquired during the first sixteen weeks of gestation are associated with fetal malformations and sequelae such as abortions, stillbirths, and congenital anomalies. Fifty percent of infants will be affected if disease occurs in the first four weeks of gestation, dropping to 10% at 20 weeks. Congenital rubella syndrome is diagnosed as given in the following Table (Centers for Disease Control. MMWR. 1990; 39(RR-15):1-18):

Diagnosis of Congenital Rubella Syndrome
Clinical
Characteristic Symptoms: Cataracts/Congenital Glaucoma; Congenital Heart Disease; Hearing Loss; Pigmentary Retinopathy
Associated Symptoms: Purpura; Splenomegaly; Jaundice; Microcephaly; Mental Retardation; Meningoencephalitis; Radiolucent Bone Disease
Laboratory
Rubella Virus Isolation
Rubella-Specific IgM Present
Rubella-Specific IgG titer that persists beyond that expected from passive transfer of maternal antibody (Rubella hemagglutination-inhibition titer is expected to decrease by one two-fold dilution per month)

Congenital rubella syndrome is classified as confirmed if a clinically compatible case is laboratory confirmed, compatible if the case is not laboratory confirmed but has any two characteristic symptoms or one characteristic and one associated symptom, and possible if the case has some compatible findings but does not meet the criteria for a compatible case. The serum of a congenitally infected infant contains actively acquired IgM antibody, and passively acquired IgG antibody; in one year IgG antibody is present. Passively acquired IgG usually disappears by 7 months of age.

It is important to determine the immune status of women of childbearing age. An IgG antibody titer indicates immunity; a susceptible woman should be immunized and is advised not to conceive for two to three months following immunization. There has been a resurgence of congenital rubella syndrome due to missed opportunities for rubella testing and vaccination (Lee SH, et al. JAMA. 1992; 267:2616-2620).

Programs have evolved to help prevent personnel working in hospitals from being sources of rubella infections (nosocomial rubella transmission). The Public Health Services Advisory Committee on Immunization Practices (ACIP) recommended in 1978 "to protect susceptible female patients and female employees, persons working in hospitals and clinics who might contract rubella from infected patients or who, if infected, might transmit rubella to pregnant patients should be immune to rubella." The American Academy of Pediatrics' Committee on Infectious Disease concurred and made similar recommendations (Pediatrics. 1980; 65:1182-1184). See also review: Bart KJ, et al. "Elimination of Rubella and Congenital Rubella from the United States," Pediatr Infect Dis. 1985; 4:14-21.

SALICYLATE, BLOOD; SCREENING, URINE

Synonyms: Aspirin, Salicylic Acid, Acetylsalicylic Acid (ASA)
SPECIMENS: Red top tube; serum may be stored in refrigerator at 4°C for at least one week. Lavender (EDTA) or green (heparin) top tubes with collection of plasma; salicylate is stable in plasma stored at 4°C for 2 months. There is no difference in salicylate values obtained from serum or plasma.
REFERENCE RANGE: Therapeutic level: 15-30mg/dL; Toxicity may appear in the range of 20-30mg/dL. At >30mg/dL, toxicity will definitely occur. Aspirin is lethal at levels >70mg/dL. Salicylate blood levels do not correlate well with degree of intoxication in chronic salicylism; Half-Life (Adults): 2-4.5 hours; Half-Life (Children): 2-3 hours; Time to Peak Plasma Level: 1-2 hours; Time to Steady State, Adults: 10.0-22.5 hours; Time to Steady State, Children: 10-15 hours. With high doses the half-life increases substantially (>15 hours). Other drugs may displace protein bound salicylate leading to increased toxicity. Salicylate doses in patients on chronic therapy may approach toxic levels. Measurement of salicylate levels in such patients are best performed just prior to the next dose.
METHOD: Urine Screen: Colorimetrically by reaction with ferric ions; Blood: colorimetric; microfluorometric procedure; HPLC; Immunoassay.
INTERPRETATION: Salicylate screens are performed on urine, and salicylate levels performed on blood in order to document ingestion and intoxication in patients suspected of salicylate overdose. Determination of drug levels and comparison of these levels with the "Done nomogram" and time of ingestion allows for a determination of the necessity of aggressive treatment. In acute salicylate toxicity, the Done nomogram (Done AK. Pediatrics. 1960; 26:800-807) can be used to predict severity of intoxication from a single determination of serum salicylate, provided that the time of ingestion is known and sufficient time has elapsed between ingestion and sampling of blood (six hours)(Atwood SJ. Pediatr Clin of N Amer. 1980; 27:871-879); the Done nomogram is shown in the next Figure:

Serum Salicylate Concentration and Expected Severity of Intoxication at Varying Intervals of Time Following Ingestion

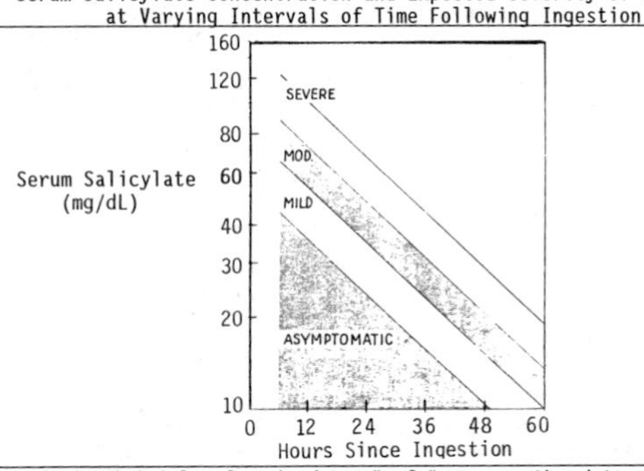

If the initial value is in a "safe" range, the determination of serum salicylate should be repeated in two to four hours. If the serum salicylate does not rise between tests, the patient is not considered to be at risk for serious intoxication. Presence of salicylate in urine may be suspected by the finding of reducing substances in the urine by a quick screening test. Other reducing substances may give a false positive test. Laboratory data suggestive of aspirin are a positive urine by clinitest (reducing substances), with a negative glucose oxidase test, and positive ferric chloride test on boiling urine.

Laboratory tests, other than serum salicylate levels, that should be used to monitor complications of salicylate toxicity are given in the next Table:

Laboratory Tests to Monitor Possible Complications of Salicylate Toxicity

Tests	Monitor Complications
Arterial Blood Gases (pH, PCO_2, HCO_3-)	Respiratory Alkalosis; Metabolic Acidosis; Hypermetabolism
Serum Electrolytes (Na^+, K^+, Cl^-, CO_2 Content)	Anion Gap Increased Due to Increased Lactic Acid, Ketones and Salicylic Acid. Inappropriate Fluid Retention: Detect the Effects of Bicarbonate Diuresis following Therapy with Bicarbonate for Acidosis
Serum and Urine Osmolality	Inappropriate Secretion of Antidiuretic Hormone (SIADH)
Blood Glucose	Blood Glucose May be Low, Normal or Elevated; Blood Glucose Should be Maintained at Elevated Levels to ensure Adequate Supply of Glucose to Brain
Prothrombin Time (PT)	If Prolonged, Vitamin K Therapy Indicated

Following acute ingestion, children quickly pass through the stage of pure respiratory alkalosis and present with some degree of metabolic acidosis. The toxic properties of salicylates result principally from direct stimulation of the respiratory center (respiratory alkalosis) and influence on metabolic pathways (metabolic acidosis). Symptoms of hypocalcemic tetany, secondary to alkalosis, may develop.

Reference: Temple AR. Pediatr. 1978; 62:873-876.

SALINE WET MOUNT FOR TRICHOMONAS, G. VAGINALIS AND CANDIDA ALBICANS

SPECIMENS AND METHOD: Obtain a specimen of vaginal discharge and place on a clean slide; mix with one or two drops of normal saline. Cover the specimen with a clean glass cover slip. Examine the specimen under low power and high dry for the active motile flagellates of Trichomonas vaginalis. A slide may also be prepared for gram stain.

INTERPRETATION: Saline wet mounts are performed in order to diagnose the cause of a vaginal discharge or a vaginitis. The usual organisms are Trichomonas vaginalis, Candida, and Gardnerella vaginalis. The most sensitive method of detecting Trichomonas is culture but it is expensive. Saline wet mount is less sensitive but inexpensive. This organism is identified on wet mount by its jerky motion. Candida may also be diagnosed by saline wet mount but potassium hydroxide preparation is the preferred method. The organism is commonly adherent to squamous cells. G. vaginalis is identified by the presence of "clue cells." These clue cells are squamous epithelial cells to which coccobacillary bacteria have become adherent. The surface of the cell is coated with these bacteria and the cell border tends to be "fuzzy" and indistinct. On gram stain they are gram negative coccobacilli. The above organisms may also be noted and reported on "pap smear."

SARCOIDOSIS PANEL

A list of laboratory tests that may assist in the diagnosis of sarcoidosis is given in the next Table (Sharma Om P. Arch Intern Med. 1983; 143:1418-1419):

Tests of Sarcoidosis

Test	Sensitivity (%)	False Positives
Kveim-Siltzbach(K-S) Test	97	Regional Enteritis Infectious Mononucleosis Chronic Lymphocytic Leukemia Nonspecific Lymphadenopathy
Serum Angiotensin-Converting Enzyme (ACE)	60-80	Leprosy Gaucher's Disease Primary Biliary Cirrhosis Silicosis Miliary Tuberculosis Lymphoma Extrinsic Allergic Alveolitis Talc Granulomatosis Asbestosis Fibrosing Alveolitis Chronic Lymphocytic Pneumonitis Coccidioidomycosis Berylliosis HIV/AIDS
Serum Lysozyme Activity	40-50	See LYSOZYME, SERUM
Transcobalamin II	50	Gaucher's Disease Lymphoproliferative Disorder Myeloma Waldenström'sMacroglobulinemia Liver Disease
Gallium Citrate GA-67 Lung Scan	High Sensitivity	Low Specificity
Bronchoalveolar Lavage	Normal: >90% Macrophages; <10% Lymphocytes; <1% PMN	Sarcoidosis: Increase in Lymphocytes

In the Kvein-Siltzbach(K-S) test, a saline suspension of human sarcoidal spleen or lymph node is injected intracutaneously; in patients with sarcoidosis, a nodule develops in two to six weeks. A biopsy specimen of the nodule demonstrates a characteristic non-caseating granulomatous reaction. The basic problem in this test is the unavailability of potent biologically active antigen for injection.

SCHILLING TEST, I, II, III

SPECIMEN: Twenty-four hour urine for assay of radioactive Co-57. Patient should be fasted and have adequate renal function.
REFERENCE RANGE: Normal >7% excretion in 24 hours.
METHOD: Co-57; Half-life = 270 days; decay mode, EC, gamma.
INTERPRETATION: The Schilling test is used in the evaluation of patients with low vitamin B12 levels. Most cases of vitamin B12 deficiency are due to inadequate absorption and not due to dietary deficiency. The Schilling urinary excretion test is used to differentiate intrinsic factor deficiency from small bowel (ileum) malabsorption of vitamin B12. In part I, vitamin B12 (Co-57) is given alone; in Part II, vitamin B12 (Co-57) plus intrinsic factor are given; in Part III, tetracycline is given for 10 days followed by vitamin B12 (Co-57) by mouth.
Part I: The patient is given vitamin B12 (Co-57) by mouth. This is followed 2 hours later by an intravenous injection of nonradioactive vitamin B12; this saturates the liver storage capacity for the vitamin; thus, the excess over that taken up by the liver is excreted in the urine. Urine is collected for 24 hours and radioactivity is measured. Interpretation of results is illustrated in the next Figure:

Results of Part I, Schilling Test

Vitamin B_{12} (Co-57) by Mouth; Collect Urine for 24 hours; Measure Percentage of Administered Radioactivity in 24 Hour Urine

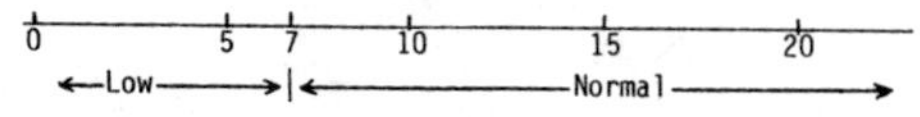

Part I Low	Part I Normal
Stomach - Intrinsic Factor Deficiency	Dietary Deficiency
a. Pernicious Anemia	Lack of B_{12} Binding Protein (Transcobalamine)
b. Gastrectomy, Total or Partial	Folate Deficiency
c. Atrophic Gastritis	
Intestine - (Ileum)	
a. Blind Loop Syndrome	
b. Idiopathic Steatorrhea	
c. Celiac Disease	
d. Tropical Sprue	
e. Crohn's Disease	
f. Small Bowel Lymphoma	
g. Pancreatic Insufficiency	
h. Helminths	
i. Drugs	
j. Others: see VITAMIN B12	

Part II: If low excretion of vitamin B12 (Co-57) is obtained, then, it is necessary to differentiate between vitamin B12 deficiency due to inadequate secretion of intrinsic factor by the stomach from an abnormality in absorption by the ileum. Vitamin B12 (Co-57) is given along with intrinsic factor. This test is done 5 to 7 days after the first test. Interpretation of results is illustrated in the next Figure:

Results of Part II, Schilling Test

Vitamin B_{12} (Co-57) Plus Intrinsic Factor by Mouth; Collect Urine for 24 Hours and Measure Percentage of Radioactivity in 24 Hour Urine

Intestine (Ileum)	Stomach-Intrinsic Factor Deficiency
a. Blind Loop Syndrome	a. Pernicious Anemia
b. Idiopathic Steatorrhea	b. Gastrectomy, Total or Partial
c. Tropical Sprue	c. Atrophic Gastritis
d. Others (see PART I)	

Part III: The third stage is performed on those patients who have low excretion of cobalt-57 labeled vitamin B12 in the urine with the first and second stage Schilling tests. The patient receives tetracycline, 250mgs, 4 times a day for 10 days prior to the test. Tetracycline is administered to reduce or alter intestinal bacteria in the possible blind loop syndrome.

Markedly improved excretion of labeled vitamin B12 after tetracycline confirms the presence of malabsorption secondary to interference with the normal absorptive process by the bacterial flora which compete for the vitamin B12.

(cont)

Schilling Test (Cont)

The Schilling Test, Parts I, II and III are summarized in the next Table:

Schilling Test, Parts I, II and III

Schilling	Nutritional Megaloblastic Anemia	Pernicious Anemia	Gastrectomy Total or Partial	Blind Loop Syndrome	Idiopathic Steatorrhea	Tropical Sprue
Dose alone (Part I)	Normal	Subnormal	Subnormal	Subnormal	50% of Patients Subnormal	Subnormal
Dose with I.F. (Part II)	Normal	Improvement	Improvement	No Improvement	No Improvement	No Improvement
Dose after antibiotics (Part II)	Normal	-	-	Improvement if the Ileum is Intact	No Improvement	Improvement in Many

Other Tests: Measurement of gastric acidity is used to identify achlorhydria. However, there are many patients with achlorhydria without pernicious anemia. Gastric fluids may be analyzed for intrinsic factor in a more direct approach. Antibodies to intrinsic factor are found in 50-60% of patients with pernicious anemia, and this finding may be supportive of the diagnosis. Urinary excretion of methylmalonic acid is increased in many patients with pernicious anemia. Infection with intestinal helminths will sometimes lead to reduced vitamin B12 absorption.

SCHLICHTER TEST (see SERUM BACTERICIDAL TEST)

SEIZURE PANEL

There are many causes of seizures; a list of etiologies of status epilepticus is given in the following Table (Bleck TP. Clin Neuropharm. 1991; 14:191-198):

Etiologies of Status Epilepticus

- Withdrawal - From anticonvulsants (e.g. phenobarbital), alcohol, other hypnosedatives (e.g. glutethimide)
- Stroke - Acute or Old
- Tumor - Primary brain tumor or metastatic
- Intracranial Infection - Brain abscess, Meningitis, Focal encephalitis (e.g. herpes simplex), Diffuse Encephalitis
- Metabolic Disorder - Hypoglycemia, Nonketotic hyperglycemia, Other
- Drug Induced - Cocaine, Theophylline, Other
- Trauma
- Postanoxic

Suggested evaluation of a first seizure is given in the following Table (Scheuer ML, Pedley TA. N Engl J Med. 1990; 323:1468-1474):

Evaluation of a First Seizure

- Physical Examination
- Infection Workup (if appropriate)
  - White Blood Cell Count, Differential
  - Cerebrospinal Fluid (CSF) Examination
  - Blood Culture
  - Other Cultures
- Serum Studies
  - Electrolytes (Sodium, Potassium, Chloride, Bicarbonate)
  - Glucose
  - Calcium
  - Magnesium
  - Tests of Hepatic and Renal Function
- Illicit Drug Screen (Blood or Urine)
- Electroencephalography (EEG)
- Neurologic Imaging Studies
  - CT Scan
  - Magnetic Resonance Imaging (MRI)

Serum chemistry abnormalities are uncommon in both adults and children with seizures. In adult patients, hypoglycemia is the most commonly identified abnormality (Nypaver MM, et al. Pediatr Emerg Care. 1992; 8:13-16; Kenney RD, Taylor JA. Pediatr Emerg Care. 1992; 8:65-66; Eisner RF, et al. Ann Emerg Med. 1986; 15:33-39; Powers RD. Ann Emerg Med. 1985; 14:416-420).

SEMEN ANALYSIS

SPECIMEN: An absolutely fresh specimen is required; thus, collection must be made at the site of examination within one hour of ejaculation. The subject is requested to abstain from sexual activity for 3 days prior to examination. The semen is collected via masturbation in a labeled dry small container. Do not refrigerate or freeze.

REFERENCE RANGE: The normal parameters for an ejaculation are given in the next Table:

Normal Parameters for an Ejaculation

Parameter	Normal
Volume	2-5 mL.
Consistency	Gelatinous, High Viscosity Initially
Color	Light Gray and Opaque
Odor	Musty
Liquefaction	Complete in 15 min.; Over 30 min. Abnormal
pH	7.2-8.0; Below 7.2 is Abnormal
Specific Gravity	About 1.033
Sperm Count	70-150 million/mL.
Sperm Motility	80% or More Active (forward migration)
Sperm Morphology	80-90% Normal Forms
Normal Heads	80% or More
Cytology	A few Crystals and Epithelial Cells

The American Society for the Study of Sterility has proposed a volume of 2-5 mL, a minimum count of 60 million/mL, and at least 60% active forms as adequate for fertility. Others in this field feel that these figures are too high and that a minimum sperm count of 20-40 million/mL with 40% or more active forms is adequate. Morphology is classified as poor when less than 60% are normal.

METHOD: Appearance, volume, color, odor, microscopic examination, sperm count, morphology with stain.

INTERPRETATION: Semen analysis is performed as part of infertility work-ups, in the evaluation of the hypothalamic-pituitary-testicular axis, and post-vasectomy.

Although sperm counts below 20-40 million/mL are abnormal, fertility may not be definitely affected until the count falls below 5 million/mL. Semen analysis may not be accurate following illness or high fevers. There are a number of causes of low sperm counts as tabulated below:

Causes of Low Sperm Counts
Exposure to Heat
High Fever
Exposure to Chemicals
Chemotherapy
Radiation
Infection
Varicocele
Retrograde Ejaculation
Obstruction of Vasa
Post-Vasectomy
Primary Testicular Failure
Klinefelter's Syndrome
Hypothalamic Lesions
Endocrine
Low Testosterone
Elevated Prolactin
Estrogen or LHRH Analogues

Following vasectomy, a vasectomy can be considered successfully only when complete azoospermia is achieved in two consecutive semen analyses, 4-6 weeks apart. Approximately 90% of men are sperm free after 6-24 ejaculations over a 2-6 month period (Alderman PM. Fertil Steril. 1989; 51:859-862). Men who have not achieved azoospermia after 6 months should be suspected of reanastomoses.

SEPTIC SHOCK TEST PANEL

Septic shock test panel is shown in the next Table:

Septic Shock Test Panel
Blood Cultures: Ideally, a total of 3-4 blood culture bottle sets (2 bottles/set, 5 mL blood/bottle) collected in a 24 hr. period, 3 different venipunctures at different sites, preferably at 1 hr. intervals. Accuracy with this protocol is greater than 90%.
Blood Gases: pH, pCO_2, pO_2
Electrolytes: Na^+, K^+, Cl^-, CO_2 Content
Lactate
Glucose
Disseminated Intravascular Coagulopathy(DIC); if clinical signs and symptoms of DIC develop, determine PT, PTT, fibrinogen level, fibrin split products (FSP), and platelet count. See DISSEMINATED INTRAVASCULAR COAGULOPATHY(DIC) PANEL

Tests for hepatic or renal dysfunction are often appropriate. See also Parrillo JE. "Pathogenetic Mechanisms of Septic Shock," N Engl J Med. 1993; 328:1471-1477.

SERUM BACTERICIDAL TEST (SCHLICHTER TEST)

SPECIMEN: Three specimens are required: two serum specimens (or other body fluid) and a bacterial isolate. Serum Specimen: Two serum specimens are obtained; one specimen is collected at the trough level just prior to the next antibiotic dose and next specimen is collected at the peak, that is, 30 minutes post I.V. infusion or one hour post I.M. or oral administration. Use red top tubes; separate serum and refrigerate or freeze. Microbiological Specimen: Request the laboratory to save the bacteriological isolate.
REFERENCE RANGE: Traditionally, a serum bactericidal titer equal to or greater than 1:8 dilution was considered significant. A study of the serum bactericidal test in infective endocarditis indicated that peak serum bactericidal titers equal to or greater than 1:64 and trough serum bactericidal titers equal to or greater than 1:32 predicted bacteriologic cure in all patients (Weinstein MP, et al. Am J Med. 1985; 78:262-269).
METHOD: Serially dilute the patient's serum with a 1:1 mixture of pooled human serum and calcium/magnesium supplemented Mueller-Hinton broth. The dilutions are inoculated with a standard suspension of organisms (5×10^5 to 10^6 colony-forming units per milliliter). The specimen is incubated; usual endpoint is 99.9 percent bacterial killing at 18 to 24 hours (Stratton CW, Reller LB. J Infect Dis. 1977; 136:187-195; Reller LB, Stratton CW. J Infect Dis. 1977; 136:196-204).
INTERPRETATION: The serum bactericidal test has been used most commonly as a guide to antibiotic choice and dosage for infective endocarditis. This test has been used in other infections such as osteomyelitis, bacteremia, urinary tract infection, septic arthritis, and bacterial pneumonia or meningitis. This test has also been used as a guide when switching from parenteral to oral antibiotics. The effectiveness of this test for guiding the adequacy of therapy in most conditions is not well established. (Review: Wolfson JS, Swartz MN. N Engl J Med. 1985; 312:968-975). The serum bactericidal test may be useful for monitoring antibiotic therapy in cystic fibrosis (Cahen P, et al. Pediatr. 1993; 91:451-455).

SERUM GLUTAMIC OXALACETIC TRANSAMINASE(SGOT)
(see ASPARTATE AMINOTRANSFERASE)

SERUM GLUTAMIC PYRUVATE TRANSAMINASE(SGPT)
(see ALANINE AMINOTRANSFERASE)

SEXUAL ASSAULT DATA

A guide to the workup of victims of sexual assault is given in the following Table (Sarles RM. Pediatr Rev. 1982; 4:93-98; Long WA, et al. Pediatrics. 1983; 72:738-740):

Sexual Assault Data Sheet

I. History
- A. Presentation in emergency room
  1. Date seen
  2. Time seen
  3. Mode of entry: police, friend, family, self-referral, other
- B. Date of assault
- C. Time of assault
- D. Circumstances of assault (including postassault activity, changes of clothing, bathing, douching. Record evidence of torn clothing, bruises, blood or semen stains)
- E. Menarche
- F. Last menstrual period
- G. Method of birth control
- H. Current medications

II. Physical Examination
- A. General appearance (include the emotional state, behavior of patient. Document areas of obvious trauma by photograph or diagram)
- B. Vital Signs; Pubertal Stage (Tanner)
- C. Evidence of trauma
- D. Description of clothing: torn, blood-stained, semen-stained, normal
- E. Description of perineum: normal, laceration, ecchymosis, bleeding
- F. Pelvic examination: vagina, cervix, uterus, adnexa, rectum

---

III. Laboratory Evaluation
- A. Wet preparation of vaginal fluid for motile sperm and T. vaginalis
- B. Vaginal washing for
  - a. Acid phosphatase
  - b. ABH agglutinogen
- C. Culture of vagina, anus, oropharynx, and urethra for GC
- D. Serologic test for syphilis
- E. Pregnancy test (pubertal females)
- F. Wood's lamp for semen
- G. Hair combing of pubis
- H. Fingernail scrapings
- I. Serum sample frozen and saved for future testing
- J. HIV Test

IV. Therapy
- A. Antibiotic prophylaxis (See CDC. 1989 Sexually Transmitted Diseases Treatment Guidelines. MMWR. 1989; 38(S-8):1-40).
- B. Tetanus toxoid as indicated according to Public Health recommendations
- C. Pregnancy prevention for pubertal females: Ovral, 4 tablets in 2 divided doses 12 hours apart.

V. Reported to police

VI. Disposition and follow-up

Reference: Sarles RM. Pediatr Rev. 1982; 4:93.

Some of the tests such as serology for syphilis and AIDS should be repeated on follow-up visits.

The laboratory testing of vaginal washings and semen stains for identification of suspects is rapidly changing. Polymorphisms among individuals were previously tested on gene products (e.g., ABO agglutinogens), whereas DNA polymorphism at the DNA level is much greater and allows for the detection of individual-specific "DNA fingerprints" (Jeffreys AJ, et al. Nature. 1985; 314:67-73). Two general techniques are being used: (1) Hybridization analysis using DNA probes to detect restriction fragment length polymorphisms, and (2) polymerase chain reaction amplification of length polymorphisms (e.g., variable number of tandem repeats, VNTR). The state of forensic DNA testing methodology is reviewed in: Sajantila A, Budowle B. Ann Med. 1991; 23:637-642. Similar testing is used for establishment of paternity.

SKELETAL MUSCLE PANEL

SPECIMEN: Red top tube, separate serum
(a) Creatine Phosphokinase (CPK) and Isoenzymes
(b) Serum Aldolase
(c) Myoglobin, Urine (Random Specimen)

Creatine Phosphokinase (CPK) is found in high concentration in skeletal muscle and the heart; CPK and aldolase are increased in the serum in patients with muscle injury from any cause, such as trauma, and in the myopathies. CPK is not generally elevated in patients with neurogenic disorders. Aldolase is not specific for skeletal muscle. It is usually not necessary to determine both CPK and aldolase. Assay of urine myoglobin is especially useful when there is massive necrosis of skeletal muscle, and the patient is at risk for the renal tubular necrosis and acute renal failure.

The conditions involving skeletal muscle that are associated with an increased in CPK are shown in the next Table:

Skeletal Muscle Conditions and Elevated Serum CPK
Rhabdomyolysis
Polymyositis
Mixed Connective Tissue Disease
Dystrophies such as Duchenne Type
Female Carriers of Muscular Dystrophy
Metabolic Myopathies: Hypothyroidism, Alcoholism, Malignant Hyperthermia

About 15 to 20 percent of patients with polymyositis have cardiac involvement; thus CPK-MB may be elevated. Almost 3/4 of patients with mixed connective tissue disease (MCTD) have muscle pain, muscle tenderness and weakness with elevated CPK and aldolase and with abnormal electromyograms consistent with inflammatory myositis. In muscular dystrophy (Duchenne's), CPK values are markedly elevated at an early age and gradually decrease with progression of the disease; both CPK-MM and CPK-MB are increased. CPK is a reliable means of identifying female muscular dystrophy carriers. The abnormality in Duchenne's and Becker's muscular dystrophy has been identified at the level of the gene so that carrier analysis may now be performed by genetic testing (Multicenter Study Group. JAMA. 1992; 267:2609-2615). In hypothyroidism, the mean value of CPK is seven times the post-treatment value.

The laboratory abnormalities in malignant hyperthermia are as follows: acidosis, hypoxemia, hyperkalemia, myoglobinemia and myoglobinuria and elevated CPK (Stehling L, Brown D. Diagnostic Medicine. 1983; 59-64). See MALIGNANT HYPERTHERMIA PANEL.

See also RHABDOMYOLYSIS PANEL.

SMOOTH MUSCLE ANTIBODY (see ANTI-SMOOTH MUSCLE ANTIBODY)

SODIUM, EXCRETION FRACTION
(see EXCRETION FRACTION OF FILTERED SODIUM)

SODIUM, SERUM

SPECIMEN: Red top tube, separate serum; or green (heparin) top tube, separate plasma.

REFERENCE RANGE: Adult: 135-148mmol/liter; Premature (cord): 116-140mmol/liter; premature: 128-148mmol/liter; newborn (cord): 126-166mmol/liter; newborn: 134-144mmol/liter; infant: 139-146mmol/liter; child: 138-145mmol/liter. The following are potentially life-threatening values and, after confirmation, preferably by repeat determination on a different specimen, should be telephoned to the responsible nursing staff, or to the responsible physician so that corrective therapy may be immediately undertaken: serum sodium: <120mmol/liter; serum sodium: >155mmol/liter. Conventional units (mEq/L) equal international units (mmol/L).

METHOD: Ion-specific electrode or flame emission photometry.

INTERPRETATION:

Hyponatremia: The causes of decreased serum sodium (hyponatremia) are shown in the next Table (Berry PL, Belsha CW. Pediatr Clin N Amer. 1990; 37:351-363):

Causes of Hyponatremia
Isotonic Hyponatremia (Plasma Osmolality 280-295)
Hyperproteinemia
Hyperlipidemia
Hypertonic Hyponatremia (Plasma Osmolality >295)
Hyperglycemia
Mannitol
Glycerol
Hypotonic Hyponatremia (Plasma Osmolality <280)

Hypotonic hyponatremia is the most common form; plasma volume estimation and urine sodium are used to diagnose the cause, as given in the following Table (DeVita MV, Michelis MF. Clin Lab Med. 1993; 13:135-148):

Diagnosis of Hypotonic Hyponatremia Based on Plasma Volume and Urine Sodium

Plasma Volume	Urine Sodium	Diagnosis
Hypovolemic	>30mmol/L	Renal Losses: Diuretic Therapy; Osmotic Diuresis; Salt-Wasting Nephropathy; Adrenal Insufficiency; Proximal Renal Tubular Acidosis; Metabolic Alkalosis; Pseudohypoaldosteronism; Cerebral Salt-Wasting
	<30mmol/L	Extrarenal Losses: Gastrointestinal; Sweat; Third Space (Burns, Ascites, Effusions)
Euvolemic	>20mmol/L	Excess ADH (SIADH, Drugs, Pain); Water Intoxication (IV Therapy, Psychogenic Water Drinking, Tap Water Enema); Reset Osmostat; Glucocorticoid Deficiency; Hypothyroidism
Hypervolemic	<10mmol/L	Edematous: Congestive Heart Failure; Cirrhosis; Nephrotic Syndrome
	>30mmol/L	Renal Failure (Acute or Chronic)

Clinical symptoms of hyponatremia depend both on the level of the serum sodium and how rapidly the decrease occurs. An acute change in Na^+ from normal range to 125mmol/liter may be associated with symptoms; a gradual change in Na^+ from normal range to 110mmol/liter may be asymptomatic.

A decrease in serum Na^+, relatively impermeable solute or an increase in extracellular water will cause water to move into cells, yielding cell swelling and intracellular hypotonicity. The symptoms of hyponatremia are primarily neurologic.

The plasma sodium concentration indicates the balance between salt and water and, by itself, gives no certain information about overall sodium deficiency or excess.

Hyponatremia occurs as a result of water retention or sodium loss or both. Beer has a sodium concentration of 1.5 to 10.0mmol/liter and dilutional hyponatremia due to beer-drinking is probably common. Bodily depletion of sodium may be due to gastrointestinal loss, e.g., diarrhea and vomiting, diuretic therapy, sweating, hydroadrenalism or a variety of renal disorders. Urinary sodium concentration may be a useful indicator of salt depletion (usually less than 20mmol/24 hours or less than 10mmol/liter on a "spot" sample). Drug-induced hyponatremia has two usual causes - diuretic induced and ADH-like action of some drugs such as chlorpropamide and carbamazepine.

Central Pontine Myelinolysis is associated with correction of hyponatremia. Female patients may be more at risk than male patients. In chronic hyponatremia, correction rate should not exceed 2.5mmol/L/hour or 20mmol/L/day (Berl T. Ann Intern Med. 1990; 113:417-419).

Hypernatremia: The causes of hypernatremia are given in the next Table (Conley SB. Pediatr Clin N Amer. 1990; 37:365-372; DeVita MV, Michelis MF, already cited):

Causes of Hypernatremia
Sodium Excess
Excess Sodium Bicarbonate
Hypertonic IV Fluids
Sodium Chloride Tablets
Ingestion of Sea Water (480mmol/L)
Improperly Mixed Formula
Aldosteronism (Primary or Secondary)
Cushing's Syndrome
Water Deficit
Central Diabetes Insipidus
Nephrogenic Diabetes Insipidus
Diabetes Mellitus
Excessive Sweating
Inadequate Access to Water
Lack of Thirst (Adipsia)
Increased Insensible Water Loss
Water Deficit in Excess of Sodium Deficit
Diarrhea
Osmotic Diuretics
Diabetes Mellitus
Obstructive Uropathy
Renal Dysplasia

A common cause of hypernatremia in pediatric cases is enteric disease; in enteric disease, hypernatremic dehydration occurs. Dehydration occurs secondary to diarrhea, vomiting, anorexia and failure of water intake.

Hypernatremia elevates serum osmolality and results in intracellular dehydration as water shifts into the extracellular space; cells shrink. The effects are mediated, pathologically, via marked brain cell shrinkage, yielding mechanical trauma intracranially. Vascular damage is extensive; venous and capillary congestion is prominent and subarachnoid and intracerebral hemorrhages occur with cortical venous thrombosis and areas of venous infarction. Neurological symptoms occur in more than 50 percent of patients. Over zealous correction of hypernatremia may lead to seizures which can be difficult to treat.

SODIUM, URINE

<u>SPECIMEN</u>: A random specimen may be used but a timed specimen, i.e., 8, 12 or 24 hour urine collection is preferred. Collect 24 hour urine as follows: Refrigerate specimen as it is collected. Instruct the patient to void at 8:00 A.M. and discard the specimen. Then collect all urine including the final specimen voided at the end of the 24 hour collection, i.e., 8:00 A.M. the next morning. Transport to lab and refrigerate specimen at 4°C.
<u>REFERENCE RANGE</u>: Dependent on intake. <u>Infant</u>: 0.3-3.5 mmol/day; <u>Child</u>: 40-180 mmol/day; <u>Adult</u>: 40-210 mmol/day.
<u>METHOD</u>: Ion-selective electrodes, flame photometry.
<u>INTERPRETATION</u>: Urinary sodium varies with intake, state of hydration, influence of drugs such as diuretics, abnormalities of renal perfusion, glomerular filtration or tubular function.

The major diagnostic value of urinary sodium determination is evaluation of patients with conditions listed in the next Table:

Major Diagnostic Value of Urinary Sodium Determination
Acute Oliguria
Hyponatremia
Volume Depletion

The level of urinary sodium in different conditions is given in the next Table:

Interpretation of Urinary Sodium

Urinary Sodium	Interpretation
0-30 mmol/liter	Extra-Renal Sodium Loss (Gastrointestinal or Sweat Loss) Prerenal Azotemia Severe Volume Depletion Edematous States (Congestive Heart Failure, Liver Disease, Nephrotic Syndrome)
20-30 mmol/liter	Acute Tubular Necrosis SIADH Adrenal Insufficiency (Addison's Disease) "Renal Salt Wasting"

Determination of urinary sodium can be extremely helpful if interpreted with knowledge of the clinical situation; this can be appreciated by the level of urinary sodium as shown in the next Table (Harrington JT, Cohen JJ. N Engl J Med. 1975; 293:1241-1243; Berry PL, Belsha CW. Pediatr Clin N Amer. 1990; 37:351-363):

Interpretation of Urinary Sodium Levels*

Diagnostic Problem	Urinary Value	Primary Diagnostic Possibilities
Acute Oliguria	Na^+, 0-10 mmol/liter	Prerenal Azotemia
	Na^+, > 30 mmol/liter	Acute Tubular Necrosis
Hyponatremia	Na^+, < 20 mmol/liter	Severe Volume Depletion; Edematous States (Congestive Heart Failure, Cirrhosis, Nephrotic Syndrome)
	Na^+, > 20 mmol/liter	SIADH; Water Intoxication; Reset Osmostat; Adrenal Insufficiency
Volume Depletion	Na^+, < 20 mmol/liter	Extra-Renal Sodium Loss (GI, Sweat, "Third" Space)
	Na^+, > 20 mmol/liter	"Renal Salt Wasting"; Adrenal Insufficiency

*For purposes of this table, it is assumed that the patient is not receiving diuretics.

<u>Acute Oliguria</u>: <u>Prerenal</u>: There is hypoperfusion of the kidney and avid reabsorption of sodium and water by the nephrons. The most important diagnostic criterion of prerenal uremia is response to treatment (Editorial, Brit Med J. June 7, 1980; 1333-1335).

Renal Parenchymal: Damaged tubules fail to reabsorb solutes and water normally. The level of urinary sodium is used to differentiate prerenal and renal causes of oliguria as illustrated in the next Figure (Espinel CH. JAMA. 1976; 236:579):

Urinary Sodium in Patients with Acute Tubular Necrosis and Patients
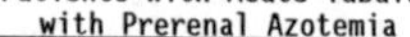
with Prerenal Azotemia

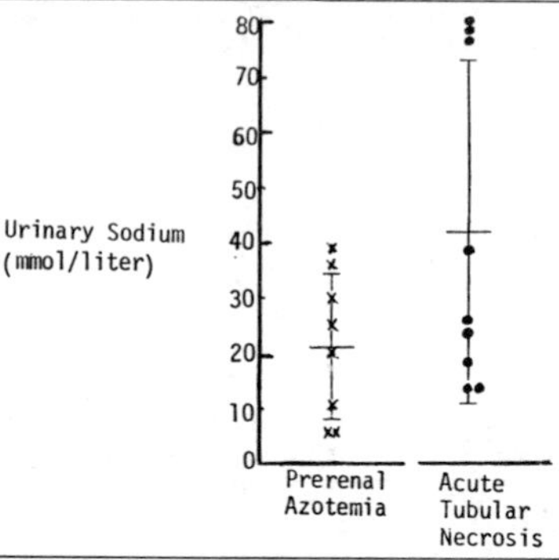

Note that when a cut-off of urinary sodium of 30 mmol/liter is used, some patients with prerenal are above that level, and some patients with acute tubular necrosis are below that level. The excretion fraction of filtered sodium is a more sensitive and specific test to differentiate prerenal azotemia from acute tubular necrosis.

Hyponatremia: The concentration of sodium in the urine tends to be proportional to the serum sodium concentration. Urinary sodium is a useful test to help differentiate causes of hyponatremia (see Table, "Causes of Hyponatremia," SODIUM, SERUM).

SOMATOMEDIN-C

Synonym: Insulin-Like Growth Factor-1 (IGF-1)

SPECIMEN: Lavender(EDTA) top tube; separate plasma into plastic tube and freeze.

REFERENCE RANGE: Reference values vary by sex and age. Values vary significantly between laboratories. Refer to the values in your laboratory. Representative published values: Female: Preadolescent 61-725ng/mL; Adolescent 113-450ng/mL; Adult 142-389ng/mL. Male: Preadolescent 66-845ng/mL; Adolescent 83-378ng/mL; adult 54-329ng/mL (Normal Reference Laboratory Values, Weekly Clinicopathological Exercises. N Engl J Med. 1992; 327:718-724).

METHOD: RIA

INTERPRETATION: The assay of somatomedin-C is useful as a surrogate measure of growth hormone (GH) activity. Accurate GH levels are difficult to obtain because: 1) levels fluctuate during drawing of the specimen (stress), 2) it may be difficult to distinguish a low normal level from GH deficiency. GH acts via stimulation of somatomedin production, assay of which provides insight into GH levels.

The measurement of somatomedin-C is used in the diagnosis and evaluation of growth disturbances, GH deficiency, the response to GH, and acromegaly. A correlation of somatomedin-C levels with clinical states is tabulated below:

Clinical Uses of Somatomedin-C Assay

Normal Serum Somatomedin-C Levels: Virtually Rules Out Deficiencies of GH Thus Eliminating Need for GH Stimulation Tests.

Low Serum Somatomedin-C Levels: Requires GH Stimulation Tests for Work-up of Subjects with Possible GH Deficiency.

High Serum Somatomedin-C Levels: Compatible with Diagnosis of Acromegaly (Clemmons DR, et al. N Engl J Med. 1979; 301:1138).

Deficiency of Growth Hormone(GH): GH deficiency is usually idiopathic. The deficiency may be isolated or it may be accompanied by a deficiency of other pituitary hormones. The prevalence of GH deficiency is about 1 child in 5,000. Low levels of somatomedin-C (IGF-1) are consistent with, but not totally diagnostic of GH deficiency. The presumptive diagnosis of GH deficiency must be confirmed with a GH stimulation test. IGF-1 levels following GH administration correlate well with clinical response to GH therapy.
Acromegaly: Elevation of GH is practically diagnostic of acromegaly.

IGF-1 levels may be low in a number of states in addition to GH deficiency. It is decreased in liver disease, malnutrition, hypothyroidism, and in some cases of dwarfism (Laron).

SPERM COUNT (see SEMEN ANALYSIS)

STONE ANALYSIS (see RENAL STONE ANALYSIS)

STOOL CULTURE AND SENSITIVITY

SPECIMEN: Rectal swab; fresh random stool in stool container or Culturette; specimen must be less than 4 hours old when delivered to the laboratory. Do not refrigerate. Blood cultures for Salmonella, Shigella and Campylobacter should be obtained in those patients with moderate to severe illness and those with high fever.
REFERENCE RANGE: Negative
METHOD: In patients with suspected Clostridium difficile colitis, stools for diagnostic toxin assays are done.
INTERPRETATION: The usual causes of bacterial diarrhea are given in the next Table:

Usual Causes of Bacterial Diarrhea
Salmonella
Shigella
Campylobacter Jejuni
Yersinia Enterocolitica
Clostridium Difficile
E. coli

Clinical dysentery is manifest by cramps, tenesmus, very frequent, small volume stool composed of blood and mucous. Dysentery is usually caused by shigella, but may be caused by campylobacter, salmonella, or yersinia.

Salmonella (nontyphoidal) is usually food borne; the most frequent sources are poultry and eggs. The incubation period is usually 12 to 36 hours (range 5-72 hours, up to 7 days). Usual symptoms are diarrhea, abdominal pain, mild fever, chills, nausea and vomiting; septicemia may occur in infants and immunocompromised patients. Salmonella can cause a wide range of sequelae including pericarditis, reactive arthritis, alkylosing spondylitis, osteomyelitis, neurological and neuromuscular diseases (Baird-Parker AC. Lancet. 1990; 336:1231-1235).

Shigella is transmitted person to person; man is the only important reservoir. Shigella is readily spread among members of a family or group; intrafamilial attack rate is up to 35%. Initial manifestations are fever and malaise followed by dysentery. Complications include dehydration, toxic megacolon, sepsis (shigella or other enterobacteriaceae), hyponatremia, or seizures (Keusch GT, Bennish ML. Pediatr Infect Dis J. 1989; 8:713-719).

Campylobacter presents like salmonella or shigella. Transmission is from direct or indirect contact with domestic or food producing animals, particularly chicken (Skirrow MB. Lancet. 1990; 336:921-923).

Yersinia may cause the following: enterocolitis, acute mesenteric adenitis, arthritis, erythema nodosa or septicemia. Enterocolitis is the most common manifestation with fever, diarrhea and abdominal pain lasting one to three weeks. The typical patient is less than five years of age. These infections are usually self-limited.

Clostridium difficile is associated with antibiotic use and colonic overgrowth occurs with cytotoxic mucosal damage (see CLOSTRIDIUM DIFFICILE TESTING).

(cont)

Stool Culture and Sensitivity (Cont)

E. coli strains may cause diarrhea. Enterotoxigenic E. coli (ETEC) produce enterotoxins and are a common cause of travelers diarrhea. Enteropathogenic E. coli (EPEC) do not produce enterotoxins or invade mucosa, but may cause prolonged diarrhea and are a cause of diarrhea outbreaks in nurseries. Enteroinvasive E. coli (EIEC) cause high fever with bloody mucous stools and a toxic appearance. Enterohemorrhagic E. coli (EHEC) cause bloody diarrhea and low grade fever and are associated with hemolytic uremic syndrome (HUS). Enteroadherent E. coli (EAEC) cause watery diarrhea, particularly in travelers (Ashkenazi S, Cleary TG. Pediatr Infect Dis J. 1991; 10:140-148).

STOOL ELECTROLYTES AND OSMOLALITY

SPECIMEN: 24 hours stool specimen
REFERENCE RANGE: Chloride: 6-17 mmol/24 hours; (6 to 17 mmol/liter); Sodium: 7.8±2.0 mmol/24 hour; Potassium: 18.2±2.5 mmol/24 hour; Measured Osmolality: should be approximately equal to that calculated from stool electrolyte values.
METHOD: Collect 24 hour stool specimen; obtain 30 gram sample, add 100 mL of water and stir for 30 minutes. It is necessary to prevent clogging of aspiration devices of electrolyte machines. Therefore, obtain 40 mL of the suspension, centrifuge, collect supernatant and set aside; then, add 40 mL of distilled water to the residue, mix, centrifuge, collect supernatant and set aside. Repeat extractions for a total of five times. Collect fluids and bring to a volume of 250 mL in a volumetric flask (Caprilli R, et al. Scan J Gastroenterol. 1978; 13:331-335). The stool specimens for osmolality and other laboratory measurement may be prefiltered (nominal cutoff 45,000 daltons)(Epstein M, Pusch AL. Clin Chem. 1983; 29:211). This will work for watery specimens.
INTERPRETATION: Stool electrolytes are measured principally to diagnose [Cl^-] losing states and to distinguish osmotic diarrhea from non-osmotic diarrhea. Chloride is increased in the feces in the following conditions: congenital chloride diarrhea(CCD); acquired chloride diarrhea (Kaplan BS, Vitullo B. J Pediatr. 1981; 99:211-214); secondary chloride-losing diarrhea (Aaronson I. Arch Dis Child. 1971; 46:479); idiopathic proctocolitis(IPC)(Caprilli R, et al. Scand J Gastroent. 1978; 13:331-335); cholera.

CCD and acquired chloride diarrhea are characterized by large concentrations of fecal chloride, metabolic alkalosis, hyponatremia, hypokalemia, hypochloremia, and almost no chloride in the urine; CCD is a more severe illness. Patients with cholera have metabolic acidosis rather than alkalosis.

The mechanism of development of hypochloremic metabolic alkalosis is illustrated in the next Figure:

Hypochloremic Metabolic Alkalosis

Renal Vein
Tubular Cell
H^+ or K^+
$Na^+ + Cl^-$
Na^+Cl^- 80%
Na^+ 20%
Normal

Renal Vein
Tubular Cell
$H^+ + K^+$ $H^+ + K^+$
$Na^+ + Cl^-$
Na^+Cl^-
Na^+
Hypochloremia

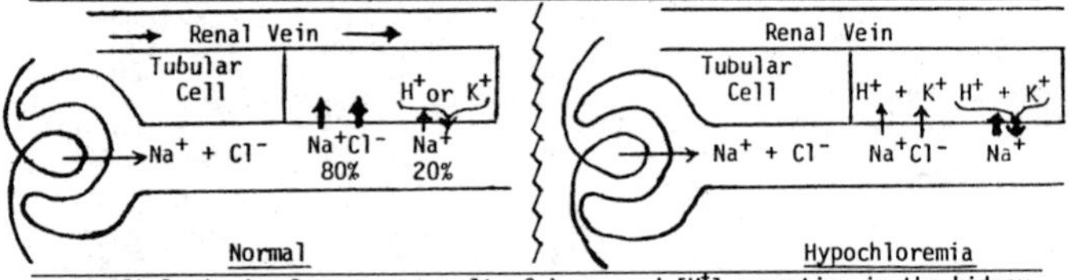

Alkalosis develops as a result of increased [H^+] excretion in the kidney. Normally 80% of the Na^+ resorbed by the kidney is resorbed accompanied by [Cl^-]. In hypochloremia, more [Na^+] must be absorbed by exchange with [K^+] and [H^+]. Metabolic alkalosis, hypokalemia, and a urine almost devoid of [Cl^-] results.

Stool electrolyte values may be used to distinguish between an osmotic and non-osmotic diarrhea. Osmotic diarrheas result from poor absorption of nutrients which are then metabolized to small organic acids by bacteria in the colon resulting in retention of water in the stool (malabsorption, lactase deficiency, rapid transit to colon, etc). The non-osmotic diarrheas result from secretion of both electrolytes and water from the small intestine (bacterial toxins, tumor hormones, certain drugs, etc). The distinction between the two types of diarrhea may be made by measuring stool osmolality and electrolytes and comparing the measured osmolality versus calculated osmolality, e.g.:

$$\text{Calculated Stool Osmolality} = 2(Na^+ + K^+)$$

A large deficit between calculated and measured osmolality indicates an osmotic component to the diarrhea (presence of unmeasured solutes).

STOOL FAT (see FECAL FAT)

STOOL LEUKOCYTES

SPECIMEN: Mucus or liquid stool preferred; rectal swabs may also be used if necessary.
REFERENCE RANGE: Negative
METHOD: A small fleck of mucus or liquid stool is carefully and thoroughly mixed with an equal amount of methylene blue stain solution (reticulocyte stain) or Wright stain and examined for the presence of leukocytes.
INTERPRETATION: This test helps to separate treatable bacterial diseases from diarrhea due to viruses, bacterial toxins or parasites; interpretation is given in the next Table (Guerrant RL, Bobak DA. N Engl J Med. 1991; 325:327-340):

Fecal Leukocytes in Diarrhea

Leukocytes Present	Leukocytes Absent
Shigella	Vibrio Cholerae
Salmonella	E. coli (Toxin-producing)
Campylobacter Jejuni	Clostridium Perfringens
E. coli (Invasive Strains)	Staphylococcus Aureus
Cytotoxic Clostridium Difficile	Bacillus Cereus
Entamoeba Histolytica	Giardia Lamblia
Ulcerative Colitis	Cryptosporidium
	Viruses (Rotavirus, Norwalk Virus, etc)

Cytotoxigenic clostridium difficile and Entamoeba Histolytica may destroy the morphology of fecal leukocytes making test interpretation difficult.

Patients with fecal leukocytes in diarrheal stools have a 70% chance of having a bacterial pathogen as the cause of infection (70% positive predictive value). Up to 90% of patients with bacterial dysentery have fecal leukocytes (90% sensitivity)(DeWitt TG. Pediatric in Review. 1989; 11:6-13).(See also Calubiran OV, et al. Hosp Phys. February 1990; 56-62).

STREPTOCOCCAL ANTIBODIES (see STREPTOZYME, ANTI-STREPTOCOCCAL-O, ASO TESTS and ANTIDESOXYRIBONUCLEASE B)

STREPTOCOCCAL (GROUP A) CULTURE AND RAPID ANTIGEN DETECTION

SPECIMEN: Obtain specimen before antimicrobial chemotherapy is started. Depress the tongue to expose the pharynx. Use a sterile cotton or dacron swab. Commercially available sterile "Culturette" may be used. Rub swab vigorously over the posterior pharynx, tonsils and tonsillar fossae; avoid the tongue, uvula, lips and buccal mucosa.
The specimen may be treated in one of two ways. Transport Culturette to the laboratory as soon as possible; if transport is delayed, refrigerate specimen; or the swab may be inoculated onto sheep blood agar plate. Sheep blood agar is the best medium for detecting Group A beta hemolytic streptococci.
REFERENCE RANGE: No beta hemolytic streptococci isolated.
METHOD: The method of inoculation of sheep blood agar is as follows:
Initially, run the swab over approximately one-sixth of the agar. Then using a sterile loop, streak the primary inoculum onto about one-half of the plate in 10 to 20 to-and-fro strokes. The plate is then turned 90° and the plate is again streaked with the same loop. "Stabs" should be put into the media to enhance hemolysis of red blood cells by streptolysin-O. Incubate at 35° to 37° overnight preferably under anaerobic conditions.

Group A beta hemolytic streptococci may be differentiated from nongroup A beta hemolytic streptococci as follows: Inoculate the organism onto a 5 percent sheep blood agar plate and place a bacitracin disk of 0.04 units of bacitracin at the center; incubate the plate for 18 to 24 hours at 35° to 37°C. Group A beta hemolytic streptococci are susceptible to 0.04 units of bacitracin while other groups of beta hemolytic streptococci are usually resistant. (cont)

Streptococcal Testing (Cont)

INTERPRETATION; The most common cause of bacterial pharyngitis is group A beta hemolytic streptococci (GABHS) and most physicians differentiate group A beta hemolytic streptococci from nonstreptococcal pharyngitis.

Patients with pharyngitis are at increased risk for development of acute rheumatic fever, acute glomerulonephritis, and toxic strep syndrome.

The frequency of isolation by age of GABHS, as a cause of pharyngitis, is given in the next Table (Levy ML, et al. Med Clin of N Amer. 1983; 67:153-172):

Frequency of Age of Isolation of Group A Beta Hemolytic Streptococci in Patients with Pharyngitis

Age	Frequency (%)
<3 years	3 to 15
Children	35 to 50
15 to 35	10 to 20
Over 35	5

Beta-hemolytic streptococci are usually susceptible to penicillin or penicillin derivatives; therefore, susceptibility studies are usually not routinely performed. If a patient is allergic to penicillin, erythromycin might be given.

Group A Streptococcal Rapid Antigen Detection: Various tests designed to rapidly identify group A beta-hemolytic streptococci have been devised utilizing ELISA, latex agglutination and other techniques. These techniques involves extraction of antigen from a throat swab and detection of this antigen with a streptococcal antibody. The specificity of such testing is greater than 90% but the sensitivity is probably only 60-70% (Medical Letter. May 3, 1991; 33[843]:40-41). One study described the sensitivity of rapid antigen detection tests to be only 41% and the sensitivity of single cultures at two days to be 72%. A two plate technique using an anaerobic trimethoprim sulfamethoxazole plate and an aerobic, 5 percent carbon monoxide sheep blood agar plate has been suggested (Wegner DL, et al. JAMA. 1992; 267:695-697). Other authors have disputed these findings, suggesting that a single culture is adequate (Kaplan EL, Amren DP. JAMA. 1992; 268:599).

STREPTOZYME

SPECIMEN: Red top tube, remove serum

REFERENCE RANGE: <100 STZ units

METHOD: Streptozyme is a commercial reagent consisting of formalin-treated sheep erythrocytes that are coated with group A streptococcal antigens, DNase, streptokinase, streptolysin O, hyaluronidase and NADase. Serum diluted 1:100 is mixed on a glass slide with a drop of Streptozyme and agglutination is observed macroscopically.

INTERPRETATION: This is a test for antibody in a patient's serum to multiple different streptococcal antigens, the extracellular product of Group A streptococcus. A significant rise of antibody titer results in a positive test. It is positive in the conditions listed in the next Table:

Positive Streptozyme Test
Post-Streptococcal Glomerulonephritis
Acute Rheumatic Fever
Streptococcal Infections of Pharynx and Skin
Bacterial Endocarditis

The antibodies measured by streptozyme increase more rapidly and appear in the blood earlier than ASO following infection with group A streptococci.

Antibodies increase 1-3 weeks after onset and decrease 8-10 weeks after uncomplicated infection. The sensitivity of the test in patients with streptococcal infection is 95 percent.

Streptozyme test is positive in 97% of patients with elevated ASO titers and 88% of patients with elevated DNase B titers.

References: Ayoub EM. Pediatr Infect Dis J. 1991; 10:S15-S19; Kotylo PK, et al. Check Sample - Immunopathology. 1987; 11:1-5; Gerber MA, et al. Pediatr Infect Dis J. 1987; 6:36-40.

SWEAT CHLORIDE (TEST FOR CYSTIC FIBROSIS)

SPECIMEN: Amount of sweat desired, 200mg; minimum amount, 100mg. The specimen can be stored in the refrigerator at 4°C for 1 week if the container is tightly sealed against evaporation; otherwise, the container may be stored in a freezer. Sweat testing is not considered accurate until the third or fourth week of life. Infants younger than six weeks may not sweat sufficiently; the back or leg may be used for sweat collection in these infants.

REFERENCE RANGE: Sweat chloride and sweat sodium ≤ 40mmol/liter. Sweat chloride >80mmol/liter or sodium >80mmol/liter is consistent with the diagnosis of cystic fibrosis. Children who have moderately elevated values of sweat chloride or sweat sodium (60 to 80mmol/liter) may have cystic fibrosis, but such values are also seen in adrenal insufficiency, glycogen storage disease type 1, fucosidosis, malnutrition/edema, nephrogenic diabetes insipidus, hypothyroidism, ectodermal dysplasia, and prostaglandin E_1 infusion (Stern RC. Pediatrics in Review. 1986; 7:276-286). Children who have intermediate sweat chloride or sodium (40 to 60mmol/liter) are more difficult to diagnose; definitive diagnosis will depend on the clinical situation. Repeat sweat testing should be considered.

During the first two or three postnatal days chloride values may be as high as 80mmol/liter and returning to normal levels on the third or fourth day. Normal adults (>21 years of age) have a range of values for chloride of 10-70mmol/liter.

A gap of more than 30mmol/liter between the sodium and chloride values indicates an error in calculation, analysis or contamination.

Sweat chloride and sweat sodium are elevated in up to 98% of persons with cystic fibrosis.

Conventional units for electrolytes in mEq/liter equal international units in mmol/liter.

METHOD: A kit for assay may be obtained from Farrall Instruments Inc., Arch Avenue and West Highway 30, Grand Island, Nebraska 68803, Tel. #308-384-1530. The method used is the standard quantitative pilocarpine iontophoresis test (Gibson LE, Cooke RE. Pediatrics. 1959; 23:545). The sweat test is done as follows: (1) Local stimulation of sweat with pilocarpine iontophoresis. Pilocarpine is a cholinomimetic; one of its parasympathomimetic effects is its potent ability to produce sweat (2) Collection of the sweat on Curity gauze pads (3) Analysis of chloride by titration; sodium and potassium by flame photometry.

INTERPRETATION: Cystic fibrosis is the most common serious genetic disease in the Caucasian population in the United States; it is transmitted as an autosomal recessive trait, with a carrier rate among Caucasians of 1 in 23 and an incidence of about 1 in 2000 live births. The carrier rate in the African American population is one in 60 to one in 100; therefore, one in 3600 to one in 10,000 African American couples is at risk; their risk is one in four; in 1967, in the African American population, there were 17 new cases of cystic fibrosis in the United States; there were 571,000 births of African American infants.

The classical clinical triad consists of malabsorption due to pancreatic insufficiency, chronic suppurative lung disease and failure to thrive. Chronic respiratory disease is the major cause of morbidity and mortality. The diagnosis of cystic fibrosis requires an elevated sweat chloride test and either a positive family history or one of the two primary manifestations of cystic fibrosis - progressive obstructive pulmonary disease or pancreatic exocrine deficiency.

(cont)

Sweat Chloride (Cont)

Sweat testing is indicated with the following conditions (Stern RC, already cited):

Indications for Sweat Testing	
Pulmonary:	1. Recurrent or nonresolving pneumonia
	2. Staphylococcal pneumonia - first episode
	3. Recurrent bronchiolitis
	4. Chronic bronchitis
	5. Radiologic evidence of bronchiectasis
	6. Lobar atelectasis - first episode
	7. Pseudomonas aeruginosa from the respiratory tract
	8. Hemoptysis in children
Gastrointestinal:	1. Meconium ileus
	2. Prolonged neonatal jaundice
	3. Malabsorption stools
	4. Failure to thrive
	5. Cirrhosis
	6. Pancreatitis
	7. Unexplained cholelithiasis in child, adolescent, or young adult
	8. Intussusception at unusual age
	9. "Meconium ileus equivalent" - unexplained intestinal obstruction in childhood
	10. Symptomatic fat-soluble vitamin (A,D,E,K) deficiency
	11. Rectal prolapse
	12. Typical appendiceal histology (after appendectomy)
Miscellaneous:	1. Family history of cystic fibrosis
	2. Clubbing
	3. Hyponatremic dehydration (without diarrhea)
	4. Unexplained metabolic alkalosis in infancy
	5. Salty taste
	6. Nasal polyps
	7. Pansinusitis
	8. Unexplained hypoproteinemia with anasarca in infancy
	9. Unexplained delayed menarche
	10. Obstructive azoospermia
	11. Parents' request

Immunoreactive Trypsinogen (IRT): Newborns with cystic fibrosis have elevated serum levels of immunoreactive trypsinogen (IRT); the feasibility of mass screening of newborns using assay of IRT is being investigated. Present testing methods have significant numbers of false positive and false negative results. Testing may improve with development of other tests (Rock MJ, et al. Pediatrics. 1990; 85:1001-1007; Lyon ICT, Webster DR. Pediatrics. 1991; 87:954-955; Rock MJ, Farrell PM. Pediatrics. 1991; 87:955-956). Methods are currently being developed to provide for prenatal diagnosis of cystic fibrosis (Johnson JP. J Pediatr. 1988; 113:957-964; Beaudet AL, Buffone GJ. J Pediatr. 1987; 111:630-633). See also review: Tizzano EF, Buchwald M. "Cystic Fibrosis: Beyond the Gene to Therapy." J Pediatr. 1992; 120:337-349).

SYNDROME OF INAPPROPRIATE ANTI-DIURETIC HORMONE(SIADH)

INTERPRETATION: In SIADH, there is excess ADH; the excess ADH causes increased absorption of water from the renal tubular lumen with increased renal sodium and urate excretion. The plasma osmolality is decreased and the urine osmolality is inappropriately increased. These characteristic plasma and urine findings provide the basis for the diagnosis. Plasma ADH levels may also be measured.

Laboratory Findings in SIADH: The laboratory findings in SIADH are given in the next Table:

Laboratory Findings in SIADH
Decreased Plasma Osmolality
Increased Urine Osmolality
Increased Plasma ADH
Hyponatremia
Renal Sodium Loss (Urine Sodium >20mmol/L)
Normal or Expanded Extracellular Fluid Volume
Low Serum Urea Nitrogen (<10 mg/dL)
Low Serum Uric Acid

The basic test for the laboratory diagnosis is the finding of decreased plasma osmolality and increased urine osmolality.

Clinical Features of SIADH: The clinical features of the syndrome are: (1) those of the underlying disease and (2) those of hyponatremia. The extracellular hyponatremia is clinically important because water moves into the body cells. Overhydration occurs in brain cells giving rise to symptoms of increased intracranial pressure - confusion, seizures and coma. Symptoms are usually G.I. and mild when serum sodium levels are above 120 mmol/L and may include anorexia, nausea and vomiting. When the concentration of serum sodium falls below 115 mmol/L, symptoms are usually CNS and finally coma. Patients with a rapid drop in serum sodium may be symptomatic at higher levels.

Causes of SIADH: The conditions associated with SIADH are predominantly pulmonary, central nervous system, or drugs. The SIADH is caused by the following conditions:

Conditions Causing SIADH
Ectopic ADH:
Lung: Oat Cell Carcinoma, Adeno- or Alveolar Carcinoma
Tumors of Pancreas and the Thymus
Infectious Disorders, e.g. TB
Meningitis
Brain Abscess
Aneurysm
Cerebral Vascular Accidents
CNS Tumors
Post-Surgical on CNS
Drugs, e.g. Vincristine

The pathogenesis of the conditions causing SIADH may be obscure. Tuberculous lung tissue may produce ADH (Vorrher et al. Ann Intern Med. 1970; 72:383).

The treatment of excessive secretion of ADH in the ectopic ADH syndrome is treatment of the underlying condition plus water restriction; or lithium or demeclocyline which inhibit arginine vasopressin sensitive adenylcylase in the collecting ducts.

SYNOVIAL FLUID ANALYSIS

SPECIMEN: Specimens: Divide specimen into 4 aliquots as follows:
(1) Red top tube for gross appearance
(2) Heparin tube for microscopic examination
(3) Sterile tube for culture
(4) Heparin tube for chemical analysis

REFERENCE RANGE: The constituents in normal synovial fluid are shown in the next Table:

Constituents of Normal Synovial Fluid

Constituent	Value
Protein	1-3g/dL
Albumin	55-70%
Hyaluronate	0.3-0.4g/dL
Glucose	70-110mg/dL
Uric Acid	
Males	2-8mg/dL
Females	2-6mg/dL
Lactate	10-20mg/dL (1-2mmol/liter)

METHOD: See specific tests. Mucin Clot (Ropes) Test: The mucin clot test reflects polymerization of hyaluronate. In the mucin clot test, a few drops of synovial fluid are added to 20mL of 5% acetic acid. Normally, a mucin clot forms within 1 minute; a poor clot indicates inflammation. This test was once one of the diagnostic criteria for rheumatoid arthritis. More accurate and objective tests are now available, to test whether fluid is inflammatory, so this test is no longer recommended (Hasselbacher P. "Arthrocentesis and Synovial Fluid Analysis." Primer on the Rheumatic Diseases, 9th ed., 1988; 55-60).

INTERPRETATION: For discussion of synovial fluid analysis, see ARTHRITIS (JOINT) PANEL.

SYPHILIS, ANTIBODIES

SPECIMEN: Red top tube, separate serum
REFERENCE RANGE: Non-reactive or negative
METHOD: There are two general types of substances used in detection of antibodies associated with syphilis: (1) antigens composed of tissue extracts (heart) and lipid; (2) antigens derived from treponema. These are used in agglutination, hemagglutination, and fluorescent antibody tests. Hemolysis and lipemia may interfere.
INTERPRETATION: There are two general approaches to the detection of antibodies produced in response to treponemal pallidum infection. Procedures utilizing antigens from heart and lipid (VRDL, RPR) detect antibodies directed against host tissues and are most useful as screening tests. Procedures utilizing antigens from treponemas (FTA-ABS, MHA-TP) are more specific and are useful in confirming the diagnosis, and in diagnosis of the disease at times when the nontreponemal antibodies are not produced (early and late disease). These tests are listed in the next Table:

Tests for Detection of Antibodies Produced in Response to Treponemal Pallidum
(1) Use of Nontreponemal Derived Substances to React with Antibody:
(a) VDRL (Venereal Disease Research Laboratory)
(b) RPR (Rapid Plasma Reagin)
(2) Use of Antigen Derived from T. Pallidum to React with Antibody:
(a) FTA-ABS (Fluorescent Treponemal Antibody-Absorption Test)
(b) MHA-TP (Microhemagglutination-Treponema Pallidum)
(c) TPI (Treponema Pallidum Immobilization)

The RPR cannot be used with cerebrospinal fluid (CSF); the VDRL can be used to test for antibody in CSF. The epidemiology, laboratory diagnosis, and treatment of syphilis is reviewed in: Hook EW, Marra CM. "Acquired Syphilis in Adults, Medical Progress." N Engl J Med. 1992; 326:1060-1069.

(1) Nontreponemal (Reaginic) Tests: The nontreponemal tests are used as screening tests and for following therapy; quantitation is done to follow the response to therapy and to detect reinfection. These tests use extracts of beef heart (cardiolipin) to give reproducible qualitative and quantitative reactions to antibody produced following infection with Treponema pallidum. It is not known whether the antibodies are produced in response to antigens in T. pallidum or to antigens resulting from the interaction of T. pallidum with tissue.

Positive tests are quantitated. Most true-positive VDRL tests are >1:8; false-positive tests are commonly <1:8. A four-fold fall in titers is expected within the first 3 months of treatment of primary and secondary syphilis. Tertiary syphilis may fail to show a decline in VDRL titers. Titers are useful in monitoring treatment of congenital syphilis. A positive serology in an infant may mean active infection, or alternately, maternally derived antibody. Rising titers indicates active infection or inadequate therapy in an infant. The CSF gives a positive VDRL in less than half the patients with neurosyphilis. A non-reactive CSF VDRL does not rule out neurosyphilis (Am J Clin Pathol. 1991; 95:397). CSF-cell counts are considered a more reliable indicator of adequate treatment. False-positive reactions occur in a variety of acute and chronic conditions as listed in the next Table:

Conditions Causing False-Positive Results Using Nontreponemal Tests for Antibodies Produced in Response to Syphilis	
Lupus Erythematosus	Miliary Tuberculosis
Other Collagen Vascular Disease	Typhus
Malaria	Related Treponemal Infections
Infectious Mononucleosis	Lyme Disease
Infectious Hepatitis	Yaws
Post-Vaccination States	Pinta
Leprosy	Pregnancy
Brucellosis	Hypergammaglobulinemia
Atypical Pneumonia	Drug Addiction

A more detailed list of both infectious and non-infectious causes of false positive nontreponemal and treponemal serological tests is given in: Hook EW, Marra CM. N Engl J Med. 1992; 326:1062. False-positive tests for syphilis, in patients with systemic lupus erythematosus are caused by anti-DNA antibodies that cross-react with cardiolipin (Koike T, et al. Clin Exp Immunol. 1984; 56:193-199).

False-negative nontreponemal results are found in early disease and in a significant number of patients with late syphilis. In these instances one of the treponemal tests (FTA-ABS, MHA-TP) may be positive. False negative "prozone"-type interference associated with high titer antibodies are sometimes seen. These are resolved by assaying a diluted specimen. AIDS patients may not develop an appropriate response to syphilitic infection.

(2) <u>Treponemal Tests: Use of antigen derived from T. pallidum</u>: These tests are most commonly used to determine whether the results of a nontreponemal test are due to syphilis or due to a condition causing false-positive results. These tests may also be used to detect syphilis in patients with negative nontreponemal test results but with clinical evidence of early or late syphilis.

The MHA-TP test is less sensitive than the FTA-ABS test in primary untreated syphilis; the sensitivity of the MHA-TP Test is 65% to 70%, and the FTA-ABS has a sensitivity of 80% or greater. However, the MHA-TP test has the following characteristics: simplicity, lower overall cost, and reliability. If the non-treponemal test is positive, then the MHA-TP test may be done for confirmation.

The treponemal tests do not indicate the patient's response to treatment and are of doubtful value in the diagnosis of active neurosyphilis. The treponemal tests tend to remain positive throughout life, even in adequately treated individuals.

The TPI test is performed by observing the immobilization of the organism by patient antibodies. It is basically of research interest since it requires a source of live organisms.

T. pallidum has not been cultivated in-vitro; it is grown intratesticularly in rabbits; thus, antigens are contaminated with rabbit tissue. False positive FTA-ABS test results have been obtained in patients with increased or abnormal globulins, with pinta, yaws and bejel, leptospirosis, Borrelia, Lyme disease, herpes, lupus erythematosus and during pregnancy. False negative results have been reported in patients with AIDS.

The Centers for Disease Control are able to perform a specific IgM antibody test by the fluorescent treponema antigen-antibody test. Rheumatoid factor is removed from the serum with a column. A positive IgM antibody is indicative of recent and ongoing infection.

The reactivity of nontreponemal and treponemal tests in untreated syphilis is given in the next Figure (Henry JB [ed.] Todd-Sanford-Davidsohn. Clinical D Diagnosis by Laboratory Methods, 16 ed., 1979. Vol. II, pg. 1890, W.B. Saunders Co., Phila. PA):

Reactivity of Nontreponemal and Treponemal Tests in Untreated Syphilis

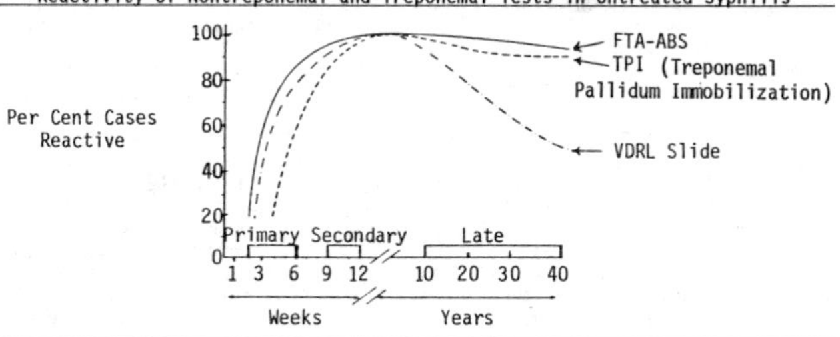

<u>Sexual Contact</u>: It is estimated that approximately 33% of partners of individuals with early syphilis will become infected (vonWerrsowetz AJ. J Vener Dis Inf. 1948; 28:132-137; Schober PC, et al. Br J Vener Dis. 1983; 59:217-219). Sexual contact during latent disease is at lower risk.

<u>Incubation Period</u>: Typically 21 days, but up to 10 weeks; during this period, there is no serological evidence of infection. T. pallidum is multiplying at the site of inoculation and is invading the lymphatics and blood stream.

Syphilis Antibodies (Cont)

Primary Stage: Appearance of lesion (chancre) at the site of inoculation; chancres usually heal spontaneously within a few weeks. At this stage, spirochetes may be detected by dark field microscopy. Fluorescent microscopy employing reaction of lesion exudates with monoclonal and polyclonal antibodies directed against the organism may be used as an alternative to dark field microscopy. The nontreponemal screening tests (VRDL, RPR) start to become reactive in early primary syphilis. The nontreponemal tests (FTA-ABS, MHA-TP) become reactive slightly sooner.

Secondary Stage: Generalized skin and mucous membrane lesions; multiple organs may be involved including the central nervous system. These manifestations disappear usually after several weeks. About 10 percent of patients with recognized secondary syphilis have clinical evidence of liver disease (Koff RS in Case Records of Massachusetts General Hospital. 1983; 309:35-44). Highest titers of VDRL and RPR are seen in secondary and early latent syphilis.

Latent Syphilis: Early studies on untreated syphilis (before 1910) revealed that approximately 25% of patients with latent syphilis had recurrent mucocutaneous lesions, usually in the first year, but up to a period of 5 years. Hence latent syphilis is arbitrarily divided into early latent (infectious) and late latent (non-infectious). The division between early and late latent syphilis is at one year (CDC definition) or at two year (WHO definition). Titers of nontreponemal tests peak in secondary or early latent syphilis. Antibody levels decline in late latent syphilis.

Tertiary (Late) Syphilis: 5 to 20 years after infection; the most frequent complications involve lesions in the central nervous and cardiovascular systems and gummatous lesions of skin, bone and viscera. Antibody levels (nontreponemal tests) decline even without treatment, and in up to 25% of untreated patients the VDRL tests become non-reactive. The more sensitive treponemal tests (FTA-ABS, MHA-TP) usually stay reactive even after successful treatment.

Congenital Syphilis: The spirochete, in a mother who has active syphilis, crosses the placenta after 18 weeks gestation and causes a spirochetemia in the fetus. Congenital syphilis occurs in less than 0.05% of all pregnancies but is increasing. Antibodies in an infant's blood can mean active infection or presence of maternal antibodies. Rising titers in the infant following delivery indicates active infection.

CNS Syphilis: Abnormal CSF findings may be observed in primary (10-20%), secondary (30-70%), and latent syphilis (10-30%). Abnormalities include increased mononuclear cells, increased globulin and protein, and reactive nontreponemal tests. In addition, organisms have been found in the CSF of 15-30% of patients with none of the above laboratory findings. Although the CSF findings resolved in 2/3 of patients by the end of the secondary stage of syphilis, the ones with the most severe findings are the ones in which neurosyphilis is likely to develop (overall 4-9% of untreated patients. The CSF-VDRL is reactive in 30-70% of patients with neurosyphilis. The use of the FTA-ABS and MHA-TP on CSF is not generally accepted (Hook EW, Marra CM. N Engl J Med. 1992; 326:1060-1069). It is said that a negative FTA-ABS on CSF is effective to exclude the diagnosis of neurosyphilis (Jaffe HW, et al. Arch Intern Med. 1978; 138:252-255). After treatment of neurosyphilis, the CSF cellular findings return to normal before the globulin and protein, and the CSF-serological tests may remain reactive for more than one year.

Syphilis and HIV: HIV infection may have a modifying effect on serological findings in syphilis. Those patients with early or asymptomatic HIV infection appear to have elevated titers to nontreponemal tests (RPR) while those in later stages of HIV infection appear to have reduced or absent titers. Persons with HIV infection but who do not have syphilis have a greater number of false positive nontreponemal tests. In response to therapy, titers appear to fall more slowly in patients with HIV infection and primary syphilis. The number of patients whose FTA-ABS and MHA-TP tests turn non-reactive is increased in HIV infected individuals. A thorough discussion of HIV and syphilis coinfections is given: Hook EW, Marra CM, already cited.

T AND B LYMPHOCYTE ENUMERATION (see LYMPHOCYTE DIFFERENTIAL)

TEGRETOL (see CARBAMAZEPINE)

TEICHOIC ACID ANTIBODY TITER

SPECIMEN: Red top tube, separate serum and refrigerate
REFERENCE RANGE: Titer less than 1:2-1:8 (variable range). Reference values and diagnostic titers may vary between laboratories.
METHOD: Counterimmunoelectrophoresis (CIE) and radial diffusion.
INTERPRETATION: Antibody to teichoic acid is increased in infection with Staphylococcus aureus. It may be of greatest interest when it is not clear if a positive blood culture is due to a true infection or due to skin contamination. The condition of a positive blood culture and elevated teichoic antibody titer is particularly significant in a patient suspected of endocarditis. Interpretation of teichoic acid antibody results is as follows (Smith-Kline Clinical Laboratories):

1. Antibodies are often negative during the first two weeks following onset of Staphylococcal infection; therefore, if Staphylococcal infection or endocarditis is a significant concern, and if antibodies are negative early, they should be repeated.
2. The incidence of antibodies relates to the seriousness of the infection.

	% Positive	% with Titer >1:2
Staphylococcal Endocarditis	87	72
Staphylococcal Bacteremia with Metastatic Infection	50	35
Uncomplicated Staphylococcal Bacteremia	8	0
Non-Bacteremic Staphylococcal Infection	58	10
Normals, Estimated	5	0

3. Occasional positive results occur in patients with other types of infections, connective tissue diseases, and neoplastic diseases. These are usually in undiluted serum only. Cross reactions may occur with H. influenzae infections.

TERMINAL DEOXYNUCLEOTIDYL TRANSFERASE (TdT)

SPECIMEN: Bone marrow; peripheral blood, lavender tube (EDTA), cerebrospinal fluid.
REFERENCE RANGE: Negative
METHOD: Bone marrow and peripheral blood are treated with 0.1M NH_4Cl to lyse erythrocytes; the mononuclear fraction is spread on a glass slide by cytocentrifuge; fix smear and maintain at room temperature in a desiccator for 48 to 72 hours. Layer rabbit antibody to TdT over the smear and incubate for 30 minutes; wash slide with phosphate buffered saline, and add fluorescein tagged antirabbit IgG (or peroxidase labelled) for 30 minutes. Wash slides and mount.
INTERPRETATION: This test is performed to aid in the subclassification of leukemias and lymphomas and to follow the progression and response to therapy. TdT is typically found in early lymphoid cells and may have diagnostic, prognostic and therapeutic significance. TdT positive cells are sensitive to vincristine and steroids while TdT negative cells are resistant to steroids (Morse EE, et al. Ann Clin Lab Sci. 1983; 13:128-132). Situations in which determination of TdT status of blasts is useful are discussed below:
ALL: Most cases of ALL (90%) are TdT positive. These include pre-T and pre B-ALL. B-ALL is negative. In remission very few if any TdT positive cells should be found in blood or CSF. Normal and regenerating bone marrow contains a large number of TdT positive blasts limiting the assessment of residual disease in bone marrow.
CML: Thirty percent of CML in blast crisis will express TdT in their blast cells.
Lymphoblastic Lymphomas: The neoplastic cells of this lymphoma express TdT.
AML: From 5-40% of AML may contain blast cells expressing TdT. This tends to be of low titer. The number of AML cases with marker expression appears to depend on the sensitivity of the reagents used and the cutoff value for TdT selected by the investigator.
Evaluation of Residual Disease: TdT is useful to follow therapy in specimens other than bone marrow (e.g., CSF, blood).

TESTOSTERONE, FREE

SPECIMEN: Red top tube, separate serum promptly, store at -20°C; or green (heparin) top tube, separate plasma and store at -20°C until assay.
REFERENCE RANGE: Adult Males: 9-30ng/dL; Females: The average free testosterone concentration in girls rises on pubertal maturation; by midpuberty, it reaches adult level (Moll GW, Rosenfield RL. J Pediatr. 1983; 102:461-464). Adult values vary with methodology; Mayo Clinic Laboratories, 0.3-1.9 ng/dL.
METHOD: Radioimmunoassay with extraction and equilibrium dialysis.
INTERPRETATION: Free testosterone measurements are useful in the evaluation of clinical states in which sex hormone binding globulin (SHBG) may be altered (liver disease, thyroid disease, androgen or estrogen imbalances, obesity). Free testosterone is the non-protein-bound fraction of circulating testosterone and is probably the physiologically active form; only 1 to 3% of plasma testosterone is in the free state. The bound testosterone is bound to a specific binding protein (SHBG); there is also non-specific binding to albumin. Usually, there is a positive linear correlation between the level of total plasma testosterone and free testosterone. However, hyperthyroidism and hyperestrogenic states, such as pregnancy and oral contraceptives, are characterized by increased SHBG, increased total testosterone and normal free testosterone. There are relatively few conditions in which assay of free testosterone is required.

In the diagnostic working-up of hirsutism, due to androgen-secreting ovarian or adrenal neoplasms, total testosterone and dehydroepiandrosterone sulfate are the most useful markers in that these neoplasms hardly ever occur without peripheral elevation of either steroid. Measurement of free testosterone is not usually necessary in diagnosis of these conditions.

The clinical significance of the use of free testosterone for the diagnosis of hirsutism is controversial. Although plasma free testosterone concentration was found to be elevated in every hirsute woman studied by Paulson JD, et al. Am J Obstet Gynecol. 1977; 128:851, other studies indicate that free testosterone is elevated in only half of hirsute women (Schwartz U, et al. Fertil Steril. 1983; 40:66-72; Wu CH. Obstet Gynecol. 1982; 60:188-194). The measurement of plasma free testosterone before and after administration of dexamethasone is said to be the most sensitive single method for detecting polycystic ovary syndrome in adolescence (Moll GW, Rosenfield RL. J Pediatr. 1983; 102:461-464).

There is some evidence that albumin-bound testosterone may also be biologically active. A new test employing ammonium sulfate precipitation called Bioavailable Testosterone is able to measure free plus albumin-bound testosterone.

TESTOSTERONE, TOTAL, SERUM

SPECIMEN: Red top tube, separate serum; store frozen; or green (heparin) top tube, separate plasma and freeze. Testosterone levels increase if serum remains in contact with RBC for a long interval.
REFERENCE RANGE: Adult male: 350-800ng/dL; in healthy men, serum testosterone remains in the normal range even with advancing age (Sparrow D, et al. J Clin Endocrinol Metab. 1980; 51:508-513). Adult female: 10-60ng/dL; Prepubertal children: <10ng/dL; Pregnancy: 75-300ng/dL.
METHOD: RIA
INTERPRETATION: This test is useful in the evaluation of hypogonadism, infertility, and impotence in the male and hirsutism and virilization in females. It may also be of use in evaluation of ambiguous genitalia, precocious puberty, and testosterone replacement therapy.

Males: Testosterone levels follow a circadian pattern with levels highest in early morning. There is a broad reference range for normal males. Testosterone is bound to sex hormone binding globulin SHBG, and to a small extent to albumin. Only the free hormone (approximately 2%) and possibly the small amount bound to albumin are biologically available. Serum testosterone is decreased in (a) testicular failure (hypergonadotropic hypogonadism) and (b) pituitary failure (hypogonadotropic hypogonadism): (see LUTEINIZING HORMONE).

There are two conditions in which male sexual dysfunction is associated with elevated serum total testosterone level; these are hyperthyroidism and syndromes of androgen resistance. Hyperthyroidism causes an increase in testosterone binding globulin and more testosterone binds to this globulin; the free testosterone is normal. When hyperthyroidism is treated successfully, testosterone binding globulin returns to normal and potency is restored (Spark RF, et al. JAMA. 1980; 243:750-755).

Syndromes of androgen resistance are also associated with increased serum testosterone; in this syndrome, there may be an abnormality of the androgen receptor or in the 5-alpha-reductase enzyme which is responsible for the conversion of testosterone to the active metabolite, dihydrotestosterone. The integrity of the adult male reproductive system may be tested by measuring parameters of ejaculation (see SEMEN ANALYSIS).

Females: Hirsutism in the female can be of adrenal or ovarian origin and is caused by excess androgens. Testosterone is a good indicator of ovarian function while dehydroepiandrosterone-sulfate (DHEA-S) is a good indicator of adrenal function. Plasma testosterone over 200ng/dL is usually indicative of an ovarian abnormality (Maroulis GB. Fertil Steril. 1981; 36:273). Levels <150 ng/dL arise from adrenal carcinoma. This can often be confirmed by demonstration of increased DHEA-S levels, an indicator of adrenal androgen production. There are many instances of exceptions to these generalizations.

The clinical causes of hirsutism are given in the next Table:

Causes of Hirsutism

Common:
- Polycystic Ovary Syndrome

Rare:
- Adrenal Origin: Cushing's Syndrome; Late-Onset or Attenuated 21-Hydroxylase Deficiency (Congenital Adrenal Hyperplasia); Androgen-Secreting Tumors
- Ovarian Origin: Hilus Cell Tumor; Androblastoma; Teratoma
- Hypothyroidism
- Acromegaly
- Hyperprolatinemia

Drugs:
- Phenytoin; Diazoxide; Minoxidil; Menopausal Mixtures Containing Androgens, Corticosteroids (rarely)

An algorithm based upon the dexamethasone-suppression test may be useful for the diagnostic approach to hirsutism (Hatch R, et al. Am J Obstet Gynecol. 1981; 140:815-830).

Stein-Leventhal Syndrome (Polycystic Ovary Syndrome): The Stein-Leventhal Syndrome is the most common hormonal cause of hirsutism. Evidence of menstrual irregularity may point towards polycystic ovarian disease. Borderline elevations of the androgens, testosterone and 17-ketosteroids are present; raised LH levels and increased LH/FSH ratios (>2.5 to 1) are consistent with this disease. Attenuated 21-hydroxylase deficiency may be clinically indistinguishable from polycystic ovary syndrome. The prevalence of the disorder is disputed with figures ranging from 1-12% of hirsute women. The diagnosis is established by comparing 17OH-progesterone levels before and after ACTH stimulation.

Galactorrhea (galactorrhea/amenorrhea syndrome) should be excluded since there is an association of hirsutism with hyperprolactinemia.

THEOPHYLLINE (AMINOPHYLLINE)

SPECIMEN: Red top tube, separate serum and refrigerate.
REFERENCE RANGE: Therapeutic: 10 to 20mcg/mL (10 to 20mg/liter). Toxicity: >20mcg/mL (20mg/liter). To convert traditional units in mg/liter to international units in micromol/liter, multiply traditional units by 5.550.

The serum half-life for different age groups is given in the next Table:

Change of Plasma Half-Life with Age

Age Group	Half-Life (Hours)
Adult (Healthy Non-Smoker)	8.7 (4 to 16)
Adult (Healthy Smoker)	4.4 (3.0 to 9.5)
Children (1 to 9)	3.7 (2 to 10)
Premature Infants	30
Term Infants (under 3 Months)	24

Factors that Alter Aminophylline Clearance (in Terms of Half-Life)

Decreased Clearance:
- Patients with cardiac decompensation or hepatic cirrhosis:
  - Half-Life: 20 to 30 hours
- Neonates and Infants
  - Half-Life: Prematures - about 30 hours
    Normal Newborns - about 24 hours
    Infants beyond Neonatal Period - gradually decreases from Newborn value to Childhood levels by year one.
- Concomitant Drug Therapy: Erythromycin, Cimetidine, Others
- Acute Febrile Illness

Increased Clearance:
- Cigarette or marijuana smokers; Half-Life: 4.4 hours
- Children one to nine years old
  - Half-Life: 3.7 hours mean (2 to 10 hours range); half-life increases with age beyond nine years to adult values at age 16

The plasma half-life is altered by the following conditions: Tobacco smoking: half-life shortened; chronic liver disease, congestive heart failure and severe obstructive pulmonary disease: half-life prolonged.

Some of the drugs that alter the metabolism of theophylline are given in the next Table (See also Committee on Drugs. Pediatr. 1992; 89:781-783; Sessler CN, Brady W. J Crit Illness. 1991; 6:1045-1054):

Drugs and Metabolism of Theophylline

Decreased Clearance (Prolonged Half-Life)	Increased Clearance (Decreased Half-Life)
Cimetidine (Tagamet)	Phenytoin
Propranol (Inderal)	Barbiturates
Allopurinol (Zyloprim and others)	
Erythromycin	
Others	

Serum concentrations of theophylline should be monitored in patients taking these drugs concurrently with theophylline, especially when these drugs are introduced or discontinued (The Medical Letter. Jan. 6, 1984. 26:1-3).

METHOD: RIA; EMIT; HPLC; SLFIA(Ames); Nephelometry(ICS); Fluorescence Polarization(Abbott).

INTERPRETATION: Theophylline is used as a bronchodilator for relief of acute asthmatic symptoms, as a "prophylactic" agent for controlling the symptoms and signs of chronic asthma and to control premature infant apnea.

Aminophylline is theophylline ethylene diamine; aminophylline and other theophylline salts can be given orally or parenterally. Theophylline can be given only orally.

Dosage of Aminophylline for Acute Therapy: For acute therapy, an intravenous loading dose of about 6mg/kg aminophylline, if ideal body weight, will usually yield a serum concentration in the range 10 to 15mcg/mL in patients who have previously taken no theophylline. The intravenous loading dose should be infused over a period of at least 30 minutes to minimize the probability of serious arrhythmias. IV aminophylline in the acute management of asthma may not provide additional bronchodilation compared to beta-adrenergic agents and steroids and side effects may be increased (Carter E, et al. J Pediatr. 1993; 122:470-476; DiGiulio GA, et al. J Pediatr. 1993; 122:464-469; Weinberger M. J Pediatr. 1993; 122:403-405; Siegel D, et al. Ann Rev Respir Dis. 1985; 132:283-286). Many clinicians no longer use this therapy for acute asthma.

A protocol for aminophylline therapy for acute asthmatic attack is given in the next Figure (Hendeles L, Weinberger M in Individualizing Drug Therapy. Taylor WJ and Finn AL, eds. Gross, Townsend and Frank, Inc. New York, NY. 1981; 1:32-65):

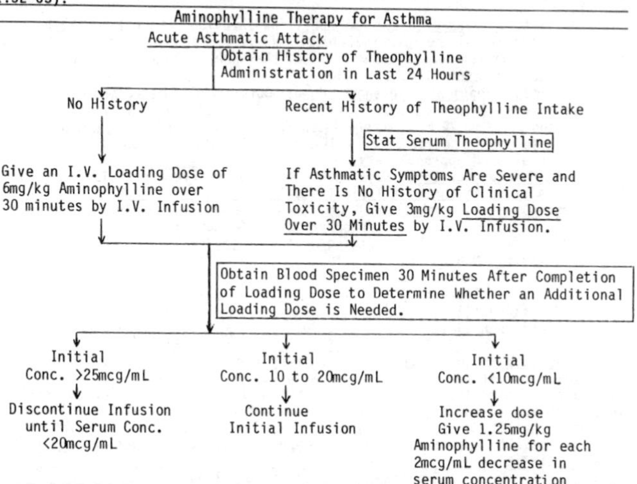

Blood Specimen: (Loading Dose in Hospital) A blood sample for aminophylline assay should be drawn prior to loading dose if the patient has a history of theophylline therapy; then obtain a blood specimen 30 minutes after completion of loading dose.

Maintenance Dose: The aminophylline infusion rate necessary to attain a serum level of 10mcg/mL for different conditions is as follows (Powell R. School of Pharmacy. University of North Carolina):

Aminophylline Infusion Rate

Adults:

Characteristic	0.5mg/kg/hr x Factor		Aminophylline Infusion Rate
Nonsmoker	0.5mg/kg/hr x 1	=	0.5 (mg/kg/hr)
Smoker (within last 3 mos > 1 pack/day) plus ages 10-16	0.5mg/kg/hr x 1.6	=	0.8 (mg/kg/hr)
CHF (edema, x-ray)	0.5mg/kg/hr x 0.4	=	0.2 (mg/kg/hr)
Pneumonia, Viral Respiratory Infection	0.5mg/kg/hr x 0.4	=	0.2 (mg/kg/hr)
Cirrhosis	0.5mg/kg/hr x 0.4	=	0.2 (mg/kg/hr)
Severe hepatic obstruction (PEER < 100L/min or PCO_2 > 45mmHg)	0.5mg/kg/hr x 0.8	=	0.4 (mg/kg/hr)

Example: The aminophylline infusion rate in a 60kg male who smokes 2 packs of cigarettes per day and currently has signs and symptoms of CHF: Aminophylline infusion rate = 0.5mg/kg/hr x 60kg x 1.6 x 0.4 = 19.2mg/hr

Theophylline (Cont)

Premature Infants and Neonates: Aminophylline Infusion Rate = 0.19mg/kg/hr
Infants 4 to 52 Weeks: Aminophylline Infusion Rate = 0.37 + (0.012 x Age in Weeks)
Children (1 to 9 Years of Age): Aminophylline Infusion Rate = 1.0mg/kg/hr

Interindividual and intraindividual differences or changes in theophylline clearance in children necessitates frequent therapeutic drug monitoring (Kubo M, et al. J Pediatr. 1986; 108:1011-1015).

Blood Specimen (Inpatient Maintenance Dose): Obtain a blood specimen 4 to 8 hours after start of maintenance I.V. infusion and 18 to 30 hours after starting the infusion and at 12 to 24 hour intervals thereafter for possible adjustment of the infusion rate until the infusion is discontinued. As a function of results, the dose may be altered appropriately.

Outpatient Maintenance Dose: Maintenance dose to achieve a peak level of 10 to 20mcg/mL is given in the next Table (Powell R):

Maintenance Dose of Theophylline or Aminophylline (mg/kg/day) Ideal Body Weight

	Theophylline (mg/kg/day)	Aminophylline (mg/kg/day)
Adult - Nonsmoker	10	12
Smoker	15	19
Cirrhotic	4	5
Children (1-8)	12-24	15-30
Children (9-12)	10-20	12.5-25
Children (12-16)	8-18	10-22.5

Blood Specimens (Outpatient Maintenance Dose): Blood specimens should be obtained at about 3 day intervals and at the time of peak concentration or about 2 to 4 hours after a scheduled dose depending on the product formulation. Concentrations of 5 to 10 mcg/ml may be effective for some patients, particularly children.

Toxicity: Toxicity usually is seen when serum levels rise above 20mcg/mL. Early signs of toxicity are nausea, vomiting and headache. Toxic reactions are listed in the next Table:

Toxic Reactions of Theophylline
Gastrointestinal: Abdominal Pain, Nausea, Vomiting
Cardiovascular: Tachycardia, Arrhythmias, Hypertension
Central Nervous system: Agitation, Irritability, Insomnia, Tremor, Seizures
Occasionally: Hyperthermia, Diuresis, Ketosis, Hyperglycemia, Hypokalemia

Theophylline may cause tachycardia and serious arrhythmias even at therapeutic serum concentrations in adults (Bittar G, Friedman HS. Chest. 1991; 99:1415-1420).

THERAPEUTIC DRUG MONITORING

Multiple Drug Doses: Drug accumulation, following multiple oral doses, is illustrated in the next Figure (The Bio-Science Handbook of Clinical and Industrial Toxicology, 1st Ed., March 1979):

Drug Concentration - Time Relationship for Multiple Oral Dose

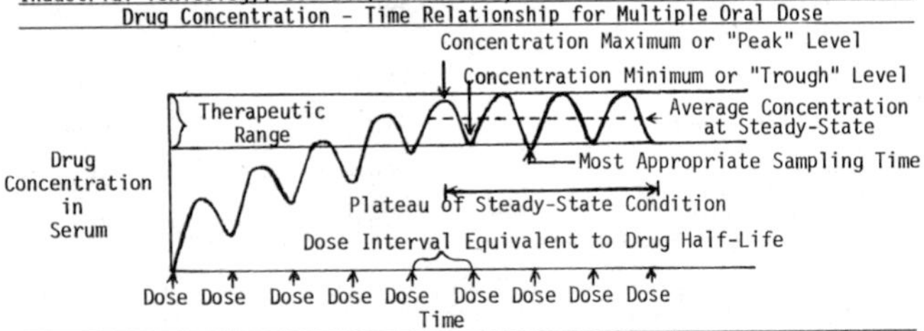

After administration of a single dose, drug levels increase and then decline. If a second dose is administered before complete elimination of the first dose, the drug levels achieved by the second dose will be higher. A third dose produces higher levels and an increased elimination.

When the rate of oral administration remains constant, the rate of elimination will increase until it is in equilibrium with the rate of administration; at this time, the steady state is established. The time required to reach the steady state is approximately five half-lives. This phenomenon occurs in reverse when drug dosing is discontinued. A blood level measured after four doses, administered at half-life intervals, will reflect approximately 94% of the ultimate steady-state drug level (Bio-Science Handbook). Approximately five half-lives are required for the nearly complete removal of the drug.

Note that in the previous Figure, drug administration causes drug concentration to fluctuate between a maximum (peak) and a minimum (trough) level. Therefore, timing of specimen in relationship to drug administration is important. Also, note that the interval between doses of the drug is equal to the drug half-life.

If the dosing interval, at steady state, is longer than the half-life of the drug, then the relationship that is obtained is shown in the next Figure:

Dosing Interval, at Steady State, Longer than Half-Life

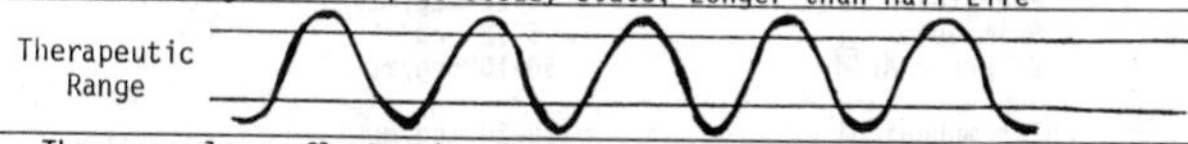

There are large fluctuations between peak and trough levels.

If the dosing interval, at steady state, is shorter than the half-life of the drug, then the relationship that is obtained is shown in the next Figure:

Dosing Interval, at Steady State, Shorter than Half-Life

Therapeutic Range

Altered Pharmacokinetics: The conditions associated with altered pharmacokinetics are listed in the next Table:

Conditions Associated with Altered Pharmacokinetics

Condition	Effects
Congestive Heart Failure	Edema of gut wall: impaired absorption Decreased liver blood flow: decreased rate of hepatic metabolism Decreased renal blood flow and decreased glomerular filtrate rate: prolongation of half-life of agents renally secreted
Abnormal Renal Function	Prolongation of half-life of agents renally secreted; increase in free drug concentration
Abnormal Liver Function	Prolongation of half-life of agents that are metabolized by the liver

The routine screening procedures recommended for elderly patients in whom use of psychotropic drugs is being considered are liver function tests and renal function tests. Since older persons have a decreased muscle mass, they have a decreased rate of creatinine production; a reduced rate of creatinine clearance may coexist with serum creatinine in the normal range (Koch-Weber J. N Engl J Med. 1983; 308:134-138).

Determination of Half-Life of a Drug: Collect at least 3 timed specimens at "appropriate" timed intervals (e.g., 30-60 minutes apart or other appropriate times) and plot log concentration versus time.

Sampling Time: Basic principles of sampling time are as follows:

Toxicity: A single stat specimen is required; determine peak levels.

Determination of Half-Life: Obtain two samples drawn at precise times after the drug is administered.

Intravenous and Intramuscular Medication: Sample one-half to one hour after administration, e.g., Gentamicin.

(cont)

Therapeutic Drug Monitoring (Cont)

<u>THERAPEUTIC RANGE</u>: Therapeutic ranges for some of the more commonly monitored drugs are given in the following Table:

Therapeutic Range

Drugs	Therapeutic Range
Aminoglycosides	
Amikacin	Peak: 10-30mcg/mL; Trough: <5mcg/mL
Gentamicin	Peak: 4-10mcg/mL; Trough: <2mcg/mL
Tobramycin	Peak: 4-10mcg/mL; Trough: <2mcg/mL
Antiepileptic Drugs	
Carbamazepine	4-12 mcg/mL
Ethosuximide	40-100mcg/mL
Pentobarbital	1-5 mcg/mL
Phenobarbital	15-40 mcg/mL
Phenytoin	10-20 mcg/mL
Primidone	5-12 mcg/mL
Valproic Acid	50-100mcg/mL
Chloramphenicol	10-20 mcg/mL
Cyclosporine	100-300mcg/mL
Digoxin	0.5-2.0ng/mL
Digitoxin	11-23ng/mL
Disopyramide	2-6mcg/mL
Lidocaine	1.2-6.0mcg/mL
Lithium	0.6-1.2mmol/liter
Procainamide	4.0-8.0mcg/mL
Quinidine	2.0-5.0mcg/mL
Theophylline	10-20mcg/mL
Tricyclic Antidepressants	
Imipramine (and Desipramine)	100-300ng/mL
Amitriptyline	75-250ng/mL
Desipramine	50-300ng/mL
Nortriptyline	50-150ng/mL
Protriptyline	50-170ng/mL
Doxepin	90-250ng/mL

THERAPEUTIC PHERESIS

Therapeutic pheresis (hemapheresis) is a procedure in which whole blood is withdrawn from the patient, a portion of the blood separated and retained (e.g., leukocytes, platelets, plasma) and the remainder retransfused into the patient.

<u>Therapeutic Leukapheresis</u>: Therapeutic leukapheresis is used to reduce the level of circulating leukocytes. It is used to remove leukocytes in newly diagnosed patients with acute leukemia, and in some patients with chronic myelogenous or lymphocytic leukemia.

<u>Therapeutic Plateletpheresis</u>: Therapeutic plateletpheresis is used to reduce the level of circulating platelets. It is used to remove platelets in patients with thrombocytosis (usually associated with myeloproliferative syndrome) who may be at risk from thrombosis. The platelet count can be reduced by approximately 50% in three to four hours.

<u>Therapeutic Plasmapheresis</u>: Plasmapheresis is used to remove a wide variety of substances. The principal indications for therapeutic plasma exchange are given in the next Table (McLeod BC. J Crit Illness. 1991: 6:487-495):

Principal Indications for Therapeutic Plasma Exchange	
Disorder	Percent of Procedures
Acute Guillain-Barré Syndrome	21.7
Myasthenia Gravis	18.0
Thrombotic Thrombocytopenic Purpura	12.1
Paraproteinemia	10.4
Systemic Lupus Erythematosus	6.6
Cryoglobinemia	4.5
Goodpasture's Syndrome	1.9
Rapidly Progressive Glomerulonephritis	1.6
Other	23.2

Recent studies have failed to demonstrate benefit from plasmapheresis in patients with polymyositis, dermatomyositis, or systemic lupus erythematosus with severe nephritis (Lewis EJ, et al. N Engl J Med. 1992; 326:1373-1379; Miller FW, et al. N Engl J Med. 1992; 326:1380-1384).

Proposed mechanisms for therapeutic benefit in various conditions are given in the following Table (Hart GK. Intensive Care World. 1990; 7:21-25):

Proposed Mechanisms for Therapeutic Benefit of Plasma Exchange		
Mechanism	Condition	Pathogenic Substance Removed
Antibody Removal	Myasthenia Gravis	Anti-Acetylcholine Receptor Ab
	Goodpasture's Syndrome	Anti-Glomerular Basement Membrane Ab
	Hemophilia	Anti-Factor VIII Ab
	Guillain-Barré Syndrome	Anti-Peripheral Nerve Myelin Ab
Immune Complex Removal	Systemic Lupus Erythematosus Rheumatoid Vasculitis	
Inflammatory Mediator Removal	Goodpasture's	Complement, Fibrinogen
	Sepsis	Endotoxin, Tumor Necrosis Factor
Immune Function Enhancement	Cancer	"Blocking Factors" Removed
Deficient Plasma Factors Added	Thrombotic Thrombocytopenic Purpura Immune Thrombocytopenia	
Substance Removal	Paraproteinemia	Paraproteins
	Hepatic Failure	"Toxins"
	Poisons	Dependent on Poison

Plasma exchange is expensive and not necessarily benign. In many of the above listed conditions, the efficacy and indications for plasma exchange have not been fully investigated. Potential complications are listed in the following Table (Hart GK, already cited):

Potential Complications of Plasma Exchange
Hemodynamic: Hypovolemia; Hypervolemia; Thrombosis; Hemorrhage
Metabolic: Citrate Toxicity; Hypocalcemia; Metabolic Alkalosis; Coagulopathy
Infection: Sepsis; Post-Transfusion Hepatitis, CMV, HIV
Hypersensitivity: FFP or Albumin Reactions; IgE Mediated Reaction to Ethylene Oxide (Used to Sterilize Components)

THROAT CULTURE FOR GROUP A BETA-HEMOLYTIC STREPTOCOCCI
(see STREPTOCOCCAL (GROUP A) CULTURE)

THROAT CULTURE FOR NEISSERIA GONORRHOEAE ONLY
(see GONORRHEA CULTURE)

THROMBIN TIME

Synonym: Thrombin Clotting Time

SPECIMEN: Blue (citrate) top tube; fill tube; separate plasma and refrigerate if necessary. Determination must be done within 2 hours from the time of collection. Cells can be separated from plasma and the specimen frozen in plastic if necessary. Do not perform test on patients who are on heparin therapy.

REFERENCE RANGE: Varies by laboratory. Clot formation in 12-18 seconds.

METHOD: Add low concentration of thrombin to citrated plasma and measure the time necessary for clot formation.

INTERPRETATION: The thrombin time is performed in order to detect hypofibrinogenemia, the presence of heparin, or other inhibitors of fibrin polymerization. It is used to aid the investigation of a prolonged partial thromboplastin time (PTT). The thrombin time is a measure of the rate of conversion of fibrinogen to fibrin and fibrin polymerization. As a result, the thrombin time has been used to monitor thrombolytic therapy. The causes of prolonged thrombin time are listed in the next Table:

Causes of a Prolonged Thrombin Time
Heparin
Severe Hypofibrinogenemia (<80mg/dL)
Inhibitors of Fibrin Polymerization
Fibrin Split Products (FSP)
Paraproteins
Dysfibrinogenemia
Uremia

This is used as a rapid semi-quantitative test for intravascular clotting and fibrinolysis.

The thrombin time is exquisitely sensitive to heparin and can be very useful in monitoring low doses. If the thrombin time is prolonged, the presence of heparin in plasma as a cause of prolongation of the thrombin time can be documented by using the thrombin time, protamine sulfate corrected.

A reptilase time may also be performed to confirm the presence of heparin. A normal reptilase time is obtained in the presence of heparin. Mixing studies with normal plasma may not correct if heparin and other inhibitors (FSP) are present. However, mixing of normal plasma will correct the thrombin time if the cause is hypofibrinogenemia.

THROMBIN TIME, PROTAMINE CORRECTED

Synonym: Heparin Neutralized Thrombin Time

SPECIMEN: Blue (citrate) top tube; fill tube; separate plasma and refrigerate if necessary. Determination must be done within 2 hours from the time of collection.

REFERENCE RANGE: Correction of prolonged thrombin time if due to heparin.

METHOD: Protamine sulfate is added to test plasma and thrombin time is determined. This test is performed in order to investigate the reason for a prolonged thrombin time. It is used to distinguish a prolongation due to the presence of heparin, from a prolongation due to other reasons.

INTERPRETATION: This test is ordered to confirm that a prolonged thrombin time is due to the presence of heparin, and not other causes. Protamine sulfate will fully correct a prolonged thrombin time due to heparin up to 2.5U/mL but will have little or no effect on a prolonged thrombin time due to hypofibrinogenemia or inhibitors of fibrin polymerization such as fibrin split products (FSP), paraproteins or dysfibrinogenemia.

THROMBOLYTIC THERAPY PANEL

USE: Thrombolytic therapy is used in the treatment of coronary artery thrombosis, proximal deep vein thrombosis, pulmonary embolism, peripheral artery thrombosis and embolism, and the unclotting of catheters and shunts. There are a number of thrombolytic agents available, each with their own unique mode of action, half-life, method of administration, complications and cost. Which of these agents is best has not yet been established in clinical trials.

THROMBOLYTIC AGENTS: Streptokinase (STK): This is a bacterial product which activates both circulating plasminogen, and plasminogen within the thrombus. Because it has a short half-life, it must be administered over a period of time. Because it is a bacterial product and all patients have experienced prior streptococcal infections, antibodies are formed against this agent within 5-6 days. After 6 months these antibody levels return to normal. Because of this, it is suggested that this agent not be reused within 6 months.

Acetylated Plasminogen-Streptokinase Activator Complex (APSAC): This is a slowly activated (following deacylation of the plasminogen) form of thrombolytic agent. Because of it's longer half-life, it may be administered as a bolus.

Urokinase (UK): This is an endogenous protein that acts on circulating plasminogen to induce a lytic state. It is not taken up by the thrombus. It is usually administered during cardiac catheterization. Because it is an endogenous protein, antibodies against it are not expected to form.

Recombinant Tissue Plasminogen Activator (rt-PA): This is a genetically engineered form of an endogenous protein. It has a greater affinity for fibrin thrombi than the other agents. It may be that this agent has better activity against "mature" cross-linked clots. rt-PA has a short half-life and must be continuously administered. Since it has weaker activity against circulating fibrinogen, heparin is administered concurrently to produce and maintain a lytic state. Since it is an endogenous protein, antibodies are not expected to be produced against it.

Complications: Complications of thrombolytic therapy include bleeding (<5%), including intracranial bleeds (<1%). Restenosis of a vessel is a major problem. Formation of antibodies against STK limits this agent's reuse. These agents can activate platelets and result in a paradoxical "rebound thrombosis" if thrombolysis is interrupted or reversed. Hypotension may be seen after STK infusion as a result of bradykinin formation.

LABORATORY TESTING: Early: A number of laboratory tests may be employed, however, their early usefulness is limited since: 1) therapy is usually completed within 60 minutes, long before the laboratory results are available, and 2) and there appears to be no clear relationship between the degree of change in laboratory parameter, and angiographically detected improvement.

However, laboratory monitoring may be useful in detecting the 10% of patients with no change in coagulation assays. Presumably in this group of patients if there is no change in coagulation assays, there is no clinical benefit to the administered agent and a change to other therapy or agent may be indicated. In the other 90% of patients, the fibrinogen is typically lowered to 80% of initial value. In this group, laboratory monitoring may be useful in standardizing future anticoagulant therapy, and evaluating hemostatic integrity prior to any surgical procedure. Bleeding complications are treated empirically. The PPT and thrombin time are useful in accessing the presence of a heparin effect. Fibrinogen levels <0.5g/dL are needed to induce thrombolysis of deep vein thrombosis (Duckert F, et al. Br Med J. 1975; 1:479-481) and coronary thrombosis (Burkart MW, et al. Am J Cardiol. 1985; 56:441-444). D-dimer assays will detect the lytic state but do not necessarily detect thrombolysis (Francis CW, et al. Circulation. 1986; 74:1027-1036).

Late: It has been suggested that the PTT should be maintained at a minimum of 45-60 seconds after use of thrombolytic agents in acute MI (Hsia J, et al. J Am Coll Cardiol. 1992; 20:31-35). The coronary arteries remained open in only 45% of patients with PTTs of less than 45 seconds at 8 and 12 hours. Thrombolysis of infarct-related coronary arteries was successful (based on 90 minute angiography) if there was a peak in creative kinase within 12 hours after the start of thrombolytic therapy (84% sensitivity) compared to a reduction in ST segment reduction (63% sensitivity). Creatine kinase activity was measured at four-hour intervals for the first 24 hours (Hohnloser SH, et al. J Am Coll Cardiol. 1991; 18:44-49). Use of heparin, dosing and monitoring, in patients following thrombolytic therapy is summarized by: Hirsh J. N Engl J Med. 1991; 324: 1565-1574.

Alternatively the thrombin time has been advocated to monitor thrombolytic therapy (Hermelin LI, Eichelberger JW. J Med Tech. 1985; 2:89-93). Results of the thrombin time are recommended to be maintained between 1.5-5 times normal control 4 hours after thrombolysis, to be rechecked every 12 hours thereafter. However, a protamine-neutralized thrombin time has to be run to distinguish prolongation due to heparin from fibrinolysis.

Other references: Marder VJ, Sherry S. N Engl J Med. 1988; 318:1512-1520; 1585-1595. Taylor GJ in Thrombolytic Therapy for Acute Myocardial Infarct. Blackwell Scientific Publications, Boston, 1992.

THROMBOSIS PANEL

There is a great need to develop diagnostic tests to identify patients at risk for thrombotic events. A number of blood protein disorders have been identified which lead to hypercoagulability. Tests being used to measure some of these proteins and tests under development to access thrombotic risk are tabulated below:

Tests for Factors Associated with Thrombotic Risk
Antithrombin III (ATIII)
Lupus Anticoagulant
Anti-cardiolipin (Anti-phospholipid Antibody)
Protein C
Protein S
Prothrombin Fragment F1.2
Native Prothrombin Antigen
Homocysteine

Only 15-20% of patients with a hypercoagulable state can be identified as having a biochemical abnormality (Furie B, Furie BC. N Engl J Med. 1992; 326:800-806). Congenital or acquired deficiency of antithrombin III is associated with thrombosis (Egeberg O. Thromb Diath Haemorrh. 1965; 13:516-530). (see ANTI-THROMBIN III). The presence of anti-phospholipid/anti-cardiolipin antibodies with lupus anticoagulant activity likewise predisposes to thrombosis and recurrent abortion (see ANTICARDIOLIPIN ANTIBODY; LUPUS ANTICOAGULANT).

The relatively recently described thrombomodulin, Protein C, Protein S system is responsible for preventing widespread thrombosis. Normal endothelium contains thrombomodulin on it's surface. Normal endothelium binds circulating activated thrombin via thrombomodulin. This thrombomodulin-bound thrombin will activate protein C. Activated protein C, (Ca), circulates and inhibits clotting by: 1) cleaving and inactivating factors V and VIII; 2) inhibiting the activity of the Va, Xa, platelet complex; 3) raising plasminogen-activator activity by neutralizing an inhibitor of tissue-plasminogen activator. Defects of the protein C/protein S system have been described and are listed in the next Table:

Defects in the Protein C/Protein S System

Defect	Remarks
Congenital:	
Partial Protein C Deficiency	Venous Thrombosis
Total Protein C Deficiency	Purpura Fulminans Neonatalis
Partial Protein S Deficiency	Thrombotic Disease
Acquired:	
Protein C Deficiency	Associated with DIC
	Acute Leukemia with Liver Dysfunction
	Early Oral Anticoagulation
	Vitamin K Deficiency
	Severe Liver Disease

Both functional and antigenic assays for proteins C and S are available. Approximately 50% of patients with protein C deficiency have normal immunological levels of protein C antigen, but decreased functional activity. Functional assay of protein S is complicated by the presence of this factor as both free protein S and protein S complexed to C4b-binding protein.

References related to thrombomodulin/Protein C/Protein S are given in: Clouse LH, Comp PC. N Engl J Med. 1986; 314:1298-1304. Comp PC, Clouse L. Lab Management. 1985; Dec:29-32. Furie B, Furie BC. N Engl J Med. 1992; 326:800-806. Griffin JH, et al. J Clin Invest. 1981; 68:1370-1373.

The peptide prothrombin fragment F1.2 is generated by the action of factor Xa on prothrombin. The F1.2 fragment consists of the amino-terminal segment of the prothrombin molecule, and increases in this fragment indicate generation of thrombin. F1.2 levels have been demonstrated to correlate with thrombotic risk in certain patient populations, and levels are decreased in those on oral anticoagulant therapy, heparin infusion, or following antithrombin III administration. Thrombotic conditions associated with an elevated F1.2 include AT III deficiency, proteins C and S deficiency, malignancy, MI, DIC, venous thrombosis, stroke, cancer, and vascular disease (Hursting M, et al. Blood. 1989; 74(suppl 1). Blood. 1990; 76(suppl 1). An ELISA method is now available for measuring F1.2 (Organon Teknika Corp). Elevated F1.2 is seen in patients with deficiencies of antithrombin III and protein C (Conway EM, et al. J Clin Invest. 1987; 80:1535-1544).

Native prothrombin antigen is the fully carboxylated form of the factor. Poorly carboxylated forms are the result of oral anticoagulant therapy. Reduction of native prothrombin antigen to the range of 12-24mcg/mL may reduce the complications of anticoagulant therapy in patients with a thrombotic tendency (Furie B, et al. Blood. 1984; 64:445-451. Furie B, et al. Blood. 1990; 75:344-349).
Review of Hypercoagulable States: See Nachman RL, Silverstein R. Ann Intern Med. 1993; 119:819-827.

THYROCALCITONIN (see CALCITONIN)

THYROGLOBULIN

SPECIMEN: Red top tube, separate serum.
REFERENCE RANGE: Up to 50ng/mL.
METHOD: RIA
INTERPRETATION: Thyroglobulin is used as a tumor marker to monitor post-treatment of patients with differentiated thyroid cancer. Thyroglobulin is the 650,000 dalton storage form of the thyroid hormones, T-4 and T-3. Serum thyroglobulin is elevated in patients with differentiated thyroid carcinoma, e.g., papillary carcinomas, follicular carcinomas and mixed papillary follicular carcinoma. Serum thyroglobulin is normal or undetectable in subjects with medullary carcinoma of the thyroid. Serum thyroglobulin is also elevated in benign conditions, e.g., non-toxic goiter, Grave's disease, subacute thyroiditis, endemic goiter. Therefore, assay of serum thyroglobulin is not useful in the diagnosis of thyroid cancer due to possible confusion with these benign conditions.

Serum thyroglobulin measurement is useful in the post-treatment follow-up of patients with differentiated thyroid cancer in that it is an indicator of tumor recurrence and metastatic disease (Hay ID, Klee GG. Endocrinol Metab Clin N Amer. 1988; 17:473). The test is more sensitive if the patients have discontinued T4 therapy and TSH stimulation is maximal. Thyroglobulin cannot be assayed in the presence of anti-thyroglobulin antibodies; anti-thyroglobulin antibodies are present in a high percent of patients with carcinoma of the thyroid (Schneider AB, Perros. J Clin Endocrinol Metab. 1978; 47:126). Interpretation is also difficult in the presence of small remnants of normal thyroid.

THYROID ANTIBODIES, ANTI-MICROSOMAL OR ANTI-THYROGLOBULIN

SPECIMEN: Red top tube, serum.
REFERENCE RANGE: Varies with laboratory and technique.
METHOD: Immunofluorescent technique on monkey tissue substrate, or agglutination techniques.
INTERPRETATION: Used in the diagnosis and classification of inflammatory and autoimmune thyroid disease, especially Hashimoto's thyroiditis. These tests are useful in detecting and confirming autoimmune thyroiditis such as Hashimoto's thyroiditis, atrophic thyroiditis, and 70-90% of Graves Disease. Both anti-microsomal and anti-thyroglobulin antibodies (anti-thyroid peroxidase) should be assayed since there are some patients which exhibit only one of the two antibodies. Very high titers of antibodies are almost always indicative of Hashimoto's thyroiditis. These autoantibody tests lack both sensitivity and specificity. Many patients with biopsy proven thyroid autoimmunity lack these antibodies. Conversely, these antibodies may be seen in patients with no obvious thyroid disease, especially in women and the aged (Peter JB. Diag Med. 1981; 19-25). They are seen in patients with other autoimmune diseases such as Sjogren's Syndrome, SLE, and rheumatoid arthritis. Conditions in which these antibodies are elevated are given in the next Table (Taylor DG, Nakamura RM. Clinical Immunology Check List-1. Am Soc Clin Path. 1977):

Thyroid Antibodies (Cont)

Clinical Diagnosis and Percentage of Patients with Thyroid Autoantibodies

Diagnosis	Thyroglobulin Antibody Titers 1:10	Thyroglobulin Antibody High Titers >1:1000	Microsomal Antibody
Hashimoto's Thyroiditis	30% - 100%	25%	80%
Subacute Thyroiditis	35% - 70%	0	40%
Grave's Disease	35% - 85%	10%	80%
Primary Myxedema	60% - 80%	"common"	80%
Nontoxic Nodular Goiter	20% - 80%	0	40%
Adenoma	10% - 30%	?	?
Carcinoma	15% - 60%	8%	2% - 20%
Juvenile Thyroiditis	30% - 60%	?	90%

The presence of these antibodies does not exclude thyroid cancer. Screening of day 2 postpartum patients for anti-microsomal and anti-thyroglobulin antibodies has been able to detect a sub-population of patients at risk for thyroid dysfunction and development of postpartum hypothyroidism (Hayslip CC, et al. Am J Obstet and Gynecol. 1988; 159:203-209).

THYROID PANEL

Abnormalities of the thyroid gland are relatively common, estimated at 2-3% of the population. Screening of individuals revealed unsuspected thyroid dysfunction in approximately 0.5% of the subjects, with increased risk in women and the elderly (Helfand M and Crapo LM. Ann Intern Med. 1990; 112:840). Important tests for thyroid disease are given in the following Table:

Important Tests for Thyroid Disease

Group I - General Screening Tests

1) Total thyroxine (T-4)
2) T-3 Resin Uptake (T-3RU)
3) Free Thyroxine Index (FTI) or Free Thyroxine Estimate (FTE)
4) Thyroid Stimulating Hormone (TSH)/Sensitive-TSH
5) T-4 by Equilibrium Dialysis (Free T4)
6) Triiodothyronine, total or free (T-3)

Group II - More Specialized Tests

7) TSH Following Thyroid Releasing Hormone (TRH)
8) Thyroid Autoantibodies (Anti-Thyroglobulin; Anti-Microsomal)
9) Anti-TSH Receptor Antibody
10) Thyroxine Binding Proteins (TBP)
11) Thyroglobulin
12) Radioactive Iodine (or Technetium) Uptake (RAIU)

The active hormones of the thyroid are thyroxine (T4), and triiodothyronine (T3). Both are produced by proteolysis of endocytosed thyroglobulin by follicular cells. Serum T4 concentrations are higher than those of T3. However, both T4 and T3 are bound to thyroid binding globulin, T4 more strongly than T3, hence there is a greater proportion of free, unbound T3 in serum. T4 is converted to the more biologically active T3 in both thyroid and in the peripheral tissues.

As mentioned above, thyroid hormones are bound to TBG, and to a lesser extent to prealbumin and albumin. There is very little free T4 and T3 present, 0.05% and 0.25% respectively.

Thyroid hormones are linked in a negative feedback mechanism with both the pituitary and hypothalamus. Decreased levels of T3 and T4 leads to increased levels of TRH and TSH, while increased levels of T3 or T4 while lead to decreased levels of TRH and TSH.

Diseases of thyroid function are usually "primary" - i.e. they are disorders of the thyroid gland. The "secondary" disorders of pituitary function are not as common. Disorders of hypothalamic origin are sometimes referred to as "tertiary." Primary thyroidal disorders can usually be fairly easily diagnosed with the tests listed above. Secondary thyroid dysfunction often requires a much more detailed workup, consultation with an endocrinologist, and often includes an evaluation of total anterior pituitary function.

The principal thyroid disorders are hyperthyroidism and hypothyroidism. The usual laboratory results observed in these conditions are summarized in the table below. Screening tests such as the FTI or the s-TSH are not always diagnostic. Additional testing employing the T3 or TRH stimulation test are sometimes required.

Once the patient has been diagnosed as either hyperthyroid or hypothyroid, additional testing of pituitary function, or antibody screening may be helpful in further classifying the disease process. The hospitalized patient and the patient under therapy represent special situations. Some thyroid testing is useful in the diagnosis and follow-up of patients with neoplastic disease. Further details are given under the respective test headings.

The usual results of thyroid function tests in various conditions is given in the following Table:

Usual Results of Thyroid Function Tests in Various Conditions*

Total T-4	T-3RU	Free T-4 or FTI**	T-3	s-TSH***	TSH Response to TRH	Condition
↓	↓	↓	↓	↑	↑	Primary hypothyroidism
↓	↓	↓	↓	↓	(-)	Secondary hypothyroidism
↑	↑	↑	↑	↓	↓	Hyperthyroidism Factitial Thyroiditis
N	N	N	↑	↓	↓	T-3 Thyrotoxicosis

* Classical laboratory findings obtained in these conditions. In cases of early or mild disease not all tests will deviate from reference range.
** FTI is calculated from measured values of Total T-4 and T-3RU.
*** s-TSH is the newer "sensitive"-TSH. Not to be confused with the older TSH which could not detect values below the normal range.

HYPERTHYROIDISM: In hyperthyroidism, total T-4, T-3 resin uptake (T-3RU), free thyroxine index (FTI) or free thyroxine are elevated. If hyperthyroidism is suspected and the results are borderline or equivocal T-3 may be done. T-3 is often elevated out of proportion to T-4. The T-3 would also be measured to document T-3 thyrotoxicosis, however, this would be rare outside of iodide deficient areas. If the results are equivocal, a sensitive-TSH assay would be useful to discriminate a euthyroid from hyperthyroid condition. This is rapidly replacing the TSH following TRH procedure which was previously favored in this circumstance. In hyperthyroidism, there is no rise in plasma TSH after TRH, and in euthyroidism, there is a rapid rise in TSH of 2 to 20 microunits/mL from the 0 time TSH value.

The three most important causes of thyrotoxicosis with increased or occasionally normal RAIU are given in the next Table:

Thyrotoxicosis with Increased or Occasionally Normal RAIU
Toxic Diffuse Goiter (Grave's Disease)
Toxic Multinodular Goiter (Plummer Syndrome)
Toxic Adenoma

Causes of hyperthyroidism with reduced RAIU are listed in the next Table:

Hyperthyroidism with Decreased RAIU
Thyroiditis
Grave's Disease with Acute Iodide Loading
Thyrotoxicosis Factitia (Surreptitious Intake of Thyroid Hormone)
Antithyroid Drugs (Propylthiouracil or Methimazole)

Thyroid antibodies (antibodies against thyroglobulin and microsomal antigen) may be present in patients with hyperthyroidism. Hyperthyroidism due to inappropriate TSH secretion is rare (approximately 150 reported cases). The majority of these are due to TSH producing pituitary adenomas while others probably represent selective pituitary resistance to thyroid hormone (Wynne AG, et al. Am J Med. 1992; 92:15-24).

HYPOTHYROIDISM: In primary hypothyroidism, total T-4, T-3 resin uptake (T-3RU) and free thyroxine index (FTI) or free thyroxine are depressed. Measurement of TSH is the most sensitive test for detecting primary hypothyroidism; TSH is markedly elevated in primary hypothyroidism. It is important to screen patients who are on lithium therapy with thyroid function tests, total T-4, free thyroxine index (FTI) and TSH, because of the high incidence, 4% to 13%, of hypothyroidism associated with lithium therapy. In thyroid screening of the newborn in Utah, the rate of hypothyroidism was one in 3,800 live births; the rate of hypopituitarism was one in 27,600 live births; the incidence of congenital absence of thyroid binding globulin was one in 41,000 live births (Buehler BA. Ann Clin Lab Science. 1983; 13:5-9).

Thyroid Panel (Cont)

Thyroid antibodies (antibodies against thyroglobulin and microsomal antigen) may be present in patients with hypothyroidism, especially when hypothyroidism is associated with thyroiditis, especially Hashimoto's thyroiditis. Significant elevation of serum cholesterol occurs in hypothyroidism. Antibody to TSH in patients with autoimmune thyroiditis may lead to a transient, reversible hypothyroidism (Takasu N, et al. N Engl J Med. 1992; 326:513-518).

In secondary (pituitary) and tertiary (hypothalamic) hypothyroidism, TSH is low. To differentiate secondary from tertiary, inject TRH (7 mcg/kg) and measure TSH at times 0, 15, 30, 45 and 60 minutes; if TSH increases then hypothyroidism is due to hypothalamic failure; if TSH remains low, then the hypothyroidism is due to pituitary failure (See THYROID STIMULATING HORMONE FOLLOWING TRH).

THYROIDITIS: In Hashimoto's thyroiditis, most patients are metabolically normal during the early stages of the disease; hypothyroidism is found in 20% to 50% and up to 15% early in the disease. High titers of the thyroid autoantibodies (antibodies against thyroglobulin and microsomal antigen) are usually present; anti-microsomal antibodies are present more often than anti-thyroglobulin antibodies (Baker BA. Am J Med. 1983; 74:941-944). In Riedel's struma, the patient is hypothyroid; autoantibodies are present as seen in Hashimoto's disease. In granulomatous thyroiditis, at least 50% of the patients are hyperthyroid, and thyroid autoantibodies are present. However, unlike classic thyrotoxicosis, RAIU is very low. The thyrotoxic state usually only lasts for a short time. In thyroiditis, the erythrocytes sedimentation rate (ESR) is elevated; cholesterol may be elevated if the patient is hypothyroid.

THYROID NODULE: Schemes for the diagnosis of thyroid nodules have been compared (Solomon DH, Keeler EB. Ann Intern Med. 1982; 96:227-231). Fine-needle aspiration biopsy has become the initial test in most patients (Mazzaferri EL. N Engl J Med. 1993; 328:553-559). A screen of thyroid function, FTI or s-TSH, is run to exclude unsuspected thyrotoxicosis, and a serum calcitonin ordered if medullary thyroid cancer is suspected.
Scan: Seventeen percent of cold nodules are malignant; 9 percent of hypofunctioning and normofunctioning (warm) nodules. Autonomously functioning hot nodules are almost never carcinomas; however, hot nodules constitute only approximately 5 percent of solitary nodules.
Confirming the Hot Nodule: Use a thyroid hormone suppression test. Perform an I-123 scan. Give 75 mcg/day liothyronine sodium (Cytomel, 25mcg/8 hours) for 7 to 10 days. Liothyronine suppresses TSH. Perform another I-123 scan. The autonomous nodule will not be suppressed by the lack of TSH stimulation but will appear as a hyperfunctioning spot on the scan. Do not perform this test in the elderly, patients with heart disease, or patients who are thyrotoxic. Thyroxine therapy may be ordered to decrease nodule size or prevent it's growth. Although there are no explicit data on the optimal degree of TSH suppression desired, it has been suggested that the TSH should be lowered to 0.2 to 0.4 microU/mL for 6 months (Mazzaferri, already cited).

THYROID CANCER: Many of the malignant tumors, especially follicular carcinomas, take up radioactive iodide. Calcitonin is often secreted by patients with medullary carcinomas in the familial form of medullary carcinoma. Tests provocative of calcitonin secretion, that is, infusion of calcium or pentagastrin or both, may be done.

EVALUATION OF THERAPY: The relative merits of using various laboratory tests to monitor thyroid replacement are not well established. Most patients taking customary doses of levothyroxine have slightly elevated T4 levels, normal or decreased T3, and suppressed s-TSH with no symptoms of hyperthyroidism (Karla J, Hart IR. Clin Biochem. 1987; 20:265; Liewendahl K, et al. Acta Endocrinol. 1987; 116:418). Recent studies suggest that some of these patients may experience adverse effects of subclinical hyperthyroidism, and advocate adjusting doses of replacement therapy by monitoring the s-TSH (Thomas SHL, et al. J Roy Col Phys London. 1990; 24:289-291; Ross DS. Mayo Clin Proc. 1988; 63:1223). A similar conclusion was reached in a 25 year retrospective study (Helfand M, Crap LM. Ann Intern Med. 1990; 113:450). The sensitive TSH may also be used to monitor anti-thyroid drug treatment in patients with hyperthyroidism (Graves' Disease)(Cho BY, et al. Clin Endocrinol. 1992; 36:585-590).

PREGNANCY: During pregnancy, TBG levels are doubled and remain so until term, after which it takes several weeks to return to normal. As a result T4 is increased. However, the FTI and s-TSH will remain in the normal range. In problem cases, a free T3 measurement (similar to a free T4), has been advocated to investigate these patients since free T3 measurements are independent of changes in TBG (Frankdyn JA, et al. Clin Chem. 1983; 29:1527). The FTI is increased in some patients with hyperemesis gravidarum. Measurement of anti-thyroid antibodies at term has been used to select a population of women at risk for postpartum hypothyroidism (Hayslip CC, et al. Am J Obstet and Gynecol. 1988; 159:203-209).

DRUGS AND THYROID TESTING: Thyroid function tests are modified by a number of drugs as shown in the partial list below. Of particular concern in regard to changes in thyroid tests in the hospitalized patient has been the effects of dopamine and glucocorticoids. Dopamine-treated patients exhibit lowered free T4 and T3, lowered s-TSH and TSH response to TRH. It has been concluded that critically ill patients with non-thyroidal illness have normal hypothalamic-pituitary-thyroid function, while those receiving dopamine have some degree of induced secondary hypothyroidism (Farber J, et al. J Clin Endocrinol Metab. 1987; 65:315). In some ill patients the FTI is increased due to impaired T4 clearance or decrease in T4 to T3 conversion. This can be the result of glucocorticoids, propananol, iodinated drugs or contrast agents, acute psychosis, and hyperemesis gravidarum (Hay ID, Klee GG. Endocrine and Metab Clinics N Amer. 1988; 17:473). A summary of some of the varied effects drugs can exert on thyroid function tests is given below:

Drugs and Their Effects on Thyroid Function Tests

Effect	Drugs
Increased TBG	Estrogen, Oral Contraceptives, Heroin, Clofibrate, Fluorouracil
Decreased TBG	Androgens, Glucocorticoids, L-Asparaginase
Increased T4	Drugs that Elevate the TBG, Extrathyroidal Thyroid Hormone
Decreased T4	Drugs that Decrease the TBG, Phenytoin, Carbamazepine, Rifampin, High Doses of Glucocorticoids, Dopamine
Decreased T3	Glucocorticoids, Propranolol, Iodine Containing Drugs and Contrast Agents, Dopamine
Increased TSH	Dopamine Antagonists, Chlorpromazine, Haloperidol, Iodine Containing Drugs
Decreased TSH	Extrathyroidal Thyroid Hormone, Glucocorticoids, Dopamine, Levodopa, Dopamine Agonists, Apomorphine, Pyridoxine

THYROID TESTING AND NON-THYROIDAL ILLNESS, SICK EUTHYROID SYNDROME: Almost all thyroid function tests are modified to some degree by illness, hence the definitive diagnosis of thyroidal illness needs to be made with care in the severely ill patient. There are a number of relevant factors: 1) thyroid function tests in the ill patient often mimic the results seen in thyroid dysfunction, 2) Hospitalized patients often receive drugs which modify the hypothalamic-pituitary-thyroidal axis, 3) The prevalence of thyroidal disease increases with age, 4) There are differences in reference ranges for young and elderly. These represent age related changes, or are the result of comorbidity in the elderly.

In a hospitalized patient, any abnormality in a screening test of thyroid function, e.g. T4, FT4E, FTI, s-TSH, does not necessarily mean thyroid disease. Many of these patients are probably euthyroid. The best advice has been to repeat the testing after recovery. In one study, the s-TSH was suppressed in 9% of hospitalized patients, and elevated in 4%. Follow-up testing revealed a return to normal range in all but 3.3% of cases (Bharkri HL, et al. Gerontology. 1990; 36:140).

Thyroid Panel (Cont)

The most common change in the sick euthyroid syndrome is a decrease in the T3 and free T3. This is due to decreased peripheral conversion of T4 to T3 and is coupled with a corresponding increase in the reverse T3. In very serious illness, the T4 may fall and is considered a very poor sign (Slag MF. JAMA. 1981; 246:2702). It has been stated that severe illness will have predictable effects on thyroid testing: the first change is a decrease in T3 and an increase in reverse T3. This initial change is followed by a decrease in total T4 and an increase in the T3RU. TSH levels remained in the normal range in this study (Felicetta JV. Postgrad Med. 1989; 85:213). However, reported values for thyroid function tests in illness are quite variable. In mild non-thyroidal illness, there has been reported an initial increase in T4, T3 decrease, and reverse T3 increase. There was also a blunted TSH response to TRH (Hennemann G, Krenning EP. Horm Res. 1987; 26:100). The highly variable effects of illness on thyroid function tests are outlined in the next Table:

The Effects of Non-Thyroidal Illness on Thyroid Function Tests

T4	Increased (Early Illness), Normal, or Decreased (Severe Illness)
T3	Consistently Decreased (Reverse T3 Increases)
Free T3	Decreased in Parallel with T3
s-TSH	Variable, some Reports no Change, Others Decrease. May Depend on Drug Use, e.g. Dopamine. Increases as Illness Resolves.
TSH following TRH	Some Reports Blunted Response. May Depend on Drugs.

Many of the highly variable effects of illness on thyroid function tests may be due to the effect of drugs. Many of the results summarized above were based on observations on unselected patients. More recent studies have emphasized the effects of dopamine and glucocorticoids on these tests (see above under DRUGS AND THYROID TESTING). Hypoalbuminemic ill patients had lower T4, T3, free T4 and FTI values compared to a normalbuminemic population which exhibited only lowering of T3 and free T3 (Desai RK, et al. Clin Chem. 1987; 33:1445).

THYROID TESTING IN THE ELDERLY: It is thought that T4 levels change little with age, but that T3 levels decline (Feit H. Clin Geriatr Med. 1988; 4:151). Approximately 2% of the untreated elderly have low s-TSH values, but normal T4 values. These individuals are thought to be euthyroid (Sawin CT, et al. Arch Intern Med. 1991; 151:165). Long range study of elderly patients with either low or high s-TSH values have been described (Parle, et al. Clin Endocrinol. 1991; 34:777). Many of these developed established thyroidal dysfunction while others appeared to remain euthyroid. Caution is urged in the interpretation of thyroidal function tests in the elderly in view of the atypical presentations of hyperthyroidism (Federman DD. Hosp Pract. 1991; 26:61) and hypothyroidism (Drinken and Nolten WE. South Med J. 1990; 83:1259) in the elderly population.

GENERAL REFERENCES: The following are a group of general references on testing of thyroid function.

Larsen PR, et al. "Revised nomenclature for tests of thyroid hormones and thyroid-related proteins in serum." J Clin Endocrinol Metab. 1987; 64:1089.

Hay ID, Klee GG. "Thyroid dysfunction." Endocrinol Metab Clin N Amer. 1988; 17:473.

de los Santos E, Mazzaferri EL. "Thyroid function tests. Guidelines for interpretation in common clinical disorders." Postgrad Med. 1989; 85:333.

Surks MI, et al. "American thyroid association guidelines for use of laboratory tests in thyroid disorders." JAMA. 1990; 263:1529-1532.

Wellby ML. "Clinical chemistry of thyroid function testing." Adv Clin Chem. 1990; 28:1.

Becker DV, et al. "Optimal Use of Blood Tests for Assessment of Thyroid Function." JAMA. 1993; 269:2736.

THYROID STIMULATING HORMONE (TSH)

SPECIMEN: Red top tube, separate serum.
REFERENCE RANGE: Adults ND-10microU/liter.
METHOD: RIA
INTERPRETATION: Has been used to (1) screen for hypothyroidism in newborn and adult populations; and (2) measured following TRH administration to access hypothalamic, pituitary, thyroid-axis function. Has been largely replaced by the "sensitive" TSH assay (s-TSH), from which it differs in that the s-TSH is able to detect TSH values below the normal range. Measurement of TSH is the most sensitive test for detecting primary hypothyroidism, a condition in which serum TSH is elevated. Now being replaced by the sensitive-TSH(s-TSH). See THYROID STIMULATING HORMONE - SENSITIVE ASSAY.

THYROID STIMULATING HORMONE (TSH) FOLLOWING TRH

SPECIMEN: Red top tube, separate serum. The test is time consuming and may be unpleasant for the patient. The test for adults is done as follows: The patient should be in a supine position. A serum specimen is obtained for serum TSH assay. Then TRH is rapidly injected IV over 30-60 seconds in a dosage of 7 mcg/kg. There is some variability in the time of drawing the serum specimens for TSH. Specimens are usually obtained at 20 and 60 minutes. If a delayed or a prolonged response is anticipated, additional specimens may be drawn at 90 and 120 minutes.
METHOD: RIA
INTERPRETATION: The TRH stimulation test is used: 1) To diagnose or exclude hyperthyroidism when equivocal screening thyroid function tests are obtained, 2) To differentiate pituitary versus hypothalamic causes of hypothyroidism (secondary versus tertiary hypothyroidism), and 3) To investigate certain psychiatric syndromes. The TRH stimulation test is not employed as often as previously in the diagnosis of hyperthyroidism since the introduction of the sensitive TSH test (Klee GG and Hay ID. J Clin Endocrinol Metab. 1987; 64:461).
Diagnosis of Hyperthyroidism: In hyperthyroidism there is an absent or blunted response to TRH stimulation due to the feedback inhibition of pituitary function by the elevated thyroid hormones. A typical time curve of a normal and hyperthyroid response is given below (Diagnostic Pathways, Damon Corporation, 1980):

Response of TSH to TRH in Hyperthyroidism

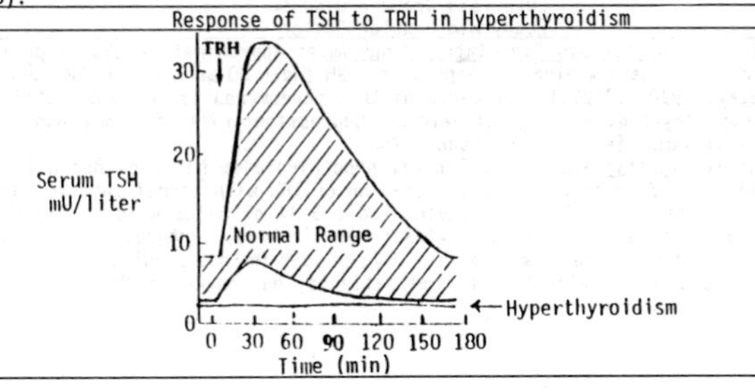

(cont)

Thyroid Stimulating Hormone (TSH) Following TRH (Cont)

Differentiation of Pituitary Versus Hypothalamic Hypothyroidism: Differentiation of Secondary (pituitary) versus Tertiary (hypothalamic) Hypothyroidism: TRH is useful in differentiating causes of secondary hypothyroidism, that is hypothyroidism due to a pituitary lesion versus hypothyroidism due to a hypothalamic lesion. In these patients, TSH is low. A control level of serum TSH is obtained; then TRH is injected intravenously and TSH is measured at 15 minute intervals as illustrated in the next Figure:

Response of TSH to TRH in a Patient with Secondary or Tertiary Hypothyroidism

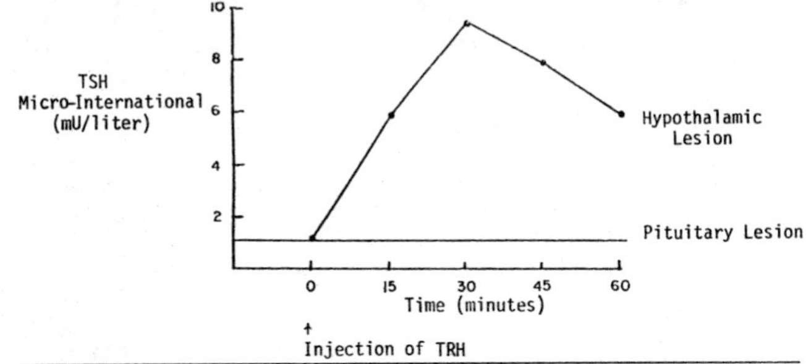

As shown in the previous illustration, serum TSH increased ten times its control level in 30 minutes. The results indicate that the patient's hypothyroidism is due to a lesion in the hypothalamus. If the serum TSH does not increase, the lesion is in the pituitary.

No response is found in the conditions listed in the next Table:

Conditions - No Response of Serum TSH to I.V. TRH
Secondary (or Pituitary) Hypothyroidism
Thyrotoxicosis
Euthyroid Grave's Disease
Euthyroid Nodular Goiter

Investigation of Certain Psychiatric Syndromes: Approximately 25% of patients with depression have a blunted response to TRH (Hein MD and Jackson IM. Gen Hosp Psychiatry. 1990; 12:232). The cause of this phenomenon is unknown, but is not due to hyperthyroidism. Range of values - Depression: rise of <5 microunits/mL; Hyperthyroidism: rise of <1 microunits/mL.

Subclinical Hypothyroidism: Subclinical hypothyroidism has been defined as an elevated TSH before or after administration of TRH with normal thyroid hormone levels. In one study, 6-7% of females and 2.5-3% of males were so classified (Staub JJ, et al. Am J Med. 1992; 92:631-642). Progression to overt hypothyroidism occurred at a rate of 5-10% per year in this population. Prolactin levels were slightly elevated in some patients, but showed marked response to TRH.

THYROID STIMULATING HORMONE (TSH) RECEPTOR ANTIBODIES

SPECIMEN: Red top tube, separate serum; store serum frozen and forward to reference laboratory in plastic tube on dry ice.
REFERENCE RANGE: Depends on method.
METHOD: Radioreceptor assay measures inhibition of TSH binding to thyroid plasma membranes or inhibition of TSH stimulated cAMP production by cell lines.
INTERPRETATION: This test is ordered to help establish the diagnosis of Grave's disease, the most common cause of hyperthyroidism, and as a predictor of relapse after therapy. Thyroid stimulating hormone(TSH) receptor antibodies are elevated in about 80 percent of patients with Grave's disease (both treated and untreated) and about 5-10 percent of patients with Hashimoto's disease (Ginsberg J, et al. Clin Endocrinol. 1983; 19:305-311; Shewring G and Rees Smith B. Clin Endocrinol. 1982; 17:173-179; de Bruin T W A, et al. Acta Endocrinologica. 1982; 100:245-251). Approximately 40-60% of patients with euthyroid Grave's disease will demonstrate these antibodies. Graves patients with pretibial dermopathy and those who require orbital decompression tend to have extremely high antibody levels. Occasionally patients with nodular thyroid disease will have borderline elevations (Hay ID, Klee GG. Endocrinol Metab Clin N Amer. 1988; 17:473).

THYROID STIMULATING HORMONE-SENSITIVE ASSAY (s-TSH)

SPECIMEN: Red top tube, separate serum.
REFERENCE RANGE: 0.38 to 6.15 microU/mL
METHOD: RIA, Fluorometric enzyme assay. Current assays are sensitive to at least 0.05 microU/mL.
INTERPRETATION: The sensitive thyroid stimulating hormone assay (s-TSH) is a relatively recent addition to the laboratory which is revolutionizing thyroid testing. This is an ultrasensitive assay in contrast to the older TSH which was unable to distinguish patient values in the normal range from those which were abnormally low. It is the feeling of some thyroid specialists that measurement of the s-TSH, complimented by an appropriate FTI measurement, represents the best and most efficient combination of blood tests for the diagnosis and follow-up of most patients with thyroid disorders (Becker DV, et al. JAMA. 1993; 269:2736).

The s-TSH is principally used to screen for thyroid disease, diagnosis of hyperthyroidism, and diagnosis of hypothyroidism. s-TSH values are elevated in hypothyroidism and decreased in hyperthyroidism. The s-TSH is used to aid in the diagnosis of equivocal hyperthyroidism and is presently favored over and currently replacing the TRH test for this purpose.

The s-TSH is also useful for the monitoring of thyroid hormone replacement therapy (Thomas SHL, et al. J Royal Coll Phys London. 1990; 24:289-291), monitoring of antithyroid therapy, and in evaluating thyroid function in critically ill patients (sick euthyroid syndrome). Dopamine or glucocorticoids will depress the s-TSH. During recovery from illness, in going from a catabolic to anabolic state, there may be a transient rise in the TSH. When recovery is complete, all blood tests of thyroid function should return to normal (Surks MI. JAMA. 1991; 266:1573).

Uses of the s-TSH are given in the next Table:

Uses of the s-TSH
Screening for Thyroid Dysfunction
Diagnosis of Hyperthyroidism (Decreased)
Diagnosis of Hypothyroidism (Elevated)
Replacement of the TRH Test
Monitoring Thyroid Replacement Therapy
Monitoring Anti-Thyroid Therapy

Articles describing the use of this recently introduced test include: Surks I, et al. "American Thyroid Association Guidelines for use of Laboratory Tests in Thyroid Disorders." JAMA. 1990; 263:1529-1532. Hay ID, et al. "American Thyroid Association Assessment of Current Free Thyroid Hormone and Thyrotropin Measurements and Guidelines for Future Clinical Assays." Clin Chem. 1991; 37:2002-2008.

THYROXINE BINDING GLOBULIN (TBG)

SPECIMEN: Red top tube, separate serum.
REFERENCE RANGE: 16-24mcg/dL (binding capacity for T-4). To convert conventional units in mcg/dL to international units in nmol/liter, multiply conventional units by 12.87.
METHOD: Radioactive T-4 is added to the serum to saturate the TBG. Unbound T-4 is removed with charcoal. The bound radioactive thyroxine is counted and reflects the TBG.
INTERPRETATION: The major use of this assay is in the identification of familial dysalbuminemic hyperthyroxinemia and other forms of protein abnormalities.

The causes of reduced concentrations of serum TBG are given in the next Table:

Causes of Reduced Thyroxine-Binding Globulin (TGB)
Congenital TBG Deficiency
Acquired:
Nephrotic Syndrome
Marked Hypoproteinemia
Hepatic Disease
Severe Acidosis
Androgenic or Anabolic Steroid Therapy
Aspirin in Large Doses

The incidence of congenital TBG deficiency is 1 in 13,000 (Dussault JH, et al. J Pediatr. 1978; 92:274-277); these patients have a normal free T-4.

The causes of increased concentration of serum TBG are given in the next Table:

Causes of Increased Thyroxine-Binding Globulin (TBG)
Congenital TBG Increase
Acquired:
Pregnancy
Oral Contraceptive
Liver Disease
Perphenazine Therapy
Hypothyroidism

Congenital TBG increase has an X-linked mode of inheritance; these patients have normal free T-4. Estimated incidence is 1 in 40,000 (Viscardi RM, et al. N Engl J Med. 1983; 309:897-899). During pregnancy, TBG levels are doubled and remain so until term; several weeks postpartum, TBG levels return to prepregnancy levels.

THYROXINE, FREE (FREE T4) BY EQUILIBRIUM DIALYSIS

SPECIMEN: Red top tube; separate serum
REFERENCE RANGE: 1.3 to 3.8ng/dL in children and adults; in infants under one month of age, normal values are higher. To convert conventional units in ng/dL to international units in pmol/liter, multiply conventional units by 12.87.
METHOD: RIA/equilibrium dialysis.
INTERPRETATION: Free thyroxine is elevated in hyperthyroidism and reduced in hypothyroidism. This test is used to confirm hypo or hyperthyroidism in individuals with abnormal or equivocal screening tests of thyroid function. This test is not necessary in the evaluation of most patients. Do not confuse this test with tests that estimate the free T4. The free T4 by equilibrium dialysis is generally too tedious and expensive a test to use as an initial screening test of thyroid dysfunction and is primarily employed to confirm or clarify abnormal or equivocal screening test results, eg, total T4, FT4E, FTI, or s-TSH.

Many technical modifications employing T4 analogs have recently been introduced in an attempt to simplify the free T4 assay. However, these have not been generally accepted as a valid measure of the free T4, and equilibrium dialysis remains the method of choice (Czako G, et al. Clin Chem. 1986; 32:108; Alexander NM. Clin Chem. 1989; 32:417). These "analog" methods should be considered as forms of the free T4 estimate-type tests (Hay ID, Klee GG. Endocrinal Metab Clin N Amer. 1988; 17:473). Attempts continue to design and evaluate better free T4 assays (Reilly CP, et al. Ann Clin Biochem. 1989; 26:517).

THYROXINE INDEX, CALCULATED (FTI)

Synonyms: Free Thyroxine Index, Free Thyroxine Estimate, FT4E, T7
SPECIMEN: Red top tube, separate serum.
REFERENCE RANGE: 0.8-2.4ng/dL. To convert conventional units in ng/dL to international units in pmol/liter, multiply conventional units by 12.87.
METHOD: RIA. The FTI is a calculated value obtained by multiplying the measured total T-4 by the T-3 resin uptake.
INTERPRETATION: The FTI is used in assessing and screening for thyroid dysfunction. The FTI reflects the serum level of the free thyroxine. The FTI is elevated in hyperthyroidism and decreased in hypothyroidism.

There are two genetic conditions in which there is an elevated FTI and normal measured levels of free T-4 in euthyroid subjects. In one type, there is an abnormal T-4 binding protein (Ruiz M, et al. N Engl J Med. 1982; 306:635-639); in the other type, there is increased binding of T-4 to immunoreactive thyroxine binding pre-albumin (TBPA) (Moses AC, et al. N Engl J Med. 1982; 306:966-969).

THYROXINE, TOTAL (T-4, TETRAIODOTHYRONINE)

SPECIMEN: Adult: Red top tube, separate serum. Newborn: In the usual procedure, blood specimens obtained by heel prick are taken on the third day after birth. Special filter paper is provided by the laboratory as illustrated in the next Figure:

Filter Paper for Blood Specimens for Thyroid Testing in the Newborn

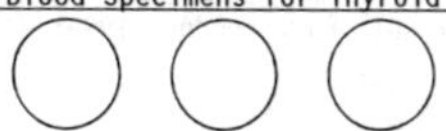

All three circles are soaked with blood; be sure blood soaks through. Patients that have values of total T-4 less than 7mcg/dL are retested by TSH. Reference: Sadler WA, Lynskey CP. Clin Chem. 1979; 25:933-938.
REFERENCE RANGE: Adults: 5-13mcg/dL; fetus: 4-16mcg/dL; amniotic fluid: 0.54 (non-detectable-1.75mcg/dL); cord: 6-17mcg/dL; 24 hours: 16-26mcg/dL; 48 hours: 12-20mcg/dL; 3-5 days: 9-20mcg/dL; 2 weeks-months: 7-15mcg/dL. To convert conventional units in mcg/dL to international units in nmol/liter, multiply conventional units by 12.87.
METHOD: RIA
INTERPRETATION: The total thyroxine (T4) is used as a measure of thyroid function. This assay gives the value of total T4, protein-bound and unbound. Only the unbound fraction is physiologically active. Thyroxine is synthesized in and released from the thyroid gland in response to TSH. When this thyroxine is released into the blood, it combines with the serum protein, thyroxine binding globulin, TBG. It also binds to a lesser extent with prealbumin and albumin. Only 0.025% of the total T4 is the biologically active, unbound T4. This unbound fraction can be determined by ordering a free T4 assay, performed by equilibrium dialysis. Because of protein binding, the total T4 is dependent on both thyroid function and the level of TBG in the blood.

The causes of an elevated T4 are hyperthyroidism and conditions which raise the TBG. Conversely, hypothyroidism and conditions which lower TBG or compete for T4 binding sites, will depress the T4. These conditions are given in the following two tables:

Causes of Elevation of Total T-4

- Hyperthyroidism
- Increased Thyroxine Binding Proteins:
  - Pregnancy
  - Drugs:
    - Estrogens
    - Clofibrate
    - Perphenazine
  - Genetic Increase of Thyroxine Binding Proteins
  - Acute Hepatitis
- Anti-T4 Antibodies

(cont)

Thyroxine Total (Cont)

The causes of decreased total T-4 are given in the next Table:

Causes of Decreased Total T-4
Hypothyroidism
Competition with T-4 for Binding Sites on Thyroxine Binding Proteins:
Phenytoin; Salicylates in High Doses; Phenylbutazone
Decreased Thyroxine Binding Proteins
Androgens; Renal Failure; Nephrosis; Malnutrition; Active Acromegaly; Corticosteroids (large doses); Major illness; Genetic Decrease of TBP

A few workers have claimed that the total T4 can be used as the primary screening test for thyroid dysfunction (Penney MD, O'Sullivan DJ. Clin Chem. 1987; 33:170). This is, however, not reliable. Because of the marked dependance of this test on thyroid binding proteins in many common conditions, most prefer to use an estimate of the free T4 (FT4E, FTI) as the screening assay. These estimations correct, in most cases, for the variability in T4 levels introduced by changes in TBG (Jatanavian RA, Braverman LE. Endocrinol. 1983; 6:493).

The T4 may also be used to follow patients undergoing thyroid hormone replacement therapy. In most instances the T4 levels in these patients have been in the high normal, or slightly elevated range (Karla J, Hart IR. Clin Biochem. 1987; 20:265; Liewendahl K, et al. Acta Endocrinol. 1987; 116:418). There is some concern that these patients may be receiving too much replacement hormone, and may be experiencing the untoward effects of cardiac arrhythmias and osteoporosis. Some clinicians are now monitoring replacement therapy with the s-TSH test (Thomas SHL, et al. J Royal Coll Phys London. 1990; 24:289-291).

T3 RESIN UPTAKE TEST, (T3RU)

Synonym: Thyroxine Binding Globulin Ratio (TBGR)
SPECIMEN: Red top tube, serum.
REFERENCE RANGE: 30-40%.
METHOD: Radioactive T3 competes with the T4 in patient serum for the binding sites on TBG. The excess radioactive T3 is then bound to a resin or other type of matrix. The measured radioactivity is then inversely proportional to the number of unoccupied binding sites.
INTERPRETATION: Used in conjunction with the T4 to calculate an estimate of the free T4, the free thyroid index (FTI). This is used as part of the screening evaluation of thyroid function. The terminology is rapidly changing in this area. Some would refer to this as a test of thyroxine binding globulin ratio (TBGR). This test is probably the most confusing and least understood laboratory measurement of thyroid function. **PRACTICALLY SPEAKING, THIS TEST IS ONLY USED IN CONJUNCTION WITH A T4 MEASUREMENT TO YIELD THE FTI, AN ESTIMATE OF THE FREE T4.**

The T3RU gives a quantitative measure of the unsaturated binding sites on TBG. The value of the T3RU is inversely proportional to the number of available thyroxine binding sites in the specimen. Hence its value depends on: 1) the level of thyroid hormones, 2) The level of TBG, and 3) The presence of drugs which may compete with thyroid hormones at their TBG binding sites. See below:

Summary of Conditions Associated with Changes in the T3RU
Elevated T3RU
Hyperthyroidism
Decreased TBG
Drugs Which Compete for T4 Binding Sites
Decreased T3RU
Hypothyroidism
Increased TBG

The conditions which are associated with changes in TBG levels are given in the next two tables. These are conditions in which binding proteins or the number of available hormone binding sites are decreased and the T3RU is elevated:

Conditions Associated with Elevation of T3RU
Hyperthyroidism
Decreased Thyroxine Binding Proteins:
Androgens
Renal Failure
Nephrosis
Malnutrition
Active Acromegaly
Corticosteroids (large doses)
Major Illness
Genetic Decrease of TBP
Competition with T-4 for Binding Sites on Thyroxine Binding Proteins:
Phenytoin
Salicylates (high doses)
Phenylbutazone

The conditions associated with decreased T3RU are given in the next Table. These are conditions in which the number of available hormone binding sites are increased.

Conditions Associated with Decreased T3RU
Hypothyroidism
Increased Thyroxine Binding Proteins
Pregnancy
Estrogen Effect, i.e., Oral Contraceptives; or Estrogens in Osteoporosis or Menopause
Clofibrate
Perphenazine
Genetic Increase of TBP
Acute Hepatitis

TOBRAMYCIN (NEBCIN)

SPECIMEN: Red top tube, separate serum and freeze. Obtain serum specimens as follows:
(1) 24-48 hours after starting therapy if loading dose is not given.
(2) 5 to 30 minutes before I.V. tobramycin (trough)
(3) 30 minutes after completion of a 30 minute I.V. infusion of tobramycin (peak)

The following times should be recorded on the laboratory requisition form and on the patient's chart:

Trough Specimen Drawn ___________(Time) (5 to 30 minutes before Tobramycin)
Tobramycin Started ___________(Time) (30 minute I.V. infusion)
Tobramycin Completed ___________(Time)
Peak Specimen Drawn ___________(Time) (30 minutes after I.V. Tobramycin)

REFERENCE RANGE: Therapeutic Range: 4-10mcg/mL; Toxic Level Peak: >12mcg/mL; Trough: <2mcg/mL; Toxic Level, Trough: >2mcg/mL.
METHOD: RIA; EMIT
INTERPRETATION: Tobramycin is an aminoglycoside antibiotic (amikacin, tobramycin, gentamicin, and kanamycin) which is frequently used in hospitals to treat patients who have serious gram-negative bacterial infections, especially septicemia and staphylococcal infections, untreatable with penicillins.

Toxicities associated with aminoglycosides are listed in the following Table (Edson RS, Terrell CL. Mayo Clin Proc. 1991; 66:1158-1164):

Aminoglycoside Toxicities	
Major:	Ototoxicity
	Vestibular
	Auditory
	Nephrotoxicity
	Neuromuscular Blockade
Minor:	Drug Fever
	Skin Rash

(cont)

Tobramycin (Cont)

The three main toxic side effects of tobramycin are ototoxicity, nephrotoxicity and neuromuscular blockage; it is important to control the dose given by monitoring peak and trough levels of the drug, particularly in patients with any degree of renal failure. As renal function declines, drug half-life increases to up to 50 hours.

To minimize risk of toxicity, it has been recommended that peak levels not exceed 10mcg/mL and that trough levels should fall between 1 and 2mcg/mL.

Tobramycin is eliminated exclusively by renal excretion; excessive serum concentrations may occur and lead to further renal impairment. The renal damage is to the renal proximal tubules and is usually reversible if discovered early. Ototoxicity is usually due to vestibular damage and is often not reversible.

Aminoglycoside antibiotics have been recently reviewed (Edson RS, Terrell CL. "The Aminoglycosides." Mayo Clin Proc. 1991; 66:1158-1164).

TORCH (ANTIBODIES TO TOXOPLASMA, RUBELLA, CYTOMEGALOVIRUS, HERPES SIMPLEX)

SPECIMEN: Red top tube, separate serum; IgG: acute and convalescent sera are required; IgM: single specimen. For congenital infections, serial sera from both the mother and infant are required.

REFERENCE RANGE: Rising specific IgG antibody titers on serial determinations indicates infection; obtain acute serum from cord blood, obtain convalescent serum at 4-6 months. Specific IgM antibody indicates active disease.

METHOD: See specific tests

INTERPRETATION: Torch tests are tests for detecting antibodies to toxoplasma, rubella, cytomegalovirus and herpes simplex. Infections of the mother in early pregnancy are associated with fetal malformations and sequelae such as abortions, stillbirths and congenital anomalies. Clinical features associated with congenital infections include the following (Editorial, Lancet. 1990; 335:1559-1561):

Clinical Features Common to Congenital Infections
Small for Gestational Age (SGA)
Purpura
Jaundice
Anemia
Chorioretinitis
Cataracts
Microphthalmia
Pneumonitis
Hepatosplenomegaly
Suspicious Rash
Maternal Infection

The presence of specific IgG antibody prior to pregnancy indicates immunity; rising IgG antibody titers and/or the presence of IgM antibody indicates active disease.

Antibodies in "normal" infants decrease during the first several months of life because they are passively acquired by placental transfer. Active infection in the neonate is indicated by antibody levels that are unchanged or increase in serial sera over several months. The absence of antibody in the mother rules out congenital infection.

The presence of IgM antibodies in the infant indicates active disease.

Many infections other than the classic TORCH infections may cause congenital infections; causes of neonatal infection and diagnostic tests are listed in the following Table (Editorial, Lancet, already cited; McMillan JA. Contemporary Pediatrics. February 1992):

Causes of and Tests for Congenital Infections	
Etiology	Tests
Toxoplasmosis	IgM; Sequential IgG; Placental Culture
Rubella	IgM; Absence of IgG excludes infection; Urine, Nasopharyngeal or CSF Culture
CMV	IgM and IgG within 3 weeks of birth; Urine, Nasopharyngeal, CSF or Blood Culture within 2 weeks of birth
Herpes Simplex	Skin lesion, CSF, Throat, or Conjunctival Culture
Hepatitis B	Maternal Hepatitis B Surface Antigen
Syphilis	VDRL (Not FTA); Neonatal level 4 times maternal level
HIV	IgG; retest every 3 months
Parvovirus B19	IgM detection
Varicella Zoster	Sequential IgG; No IgM available
Enterovirus	Nasopharyngeal or rectal culture

The diagnostic evaluation of congenital infections should be based on the clinical presentation.

TOTAL IRON BINDING CAPACITY (see IRON BINDING CAPACITY, TOTAL)

TOURNIQUET TEST

SPECIMEN: This test is done at the bedside.
REFERENCE RANGE AND METHOD: Inspect arm for petechiae. Inflate a blood pressure cuff on the arm to a pressure midway between systolic and diastolic pressures. After 5 minutes, the cuff is deflated and the arm distal to the cuff inspected for the appearance of new petechiae over the next 5 minutes. The grading is as follows: 4+ = confluent petechiae; 3+ = non-confluent petechiae; 2+ = large number of petechiae; 1+ = minimal change.
INTERPRETATION: This test is performed when there is a suspected defect in platelet-endothelial wall interaction. A positive tourniquet test is seen in the conditions listed in the next Table:

Positive Tourniquet Test
Primary Vascular Abnormalities:
Scurvy
Collagen Vascular Disease
Decreased Number of Platelets
Qualitative Platelet Disorder

A positive tourniquet test and a normal bleeding time is suggestive of a primary vascular abnormality, that is, scurvy or collagen vascular disease.

TOXICOLOGY PANEL

A suggested stat toxicology panel is listed in the next Table (Modified from Saxena K. Res and Staff Physician. 1983; 47-57):

Stat Toxicology Tests
Salicylate
Alcohol
Barbiturate
Acetaminophen
Iron
Volatile Screen: Methanol, Ethylene Glycol, Isopropanol
Methemoglobin Determination in Nitrate or Nitrite Ingestion
Carbon Monoxide
Morphine (Opiates, Opioids)
Benzodiazepines
Cocaine
PCP
Tricyclic Antidepressants

Some substances will be missed because the techniques are insensitive or the interval between administration and screening is too long.

Tricyclic Antidepressants: A QRS duration of 100msec or more suggests the possibility of a serious overdose of tricyclic antidepressants.

See DRUG SCREEN for additional details.

TOXICOLOGY SCREEN (see DRUG SCREEN)

TOXOPLASMOSIS TESTING

SPECIMEN: Culture: Placenta, Lung, Other Tissue, Bronchoalveolar Lavage Fluid; Histopathologic Examination: Lung or Other Tissue, Bronchoalveolar Lavage Fluid; Antibody Tests: Serum; Antigen Tests: Amniotic Fluid, Serum, CSF.

REFERENCE RANGE: Negative

METHODS: Culture: Toxoplasma gondii may be isolated in tissue culture or in mice. Infection is confirmed by microscopic examination. This process is difficult and prolonged. Histopathologic Exam: Identification of tachyzoites by hematoxylin and eosin, eosin/methylene blue fast staining, or Giemsa staining. Antibody Tests: Indirect Immunofluorescence; ELISA; Direct Agglutination Test; Complement Fixation. Antigen Tests: ELISA; Dot Immunobinding Assay (Review: Koskiniemi M, et al. AJDC. 1989; 143:724-728).

INTERPRETATION: Toxoplasma gondii is a protozoan obligate intracellular parasite. The sources of toxoplasma are uncooked meat and domestic animals. Toxoplasma infection is asymptomatic in 80-90 percent of cases; symptomatic patients may have cervical lymphadenopathy, fever, sore throat, malaise, rash, or hepatosplenomegaly. Immunodeficient patients may have severe or fatal toxoplasmosis. Patients with AIDS may have systemic infection, pneumonia, chorioretinitis, or CNS manifestations such as abscesses, encephalopathy or encephalitis. Toxoplasmosis in the pregnant female may cause spontaneous abortion or disease in the infant with symptoms developing after birth (congenital toxoplasmosis).

The usual changes in IgM and IgG antibodies in response to T. gondii infection are as follows: IgM antibodies appear within one week, peak in three to four weeks and disappear in three to four months. IgG antibodies appear in about three weeks, peak in two months with the immunofluorescent test. Newborns and immunodeficient patients may not demonstrate an IgM antibody response. IgM antibodies that form may persist for more than a year. Antinuclear antibodies and rheumatoid factor cause false-positive results with the indirect immunofluorescence assay for IgM. These problems make interpretation of toxoplasma serology difficult (Koskiniemi M, et al, already cited; Frenkel JK. AJDC. 1990; 144:956-957). IgA antibody testing is promising as a more sensitive and specific method of diagnosing congenital and acute toxoplasmosis (Decoster A, et al. Lancet. November 12, 1988. 1104-1107).

Pregnant Female: Transplacental transmission of toxoplasma to the fetus occurs if the mother first acquires acute toxoplasmosis during pregnancy. It does not repeat in subsequent pregnancies. Fetal parasitemia and generalized disease develop in about 50% of such pregnancies. In 10% of cases, the mother shows evidence of acute gestation illness. When the maternal infection occurs in the third trimester, the fetal infection rate is high, but the disease in the newborn is mild or subclinical. When the maternal infection occurs in the second trimester, fetal disease is devastating.

A positive toxoplasma IgG test in early pregnancy indicates that the patient was exposed to toxoplasmosis in the past and is now immune. A positive IgM test may indicate active disease. Serial testing every three months demonstrating four-fold rise in IgG titer or the appearance of IgM antibodies may indicate active infection.

Newborn: Detection of active infection in the newborn is more difficult in that the immunofluorescence test for IgM antibody on cord blood is not sensitive; most neonates who have been infected in-utero are born without detectable IgM antibody, and IgG antibody may be present in the neonate by passive transfer from immune mother. The laboratory diagnosis in the neonate of acquired toxoplasmosis in-utero is done by demonstrating a rising IgG titer over a several month period.

Immunocompromised Host: Toxoplasmosis in immunocompromised hosts may represent primary infection or reactivation. These patients make immunoglobulins unreliably and may not seroconvert. In such cases, histologic and microbiologic criteria may be more important (Pomeroy C, Filice GA. Clin Infect Dis. 1992; 14:863-870).

There is no single test or combination of tests that is diagnostic of toxoplasmosis in all clinical situations. Diagnostic evaluation must take into consideration the clinical situation and limitations of the available tests (Kaufman L. Clin Infect Dis. 1992; 14(Suppl 1):S23-S29; Hohlfeld P, et al. J Pediatr. 1989; 115:765-769; Sever JL, et al. Pediatr. 1988; 82:181-192; Daffos F, et al. N Engl J Med. 1988; 318:271-275).

TRANSFERRIN

SPECIMEN: Red top tube, separate serum; avoid hemolysis

REFERENCE RANGE: 204-306mg/dL. There is a diurnal variation in transferrin levels, with higher levels found in the A.M. To convert conventional units in mg/mL to international units in g/liter, multiply conventional units by 0.01.

METHOD: Immunoassay. Gel diffusion or light scattering

INTERPRETATION: Transferrin levels are primarily measured in the work-up of anemias. Transferrin levels can also be assessed by measurement of the total iron binding capacity (TIBC). Transferrin is elevated in the serum in the conditions listed in the next Table:

Causes of Elevated Serum Transferrin
Iron Deficiency Anemia
Oral Contraceptives
Late Pregnancy
Viral Hepatitis

Transferrin is decreased in the serum in the conditions listed in the next Table:

Causes of Decreased Serum Transferrin
Chronic Infections (IL-1 Dependent)
Malignancy
Iron Poisoning
Hemolytic Anemia (sometimes Normal)
Hemochromatosis (sometimes Normal)
Nephrosis
Kwashiorkor
Thalassemia
Following Treatment of B-12 or Folate Deficiency

Sialic acid deficient forms of transferrin appear in the serum during alcohol abuse (Stibler H, Borg S. Alcoholism: Clin Exp Res. 1986; 10:61; Behrens UJ, et al. ibid. 1988; 12:427).

See related topics, IRON; IRON BINDING CAPACITY, TOTAL.

TRICYCLIC ANTIDEPRESSANTS

Synonyms: Amitriptyline, Nortriptyline, Imipramine (Tofranil), Desipramine, Protriptyline, Doxepin and Desmethyldoxepin

SPECIMEN: Urine: 50mL random, may be used for qualitative screening. Blood: The prerequisites for therapeutic monitoring are: 1) dosage must be stable for at least ten days and 2) dosage cannot be changed or missed. Specimens for monitoring therapeutic response should be drawn just prior to the first dose of the day or in the morning if single bedtime doses are taken. These sampling times are at least six hours, but preferably 10-15 hours after the last dose. Specimens for toxic overdose are obtained anytime. Blood: Green (heparin) top tube, remove plasma from red cells within two hours. There is a substance in some vacutainer tubes that displaces tricyclic antidepressants (TCAs) from plasma protein. The TCAs may then enter the red blood cells yielding artifactually low plasma values. Serum separator tubes may not be used. Urine: 50mL random urine.

REFERENCE RANGE:

Therapeutic plasma ranges for TCAs are given in the next Table:

Therapeutic Plasma Ranges for Tricyclic Antidepressants

Drug	Active Metabolite	Therapeutic Ranges (ng/mL)	Half-Life (hours)
Amitryptyline (Elavil Endep and Triavil)	Nortriptyline	120-250	17-40
Nortriptyline (Aventyl)		50-150	18-93
Imipramine (Tofranil and Presamine)	Desipramine	150-250	9-24
Desipramine (Norpramin and Pertofrane)		150-250	14-76
Doxepin (Sinequan and Adapin)	Desmethyldoxepin	150-300	8-25

Although there is individual variation in response to these drugs, the frequency of adverse effects increases sharply at levels above the therapeutic range. The therapeutic index is quite low.

Therapeutically effective concentrations are approximately 200ng/mL except for nortriptyline (100ng/mL). A level of tricyclics above 600ng/mL may produce serious side effects and extremely toxic concentrations exceed 1000ng/mL. Toxic levels are lower for nortriptyline and protriptyline. These levels include the parent compounds and active metabolite.

METHOD: Qualitative screens of urine use thin layer chromatography. Immunoassays are semiqualitative since they lack specificity and cross-react with related drugs and metabolites. Gas liquid chromatography (GLC) and high-pressure liquid chromatography (HPLC) are used to definitely identify and quantitate these drugs. The method used must measure the active metabolite for amitriptyline (nortriptyline) and the active metabolite for imipramine (desipramine).

INTERPRETATION: Levels of TCAs are determined as part of therapeutic monitoring therapy and compliance and in cases of intentional overdose. Patients with renal and hepatic therapy, the young and old, and those on additional drugs may in particular require more careful monitoring. TCA drugs are used primarily for the treatment of acute episodes of endogenous depression and for long term treatment to prevent recurrence and relapse.

The TCAs most frequently prescribed are amitriptyline, imipramine and doxepin; these same drugs are most frequently involved in suicidal attempts (Priest RG, et al. J Intern Med. 1980; 8:8-13). At levels greater than 1000ng/mL these drugs are toxic to the nervous system causing grand mal seizures and respiratory arrest, and to the heart causing ventricular arrhythmias, e.g., ventricular tachycardia and fibrillation. Orthostatic hypotension may develop. Total tricyclic concentrations reflect the severity of overdose; however, the best correlation is with an ECG finding of QRS widening. In the absence of severe impairment of myocardial performance, depressed patients with preexisting heart disease can be effectively treated with TCAs without an adverse effect on ventricular rhythm or hemodynamic function (Veith RC, et al. N Engl J Med. 1982; 306:954-959).

TCAs may also cause vasospasm in some patients, and the vasospasm may be dose related (Appelbaum P, Kapoor W. Am J Psychiatry. 1983; 140:913-915). The most common adverse effects of tricyclic antidepressants are anticholinergic, e.g., dry mouth, decreased gastrointestinal motility, mydriasis, urinary hesitancy or retention and tachycardia. Treatment of mild toxicity is supportive; moderate and severe overdoses may require bicarbonate infusion respiratory support, anticonvulsants, physostigmine and beta-blockers. Cardioversion and pacing may be necessary.

Withdrawal symptoms (nausea, dizziness, headache, increased perspiration and salivation) have been reported after stopping treatment with tricyclics (Dilsaver SC, et al. Am J Psychiatry. 1983; 140:249); dosage should be tapered when tricyclics are discontinued.

Lack of therapeutic response may be due to doses that are too low or too high, or to an unresponsive type of depression or non-compliance. A serum assay should be done, and the dose adjusted appropriately; a therapeutic response will be obtained in an additional 25-30 percent of the patients.

Patients with low serum albumin have a higher percentage of free to bound drug ratio (protein binding, 90 percent) and thus therapeutic or toxic reactions occur at a lower serum level. The half-life of the tricyclics is accelerated by barbiturates and smoking; the half-life is decreased by corticosteroids, phenothiazines, haloperidol and disulfiram.

TRIGLYCERIDES

SPECIMEN: Fasting specimen, red top tube, separate serum.
REFERENCE RANGE: Age adjusted normal values for serum triglycerides are given in the following Table:

Age Adjusted Normal for Serum Triglycerides

Age	Triglycerides (mg/dL)
0-19	10-140
20-29	10-140
30-39	10-150
40-49	10-160
50-59	10-190

To convert conventional units in mg/dL to international units in mmol/L, multiply conventional units by 0.01129.

METHOD: Breakdown of triglycerides to glycerol plus fatty acids via enzymatic or chemical hydrolysis. Glycerol release is then assayed enzymatically.

INTERPRETATION: Triglyceride (TG) levels vary markedly following a meal. A fasted sample (12-14 hours) must be used for accurate cardiac risk assessment and calculation of LDL and VLDL. There is about a 25% within-individual variation in TG levels (Cooper GR, et al. JAMA. 1992; 267:1652-1660). Triglycerides are insoluble in blood, and thus they do not circulated free in the serum but are transported as chylomicrons and pre-beta lipoproteins (very low density lipoproteins (VLDL). 80% of chylomicrons are triglycerides; 55% of pre-beta lipoproteins are triglycerides. It takes 10-12 hours to clear the blood of chylomicrons after a meal; peak lipidemia is reached in 3 to 5 hours and persists for another 6 to 8 hours.

Triglyceride levels of 250-500mg/dL are considered borderline hypertriglyceridemic, while levels >500mg/dL are definitely hypertriglyceridemic (Expert Panel, National Cholesterol Education Program. Arch Intern Med. 1988; 148:36-39). TG levels in the range 250-750mg/dL are usually due to increased in VLDL (Type IV hyperlipoprotein), while those >750mg/dL are due to elevations in VLDL and/or chylomicrons (Type V or rarely Type I hyperlipoproteinemia). TG levels greater than 1000mg/dL are associated with pancreatitis. In most studies elevated TG level is not an independent risk factor for coronary heart disease, however, the Framingham Heart Study reported it as an independent predictor in women. At present the role of elevated TG as an independent risk factor for cardiovascular disease remains controversial. A NIH consensus conference concluded that current evidence did not allow one to conclude causality between high TG levels and CHD (NIH Consensus Conference. JAMA. 1993; 269:505-510). Publications suggesting such a link are: Bainton D, et al. Br Heart J. 1992; 68:60-66; Castelli WP. Am J Cardiol. 1992; 70:3H-9H; Manninen V, et al. Circulation. 1992; 85:37-45). In a few studies, TG appeared to contribute to risk in subgroups in association with low HDL and LDL (Criqui MH, et al. N Engl J Med. 1993; 328:1220-1225).

Triglycerides (Cont)

The triglyceride level is used to calculate the LDL cholesterol. This is based on the fact that VLDL cholesterol is approximately one-fifth that of the triglyceride. The formula is:

$$\text{LDL cholesterol} = \text{Total serum cholesterol} - \left(\frac{TG}{5} + \text{HDL cholesterol}\right)$$

In some reports, 6.25 is used instead of 5. This formula is only valid for triglyceride levels <400mg/dL.

The causes of elevation of serum triglycerides are as follows:

Causes of Elevation of Serum Triglycerides
Non-fasting Specimen
Primary Hyperlipoproteinemia
(Fredrickson Types I, IIb, III, IV, V)
Secondary Hyperlipoproteinemias:
Lipid Infusion in Hospitalized Patients
Diabetes Mellitus
Diabetic Ketoacidosis
Acute Alcoholism
Oral Contraceptives
Nephrotic Syndrome
Chronic Renal Failure
Steroids
Acute Pancreatitis
Gout
Gram Negative Infections
Glycogen Storage Disease

Secondary hyperlipoproteinemias with elevation of serum triglycerides are very common, e.g., diabetes mellitus, acute alcoholism, nephrotic syndrome and use of oral contraceptives.

In most women taking oral contraceptive steroids there is a small but definite increase in serum lipids. However, in about 1/3 of the women taking oral contraceptive steroids, the triglyceride level exceeds the upper limit of normal. The increase in the triglyceride is related to the estrogen but not the progesterone content of the contraceptive drug. Patients on oral contraceptives who are most "susceptible" to hyperlipemia are the following: obese patients, patients showing evidence of carbohydrate intolerance, patients having pre-existing hyperlipemia, and patients having a family history of hypertriglyceridemia. Caution should be exercised in the use of contraceptive agents and other estrogenic compounds in patients with pre-existing hyperlipemia, e.g., women with Type I, IV and Type V hyperlipoproteinemia, because elevated plasma triglyceride may trigger an attack of pancreatitis.

TRIIODOTHYRONINE (T-3)

SPECIMEN: Red top tube, separate serum.
REFERENCE RANGE: Cord blood 10-90 ng/dL (mean = 58); Full term 1-3 days 89-405 ng/dL (mean = 273); One week 91-300 ng/dL (mean = 190); 1-12 months 85-250 ng/dL (mean = 175); Prepubertal 119-218 ng/dL (mean = 168); Pubertal and Adult 55-170 ng/dL (mean = 105). To convert conventional units in mg/dL to international units in mmol/L, multiply conventional units by 0.01536.
METHOD: RIA
INTERPRETATION: The T-3 is measured to aid in the diagnosis of thyroid disorders, especially hyperthyroidism and T-3 thyrotoxicosis. The T-3 is increased in hyperthyroidism and decreased in hypothyroidism. Although the T-3 may be useful in the diagnosis of hyperthyroidism, it has a low sensitivity in the diagnosis of hypothyroidism. Measurement of the T-3 may be useful when other blood tests of thyroid function are equivocal. Early in the course of hyperthyroidism, the serum T-3 may be elevated before the T-4. Patients with T-3 thyrotoxicosis have signs and symptoms of hyperthyroidism, an elevated T-3, but normal T-4 and free T-4. This relatively rare disorder appears to be more common in areas of iodine deficiency. In sick euthyroid patients there is reduced conversion of T-4 to T-3 and an increase in the inactive metabolite, reverse T-3. T-3 is reduced in sick euthyroid patients and may be normal in sick hyperthyroid patients. Endogenous anti-T-3 antibodies may cause either spurious elevated or lowered serum T-3 values (De Baets M, et al. Clin Chim Acta. 1982; 118:293-301).

TRYPSIN/TRYPSINOGEN, IMMUNOREACTIVE, SERUM

SPECIMEN: Red top tube, separate serum; store sera at -20°C until assay. Assay may be done on dried blood-spots collected for neonatal screening of other inborn errors. Dried blood specimens stored at room temperature lose half their immunoreactivity over a period of three months (Heeley AF, et al. Arch Dis Childhood. 1982; 57:18-21).
REFERENCE RANGE: Adult: 20 to 80 mcg/L; reference range dependent on kit used. For cystic fibrosis screening: abnormal >150 mcg/L days 1-4; >120 mcg/L on remeasurement (mean age 38 days - Hammond KB, et al. N Engl J Med. 1991; 325:769-774).
METHOD: Radioimmunoassay. The assay usually employs trypsin which is bound to and inhibited by a protease inhibitor. The antibodies crossreact with and measure trypsinogen which has been released into the circulation.
INTERPRETATION: Immunoreactive serum trypsin is useful in the diagnosis and evaluation of cystic fibrosis, acute pancreatitis, and chronic pancreatitis. Immunoreactive trypsin(IRT) is actually circulating trypsinogen, a precursor of trypsin which is synthesized in the pancreas.
Cystic Fibrosis: The concentration of immunoreactive trypsin(IRT) is increased in the blood of infants with cystic fibrosis(CF) during the first few months of life; it is thought that in these patients, the pancreatic duct is progressively blocked with initial "back-leakage" of acinar contents into the plasma. The level of IRT subsequently decreases to values below normal. The elevation of IRT is characteristic of newborn CF, whether or not they have residual exocrine pancreatic function, e.g., measurable stool trypsin activity (Crossley JR, et al. Clin Chem Acta. 1981; 113:111-121).

The blood spot IRT assay performed in the neonatal period is an excellent screening test for detection of cystic fibrosis (Wilchen B, et al. J Pediatrics. 1983; 102:383-387). The incidence of false negative test results is 10% to 25%; the incidence of false positive test results is 0.1% to 0.3%. The IRT test may be particularly useful during the first month of life when the results of the sweat test are not reliable. (For Discussion see Ad Hoc Committee Task Force on Neonatal Screening. Cystic Fibrosis Foundation: Neonatal Screening for Cystic Fibrosis: Position Paper. Pediatrics. 1983; 72:741-745; Holtzman NA. Pediatrics. 1984; 73:98-99; Farrell PM. Pediatrics. 1984; 73:115-117). The feasibility and efficacy of serial IRT measurements on dried blood spots in a large state-wide cystic fibrosis screening program has been reported (Hammond KB, et al. N Engl J Med. 1991; 325:769-774).
Chronic Pancreatitis with Steatorrhea: IRT is decreased in patients with chronic pancreatitis with steatorrhea; it is normal in patients with chronic pancreatitis without steatorrhea and in patients with steatorrhea but with normal pancreatic function (Jacobson DG, et al. N Engl J Med. 1984; 310:1307-1309).
Acute Pancreatitis: Serum immunoreactive trypsin is a sensitive and specific indicator of pancreatic damage. Levels increase 5-10 times normal in acute pancreatitis and remain elevated for 4-5 days. However, although it is quite specific for pancreatitis, it is a radioimmunoassay not available on a 24 hour basis, and hence does not present any clear advantage over serum amylase and lipase in the routine diagnosis of acute pancreatitis.

TUMOR MARKER PANEL (BLOOD)

Synonym: Cancer Marker Panel

Tumor markers are used to help direct or monitor therapy, aid in arriving at a diagnosis, establish prognosis, predict sensitivity of the malignancy to therapy and study the biology of tumors. The principle use of tumor markers is to assess therapy and to detect tumor recurrences. Rarely, tumor markers may be used to screen for malignancies in selected populations, although in most cases, measurement of tumor markers cannot be used in screening programs.

Tumor markers are most often measured in blood, although they may also be detected in body fluids. Tissue and cell preparations may be used to help establish a diagnosis (see TUMOR MARKER PANEL, TISSUE).

The classical tumor markers have been tumor associated antigens (oncofetal antigens), enzymes associated with malignancies, and hormones associated with malignancies. To this list must now be added cytogenetic markers, molecular markers (oncogene and suppressor gene), and immune system markers. The types of tumor markers used in medicine and one example of each is given in the next Table:

Tumor Markers Panel (Blood) (Cont)

Tumor Markers
Tumor Associated Antigens (Oncofetal Antigens)
Carcinoembryonic Antigen (CEA) - Colon Cancer
Enzymes
Acid Phosphatase - Prostatic Cancer
Hormones
HCG - Choriocarcinoma
Immune System Markers
Monoclonal Ig - Multiple Myeloma
Cytogenetic Markers
Philadelphia Chromosome - CML
Molecular Markers (Oncogenes)
c-myc Amplication - Neuroblastoma

Problems associated with the use of tumor markers include:

1) They are not sensitive - the malignancy is often metastatic or has reached a large size before the tumor marker is elevated.
2) They are not specific - elevations are also seen in non-malignant disease.
3) Not all of a class of tumors will produce the antigen or tumor marker.

Tumor Associated Antigens: Tumor associated antigens (oncofetal antigens), are cellular antigens found during fetal development, and associated with various tumors. They constitute the largest group of tumor markers measured in blood and other fluids.

Commonly measured tumor associated antigens are given in the next Table:

Commonly Measured Tumor Associated Antigens

Antigen	Associated Tumor
Alpha-Fetoprotein	Hepatoma Germ Cell Neoplasms
Carcinoembryonic Antigen (CEA)	Colon Cancer Breast Cancer Many Others
Cancer Antigen (CA) 125	Ovarian Cancer
Cancer Antigen (CA) 15-3	Breast Cancer
Cancer Antigen (CA) 19-9	Pancreatic Cancer
Neuron-Specific Enolase	Endocrine Neoplasms, Seminoma
Prostate-Specific Antigen (PSA)	Prostate Cancer

Both CEA and CA 15-3 have been found to be helpful in monitoring therapy of breast cancer (Dnistrian AM, et al. Clin Chem Acta. 1991; 200:81-93). Serum neuron-specific enolase (NSE) has been used to follow therapy in patients with seminoma. It is claimed to have approximately the same sensitivity and specificity as HCG (Fossa SD, et al. Br J Cancer. 1992; 65:297-299). Prostate-specific antigen (PSA) appears to be a more specific and sensitive marker for prostatic carcinoma than either acid phosphatase or beta-microseminoprotein (Abrahamsson PA, Lilja H. Andrologia. 1990; 22 suppl. 1:122-131).

Enzymes: A list of some enzymes that have been used to monitor cancers is given in the next Table:

Enzymes As Tumor Markers
Acid Phosphatase
Alkaline Phosphatase (Placenta)
Lactate Dehydrogenase(LDH)
Gamma Glutamyl Transpeptidase
Cathepsin-D
Muramidase (Lysozyme)
Creatine Phosphokinase - "Brain" Isoenzyme(CPK-BB)
Galactosyl Transferase Isoenzyme II(GT-II)
Beta-Glucuronidase
Sialytransferase
Terminal Deoxynucleotidyl Transferase
Ribonuclease
Neuron-Specific Enolase
Histaminase (Medullary Carcinoma of the Thyroid)
Amylase
Cystine Aminopeptidase

Those enzymes marked in bold are discussed under their respective titles elsewhere in the text. The most commonly employed of these tests has been the prostatic acid phosphatase, a test which is being rapidly replaced by the prostate-specific antigen test. In general, tumor associated enzyme activities are the least specific of the various types of tumor markers.

Hormones: Elevation of hormones is associated with certain malignancies. Some examples are given in the next table:

Hormone Elevations Associated with Neoplasia	
HCG	Germ Cell Tumors
	Gestational Trophoblastic Disease
Insulin	Pancreatic Islet Neoplasms
	Retroperitoneal Tumors
Calcitonin	Medullary Carcinoma of Thyroid
Erythropoietin	Renal Cell Carcinoma
ACTH	Small Cell Carcinoma, Carcinoid
ADH	Undifferentiated Small Cell Carcinoma
PTH	Squamous Cell Carcinoma
PTH-Like Peptide	Many Malignancies

This list is not all-inclusive. In addition, there are many examples of hormone production by neoplasms of endocrine tissues (eg, steroids by adrenal cortical neoplasm; gastrin by islet cell tumors; catecholamines by adrenal medulla neoplasms, etc).

Screening for Malignancy: In general, tumor markers are not useful in screening for malignancy. Exceptions are: 1) calcitonin measurements in screening family members at risk for medullary carcinoma of the thyroid, 2) HCG measurements in postpartum women at risk for choriocarcinoma following delivery of a molar pregnancy, and 3) alpha fetoprotein in screening for hepatocellular carcinoma in high prevalence areas, e.g., China (Xu KL in Fishman W (ed); Oncodevelopmental Markers, Academic Press, San Diego. 1983; 395-399). The use of the prostate-specific antigen (PSA) test in the screening of prostatic carcinoma is controversial at present. One study has recommended the use of PSA test in conjunction with rectal examination to screen for prostatic carcinoma (Calalona WJ, et al. NEJM. 1991; 324:1156-1161). However, others have stated that the value of the PSA test for screening remains unclear (Gerber GS, Chodak GW. J Natl Cancer Inst. 1991; 83:329-325).

TUMOR MARKER PANEL (TISSUE)

Tumor markers are determined on removed tissues and cells to establish or confirm a diagnosis, subclassify a lesion, assess the prognosis, help select therapy, and to study the biology of the process.

Immunocytochemistry of Tumor Markers: Tumor markers in anatomical pathology are usually performed to establish a diagnosis on a surgical specimen that is difficult to classify by conventional light microscopy. Tumor markers are usually assessed by immunocytochemistry. The conventional approach is to use a screen of tumor markers to classify a lesion into the broad categories of carcinoma, melanoma, lymphoma and sarcoma as illustrated below:

Immunohistochemical Staining of Malignant Tissues

	Cytokeratin	LCA	S-100 Protein	Vimentin	Desmin
Carcinoma	+	-	-	-	-
Melanoma	-	-	+	+	-
Lymphoma	-	+	-	+/-	-
Sarcoma	-	-	-	+	+/-

LCA = Leukocyte Common Antigen. Adapted from: Roche Diagnostic Oncology

In some instances, measurement in serum or urine of some tumor markers may help elucidate or confirm the identify of a malignancy. Serum tumor markers commonly employed in the workup of malignancies with an unknown site of origin are outlined in the tables below (Modified from Bitran JD, Ultmann JE. Dis-Mon. 1992: 38:213-260):

Tumor Markers Panel (Tissue) (Cont)

Malignancies with Unknown Primary Site: Serum and Urine Tumor Markers	
Tumor Marker	**Possible Primary**
CEA	Colon, Lung, Breast, Cervix
AFP	Liver, Germ Cell
HCG	Gestational Trophoblastic Disease, Germ Cell, Colon and Lung (infrequent)
Acid Phosphatase, PSA	Prostate
Catecholamines, VMA, HMA	Neuroblastoma, Malignant Pheochromocytoma
CA 125	Ovary (non-mucinous carcinomas)
CA 15-3	Breast
CA 19-9	Pancreas, Stomach, Hepato-biliary, Colon (occasionally)

The diagnostic and treatment aspects of the patient who presents with a malignancy of unknown primary site are reviewed (Hainsworth JD, Greco FA. N Engl J Med. 1993; 329:257-263).

Newer techniques and reagents are constantly being introduced to diagnose and classify neoplasms. Some of these include oncogene and suppressor gene expression, and assays of gene mutation and rearrangements.

More detailed subclassification and confirmation in difficult cases can be made by use of a variety of other tumor markers as listed below:

Malignancies with Unknown Primary Site: Tests Performed on Tissue	
Tumor Markers Present	**Possible Primary**
Carcinoma:	
HCG	Germ Cell Tumors/Choriocarcinoma
AFP	Germ Cell Tumors/Hepatoma
Acid Phosphatase/PSA	Prostate
α-1 Antitrypsin	Hepatoma
Neuron-Specific Enolase	Undifferentiated Small Cell Carcinoma and Other Neural-Crest
S-100	Melanoma/Nervous System
Estrogen/Progesterone Receptor	Breast/Uterus/Ovary/Fallopian Tube/ Colon/Melanoma (rare)
Lactoferrin/Lactalbumin/Casein	Breast
Ferritin	Hepatoma
Bombesin	Undifferentiated Small Cell
Leu-7	Melanoma and Others (Small Cell, Lymphoma)
Melanoma Specific Ab-HMB45	Melanoma
Chromogranin	Neuroendocrine Neoplasms
Thyroglobulin	Follicular Thyroid Carcinoma
Calcitonin	Medullary Thyroid Carcinoma
Sarcoma:	
Desmin/Myosin/Myoglobin	Rhabdomyosarcoma/Leiomyosarcoma
Factor VIII	Angiosarcoma/Kaposi's Sarcoma
Myelin Basic Protein	Schwanoma/Neurilemmoma
Glial Fibrillary Acidic Protein	Glioma
Leukemia/Lymphoma:	
Alkaline Phosphatase	Leukemia/Lymphoma
Terminal Deoxynucleotidyl Transferase	
Cell Surface Markers	
Surface Ig	
Light Chain Analysis	
Immunoglobulin Gene Rearrangements	

(cont)

Malignant Mesothelioma Versus Adenocarcinoma: Immunocytochemistry does not allow total confirmation of a diagnosis of malignant mesothelioma. The problem of distinguishing malignant mesothelioma from adenocarcinoma has been approached by application of a battery of immunochemical stains. Representative results are tabulated below (Delahaze M, et al. J Pathol. 1991; 165:137-143):

Positive Staining of Malignant Mesothelioma and Adenocarcinoma in Effusions

Marker	% Positive	
	Malignant Mesothelioma	Adenocarcinoma
CEA	83%	100%
EMA	4%	45%
MOC 31	8%	58%
OV 632*	91%	9%

*The reported percentages are for non-ovarian carcinoma. One hundred percent of ovarian cancer stains with OV 632. Epithelial membrane antigen is abbreviated EMA.

In a similar study, reactivity to anti-CEA, EMA, and B72.3 was shown to be sufficient to accurately characterize 97% of effusions (Frisman DM, et al. Mod Pathol. 1993; 6:179-184).

Molecular Markers (Oncogenes): Oncogenes are typically growth-related factors that are either: 1) surface receptors of growth factors; 2) cytoplasmic proteins involved in signal transduction; 3) nuclear proteins involved in cell division and gene expression. Aberrant oncogene expression can result from: 1) mutation; 2) duplication; 3) translocations. Examples of molecular markers being measured in medical practice are given below:

Molecular Markers Associated with Neoplasia

Neoplasm	Marker
Breast Cancer	erbB2/HER2/neu
Neuroblastoma	c-myc
Colon Cancer	Loss of p53

Cytogenetic Markers: A number of associations between cytogenetic changes and specific neoplasms exist. These were classically detected by karyotyping. However, many of these are now being detected with the polymerase chain reaction (PCR) followed by gene probes. Examples of cytogenetic markers and associated neoplasms are tabulated below:

Some Cytogenetic Markers and Neoplasms

Neoplasm	Marker
Chronic Myelogenous Leukemia (CML)	Philadelphia Chromosome, t(9;22); (bcr/abl fusion)
Acute Non-Lymphocytic Leukemia (ANLL)	t(8;28)
Acute Apromyelocytic Leukemia (APL)	t(15;17)
Follicular B-cell Lymphoma	t(14;18)(q32;q21)(BCL-2)
Colon Cancer	Deletion 17p and 18q

General References:

Jacobs EL, Haskell CM. Curr Probl Cancer. 1991; 15:299-360.
Beastall GH, et al. Ann Clin Biochem. 1991; 28:5-18.
Stener K. Scand J Clin Lab Invest Suppl. 1991; 206:6-11.
Fenoglio-Preiser CM. Cancer. 1992; 69(Suppl 6):1607-1632.

TYPE AND CROSS **(see ABO AND RH TYPE)**

UREA NITROGEN, BLOOD (BUN)

SPECIMEN: Red top tube, separate serum; or green (heparin) top tube, separate plasma.
REFERENCE RANGE: 8-18mg/dL. To convert conventional units in mg/dL to international units in mmol/liter, multiply conventional units by 0.3570.
METHOD: Enzymatic Degradation:

$$\text{Urea} + H_2O \xrightarrow{\text{Urease}} 2NH_3 + CO_2$$

$$NH_3 + \text{Alpha-Ketoglutarate} + NADH + H^+ \xrightarrow[\text{Dehydrogenase}]{\text{Alpha-Ketoglutarate}} \text{L-Glutamate} + NAD^+$$

Some methods use an ammonia-sensitive electrode.
INTERPRETATION: Urea is increased in the serum in conditions listed in the next Table:

Causes of Elevation of Serum Urea
Renal Disease
Prerenal Azotemia
G.I. Hemorrhage
G.I. Obstruction
Shock
Tissue Necrosis
Third Degree Burns
Fever, Protracted
Dehydration
Diarrhea
Diabetic Coma
Congestive Heart Failure
Renal Artery Stenosis
Addison's Disease
Steroid Therapy
High Protein Diet
Post-Renal (Renal Vein Thrombosis, Urinary Tract Obstruction)
Tetracycline (Results in Net Catabolism)

The BUN is increased in both renal and prerenal azotemia. Serum creatinine and creatinine clearance are needed to differentiate pre-renal and renal azotemia. See RENAL FAILURE, ACUTE, PANEL.

The three general causes of an increased serum urea concentration are decreased glomerular filtration rate, an increased load of urea for excretion from the diet or tissue metabolism, and an increased tubular reabsorption of urea.

Causes of raised urea concentrations in a hospital population are given in the next Table (Morgan DB, Carver ME, Payne RB. Brit Med J. 1977; 2:929-932):

Causes of Elevated Serum Urea in a Hospital Population

Cause	Incidence (Percent)
Congestive Heart Failure	36
Dehydration	12
Post-operation	6
Hypotension	3
Acute Renal Failure	2
Chronic Renal Failure	3
Increased Urea Load	2
Obstructive Renal Disease	1
Combined Causes	9
Unclassified	26

Urea is decreased in the serum in conditions listed in the next Table:

Causes of Decrease in Serum Urea
SIADH
Liver Disease
Overhydration
Anabolic Hormones
Malnutrition
Normal Pregnancy

Review: Lyman JL. Emerg Med Clin N Amer. 1986; 4:223-233.

URIC ACID

SPECIMEN: Red top tube, separate serum; stable for 3 days at room temperature.
REFERENCE RANGE: Adult Females: 2.5-6.2mg/dL; Adult Males: 3.5-8.0mg/dL; One Month to 12 years: 2.0-7.0mg/dL. To convert conventional units in mg/dL to international units in micromol/liter, multiply conventional units by 59.48. Fasting sample (4 hour) preferred.
METHOD: Enzymatic (uricase).
INTERPRETATION: There are four main causes of hyperuricemia; tests used to differentiate between these causes is given in the following Table (Emmerson BT. Lancet. 1991; 337:1461-1463):

Tests for Causes of Hyperuricemia

Test (Normal Range)	Excess Purine Consumption	1° Urate Under-Excretion; NL Renal Function	Renal Disease (1° or 2°)	Urate Overproduction
Creatinine Clearance (100-130mL/min)	Normal	Normal	Greatly ↓	Normal
Urate Clearance (6-11mL/min)	Normal	Greatly ↓	Greatly ↓	NL or ↑
24 hr Urinary Urate Excretion, Low-Purine Diet (2-4mmol/24h)	Normal	Greatly ↓ (<2mmol/24h)	Greatly ↓	Greatly ↑ (>4.5mmol/24h)
Fall in Plasma Urate on Low-Purine Diet (<0.06mmol/L)	↑	NL or ↑	↑	NL or ↓
Fall in Urine Urate on Low-Purine Diet (<1.2mmol/24h)	↑	Normal	↑	NL or ↓

Uric acid is increased in the serum in the conditions listed in the next Table:

Causes of Increased Serum Uric Acid

Gout	Tissue Necrosis
Dehydration	Chemotherapy
Acute Inflammation	Malnutrition of All Types
Hematologic Conditions:	Therapeutic Radiation
Leukemia	Alcohol
Lymphoma	Lead Poisoning
Hemolytic Anemia	Glycogen Storage Dis., Type I
Megaloblastic Anemia	Lactic Acidosis
Infectious Mononucleosis	Toxemia of Pregnancy
Polycythemia Vera	Psoriasis (Active)
Chronic Renal Disease (Renal Failure)	Lesch-Nyhan Syndrome
Drug Induced:	Down Syndrome
Thiazides	Hypothyroidism
Salicylates (Low Dose)	Hypoparathyroidism
Pyrazinamide (PZA)	
Ethambutol	
Nicotinic Acid	
Cytotoxics	

Increased concentration of uric acid is found in gout, in conditions involving increased cellular destruction, such as leukemia and lymphoma. In treating leukemia, uric acid must be monitored because massive destruction of cancer cells by cytotoxic agents causes release of large amounts of uric acid. In conditions whereby increased lactic acid is produced, the lactic acid and uric acid compete for renal excretory sites and serum uric acid is increased. Hyperuricemia appears to be associated with risk of coronary heart disease (Brand FEN, et al. Am J Epidemiol. 1985; 121:11-18; Frohlich ED. JAMA. 1993; 270:378-379).

Gout: All patients with gout have hyperuricemia. However, all patients with an elevated serum uric acid do not have gout. Males with uric acid levels >7.0 mg/dL and females >6.0mg/dL should be followed for long term renal complications of elevated blood and urine uric acid levels. Using 7.0 mg/dL as cutoff for men, serum uric acid has a sensitivity of 90% and a specificity of 95% for the diagnosis of gout.

Uric Acid (Cont)

Uric acid is decreased in the serum in conditions listed in the next Table:

Causes of Decreased Serum Uric Acid
Severe Alcoholism with Liver Disease
Chronic Debilitating Disease
Renal Tubular Defects (Fanconi Syndrome)
Salicylates (High Dose)
Corticosteroids
Ascorbic Acid (High Dose)
Allopurinol
Probenecid
X-Ray Contrast Media
Glyceryl Guaiacolate
Wilson's Disease
Hemochromatosis
Cystinosis
Xanthine Oxidase Deficiency
Biliary Obstruction
SIADH
Acute Intermittent Porphyria
Galactosemia

URINALYSIS ROUTINE

SPECIMEN: 10mL fresh urine
REFERENCE RANGE: Color, Straw; Turbidity, clear; Sp.Gr. 1.001-1.020. Dipsticks, pH 4.5-7.5; Protein, Negative; Sugar, Negative; Acetone, Negative; Bile, Negative; Hemoglobin, Negative; Nitrite, Negative; Leukocyte Esterase, Negative; Urobilinogen, Present. Positive dipstick tests are confirmed as follows: Protein, Sulfosalicylic acid; Coomassie blue dye binding method; Sugar, Clinitest; Acetone, Acetest; Bilirubin, Ictotest; Nitrite and Leukocyte Esterase: negative. Microscopic: WBC, 0-2 hpf; RBC, 1-2 hpf; Cast, 0-1 hyaline, occasional granular, lpf; hpf = high powered field; lpf = low powered field. Bacteria, rare.
METHOD: Appearance; Sp.Gr. by Refractometer; Dipstick; Microscopic.
INTERPRETATION: Urinalysis is performed as a routine health screen, in the evaluation of patients with suspected renal dysfunction, urinary tract disease, urinary tract infection, diabetes, prenatal evaluations, and many other conditions. Routine urinalysis requires a random specimen; an early morning specimen is preferred. With specimens for routine analysis, avoid preservatives and refrigeration. Analyze specimen within two hour after collection. If a urine specimen is left standing at room temperature for 5 hours, it will become alkalinized; it is not suitable for culture; the erythrocytes, if present, will decompose and urinary casts will disintegrate.

If immediate examination of urine is impractical, refrigeration is the preferred form of preservation; refrigeration does not affect the appearance of casts; it does not affect the urine osmolality and does not destroy erythrocytes.

The normal ranges for urine volume are as follows:

Age	Volume
Pediatrics	Normal: 2-5mL/kg/hr
	Oliguric: <1mL/kg/hr
Adult	1000-1600mL/day

Routine urinalysis usually consists of gross observation of the specimen, use of dipsticks and microscopic analysis.

APPEARANCE: Cloudiness due to phosphates (alkaline urine) and urates (acid urine) is normal. Cloudiness and color in urine may be associated with the conditions given in the next Table:

Color and Cloudiness in Urine
Blood
Myoglobin, Hemoglobin
Leukocytes (Infection)
Mucous (Normal or Inflammation)
Urobilinogen (Hemolytic Anemia, Liver Disease)
Conjugated Bilirubin (Liver Disease, Biliary Tract Obstruction)
Various Drugs
Certain Foods

A few important but relatively rare diseases are associated with changes in the color of urine. Some of the porphyrias cause excretion of porphobilinogen. This results in a red urine upon standing. In alkaptonuria, homogentisic acid is excreted which turns brown-black on standing. Many drugs and certain foods will impart a variety of colors to urine.

SPECIFIC GRAVITY: Urea (20 percent), chloride (25 percent), sulfate and phosphate contribute most to normal specific gravity urine. The measurement of specific gravity is done using a hydrometer (urinometer); other measurements that reflect specific gravity are measurement of refractive index using a refractometer and measurement of osmolality using an osmometer and the measurement of specific gravity using dip-stick test strips. The specific gravity may range from 1.003 to 1.030 depending on fluid intake. A specific gravity of 1.023 or more indicates normal urine concentrating ability. An inappropriately low specific gravity may indicate a tubular defect of concentrating ability as in sickle cell disease or diabetes insipidus. Common causes of greatly elevated specific gravity measurements include excretion of mannitol and radiographic contrast agents.

MULTIPLE REAGENT STRIP: The multiple reagent strip is dipped into the urine and read; the color indicates the pH range or the presence (concentration dependent) of the substance being tested. pH: Urine is normally acid; the distal tubular cells exchange $[H^+]$ for sodium of the glomerular filtrate and the urine becomes acid. The usual pH of urine is pH = 6. The pH of the urine reflects acid-base status, e.g., acid urine in acidosis and alkaline urine in alkalosis. However, the urine is alkaline in renal tubular acidosis; in this condition, the tubular ability to form ammonia and exchange $[H^+]$ for $[Na^+]$ is defective. An alkaline urine may indicate a defect in tubular function, or may be seen in certain urinary tract infections such as *Proteus* in which the organism generates a large amount of ammonia. If this organism contaminates the urine, and the urine is left standing for a prolonged period, the urine may be artifactually alkaline.

PROTEIN: Protein is found in the urine in the conditions listed in the next Table:

Protein in Urine
Nephrotic Syndrome
Nephritic Syndrome
Toxic Nephropathies
Renal Tubular Diseases
Nephrosclerosis
Polycystic Kidney Disease
Severe Venous Congestion of Kidney
Pyelonephritis
Pre-eclampsia of Pregnancy
Postural Proteinuria
Other Conditions: Multiple Myeloma, Hemorrhage Salt Depletion and Febrile Illness

The upper limit of normal for urine protein is about 150mg/24 hr. The protein lost in greatest amount in renal disease is albumin; however, when severe glomerular damage is present, larger proteins may be found in the urine. With glomerular damage, proteins with molecular weight greater than 60,000 appear in the urine. With tubular damage, low molecular weight proteins, which are normally absorbed in the proximal tubule appear (ß-2-microglobulin, lysozyme).

Urinalysis (Cont)

Bence-Jones protein is present in the urine in the absence of a serum paraprotein in about 20 percent of proven cases of multiple myeloma (Hobbs JR. Essays in Med Biochem. 1975; 1:105). The test material on reagent strips is not as sensitive to globulins nor to Bence-Jones protein as to albumin.

The nephrotic syndrome is by far the most common cause of a decrease in serum immunoglobulin levels. IgG and IgA concentrations are decreased and IgM concentration is usually normal. The synthesis of alpha-2 macroglobulin is switched on; this may be due to albumin loss and decreased oncotic pressure in the sinusoids; these proteins are not lost in the urine and accumulate in the blood due to their high molecular weight. The increased beta-lipoprotein (low density lipoprotein) accounts for the increase in serum cholesterol in the nephrotic syndrome.

GLUCOSE: Normally, no detectable glucose is present in the urine; glucose is elevated in the urine when glucose is elevated in the serum or when there is a lower renal threshold for glucose. The conditions in which glucose is found in the urine are given in the next Table:

Excess Glucose in Urine
Diabetes Mellitus
Other Endocrine Disorders: Cushing's Syndrome, Hyperadrenocorticism, Acromegaly, Pheochromocytoma, Hyperthyroidism
Pancreatic Disorders: Hemochromatosis; Pancreatitis; Carcinoma of the Pancreas
Central Nervous System Disorders
Disturbances of Metabolism
Burns, Infection, Fractures, Myocardial Infarction, Uremia, Liver Disease, Glycogen Storage Diseases, Obesity
Drugs: Thiazides, Corticosteroids, ACTH, Oral Contraceptives
IV Fluids
Renal Tubular Dysfunction
Pregnancy

The test material on the reagent strip (glucose oxidase) is specific for glucose. Other sugars in the urine will not give a positive test. Ascorbic acid (vitamin C) inhibits the test.

The renal threshold for glucose is a serum concentration of 180mg/dL; however, there is wide individual variation in the renal threshold for glucose. The test strip is positive at urine levels of 75-300mg/dL. Patients with significant glycosuria have increased specific gravity due to increased glucose in the urine.

KETONES: The test material on reagent strips measures acetone and acetoacetic acid but does not detect beta-hydroxybutyric acid; the strips detect 5 to 10mg/dL of ketones in urine. Ketonuria is found in conditions listed in the next Table:

Ketonuria
Uncontrolled Diabetes Mellitus
Non-Diabetic Ketonuria:
Acute Febrile Illnesses in Children
Toxic States with Vomiting or Diarrhea
Alcoholics
Starvation
Some Weight Reducing Diets

BLOOD: The test reagent strip material detects the heme in hemoglobin and myoglobin, it detects 0.05 to 0.3mg hemoglobin/dL of urine; 0.3mg hemoglobin is equivalent to 10 lysed red blood cells per microliter. When red blood cells enter the urine, hemolysis usually occurs, and free hemoglobin is released in the urine. Hemoglobin can be found in the urine as a result of intravascular hemolysis, or as a result of urinary tract bleeding and subsequent to hemolysis in the urine. False positives can be due to oxidizing compounds used as disinfectants; false negatives can be due to ascorbic acid or a urine pH <5.1.

Conditions in which hemoglobin appear in the urine are given in the next Table:

Hemoglobin in the Urine

- Hemolytic Anemias
- Renal Disease
  - Glomerulonephritis
  - Lupus Nephritis
  - Calculi
  - Tumor
  - Acute Infection
  - Tuberculosis
  - Infarction
  - Renal Vein Thrombosis
  - Trauma
  - Hydronephrosis
  - Polycystic Kidney
  - Acute Tubular Necrosis
  - Malignant Nephrosclerosis
- Lower Urinary Tract Disease
  - Acute and Chronic Infection
  - Calculus
  - Tumor
  - Stricture

The most common causes of hematuria are stones (20%) malignant neoplasm (15%), uretherotrigonitis, bacterial infection (10%), prostatic hypertrophy (10%) and glomerulonephritis (Abuelo JG. Arch Intern Med. 1983; 143:967-970). A detailed discussion of the causes of hematuria and workup of such patients is given in: Sutton JM. "Evaluation of Hematuria in Adults." JAMA. 1990; 263:2475-2480.

BILIRUBIN: The bilirubin that appears in the urine is bilirubin diglucuronide. The test material in the reagent strip is sensitive from 0.2-0.4mg/dL. The conditions in which positive tests are obtained are liver disease, obstructive biliary tract disease and the congenital hyperbilirubinemias, e.g., Dubin-Johnson and Rotor types.

LEUKOCYTE ESTERASE: The number of leukocytes in urine is indicated by a color reaction on the reagent strip. It exploits the esterase activity of leukocytes. Leukocytes liberate esterase which acts on the substrate on the dipstick. The substrate is an indoxyl carbonic acid ester from which indoxyl is liberated; indoxyl is unstable and oxidizes to the blue compound, indigo when exposed to atmospheric oxygen. The presence of leukocytes indicates pyelonephritis or inflammation involving other structures in the urinary tract.

NITRITE: Nitrite indicates the presence of bacteria which have reduced urine nitrate to nitrite. Some organisms which infect the urinary tract may not be associated with nitrite production. There should be a 4 hour period between voidings in order to insure enough time for bacterial nitrite production. This test is evaluated in conjunction with the microscopic examination, leukocyte esterase, and culture, to diagnose bacterial infection.

MICROSCOPIC EXAMINATION OF URINE: The following values obtained on examination of sediment of a centrifuged urine specimen are considered normal: white blood cells, 0 to 2/hpf; hyaline casts, rare to 1/lpf; red blood cells 0 to 2/hpf, (hpf = high power field; lpf = low power field). The number of casts present are counted per low power microscopic field (10x objective lens).

The presence of increased leukocytes in urine is probably a better indicator of infection than observation of bacteria in the urine (usually <5/hpf). Numbers of leukocytes (5-50/hpf) are consistent with localized acute or chronic inflammation. Leukocytes >50/hpf are consistent with acute infection. Crystals are commonly found in urine and usually represent normal constituents.

URINE CULTURE

SPECIMEN: "Clean Catch" Method: The patient should be supplied with a sterile urine collection cup and instructed as follows: First morning specimen is preferred. 1. Wash hands thoroughly. 2. Wash penis or vulva using downward strokes. 3. Start to urinate directly into toilet or bedpan - stop - position container and take urine sample. 4. Screw cap on securely without touching the inside rim. 5. Keep specimen refrigerated. The urine should be transported to the laboratory and refrigerated until processed. Urine specimens must be cultured within two hours (preferably within one hour) of collection. If a specimen cannot be cultured immediately, it may be held in a refrigerator (not frozen) for up to 48 hours.
Catheterized Specimen: Catheterization of the urinary bladder is not benign; a single, short-time catheterization causes bacteriuria in 1% to 5% of patients. This risk may be higher in patient who have urinary tract abnormalities (Klein RS. Mayo Clin Proc. 1979; 54:412). Indwelling bladder catheters account for more than 500,000 nosocomial infections per year in United States hospitals. Apparently, sealed catheter systems can reduce infection and mortality among hospitalized patients (Platt R, et al. Lancet. 1983; 1:893). Bacteria gain entry into the catheterized bladder by two routes: migration from the collection bag or the catheter drainage tube junction within the catheter lumen, or they may ascend into the periurethral mucous sheath outside the catheter. Migration of bacteria extraluminally in the periurethral space is the major pathway for entry into the bladder and meatal colonization by bacteria is a major risk factor (Garibaldi RA, et al. N Engl J Med. 1980; 303:316-318).

Collection of urine in children less than three years of age is a particularly difficult problem. In one way, urine is collected by use of a bag with adhesive placed over the genitalia; however, this leads to a high level of contamination and results obtained by this method are difficult or impossible to interpret. Straight catheterization with proper preparation yields reliable culture results. Suprapubic aspiration is another method for the collection of urine and may be useful for obtaining uncontaminated urine specimens in children. This procedure minimizes the risk of introducing bacteria into an uninfected bladder and any growth indicates infection.
REFERENCE RANGE: Formerly, a bacteria count of 10^5 or more organisms per mL of urine indicated the presence of infection. That number has been revised downward (Stamm WE, et al. N Engl J Med. 1982; 307:463-468; Stamm WE, et al. ibid. 1981; 304:956; Stamm WE, et al. ibid. 1980; 303:409). Lower counts are considered significant in symptomatic patients, in patients who have urological problems, and in patients who have undergone urological procedures. On suprapubic aspiration, 150 or more bacteria per mL is a significant count in bladder urine.
METHOD: For bacteriologic studies, an unspun, first morning, midstream, clean catch specimen preceded by cleaning of the external genitalia, is used for culture, for identification, and for antibiotic sensitivity.

Most laboratories estimate the bacterial count in urine by streaking the surface of the plates with wire loops calibrated to deliver 0.01 or 0.001mL. Other techniques include the filter paper and dip slide methods (Cohen SN, Kass EH. N Engl J Med. 1967; 277:176).

A presumptive diagnosis of bacteriuria may be made by microscopic examination of the urine (Farrar WE. Med Clin of N Amer. 1983; 67:187-201). Unspun Specimen: One or more bacteria per oil immersion field of the unstained centrifuged sediment. Spun specimens yield a higher positive bacterial detection rate than unspun specimens on microscopic examination (Wallach J. JAMA. 1982; 248:1509). A 12mL aliquot of urine is centrifuged at 1500rpm for three minutes at 400rcf (relative centrifugal force); the supernatant is decanted and the sediment is resuspended in 1mL of urine. The sediment is observed on a slide. Examine the sediment under high power and look for rods and cocci; rods are relatively easy to identify. Twenty or more bacteria per high power field may indicate urinary tract infection. Presence of increased leukocytes is probably more indicative of urinary tract infection, especially in a possibly contaminated specimen.
Gram Stain: Allow a drop of urine to air-dry on a microscopic slide; heat fix and stain with Gram stain or methylene blue stain (Wright's stain); Gram stain procedure is described; see GRAM STAIN.

UROBILINOGEN, URINE

SPECIMEN: Fresh random urine, 20mL; any clean container, no preservatives; do not expose specimen to light.
REFERENCE RANGE: 0.5 to 1.0 Ehrlich Units
METHOD: Specimens are screened with Urobilistix. If greater than 1.0 Ehrlich units, Watson-Schwartz semi-quantitative test is run.
INTERPRETATION: This test is used as a liver function test and for the differential diagnosis of obstructive jaundice versus hemolytic disease of the newborn; urobilinogen is increased in the urine in hemolytic anemias but is decreased in obstructive liver disease, especially in patients with complete obstruction. The test material on the reagent strips is sensitive to 0.1 Ehrlich unit per dL of urine. The test material (p-dimethylamino-benzaldehyde) is not specific for urobilinogen; it reacts with substances in the Watson-Schwartz reaction, e.g., porphobilinogen and the drugs sulfisoxazole and p-amino-salicylic acid. Porphobilinogen and urobilinogen can be distinguished in the Watson-Schwartz test by a solvent extraction step following reaction with Ehrlich's reagent (para-dimethylaminobenzaldehyde). Increased urobilinogen is found in the conditions listed in the Table:

Increased Urobilinogen in Urine
Hemolytic Anemia
Liver Disease

UROPORPHYRINOGEN I SYNTHETASE

Synonym: Porphobilinogen deaminase
SPECIMEN: 1mL blood in EDTA or Heparin
REFERENCE RANGE: Females: 8.1-16.8nmol/sec/liter; males: 7.9-14.7nmol/sec/liter; indeterminate: 6.0-8.0nmol/sec/liter; definitive acute intermittent porphyria: less than 6.0nmol/sec/liter.
METHOD: Measurement of enzyme activity. Determination of the rate of synthesis of uroporphyrin from porphobilinogen.
INTERPRETATION: Measurement of erythrocyte uroporphyrinogen I synthetase is useful in the diagnosis of patients with, or carriers suspected of having the trait for acute intermittent porphyria (AIP). Patients with the disease may have elevated porphobilinogen levels during acute attacks. Decreased uroporphyrinogen I synthetase activity may be found in erythrocytes of carriers of the genetic defect of AIP in the absence of clinical or chemical manifestations of the disease.

VALIUM (see DIAZEPAM)

VALPROIC ACID (VPA, DEPAKENE)

SPECIMEN: Red top tube, separate serum
REFERENCE RANGE: Therapeutic: 50-100mcg/mL (350-700micromol/L); Time to Obtain First Specimen (Steady State): 2-4 days after starting therapy; Half-Life: 4-15 hours; Trough: just before next dose.
METHOD: EMIT; GLC; Fluorescence Polarization (Abbott)
INTERPRETATION: Valproic acid is used in the treatment of generalized tonic-clonic, myoclonic and atonic seizures as well as mixed generalized seizure disorders. Valproic acid is as effective as carbamazepine for the treatment of secondarily generalized tonic-clonic seizures but is not as effective for treatment of complex partial epilepsy (Mattson RH, et al. N Engl J Med. 1992; 327:765-771). Valproic acid is an alternative therapy for absence (petit mal) seizures. Dosage guidelines are shown in the next Table:

Dosage and Blood Specimen Guidelines

Dosage	Blood Specimens
Adults: 15 to 45mg/kg/day (Usual: 1-3 grams/day Children: 15 to 100mg/kg/day Usual: 15-60mg/kg/day Dosage 1-3 times/day	Two to four days following initiation of therapy; obtain blood specimens two hours post-drug (peak) and just before next dose

Valproic Acid (Cont)

Valproic acid serum levels show little correlation between concentration and clinical effect. Target range of 50-100mcg/mL is only a guide to therapy, routine monitoring is unnecessary. Valproic acid serum levels are obtained for the following purposes (Editorial, Lancet. November 26, 1988; 1229-1231):

(a) Identify noncompliance
(b) Establish therapeutic failure (concentration consistently above 150mcg/mL)
(c) Signs of toxicity
(d) Dosage adjustments (decreased seizure control, addition or withdrawal of other antiepileptic drugs)

The most common side effect of valproic acid treatment is gastrointestinal upset; this may be avoided by taking valproate with food; it may be decreased by using enteric-coated tablets. Other side effects include pancreatitis, weight gain, hair loss, tremor, and thrombocytopenia.

The most serious adverse effect is potentially fatal hepatotoxicity. Highest risk is in children under the age of 2 years; most fatalities occur in the first 6 months of therapy; but some have occurred up to 2 years after initiation of therapy. Although it is recommended that liver function tests be performed prior to initiation of therapy and frequently thereafter, biochemical tests may not be abnormal in some cases of hepatotoxicity. Nonspecific findings such as seizures, malaise, weakness, lethargy, anorexia, and vomiting may be signs of hepatotoxicity (Depakene package insert, Abbott Laboratories, 1993). Clinical monitoring is superior to routine laboratory monitoring for identifying patients with hepatotoxicity (Pellock JM, Willmore LJ. Neurology. 1991; 41:961-964).

Pregnant women taking valproic acid have an increased incidence of congenital anomalies including neural tube defects. (Review of adverse effects: Pellock JM. Pediatr Clin N Amer. 1989; 36:435-448).

VANCOMYCIN

SPECIMEN: Red top tube, separate serum. Peak sample should be drawn 30 minutes after completion of a 60 minute infusion. Trough sample should be drawn immediately prior to the next dose. CSF may also be tested.

REFERENCE RANGE: Therapeutic Levels: Peak: 25-40 micrograms/mL; Trough: <5-10 micrograms/mL.

METHOD: Enzyme Immunoassay, Radioimmunoassay, Others.

INTERPRETATION: Vancomycin is active against most gram-positive organisms but generally not gram-negative organisms, fungi, or yeast.

Vancomycin is bactericidal against the gram-positive organisms: Staphylococcus aureus and Staphylococcus epidermidis, both methicillin-susceptible and methicillin-resistant, diphtheroids, hemolytic streptococci, pneumococci, clostridium and viridans streptococci. It is bacteriostatic against enterococcus (Levine JF. Med Clin N Amer. 1987; 71:1135-1145).

Parenteral vancomycin is the drug of choice for treatment of methicillin-resistant staphylococcal infections; it can also be used to treat other gram-positive infections in patients allergic to penicillin. Oral vancomycin is the drug of choice for C. difficile enterocolitis.

High-risk patients for staphylococcal infections are patients with diabetes mellitus, peripheral vascular disease or burn wounds, prosthetic valve endocarditis, infected CSF shunts and granulocytopenic children with cancer, hemodialysis patients, peritoneal dialysis patients.

IV vancomycin is used for prophylaxis of bacterial endocarditis in penicillin-allergic adults and children with congenital heart disease, rheumatic or other acquired valvular heart disease, prosthetic or graft heart valves, idiopathic hypertrophic subaortic stenosis or mitral valve prolapse who undergo dental procedures that are likely to cause gingival bleeding, minor upper respiratory tract surgery or instrumentation. IV vancomycin is also used for prophylaxis in penicillin-allergic adults and children undergoing GI, biliary, or genitourinary tract surgery or instrumentation who are at risk of developing enterococcal endocarditis.

Vancomycin is administered by slow IV infusion (usually over a minimum of 1 hour) or orally. The dosage must be decreased in patients with renal impairment.

Adverse effects of vancomycin are given in the following Table (Wilhelm MP. Mayo Clin Proc. 1991; 66:1165-1170):

Adverse Effects of Vancomycin
Infusion-Related Side Effects
"Red Man" Syndrome
"Pain and Spasm" Syndrome
Hypotension
Thrombophlebitis
Ototoxicity: Hearing loss is rare, often irreversible, and associated with toxic drug levels. Potentiated by coadministration of aminoglycosides.
Nephrotoxicity: Rare, often reversible. Potentiated by coadministration of aminoglycosides.
Hematologic: Neutropenia, thrombocytopenia
Hypersensitivity Reactions: Drug fever or allergy

Red Man Syndrome: Rapid IV infusion of vancomycin may cause pruritis, flushing of the face, neck, and torso ("red neck" or "red man syndrome"). Rapid bolus injection can also cause pain and muscle spasms of the chest and back. Life-threatening hypotension and cardiac arrest have occurred. The effect of vancomycin is usually due to a decrease in strength of contraction of the heart (negative inotropic effect) and vasodilation effect produced, in part, by histamine release. The phenomena may be an anaphylactoid reaction. Steroids and antihistamines as well as fluid resuscitation are used to treat more severe acute episodes. These reactions can be minimized in most patients if the drug is infused slowly (Levy M, et al. Pediatr. 1990; 86:572-580).

VANILLYLMANDELIC ACID (VMA), URINE

SPECIMEN: Add 25mL of 6N HCl to container prior to collection. Caution must be taken to avoid skin burns. 24 hour urine: Instruct the patient to void at 8:00 A.M. and discard the specimen. Then, collect all urine including the 8:00 A.M. specimen at the end of the 24 hour collection period. Refrigerate jug as each specimen is collected. Following collection, add 6N HCl to pH 1 to 2 (Do not use boric acid). A pH <3 is cause for specimen rejection. Forward a 100mL aliquot for analysis.

REFERENCE RANGE: Adults: 2.0-7.0mg/24 hours, less than 8mg/24 hour. The results may also be reported as the ratio: mg VMA/g creatinine.

Age	VMA (mg/gm Creatinine)
1-12 months	1.40-15.0
1- 2 years	1.25- 8.0
2- 5 years	1.50- 7.5
5-10 years	0.50- 6.0
10-15 years	0.25- 3.25

To convert conventional units in mg/gm creatinine to international units in micromol/gm creatinine, multiply conventional units by 5.046.

METHOD: Methods of VMA measurement continue to evolve. Spectrophotometric methods suffered interference from dietary substances and drugs. Extraction procedures and measurement at multiple wavelengths solved some of these problems. More recently HPLC is being employed in some laboratories. Reference range depends on the method used.

INTERPRETATION: This test is used in the evaluation of suspected pheochromocytoma and neuroblastoma, conditions in which most patients have elevations of the VMA, 90 and 75% respectively. The VMA may also be elevated as a result of diet or drugs (see lists in Spilker, et al. Ann Clin Lab Sci. 1983; 13:16-19; Stein PP, Black HR. Medicine. 1991; 70:46-66). The VMA may also be elevated in cases of ganglioneuroblastoma, retinoblastoma, carcinoid and carotid body tumors. Laboratory diagnosis and screening for pheochromocytoma usually involves measurement of more than one test of catecholamines or their metabolites, i.e. VMA, metanephrines, plasma catecholamines or urine free catecholamines. Diagnosis of pheochromocytoma is discussed further under PHEOCHROMOCYTOMA SCREEN.

VDRL (VENEREAL DISEASE RESEARCH LABORATORY)

Synonym: Nontreponemal test for syphilis; Reaginic test for syphilis
SPECIMEN: Red top tube, serum
METHOD: Agglutination
REFERENCE RANGE: Nonreactive.
INTERPRETATION: The VDRL is employed in the screening of patients for syphilis, the monitoring of the response to therapy, the detection of CNS involvement, and as an aid in the diagnosis of congenital syphilis.
Screening: The VDRL is a nontreponemal serological test for syphilis. It and the RPR are the two most commonly used nontreponemal tests for screening patients for syphilis. Following a positive (reactive) screening result, this finding must be confirmed with a more specific and more tedious treponemal test such as the FTA-ABS or MHA-TP. Nontreponemal tests such as the VDRL become reactive in early primary syphilis, the titers peak in secondary and/or early latent syphilis, and the titers decrease in late syphilis. In fact, the VDRL is non-reactive in approximately 25% of patients with late syphilis. The VDRL may not be the initial test of choice in very early or late syphilis, and one of the treponemal tests (FTA-ABS or MHA-PT) is preferable under those circumstances.

There are a lot of causes of false positive VDRL tests, which usually are reactive at low dilutions (<1:8). Most of these are due to infectious organisms (bacterial and viral), inflammatory conditions, connective tissue disease, and various gammopathies. An exhaustive list of the causes of false positive tests is given in: Hook EW, Marra CM. N Engl J Med. 1992; 326:1006-1069.
Congenital Syphilis: Fetuses of mothers with syphilis can contract the disease after the 18th week of pregnancy. When such an infant is born, the question is if the infant's reactive serology is due to active infection of the infant or passively acquiring the antibody from the mother. Titers are informative in such a situation. If infant titers decrease, the interpretation is passively acquired antibodies. If titers increase, it indicates active infection. Tests for IgM nontreponemal antibodies are available in some Public Health referral laboratories for the same purpose.
Tests on CSF: Syphilis can cause changes in CSF which include a mononuclear pleocytosis, elevated protein and globulin, and reactive CSF-VDRL. The RPR should not be performed on CSF and use of treponemal tests (FTA-ABS, MHA-TP) on CSF is controversial. A negative CSF-VDRL does not exclude CNS involvement and up to 25% of patients with neurosyphilis had a negative VDRL.

In the CSF, the pleocytosis is observed before the VDRL. The CSF abnormalities usually resolve (2/3 of patients) by the end of the secondary stage of syphilis. However, the patients with the more marked abnormalities are the ones at highest risk for development of neurosyphilis. Following successful treatment the CSF pleocytosis is the first to resolve followed by the return of the elevated protein and globulin to normal. The reactive CSF-VDRL titers may persist for longer than one year.
Monitoring Therapy: VDRL titers may be followed to assess the success of therapy. Following therapy the VDRL should decline by two or more serial dilutions. Titers decline more rapidly in patients in the earlier stages of infection as compared to later stages, and are slower to decline in patients with higher titers and in patients with more than a single syphilitic infection.

VIRAL LABORATORY DIAGNOSIS

<u>SPECIMENS</u>: Specimens for viral culture and for antibodies should be collected as indicated in the next Figure:

Specimens for Viral Culture and Serology (Antibodies)

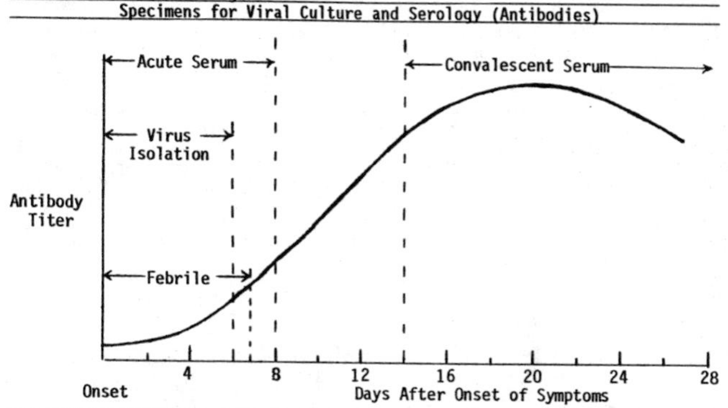

<u>Specimens for Viral Antibodies</u>: Collect both acute and convalescent blood specimens; the convalescent specimen is obtained 7 to 30 days after the acute serum is collected. Use red top tubes; separate serum and refrigerate.

<u>INTERPRETATION</u>: Virologic tests include viral culture, antigen identification (e.g. enzyme immunoassay, fluorescent antibody; others), electron microscopy, or detection of viral nucleic acids using DNA probes. Serologic tests includes demonstration of virus-specific IgM antibody or antiviral IgG antibody using acute and convalescent sera. The preferred method for diagnosis of various viral infections is given in the following Table (Overall JC. Pediatric Rev. 1993; 14:251-261):

Preferred Methods for Viral Laboratory Diagnosis

Method	Viruses
Cell Culture	Adenoviruses: Respiratory, Eye
	Cytomegalovirus (CMV)
	Enteroviruses: Coxsackie, Echo, Polio
	Herpes Simplex Virus
	Influenza Viruses
	Measles
	Mumps
	Parainfluenza Viruses
	Varicella Zoster Virus
Antigen Detection	Adenoviruses: Enteric
	Hepatitis B, D
	Human Immunodeficiency Virus (HIV)
	Influenza A and B
	Measles
	Respiratory Syncytial Virus (RSV)
	Rotavirus
Serology (*indicates that virus-specific IgM antibody test available)	Arboviruses
	Epstein-Barr Virus (EBV)*
	Hepatitis A*, B*, C, D*
	Human Herpes Virus Type 6 (HHV-6)*
	Human Immunodeficiency Virus (HIV)
	Measles*
	Mumps*
	Parvovirus B19*
	Rubella*

VISCOSITY, SERUM OR WHOLE BLOOD

SPECIMEN: Red top tube, separate serum; heparinized for whole blood measurements.
REFERENCE RANGE: <1.8 centipoises range 1.4-1.8 relative to water; symptoms occur when viscosity >4 centipoises.
METHOD: Viscosity is measured using a cone rotated in the sample; the torque necessary to overcome the viscous resistance of the serum is measured in centipoises.
INTERPRETATION: Viscosity measurements are performed on serum to evaluate hyperviscosity syndromes most commonly associated with Waldenström's macroglobulinemia and less commonly multiple myeloma. This syndrome is sometimes seen in rheumatoid arthritis, SLE, and hyperfibrinogenemia. Viscosity measurements are usually performed on whole blood because of hyperviscosity in the newborn period. Viscosity varies as a function of hematocrit level, red cell size and deformability, plasma proteins, fibrinogen concentrations and presence of abnormal plasma proteins. Hyperviscosity syndromes are frequently observed in Waldenström's macroglobulinemia and sometimes multiple myeloma when the serum concentration of immunoglobulins is high.

Hyperviscosity of blood may be due not only to elevated serum immunoglobulins, but also to increased number of cells (polycythemia or leukemia), or to increased resistance of cells to deformation (sicklemia or spherocytosis).

Diseases associated with hyperviscosity are given in the next Table:

Diseases Associated With and Causes of Hyperviscosity

Diseases	Causes of Hyperviscosity
Monoclonal Gammopathies especially in Waldenström	Increased Protein
Polycythemia; Leukemia; Post Splenectomy	Increased Number of Cells
Sickle Cell Anemia; Other Hemoglobinopathies; Pyruvate Kinase Deficiency; Burr-Cell Formation; Hereditary Spherocytosis	Increased Resistance of Cells to Deformation

Patients with hyperviscosity may present with mucous membrane hemorrhages. In-vivo, the increased viscosity may result in a decreased flow rate of blood and ischemic changes in the tissues with secondary necrosis and hemorrhage. The most common cause of hyperviscosity in adults is Waldenström's macroglobulinemia.

Neonatal polycythemia has been associated with an increased risk of neurologic and motor abnormalities (Black VD, et al. Pediatrics. 1982; 69:426-431). There is decreased cerebral blood flow and cardiovascular, respiratory sequelae, necrotizing enterocolitis and even acute renal failure (Herson VC, et al. J Pediatrics. 1982; 100:137-139). Hyperviscosity occurs in up to five percent of all neonates; four percent had polycythemia while one percent had hyperviscosity without polycythemia (Wirth FH, et al. Pediatrics. 1979; 63:833). At birth, the hematocrit is relatively high; viscosity of the blood increases as the hematocrit rises. The relationship between viscosity and hematocrit is shown in the next Figure (Barum RS. J Pediatrics. 1966:69:975):

Viscosity versus Hematocrit

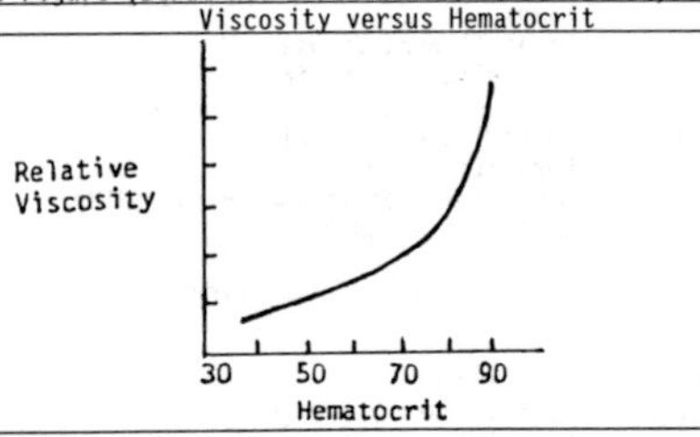

Viscosity increases linearly with peripheral venous hematocrits up to 65 percent; then the curve becomes exponential. A recommendation for screening hematocrits in the neonates is given in the next Figure (Ramamurthy RJ, Brans YW. Pediatrics. 1981; 68:168-174):

Hematocrit and Exchange Transfusions

Capillary Hematocrit
↓ > 69 Percent
Peripheral Venous Hematocrit
↓ > 64 Percent
Insert Umbilical Vein Catheter
↓ > 62 Percent
or
Viscosity at 11.5 sec. Shear Rate > 14.0 Shear Rate
Consider Partial Exchange Transfusion

Kamamurthy and Brans (Pediatrics. 1981; 68:168-174) recommend screening of all neonates by measuring capillary hematocrit followed by peripheral venous hematocrit levels in cases in which the capillary hematocrit level is greater than 69 percent. If the peripheral venous hematocrit is greater than 64 percent, an umbilical vein catheter should be inserted and the umbilical vein hematocrit level and, preferably, viscosity should be determined. If the umbilical vein hematocrit level is greater than 62 percent and/or viscosity at 11.5/sec. shear rate is greater than 14.0 cps, then partial exchange transfusion should be considered. The usual volume for exchange is determined by the equation:

$$\frac{\text{Current Venous Hematocrit} - \text{Desired Venous Hematocrit}}{\text{Current Venous Hematocrit}} \times \text{Weight(kg)} \ \frac{\text{(80mL)}}{\text{kg}}$$

(Black VD, Lubchenco LO. Pediatr Clin of N Amer. 1982; 29:1137-1148).

A cord blood hematocrit greater than 56 percent is associated with a markedly increased risk of neonatal polycythemia (Shohat M, et al. Pediatrics. 1984; 73:7-10).

VITAMIN A (RETINOL)

SPECIMEN: Overnight fast. Red top tube, separate serum. Protect from light. Forward specimen frozen in plastic vial on dry ice to reference laboratory.
REFERENCE RANGE: Vitamin A: 30-95mcg/dL; 125-400 IU/dL. To convert conventional units in mcg/dL to international units in micromol/Liter multiply conventional units by .03491.
METHOD: Multiple methodologies. High performance liquid chromatograph(HPLC) preferred.
INTERPRETATION: Vitamin A measurements are useful in cases of suspected malnutrition, vitamin A deficiency due to various causes determining the severity of malabsorption, and in cases of hypervitaminosis A syndrome due to excessive dietary intake. The provitamin beta-carotene levels reflect dietary intake, while vitamin A (retinol) levels reflect both intake and liver storage. Vitamin A retinols direct normal differentiation of epithelia and affect the plasma membrane. Vitamin A is a fat soluble vitamin; it is decreased in the conditions listed in the next Table:

Causes of Decreased Vitamin A

(1) Fat Malabsorption Syndromes
- A. Deficiency of Pancreatic Digestive Enzymes:
  - Chronic Pancreatitis
  - Cystic Fibrosis
  - Pancreatic Carcinoma
  - Pancreatic Resection
- B. Impairment of Intestinal Absorption:
  - Celiac Disease (Gluten-Sensitive Enteropathy)
  - Crohn's Disease
  - Rare Causes:
    - Tropical Sprue
    - Abetalipoproteinemia
    - Familial Hypobetalipoproteinemia
    - Lymphangiectasis
    - Intestinal Lipodystrophy
    - Amyloidosis
    - Lymphoma
    - Surgical Loss of Functional Bowel
    - Mineral Oil Laxative Use
- C. Deficiency of Bile:
  - Extrahepatic Bile Duct Obstruction
  - Intrahepatic Disease
  - Cholecystocolonic Fistula

(2) Dietary Deficiency

(3) Other
- A. Plasmapheresis
- B. Liver Disease
- C. Fever and Chronic Infection

The incidence of vitamin A deficiency in Crohn's disease is 11 percent (Driscoll RH, Rosenberg IH. Med Clin of North America. 1978; 62:185-201).

Vitamin A deficiency may lead to blindness and epithelial metaplasia with replacement of mature differentiated cells of various epithelial tissue by squamous and keratinizing cells. There is increased respiratory infections, plugging of salivary and pancreatic ducts and increased incidence of bladder stones.

Increased vitamin A is associated with excessive ingestion, or impaired disposal caused by diabetes mellitus, chronic nephritis or myxedema. The effects of excess vitamin A are as follows: acute overdose: elevation of intracranial pressure and skin desquamation; chronic excess: increased intracranial pressure leading to hydrocephalus, skin desquamation and other dermatological findings; bone pain.

"Retinoids" is a generic term that includes all natural and synthetic analogues of vitamin A(retinol). Retinoids are used in the treatment of acne and psoriasis. Vitamin A has been found to reduce the growth of chemically-induced tumors, especially those of epithelial origin.

The provitamin beta-carotene is obtained in the diet. It is an anti-oxidant and has been associated with a protective anti-cancer effect (Stahelin HB, et al. "B-Carotene and Cancer Prevention: the Basal Study." Am J Clin Nutr. 1991; 53:265S-269S). Vegetables, especially carrots, sweet potatoes, squash and spinach, are rich in beta-carotene. Reference values for beta-carotene: 50-200mcg/dL.

VITAMIN B-6 (PYRIDOXAL PHOSPHATE)

SPECIMEN: Lavender(EDTA) top tube; separate plasma and freeze. Protect from light.
REFERENCE RANGE: 5-24ng/mL.
METHOD: Enzymatic assay using B-6-dependent apoenzymes (e.g., tyrosine decarboxylase); spectrophotometry; HPLC.
INTERPRETATION: This test may be of use in evaluating suspected deficiency of vitamin B-6. Sphingosine synthesis (nerve), heme synthesis (RBC), and transaminase activity are among the many processes which are vitamin B6 dependent. Vitamin B-6 comprises a group of water-soluble vitamins which includes pyridoxine, pyridoxal and pyridoxamine. The vitamin B-6 group functions as coenzymes participating in the metabolism of amino acids and the breakdown of glycogen to glucose-1-phosphate.
Deficiency: Conditions associated with vitamin B-6 deficiency are given in the next Table:

Conditions Associated with Vitamin B-6 Deficiency
Chronic Alcoholism
Malnutrition
Malabsorption
Pregnancy
Gestational Diabetes
Drugs: Isoniazid, L-dopa, Oral Contraceptives

Excess: There is a health fad of taking large doses of vitamin B-6. Vitamin B-6 has been used to promote muscle growth and relieve premenstrual swelling. However, nerve damage may occur with difficulty in walking, and loss of feeling in arms and legs. Recovery takes one to two years and may not be complete.

VITAMIN B12 (COBALAMIN)

SPECIMEN: Red top tube, separate and freeze serum in a plastic tube. Protect from light. Fasting specimen preferred. This test should not be done in patients who have recently received radioisotopes therapeutically or diagnostically (Schilling test). Draw before B12 therapy or transfusions.
REFERENCE RANGE: 200-800pg/mL; indeterminate, 100-200pg/mL; early vitamin B12 deficiency, 100-150pg/mL; deficiency, <100pg/mL. To convert conventional units in pg/mL to international units in pmol/liter, multiply conventional units by 0.7378.
METHOD: RIA.
INTERPRETATION: Vitamin B12 measurements are useful in the diagnosis and evaluation of vitamin B12 deficiency, macrocytic anemias, megaloblastic anemias, neurological diseases, folate deficiencies, malabsorption and GI disorders, rare inherited conditions, neutropenia with hypersegmented nuclei, and evaluation of certain hematological disorders. Vitamin B12 measurements may be made following oral ingestion of radioactivity labeled B12 (Schilling Test) in order to evaluate intrinsic factor production and intestinal disorders. Vitiligo and autoimmune hypothyroidism are sometimes associated with pernicious anemia, and vitamin B12 measurements may be useful in these conditions also. Vitamin B12 measurements are often performed simultaneously with folate levels. Causes of decreased vitamin B12 are given in the next Table:

Causes of Decreased Serum Vitamin B12
(1) Inadequate Diet: Strict Vegetarianism
(2) Inadequate Absorption:
Deficient or Defective Intrinsic Factor:
Pernicious Anemia
Total Gastrectomy; Gastritis; Gastric carcinoma
Small Bowel Disease:
Ileal Resection or Bypass; Blind Loop Syndrome with Abnormal Gut Flora; Malabsorption; Coeliac Disease; Tropical Sprue; Crohn's Disease
Pancreatic Insufficiency (Chronic Pancreatitis, Cystic Fibrosis)
(3) Interference with Vitamin B12 Absorption:
Drugs: Neomycin, Metformin, Colchicine, Ethanol, p-Aminosalicylic Acid
Bacterial Overgrowth
Fish Tapeworm (Diphyllobothrium Latum)
(4) Rare Congenital Disorders: Orotic Aciduria; Transcobalamin II Deficiency; Defective Intrinsic Factor Production
(5) Other:
Anticonvulsants
Dietary Folic Acid Deficiency (Occasionally)
Multiple Myeloma

Subjects who eat no food of animal origin are liable to develop B12 deficiency; the largest group of vegetarians are religious Hindus. Ancillary evidence of vitamin B12 deficiency include macrocytosis, hypersegmented neutrophils, elevations in serum methylmalonic acid and homocysteine, atrophic gastritis and a positive Schilling test. Some B12 is excreted in the bile only to be resorbed in the small intestine. With defective enterohepatic circulation (small bowel disease), stores may be depleted.

There appear to be a number of elderly patients (25% in one study) with vitamin B12 levels in the 100-300ng/mL range who appear to benefit from cobalamin (B12) therapy as evidenced by a fall in methylmalonic acid levels and a fall in MCV (Pennypocker LC, et al. J Am Geriatr Soc. 1992; 40:1197-1204).

The causes of increased serum vitamin B12 are given in the next Table (Fairbanks VF, Elveback LR. Mayo Clin Proc. 1983; 58:135-137):

Causes of Increased Serum Vitamin B12
Acute Hepatitis
Myeloproliferative Diseases
Acute and Chronic Granulocytic Leukemia
Myelomonocytic Leukemia
Polycythemia Vera
Leukocytosis
Oral Contraceptives
Uremia

VITAMIN C (ASCORBIC ACID)

SPECIMEN: Two gray (oxalate) top tubes, separate plasma and freeze plasma. Serum, may also be used.
REFERENCE RANGE: 0.2-2.0mg/dL. Acceptable levels of vitamin C are 0.30mg/dL in plasma; >15mg/dL in leukocytes. To convert conventional units in mg/dL to international units in micromol/liter, multiply conventional units by 56.78.
METHOD: Ascorbic acid is oxidized to dehydroascorbic acid which is coupled with 2,4-dinitrophenylhydrazine to form a 2,4-dinitrophenylhydrazone which is rearranged in H_2SO_4 to a compound that is measured at 515nm (Carr RS, et al. Anal Chem. 1983; 55:1229-1232). Fluorometric methods are also available.
INTERPRETATION: Vitamin C measurements are useful in the diagnosis of vitamin C deficiency (scurvy), and evaluation of possible vitamin C deficiency related growth and bleeding disorders. Although vitamin C levels can be measured on blood, plasma, urine and white cell preparations, plasma levels are the most convenient. Plasma levels are more sensitive than whole blood levels to deficiency states. White cell levels closely approximate body vitamin stores but require larger specimens. Urine levels may vary as a result of recent dietary intake.

Vitamin C is a water soluble vitamin which is necessary for the preservation of capillary integrity; it may function to maintain normal venous endothelium. Vitamin C is involved in the hydroxylation of proline to hydroxyproline which occurs in collagen. Thus, vitamin C is essential for collagen synthesis; in addition, it is also associated with the metabolism of the mucopolysaccharide ground substance of connective tissue. Ascorbic acid is also a soluble antioxidant and may have a role in limiting oxidizing and free radical tissue damage.

Severe ascorbic acid deficiency (scurvy) is characterized by connective tissue changes with delayed wound healing, poor scar strength, edema, hemorrhage, and bone weakness, sometimes leading to fractures. In addition, there may be anemia relating to the role of ascorbic acid in absorption of dietary iron and folic acid metabolism.

Vitamin C needs are increased and a tendency towards deficiency may be associated with smoking and alcohol use, obesity, stress, infection, diabetes, chronic inflammatory disease, excessive iron intake and oral contraceptive use. Long-stay (years) psychiatric patients are at increased risk for development of vitamin C deficiency (Thomas SJ, et al. J Plant Foods. 1982; 4:191-197).

Ascorbic acid is a strong reducing agent. This property results in interferences with a number of common laboratory determinations as given in the next Table (Woolliscroft JA. Disease-A-Month. 1983; 5:29):

Interference by Megadoses of Vitamin C with Laboratory Determinations

Test	Result
Multistix Test for Blood in Urine	False Negative
Hemoccult Test for Blood in Stool	False Negative or Delay
Urine Glucose: Test-Tape, Clinistix or Labstix	False Negative
Blood Tests:	
Glucose	Elevated
Uric Acid	Elevated
Cholesterol	Elevated

It is recommended that patients refrain from ascorbic acid supplementation for 48-72 hours prior to testing for occult blood. Ascorbic acid interferes with those tests that are based on oxidation-reduction reactions, that is, those with reactions that produce hydrogen peroxide including tests for glucose, uric acid and cholesterol.

VITAMIN D (25-HYDROXYVITAMIN D)

Synonyms: Cholecalciferol; 25-hydroxycalciferol
SPECIMEN: Red-top tube, separate serum or green (heparin) top tube; separate plasma; place plasma in plastic vial and send frozen on dry ice.
REFERENCE RANGE: Deficiency: <15ng/mL; Winter: 14 to 42ng/mL; Summer: 15-80ng/mL. Values obtained during the summer are usually higher than those obtained in the winter. Levels in newborns are lower than children, children are lower than adults. To convert conventional units in ng/dL to international units in nmol/liter, multiply conventional units by 2.599.
METHOD: HPLC: Competitive Protein Binding.
INTERPRETATION: Vitamin D levels are useful in the workup of patients with suspected poor dietary intake and malnutrition, patients with bone disease thought secondary to vitamin D deficiency, and patients with hypercalcemia and suspected elevated vitamin D. 25-Hydroxyvitamin D is synthesized in the liver; the steps in the synthesis of the biologically active form of vitamin D (1,25 DiOH-cholecalciferol) are shown in the next Figure:

Synthesis of Biologically Active Vitamin D

7-Dehydro-cholesterol	Skin → U.V.Light	Chole-calciferol (CC)	Liver → +25-OH	25-OH-CC	Kidney → +1-OH PTH	1,25 DiOH-CC (calcitriol)

Vitamin D_3 (cholecalciferol) is obtained in the diet or is synthesized in the skin from 7-dehydrocholesterol; vitamin D_3 is converted to 25-OH vitamin D in the liver and finally converted to 1,25-$(OH)_2$ vitamin D (calcitriol) in the kidney. 1,25-$(OH)_2$ vitamin D has 100 times the activity of 25(OH) vitamin D. Vitamin D stimulates calcium absorption, regulates enterocyte, osteoblast and hematopoietic cell function, and suppresses parathyroid gland activity. Surrogate markers of vitamin D activity includes serum calcium and phosphate measurement. The causes of 25-hydroxyvitamin D deficiency are listed in the next Table:

Causes of 25-Hydroxyvitamin D Deficiency

Cause	Mechanism
Lack of Sunlight	Failure to Convert 7-Dehydrocholesterol to Cholecalciferol
Poor Diet	Lack of D_2, (Ergocalciferol) in Diet
Malabsorption	Failure to Absorb Fat Soluble Vitamins and Loss of 25-(OH) Because of its Enterohepatic Circulation
Anticonvulsant Therapy	Induction of Hepatic Microsomal Enzymes and Subsequent Inactivation of 25-(OH)CC
Others: Liver Disease, Hyperthyroidism, Diabetes, Rheumatoid Arthritis, Nephrotic Syndrome	Reduced Vitamin D Metabolites

Elevated levels of vitamin D may be seen with excessive sun exposure, excessive dietary intake, sarcoidosis, and some cases of lymphoma. Excessive vitamin D causes hypercalcemia, hypercalciuria, renal stones, extraskeletal ossification and renal disease. 25-Hydroxyvitamin-D is increased in 25-OH-vitamin-D deficient patients after even brief exposure to ultraviolet radiation (Adams JS. N Engl J Med. 1982; 306:722-725).

VITAMIN E (TOCOPHEROL)

SPECIMEN: Overnight fast. Red top tube, separate serum. Protect from light. Forward specimen frozen in plastic vial on dry ice to reference laboratory.
REFERENCE RANGE: 0.550-1.750mg/dL. To convert conventional units in mg/dL to international units in micromol/liter, multiply conventional units by 23.22.
METHOD: High performance liquid chromatography(HPLC), or fluorometric assay.
INTERPRETATION: Vitamin E measurements are most useful in workup of newborns with hemolytic disease or neurological disease. Vitamin E (tocopherol) is a fat-soluble vitamin associated with polyunsaturated fatty acids and is present in most conventional diets. There are relatively few conditions that respond to vitamin E therapy (Roberts HJ. JAMA. 1981; 246:129-131). Nutritional inadequacy or frank deficiency of vitamin E is found only in patients with various genetic or acquired diseases, with the exception of premature children in whom the deficiency may be iatrogenic. Very low vitamin E levels are associated with abetalipoproteinemia. There is some interest in the possible role of vitamin E in the prevention of atherosclerosis. Vitamin E is a fat-soluble antioxidant that is incorporated into lipoproteins and may prevent their oxidation and uptake by atheromata. There is also some evidence that vitamin E may help prevent development of cataracts (Knekt P, et al. BMJ. 1992; 305:1392-1394).

Rich sources of vitamin E are the vegetable oils, soybean, corn, cottonseed and safflower oils. The average daily diet contains 8 to 11 mg of alpha-tocopherol; the recommended dietary allowance of vitamin E is 8 mg of alpha-tocopherol for women and 10 mg for men.

The possible uses of vitamin E are listed in the next Table (Oski FA. N Engl J Med. 1980; 303:454-455; Bieri JG, et al. ibid. 1983; 308:1063):

Possible Uses of Vitamin E

Category	Examples
Premature Infants	Hemolytic Anemia of Low-Birth Weight Infants
Correct a Deficiency State e.g., Malabsorption Disorders	Neuropathic and Myopathic Abnormalities
	Hereditary Abetalipoproteinemia: Retinitis Pigmentosa, Myopathy and Cerebellar Dysfunction
	Modest Shortening of Red-Cell Life Span in Cystic Fibrosis
	Hyperaggregability of Platelets in Patients with Biliary Atresia
Anti-Oxidant	Infants Exposed to Prolonged Oxygen Administration in the Treatment of Respiratory-Distress Syndrome: Retrolental Fibrodysplasia; Bronchopulmonary Dysplasia
	Counter Cardiotoxic Effects of the Chemotherapeutic Agent Doxorubicin
Defense Against Free Radicals(Hematolog. Disorder)	Hereditary Hemolytic Anemias due to Deficiency of Glutathione Synthetase

High dose vitamin E does not decrease the rate of chronic hemolysis in glucose-6-phosphate dehydrogenase deficiency (Johnson GJ, et al. N Engl J Med. 1983; 308:1014-1017).

Vitamin E is decreased in conditions causing fat malabsorption (see FAT MALABSORPTION PANEL). The most common cause of vitamin E deficiency is cystic fibrosis. Neurologic abnormality may be associated with vitamin E deficiency and malabsorption. In children, neurologic abnormalities can be detected by clinical examination only after 18 to 24 months of malabsorption; in adults, neurologic abnormalities can be detected after 10 to 20 years of fat malabsorption reflecting vitamin E stores (Sokol RJ, et al. Gastroenterology. 1983; 85:1172-1182; Sokol RJ. Ann Intern Med. 1984; 100:769).

High dose vitamin E is associated with a wide variety of disorders (Roberts HJ. JAMA. 1981; 246:129-131); the more serious ones are as follows: thrombophlebitis; pulmonary embolism; hypertension; severe fatigue; gynecomastia in men and women; breast tumors. Cohen MH (N Engl J Med. 1983; 289:980) described the onset of several episodes of fatigue following vitamin E therapy with relief following withdrawal of the vitamin.

VITREOUS HUMOR, POSTMORTEM

The constituents in the vitreous humor are relatively stable as compared to blood values; therefore, analysis of vitreous humor may be done at postmortem to reflect certain body constituents during life.

The changes in vitreous humor, postmortem, are shown in the next Table:

Examination of Vitreous Humor

Substance	Comment
BUN	Constant for 30 hours
Creatinine	Changes slightly
Sodium	Constant for 30 hours
Calcium	Changes slightly
Magnesium	Changes slightly
Chloride	Constant for 30 hours
Glucose	Decreases

VOLATILES (SCREENING TESTS)

SPECIMEN: Red top tube, separate serum and store at 4°C.
REFERENCE RANGE: Serum osmolality: 285-295mosm/liter; pH 7.4; Anion gap: 15-20mmol/liter; Acetone: negative
METHOD: Osmolality; pH; electrolytes (calculate anion gap); acetone.
INTERPRETATION: There are simple screening tests that may be done to reflect the presence of methanol, ethylene glycol or isopropanol; these are illustrated in the next Figure:

Screening Tests for Volatiles

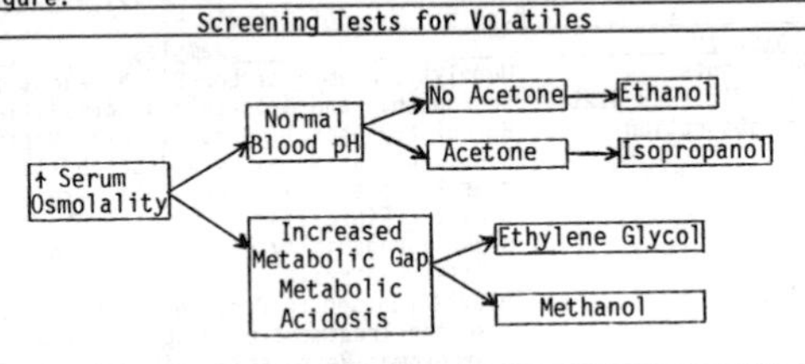

The contribution of each 100mg/dL of the volatiles to the serum osmolality is given in the next Table:

Volatiles and Osmolality

Volatile (100mg/dL)	Osmolality Increased (mosm/liter)
Ethanol	22
Isopropanol	17
Ethylene Glycol	16
Methanol	31

The volatiles increase serum osmolality between 16 and 31 mosm/liter thus increasing the measured serum osmolality from 285-295mosm/liter to more than 300mosm/liter. Methanol and ethylene glycol are metabolized to formic acid and oxalic acid respectively. They will result in a metabolic acidosis. Isopropyl alcohol is metabolized to acetone.

vonWILLEBRAND DISEASE PROFILE

von Willebrand's disease is the most common of the inherited bleeding disorders. The disease can be missed because many of these patients are asymptomatic unless they experience major trauma or surgery. von Willebrand's disease results from quantitative or qualitative defects in von Willebrand factor (vWF), a high molecular weight protein formed in endothelium and megakaryocytes, and stored in platelet granules. vWF forms a complex with factor VIII and circulates as a series of high molecular weight multimers. The following laboratory tests may be useful in the diagnosis of suspected von Willebrand's disease:

Tests for Von Willebrand's Disease
Bleeding Time
Platelet Count
PTT
Factor VIII
Factor VIII: Antigen
Platelet Aggregation
vWF: Ristocetin Cofactor Activity
vWF: Antigen
Electrophoresis of vWF (crossed immunoelectrophoresis, SDS agarose gel)

There are three major types of hereditary von Willebrand's disease, and in addition a platelet type and an acquired form. In general, Type I is characterized by a reduced level of vWF, type II by abnormalities in multimer composition, and type III by absence of vWF in cells and plasma and severe symptoms. A platelet type (Pseudo-von Willebrand's disease) is manifest by platelet receptors with increased affinity for vWF resulting in vWF clearance and thrombocytopenia. Acquired von Willebrand's disease is seen in some patients with autoimmune or lymphoproliferative diseases.

Type I von Willebrand's disease constitutes approximately 85% of all von Willebrand's disease. Typical laboratory results are given in the next Table:

Typical Laboratory Results in von Willebrand's Disease Type I

Test	Result
Bleeding Time	Increased
Platelet Count	Normal
PTT	Increased
Factor VIII	Decreased
vWF	Decreased
Ristocetin Platelet Aggregation	Negative
Electrophoresis	Normal Pattern

Factor VIII levels increase following blood component therapy to levels in excess of that expected. vWF and factor VIII levels increase to normal in response to estrogen or pregnancy.

von Willebrand Disease Type II is characterized by abnormal bleeding times and abnormal vWF multimeric patterns on electrophoresis. A large number of subtypes with a plethora of multimeric patterns have been described, the details of which are beyond the level of this discussion.

Variants of von Willebrand's disease continue to be described. The laboratory features of these variants can be found in a number of publications (Bowie EJ. ASCP Check Sample, Thrombosis and Hemostasis, 1986; vol 8(3); Weiss HJ, et al. J Lab Clin Med. 1983; 101:411-425; Ruggeri AM, et al. J Clin Invest. 1982; 70:1124-1127; Battle J, et al. Am J Hematol. 1986; 21:177-188; Zimmerman TS, et al. J Clin Invest. 1986; 77:947-951).

von WILLEBRAND FACTOR ANTIGEN (VWF:Ag)

Synonym: VIII R:Ag
SPECIMEN: Blue (citrate) top tube.
METHOD: Immunoelectrophoresis. Levels are compared to a pool of normal plasma.
REFERENCE RANGE: Varies by blood group and age. Blood group O patients have lower levels than other groups. Levels tend to increase with age. Must be combined with other tests to interpret correctly in the 30-70% range. Pregnancy and estrogen may raise levels.
INTERPRETATION: Used in the diagnosis of von Willebrand's disease. In the most common type of von Willebrand's disease, type I (85% of patients affected), VWF is quantitatively decreased. In type II, VWF levels may be quantitatively within normal range, however there is a qualitative abnormality in the distribution of high molecular weight multimers. See von WILLEBRAND'S DISEASE PROFILE.

von WILLEBRAND FACTOR (VWF: Rcof)

Synonyms: vWF assay; Ristocetin cofactor, vWF: Rcof, VW cofactor assay
SPECIMEN: Blue-top tube (citrate)
METHOD: Platelet aggregation with ristocetin. Patient plasma is incubated with a standardized suspension of formalin-fixed platelets and ristocetin.
REFERENCE RANGE: 70-150% of normal mean. May vary among laboratories. Blood group O patients have lower levels.
INTERPRETATION: von Willebrand factor (VWF) is assayed to aid in the diagnosis of von Willebrand's disease and to distinguish this disease from hemophilia A. In the most common type of von Willebrand's disease, type I (85% of patients) there is a corresponding decrease in factor VIII, vWF: Ag, and vWF: Rcof. Levels of VWF: Rcof increase in pregnancy and with estrogen. Interpretation should be made in conjunction with other tests. See von WILLEBRAND'S DISEASE PROFILE.

WATSON-SCHWARTZ (SCREEN FOR PORPHOBILINOGEN)

SPECIMEN: Fresh random urine; keep specimen protected from light during transit to lab by wrapping in foil; freeze until ready for assay.
REFERENCE RANGE: Negative
METHOD: Reaction with Ehrlich's Reagent, p-dimethylaminobenzaldehyde. Porphobilinogen, urobilinogen and substituted indoles all react with p-dimethylaminobenzaldehyde to give a red color. If a red color is obtained, the compounds are identified on the basis of differential solubility; porphobilinogen is soluble in aqueous solvents while urobilinogen is soluble in organic solvents, e.g., chloroform and butanol.
INTERPRETATION: The Watson-Schwartz test is a rapid qualitative screening test for urine porphobilinogen, a substance found in the urine during an acute attack of certain types of porphyrias. A positive result should be confirmed by further testing (See PORPHOBILINOGEN). Porphobilinogen is increased in the conditions listed in the next Table:

Increase in Porphobilinogen
Acute Intermittent Porphyria
Variegate Porphyria
Hereditary Coprophorphyria

Forms of porphyria in which porphyrin precursors (e.g. porphobilinogen) are increased are characterized by neurological symptoms, while those in which excess porphyrins are produced experience cutaneous photosensitivity. Porphobilinogen is not increased in porphyria cutanea tarda or lead poisoning.

Urine urobilinogen is increased in the conditions listed in the next Table:

Increase in Urine Urobilinogen
Hemolytic Anemias
Liver Disease

Hemolytic disease leads to increased breakdown of heme while in hepatitis there is decreased hepatic uptake of urobilinogen and excretion in the bile.

WHITE BLOOD CELL COUNT (WBC) AND DIFFERENTIAL

SPECIMEN: Lavender (EDTA) top tube or microtube containing EDTA. The specimen is stable for WBC counts for up to 24 hours in the refrigerator.

Prepare blood film as follows: Place small drop of blood from needle on slide. Allow the blood to spread at junction of spreader slide, and then push the spreader slide, at a 45° angle, smoothly and quickly. Allow the film to air dry. Differential counts should be prepared on fresh specimens if possible (<2 hours old).

METHOD: Manual Methods: Total leukocyte count and absolute eosinophil count by hemocytometer.

Manual Differential Counts: The smear is scanned under low-power magnification for an area of reasonable cell distribution, and then examined under oil immersion magnification; the percent distribution of the various types of leukocytes, in a sample of 100-200 cells, is obtained. Neutrophils tend to distribute at the periphery of the smear and lymphocytes in the center hence counts are performed by moving from one lateral border to the other. In areas where the smear is thick, WBC may be misclassified.

Automated Methods: Counts are by change in impedance or interruption of a laser light beam. RBC are removed by lysing solutions. Some lysing solutions lyse the WBC leaving bare nuclei, other solutions shrink the WBC to 50-70% of original size. Differential counts by automated methods are by cytochemical staining, low and high angle light scattering, and analysis of volume. Image analysis systems have been used in the past, and new techniques continue to be developed.

TOTAL WHITE BLOOD COUNT:

REFERENCE RANGE: White cells (thousands/cu mm), white cell groups (percentage) with age, are listed in the next Table:

White Cells (Thousands) and White Cell Groups (Percentage) with Age; Mean Values

White Cell	Age							
	Birth	2 Days	14 Days	3 Mos	1 Yr.	4 Yrs.	8-21 Yrs.	Adult
WBC x 10^3/cumm (range)	15 (9-30)	21	11 (5-20)	9.5	9.0 (6-18)	8.0 (5-15)	8.0 (4.5-13.5)	7.5 (4.5-11)
White Cell Groups (%)								
Neutrophils	45	55	36	35	40	50	60	60
Lymphocytes	30	20	53	55	53	40	30	32
Monocytes	12	15	8	7	5	8	8	4
Eosinophils	2	4	2	2	1	1	1	3
Basophils	1	1	1	1	1	1	1	1
Immature White Cells	10	5	-	-	-	-	-	-

Overall ranges: Segmented neutrophils 50-70%; Band neutrophils 0-5%; Lymphocytes 20-40%; Monocytes 1-6%; Eosinophils 1-5%; Basophils 0-1%.

Reference ranges vary among studies, a reflection probably in observer bias and the variation inherent in manual methods. Reference values vary also among ethnic groups with a lower limit of normal in black patients due to decreased neutrophils. In certain conditions, WBC counts have to be corrected for nucleated RBC when automated counts are performed.

INTERPRETATION: The white blood cell (WBC) count is useful in the diagnosis and management of infection, inflammatory disorders, hematopoietic malignancies, evaluation of myelopoietic disorders, drug effects, and response to various cytotoxic agents. The differential count is performed to acquire further information concerning the above states and enables one to arrive at values for the absolute value of discreet WBC cell populations. Absolute values for individual cell populations are obtained from a combination of the WBC count and the % of each cell type from the differential. Interpretation of the various changes in populations of individual types of white cells are given below.

White Blood Cell Count (Cont)

NEUTROPHILS:

REFERENCE RANGE: (Altman PL, Dittmer DS eds., Blood and Other Body Fluids, Bethesda, Maryland 1961, Fed Am Soc Exp Biol):

	Polymorphonuclear Neutrophils		
	Thousands/cu mm (Range)		
Age	Total	Segmented	Band
Birth	11.0 (6-26)	9.4	1.6
12 Hours	15.5 (6-28)	13.2	2.3
24 Hours	11.5 (5-21)	9.8	1.7
1 Week	5.5 (1.5-10.0)	4.7	0.8
4 Weeks	3.8 (1.0-9.0)	3.3	0.5
1 Year	3.5 (1.5-8.5)	3.2	0.3
10 Years	4.4 (1.8-8.0)	4.2	0.2
20 Years	4.4 (1.8-7.7)	4.2	0.2

INTERPRETATION: Conditions associated with an increase in neutrophilic leukocytes are given in the next Table:

Conditions Associated with Increased Neutrophilic Leukocytes

Physiologic:
Newborn; Pregnancy; Delivery; Emotional Disturbances; Crying (infant), Nausea and Vomiting; Strenuous Physical Exercise; Ultraviolet Light; Cold; Severe Stress; Heat

Pathologic:
Acute Infections: Bacterial; Some Viral; Mycotic; Spirochetal; Rickettsial and Parasitic Infections
Acute Inflammatory Disorders: Acute Rheumatoid Arthritis; Rheumatic Fever; Vasculitis; Myositis; Hypersensitivity Reactions
Metabolic Disturbances: Uremia; Diabetic Acidosis; Eclampsia; Thyroid Storm
Hematologic Disorders: After Hemorrhage; Hemolytic Anemias; Leukemias; and Myeloproliferative Disorders
Tissue Necrosis: Burns; Myocardial Infarction; Gangrene; Malignant Neoplasia
Drugs and Toxins: Heparin; Digitalis; Epinephrine; Lithium; Histamines; Corticosteroids
Allergies

Conditions associated with a decrease in neutrophilic leukocytes are given in the next Table:

Conditions Associated with Decreased Neutrophilic Leukocytes

Infections:
Bacterial: Typhoid Fever; Paratyphoid; Brucellosis; and Septicemia (Mainly Gram Negative)
Viral: Hepatitis; Infectious Mononucleosis; Measles; Rubella; Influenza; Chicken Pox; Colorado Tick Fever
Other: Protozoa (Especially Malaria); Overwhelming Infections of any Kind

Myeloid Hypoplasia: Aplastic Anemia; Vitamin B12 and Folic Acid Deficiency; Agranulocytosis; Space-Occupying Bone Marrow Lesions

Chemical and Physical Agents: Bone Marrow Depressants (Radiation, Cytotoxic Drugs, Benzene); Drug Reactions, e.g., Chloramphenicol, Phenothiazines

Other Conditions: Collagen-Vascular Diseases, especially Lupus Erythematosus; Rheumatoid Arthritis; Infectious Mononucleosis; Hypersplenism, e.g., Liver Disease, Storage Diseases

LYMPHOCYTES:

REFERENCE RANGE: The mean values and reference ranges in thousands per cu mm, with age, are as follows: Birth: 5.5 (2.0-11.0); 1 Week: 5.0 (2.0-17); 4 Weeks: 6.0 (2.5-16.5); 6 Months: 7.3 (4.0-13.5); 1 Year: 7.0 (4.0-10.5); 2 Years: 6.3 (3.0-9.5); 4 Years: 4.5 (2.0-8.8); 6 Years: 3.5 (1.5-7.0); 10 Years: 3.1 (1.5-6.5); 16 Years: 2.8 (1.2-5.2); 21 Years: 2.5 (1.0-4.8).

INTERPRETATION: Conditions associated with an increase in lymphocytes are given in the next Table:

Conditions Associated with Lymphocytosis

- Infections:
  - Viral: Hepatitis; Infectious Mononucleosis; Cytomegalovirus Infections; Herpes Zoster and H. Simplex; Mumps; Chicken Pox; Viral Pneumonia; Measles
  - Bacterial: Pertussis; Brucellosis; Typhoid and Paratyphoid; Tuberculosis (Occasionally); Secondary and Congenital Syphilis
  - Other: Toxoplasmosis, Infectious Lymphocytosis
- Chronic Inflammatory Conditions:
  - Ulcerative Colitis
  - Immune Diseases: Serum Sickness, Idiopathic Thrombocytopenic Purpura
- Metabolic:
  - Hypoadrenalism
  - Hyperthyroidism (Occasionally)
- Blood Diseases:
  - Lymphocytic Leukemia (Acute and Chronic); Aplastic Anemia; Agranulocytosis; Heavy-Chain Disease; Multiple Myeloma; Felty's Syndrome; Banti's Syndrome

Conditions associated with a decrease in lymphocytes are given in the next Table:

Conditions Associated with Decreased Lymphocytes

- Acute Infections and Illnesses: Associated with Increased Plasma Corticosteroids; ACTH; Epinephrine
- Increased Corticosteroids: Cushing's Syndrome; Corticosteroid Therapy
- Immunodeficiency Syndromes: Congenital Defects of Cell Mediated Immunity; Immunosuppressive Medication; AIDS
- Defects of Lymphatic Circulation: Intestinal Lymphangiectasia; Thoracic Duct Drainage; Disorders of Intestinal Mucosa
- Severe Debilitating Diseases: Miliary Tuberculosis; Hodgkin's Disease; Lupus Erythematosis; Terminal Carcinoma; Renal Failure

MONOCYTES:

REFERENCE RANGE: The mean values and reference ranges in the thousands per cu mm, with age, are as follows: Birth: 1.0 (0.40-3.1); 12 Hours: 1.2 (0.4-3.6); 24 Hours to 1 Week: 1.10 (0.2-3.1); 2 Weeks: 1.0 (0.2-2.4); 4 Weeks: 0.7 (0.15-2.0); 2 Months: 0.65 (0.1-1.8); 4 Months: 0.6 (0.1-1.5); 12 Months: 0.5 (0.05-1.1); 10 Years: 0.35 (0-0.8); Adult: 0.3 (0-0.8).

INTERPRETATION: Conditions associated with an increase in monocytes are given in the next Table:

Conditions Associated with Increased Monocytes

- Infections:
  - Bacterial Infections: Tuberculosis; Subacute Bacterial Endocarditis; Syphilis; Brucellosis; Listeria
  - Viral Infections: Hepatitis; Mumps
  - Parasitic Diseases: Malaria; Kala-Azar
  - Other: Rickettsial Infections; Mycotic; Protozoal Infections
- Hematologic Disorders:
  - Preleukemic States:
  - Leukemia: Chronic Myelomonocytic; Acute Monocytic; Chronic Myelogenous
  - Lymphomas: Hodgkin's Disease and Non-Hodgkin Lymphomas
  - Histiocytic Medullary Reticulosis
  - Myeloproliferative Disorders: Myelosclerosis; Agnogenic Myeloid Aplasia; Polycythemia Vera
  - Hemolytic Anemias
- Collagen-Vascular Disease: Periarteritis Nodosa; Lupus Erythematosus; Rheumatoid Arthritis
- Other: Ulcerative Colitis; Regional Ileitis; Cirrhosis; Malignancies; Hand-Schüller-Christian Disease

White Blood Cell Count (Cont)

EOSINOPHILS:

REFERENCE RANGE: The mean values and reference ranges in thousands per cu mm, with age, are as follows: Birth: 0.4 (0.02-0.85); 24 Hours: 0.45 (0.05-1.00); 4 Weeks: 0.30 (0.07-0.90); 6 Months: 0.30 (0.07-0.75); 1 Year: 0.30 (0.05-0.70); 10 Years: 0.20 (0-0.60); 21 Years: 0.20 (0-0.45)

INTERPRETATION: The causes of increased blood eosinophils are listed in the next Table (Wolfe MS. Infect Dis Clin N Amer. 1992; 6:489-502; Nutman TB, et al. Allergy Proc. 1989; 10:33-46, 47-62):

Causes of Increased Blood Eosinophils
Parasitic Diseases: Trichinosis; Visceral Larva Migrans (Toxocara Canis or T. Cati); Ascaris Pneumonia; Strongyloides Filariasis; Schistosomiasis; Microfilariae (Causes Tropical Pulmonary Eosinophilia)
Allergic Diseases: Asthma; Seasonal Rhinitis (Hay Fever)
Skin Disorders: Atopic Dermatitis; Eczema; Acute Urticarial Reactions; Pemphigus; Dermatitis Herpetiformis
Pulmonary Eosinophilias: Loeffler's Syndrome; Pulmonary Infiltrate with Eosinophilia (PIE Syndrome); Hypersensitivity Pneumonitis [e.g. Allergic Bronchopulmonary Aspergillosis(ABPA)]; Tropical Pulmonary Eosinophilia (caused by Microfilariae)(Review of Eosinophilic Pneumonia: Umeki S, Arch Intern Med. 1992; 152:1913-1919)
Collagen Vascular Diseases: Dermatomyositis; Progressive Systemic Sclerosis; Eosinophilic Fascitis; Hypersensitivity Vasculitis
Malignancies: Ovarian Carcinoma; Epidermoid Carcinoma (Cervix, Uterus, Penis, Lip, Tongue); Villous Carcinoma of the Bladder; Carcinoma of Lung; Metastasis from Vulvar or Penile Carcinoma; Adenocarcinoma of Colon
Immunodeficiency: Wiskott-Aldrich Syndrome; Hyperimmunoglobulinemia E; IgA Deficiency; Nezelof Syndrome
Hematologic Disorders: Polycythemia Vera; Pernicious Anemia; Myelofibrosis; Myeloid Metaplasia; Chronic Myelogenous Leukemia
Drugs: Arsenicals; Chlorpromazine (Phenothiazines); Gold; Iodides; Nitrofurantoin; Para-aminosalicylic Acid; Ampicillin; Phenytoin; Streptomycin; Sulfonamides
Hypereosinophilic Syndrome(HES) - Liesveld JL, Abboud CN. "State of the Art: The Hypereosinophilic Syndromes." Blood Reviews. 1991; 5:29-37).
Fungal Infections: Coccidioidomycosis; Histoplasmosis; Allergic Bronchopulmonary Aspergillosis (ABPA)
Other: Wegener's Granulomatosis; Eosinophilia-Myalgia Syndrome; Inflammatory Bowel Disease

BASOPHILS:

REFERENCE RANGE: The mean values and reference ranges, in thousands per cu mm, with age, are as follows: Birth: 0.10 (0-0.65); 12 Hours: 0.10 (0-0.50); 24 Hours: 0.10 (0-0.30); 1 Week to 8 Years: 0.05 (0-0.20); 9 Years to Adult: 0.04 (0-0.20).

INTERPRETATION: The conditions causing basophilia are listed in the next Table:

Causes of Basophilia
Chronic Hypersensitivity Reactions: to Food, Drugs or Inhalants
Myeloproliferative Disorders: Polycythemia Vera, Chronic Granulocytic Leukemia, and Basophilic Leukemia
Mast Cell Disease: Urticaria Pigmentosa
Others:
Chronic Hemolytic Anemias
Ulcerative Colitis
Myxedema

WHOLE BLOOD CLOT LYSIS (see CLOT LYSIS TIME)

WHOLE BLOOD CLOTTING TIME (see CLOTTING TIME)

XYLOSE TOLERANCE TEST

Synonym: Xylose Absorption Test

SPECIMEN: Adult patient is prepared as follows: Nothing by mouth except water after midnight; the patient empties their bladder between 8:00 A.M. and 9:00 A.M. and then takes 25g of D-xylose (or significantly lower dose [5g] to avoid GI discomfort) dissolved in 250mL of tap water. Then, the patient takes an additional 750mL of water the time is recorded after the patient finishes drinking the water. All urine is collected without preservation for five hours; keep urine refrigerated during collection. Mix the specimens and measure and record volume. Freeze until time of assay.

Children (10kg weight to 9 years old): The child takes nothing by mouth for six hours. Between 8:00 A.M. and 10:00 A.M., draw 1mL of blood into a red top tube. The patient is given 5.0gm of D-xylose dissolved in 100-200mL of water. The time is recorded. After exactly one hour, obtain 1mL of blood in a red top tube. Separate serum from both tubes and freeze in plastic vials.

Infants (<10kg body weight): The child takes nothing by mouth for 4 hours. Between 8:00 A.M. and 10:00 A.M., draw 1mL of blood into a red top tube. The patient is given 5% solution of D-xylose (10mL/kg). The time is recorded. After exactly one hour, obtain 1mL of blood into a red top tube. Separate serum from both tubes and freeze in plastic vials.

REFERENCE RANGE: Urine: 5g of D-xylose in 5 hour urine collection with 25g dose (1.2g/5 hours with the 5g dose). Blood: Children (10kg weight to 9 years old): ≥20mg/dL at one hour. Infants (≤10kg body weight): ≥ 15mg/dL at one hour. To convert conventional units in mg/dL to international units in mmol/liter, multiply conventional units by 0.06661.

METHOD: Reaction of D-xylose with phloroglucinol (Clin Chem. 1979; 25:1440-1443). Destruction of glucose followed by o-toluidine.

INTERPRETATION: This test is used to access proximal small bowel function. It is performed to differentiate malabsorption caused by intestinal problems versus malabsorption caused by pancreatic insufficiency. Pancreatic enzymes are not required for absorption of xylose. D-xylose is absorbed chiefly from the upper small intestine, especially the jejunum. The mechanism for absorption is by diffusion facilitated by a carrier system. D-xylose is not metabolized and is excreted by the kidney.

The causes of decreased absorption of D-xylose are given in the next Table:

Causes of Decreased Absorption of D-xylose
Celiac Disease (Coeliac Disease, Nontropical Sprue, Gluten-Sensitive Enteropathy)
Other Causes:
Tropical Sprue
Abetalipoproteinemia
Lymphangiectasis
Intestinal Lipodystrophy
Amyloidosis
Lymphoma
Intestinal
Scleroderma and Other "Collagen" Diseases
Whipple's Disease
Surgical Removal of Jejunum

The test requires normal renal function and tends to yield a lower reference interval in the elderly. Vomiting, gastric stasis and bacterial overgrowth will lead to reduced values. Falsely reduced values may also result from medications such as diuretics, non-steroidal anti-inflammatory agents, and certain antibiotics.

ZARONTIN (see ETHOSUXIMIDE)

ZINC PROTOPORPHYRIN

Synonym: Free erythrocyte protoporphyrin (FEP).
SPECIMEN: Lavender (EDTA) top tube; heparinized whole blood; zinc protoporphyrin (ZPP) is stable for 4 days at room temperature. Protect from light.
REFERENCE RANGE: <35mcg/dL. Elevated in iron deficiency: typical values 35-600mcg/dL. Increased with lead levels >35mcg/dL. **Now not considered sensitive enough to exclude lead poisoning.**
METHOD: Hematofluorometer; extraction followed by HPLC
INTERPRETATION: Zinc protoporphyrin is increased in the conditions listed in the next Table. Zinc protoporphyrin levels are useful in the diagnosis of iron deficiency anemia, lead poisoning (see caution below), and erythropoietic protoporphyria.

Increase in Zinc Protoporphyrin
Lead Poisoning
Iron Deficiency Anemia
Anemia of Chronic Disease
Erythropoietic Protoporphyria

Lead Poisoning: The measurement of zinc protoporphyrin(ZPP) in red blood cells was often used as a screening test for lead poisoning. It is now believed that children exhibit neurological and developmental deficits at blood lead levels of 10-15mcg/dL. As a result the CDC has lowered the acceptable blood lead level in children from 25mcg/dL to 10mcg/dL (October 1991). However, zinc protoporphyrin elevation is not seen until lead levels reach at least 15-20mcg/dL. **As a result zinc protoporphyrin is not sensitive enough to detect toxic levels of lead (Turk DS, et al. N Engl J Med. 1992; 326:137-138).**

Lead inhibits the enzyme ferrochelatase, thus leading to the accumulation of "free" protoporphyrin(FPP); FPP is in fact not "free" but in the erythrocyte, chelates zinc, and forms zinc protoporphyrin. The level of red blood cell zinc protoporphyrin(ZPP) is directly related to the level to lead in the blood. Abnormal fingerstick lead measurements should be confirmed with a venous blood specimen. Alternate methods of screening for lead poisoning include direct measurement of blood lead levels (recommended test) or ALA-F inhibition (inhibited in the 10-15mcg/dL blood lead level range - not quite sensitive enough).
Iron Deficiency: The measurement of ZPP is useful in patients with iron deficiency anemia, in patients who have received iron therapy, and in the differential diagnosis of iron deficiency anemia from that of beta-thalassemia. Normal ZPP is <35 ± 50mcg/dL RBC; levels greater than 100mcg/dL indicate overt iron deficiency anemia (Cook JD, Seminars in Hematology. 1982; 19:6-18). In microcytic anemia due to iron deficiency, ZPP is increased, in thalassemia it is normal. Because of the 120 day lifetime of the RBC, ZPP levels change slowly, hence in the recently iron deficient patient who has received iron therapy, the ZPP will remain increased for several weeks. In general, ferritin levels will reflect the iron deficient state better than the ZPP. However, ferritin is an acute phase reactant and will change with acute inflammation, while the ZPP will not.

The ZPP-hemoglobin ratio (mcg protoporphyrin/g hemoglobin) is a sensitive index of iron deficiency erythropoiesis. The mean protoporphyrin-hemoglobin ratio is 16 with an upper limit of 32 (Labbe RF, et al. Clin Chem. 1979; 25:87-92). Transferrin saturation is erratically affected following iron therapy and acute viral illnesses. The ZPP:hemoglobin ratio is not affected by these conditions (Thomas WJ, et al. Blood. 1977; 49:455-462).

The serum ZPP level may be artifactually elevated in hyperbilirubinemia and after riboflavin or multivitamin therapy. These levels may be artifactually lowered if the specimen is inadequately oxygenated.
Erythropoietic Protoporphyria: In this condition the defect appears to be at the ferrochetalase step of iron insertion into the protoporphyrin. Protoporphyrin accumulates in the RBC and is also excreted in the bile. As a result, elevated ZPP levels are found in RBC and porphyrins in the stool. The disease is manifest as a photosensitive dermatosis.

ZINC

SPECIMEN: May vary by laboratory. Specimen collection may be difficult because of the possibility of contamination. Some laboratories employ acid washed polypropylene containers. The following is a procedure suggested by Mayo Medical Laboratories: Equipment for venipuncture: One disposable plastic syringe, two appropriately cleansed Sarstedt® syringes and one Monoject® needle.

1. Draw 3mL of blood through the Monoject® needle into a regular disposable plastic syringe. The purpose of this is to rinse the needle.
2. Discard the blood and the syringe.
3. Utilizing the Monoject® needle already in place, slowly draw the required volume of blood into the first Sarstedt® syringe. Cap the syringe.
4. After adequate clotting, centrifuge the specimen in the Sarstedt® syringe.
5. After centrifugation, pour (do not transfer with pipette) the serum into the second Sarstedt® syringe (5mL adequate for multiple requests).
6. Cap the second syringe and ship at room temperature.

Reference: Moody JR, Lindstrom RM: Selection and cleaning of plastic containers for storage of trace element samples. Anal Chem. 1977; 49:2264).
REFERENCE RANGE: 60-130mcg/dL. To convert conventional units in mcg/dL to international units in micromol/liter, multiply conventional units by 0.1530.
METHOD: Atomic absorption spectrometer; anodic stripping voltammetry
INTERPRETATION: Zinc is the second most common trace element in the body; there are over 90 zinc metalloenzymes. Causes of decreased serum levels of zinc are given in the next Table:

Causes of Decreased Serum Levels of Zinc
Dwarfism
Acrodermatitis Enteropathica
Acute Tissue Injury
Chronic Liver Disease
Sickle Cell Disease
Some Cancers, Especially Carcinoma of the Bronchus
Parenteral Nutrition for Several Weeks
Hypoalbuminemia
Stress
Infection
Pregnancy and Oral Contraceptives
Malnutrition
Pica
Burns

Subnormal serum zinc levels are found in patients with cancer of the lung and colon but usually not in association with other tumors. Zinc deficiency is often associated with delayed wound healing.

In deficiency states, urine measurements are confounded by specimen contamination and certain disease states that increase zinc excretion. There are a number of variables in the determination of serum zinc levels; these variables include: (1) contamination of samples; (2) serum albumin concentration; and (3) prolonged occlusion of blood vessels which may occur during phlebotomy producing localized increases in serum zinc levels.

Another approach involves the determination of serum zinc levels following ingestion of a challenge dose of zinc: Zinc Tolerance Test (Sullivan JF, et al. J Lab Clin Med. 1979; 93:485-492; Capel ID, et al.. Clin Biochem. 1982; 15:257-260). In this test, the subject ingests 200mg of zinc sulfate (equivalent to 50mg of the metal) in a small quantity of aqueous solution after an overnight fast. Blood is drawn before and 2, 4 and 6 hours after ingestion.

Zinc intoxication may result from industrial exposure, accidental ingestion (acidic beverages and foods in zinc containers), and overdosage of zinc supplements. In overdose situations, assays may be made on either blood or urine specimens.

NOTES

NOTES

NOTES

NOTES

NOTES

NOTES

NOTES

NOTES